STEM CELLS IN CRANIOFACIAL DEVELOPMENT AND REGENERATION

STEM CELLS IN CRANIOFACIAL DEVELOPMENT AND REGENERATION

Edited by

GEORGE T.-J. HUANG, D.D.S., M.S.D., D.Sc.
Professor, Department of Bioscience Research
College of Dentistry
The University of Tennessee Health Science Center
Memphis, Tennessee

IRMA THESLEFF, D.D.S., Ph.D.
Professor, Developmental Biology Program
Institute of Biotechnology
University of Helsinki
Helsinki, Finland

WILEY-BLACKWELL
A JOHN WILEY & SONS, INC., PUBLICATION

Wiley Blackwell is an imprint of John Wiley & Sons, formed by the merger of Wiley's global Scientific Technical and Medical business with Blackwell Publishing.

Published by John Wiley & Sons, Inc., Hoboken, New Jersey.
Published simultaneously in Canada.

For general information on our other products and services or for technical support, please contact our Customer Care Department within the United States at (800) 762-2974, outside the United States at (317) 572-3993 or fax (317) 572-4002.

Wiley also publishes its books in a variety of electronic formats. Some content that appears in print may not be available in electronic books. For more information about Wiley products, visit our web site at www.wiley.com.

Library of Congress Cataloging-in-Publication Data:

Stem cells in craniofacial development and regeneration / edited by George T.-J. Huang, Irma Thesleff.
p. cm.
Includes bibliographical references and index.
ISBN 978-1-118-27923-6 (cloth)
1. Regeneration (Biology). 2. Bone regeneration. 3. Guided bone regeneration. 4. Stem cells. I. Huang, George T.-J. II. Thesleff, Irma.
QH499.S816 2013
571.8′89–dc23

2012028577

Printed in Singapore

10 9 8 7 6 5 4 3 2 1

CONTENTS

PART III STEM CELL–MEDIATED CRANIOFACIAL TISSUE BIOENGINEERING

CONTRIBUTORS

John Abramyan, Ph.D., Postdoctoral Fellow, Department of Oral Health Sciences, Life Sciences Institute, University of British Columbia, Vancouver, British Columbia, Canada

Barbara D. Boyan, Ph.D., Professor, Price Gilbert, Jr. Chair in Tissue Engineering, Wallace H. Coulter Department of Biomedical Engineering, Georgia Institute of Technology, Atlanta, Georgia

Yang Chai, D.D.S., Ph.D., George and MaryLou Boone Professor of Craniofacial Biology, Center for Craniofacial Molecular Biology, Herman Ostrow School of Dentistry, University of Southern California, Los Angeles, California

Fa-Ming Chen, Ph.D., D.D.S., Professor and Dental Surgeon, Department of Periodontology and Oral Medicine, and Translational Research Team, School of Stomatology, Fourth Military Medical University, Shaanxi, China

Mo Chen, Ph.D., Associate Research Scientist, Center for Craniofacial Regeneration, Columbia University Medical Center, New York, New York

YiPing Chen, Ph.D., Professor, Department of Cell and Molecular Biology, Tulane University, New Orleans, Louisiana

Choko Cho, Ph.D., Postdoctoral Fellow, Center for Craniofacial Regeneration, Columbia University Medical Center, New York, New York

Michael T. Chung, B.S., Hagey Laboratory for Pediatric Regenerative Medicine, Department of Surgery, Plastic and Reconstructive Surgery Division, Stanford University School of Medicine, Stanford, California

Gang Ding, D.D.S., Ph.D., Associate Professor, Molecular Laboratory for Gene Therapy and Tooth Regeneration, Capital Medical University School of Stomatology, Beijing, China

Mildred C. Embree, D.M.D., Ph.D., Assistant Professor, Center for Craniofacial Regeneration, Columbia University Medical Center, New York, New York

Jifan Feng, Ph.D., Craniofacial Development and Stem Cell Biology, Biomedical Research Centre, and MRC Centre for Transplantation, Dental Institute, King's College, London, UK; Postdoctoral Research Associate, Center for Craniofacial Molecular Biology Herman Ostrow School of Dentistry, University of Southern California, Los Angeles, California

Brian L. Foster, Ph.D., National Institute of Arthritis and Musculoskeletal and Skin Diseases, National Institutes of Health, Bethesda, Maryland

Jonathan A. Garlick, D.D.S., Ph.D., Professor, Division of Cancer Biology and Tissue Engineering, Department of Oral and Maxillofacial Pathology, School of Dental Medicine, Tufts University, Boston, Massachusetts

Jeffrey Gimble, M.D., Ph.D., Professor, Stem Cell Biology Laboratory, Pennington Biomedical Research Center, Louisiana State University System, Baton Rouge, Louisiana

Rolando Gittens, Wallace H. Coulter Department of Biomedical Engineering, Georgia Institute of Technology, Atlanta, Georgia

Sander Grefte, Ph.D., Postdoctoral Researcher, Department of Biochemistry, Radboud University Nijmegen Medical Centre, Nijmegen, The Netherlands

Stan Gronthos, Ph.D., Professor, Mesenchymal Stem Cell Laboratory, Department of Haematology, University of Adelaide, Adelaide, South Australia, Australia

Monica Grova, B.S., Hagey Laboratory for Pediatric Regenerative Medicine, Department of Surgery, Plastic and Reconstructive Surgery Division, Stanford University School of Medicine, Stanford, California

Shuping Gu, D.D.S., Ph.D., Assistant Research Professor, Department of Cell and Molecular Biology, Tulane University, New Orleans, Louisiana

Christopher D. Hermann, Ph.D., Wallace H. Coulter Department of Biomedical Engineering, Georgia Institute of Technology, Atlanta, Georgia

Kyle J. Hewitt, Ph.D., Program in Cell, Molecular and Developmental Biology, Sackler School of Graduate Biomedical Sciences, Tufts University, Boston, Massachusetts

Joshua W. K. Ho, Ph.D., Instructor, Division of Genetics, Department of Medicine, Brigham and Women's Hospital, Harvard Medical School, Boston, Massachusetts

Matthew P. Hoffman, B.D.S., Ph.D., National Institute of Dental and Craniofacial Research, National Institutes of Health, Bethesda, Maryland

George T.-J. Huang, D.D.S., M.S.D., D.Sc., Professor, Director for Stem Cells and Regenerative Therapies, Department of Bioscience Research, College of Dentistry, The University of Tennessee Health Science Center, Memphis, Tennessee

Jeong S. Hyun, M.D., Postdoctoral Research Fellow, Hagey Laboratory for Pediatric Regenerative Medicine, Department of Surgery, Plastic and Reconstructive Surgery Division, Stanford University School of Medicine, Stanford, California

John A. Jansen, D.D.S., Ph.D., Professor, Department of Biomaterials, Radboud University Nijmegen Medical Centre, Nijmegen, The Netherlands

Andrew H. Jheon, D.D.S., Ph.D., Assistant Adjunct Professor, Department of Orofacial Sciences and Program in Craniofacial and Mesenchymal Biology, University of California–San Francisco, San Francisco, California

Nan Jiang, D.D.S., Center for Craniofacial Regeneration, Columbia University Medical Center, New York, New York

Yan Jin, Ph.D., D.D.S., Professor, Research and Development Center for Tissue Engineering, Fourth Military Medical University, Shaanxi, China

Maria Jussila, M.Sc., Developmental Biology Program, Institute of Biotechnology, University of Helsinki, Helsinki, Finland

Emma Juuri, D.D.S., M.Sc. Developmental Biology Program, Institute of Biotechnology, University of Helsinki, Helsinki, Finland

L. Keller, INSERM UMR 1109, Team "Osteoarticular and Dental Regenerative NanoMedicine," Strasbourg, France; and Faculté de Chirurgie Dentaire, Université de Strasbourg, Strasbourg, France

Robert G. Kelly, Ph.D., Investigator, Developmental Biology Institute of Marseilles–Luminy, Aix–Marseille University, Marseille, France

Ophir D. Klein, M.D., Ph.D., Associate Professor, Departments of Orofacial Sciences and Pediatrics, Program in Craniofacial and Mesenchymal Biology, and Institutes for Human Genetics and Regeneration Medicine, University of California–San Francisco, San Francisco, California

Kimi Kong, Ph.D., Associate Research Scientist, Center for Craniofacial Regeneration, Columbia University Medical Center, New York, New York

S. Kuchler-Bopp, Ph.D., Chargée de recherche, INSERM UMR 1109, Team "Osteoarticular and Dental Regenerative NanoMedicine," Strasbourg, France; and Faculté de Chirurgie Dentaire, Université de Strasbourg, Strasbourg, France

Chang Hun Lee, Ph.D., Associate Research Scientist, Center for Craniofacial Regeneration, Columbia University Medical Center, New York, New York

Christopher S. D. Lee, Ph.D., Wallace H. Coulter Department of Biomedical Engineering, Georgia Institute of Technology, Atlanta, Georgia

Hervé Lesot, Ph.D., Investigator, INSERM UMR 1109, Team "Osteoarticular and Dental Regenerative NanoMedicine," Strasbourg, France; and Faculté de Chirurgie Dentaire, Université de Strasbourg, Strasbourg, France

Yi Liu, D.D.S., Ph.D., Associate Professor, Molecular Laboratory for Gene Therapy and Tooth Regeneration, Capital Medical University School of Stomatology, Beijing, China

David D. Lo, M.D., Postdoctoral Research Fellow, Hagey Laboratory for Pediatric Regenerative Medicine, Department of Surgery, Plastic and Reconstructive Surgery Division, Stanford University School of Medicine, Stanford, California

Isabelle M. A. Lombaert, Ph.D., National Institute of Dental and Craniofacial Research, National Institutes of Health, Bethesda, Maryland

Michael T. Longaker, M.D., M.B.A., Deane P. and Louise Mitchell Professor, Hagey Laboratory for Pediatric Regenerative Medicine, Department of Surgery, Plastic and Reconstructive Surgery Division, Stanford University School of Medicine, Stanford, California; and Institute for Stem Cell Biology and Regenerative Medicine, Stanford University, Stanford, California

Richard L. Maas, M.D., Ph.D., Professor, Division of Genetics, Department of Medicine, Brigham and Women's Hospital, Harvard Medical School, Boston, Massachusetts

Jeremy J. Mao, D.D.S., Ph.D., Professor and Zegarelli Endowed Chair, Center for Craniofacial Regeneration, Columbia University Medical Center, New York, New York

Avital Mendelson, Ph.D., Postdoctoral Fellow, Center for Craniofacial Regeneration, Columbia University Medical Center, New York, New York

Frederic Michon, Ph.D., Academy Fellow, Institute of Biotechnology, Developmental Biology Program, University of Helsinki, Helsinki, Finland

Daniel T. Montoro, B.S., Life Science Research Associate, Hagey Laboratory for Pediatric Regenerative Medicine, Department of Surgery, Plastic and Reconstructive Surgery Division, Stanford University School of Medicine, Stanford, California

Christian Morsczeck, M.Sc., Ph.D., Private lecturer, Department of Cranio- and Maxillofacial Surgery, University Hospital Regensburg, Regensburg, Germany

Misako Nakashima, D.D.S., Ph.D., Investigator, Department of Dental Regenerative Medicine, Center of Advanced Medicine for Dental and Oral Diseases, National Center for Geriatrics and Gerontology, Research Institute, Obu, Aichi, Japan

Hemin Nie, Ph.D., Postdoctoral Fellow, Center for Craniofacial Regeneration, Columbia University Medical Center, New York, New York

Francisco H. Nociti, Jr., D.D.S., Ph.D., National Institute of Arthritis and Musculoskeletal and Skin Diseases, National Institutes of Health, Bethesda, Maryland; State University of Campinas School of Dentistry, Piracicaba, São Paulo, Brazil

Daniel J. O'Connell, Ph.D., Postdoctoral Fellow, Division of Genetics, Department of Medicine, Brigham and Women's Hospital, Harvard Medical School, Boston, Massachusetts

Rene Olivares-Navarrete, D.D.S., Ph.D., Senior Research Scientist, Wallace H. Coulter Department of Biomedical Engineering, Georgia Institute of Technology, Atlanta, Georgia

Carolina Parada, D.M.D, Ph.D, Center for Craniofacial Molecular Biology, Herman Ostrow School of Dentistry, University of Southern California, Los Angeles, California

Adelina Plachokova, D.D.S., Ph.D., Assistant Professor, Department of Biomaterials, Radboud University Nijmegen Medical Centre, Nijmegen, The Netherlands

Maryam Rezai Rad, Department of Comparative Biomedical Sciences, Louisiana State University School of Veterinary Medicine, Baton Rouge, Louisiana

David P. Rice, B.D.S., F. Orth., Ph.D., Professor, Department of Orthodontics, Institute of Dentistry, University of Helsinki, and Oral and Maxillofacial Diseases, Helsinki University Central Hospital, Helsinki, Finland

Ritva Rice, Ph.D., Postdoctoral Researcher, Developmental Biology Program, Institute of Biotechnology, University of Helsinki, Helsinki, Finland

Joy M. Richman, Ph.D., D.M.D., Professor, Department of Oral Health Sciences, Life Sciences Institute, University of British Columbia, Vancouver, British Columbia, Canada

Zvi Schwartz, D.M.D., Ph.D., Professor, Wallace H. Coulter Department of Biomedical Engineering, Georgia Institute of Technology, Atlanta, Georgia

Kerstin Seidel, Ph.D., Postdoctoral Fellow, Department of Orofacial Sciences and Program in Craniofacial and Mesenchymal Biology, University of California–San Francisco, San Francisco, California

Yulia Shamis, Ph.D., Program in Cell, Molecular and Developmental Biology, Sackler School of Graduate Biomedical Sciences, Tufts University, Boston, Massachusetts

Paul T. Sharpe, Ph.D., Professor, Craniofacial Development and Stem Cell Biology, Biomedical Research Centre and MRC Centre for Transplantation, Dental Institute, King's College, London, UK

Songtao Shi, D.D.S., Ph.D., Professor, Center for Craniofacial Molecular Biology, Herman Ostrow School of Dentistry, University of Southern California, Los Angeles, California

Martha J. Somerman, D.D.S., Ph.D., National Institute of Arthritis and Musculoskeletal and Skin Diseases, National Institutes of Health, Bethesda, Maryland

Irma Thesleff, Ph.D., Professor, Developmental Biology Program, Institute of Biotechnology, University of Helsinki, Helsinki, Finland

Paul A. Trainor, Ph.D., Investigator, Stowers Institute for Medical Research, Kansas City, Missouri; and Professor, Department of Anatomy and Cell Biology, University of Kansas Medical Center, Kansas City, Kansas

Takashi Tsuji, Ph.D., Professor, Research Institute for Science and Technology, and Department of Biological Science and Technology, Faculty of Industrial Science and Technology, Tokyo University of Science, Noda, Chiba, Japan; and Organ Technologies Inc., Tokyo, Japan

Johannes W. Von den Hoff, Ph.D., Assistant Professor, Department of Orthodontics and Craniofacial Biology, Radboud University Nijmegen Medical Centre, Nijmegen, The Netherlands

X. Frank Walboomers, Ph.D., Associate Professor, Department of Biomaterials, Radboud University Nijmegen Medical Centre, Nijmegen, The Netherlands

Derrick C. Wan, M.D., Assistant Professor, Hagey Laboratory for Pediatric Regenerative Medicine, Department of Surgery, Plastic and Reconstructive Surgery Division, Stanford University School of Medicine, Stanford, California

Lei Wang, Ph.D., D.D.S., Postdoctoral Fellow, Center for Craniofacial Molecular Biology, Herman Ostrow School of Dentistry, University of Southern California, Los Angeles, California; and Faculty of Department of Oral and Maxillofacial Surgery, School of Stomatology, Fourth Military Medical University, Shaanxi, China

Songlin Wang, D.D.S., Ph.D., Professor, Molecular Laboratory for Gene Therapy and Tooth Regeneration, Capital Medical University School of Stomatology, Beijing, China

Fulan Wei, D.D.S., Ph.D., Lecturer, Molecular Laboratory for Gene Therapy and Tooth Regeneration, Capital Medical University School of Stomatology, Beijing, China

John A. Whitlock, Ph.D., Postdoctoral Fellow, Department of Oral Health Sciences, Life Sciences Institute, University of British Columbia, Vancouver, British Columbia, Canada

Xing Yan, D.D.S., Ph.D., Postdoctoral Fellow, Department of Endodontics, Boston University Henry M. Goldman School of Dental Medicine, Boston, Massachusetts, and Associate Chief Physician, Department of Stomatology, Beijing Friendship Hospital (Second Clinical School), Capital Medical University, Beijing, China

Fang Yang, Ph.D., Assistant Professor, Department of Biomaterials, Radboud University Nijmegen Medical Centre, Nijmegen, The Netherlands

Shaomian Yao, Ph.D., Department of Comparative Biomedical Sciences, Louisiana State University School of Veterinary Medicine, Baton Rouge, Louisiana

Na Yu, D.D.S., Department of Biomaterials, Radboud University Nijmegen Medical Centre, Nijmegen, The Netherlands

Ying Zheng, D.D.S., Ph.D., Postdoctoral Fellow, Center for Craniofacial Regeneration, Columbia University Medical Center, New York, New York

Jian Zhou, D.D.S., Ph.D., Postdoctoral Fellow, Center for Craniofacial Regeneration, Columbia University Medical Center, New York, New York

Xiao-Ying Zou, D.D.S., M.D., Postdoctoral Fellow, Department of Endodontics, Boston University Henry M. Goldman School of Dental Medicine, Boston, Massachusetts, and Department of Cariology, Endodontology, and Operative Dentistry, School and Hospital of Stomatology, Peking University, Beijing, China

PREFACE

This book project was motivated by the need to present in one place current knowledge on the regulation of normal development of craniofacial tissues, and on the characteristics of tissue-specific stem cells and their potential use in bioengineering/regeneration of craniofacial tissues and organs. It has become obvious that knowledge of the mechanisms of normal development will be essential when tissues and organs are attempted to generate from stem and progenitor cells. In particular, developmental biology research has unraveled the key roles of cell–cell interactions in all developmental processes, and identified specific signal molecules as the molecular mediators of these interactions. These same signals are the main tools in guiding stem cell proliferation and differentiation in the process of tissue regeneration via bioengineering technologies.

In recent years there have been huge advances in stem cell biology and in characterization of pluripotent stem cells and tissue-specific stem cells. The discovery of reprogramming differentiated cells to pluripotent stem cells has opened the possibility of using the patient's own cells for a variety of biomedical applications. Various adult stem cells have also been tested for their tissue regeneration potential. At the same time, major strides have been made in the field of tissue engineering. Engineered organs have been transplanted into patients to restore damaged ones. Strategies and study models for engineering and regenerating craniofacial tissues and organs, including teeth, have also shed light on their future clinical applications.

In the first part of the book, there are nine chapters summarizing the current knowledge on developmental mechanisms involved in selected craniofacial tissues and organs. During embryogenesis, the morphogenesis and cell differentiation are intimately linked. The complex shapes of organs, as well as the specialized cell types, are generated in concert step-by-step from progenitor cells.

The second part elaborates on stem cells and their niches. It covers the general area of stem cells, including embryonic stem cells and induced pluripotent stem cells. The physiological renewal and regeneration of tissues is based on stem cells. Postnatal stem cells of various tissue origins are reviewed with an emphasis on their potential application for craniofacial tissue regeneration. Tissue-specific stem cells, such as salivary gland stem cells and tooth stem cells, have been identified and characterized in craniofacial tissues. The details on stem cells and their differentiation are best known in continuously renewing tissues such as bone. However, stem cells are also present in adult permanent teeth, for example, pulp tissue, functioning as the source of replacement odontoblasts to form new dentin.

The third part gives an overview of ongoing research on bioengineering of craniofacial tissues, including bone, muscle, dental tissues, periodontal tissues, and teeth. The use of scaffolds, growth factors, and stem cells are the key elements for engineered tissue regeneration.

In the case of teeth, one scenario is to grow new teeth from progenitor cells by applying knowledge of the mechanisms of their normal development. The regeneration of many types of craniofacial tissues has been tested and has achieved success in small and large animals. Some of the regeneration technologies are being studied in clinical trials. It appears inevitable that tissue regeneration and regenerative medicine will become a mainstream medical practice in the near future.

We are very grateful to have such group of authors reviewing the latest work in their fields, including their own work, and sharing their expert views on future possibilities and challenges. Everyone we asked agreed to contribute to the book. All are respected specialists in their fields. We are thankful to all of them for writing the excellent chapters and we are extremely happy with the end result. We hope that students, as well as scientists in the field, young and advanced, will find this book useful.

George Huang
Irma Thesleff

PART I

DEVELOPMENT AND REGENERATION OF CRANIOFACIAL TISSUES AND ORGANS

1

MOLECULAR BLUEPRINT FOR CRANIOFACIAL MORPHOGENESIS AND DEVELOPMENT

PAUL A. TRAINOR

Stowers Institute for Medical Research, Kansas City, Missouri, and University of Kansas Medical Center, Kansas City, Kansas

1 INTRODUCTION

The vertebrate head is a sophisticated assemblage of cranial specializations, including the central and peripheral nervous systems and viscero-, chondro-, and neurocraniums, and each must be properly integrated with musculature, vasculature, and connective tissue. Anatomically, the head is the most complex part of the body, and all higher vertebrates share a common basic plan or craniofacial blueprint that is established during early embryogenesis. This process begins during gastrulation and requires the coordinated integration of each germ layer tissue (i.e., ectoderm, mesoderm, and endoderm) and its derivatives in concert with the precise regulation of cell proliferation, migration, and differentiation for proper craniofacial development (Figs. 1 and 2). For example, the appropriate cranial nerves must innervate the muscles of mastication, which, via tendon attachment to the correct part of the mandible, collectively articulate jaw opening and closing. In addition, each of these tissues must be sustained nutritionally and remain oxygenated and thus are intimately associated with the vasculature as part of a fully functioning oral apparatus.

Given this complexity, it is not surprising that a third of all congenital defects affect the head and face (Gorlin et al., 1990). Improved understanding of the etiology and pathogenesis of head and facial birth defects and their potential prevention or repair depends on a thorough appreciation of normal craniofacial development. But what are the signals and mechanisms that establish each of these individual cells and tissues and govern their differentiation and integration? In this chapter specification of the

Stem Cells in Craniofacial Development and Regeneration, First Edition.
Edited by George T.-J. Huang, Irma Thesleff.

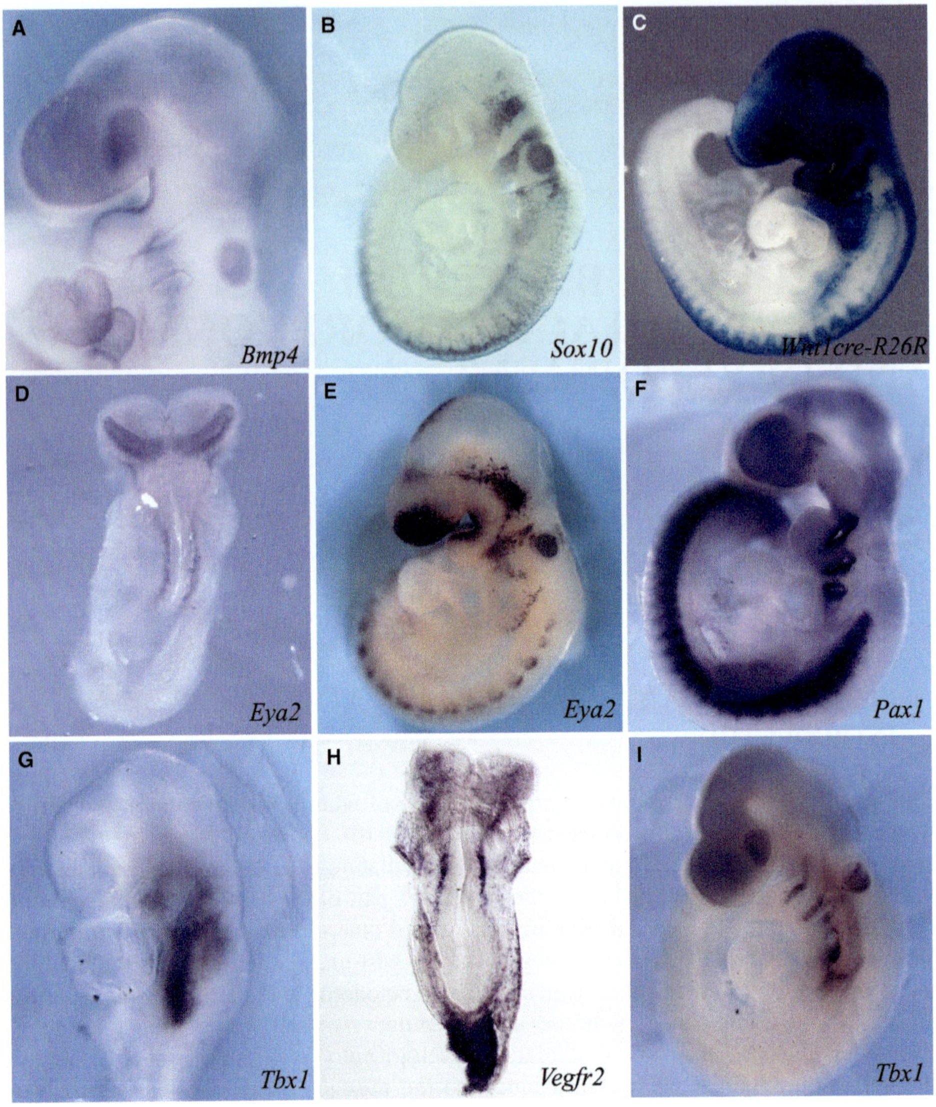

FIGURE 1 Specification of ectoderm, neural crest, placodes, mesoderm, and endoderm. In situ hybridization (A, B, D–I) or *lacZ* staining (C) of E8.5–9.5 mouse embryos as indicators of differentiation of ectoderm (A, *Bmp4*), neural crest cells (B, *Sox10*; C, *Wnt1cre-R26R*), ectodermal placodes (D and E, *Eya2*), endoderm (F, *Pax1*), mesoderm (G and I, *Tbx1*), and endothelial cells (H, *Vegfr2*).

major cell lineages, tissues, and structures that establish the blueprint for craniofacial development is described, as well as the interactions and integration that are essential for normal functioning throughout embryonic as well as adult life.

Craniofacial development begins during gastrulation, which is the process that generates a triploblastic organism. During gastrulation, cells from the epiblast (embryonic ectoderm) are allocated to three definitive germ layers: ectoderm, mesoderm, and endoderm. Formation of the mesoderm and endoderm is accomplished by morphogenetic

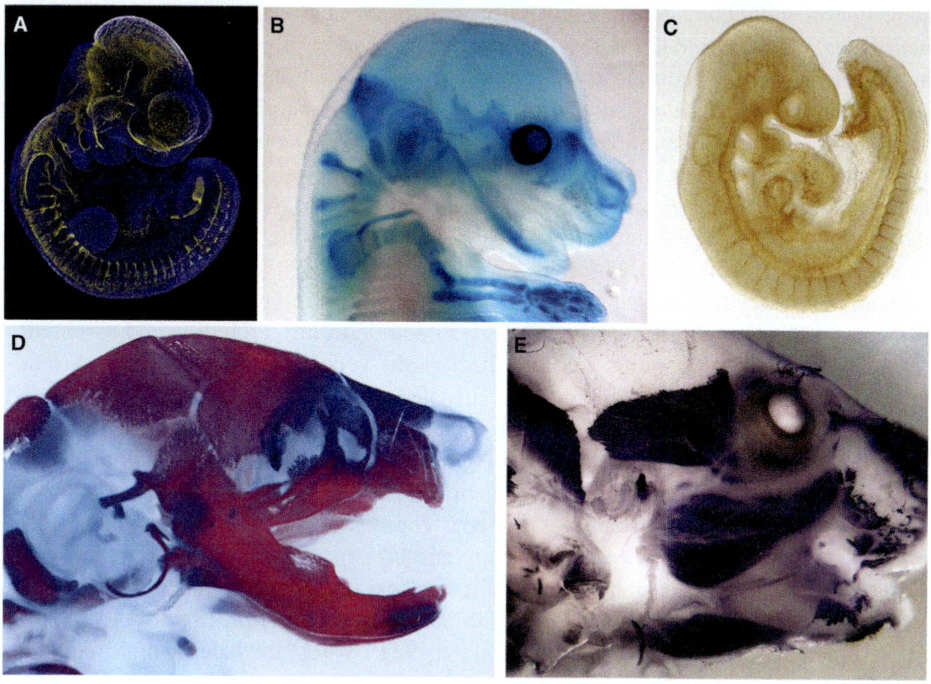

FIGURE 2 Formation of the nervous system, skeleton, musculature, and vasculature. Immunostaining (A, C, and E) and histochemical stainining (B and D) as indicators of formation of the peripheral nervous system (A, E10.5, Tuj1), cartilage (B, E15.5, alcian blue), vasculature (C, E9.5, PECAM), skeletal bone and cartilage (E18.5, alizarin red/alcian blue), and muscle (E18.5, MHC).

cell movement that comprises ingression of epiblast cells through the primitive streak (a site of epithelial–mesenchyme transition), followed by organization of ingressed mesoderm progenitors into a mesenchymal layer and incorporation of the endoderm progenitors into a preexisting layer of visceral endoderm (Arkell and Tam, 2012). Notably, a general axial registration exists between these progenitor germ layer tissues as they are established and influences their differentiation (Trainor and Tam, 1995a). These relationships and the tissue boundaries they create are often maintained throughout embryogenesis and into adult life and are critically important for proper vertebrate head and facial function. Thus, gastrulation and generation of the three germ layers create the principal building blocks of the head and face (Arkell and Tam, 2012). The ensuing morphogenetic movements that bring these tissue components to their proper place in the body plan establish the initial blueprint. Subsequent morphogenetic events continue to build on this scaffold until the fully differentiated structures emerge that define the head and face.

2 ECTODERM: NEURAL INDUCTION

Motor coordination, sensory perception, and memory all depend on precise, complex cell connections that form between distinct nerve cell types within the central nervous system. Development of the central nervous system occurs in several steps. The first step, *neural induction*, generates a uniform sheet of neuronal progenitors called the *neural plate*. Neural induction is followed by *neurulation*, a process in which the two halves of the neural plate are transformed into a hollow tube. Neurulation is accompanied by *regionalization* of the neural tube anterior–posteriorly into the brain and spinal cord and dorsoventrally into neural crest cells and numerous classes of sensory and motor neurons. Proper development of the central nervous system requires finely balanced control of cell specification and proliferation, which is achieved through the complex interplay of multiple signaling pathways. Bone morphogenetic proteins (BMPs), retinoic acid (RA), fibroblast growth factors (FGFs) and Hedgehog (Hh) proteins are a few key factors that interact to pattern the developing central nervous system.

Neural induction constitutes the first step in ectoderm differentiation and essentially resolves ectoderm progenitors into neuroectoderm versus surface ectoderm. A landmark experiment in amphibian embryos revealed that differentiation of uncommitted ectoderm into neuroectoderm depended on the underlying mesoderm (Spemann and Mangold, 1924). Transplantation of this mesoderm, called the *blastopore lip*, or *Spemann's organizer*, induced the formation of a duplicated body axis, including an almost complete second nervous system. The discovery of a number of secreted molecules expressed by the organizer in amphibian and avian embryos provides a molecular mechanism underpinning the process of neural induction. The most important molecules include noggin (Lamb et al., 1993), chordin (Sasai et al., 1994), and follistatin (Hemmati-Brivanlou et al., 1994), which mediate neural induction by binding to and inhibiting a subset of bone morphogenetic proteins (BMPs) (reviewed by Sasai and De Robertis, 1997). Each of these secreted factors has potent neural-inducing ability when added to *Xenopus* ectodermal explants and mimics the capacity of the organizer to induce and pattern a secondary axis. Interestingly, during *Xenopus* gastrulation, *Bmp4* expression is repressed by signals from the organizer in the portion of the ectoderm fated to become the neural plate (Fainsod et al., 1994). Therefore, inhibition of BMP signaling represses epidermal fate and induces neural differentiation. Consistent with this idea, single ectoderm cells taken from gastrula-stage *Xenopus* embryos and cultured in the absence of any additional factors (e.g., BMP4) will differentiate into neural tissue. This prompted the idea of a "default model" for neural induction in which ectodermal cells, by default, adopt a neural fate when removed from the influence of extracellular signals during gastrulation (Wilson and Hemmati-Brivanlou, 1995, 1997). However, difficulties arose when attempts were made to extrapolate this model to amniotes and mammals.

In chick embryos, the organizer (Hensen's node) expresses the BMP inhibitors *Noggin* and *Chordin*, yet neither *Noggin* nor *Chordin* induces neural cell differentiation in avian embryos (Streit et al., 1998). Furthermore, their temporal expression does not coincide with neural induction (Streit and Stern, 1999b). In addition, a neural plate still forms in chick, frog, zebrafish, and mouse embryos, despite surgical removal of the organizer (Wilson et al., 2001), and gene-targeting experiments in mouse have shown that neural differentiation occurs in the absence of BMP inhibitors, arguing that BMP signaling is not required for neural induction (Matzuk et al., 1995; McMahon et al., 1998; Bachiller et al., 2000). The evolution of fundamentally different molecular mechanisms for specifying neural fate in amniotes versus anamniotes seems unlikely, and in agreement with this, the avian organizer can substitute for the *Xenopus* blastopore lip (Kintner and Dodd, 1991). Avian neural induction appears to be initiated by FGF signals emanating from the precursors of Hensen's node (Streit et al., 2000;

Wilson et al., 2000). *Fgf8* is expressed during gastrulation in the anterior of the primitive streak, including the node; however, its expression is downregulated as the node begins to lose its neural-inducing ability. Consistent with this, inhibition of FGF signaling downregulates the expression of neural plate markers (Streit et al., 2000). Thus, one possible function for FGF signaling may be to attenuate BMP signaling in prospective neural cells. In support of this idea, inhibition of FGF results in maintenance of *Bmp4* and *Bmp7* expression, both of which are normally downregulated in epiblast cells of prospective neural character. This implies a role for FGF in repressing BMP signaling. Thus, as in *Xenopus*, acquisition of neural fate requires the repression of *Bmp* activity, while epidermal cell fate requires maintenance of *Bmp* expression (Fig. 1A). However, neither FGF signaling alone or in combination with BMP antagonists is sufficient for the induction of *Sox2* or later neural markers (Harland, 2000; Streit et al., 2000; Wilson et al., 2000). WNT proteins are one of the additional signals required for the regulation of neural versus epidermal fates (Wilson et al., 2001). In chick embryonic ectoderm, lateral or prospective epidermal tissue expresses *Wnt3* and *Wnt8*, whereas medial or prospective neural tissue does not. The lack of exposure to WNT signaling in the medial ectoderm permits *Fgf8* expression, which in turn represses BMP signaling, specifying neural fate. Conversely, high levels of WNT signaling in lateral epiblast cells inhibit FGF signaling, allowing for BMP activity, which in turn directs cells to an epidermal fate (Wilson et al., 2001).

Thus, vertebrate neural induction involves the coordinated interaction of three different signaling pathways—FGFs, BMPs, their associated antagonists, and WNTs—all of which play significant but distinct roles in the differentiation of neural versus epidermal fate. Notably, a key conserved feature among vertebrates is the exclusion of *Bmp* expression from the neural-induced territory.

3 ECTODERM: NEURULATION

Neural induction is followed by neurulation, the process by which the neural plate is transformed into a hollow neural tube. In amphibians, mice, and chicks, the neural tube forms through uplifting of the two halves of the neural plate and their fusion at the dorsal midline. In contrast, in fish, formation of the neurocele occurs via cavitation of the neural plate. The neural tube then becomes partitioned via differential cell proliferation into a series of swellings and constrictions that define the major compartments of the adult brain: forebrain (prosencephalon), midbrain (mesencephalon), and hindbrain (rhombencephalon). The forebrain becomes further regionalized anteriorly into the telencephalon and posteriorly into the diencephalon. The telencephalon develops into the cerebral hemispheres, and the diencephalon gives rise to the thalamic and hypothalamic brain regions. Similar to the forebrain, the hindbrain becomes subdivided further. The anterior portion forms the metencephalon, which gives rise to the cerebellum, the specific part of the brain responsible for coordinating movements, posture, and balance. The posterior portion forms the myelencephalon, which generates the medulla oblongata, the nerves of which regulate respiratory, gastrointestinal, and cardiovascular movements. In contrast to the forebrain and hindbrain, the midbrain is not subdivided further. However, the lumen of the midbrain gives rise to the cerebral aqueduct.

An important question relates to how cells in the neural plate become regionalized and specified into forebrain, midbrain, hindbrain, and spinal cord domains, since immediately following induction, the neural plate is assumed to have a uniformly rostral character. Is there a mechanism that can account for the anterior–posterior specification of individual cells along the entire neural axis? In chick embryos, medial epiblast cells in blastula-stage embryos generally express *Sox2, Sox3, Otx2*, and *Pax6*, a combination of markers characteristic of the forebrain. In addition, these cells do not express *En1/2, Krox20*, or *Hoxb8*, which are typical markers of the midbrain, hindbrain, and spinal cord, respectively (Wilson et al., 2001). Thus, initially, neural progenitors possess a rostral "forebrain-like" character which implies that midbrain, hindbrain, and spinal cord characteristics are generated by subsequent reprogramming.

Posteriorizing the early neuroepithelium, at least in chick embryos, involves the convergent actions of FGF signaling with graded concentration-dependent WNT signals to specify cells of the caudal forebrain (*Otx2*$^+$, *Pax6*$^+$), midbrain (*Otx2*$^+$, *En1*$^+$), rostral hindbrain (*Gbx2*$^+$, *Krox20*$^+$, *Pax6*$^+$), and caudal hindbrain (*Krox20*$^+$, *Gbx2*$^-$, *Pax6*$^-$) character (Nordstrom et al., 2002). Higher concentrations of WNT signals induce progressively more caudal character in the neural tube, while conversely,caudal neural cells grow in vitro, in the absence of WNT signaling. *Hox* genes also play important roles in establishing regional cell identity in the hindbrain and spinal cord, and this is achieved via opposing gradients of retinoic acid and FGF signaling (Bel-Vialar et al., 2002).

Interestingly, the progenitor cells for the forebrain, midbrain, and hindbrain are allocated during gastrulation in an anterior-to-posterior order; however, the relative size of each progenitor domain does not correlate with the final size of each region of the brain. In fact, the forebrain has undergone a disproportionate expansion during neurulation, which is underscored by the wide area covered by lineage-traced clones in ectoderm-fate mapping experiments (Cajal et al., 2012). This may underpin the vulnerability of the forebrain to developmental errors, which often leads to head truncation and raises the question of what triggers the initiation of head induction.

4 HEAD INDUCTION

The initiation of head formation depends on signaling centers juxtaposed with the progenitor tissues of the head. The anterior visceral endoderm (AVE) forms initially at the distal end of an embryonic day (E) 5.5 gastrulating embryo and then migrates to the prospective anterior of the embryo by E6.0, where it has a lasting impact on the differentiation and morphogenesis of epiblast-derived tissues into head structures (Arkell and Tam, 2012). WNT and Nodal pathway inhibitors secreted from AVE inhibit posterior development of the adjacent embryonic tissue, thus defining its anterior character. Fate-mapping studies have shown that the lateral frontonasal prominence, telencephalon, and diencephalon progenitor regions of the mouse embryo are devoid of active WNT signaling (Lewis et al., 2008) and that the lack of WNT signaling activity in this region might be required for normal head development. Consistent with this, *Dkk1*-knockout mice display ectopic and elevated WNT signaling activity in the head primordia and lack head structures anterior to the midbrain at birth (Mukhopadhyay et al., 2001). These defects can be reversed by reducing the levels of WNT3 activity (Lewis et al., 2008) or by genetic suppression of LRP6 coreceptor (MacDonald et al., 2004). The demonstration of genetic interactions between *Dkk1, Wnt3*, and *Lrp6* provides compelling evidence that stringent regulation of canonical WNT signaling levels is necessary for head formation. Furthermore, the rostral parts of the brain and the head are differentially more sensitive to canonical WNT signaling, and their development is contingent on negative modulation of WNT activity. Thus, AVE-mediated WNT signaling is a critical regulator of head induction.

5 ECTODERM: NEURAL CREST CELLS

5.1 Induction of Neural Crest Cell Formation

In addition to being regionalized anterioposteriorly, the neuroectoderm is also patterned dorsoventrally. During neural induction and neurulation, a vertebrate-specific cell type known as the *neural crest* is born at the interface between the nonneural ectoderm (presumptive epidermis–surface ectoderm) and the dorsal region of the neural plate a region commonly referred to as the *neural plate border*. Cell lineage tracing has indicated that both neuroepithelium and surface ectoderm give rise to neural crest cells (Selleck and Bronner-Fraser, 1995), although the vast majority come from the neuroepithelium. Explants of neural plate, do not endogenously generate neural crest cells. Therefore, neural crest cell induction is a multistep process, requiring contact-mediated interactions between nonneural (i.e., the surface ectoderm or paraxial mesoderm) and neural tissues (neural plate) (Rollhauser-ter Horst, 1977; Moury and Jacobson, 1990; Selleck and Bronner-Fraser, 1995). In frog and fish embryos a precise level of BMP signaling was considered central to neural crest cell induction (Mayor et al., 1995, Morgan and Sargent, 1997). Moreover, the underlying mesoderm is thought to regulate the levels by secreting BMP inhibitors that help to define low, intermediate, and high localized levels of BMP4/7 activity, which induce the overlying neural plate, neural crest, and surface ectoderm, respectively (Fig. 1B) (Marchant et al., 1998).

However, more recently it was argued that WNT signaling from the surface ectoderm drives neural crest cell formation in avian and fish embryos (Garcia-Castro et al., 2002; Lewis et al., 2004). Furthermore, FGF signaling from the underlying mesoderm has also been shown to be capable of independently inducing neural crest cell formation in frog embryos (Monsoro-Burq et al., 2003) such that WNT and FGF signaling may operate in parallel but independent pathways (Monsoro-Burq et al., 2005). Although the BMP, FGF, and Wnt signaling pathways have each been identified in species-specific contexts as key factors governing neural crest induction, the limited temporal separation between neural induction and neural crest cell formation in avian, frog, and fish embryos, and the reiterated use of the same signaling pathways, have contributed to conflicting results and difficulties in establishing the true pathways regulating neural crest cell formation.

Recently, it was provocatively proposed that neural crest cells in avian embryos are specified by *Pax7* during early gastrulation, which is much earlier than previously thought (Basch et al., 2006). Interestingly, *Pax* gene involvement in neural crest cell formation has also been observed in *Xenopus* (Maczkowiak et al., 2010), but this process does not appear to be conserved in mammals. Although *Pax3*- and *Pax7*-mutant mouse embryos exhibit craniofacial malformations, neural crest cell formation is not abrogated. Thus, although BMP, WNT, FGF, and Pax signaling have each been identified as key regulators of neural crest cell formation in diverse species, such as avians, fish, and amphibians, there is no conclusive evidence that supports an absolute role for these factors in mammalian neural crest cell induction (Crane and Trainor,

2006). Instead, mouse knockouts imply that each of these signaling pathways are more important for promoting neural crest cell survival and for specifying cell-type specification and differentiation (reviewed by Crane and Trainor, 2006). Therefore, the signals and switches governing mammalian neural crest cell formation remain to be identified.

5.2 Delamination of Neural Crest Cells

Initially, neural crest cells are integrated within the neural plate, where they are morphologically indistinguishable from other neuroepithelial cells. In response to inductive signals, neuroepithelial cells undergo an epithelial-to-mesenchymal transformation to form neural crest cells, which then delaminate from the neural plate and migrate extensively throughout the embryo (Fig. 1B and C), generating a remarkably diverse array of cell and tissue types, ranging from neurons and glia to bone and cartilage, among many others (Fig. 2A, B, and D).

The delamination of neural crest cells from the neural tube requires significant cytoarchitectural and cell adhesive changes and typically is recognized by the activity of members of the *Snail* transcription factor gene family. *Snail1*, for example, demarcates neural crest cells in mouse embryos (Sefton et al., 1998). Snail1 and Snail2 can directly repress the cell adhesion molecule, *E-cadherin*, by binding to its promoter, which is thought to facilitate delamination and cell migration (Cano et al., 2000; Bolos et al., 2003). However, in contrast to avians, fish, and amphibians, loss-of-function analyses of *Snail1* and *Snail2* either individually or in combination, do not inhibit neural crest cell induction and delamination in mice (Jiang et al., 1998; Murray and Gridley, 2006). To date, only mutations in *Zfhx1b*, which is also known as *Smad-interacting protein 1* (*SIP*1) or *Zeb2*, have been shown to affect neural crest cell formation and delamination in mammalian embryos (Van de Putte et al., 2003). *Zfhx1b*-knockout mice do not develop postotic vagal neural crest cells, and the delamination of more anterior cranial neural crest cells is perturbed. This is due to the persistent expression of E-cadherin throughout the epidermis and neural tube, as *Zfxh1b* is a direct repressor of E-cadherin. Hence, appropriate regulation of cell adhesion is critical for formation, EMT, and subsequent delamination and migration of mammalian neural crest cells.

5.3 Migration and Differentiation of Neural Crest Cells

Neural crest cells are born in a progressive rostrocaudal order along nearly the entire length of the neuraxis and, based on their axial level of origin, can be classified into at least four distinct axial groups: cranial, cardiac, vagal, and trunk, each of which exhibits specific migration pathways and differentiation capacities (Fig. 1C). The cranial neural crest demonstrates astonishing multipotentiality, giving rise to the majority of the bone and cartilage of the head and face, as well as to nerve ganglia, smooth muscle, connective tissue, and pigment cells. The remarkable capacity of neuroectoderm-derived neural crest cells to differentiate into both neuronal and mesenchymal derivatives has led to the neural crest being described as the fourth germ layer (Hall, 1999). An important feature that distinguishes the cranial neural crest from the trunk and other axial populations of neural crest cells is their ability to differentiate into mesenchymal tissues. The evolutionary significance of cranial neural crest cells has been well documented. Synonymous with the "new head" (Northcutt and Gans, 1983) and jaw formation, cranial neural crest cells carry species-specific programming information that is integral to craniofacial development, evolution, variation, and disease (Noden, 1983a; Trainor and Krumlauf, 2001; Schneider and Helms, 2003; Trainor, 2003a; Trainor et al., 2003; Noden and Trainor, 2005).

There are several mechanisms that could account for the ability of neural crest cells to differentiate into a diverse array of cell types and tissues. Neural crest cells could comprise a heterogeneous mixture of progenitor cells, with each progenitor giving rise to a distinct cell type within the body. This would require some degree of neural crest cell specification prior to their emigration from the neural tube and would be largely dependent on intrinsic signals regulating their development. Alternatively, neural crest cells could be multipotent, with their differentiation into distinct cell types being dependent on extrinsic signals emanating from the tissues they contact during their migration. The question of extrinsic versus intrinsic specification of neural crest cells and the appropriateness of their classification as true stem cells or progenitor cells has been addressed extensively elsewhere (Trainor and Krumlauf, 2001; Trainor, 2003b; Trainor et al., 2003; Crane and Trainor, 2006). Suffice it to say that neural crest cells comprise a heterogeneous migratory cell population and are governed by both intrinsic and extrinsic cues. The remarkable lineage potential, together with a limited capacity for self-renewal that persists even into adult life, demonstrates that neural crest cells exhibit some of the key hallmarks of stem and progenitor cells, even though neural crest cells are only generated transiently during embryogenesis. Much of the focus on neural crest cells today revolves around their stem cell–like characteristics and potential for use in regenerative medicine (Crane and Trainor, 2006; Achilleos and Trainor, 2012).

6 ECTODERM: PLACODES

6.1 Induction of Placode Formation

The vertebrate head contains numerous sense organs, including the nose, eyes, ears, and tongue, as well as the peripheral sensory nervous system that serves to relay the sensory information of touch, smell, taste, sound, and sight to the central nervous system as well as to provide autonomic control over the muscles of the body. The cranial sensory structures arise at least in part from ectodermal thickenings called *placodes* (von Kupffer, 1891; Webb and Node, 1993), discrete areas of thickened nonneural or surface ectoderm that form in characteristic positions in the head of vertebrate embryos and are comprised of specialized epithelial cells (Le Douarin et al., 1986).

Placodogenesis begins around gastrulation with subdivision of the nonneural ectoderm into preplacodal ectoderm and surface ectoderm. Ectoderm cells that are not incorporated into the neural plate or placodes give rise to the surface ectoderm or epidermis of the skin. The preplacodal ectoderm is located in the anterior of the embryo and is initially competent to form any of the cranial placodes. However, interactions with underlying tissues segregates the preplacodal ectoderm into discrete placodes or territories with distinct fates.

These include the adenohypophyseal placode that forms Rathke's pouch and eventually the adenohypophysis (the anterior lobe of the pituitary gland), which is of central importance to the hormonal control of body function and contains six types of endocrine secretory cells: corticotropes, melanotropes, gonadotropes, thyrotropes, lactotropes, and somatotropes (Couly and Le Douarin, 1985). The olfactory placode forms the olfactory and vomeronasal organs (Mendoza et al., 1982) and gives rise to mucus-producing cells, secretory support cells, and primary sensory cells that migrate into the forebrain to become gonatotropin-releasing secreting neurons; it is the only placode to generate glia such as Schwann cells (Couly and Le Douarin, 1985). The lens forms the lens vesicle, which generates the crystalline- accummulating cells of the lens (McAvoy, 1980a,b). The ophthalmic and trigeminal placodes combine to form the trigeminal, which gives rise to neuronal precursors and together with neural crest cells forms the sensory neurons of the trigeminal ganglia, which monitor somatosensory information (touch, temperature, pain) in the oral cavity and rostral part of the face (Noden, 1980a,b; D'Amico-Martel and Noden, 1983). The otic placode develops initially into the otic vesicle and then generates the inner ear and sensory neurons of the vestibulocochlear ganglion (Torres and Giráldez, 1998). The inner ear contains many different specialized epithelial cells, including endolymph-producing secretory cells, supporting cells, and mechanosensory hair cells (Muller and Littlewood-Evans, 2001). The epibranchial placodes which are aligned with the branchial arches between adjacent pouches give rise to neuronal precursors that form the sensory neurons of the distal ganglia of the facial (geniculate ganglion) glossopharyngeal (petrosal ganglion) and vagal nerves (nodose ganglia).

Several models have been proposed to describe induction of the preplacodal ectoderm, including the delay, gradient, neural plate border state, and binary competence models (Schlosser, 2006a). Nonetheless, induction of the preplacodal ectoderm is dependent on cooperation among the WNT, FGF, and BMP signaling pathways (Litsiou et al., 2005). Active FGF signaling can induce proneural gene expression (*Sox3* and *Erni*) in naive ectoderm of chick embryos (Streit and Stern, 1999a) and thus promote neural versus nonneural character in ectodermal cells. Later, FGF signaling from the head mesoderm (in chick) or neural plate (in *Xenopus*) induces preplacodal marker expression (Brugmann et al., 2004; Litsiou et al., 2005; Streit, 2007). FGF signals have also been shown to induce formation of the posterior placodal area (progenitors for otic and epibranchial placodes) (Ladher et al., 2010). Transient activation of BMP signaling is also required to establish preplacodal competence in nonneural ectoderm cells. Once competence is established, inhibition of BMP signaling along with active FGF signaling induces pleplacodal ectoderm development within this zone of competence (Kwon et al., 2010)

It has long been debated whether cranial placodes arise through subdivision of a common primordium or form as individual distinct thickenings in various positions of the head (Northcutt and Brandle, 1995; Baker and Bronner-Fraser, 2001; Begbie and Graham, 2001). Fate-mapping experiments in teleost, amphibian, and amniote embryos suggest that all placodes originate from preplacodal ectoderm, which lies between the neural ectoderm and surface ectoderm (epidermis) during neurulation and gastrulation. Consistent with this, transcription factors of the *Six* and *Eya* gene families are expressed initially throughout the preplacodal domain and continue to be active in some or all cranial placodes (Fig. 1D and E). In fact, *Six1* has been shown to promote generic placodal fate in early *Xenopus* embryos (Brugmann et al., 2004). Furthermore, when prospective olfactory or lens placode is replaced by prospective otic placodes, or vice versa, the donor ectoderm adopts the fate of its new location (Yntema, 1933). This suggests that the preplacodal ectoderm has a bias or plasticity for generic placode development. The individualization of different placodes from the preplacodal ectoderm involves subdivision, and inherent within this process are mechanisms to establish groups of cells with unique identities and keep them segregated from each other through the formation of stable boundaries. This type of compartmentalization and segregation may be analogous to the formation of rhombomeres in the hindbrain, which involves differential gene expression, differential cell adhesion, cell movement, and cellular plasticity (Trainor and Krumlauf, 2000). Although placodes are known to be distinguished by differential gene expression, nothing is known regarding differential cell adhesion, which could mediate compartmentalization of the preplacodal ectoderm, nor is it known whether molecular subdivision of the preplacodal ectoderm constitutes lineage-restricted compartments before individual placodes segregate.

6.2 Regulation of Cranial Placode Patterning

Little is known about the molecular control of cranial placode development; however, studies in chicken, frog, fish, and mouse embryos collectively suggest the involvement of a genetic hierarchy. *SIX1* and *Eya1/2* gene expression is currently recognized as the first true markers of the preplacodal ectoderm (Fig. 1D and E) (Streit, 2007) *Six1* and *Eya2* have also been shown to promote preplacodal-specific fate while suppressing neural and neural crest fates from the neural plate border cells (Brugmann et al., 2004; Christophorou et al., 2009). Mutations in human *SIX1* gene have been linked to branchio-oto renal (BOR) syndrome, which results in hearing loss along with malformations of neck and kidney. Mice heterozygous for *Eya1* phenocopy the BOR syndrome defects, while mice homozygous for *Eya1* or *Six1* show severe defects in inner ear formation (smaller otic vescicle, lack of vestibule–cochlear ganglion), and loss or reduction in the number of trigeminal and epibranchial ganglia neurons (Abdelhak et al., 1997; Ozaki et al., 2004).

SoxB1 (*Sox2* and *Sox3*) family genes seem to be important for the maintenance of placodal cell progenitors by preventing premature differentiation. Expression of *Pax2/5/8, Pax3/7*, and *Pax6* confers a more placode-specific identity; *Pax2/5/8* expression marks the otic placode, *Pax2/8* expression also marks the epibranchial placode (Bouchard et al., 2004; Ohyama et al., 2006) *Pax3/7* expression marks the trigeminal placode, and *Pax6* expression marks the olfactory and the lens placode (Baker and Bronner-Fraser, 2001; Schlosser, 2006b). *Pax2* and *Pax8* have crucial but redundant functions in otic placode development (Burton et al., 2004) *Pax2*-mutant mice exhibit severe defects in sensory organ formation and neurogenesis from the inner ear (Baker and Bronner-Fraser, 2000), while *Pax3* mutants display severe hypoplasticity in various cranial ganglia. Proneural genes *Ngn1* and *Ngn2* are involved in the formation of the neurogenic placodes and regulate the expression of downstream neuronal determination genes. *Ngn1* is extremely crucial for trigeminal ganglion formation, and *Ngn1* mutants fail to generate the proximal set of ganglia of cranial nerves VII, IX, and X (Ma et al., 1998, 2000) *Ngn2* mutants also display defects in formation of distal cranial sensory ganglia of cranial nerves VII, IX, and X (Fode et al., 1998).

6.3 Neural Crest and Placode Interactions

The peripheral nervous system (PNS) comprises all the neurons and glia of the body, except for those within the brain and spinal cord, and can be divided into somatic, autonomic, and enteric networks, depending on their specific functions. The PNS receives external stimuli, coordinates body movements, and is responsible for functions that are not under conscious control.

In the vertebrate head, the PNS is derived from ectodermal placodes and neural crest cells (Fig. 2A). Proper migration of the neural crest is essential to form a functional PNS, and tissue microenvironments play critical roles in dynamically regulating neural crest cell migration. Recent evidence demonstrates that cranial neural crest cells can organize placodally derived neurons (Schwarz et al., 2008), and a common feature of chick trigeminal and epibranchial ganglia is the expression of N-cadherin and Robo2 on placodal neurons and Slit1 on neural crest cells (Shiau and Bronner-Fraser, 2009). N-cadherin localizes to intercellular adherens junctions between placodal neurons during ganglion assembly, and depletion of N-cadherin causes loss of proper ganglion coalescence, similar to that observed after loss of Robo2, suggesting that the two pathways might intersect. Blocking or augmenting Slit-Robo signaling modulates N-cadherin protein expression on the placodal cell surface, concomitant with alteration in placodal adhesion. Coexpression of N-cadherin with dominant-negative Robo abrogates the Robo2 loss-of-function phenotype of dispersed ganglia, whereas loss of N-cadherin reverses the aberrant aggregation induced by increased Slit-Robo expression. Thus, N-cadherin acts in concert with Slit-Robo signaling in mediating the placodal cell adhesion required for proper gangliogenesis (Shiau et al., 2008; Shiau and Bronner-Fraser, 2009).

After completion of neural crest cell migration and integration with placode progenitors, the cranial ganglia normally extend axons toward the olfactory, ophthalmic, and distal branchial arch and cardiac tissues. However, when the neural crest cell number is decreased in avian embryos through hindbrain extirpation, axons become mispositioned (Begbie and Graham, 2001), further demonstrating the importance of proper integration between progenitor cells and tissues for normal PNS development.

7 MESODERM: MUSCLE

7.1 Mesoderm Formation and Patterning

Mesoderm is, by definition, the middle layer of the embryo and forms only in triploblastic organisms during the process of gastrulation. A key event in the establishment of the body plan during vertebrate embryogenesis is the regionalization of the mesoderm into axial, paraxial, and lateral compartments, together with their sequential allocation to the head, heart, and trunk along the anterior–posterior body axis (Tam and Beddington, 1987; Tam, 1989). The cranial mesoderm lies adjacent to the developing brain and stretches from the forebrain to the primitive ear. Unlike the paraxial mesoderm in the trunk, the cranial mesoderm lacks any overt signs of segmentation. The diversity of cell lineages that arise from head paraxial mesoderm has been well documented, through transplantation and labeling studies (Noden, 1983b; Evans and Noden, 2006). These studies identified progenitors for endothelial cells, smooth muscles, and a wide range of connective tissues, including cartilage, bone, and skeletal muscles within paraxial mesoderm.

The progenitors of the axial mesoderm ingress through the anterior segment of the primitive streak and extend along the embryonic midline by convergent extension to reach the entire length of the body axis. The resulting midline structure underlies the brain and spinal cord and is given different names according to its position along the anterior–posterior axis. The axial mesoderm that underlies the forebrain is the prechordal plate, that which associates with the rest of the brain is the anterior notochord, and the segment underneath the spinal cord is the notochord (Kinder et al., 2001).

Cranial paraxial mesoderm together with prechordal mesoderm give rise to approximately 60 distinct skeletal muscles in the head, which are used to facilitate food intake, move the eyeball, provide facial expressions, and aid in speech (Fig. 2E) (Wigmore and Evans, 2002). Traditionally, craniofacial skeletal muscles are cataloged as four distinct populations: extraocular, branchial, laryngoglossal, and axial, which includes epaxial and hypaxial muscle groups that elevate, depress, and rotate the head. Mammals are endowed additionally with an extensive set of superficial facial muscles that allow fine movements of lips, eyelids, and cheeks, and with specialized pharyngeal constrictors. Extraocular muscles (EoMs) move and maintain the rotational stability of the eye, with additional accessory ocular muscles involved in protecting the cornea. The basic pattern of six EOMs is shared among all vertebrate classes and has metabolic and fiber-type composition distinct from that of most branchial and trunk muscles (Wigmore and Evans, 2002). Transplantation (Wachtler et al., 1984) and retroviral mapping (Evans and Noden, 2006) studies show that prechordal plate cells contribute to formation of the dorsal, ventral, and medial rectus and ventral oblique muscles, all of which are innervated by the oculomotor (III) nerve. Whether prechordal plate cells are the exclusive source of myoblasts for these EOMs or share this fate with paraxial mesodermal populations remains unresolved.

Branchial arch muscles are those associated with jaw, hyoid, and caudal branchial skeletal structures and their homologs. Historically, these muscles were thought to have evolved from an iterative set of serially homologous gill muscles, which together with gill skeletal elements constituted the branchiomeric apparatus (Edgeworth, 1935). The muscles that elevate (move rostrally and dorsally) the larynx and root of the tongue in mammals are functionally grouped as the suprahyoid musculature. Unique to mammals are muscles associated with external ear movement and facial expression.

Tongue (glossal) and laryngeal structures are relatively recent evolutionary adaptations of the craniofacial musculoskeleton, appearing coincident with terrestrial amphibian species (Edgeworth, 1935). Laryngoglossal muscles function to lower (move caudally) the larynx and root of the tongue, and are innervated by both hypoglossal (XII) and cervical nerves.

It is well accepted that distinct developmental programs control skeletal muscle formation in the head and in the trunk (Mootoosamy and Dietrich, 2002). *Pax3* is not expressed in the head mesoderm, and muscle myopathies are known to be differentially linked to a specific cranial or trunk region (Emery, 2002). Head mesoderm myogenesis depends on the head environment, and signaling molecules that trigger somitic myogenesis suppress muscle formation in the head (Mootoosamy and Dietrich, 2002; Tzahor et al., 2003). Thus, the head mesoderm is a distinct mesodermal tissue. Similarly, within the head musculature, eye muscles differ from branchiomeric muscles, and there is evidence that branchiomeric muscle development varies among the branchial arches.

Head mesoderm patterning relies on interconnected molecular networks, and three discrete phases can be distinguished. The first is characterized by activation of *Pitx2* in the anterior and *Tbx1* in the posterior head mesoderm (Fig. 1G). This relies on the absence of retinoic acid (RA) in the anterior head mesoderm, and the reduction of RA plus initiation of FGF signaling in the posterior head mesoderm. In the second phase, the anterior pattern is refined by *Alx4* and *MyoR* activation in response to rising BMP and FGF levels. BMP also sets the anterior boundary of *Tbx1* expression. In the posterior domain, increasing FGF levels reinforce *Tbx1* expression and determine the posterior boundary of *Pitx2* and *Alx4* expression. In the third phase, FGF signals spread along the pharynx, driving the anterior extension of *Tbx1* and the posterior extension of *MyoR* expression in combination with further reduction of RA. This leads to the final pattern of combinatorial marker gene expression, with *Pitx2* labeling the precursors of the extraocular and mandibular arch musculature, *MyoR* and *Tbx1* labeling the precursors of all branchiomeric muscles (Fig. 1I), and all three markers labeling the region that also contributes to the outflow tract of the heart (Bothe and Dietrich, 2006; Bothe et al., 2011).

7.2 Neural Crest and Mesoderm Interactions

The ability of neural crest cells to regulate cranial muscle development has been known from neural crest extirpations which were shown to disrupt jaw muscle architecture in amphibian embryos (Olsson et al., 2001; Ericsson and Olsson, 2004; Ericsson et al., 2004). Consistent with this, the musculoskeletal anatomy of the second arch (i.e., hyoid) can be transformed into that of the first arch (i.e., mandibular) simply by exchanging premigratory first and second arch neural crest in avians (Noden, 1983a). Similarly, when *Hoxa2*, a gene normally expressed in neural crest mesenchyme and required for second arch identity, is expressed ectopically throughout the jaw primordia of either *Xenopus* or chick embryos, jaw muscle morphology is transformed homeotically (Grammatopoulos et al., 2000; Pasqualetti et al., 2000). Furthermore, zebrafish with defects in cranial neural crest development exhibit jaw muscle differentiation anomalies (Schilling et al., 1996). In addition, avian chimeras in which quail neural crest cells are transplanted into duck hosts result in ducks with muscles that resembled the shapes of those found in quail, even though these muscles were derived entirely from the duck host (Tokita and Schneider, 2009). *Tcf4* and *Scx* are dynamically expressed in jaw muscle connective tissues and precursor cells, and these genes are directly regulated by neural crest mesenchyme in a species-specific fashion. By executing autonomous molecular programs, neural crest–derived skeletal and muscular connective tissues convey species-specific patterning information to the jaw muscles. Thus, neural crest mesenchyme is the source of species-specific patterning that directs and integrates musculoskeletal development (Tokita and Schneider, 2009).

Evolutionary diversity in jaw muscle morphology can arise by transposition of attachment sites on skeletal elements, changes in muscle shape, increases or decreases in individual muscle size, and/or modifications in the number of muscles comprising a given complex. Neural crest cells mediate the first two processes, and in so doing, play a fundamental mechanistic role in establishing species-specific muscle morphology. The capacity of neural crest cells to orchestrate species-specific genetic programs, and as a consequence to implement muscle pattern across species via its connective tissue derivatives, provides a potent mechanism to explain how the musculoskeletal system remains structurally and functionally integrated during the course of vertebrate evolution (Fig. 2E) (Tokita and Schneider, 2009).

8 MESODERM: ENDOTHELIAL CELLS

8.1 Vasculogenesis and Angiogenesis

The vascular system is crucial for the normal health and development of the growing embryo and for adult tissue homeostasis, but it is particularly important for proper craniofacial morphogenesis (Fig. 2C). The vasculature has long been known to be critical for meeting the metabolic demands of the tissues by supporting gas exchange and supplying nutrients. Deficiency in either the timing or extent of formation of the vascular system has been proposed as an underlying etiologic mechanism in the pathogenesis of first-arch anomalies and cleft palate (McKenzie, 1958). The concept that primary embryonic vascular deficiency can result in congenital anomalies has received considerable experimental support. For example, interfering with the avian carotid arterial supply results in severe craniofacial anomalies, including anencephaly, anophthalmia, microphthalmia, maloccluded mandibles, and other beak deformities (Vogel and Mc, 1952).

Thus, proper spatiotemporal formation and remodeling of the vascular system is one of the most critical events to occur during embryogenesis, and classically, this process can be divided into distinct phases, the earliest of which is initiated during gastrulation and coincides with the onset of head induction and morphogenesis. At this time a subset of mesoderm cells differentiate into endothelial cells (Fig. 1H). These endothelial cells form clusters which subsequently become connected to each other to form a network of vessels. This initial vascular network is termed the *primary capillary plexus* (Risau and Flamme, 1995), and the process by which it occurs is called *vasculogenesis*.

As the embryo grows, the primary capillary plexus expands by forming additional branches while continually remodeling existing networks. This process is known as *angiogenesis*, and two distinct mechanisms have been proposed for the formation of additional vessel branches: sprouting and nonsprouting. The *sprouting process* requires endothelial cells that are already part of a continuous vessel to transform their shape and invade nearby tissue, thereby establishing an additional vascular channel from a preexisting channel. In contrast, in the *nonsprouting process*, cells or tissues surrounding an existing vessel intercept or invade the vessel, splitting one vascular channel into two. The branching that occurs during angiogenesis via sprouting and nonsprouting mechanisms is augmented by the incorporation of nonendothelial cells, which add structural and functional complexity to the primitive vasculature. For example, neural crest–derived smooth muscle cells infiltrate the vascular channels and provide rigid integrity and contractility to the maturing vascular network.

Thus, endothelial cells are integral to, and the primary cell type involved in, the initial phases of vascular development. Endothelial cells can be generated by two different processes, differentiation and expansion, and it is known that endothelial cells cultured in vitro will spontaneously form structures that resemble the primary plexus in vivo (Folkman and Haudenschild, 1980). Endothelial cells within the embryo are derived from mesoderm and begin to differentiate in the mouse between E7.5 and 8.0 (Fig. 1H). Vascular endothelial growth factor (VEGF) signaling appears to be a critical regulator of this process, since *Vegfr2* (*Kdr*)$^{-/-}$-mutant embryos exhibit complete agenesis of all endothelial and blood cells (Shalaby et al., 1995). Furthermore, *Vegfr2* (*Kdr*)$^{-/-}$ ES cells do not contribute to either endothelial or hematopoietic cells in chimeric mice (Shalaby et al., 1997). By contrast, *Vegfr1*$^{-/-}$-mutant embryos exhibit a drastically increased number of endothelial cells (Fong et al., 1999), which leads to perturbed patterning of the primary plexus. Thus, a precise number of endothelial cells appears to be essential for producing the network that constitutes the primary capillary plexus, and this process is driven largely by VEGFR signaling.

The lack of endothelial and hematopoietic cells in *Vegfr2* (*Kdr*)$^{-/-}$ embryos raised a question as to the existence of a precursor called the *hemangioblast*, with the potential to give rise to both endothelial and hematopoietic lineages, or whether, in fact, these lineages are derived from distinct mesodermal subpopulations. Support for the hemangioblast concept came initially from ES cell differentiation assays which indicated that the blast colony–forming cell (BL-CFC) is a progenitor with both vascular and hematopoietic potential (Kennedy et al., 1997). Embryo-derived hemangioblasts have now been definitively identified and first appear during the midstreak stage of gastrulation and peak in number during the neural plate stage (Huber et al., 2004). The hemangioblast constitutes a subpopulation of mesoderm cells that coexpress Vegfr2 (Kdr) and brachyury (T).

8.2 Vasculature and Muscle Interactions and Integration

Hemangioblasts move invasively within the mesoderm and neural tissues and vascularization of the central nervous system occurs by angiogenic sprouting of endothelial cells (Noden, 1990, 1991). Little attention has, however, focused on the formation of intramuscular vascular channels during craniofacial myogenesis, even though muscles are one of the most vascularized structures in the head. Both angiogenic and vasculogenic processes are involved in forming the intramuscular plexus, and interestingly, the capillary network associated with mature slow (oxidative) skeletal muscle is more tortuous and of greater density than that associated with fast (glycolytic) muscles (reviewed by Noden, 1989). Capillary density changes continuously in response to dynamic external stimuli. This raises the question of whether any particular or specific vascular pattern is present during the initiation of myogensis that precedes the mature pattern associated with distinct fiber types.

In avian embryos, as myogenic cells condense to form primary myotubes, these tissues become almost but not completely devoid of endothelial cells (Ruberte et al., 2003). Subsequently, a dense vascular plexus forms around the perimeter of all head muscles; this is only transient, however. Concomitant with disappearance of the plexus is an increase in intramuscular endothelial cells, endothelial cords, and local patent channels. Interestingly, embryonic blood vessels show similar densities and spatial organization within slow (oculorotatory), fast (mandibular depressor), and most mixed (jaw adductor complex) muscles. This indicates that, at least initially, intramuscular blood vessel development occurs independent of the myotube composition of avian head muscles, during embryonic development. The quantitative and qualitative differences seen in mature fast and slow muscles must arise at late development stages during secondary myotube formation. Whether these changes reflect selective growth or loss of embryonic capillaries is not known, nor have the factors that direct these changes been identified (Ruberte et al., 2003). Of note however, is the fact that VEGF protein is found abundantly in head muscles in the avian embryos. Moreover, intramuscular VEGF protein is localized in myotubes, not in endothelial cells. This suggests that VEGF may act as a paracrine factor secreted by myotubes to modulate remodeling of the endothelium. It will be interesting in the future to explore the spatiotemporal activity of VEGF signaling during muscle development to determine any correlation with the selective growth or loss of embryonic capillaries, because it is well known that other differentiating cells and organs signal to the endothelium to shape vascular architecture and network formation.

Conversely, recent evidence suggests that the endothelial cells themselves play a critical role in early embryonic patterning and organ differentiation by reciprocal signaling to nearby tissues. This type of crosstalk also takes place between the developing vascular and nervous systems.

8.3 Vasculature and Neural Interactions and Integration

The vascular and nervous systems are two precisely patterned networks that develop in close proximity to each other, and frequently the two cell types share similar migratory paths. Endothelial cells have been shown to provide neurotrophic factors that promote the recruitment, growth, and survival of neural precursors (Leventhal et al., 1999). Endothelial cells also maintain a microenvironment that allows for active neurogenesis, particularly throughout adult life (Palmer et al., 2000), demonstrating a role for endothelial signaling in neural development during both early and late development. Some of the signaling pathways used during vasculogenesis, such as VEGF, have recently been implicated in patterning the nervous system. Conversely, axon guidance molecules, which are widely expressed in multiple cell types, also play important roles in angiogenesis. However, it is not known how interdependent the neuronal and vascular networks are during their formation and establishment.

Neurovascular disease encompasses complex disorders of the brain, spinal cord, and blood vessels and is a major cause of embryonic lethality and adult disability. Hence, it is important to understand the mechanisms regulating proper development of the nervous and vascular networks, which are functionally interdependent. Arteries supply neurons with oxygenated blood, and nerves control vessel dilation and contraction. The nervous and vascular systems are both exquisitely branched, and additional similarities between neural and vascular networks are also evident at the cellular level. Within the nervous system, neurons explore their surroundings using *growth cones*. The vascular system equivalent, known as *tip cells*, are specialized cells located at the front of navigating blood vessels (Gerhardt et al., 2003). Blood vessels and nerves often run in parallel, and neuronal and vascular morphogenesis is also tightly interwoven at the molecular level. At least four major axon guidance molecule families (Eph/ephrin, neuropilin/semaphorins, Slit/Robo, and netrin), which are widely expressed in multiple cell types, have been implicated in angiogenesis. Conversely, some of the same signaling pathways as those used during vasculogenesis have recently been implicated in patterning the nervous system. For example, VEGF signaling is a well-known promoter of angiogenesis. Endothelial tip cells migrate toward higher concentrations of VEGF, and stalk cells proliferate in response to high VEGF concentrations (Gerhardt et al., 2003). VEGF signaling can also mediate cortical neuron proliferation in vitro and plays an important role in maintaining neural progenitor proliferation (Zhu et al., 2003).

Although much is known about the function of VEGF signaling during vascular development, the role of VEGF signaling in patterning and morphogenesis of the peripheral nervous system during early embryogenesis has not yet been thoroughly explored. Furthermore, it is not known whether endothelial cells and the vascular network influences neural crest development or whether VEGF signaling plays a critical role in patterning neural crest cell formation, migration, and/or differentiation.

However, in a recent screen for novel factors involved in neural crest cell induction, at least eight of the genes identified had previously been implicated in endothelial cell development (Gammill and Bronner-Fraser, 2002). This includes factors involved in VEGF production and signaling (e.g., *ORP150* or *neuropilin 2a1*) as well as proteins important for endothelial cell migration (such as *laminin* α5 and γ1). ORP150 is an ER chaperone whose function is required for VEGF secretion, and neuropilin 2 is an isoform-specific VEGF receptor. Thus, the screen identified factors that both produce and respond to VEGF. Moreover, VEGF is expressed in tissues that could affect neural crest development, including the headfolds, neural tube, and cephalic mesenchyme of E8.5 to 9.0 mouse embryos (Miquerol et al., 1999), and VEGF-mutant embryos exhibit poorly developed branchial arches (Ferrara et al., 1996). Collectively, this implies that endothelial cells and neural crest cells may employ similar developmental programs and be interdependent during early embryogenesis, which presages the integration, interactions, and interdependency of the neurovascular systems during embryogensis and adulthood.

Thus, it will be important to explore a role for endothelial cells and vasculature during neural crest cell formation, migration, and differentiation. It seems likely that endothelial cell networks will be essential for neural crest cell survival, but perhaps also for cell differentiation and, consequently, for proper establishment of sensory neuronal networks in the peripheral nervous system. It remains to be determined whether there is any interdependency of endothelial cells and neural crest cells during formation and patterning of mature neuronal networks, but one could imagine that progenitor cell interdependency combined with shared signaling cascades could facilitate functional integration between the neuronal and vascular systems, both of which are essential for organism survival.

9 ENDODERM: ORAL CAVITY

In contrast to mesoderm, a layer of endoderm cells called the *visceral endoderm* is already present prior to gastrulation. Descendants of the visceral endoderm contribute to the anterior and posterior segments of the embryonic gut (Kwon et al., 2008), however the ultimate fate of these cells in the digestive tract is not known, as the bulk of gut (definitive) endoderm cells are recruited from the embryonic ectoderm through the anterior end of the primitive streak during gastrulation. Cells destined for the upper digestive tract (the foregut) congregate to the anterior region of the endoderm layer (the anterior definitive endoderm, underlying the cranial and heart mesoderm and the prospective brain domains in the ectoderm (Fig. 1F) (Tam et al., 2007). Through a concerted movement, the endoderm forms the lining of the embryonic foregut and the associated organs during head morphogenesis (Tremblay and Zaret, 2005; Franklin et al., 2008). The contribution of the endoderm to head development is poorly understood compared to ectoderm and mesoderm. Nonetheless, formation of the three germ layer derivatives henceforth completes the building blocks of the head. Later events will continue to build on this scaffold until the fully differentiated head structures emerge. For example, neural crest cells differentiate into cartilage only in the presence of pharyngeal endoderm, whereas ectoderm is crucial for dermal bone differentiation (Le Douarin, 1982). Furthermore, the foregut endoderm of the avian neurula displays a regional activity essential in specifying the identity and orientation of the neural crest–derived bones forming the vertebrate facial skeleton (Couly et al., 2002). The nature of the signal(s) arising from the endoderm is so far unknown, but the signals are active at the early neurula stage, which is well before the endoderm contributes to formation of the pharyngeal pouches.

10 CONCLUSIONS AND PERSPECTIVES

Proper craniofacial development begins during gastrulation and requires the coordinated integration of each germ layer tissue (ectoderm, mesoderm, and endoderm) and its derivatives in concert with the precise regulation of cell proliferation, migration, and differentiation to generate a fully functioning head. The head must house and protect the brain as well as contain the majority of the sense organs, while the face is essential for individual identity and communication, as it conveys feelings, emotions, recognition, and sense of self.

In all higher vertebrates the facial prominences from which the head and face are derived share a common basic plan or blueprint. With respect to the pharyngeal arches, for example, the central core region of each arch is composed of cranial mesoderm, which will ultimately generate the branchiomeric musculature and vasculature (Noden, 1982, 1983b; Trainor et al., 1994). The mesodermal cores are enveloped by neural crest cells, which generate most of the bone, cartilage, and connective tissue in the head and face (Trainor and Tam, 1995b). The pharyngeal arches are then lined internally by the endoderm and externally by the ectoderm and are segregated by a reiterated series of pouches where the ectoderm and endoderm contact each other. A general axial registration exists between neural crest cells and mesoderm and ectoderm that persists during their migration and differentiation (Noden, 1991; Trainor and Tam, 1995b). These relationships and the tissue boundaries they create are maintained throughout development (Kontges and Lumsden, 1996). Moreover, the neural crest–derived connective tissue mesenchyme provides the cues necessary to direct the distribution and alignment of the mesoderm-derived myoblasts (Noden, 1983a). This congruence and axial registration include the cranial motor nerves and precursors of epipharyngeal placodes (D'Amico-Martel and Noden, 1983; Baker and Bronner-Fraser, 2001), which will innervate specific craniofacial muscles. These coordinated interactions are essential for generating a fully functioning jaw and indicate that the registration between various tissues in the head during early embryogenesis is critical for establishing the blueprint or foundations of vertebrate craniofacial development.

Defects in the formation, proliferation, migration, and differentiation of neural crest cells give rise to craniofacial malformations, which account for approximately one-third of all congenital birth defects. Depending on which phase of neural crest development is affected, vastly distinct craniofacial anomalies can arise. For example, Treacher–Collins syndrome, which is characterized by severe visceroskeletal hypoplasia, is caused by defects in neural crest cell formation and proliferation. In contrast, in craniosynostosis the sutural mesenchyme separating the calvarial bones ossifies prematurely, and this is considered to be a defect in neural crest cell differentiation but can also occur due to the disruption of neural crest mesoderm boundaries (Merrill et al., 2006). To develop therapeutic avenues for minimizing or preventing craniofacial anomalies, it is essential to understand the precise etiology and pathogenesis of individual malformation syndromes. This requires a thorough understanding of the normal developmental events that induce neural crest cells to form, maintain their survival, guide their migration, and influence their differentiation during embryogenesis.

Neural crest cell development is regulated by a combination of intrinsic cell autonomous signals acquired during their formation, balanced with extrinsic signals from tissues with which the neural crest cells interact during their migration and differentiation. Craniofacial anomalies are not, however, always the consequence of defects autonomous or intrinsic to the neural crest. Abnormal neural crest cell patterning can also arise secondarily as a consequence of cell nonautonomous or extrinsic defects in the mesoderm, ectoderm, and endoderm tissues with which the neural crest cells interact (Trainor and Krumlauf, 2001). In fact, the ectoderm, endoderm, and mesoderm tissues each play critical roles in regulating neural crest cell patterning, and reciprocal interactions between each of these tissues are absolutely critical for normal craniofacial development (Trainor and Krumlauf, 2000). Current analyses of craniofacial development must therefore be cognizant of the fact that not all craniofacial anomalies arise through defects intrinsic to the neural crest. Furthermore, it is vital to investigate the molecular and cellular nature of reciprocal interactions between all of these tissues during embryonic development to better understand the etiology and pathogenesis of congenital craniofacial malformations and to better design therapeutic avenues for prevention and repair, such as in the form of tissue engineering.

Acknowledgments

I am extremely grateful to Kimberly Inman, Kristin Melton, Margot Leroux-Berger, Naomi Butler, and Lisa Sandell for providing the figures.

REFERENCES

Abdelhak S, Kalatzis V, Heilig R, Compain S, Samson D, Vincent C, Weil D, Cruaud C, Sahly I, Leibovici M, et al. (1997). A human homologue of the *Drosophila* eyes absent gene underlies branchio-oto-renal (BOR) syndrome and identifies a novel gene family. *Nat Genet* 15(2):157–164.

Achilleos A, Trainor PA (2012). Neural crest stem cells: discovery, properties and potential for therapy. *Cell Res* 22(2):288–304.

Arkell RM, Tam PPL (2012). Signalling and transcriptional activity: initiating head development in mouse embryos: integrating. Open Biol 2(120030).

Bachiller D, Klingensmith J, Kemp C, Belo JA, Anderson RM, May SR, McMahon JA, McMahon AP, Harland RM, Rossant J, et al. (2000). The organizer factors Chordin and Noggin are required for mouse forebrain development. *Nature* 403(6770):658–661.

Baker CV, Bronner-Fraser M (2000). Establishing neuronal identity in vertebrate neurogenic placodes. *Development* 127(14):3045–3056.

Baker CV, Bronner-Fraser M (2001). Vertebrate cranial placodes: I. Embryonic induction. *Dev Biol* 232(1):1–61.

Basch ML, Bronner-Fraser M, Garcia-Castro MI (2006). Specification of the neural crest occurs during gastrulation and requires *Pax7*. *Nature* 441(7090):218–222.

Begbie J, Graham A (2001). Integration between the epibranchial placodes and the hindbrain. *Science* 294(5542):595–598.

Bel-Vialar S, Itasaki N, Krumlauf R (2002). Initiating Hox gene expression: in the early chick neural tube differential sensitivity to FGF and RA signaling subdivides the HoxB genes in two distinct groups. *Development* 129:5103–5115.

Bolos V, Peinado H, Perez-Moreno MA, Fraga MF, Esteller M, Cano A (2003). The transcription factor Slug represses E-cadherin expression and induces epithelial to mesenchymal transitions: a comparison with Snail and E47 repressors. *J Cell Sci* 116(3): 499–511.

Bothe I, Dietrich S (2006). The molecular setup of the avian head mesoderm and its implication for craniofacial myogenesis. *Dev Dyn* 235(10):2845–2860.

Bothe I, Tenin G, Oseni A, Dietrich S (2011). Dynamic control of head mesoderm patterning. *Development* 138(13):2807–2821.

Bouchard M, Souabni A, Busslinger M (2004). Tissue-specific expression of Cre recombinase from the *Pax8* locus. *Genesis* 38(3):105–109.

Brugmann SA, Pandur PD, Kenyon KL, Pignoni F, Moody SA (2004). *Six1* promotes a placodal fate within the lateral neurogenic ectoderm by functioning as both a transcriptional activator and repressor. *Development* 131(23):5871–5881.

Burton Q, Cole LK, Mulheisen M, Chang W, Wu DK (2004). The role of *Pax2* in mouse inner ear development. *Dev Biol* 272(1):161–175.

Cajal M, Lawson KA, Hill B, Moreau A, Rao J, Ross A, Collignon J, Camus A (2012). Clonal and molecular analysis of the prospective anterior neural boundary in the mouse embryo *Development* 139(2):423–436.

Cano A, Perez-Moreno MA, Rodrigo I, Locascio A, Blanco MJ, del Barrio MG, Portillo F, Nieto MA (2000). The transcription factor snail controls epithelial–mesenchymal transitions by repressing E-cadherin expression. *Nat Cell Biol* 2(2):76–83.

Christophorou NA, Bailey AP, Hanson S, Streit A (2009). Activation of *Six1* target genes is required for sensory placode formation. *Dev Biol* 336(2):327–336.

Couly GF, Le Douarin NM (1985). Mapping of the early neural primordium in quail-chick chimeras: I. Developmental relationships between placodes, facial ectoderm, and prosencephalon. *Dev Biol* 110(2):422–439.

Couly G, Creuzet S, Bennaceur S, Vincent C, Le Douarin NM (2002). Interactions between Hox-negative cephalic neural crest cells and the foregut endoderm in patterning the facial skeleton in the vertebrate head. *Development* 129(4):1061–1073.

Crane JF, Trainor PA (2006). Neural crest stem and progenitor cells. *Annu Rev Cell Dev Biol* 22:267–286.

D'Amico-Martel A, Noden DM (1983). Contributions of placodal and neural crest cells to avian cranial peripheral ganglia. *Am J Anat* 166(4):445–468.

Edgeworth F (1935). *The Cranial Muscles of Vertebrates*, Cambridge University Press, Cambridge, UK.

Emery AE (2002). The muscular dystrophies. *Lancet* 359(9307):687–695.

Ericsson R, Olsson L (2004). Patterns of spatial and temporal visceral arch muscle development in the Mexican axolotl (*Ambystoma mexicanum*). *J Morphol* 261(2):131–140.

Ericsson R, Cerny R, Falck P, Olsson L (2004). Role of cranial neural crest cells in visceral arch muscle positioning and morphogenesis in the Mexican axolotl, *Ambystoma mexicanum*. *Dev Dyn* 231(2):237–247.

Evans DJ, Noden DM (2006). Spatial relations between avian craniofacial neural crest and paraxial mesoderm cells. *Dev Dyn* 235(5):1310–1325.

Fainsod A, Steinbeisser H, De Robertis EM (1994). On the function of BMP-4 in patterning the marginal zone of the *Xenopus* embryo. *EMBO J* 13(21):5015–5025.

Ferrara N, Carver-Moore K, Chen H, Dowd M, Lu L, O'Shea KS, Powell-Braxton L, Hillan KJ, Moore MW (1996). Heterozygous embryonic lethality induced by targeted inactivation of the VEGF gene. *Nature* 380(6573):439–442.

Fode C, Gradwohl G, Morin X, Dierich A, LeMeur M, Goridis C, Guillemot F (1998). The bHLH protein Neurogenin 2 is a determination factor for epibranchial placode-derived sensory neurons. *Neuron* 20(3):483–494.

Folkman J, Haudenschild C (1980). Angiogenesis in vitro. *Nature* 288(5791):551–556.

Fong GH, Zhang L, Bryce DM, Peng J (1999). Increased hemangioblast commitment, not vascular disorganization, is the primary defect in *flt-1* knock-out mice. *Development* 126(13):3015–3025.

Franklin V, Khoo PL, Bildsoe H, Wong N, Lewis S, Tam PP (2008). Regionalisation of the endoderm progenitors and morphogenesis of the gut portals of the mouse embryo. *Mech Dev* 125(7):587–600.

Gammill LS, Bronner-Fraser M (2002). Genomic analysis of neural crest induction. *Development* 129(24):5731–5741.

Garcia-Castro MI, Marcelle C, Bronner-Fraser M (2002). Ectodermal Wnt function as a neural crest inducer. *Science* 297(5582):848–851.

Gerhardt H, Golding M, Fruttiger M, Ruhrberg C, Lundkvist A, Abramsson A, Jeltsch M, Mitchell C, Alitalo K, Shima D, et al. (2003). VEGF guides angiogenic sprouting utilizing endothelial tip cell filopodia. *J Cell Biol* 161(6):1163–1177.

Gorlin RJ, Cohen MM, Levin LS (1990). *Syndromes of the Head and Neck*, Oxford University Press, Oxford, UK.

Grammatopoulos GA, Bell E, Toole L, Lumsden A, Tucker AS (2000). Homeotic transformation of branchial arch identity after *Hoxa2* overexpression. *Development* 127(24):5355–5365.

Hall BK (1999). *The Neural Crest in Development and Evolution*, Springer-Verlag, New York.

Harland R (2000). Neural induction. *Curr Opin Genet Dev* 10(4):357–362.

Hemmati-Brivanlou A, Kelly OG, Melton DA (1994). Follistatin, an antagonist of activin, is expressed in the Spemann organizer and displays direct neuralizing activity. *Cell* 77(2):283–295.

Huber TL, Kouskoff V, Fehling HJ, Palis J, Keller G (2004). Haemangioblast commitment is initiated in the primitive streak of the mouse embryo. *Nature* 432(7017):625–630.

Jiang R, Lan Y, Norton CR, Sundberg JP, Gridley T (1998). The Slug gene is not essential for mesoderm or neural crest development in mice. *Dev Biol* 198(2):277–285.

Kennedy M, Firpo M, Choi K, Wall C, Robertson S, Kabrun N, Keller G (1997). A common precursor for primitive erythropoiesis and definitive haematopoiesis. *Nature* 386(6624):488–493.

Kinder SJ, Tsang TE, Wakamiya M, Sasaki H, Behringer RR, Nagy A, Tam PP (2001). The organizer of the mouse gastrula is composed of a dynamic population of progenitor cells for the axial mesoderm. *Development.* 128(18):3623–3634.

Kintner CR, Dodd J (1991). Hensen's node induces neural tissue in *Xenopus* ectoderm: implications for the action of the organizer in neural induction. *Development* 113(4):1495–1505.

Kontges G, Lumsden A (1996). Rhombencephalic neural crest segmentation is preserved throughout craniofacial ontogeny. *Development* 122(10):3229–3242.

Kwon GS, Viotti M, Hadjantonakis AK (2008). The endoderm of the mouse embryo arises by dynamic widespread intercalation of embryonic and extraembryonic lineages. *Dev Cell* 15(4):509–520.

Kwon HJ, Bhat N, Sweet EM, Cornell RA, Riley BB (2010). Identification of early requirements for preplacodal ectoderm and sensory organ development. *PLoS Genet* 6(9).

Ladher RK, O'Neill, P, Begbie J (2010). From shared lineage to distinct functions: the development of the inner ear and epibranchial placodes. *Development* 137(11):1777–1785.

Lamb TM, Knecht AK, Smith WC, Stachel SE, Economides AN, Stahl N, Yancopolous GD, Harland RM (1993). Neural induction by secreted polypeptide noggin. *Science* 262:713–718.

Le Douarin NM (1982) *The Neural Crest*, Cambridge University Press, Cambridge, UK.

Le Douarin N, Fontaine-Perus J, Couly G (1986). Cephalic ectodermal placodes and neurogenesis. *Trends Neurosci* 9:175–180.

Leventhal C, Rafii S, Rafii D, Shahar A, Goldman SA (1999). Endothelial trophic support of neuronal production and recruitment from the adult mammalian subependyma. *Mol Cell Neurosci* 13(6):450–464.

Lewis JL, Bonner J, Modrell M, Ragland JW, Moon RT, Dorsky RI, Raible DW (2004). Reiterated Wnt signaling during zebrafish neural crest development. *Development* 131(6):1299–1308.

Lewis SL, Khoo PL, De Young RA, Steiner K, Wilcock C, Mukhopadhyay M, Westphal H, Jamieson RV, Robb L, Tam PP (2008). *Dkk1* and *Wnt3* interact to control head morphogenesis in the mouse. *Development* 135(10):1791–1801.

Litsiou A, Hanson S, Streit A (2005). A balance of FGF, BMP and WNT signalling positions the future placode territory in the head. *Development* 132(18):4051–4062.

Ma Q, Chen Z, del Barco Barrantes I, de la Pompa JL, Anderson DJ (1998). *Neurogenin1* is essential for the determination of neuronal precursors for proximal cranial sensory ganglia. *Neuron* 20(3):469–482.

Ma Q, Anderson DJ, Fritzsch B (2000). *Neurogenin 1* null mutant ears develop fewer, morphologically normal hair cells in smaller sensory epithelia devoid of innervation. *J Assoc Res Otolaryngol* 1(2):129–143.

MacDonald BT, Adamska M, Meisler MH (2004). Hypomorphic expression of *Dkk1* in the doubleridge mouse: dose dependence and compensatory interactions with *Lrp6*. *Development* 131(11):2543–2552.

Maczkowiak F, Mateos S, Wang E, Roche D, Harland R, Monsoro-Burq AH (2010). The Pax3 and Pax7 paralogs cooperate in neural and neural crest patterning using distinct molecular mechanisms, in *Xenopus laevis* embryos. *Dev Biol* 340(2):381–396.

Marchant L, Linker C, Ruiz P, Guerrero N, Mayor R (1998). The inductive properties of mesoderm suggest that the neural crest cells are specified by a BMP gradient. *Dev Biol* 198(2):319–329.

Matzuk MM, Lu N, Vogel H, Sellheyer K, Roop DR, Bradley A (1995). Multiple defects and perinatal death in mice deficient in follistatin. *Nature* 374(6520):360–363.

Mayor R, Morgan R, Sargent MG (1995). Induction of the prospective neural crest of *Xenopus*. *Development* 121(3):767–777.

McKenzie J (1958). The first arch syndrome. *Arch Dis Child* 33(171):477–486.

McAvoy JW (1980a). Beta- and gamma-crystallin synthesis in rat lens epithelium explanted with neural retinal. *Differentiation* 17(2):85–91.

McAvoy JW (1980b). Induction of the eye lens. *Differentiation* 17(3):137–149.

McMahon JA, Takada S, Zimmerman LB, Fan CM, Harland RM, McMahon AP (1998). Noggin-mediated antagonism of BMP signaling is required for growth and patterning of the neural tube and somite. *Genes Dev* 12(10):1438–1452.

Mendoza AS, Breipohl W, Miragall F (1982). Cell migration from the chick olfactory placode: a light and electron microscopic study. *J Embryol Exp Morphol* 69:47–59.

Merrill AE, Bochukova EG, Brugger SM, Ishii M, Pilz DT, Wall SA, Lyons KM, Wilkie AO, Maxson RE, Jr (2006). Cell mixing at a neural crest–mesoderm boundary and deficient ephrin-Eph signaling in the pathogenesis of craniosynostosis. *Hum Mol Genet* 15(8):1319–1328.

Miquerol L, Gertsenstein M, Harpal K, Rossant J, Nagy A (1999). Multiple developmental roles of VEGF suggested by a *LacZ*-tagged allele. *Dev Biol* 212(2):307–322.

Monsoro-Burq AH, Fletcher RB, Harland RM (2003). Neural crest induction by paraxial mesoderm in *Xenopus* embryos requires FGF signals. *Development* 130(14):3111–3124.

Monsoro-Burq AH, Wang E, Harland R (2005). *Msx1* and *Pax3* cooperate to mediate FGF8 and WNT signals during *Xenopus* neural crest induction. *Dev Cell* 8:167–178.

Mootoosamy RC, Dietrich S (2002). Distinct regulatory cascades for head and trunk myogenesis. *Development* 129(3):573–583.

Morgan R, Sargent MG (1997). The role in neural patterning of translation initiation factor eIF4AII; induction of neural fold genes. *Development* 124:2751–2760.

Moury JD, Jacobson AG (1990). The origins of neural crest cells in the axolotl. *Dev Biol* 141(2):243–253.

Mukhopadhyay M, Shtrom S, Rodriguez-Esteban C, Chen L, Tsukui T, Gomer L, Dorward DW, Glinka A, Grinberg A, Huang SP, et al. (2001). *Dickkopf1* is required for embryonic head induction and limb morphogenesis in the mouse. *Dev Cell* 1(3):423–434.

Muller U, Littlewood-Evans A (2001). Mechanisms that regulate mechanosensory hair cell differentiation. *Trends Cell Biol* 11(8):334–342.

Murray SA, Gridley T (2006). Snail family genes are required for left–right asymmetry determination, but not neural crest formation, in mice. *Proc Natl Acad Sci USA* 103(27):10300–10304.

Noden DM (1980a). Somatotopic and functional organization of the avian trigeminal ganglion: an HRP analysis in the hatchling chick. *J Comp Neurol* 190(3):405–428.

Noden DM (1980b). Somatotopic organization of the embryonic chick trigeminal ganglion. *J Comp Neurol* 190(3):429–444.

Noden DM (1982). Patterns and organization of craniofacial skeletogenic and myogenic mesenchyme: a perspective. *Prog Clin Biol Res* 101:167–203.

Noden D (1983a). The role of the neural crest in patterning of avian cranial skeletal, connective, and muscle tissues. *Dev Biol* 96:144–165.

Noden DM (1983b). The embryonic origins of avian cephalic and cervical muscles and associated connective tissues. *Am J Anat* 168(3):257–276.

Noden DM (1989). Embryonic origins and assembly of blood vessels. *Am Rev Respir Dis* 140(4):1097–1103.

Noden DM (1990). Origins and assembly of avian embryonic blood vessels. *Ann NY Acad Sci* 588:236–249.

Noden DM (1991). Cell movements and control of patterned tissue assembly during craniofacial development. *J Craniofac Genet Dev Biol* 11(4):192–213.

Noden DM, Trainor PA (2005). Relations and interactions between cranial mesoderm and neural crest populations. *J Anat* 207(5):575–601.

Nordstrom U, Jessell TM, Edlund T (2002). Progressive induction of caudal neural character by graded Wnt signaling. *Nat Neurosci* 5(6):525–532.

Northcutt RG, Brandle K (1995). Development of branchiomeric and lateral line nerves in the *axolotl*. *J Comp Neurol* 355(3):427–454.

Northcutt RG, Gans C (1983). The genesis of neural crest and epidermal placodes: a reinterpretation of vertebrate origins. *Q Rev Biol* 58(1):1–28.

Ohyama T, Mohamed OA, Taketo MM, Dufort D, Groves AK (2006). Wnt signals mediate a fate decision between otic placode and epidermis. *Development* 133(5):865–875.

Olsson L, Falck P, Lopez K, Cobb J, Hanken J (2001). Cranial neural crest cells contribute to connective tissue in cranial muscles in the anuran amphibian, *Bombina orientalis*. *Dev Biol* 237(2):354–367.

Ozaki H, Nakamura K, Funahashi J, Ikeda K, Yamada G, Tokano H, Okamura HO, Kitamura K, Muto S, Kotaki H, et al. (2004). *Six1* controls patterning of the mouse otic vesicle. *Development* 131(3):551–562.

Palmer TD, Willhoite AR, Gage FH (2000). Vascular niche for adult hippocampal neurogenesis. *J Comp Neurol* 425(4):479–494.

Pasqualetti M, Ori M, Nardi I, Rijli FM (2000). Ectopic *Hoxa2* induction after neural crest migration results in homeosis of jaw elements in *Xenopus*. *Development* 127(24):5367–5378.

Risau W, Flamme I (1995). Vasculogenesis. *Annu Rev Cell Dev Biol* 11:73–91.

Rollhauser-ter Horst J (1977). Artificial neural induction in amphibia. I. Sandwich explants. *Anat Embryol (Berl)* 151:309–316.

Ruberte J, Carretero A, Navarro M, Marcucio RS, Noden D (2003). Morphogenesis of blood vessels in the head muscles of avian embryo: spatial, temporal, and VEGF expression analyses. *Dev Dyn* 227(4):470–483.

Sasai Y, De Robertis EM (1997). Ectodermal patterning in vertebrate embryos. *Dev Biol* 182(1):5–20.

Sasai Y, Lu B, Steinbeisser H, Geissert D, Gont LK, De Robertis EM (1994). *Xenopus* chordin: a novel dorsalizing factor activated by organizer-specific homeobox genes. *Cell* 79(5):779–790.

Schilling TF, Piotrowski T, Grandel H, Brand M, Heisenberg CP, Jiang YJ, Beuchle D, Hammerschmidt M, Kane DA, Mullins MC, et al. (1996). Jaw and branchial arch mutants in zebrafish I: Branchial arches. *Development* 123:329–344.

Schlosser G (2006a). Induction and specification of cranial placodes. *Dev Biol* 294(2):303–351.

Schlosser G (2006b). Induction and specification of cranial placodes. *Dev Biol* 294(2):303–351.

Schneider RA, Helms JA (2003). The cellular and molecular origins of beak morphology. *Science* 299(5606):565–568.

Schwarz Q, Vieira JM, Howard B, Eickholt BJ, Ruhrberg C (2008). *Neuropilin 1* and *2* control cranial gangliogenesis and axon guidance through neural crest cells. *Development* 135(9):1605–1613.

Sefton M, Sanchez S, Nieto MA (1998). Conserved and divergent roles for members of the Snail family of transcription factors in the chick and mouse embryo. *Development* 125(16):3111–3121.

Selleck MA, Bronner-Fraser M (1995). Origins of the avian neural crest: the role of neural plate–epidermal interactions. *Development* 121(2):525–538.

Shalaby F, Rossant J, Yamaguchi TP, Gertsenstein M, Wu XF, Breitman ML, Schuh AC (1995). Failure of blood-island formation and vasculogenesis in *Flk-1*-deficient mice. *Nature* 376(6535):62–66.

Shalaby F, Ho J, Stanford WL, Fischer KD, Schuh AC, Schwartz L, Bernstein A, Rossant J (1997). A requirement for *Flk1* in primitive and definitive hematopoiesis and vasculogenesis. *Cell* 89(6):981–990.

Shiau CE, Bronner-Fraser M (2009). N-cadherin acts in concert with Slit1-Robo2 signaling in regulating aggregation of placode-derived cranial sensory neurons. *Development* 136(24):4155–4164.

Shiau CE, Lwigale PY, Das RM, Wilson SA, Bronner-Fraser M (2008). Robo2-Slit1 dependent cell-cell interactions mediate assembly of the trigeminal ganglion. *Nat Neurosci* 11(3):269–276.

Spemann H, Mangold H (1924). Über Induktion von Embryonalanlagen durch Implantation artfremder Organisatoren. *Arch Mikrosk Anat Entw Mechan Wilh Roux Arch Entw Mech Organ* 100:599–638.

Streit A (2007). The preplacodal region: an ectodermal domain with multipotential progenitors that contribute to sense organs and cranial sensory ganglia. *Int J Dev Biol* 51(6–7):447–461.

Streit A, Stern CD (1999a). Establishment and maintenance of the border of the neural plate in the chick: involvement of FGF and BMP activity. *Mech Dev* 82(1–2):51–66.

Streit A, Stern CD (1999b). Neural induction. A bird's eye view. *Trends Genet* 15(1):20–24.

Streit A, Lee KJ, Woo I, Roberts C, Jessell TM, Stern CD (1998). Chordin regulates primitive streak development and the stability of induced neural cells, but is not sufficient for neural induction in the chick embryo. *Development* 125(3):507–519.

Streit A, Berliner AJ, Papanayotou C, Sirulnik A, Stern CD (2000). Initiation of neural induction by FGF signalling before gastrulation. *Nature* 406(6791):74–78.

Tam PPL (1989). Regionalization of the mouse embryonic ectoderm: allocation of prospective ectodermal tissues during gastrulation. *Development* 107:55–67.

Tam PPL, Beddington RSP (1987). The formation of mesodermal tissues in the mouse embryo during gastrulation and early organogenesis. *Development* 99:109–126.

Tam PP, Khoo PL, Lewis SL, Bildsoe H, Wong N, Tsang TE, Gad JM, Robb L (2007). Sequential allocation and global pattern of movement of the definitive endoderm in the mouse embryo during gastrulation. *Development* 134(2):251–260.

Tokita M, Schneider RA (2009). Developmental origins of species-specific muscle pattern. *Dev Biol* 331(2):311–325.

Torres M, Giráldez F (1998). The development of the vertebrate inner ear. *Mech Dev* 71:5–21.

Trainor P (2003a). Development: the bills of qucks and duails. *Science* 299(5606):523–524.

Trainor PA (2003b). Making headway: the roles of Hox genes and neural crest cells in craniofacial development. *Sci World J* 3:240–264.

Trainor P, Krumlauf R (2000). Patterning the cranial neural crest: Hindbrain segmentation and *Hox* gene plasticity. *Nat Rev Neurosci* 1:116–124.

Trainor PA, Krumlauf R (2001). Hox genes, neural crest cells and branchial arch patterning. *Curr Opin Cell Biol* 13(6):698–705.

Trainor PA, Tam PP (1995a). Cranial paraxial mesoderm and neural crest cells of the mouse embryo: co-distribution in the craniofacial mesenchyme but distinct segregation in branchial arches. *Development* 121(8):2569–2582.

Trainor PA, Tam PPL (1995b). Cranial paraxial mesoderm and neural crest of the mouse embryo-codistribution in the craniofacial mesenchyme but distinct segregation in the branchial arches. *Development* 121(8):2569–2582.

Trainor PA, Tan SS, Tam PP (1994). Cranial paraxial mesoderm: regionalisation of cell fate and impact on craniofacial development in mouse embryos. *Development* 120(9):2397–2408.

Trainor PA, Melton KR, Manzanares M (2003). Origins and plasticity of neural crest cells and their roles in jaw and craniofacial evolution. *Int J Dev Biol* 47(7–8):541–553.

Tremblay KD, Zaret KS (2005). Distinct populations of endoderm cells converge to generate the embryonic liver bud and ventral foregut tissues. *Developmental Biology* 280(1):87–99.

Tzahor E, Kempf H, Mootoosamy RC, Poon AC, Abzhanov A, Tabin CJ, Dietrich S, Lassar AB (2003). Antagonists of Wnt and BMP signaling promote the formation of vertebrate head muscle. *Genes Dev* 17(24):3087–3099.

Van de Putte T, Maruhashi M, Francis A, Nelles L, Kondoh H, Huylebroeck D, Higashi Y (2003). Mice lacking *Zfhx1b*, the gene that codes for Smad-interacting protein–1, reveal a role for multiple neural crest cell defects in the etiology of Hirschsprung disease–mental retardation syndrome. *Am J Hum Genet* 72(2):465–470.

Vogel FS, Mc CJ (1952). Anomalies of major cerebral arteries associated with congenital malformations of the brain, with special reference to the pathogenesis of anencephaly. *Am J Pathol* 28(4):701–723.

von Kupffer C (1891). The development of cranial nerves of vertebrates. *J Comp Neurol* 1:246–264.

Wachtler F, Jacob HJ, Jacob M, Christ B (1984). The extrinsic ocular muscles in birds are derived from the prechordal plate. *Naturwissenschaften* 71(7):379–380.

Webb JF, Node DM (1993). Ectodermal placodes: contributions to the development of the vertebrate head. *Am Zool* 33:434–447.

Wigmore PM, Evans DJ (2002). Molecular and cellular mechanisms involved in the generation of fiber diversity during myogenesis. *Int Rev Cytol* 216:175–232.

Wilson PA, Hemmati-Brivanlou A (1995). Induction of epidermis and inhibition of neural fate by *Bmp-4*. *Nature* 376(6538):331–333.

Wilson PA, Hemmati-Brivanlou A (1997). Vertebrate neural induction: inducers, inhibitors, and a new synthesis. *Neuron* 18(5):699–710.

Wilson SI, Graziano E, Harland R, Jessell TM, Edlund T (2000). An early requirement for FGF signalling in the acquisition of neural cell fate in the chick embryo. *Curre Biol* 10(8):421–429.

Wilson S, Rydstrom A, Trimborn T, Willert K, Nusse R, Jessell TM, Edlund T (2001). The status of Wnt signalling regulates neural and epidermal fates in the chick embryo. *Nature* 411(6835):325–330.

Yntema CL (1933). Experiments on determination of the ear ectoderm in the embryo of *Amblystoma punctatum*. *J Exp Zool* 65:317–357.

Zhu Y, Jin K, Mao XO, Greenberg DA (2003). Vascular endothelial growth factor promotes proliferation of cortical neuron precursors by regulating E2F expression. *FASEB J* 17(2):186–193.

2

CRANIAL NEURAL CREST CELLS IN CRANIOFACIAL TISSUES AND ORGANS

CAROLINA PARADA AND YANG CHAI
Center for Craniofacial Molecular Biology, Herman Ostrow School of Dentistry, University of Southern California, Los Angeles, California

1 INTRODUCTION

Craniofacial development involves the fine molecular regulation of interactions among diverse cells and tissues. An evolutionary feature that defines the vertebrate craniofacial region is the emergence of neural crest cells (NCCs), which interact with endoderm, ectoderm, and mesoderm during development of the face and skull. The vertebrate neural crest is a multipotent and migratory cell population that gives rise to a wide array of differentiated cell types (Bronner-Fraser, 1993).

During craniofacial development, cranial neural crest cells (CNCCs) migrate ventrolaterally and establish contact with the pharyngeal ectoderm and endoderm. Subsequently, the CNCC-derived mesenchyme proliferates and forms a series of discrete swellings known as *branchial arches* (BAs) (Fig. 1A). CNCCs from rhombomeres 1 to 3 in the forming posterior midbrain and anterior hindbrain migrate into the first BA (Bronner-Fraser, 1993; Chai et al., 2000). Once in the BAs, a Hox-negative distal-less homeobox (Dlx) code provides the CNCCs with patterning information and intraarch polarity along the dorsoventral (DV)/proximodistal axis. This polarity is essential for establishment of the maxillary and mandibular prominences from the first BA and is established through the differential expression of Dlx genes. Thus, the subdivision of the first BA is achieved primarily with two Dlx combinations: *Dlx1/2* for the maxillary and *Dlx1/2/5/6* for the mandibular process (Depew et al., 2005; Jeong et al., 2008; Minoux and Rijli, 2010). The maxillary and mandibular prominences, together with

Stem Cells in Craniofacial Development and Regeneration, First Edition.
Edited by George T.-J. Huang, Irma Thesleff.

the frontonasal prominence, constitute the primitive face and surround the primordial mouth, also known as the *stomodeum*. The frontonasal prominence contributes significantly to formation of the nose and upper lip. The maxillary process gives rise to part of the upper lip, the maxillary bone, and the secondary palate, whereas the mandibular prominence forms the mandible and part of the tongue (Moore and Persaud, 2008). Maxillary and mandibular CNCCs contribute significantly to the development of upper and lower teeth, respectively. Here we review the current concepts regarding the contribution of CNCCs in palate, tooth, and tongue development.

2 PALATE DEVELOPMENT

The mammalian palate forms from two primordia: the primary and secondary palates. The primary palate, which shares its embryological origin with the lip, represents only a small part of the adult hard palate. The development of the secondary palate begins with the appearance of palatal primordia from the maxillary process (Ferguson, 1988). The palatal shelves grow vertically along the two sides of the tongue. Once the mandible starts growing in length and the tongue descends, the palatal shelves reorient to acquire a horizontal position. This movement is also known as *palatal shelf elevation*. The bilateral palatal shelves grow toward each other and establish contact at the midline, where a midline epithelial seam (MES) is formed (Gato et al., 2002). For the mesenchyme to become continuous, the MES must degenerate, and it has been proposed to achieve this process through programmed cell death (Cuervo and Covarrubias, 2004) and migration to the oral or nasal side of the palate (Jin and Ding, 2006) (Fig. 1B). Next, ossification occurs in the anterior two-thirds of the palate to form the hard palate, whereas the posterior third develops into the soft palate without ossification. The complexity of palatogenesis and its molecular regulation is evident by the frequent incidence of palate malformations in humans.

2.1 Patterning of the CNCC-Derived Palatal Mesenchyme Along the Anterior–Posterior Axis

The palatal shelves are composed of CNCC-derived mesenchyme covered by pharyngeal ectoderm-derived epithelium (Chai et al., 2000). Using the *Wnt1-Cre; R26R* model, previous studies have shown that more than 90% of palatal mesenchymal cells are derived from CNCCs, whereas the remaining cells originate from the mesoderm (Chai et al., 2000; Ito et al., 2003). A number of recent studies have demonstrated that the CNCC-derived palatal mesenchyme is not homogeneous. This finding has been a fundamental advance in our understanding of the molecular basis of normal and abnormal palate development. Here we review the most significant studies related to CNCC patterning during palate development.

The CNCC composing the developing palate are tightly patterned in the anterior–posterior (AP) and oronasal (ON) axes (Fig. 1C and D and FaceBase.org). The AP pattern is defined by the differential expression of *Barx1, Efnb1, Fgf10, Fgfr2, Meox2, Mn1, Msx1, Pax9, Shox2*, and *Tbx22* (Fig. 1C). Before reorientation, *Efnb1, Msx1*, and *Shox2* are expressed in the anterior region of the palate. Deficiency of *Efnb1, Msx1*, or *Shox2* results in clefts of the secondary palate with different phenotypic characteristics. In *Efnb1*- and *Msx1*-null mice the cleft palate is complete,

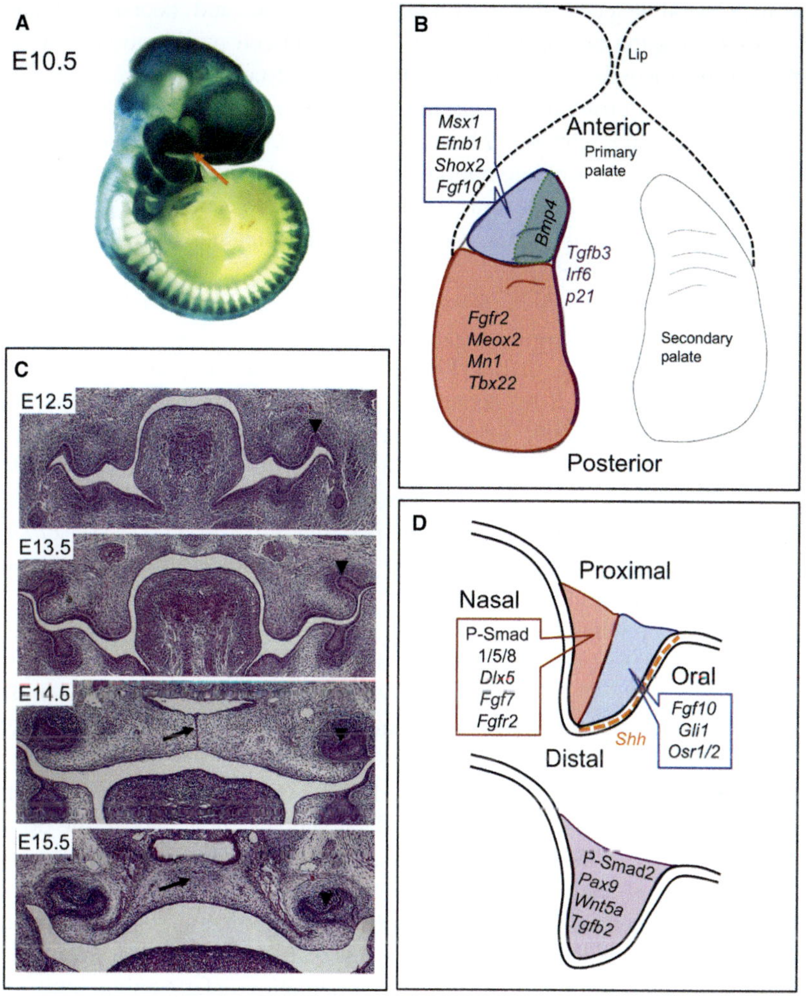

FIGURE 1 CNCC contribution to palate development. (A) CNC-derived cells (blue) populate the frontonasal process and the branchial arches. Arrow indicates the maxillary process of the first branchial arch. (B) Anatomy of palatogenesis from the appearance of the palatal primordia at E12.5 to the disappearance of the midline palatal epithelium in the fusion process of the palatal shelves at E15.5. The arrow indicates the midline epithelial seam at E14.5 and an equivalent region at E15.5, where the mesenchyme is continuous. Arrowheads point to tooth germs at different stages. (C and D) Patterning of the developing murine palate at E13.5. Genes expressed in the anterior–posterior (C) and oral–nasal (D) axes are depicted.

although the defect in mesenchymal proliferation is restricted to the anterior region, whereas in *Shox2*-null mice the cleft affects only the anterior palate (Satokata and Maas, 1994; Hilliard et al., 2005; Yu et al., 2005; Bush and Soriano, 2009). Interestingly, the expression of *Msx1* and *Shox2* appears to be dependent on BMP signals (Bush and Jiang, 2012). *Bmp4* expression is also restricted to the anterior region of the palatal mesenchyme. Using a transgenic approach, Zhang and co-workers (2002) expressed a *Bmp4* transgene ectopically in *Msx1* mutant palatal mesenchyme, and its expression rescued the cleft palate phenotype and neonatal lethality. In addition, the expression pattern of *Shh* and *Bmp2* and cell proliferation were restored in the palatal mesenchyme. Therefore, BMP4 appears to circumvent a requirement for *Msx1* and to function upstream of *Shh* and *Bmp2* to regulate epithelial–mesenchymal interactions during palatogenesis (Zhang et al., 2002). This study also demonstrated that Shh derived from the midline epithelial cells is able to activate *Bmp2* expression in the mesenchyme, which in turn stimulates cell proliferation. Therefore, *Msx1* regulates a genetic hierarchy encompassing BMP and Shh signals, which controls the growth of the anterior palate.

In contrast, the expression of *Barx1, Fgfr2, Meox2, Mn1, Pax9*, and *Tbx22* is restricted to the posterior region of the palatal shelves (Peters et al., 1998; Hilliard et al., 2005; Bush and Jiang, 2012) (Fig. 1C). *Mn1*-null mice also exhibit complete cleft palate, which is caused by defects in the middle and posterior regions of the palatal shelves (Liu et al., 2008). Loss of *Pax9* leads to malformed palatal shelves and cleft palate (Peters et al., 1998). *Tbx22*-null mice exhibit cleft palate, with variable severity from complete cleft palate to decreased palatal shelf expansion to submucous cleft palate (Bush et al., 2002). The phenotypes of *Efnb1*-, *Msx1*-, *Mn1*-, *Pax9*-, and *Tbx22*-null mice suggest that although there is a differential pattern of expression of these genes, there is also interdependence of the anterior and posterior regions of the palate during development.

AP patterning of the palatal CNCC-derived mesenchyme is also related to reorientation processes. This drastic movement from a vertical to a horizontal position is not very well understood, but a heterogeneous mechanism along the AP has been proposed (Ferguson, 1988). Coleman suggested that the anterior part of the palatal shelves elevate by rotation, whereas the posterior and middle regions change their orientation by a remodeling mechanism in which the palatal shelves "flow" above the dorsum of the tongue (Coleman, 1965; Bush and Jiang, 2012). The latter mechanism has recently been supported by an analysis of *Zfhx1a*$^{-/-}$ mice, which exhibit a delay in palatal shelf reorientation (Jin et al., 2008). The midline edge epithelium of post-reorientation palatal shelves in the medial-to-posterior regions of the palate may correspond to the nasal rather than the distal region of the pre-reorientation palatal epithelium.

At later stages, the AP heterogeneity may provide differential regulatory mechanisms for the fusion of the anterior versus posterior region of the palate (Chai and Maxson, 2006). For example, MES cells begin to undergo apoptosis at different times during palatal fusion, depending on their location. Apoptosis of MES cells is triggered by palatal shelf contact in the anterior region, whereas it is initiated before any contact between the opposing shelves in the posterior region (Cuervo and Covarrubias, 2004). This difference in mechanism may be the result of differential molecular signals in the palatal mesenchyme along the AP axis that instruct different fates to the palatal epithelium (Ferguson et al., 1984). Another example of AP patterning of the palatal epithelium is the development of rugae, which are metameric epithelial thickenings on the oral surface of the palate that appear in a specific sequence and establish a rugae growth zone (Welsh and O'Brien, 2009). Interestingly, it has been hypothesized that the palatal mesenchyme underlying the rugae growth zone may proliferate faster than both the anterior and posterior regions of the palate (Bush and Jiang, 2012). A number of studies have demonstrated that constant and reciprocal interactions between the palatal epithelium and the CNCC-derived mesenchyme are responsible for establishing this heterogeneity along the AP axis and are crucial for normal palatal outgrowth (Zhang et al., 2002; Rice et al., 2004).

2.2 Patterning of the CNCC-Derived Palatal Mesenchyme Along the Oral–Nasal Axis

For the *ON axis*, the oral side corresponds to the side closer to the oral cavity after palatal shelf reorientation. Some researchers utilize the term *medial–lateral axis* as a synonym for *ON axis*. We suggest using *oral–nasal axis* because a medial–lateral axis is no longer applicable following palatal shelf reorientation. Before reorientation, the BMP mediators phospho-Smad1/5/8 are restricted to the nasal side of the palatal mesenchyme, together with *Dlx5* and *Fgf7*. In contrast, *Fgf10* and *Gli1* are expressed in the oral side of the palatal mesenchyme, and *Shh* is detectable only in the oral side of the palatal epithelium (Fig. 1D). Strikingly, inactivation of *Dlx5* leads to an expansion of the oral region of the palatal shelf (Han et al., 2009). Loss of *Dlx5* results in downregulation of *Fgf7* and expanded *Shh* expression from the oral to the nasal side of the palatal shelf. This expanded Shh signaling is sufficient to restore palatal expansion and fusion in mice with compromised palatal mesenchymal cell proliferation, such as *Msx1*-null mutants. Exogenous *Fgf7* inhibits *Shh* signaling and reverses the CNCC proliferation rescue in *Msx1/Dlx5* double-knockout palatal mesenchyme. Thus, *Dlx5*-regulated FGF7 signaling inhibits the expression of *Shh*, which in turn controls the fate of CNCCs through tissue–tissue interactions (Han et al., 2009). In addition, a positive-feedback loop between *Shh* and *Fgf10* has been suggested (Lan and Jiang, 2009). The expression domains of these two factors are restricted to the oral regions of the epithelium and mesenchyme, respectively, and may drive the outgrowth of the palatal shelves (Lan and Jiang, 2009).

In addition, the ON pattern is also established through a gradient of expression of *Osr1* and *Osr2* in the palatal shelf, with *Osr1* restricted to the oral side and *Osr2* expressed strongly in the oral mesenchyme and weakly in the nasal mesenchyme (Lan et al., 2004, and our unpublished data) (Fig. 1D). Significantly, mutation of the *Osr2* gene results in compromised development of the nasal aspect of the palatal shelf and retards palatal shelf elevation (Lan et al., 2004). The expression of *Fgfr2* is restricted to the nasal aspect of the developing palatal shelf and may also help define the ON axis (Hosokawa et al., 2009; Snyder-Warwick et al., 2010) (Fig. 1D). In *Fgfr2*-null mice, proliferation is altered in both palatal epithelium and mesenchyme, and reorientation is defective (Rice et al., 2004). Moreover, a gain-of-function mutation in *Fgfr2* leads to increased mesenchymal cell proliferation in the oral side of the palatal shelf and delayed reorientation, resulting in cleft palate (Snyder-Warwick et al., 2010). The phenotypes of both *Osr2*- and *Fgfr2*-mutant mice suggest the possibility that the correct ON pattern is needed for the reorientation process. It is still unknown if this requirement is related to differential proliferation patterns or to extracellular matrix (ECM) production and remodeling along the ON axis. At the newborn stage, the mesenchymal heterogeneity of the nasal and oral regions of the palate is morphologically subtle, with palatine bone in the nasal region and soft tissue in the oral region, which reflects the early expression of oral and nasal markers (Han et al., 2009). The expression domain of BMP-associated Smads correlates with the osteogenic region of the palate, consistent with a BMP function in inducing bone formation (Kamiya and Mishina, 2011). ON heterogeneity is also obvious in the epithelium of newborn mice, with pseudostratified, ciliated, columnar epithelia covering the nasal side of the palatal shelf, and stratified keratinizing squamous epithelia covering the oral side (Han et al., 2009). Collectively, these studies indicate that correct patterning of the CNCC-derived mesenchyme in the palatal shelves along the ON and AP axes is essential for the formation of normal palatal structures.

2.3 TGFβ Signaling in Postmigratory Cranial Neural Crest Cells During Palatogenesis: Insights from Animal Models

Palate development relies on a tightly regulated network that includes the activity of secreted proteins and their signaling pathways, cell-surface receptors, ECM components, and transcription factors. The transforming growth factor beta (TGFβ)) superfamily is a critical part of this network, as deficiency of its members leads to malformations affecting the craniofacial region, particularly the palate, in mouse models. This superfamily includes activin, TGFβ, and BMP cytokines (Massague, 2000). TGFβ/BMP ligands trigger intracellular cascades by binding to cell-surface receptor serine/threonine kinases. Upon ligand binding, the type II receptor phosphorylates the type I receptor, which in turn activates a group of transcriptional coactivators called Smads, specifically Smad2/3 for TGFβ ligands and Smad1/5/8 for BMPs. These Smads associate with Smad4 (common Smad), forming a complex that translocates into the nucleus, where it regulates the transcription of downstream targets (Massague, 2000). Using the *Wnt1-Cre* and *Osr2-Cre* systems, the functions of diverse members of the TGFβ/BMP signaling pathways in CNCCs have been evaluated. For example, *Tgfbr2* conditional gene ablation in CNCCs leads to a complete cleft secondary palate with 100% phenotype penetrance. Disruption of TGFβ signaling does not affect CNCC migration adversely. Instead, cleft palate in *Tgfbr2*-mutant mice results from a cell proliferation defect within the CNCC-derived palatal mesenchyme (Ito et al., 2003). Reduction in proliferation appears to be associated with downregulation of *Fgf9* and *Pitx2* expression (Iwata et al., 2012b). Both *Fgf9*- and *Pitx2*-null mice exhibit cleft palate. Interestingly, FGF9 protein is able to rescue the proliferation defect observed in *Wnt1-Cre;Tgfbr2* palates in vitro (Iwata et al., 2012b). In addition, p38 MAPK is overactivated in *Tgfbr2* mutant palates, suggesting that activation of alternative pathways mediating TGFβ signals has a significant impact on CNCC fate during palatogenesis (Iwata et al., 2012a).

Specific inactivation of *Bmpr1a* in CNCC-derived mesenchyme (*Osr2-Cre;Bmpr1a*$^{fl/fl}$) causes reduced cell proliferation in the primary and anterior secondary palate, resulting in partial cleft of the anterior palate at birth. In these mutant mice, *Msx1* and *Fgf10* expression is downregulated in the anterior palate mesenchyme, and expression of *Shh* is reduced in the overlying palatal epithelium. These findings indicate that Bmp signaling regulates mesenchymal–epithelial interactions during palatal outgrowth. Additionally, development of the palatal processes of the maxilla is deficient, and formation of the palatal processes of the palatine is significantly delayed, leading to a submucous cleft of the hard palate in *Osr2-Cre;Bmpr1a*$^{fl/fl}$ mice (Baek et al., 2011). Similarly, mice lacking *Alk2* in the neural crest (*Wnt1-Cre;Alk2*$^{fl/fl}$) display multiple craniofacial defects, including cleft palate and a hypotrophic mandible (Dudas et al., 2004). Inactivation of *Smad4* in the CNCC-derived mesenchyme in *Wnt1-Cre;Smad4*$^{fl/fl}$ mice demonstrates that Smad4 is not required for the migration of CNCC but that embryonic development is arrested at embryonic day E11.5–E12.5. *Wnt1-Cre;Smad4* $^{fl/fl}$ embryos display underdevelopment of the first BA and failure of fusion in the middle of the frontonasal process and in the middle of the mandibular process of the first BA. The defects in lateral development of the first BA are more severe than those in anterior–posterior development in *Wnt1-Cre;Smad4* $^{fl/fl}$ embryos, due to a dramatic increase in the number of apoptotic cells in the CNCC-derived mesenchyme (Ko et al., 2007). In these mutant embryos, the analysis of Smad4 function in palatogenesis is also impeded by the early arrested development. To overcome this issue, *Osr2-Cre;Smad4* embryos have been generated and they display complete cleft palate due to a significant reduction in mesenchymal cell proliferation (our unpublished data). These studies underscore the requirement for TGFβ/BMP signaling pathways in the regulation of CNCC fate during palatogenesis. Their modulation offers a therapeutic alternative for palatal malformations.

3 TOOTH DEVELOPMENT

Teeth are one of the key innovations in vertebrates. The emergence of teeth in vertebrates is closely associated with the appearance of NCCs during evolution. Dental CNCC-derived mesenchyme interacts actively with the oral epithelium during tooth morphogenesis and originates most of the dental tissues and the periodontium, including the alveolar bone (Mitsiadis and Sharpe, 2004; Tucker and Sharpe, 2004). In this section we review the current and most important aspects of patterning of the CNCC-derived mesenchyme during tooth development, the contribution of CNCCs to dental tissues, and the stem cell property of CNCCs.

3.1 Patterning of the CNCC-Derived Mesenchyme During Tooth Initiation

Once the first BA is formed, the oral epithelium thickens to develop the dental lamina, which is the first morphological evidence of tooth development. Studies using rodent models have suggested that the ectoderm-derived dental epithelium possesses the competency to initiate and instruct tooth development at very early stages. However, a recent study using axolotls proposes a leading role for the CNCCs over the epithelia in tooth initiation and morphogenesis. From an evolutionary point of view, this finding suggests that an essential factor in tooth evolution was the odontogenic capacity of CNCCs, regardless of the origin of the epithelium (Soukup et al., 2008).

Dental CNCC-derived mesenchyme is patterned along the distal–proximal, oral–aboral (rostral–caudal), dorso–ventral (maxillary–mandibular arch), and lingual–buccal axes in a complex fashion that requires the action of the epithelium (Chai and Maxson, 2006) (Fig. 2). Before tooth initiation, the mandibular ectoderm generally can be divided into the proximal domain, which expresses *Fgf8* and is the origin of molars, and the distal domain, which expresses *Bmp4* and gives rise to incisors. FGF and BMP act antagonistically to restrict *Barx1* and *Dlx2* expression to the proximal domain of the first BA mesenchyme and *Msx1, Msx2*, and *Alx4* expression to the distal domain, respectively (Chai and Maxson, 2006) (Fig. 2). FGF signals also influence establishment of the oral–aboral axis through the regulation of *Lhx* and *Gsc* expression. Postmigratory CNCCs located in the oral (rostral) region of the mandibular arch express *Lhx6* and *Lhx7* (Grigoriou et al., 1998). In the aboral

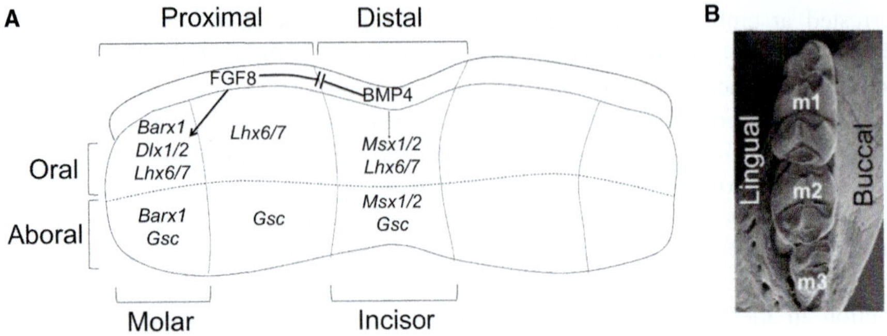

FIGURE 2 Patterning of the tooth fields in mice. The first branchial arch can be divided into proximal–distal (A), oral–aboral (A), and buccal–lingual (not depicted) axes, which will be reflected in the adult dentition (B). Differential expression of a number of genes is depicted along the aforementioned axes.

(caudal) region, *Gsc* is expressed in an *Lhx6/7*-negative area (Fig. 2) (Tucker et al., 1999). This subdivision along the rostral–caudal axis is established by FGF8 from the ectoderm and is achieved by the differential response of CNCCs to the FGF8 signal according to their proximity to it (Tucker et al., 1999).

Surprisingly, $Lhx6^{-/-}$; $Lhx7^{-/-}$ mice only lack molar teeth, whereas they have normal incisors that are flanked in the maxilla by a supernumerary pair of incisor-like teeth (Denaxa et al., 2009). This finding is noteworthy because the dentition is also patterned in a different fashion in the maxilla and mandible. As mentioned before, the Dlx code defines maxillary and mandibular structures, including teeth. Maxillary molars in $Dlx1^{-/-}$; $Dlx2^{-/-}$ mice fail to develop, whereas all the other teeth appear normal (Thomas et al., 1997). In contrast, maxillary molars are unaffected in *activin-βA*-mutant mice that lack all other teeth (Ferguson et al., 1998). The lack of a maxillary molar defect in *activin-βA* mice suggests that the molecular regulation of maxillary molar development is different from that of other teeth. Thus, *Dlx1/2* may be a crucial element for the establishment of a maxillary molar domain within the CNCC-derived mesenchyme (Fig. 2) (Miletich and Sharpe, 2004).

3.2 CNCC and Tooth Morphogenesis

The establishment of axis patterning occurs prior to and simultaneously with formation of the tooth bud. During tooth bud formation, the epithelial thickenings invaginate into the underlying CCNC-derived mesenchyme, which condenses around these epithelial buds. At this stage, the odontogenic potential switches from the epithelium to the mesenchyme (Kollar and Baird, 1969). Recent results demonstrate that the mesenchymal condensation that subsequently drives tooth formation is induced by epithelial FGF8 and SEMA3F. Mesenchymal compaction induces changes in cell shape and, consequently, expression of the odontogenic transcription factors *Pax9, Msx1*, and *Bmp4* (Mammoto et al., 2011). The expression of *Pax9* and *Msx1* in the CNCC-derived mesenchyme is pivotal for the progression of tooth development, and they regulate the maintenance of mesenchymal *Bmp4* expression, which ultimately drives morphogenesis of the dental organ (Peters and Balling, 1999). Deficiency of *Msx1* or *Pax9* in mice causes arrested tooth development at the bud stage, which is associated with downregulation of *Bmp4* expression in the mesenchyme (Satokata and Maas, 1994; Peters et al., 1998; Zhang et al., 2002). Similarly, deficiency of *Bmprla* in CNCCs leads to an arrest of tooth development at the bud or early cap stage. Defective tooth development is accompanied by the downregulation of BMP-responsive genes and decreased cell proliferation levels in the dental mesenchyme (Li et al., 2011b). In addition, the Bmp4–Msx1 pathway appears to propagate mesenchymal activation for sequential tooth induction and to pattern the mouse dentition along the lingual–buccal axis. An abnormal enlargement of the expression domain of *Bmp4* and its feedback activator *Msx1* leads to supernumerary teeth lingual to molars in *Osr2*-deficient mice (Zhang et al., 2009). This finding is associated with an expansion of the odontogenic field. Recent results have shown that Osr2 acts downstream of Pax9 and interacts with both Msx1 and Pax9 to pattern the tooth developmental field (Zhou et al., 2011).

3.3 CNCC and Dental Stem Cells

After morphogenesis is complete, CNCCs in the teeth differentiate and their cell fate is controlled by the context-dependent integration of extrinsic and intrinsic signals. CNCCs give rise to odontoblasts, dentine matrix, pulp tissue, cementum, and periodontal ligaments of the teeth in the adult dentition (Chai et al., 2000). During dentinogenesis, CNCC-derived odontoblasts secrete predentin and dentin following terminal differentiation (Chai et al., 2000; Ruch et al., 1995). Odontoblast terminal differentiation is controlled in part by the inner enamel epithelium and is also supported by matrix-mediated interactions (Ruch et al., 1995; Thesleff et al., 2001). Several studies have shown that the TGFβ superfamily and WNT contribute to odontoblast terminal differentiation (Oka et al., 2007; Lohi et al., 2010). Exogenous TGFβ1, BMP2, BMP4, and BMP7 can induce odontoblast differentiation and dentin formation in dental papilla cells in vitro (reviewed by Li et al., 2011a). Furthermore, inhibition of TGFβ signaling in *Wnt1-Cre;Tgfbr2*$^{fl/fl}$ mice results in abnormal dentin formation (Oka et al., 2007). Accordingly, ablation of *Smad4* in the dental mesenchyme leads to a defect in odontoblast differentiation. Instead of dentin formation, ectopic bone-like structures develop in *Osr2-Cre;Smad4*$^{fl/fl}$ mice (Li et al., 2011a), similar to that observed in transgenic mice, in which *Runx2* is overexpressed in odontoblasts (Li et al., 2011c). In *Osr2-Cre;Smad4*$^{fl/fl}$ mice, the canonical WNT signaling pathway is upregulated. Thus, proper CNCC fate choice during odontogenesis appears to be dependent on the interplay between TGFβ/BMP and other signaling pathways, including WNT (Li et al., 2011a).

Consistent with the studies described above, postmigratory CNCCs maintain mesenchymal stem cell characteristics and are able to differentiate into specific cell types according to environmental conditions. Using the *Wnt1-Cre;R26R* mouse model, Zhao and co-workers (2006) isolated pure populations of CNCCs, which are able to give rise to neurons, glial cells, osteoblasts, and other cell types, depending on the culture conditions. The osteogenic differentiation potential of postmigratory CNCCs has been demonstrated as well as the influence of BMP and TGFβ signaling pathways in this process (Chung et al., 2009). Moreover, transplantation of postmigratory CNCCs together with tooth germs supported normal tooth development and the formation of periodontal tissues, including alveolar bone. The CNCCs appear to have specific effects because bone marrow mesenchymal stem cells transplanted together with tooth germs fail to support tooth development (Chung et al., 2009).

Strong evidence supports the existence of stem cell populations in adult teeth. Mouse incisors grow continuously throughout life, supported by the division of dental epithelial stem cells that reside in the cervical loop region. TGFβ signaling-mediated mesenchymal–epithelial interactions control dental epithelial stem cells (Zhao et al., 2011). Deficiency of TGFβ type 1 receptor in the CNCC-derived dental mesenchyme affects the proliferation of transit-amplifying cells and the maintenance of dental epithelial stem cells. Incisors of *Wnt1-Cre;Alk5*$^{fl/fl}$ mice lose their ability to continue growing in vitro, which is due to downregulation of FGF signals in the dental mesenchyme (Zhao et al., 2011). On the other hand, in the dental pulp, mesenchymal stem cells (MSCs) are reported to reside in a perivascular niche (Shi and Gronthos, 2003; Feng et al., 2011). Feng and co-workers have recently shown that some pericytes differentiate into odontoblasts during tooth growth and also in response to damage in vivo (Feng et al., 2011). In addition to the dental pulp stem cells (Miura et al., 2003), other MSC populations have been isolated from human CNCC-derived dental tissues such as the periodontal ligament (Seo et al., 2005) and the dental follicle (Morsczeck et al., 2005). The presence of MSCs with a CNCC origin is promising in the field of regenerative medicine, not only for repairing missing or defective dental tissues but also for use in the treatment of diverse life-threatening diseases.

4 TONGUE DEVELOPMENT

The tongue is a complex and multifunctional organ composed of a muscular core surrounded by connective tissue and covered in papillae and taste buds. The tongue is derived from all BAs, with each contributing different components. The development of the tongue begins with the formation of a medial triangular elevation on the floor of the pharynx anlagen called the *median lingual swelling*, which is composed of CNCC-derived mesenchyme from the first BA. Next, lateral lingual swellings form on each side of the median tongue bud. These lateral swellings increase in volume, fuse with each other, and overgrow the medial lingual swelling. The merged lateral lingual swellings develop into the anterior two-thirds of the tongue. The fibrous CNCC-derived lingual septum is the fusion site of those lateral swellings. Two other outgrowths, the copula and the hypopharyngeal eminence, develop caudal to the foramen cecum from the third BA and will constitute the posterior third of the tongue. As development proceeds, the copula is progressively overgrown by the hypopharyngeal eminence and disappears. Consequently, the posterior third of the tongue forms from the rostral part of the hypopharyngeal eminence (Moore and Persaud, 2008). Finally, the early tongue undergoes rapid enlargement and differentiates into a muscular organ (Huang et al., 1999).

4.1 Hybrid Origin of Tongue Mesenchyme: Morphogenesis and Tongue Muscle Formation

The tongue mesenchyme has a hybrid origin. Tongue connective tissue is derived from CNCCs, whereas most of the tongue muscles originate from myoblasts that have migrated from the occipital somites (Noden and Francis-West, 2006) (Fig. 3). The intimate relationship between these cell lineages suggests that reciprocal interactions between CNCCs and myogenic cells occur during tongue development. Recent results have demonstrated that CNCCs populate the tongue primordium before the invasion of

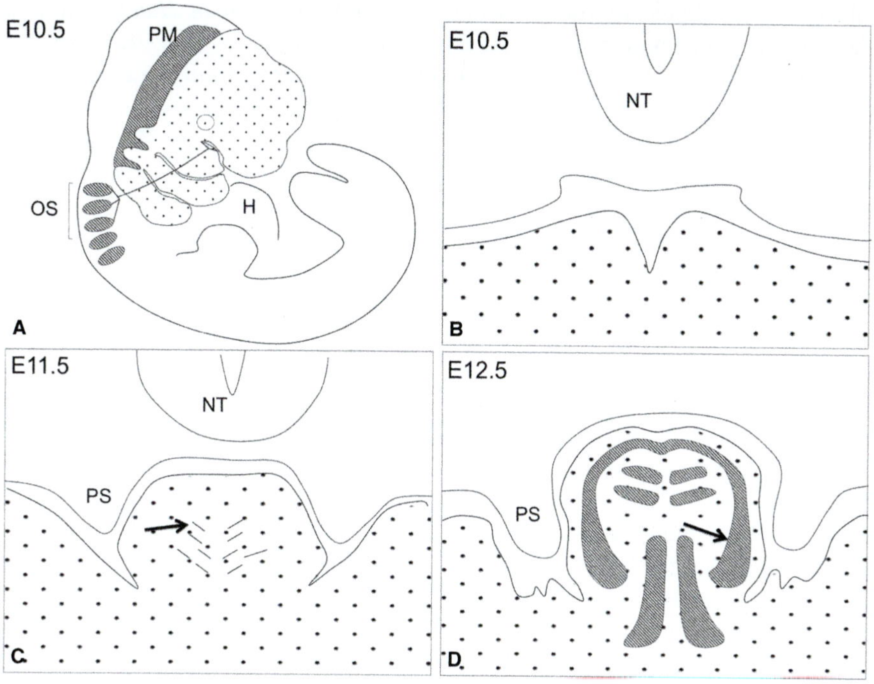

FIGURE 3 Early development of the tongue. Close interaction between CNCC-derived mesenchyme (dots) and occipital somites migrating myoblasts (stripes) directs tongue myogenesis. The arrows in C and D point to myoblasts and show the influence of CNCC-derived mesenchyme on muscle cells. A, E10.5 mouse embryo, B,C,D, higher magnifications of forming tongue in E10.5, E11.5 and E12.5 embryos. NT, neural tube; OS, occipital somites; PS, palatal shelf.

myogenic progenitors in mouse embryos (Han et al., 2012) (Fig. 3A and B), suggesting that CNCCs are the cell type initiating and directing tongue development. Although chick/quail recombination experiments have previously indicated that CNCCs surround the myogenic cell lineage at a very early stage but do not penetrate into the myogenic core (Noden, 1986; Noden and Francis-West, 2006), other reports demonstrate that secreted factors from CNCCs are able to regulate myogenic proliferation and other processes during tongue development (Hosokawa et al., 2010). Based on our work and that of others, we hypothesize that CNCC-derived mesenchyme has two main functions during tongue myogenesis: (1) it acts as a scaffolding structure for the organization of migrating myoblasts into the myogenic core, and (2) it operates as a source of molecular instruction to direct survival, proliferation, and differentiation of myogenic progenitors. Here we review the influence of CNCC-derived mesenchyme on tongue myogenesis as well as lingual morphogenesis and gustatory papillae formation.

CNCCs and mesodermal cells that colonize the first BA and consequently a great part of the tongue primordium do not express Hox genes (Couly et al., 1998). *Dlx1, 2, 5*, and *6* are expressed in the mandibular process, where the anterior two-thirds of the tongue develop, and are essential for tongue patterning (Depew et al., 2005; Heude et al., 2010). *Hand2* also plays an important function in the establishment of proximal–distal patterning of the lower jaw in mouse embryos. This occurs through a negative-feedback loop in which *Hand2* represses *Dlx5* and *Dlx6* expression in the distal arch mesenchyme following *Dlx5/6*-mediated induction of *Hand2* expression in the same region. Interestingly, failure to inhibit distal *Dlx5* and *Dlx6* expression leads to the absence of lateral lingual swelling expansion, from which the tongue arises, resulting in aglossia. Thus, *Hand2* seems to determine a distal mandibular arch domain that is favorable for lower jaw development, including the induction of tongue morphogenesis (Barron et al., 2011). Recently, Heude and co-workers have demonstrated that craniofacial myogenesis is also dependent on *Dlx5/6* expression by CNCCs. Inactivation of *Dlx5* and *Dlx6* in the mouse results in the loss of jaw muscles and compromised tongue development (Heude et al., 2010). Because *Dlx5/6* are not expressed by myogenic cells, this result indicates that *Dlx5/6*-positive CNCCs play an instructive role in muscle formation. The alteration in muscularization is not necessarily a consequence of the loss of mandibular identity in *Dlx5/6*-deficient mice, because masticatory muscles are still present in endothelin receptor type A–null mice ($EdnRA^{-/-}$), which display a jaw transformation similar to that of *Dlx5/6*-deficient mice. The intrinsic muscles of the tongue and sublingual muscles are altered significantly in *Dlx5/6*-deficient mice: the genioglossus and the geniohyoid are absent, and other intrinsic muscles of the tongue are reduced and disorganized. Interestingly, limb and trunk muscles in *Dlx5/6*-mutant mice are not affected, indicating a specific function of Dlx genes in tissue–tissue interactions involving neural crest derivatives. *Dlx5/6* expression by CNCCs is necessary for the occurrence of interactions between CNCC-derived mesenchyme and mesoderm, which result in myogenic determination, differentiation, and patterning. However, these interactions are not required for the early migration and specification of myogenic progenitors.

Thus, the importance of Dlx genes in tongue development and evolution is twofold: They establish the dorsoventral pattern of the first BA and indirectly that of tongue, and they regulate myogenic determination and differentiation processes, including those affecting the tongue myogenic core. From an evolutionary perspective, Dlx genes might have had a central role in coordinating the development of the oral skeletomuscular apparatus at the origin of active predation in vertebrates (Heude et al., 2010). Morphological innovations led to the emergence of the *new head* in vertebrates. The new head is considered a neomorphism in which the main component is the NCCs, thought to be unique to vertebrates (Northcutt, 2005). Besides the NCC-derived structures, the new head also contained a muscularized oral structure that allowed the transition from filter feeding to active predation in a vertebrate ancestor. The feeding mechanism is clearly an important factor that determines the success of adaptation of vertebrates to their environment. The tongue plays a major role in feeding, together with other organs within and close to the oral cavity, particularly in tetrapods (Iwasaki, 2002).

Hedgehog (Hh) signaling activation in NCCs also exerts profound effects on craniofacial structures that are strictly limited to neural crest–derived elements. Removal of Hh signaling in NCCs in *Wnt1-Cre;Smo*$^{n/c}$ embryos leads to a severe disruption of tongue development first detectable at E10.75, before the tongue starts to grow visibly in wild-type embryos (Jeong et al., 2004). At this stage, streams of *Myf5*- and *MyoD*-expressing myogenic progenitors arise from the somites and accumulate around the midline of the mandibular arch, where the tongue will form. These prospective tongue muscle cells are not detectable in *Wnt1-Cre;Smo*$^{n/c}$ embryos. Interestingly, the position of the tongue appears to have been specified in the epithelium of *Wnt1-Cre;Smo*$^{n/c}$ embryos at E11.5, as judged by the expression of *Shh*, a marker for tongue epithelium.

Thus, Hh signaling in the CNCC-derived mesenchyme might be involved in transmitting information from the epithelium to the myogenic progenitors to coordinate tongue formation. Tongue defects and other malformations in *Wnt1-Cre;Smo*$^{n/c}$ mutants are proposed to be dependent on alteration of Fox gene expression (Jeong et al., 2004).

4.2 Functional Significance of TGFβ and FGF Signaling Pathways from the CNCCs on Tongue Myogenesis

TGFβ family members have a critical function in regulating skeletal muscle development through tissue–tissue interactions. In the head, BMPs act to repress skeletal muscle differentiation. Myogenic differentiation of the cranial paraxial mesoderm is initiated following CNCC production of BMP inhibitors such as Noggin and Gremlin (Tzahor et al., 2003). Specific deletions of TGFβ pathway members in NCCs lead to severe tongue defects, consistent with the conclusion that CNCCs influence tongue morphogenesis and tongue muscle formation and organization. For example, loss of *Tgfbr2* in CNCCs results in microglossia with defects in CNCC-derived connective tissue and tongue muscle development (Hosokawa et al., 2010). Similarly, *Wnt1-Cre;Alk5*$^{fl/fl}$ mice exhibit severe microglossia (our unpublished results). *Wnt1-Cre;Erk2*$^{fl/fl}$ mice exhibit microglossia in combination with micrognathia and cleft palate. This phenotype is worsened by the homozygous inactivation of the Erk1 gene (*Erk1*$^{-/-}$; *Wnt1-Cre;Erk2*$^{fl/fl}$), which results in absence of the tongue (Newbern et al., 2008).

Our recent study addressed the molecular and cellular mechanisms underlying the microglossia in *Wnt1-Cre;Tgfbr2*$^{fl/fl}$ mice (Hosokawa et al., 2010). The reduction in tongue size in *Tgfbr2*-mutant mice is due to a decrease in proliferation activity of myogenic cells, which is associated with downregulation of *Fgf10* expression. In contrast, proliferation is unaffected in CNCC cells in *Tgfbr2*-mutant mice. Exogenous FGF10 reverses the reduction of tongue muscle cell number in *Tgfbr2* mice in vitro, which suggests a noncell autonomous activity because FGF10 is expressed only by CNCCs. A function for FGF10 during tongue development is consistent with previous studies showing that FGF ligands and their receptors are required for skeletal muscle formation (Flanagan-Steet et al., 2000). For example, the viral expression of truncated FGF type 1 receptor causes a skeletal muscle defect in the chick limb (Flanagan-Steet et al., 2000). *Fgf4* and *Fgf8* are expressed in muscle and tendon boundary regions during limb development, suggesting a potential role for the FGF signaling pathway in muscle and tendon interactions (Eloy-Trinquet et al., 2009). In addition, FGFs also function in a cell-autonomous manner in the myogenic cells of limb muscles (Floss et al., 1997) and tongue (Han et al., 2012). In wild-type mice, FGF6 is upregulated after skeletal muscle injury, implying that FGF6 is crucial for muscle repair and regeneration. Moreover, *Fgf6*$^{-/-}$ mutant mice have a severe regeneration defect, with fibrosis and myotube degeneration. The number of *MyoD*- and myogenin-expressing activated satellite cells after injury is reduced significantly in these mice. This reduction is not caused by a reduced pool of quiescent satellite cells but presumably by a lack of activation or proliferation. Thus, FGF6 appears to be a critical component of the muscle regeneration machinery in mammals, possibly by stimulating or activating satellite cells (Floss et al., 1997). In tongue, ablation of *Smad4* in myogenic progenitors results in defects of myogenic differentiation and myoblast fusion during embryonic development, associated with a significant reduction in the expression of *Fgf6* and *Fgfr4*. Interestingly, FGF6 is able to rescue the tongue myoblast fusion defect in *Smad4*-mutant mice (Han et al., 2012). These studies suggest that the TGFβ and FGF signaling pathways cross-talk to regulate tongue myogenesis.

4.3 Influence of CNCC-Derived Tongue Mesenchyme in Gustatory Papillae Development

Another physiologically crucial component of the vertebrate tongue is the taste buds. The majority of taste buds in mammals reside on the tongue in epithelial–mesenchymal specializations termed *gustatory papillae*. There are four types of papillae present in the mammalian tongue: fungiform papillae, filiform papillae, circumvallate papillae, and foliate papillae (Zhang and Oakley, 1996). Taste buds are concentrated on three types of lingual papillae—fungiform, circumvallate, and foliate—and they are absent from filiform papillae. In the rodent, the smaller fungiform papillae are located on the anterior tongue. The larger bilateral foliate papillae and a single midline circumvallate papilla are located on the posterior tongue (Petersen et al., 2011). All papillae initially form as epithelial thickenings that evaginate, undergo complex morphogenesis, and acquire a mesenchymal core in a process that is common to other ectodermal appendages (Farbman and Mbiene, 1991). However, the molecular regulation of papilla development is different along the AP axis of the tongue. These differences might be related to the dissimilar embryonic origins of the anterior and posterior regions of the tongue. The anterior tongue epithelium is probably derived from ectoderm, whereas the posterior epithelium appears to have an endodermal origin (Zhang and Oakley, 1996). Expression of key signaling factors such as SHH, BMPs, epidermal growth factor (EGF), and WNTs has been identified in the epithelium during papillae development (Petersen et al., 2011).

The molecular network in the mesenchymal component of the papillae is not well defined. Based on studies using the *Wnt1-Cre;R26R* system, CNCC-derived mesenchyme extends into the taste papilla but does not contribute to taste buds or to papillary epithelium (Chai et al., 2000; Thirumangalathu et al., 2009). Despite indications that mesenchyme-derived factors are involved in lingual papillae development (Barlow, 2003), only two have been identified to date: follistatin in fungiform papilla development and FGF10 in circumvallate papilla development (Beites et al., 2009; Petersen et al., 2011). Loss of follistatin in mice leads to increased activity and overexpression of epithelial *Bmp7*, which is coherent with BMP7-positive autoregulation in papillae development. Disruption of a follistatin-BMP7 regulatory circuit results in abnormalities in taste papillae patterning. Innervation and maintenance of taste buds and their progenitor cells are also dependent on the correct balance of BMP signaling. This model supports reciprocal epithelial–mesenchymal interactions during fungiform papilla development because follistatin is only derived from the mesenchyme and BMP7 from the epithelium (Beites et al., 2009). More recently, Klein and co-workers identified FGF10 as the first inductive factor from the CNCC-derived mesenchyme for taste papilla development, specifically required for circumvallate papilla formation (Petersen et al., 2011). Consistent with a role in taste papilla induction, *Fgf10* mRNA is localized directly underneath the developing circumvallate placode. A balance between Sprouty (Spry) genes and *Fgf10*, which respectively antagonize and trigger FGFR signaling, controls the number of circumvallate papilla. Loss of *Spry2* results in duplication of the circumvallate papilla as a result of an increase in the size of the placode progenitor field, and *Spry1*$^{-/-}$; *Spry2*$^{-/-}$ embryos had multiple circumvallate papillae. These findings suggest redundancy of Sprouty genes in regulating the number of progenitor for papillae development. In contrast, deficiency of *Fgf10* causes a lack of the circumvallate papilla. Thus, FGF signaling may also be important evolutionarily, helping to determine the papilla progenitor field size in vertebrates (Petersen et al., 2011).

5 CONCLUSIONS AND PERSPECTIVES

A comprehensive understanding of the regulatory mechanisms that control the development of craniofacial organs and the functional significance of CNCCs during tooth, palate, and tongue morphogenesis will provide the opportunity to strive toward an ideal regeneration approach in craniofacial abnormalities such as tooth agenesis and cleft lip and palate as well as in tongue cancer. This approach will probably be based on stem cell and tissue-engineering technologies, using isolated CNCC-derived stem cells under the precise control of the proper molecular regulatory mechanisms, and will ultimately generate functional craniofacial organs.

REFERENCES

Baek J-A, Lan Y, Liu H, Maltby KM, Mishina Y, Jiang R (2011). Bmpr1a signaling plays critical roles in palatal shelf growth and palatal bone formation. *Dev Biol* 350:520–531.

Barlow LA (2003). Toward a unified model of vertebrate taste bud development. *J Comp Neurol* 457:107–110.

Barron F, Woods C, Kuhn K, Bishop J, Howard MJ, Clouthier DE (2011). Downregulation of *Dlx5* and *Dlx6* expression by *Hand2* is essential for initiation of tongue morphogenesis. *Development* 138:2249–2259.

Beites CL, Hollenbeck PLW, Kim J, Lovell-Badge R, Lander AD, Calof AL (2009). Follistatin modulates a BMP autoregulatory loop to control the size and patterning of sensory domains in the developing tongue. *Development* 136:2187–2197.

Bronner-Fraser M (1993). Neural crest cell migration in the developing embryo. *Trends Cell Biol* 3:392–397.

Bush JO, Jiang R (2012). Palatogenesis: morphogenetic and molecular mechanisms of secondary palate development. *Development* 139:231–243.

Bush JO, Soriano P (2009). Ephrin-B1 regulates axon guidance by reverse signaling through a PDZ-dependent mechanism. *Genes Dev* 23:1586–1599.

Bush JO, Lan Y, Maltby KM, Jiang R (2002). Isolation and developmental expression analysis of *Tbx22*, the mouse homolog of the human X-linked cleft palate gene. *Dev Dyn* 225:322–326.

Chai Y, Maxson RE (2006). Recent advances in craniofacial morphogenesis. *Dev Dyn* 235:2353–2375.

Chai Y, Jiang X, Ito Y, Bringas P, Han J, Rowitch DH, Soriano P, McMahon AP, Sucov HM (2000). Fate of the mammalian cranial neural crest during tooth and mandibular morphogenesis. *Development* 127:1671–1679.

Chung I-H, Yamaza T, Zhao H, Choung P-H, Shi S, Chai Y (2009). Stem cell property of postmigratory cranial neural crest cells and their utility in alveolar bone regeneration and tooth development. *Stem Cells* 27:866–877.

Coleman RD (1965). Development of the rat palate. *Anat Rec* 151:107–117.

Couly G, Grapin-Botton A, Coltey P, Ruhin B, Le Douarin NM (1998). Determination of the identity of the derivatives of the cephalic neural crest: incompatibility between Hox gene expression and lower jaw development. *Development* 125:3445–3459.

Cuervo R, Covarrubias L (2004). Death is the major fate of medial edge epithelial cells and the cause of basal lamina degradation during palatogenesis. *Development* 131:15–24.

Denaxa M, Sharpe PT, Pachnis V (2009). The LIM homeodomain transcription factors Lhx6 and Lhx7 are key regulators of mammalian dentition. *Dev Biol* 333:324–336.

Depew MJ, Simpson CA, Morasso M, Rubenstein JLR (2005). Reassessing the Dlx code: the genetic regulation of branchial arch skeletal pattern and development. *J Anat* 207:501–561.

Dudas M, Sridurongrit S, Nagy A, Okazaki K, Kaartinen V (2004). Craniofacial defects in mice lacking BMP type I receptor Alk2 in neural crest cells. *Mech Dev* 121:173–182.

Eloy-Trinquet S, Wang H, Edom-Vovard F, Duprez D (2009). Fgf signaling components are associated with muscles and tendons during limb development. *Dev Dyn* 238:1195–1206.

Farbman AI, Mbiene JP (1991). Early development and innervation of taste bud–bearing papillae on the rat tongue. *J Comp Neurol* 304:172–186.

Feng J, Mantesso A, De Bari C, Nishiyama A, Sharpe PT (2011). Dual origin of mesenchymal stem cells contributing to organ growth and repair. *Proc Natl Acad Sci USA* 108:6503–6508.

Ferguson CA, Tucker AS, Christensen L, Lau AL, Matzuk MM, Sharpe PT (1998). Activin is an essential early mesenchymal signal in tooth development that is required for patterning of the murine dentition. *Genes Dev* 12:2636–2649.

Ferguson MW (1988). Palate development. *Development* 103:(Suppl): 41–60.

Ferguson MW, Honig LS, Slavkin HC (1984). Differentiation of cultured palatal shelves from alligator, chick, and mouse embryos. *Anat Rec* 209:231–249.

Flanagan-Steet H, Hannon K, McAvoy MJ, Hullinger R, Olwin BB (2000). Loss of FGF receptor 1 signaling reduces skeletal muscle mass and disrupts myofiber organization in the developing limb. *Dev Biol* 218:21–37.

Floss T, Arnold H-H, Braun T (1997). A role for FGF-6 in skeletal muscle regeneration. *Genes Dev* 11:2040–2051.

Gato A, Martinez ML, Tudela C, Alonso I, Moro JA, Formoso MA, Ferguson MWJ, Martínez-Alvarez C (2002). TGF-β-induced chondroitin sulphate proteoglycan mediates palatal shelf adhesion. *Dev Biol* 250:393–405.

Grigoriou M, Tucker AS, Sharpe PT, Pachnis V (1998). Expression and regulation of *Lhx6* and *Lhx7*, a novel subfamily of LIM homeodomain encoding genes, suggests a role in mammalian head development. *Development* 125:2063–2074.

Han D, Zhao H, Parada C, Hacia JG, Bringas P, Chai Y (2012). A TGFβ-Smad4-Fgf6 signaling cascade controls myogenic differentiation and myoblast fusion during tongue development. *Development* 139:1640–1650.

Han J, Mayo J, Xu X, Li J, Bringas P, Maas RL, Rubenstein JLR, Chai Y (2009). Indirect modulation of Shh signaling by Dlx5 affects the oral–nasal patterning of palate and rescues cleft palate in *Msx1*-null mice. *Development* 136:4225–4233.

Heude A, Bouhali K, Kurihara Y, Kurihara H, Couly GR, Janvier P, Levi G (2010). Jaw muscularization requires Dlx expression by cranial neural crest cells. *Proc Nat Acad Sci USA* 107:11441–11446.

Hilliard SA, Yu L, Gu S, Zhang Z, Chen YP (2005). Regional regulation of palatal growth and patterning along the anterior–posterior axis in mice. *J Anat* 207:655–667.

Hosokawa R, Deng X, Takamori K, Xu X, Urata M, Bringas P, Chai Y (2009). Epithelial-specific requirement of FGFR2 signaling during tooth and palate development. *J Exp Zool B* 312B:343–350.

Hosokawa R, Oka K, Yamaza T, Iwata J, Urata M, Xu X, Bringas Jr P, Nonaka, K, Chai Y (2010). TGF-[beta] mediated FGF10 signaling in cranial neural crest cells controls development of myogenic progenitor cells through tissue–tissue interactions during tongue morphogenesis. *Dev Biol* 341:186–195.

Huang R, Zhi Q, Izpisua-Belmonte J-C, Christ B, Patel K (1999). Origin and development of the avian tongue muscles. *Anat Embryol* 200:137–152.

Ito Y, Yeo JY, Chytil A, Han J, Bringas P, Nakajima A, Shuler CF, Moses HL, Chai Y (2003). Conditional inactivation of *Tgfbr2* in cranial neural crest causes cleft palate and calvaria defects. *Development* 130:5269–5280.

Iwasaki S-I (2002). Evolution of the structure and function of the vertebrate tongue. *J Anat* 201:1–13.

Iwata J-I, Hacia JG, Suzuki A, Sanchez-Lara PA, Urata M, Chai Y (2012a). Modulation of noncanonical TGF-β signaling prevents cleft palate in *Tgfbr2* mutant mice. *J Clin Investig* 122:873–885.

Iwata J-I, Tung L, Urata M, Hacia JG, Pelikan R, Suzuki A, Ramenzoni L, Chaudhry O, Parada C, Sanchez-Lara PA, et al. (2012b). Fibroblast growth factor 9 (FGF9)–pituitary homeobox 2 (PITX2) pathway mediates transforming growth factor β (TGFβ) signaling to regulate cell proliferation in palatal mesenchyme during mouse palatogenesis. *J Biol Chem* 287:2353–2363.

Jeong J, Mao J, Tenzen T, Kottmann AH, McMahon AP (2004). Hedgehog signaling in the neural crest cells regulates the patterning and growth of facial primordia. *Genes Dev* 18:937–951.

Jeong J, Li X, McEvilly RJ, Rosenfeld MG, Lufkin T, Rubenstein JLR (2008). Dlx genes pattern mammalian jaw primordium by regulating both lower jaw-specific and upper jaw-specific genetic programs. *Development* 135:2905–2916.

Jin J-Z, Ding J (2006). Analysis of cell migration, transdifferentiation and apoptosis during mouse secondary palate fusion. *Development* 133:3341–3347.

Jin J-Z, Li Q, Higashi Y, Darling D, Ding J (2008). Analysis of *Zfhx1a* mutant mice reveals palatal shelf contact-independent medial edge epithelial differentiation during palate fusion. *Cell Tissue Res* 333:29–38.

Kamiya N, Mishina Y (2011). New insights on the roles of BMP signaling in bone: a review of recent mouse genetic studies. *BioFactors* 37:75–82.

Ko SO, Chung IH, Xu X, Oka S, Zhao H, Cho ES, Deng C, Chai Y (2007). Smad4 is required to regulate the fate of cranial neural crest cells. *Dev Biol* 312:435–447.

Kollar EJ, Baird GR (1969). The influence of the dental papilla on the development of tooth shape in embryonic mouse tooth germs. *J Embryol Exp Morphol* 21:131–148.

Lan Y, Jiang R (2009). Sonic hedgehog signaling regulates reciprocal epithelial–mesenchymal interactions controlling palatal outgrowth. *Development* 136:1387–1396.

Lan Y, Ovitt CE, Cho E-S, Maltby KM, Wang Q, Jiang R, (2004). Odd-skipped related 2 (*Osr2*) encodes a key intrinsic regulator of secondary palate growth and morphogenesis. *Development* 131:3207–3216.

Li J, Huang X, Xu X, Mayo J, Bringas P, Jiang R, Wang S, Chai Y (2011a). SMAD4-mediated WNT signaling controls the fate of cranial neural crest cells during tooth morphogenesis. *Development* 138:1977–1989.

Li L, Lin M, Wang Y, Cserjesi P, Chen Z, Chen Y (2011b). Bmprla is required in mesenchymal tissue and has limited redundant function with Bmprlb in tooth and palate development. *Dev Biol* 349:451–461.

Li S, Kong H, Yao N, Yu Q, Wang P, Lin Y, Wang J, Kuang R, Zhao X, Xu J, et al. (2011c). The role of runt-related transcription factor 2 (Runx2) in the late stage of odontoblast differentiation and dentin formation. *Biochem Biophys Res Commun* 410:698–704.

Liu W, Lan Y, Pauws E, Meester-Smoor MA, Stanier P, Zwarthoff EC, Jiang R (2008). The Mn1 transcription factor acts upstream of Tbx22 and preferentially regulates posterior palate growth in mice. *Development* 135:3959–3968.

Lohi M, Tucker AS, Sharpe PT (2010). Expression of *Axin2* indicates a role for canonical Wnt signaling in development of the crown and root during pre- and postnatal tooth development. *Dev Dyn* 239:160–167.

Mammoto T, Mammoto A, Torisawa Y-S, Tat T, Gibbs A, Derda R, Mannix R, deBruijn M, Yung CW, Huh D, et al. (2011). Mechanochemical control of mesenchymal condensation and embryonic tooth organ formation. *Dev Cell* 21:758–769.

Massague J (2000). How cells read TGF-[beta] signals. *Nat Rev Mol Cell Biol* 1:169–178.

Miletich I, Sharpe PT (2004). Neural crest contribution to mammalian tooth formation. *Birth Defects Res C* 72:200–212.

Minoux M, Rijli FM (2010). Molecular mechanisms of cranial neural crest cell migration and patterning in craniofacial development. *Development* 137:2605–2621.

Miura M, Gronthos S, Zhao M, Lu B, Fisher LW, Robey PG, Shi S (2003). SHED: stem cells from human exfoliated deciduous teeth. *Proc Nat Acad Sci USA* 100:5807–5812.

Moore KL, Persaud TVN (2008). *The Developing Human*, W.B. Saunders, Philadelphia.

Morsczeck C, Götz W, Schierholz J, Zeilhofer F, Kühn U, Möhl C, Sippel C, Hoffmann KH (2005). Isolation of precursor cells (PCs) from human dental follicle of wisdom teeth. *Matrix Biol* 24:155–165.

Newbern J, Zhong J, Wickramasinghe RS, Li X, Wu Y, Samuels I, Cherosky N, Karlo JC, O'Loughlin B, Wikenheiser J et al. (2008). Mouse and human phenotypes indicate a critical conserved role for ERK2 signaling in neural crest development. *Proc Nat Acad Sci USA* 105:17115–17120.

Noden DM (1986). Patterning of avian craniofacial muscles. *Dev Biol* 116:347–356.

Noden DM, Francis-West P (2006). The differentiation and morphogenesis of craniofacial muscles. *Dev Dyn* 235:1194–1218.

Northcutt R (2005). The new head hypothesis revisited. *J Exp Zool B* 304B:274–297.

Oka S, Oka K, Xu X, Sasaki T, Bringas Jr P, Chai Y (2007). Cell autonomous requirement for TGF-β signaling during odontoblast differentiation and dentin matrix formation. *Mech Dev* 124:409–415.

Peters H, Balling R (1999). Teeth: where and how to make them. *Trends Genet* 15:59–65.

Peters H, Neubüser A, Kratochwil K, Balling R (1998). *Pax9*-deficient mice lack pharyngeal pouch derivatives and teeth and exhibit craniofacial and limb abnormalities. *Genes Dev* 12:2735–2747.

Petersen CI, Jheon AH, Mostowfi P, Charles C, Ching S, Thirumangalathu S, Barlow LA, Klein OD (2011). FGF signaling regulates the number of posterior taste papillae by controlling progenitor field size. *PLoS Genet* 7: e1002098.

Rice R, Spencer-Dene B, Connor EC, Gritli-Linde A, McMahon AP, Dickson C, Thesleff I, Rice DPC (2004). Disruption of Fgf10/Fgfr2b-coordinated epithelial–mesenchymal interactions causes cleft palate. *J Clin Investig* 113:1692–1700.

Ruch J, Lesot H, Bègue-Kirn C (1995). Odontoblast differentiation. *Int J Dev Biol* 39:51–68.

Satokata I, Maas RL (1994). Msx1 deficient mice exhibit cleft palate and abnormalities of craniofacial and tooth development. *Nat Genet* 6:348–356.

Seo B-M, Miura M, Sonoyama W, Coppe C, Stanyon R, Shi S (2005). Recovery of stem cells from cryopreserved periodontal ligament. *J Dent Res* 84:907–912.

Shi S, Gronthos S (2003). Perivascular niche of postnatal mesenchymal stem cells in human bone marrow and dental pulp. *J Bone Miner Res* 18:696–704.

Snyder-Warwick AK, Perlyn CA, Pan J, Yu K, Zhang L, Ornitz DM (2010). Analysis of a gain-of-function FGFR2 Crouzon mutation provides evidence of loss of function activity in the etiology of cleft palate. *Proc Natl Acad Sci USA*.

Soukup V, Epperlein H-H, Horacek I, Cerny R (2008). Dual epithelial origin of vertebrate oral teeth. *Nature* 455:795–798.

Thesleff I, Keranen S, Jernvall J (2001). Enamel knots as signaling centers linking tooth morphogenesis and odontoblast differentiation. *Adv Dent Res* 15:14–18.

Thirumangalathu S, Harlow DE, Driskell AL, Krimm RF, Barlow LA (2009). Fate mapping of mammalian embryonic taste bud progenitors. *Development* 136:1519–1528.

Thomas BL, Tucker AS, Qui M, Ferguson CA, Hardcastle Z, Rubenstein JL, Sharpe PT (1997). Role of *Dlx-1* and *Dlx-2* genes in patterning of the murine dentition. *Development* 124:4811–4818.

Tucker A, Sharpe P (2004). The cutting-edge of mammalian development; how the embryo makes teeth. *Nat Rev Genet* 5:499–508.

Tucker AS, Yamada G, Grigoriou M, Pachnis V, Sharpe PT (1999). Fgf-8 determines rostral–caudal polarity in the first branchial arch. *Development* 126:51–61.

Tzahor E, Kempf H, Mootoosamy RC, Poon AC, Abzhanov A, Tabin CJ, Dietrich S, Lassar AB (2003). Antagonists of Wnt and BMP signaling promote the formation of vertebrate head muscle. *Genes Dev* 17:3087–3099.

Welsh IC, O'Brien TP (2009). Signaling integration in the rugae growth zone directs sequential SHH signaling center formation during the rostral outgrowth of the palate. *Dev Biol* 336:53–67.

Yu L, Gu S, Alappat S, Song Y, Yan M, Zhang X, Zhang G, Jiang Y, Zhang Z, Zhang Y, et al. (2005). *Shox2*-deficient mice exhibit a rare type of incomplete clefting of the secondary palate. *Development* 132:4397–4406.

Zhang C, Oakley B (1996). The distribution and origin of keratin 20-containing taste buds in rat and human. *Differentiation* 61:121–127.

Zhang Z, Song Y, Zhao X, Zhang X, Fermin C, Chen Y (2002). Rescue of cleft palate in *Msx1*-deficient mice by transgenic *Bmp4* reveals a network of BMP and Shh signaling in the regulation of mammalian palatogenesis. *Development* 129:4135–4146.

Zhang Z, Lan Y, Chai Y, Jiang R (2009). Antagonistic actions of *Msx1* and *Osr2* pattern mammalian teeth into a single row. *Science* 323:1232–1234.

Zhao H, Bringas P, Chai Y (2006). An in vitro model for characterizing the post-migratory cranial neural crest cells of the first branchial arch. *Dev Dyn* 235:1433–1440.

Zhao H, Li S, Han D, Kaartinen V, Chai Y (2011). Alk5-mediated transforming growth factor β signaling acts upstream of fibroblast growth factor 10 to regulate the proliferation and maintenance of dental epithelial stem cells. *Mol Cell Biol* 31:2079–2089.

Zhou J, Gao Y, Zhang Z, Zhang Y, Maltby KM, Liu Z, Lan Y, Jiang R (2011). Osr2 acts downstream of Pax9 and interacts with both Msx1 and Pax9 to pattern the tooth developmental field. *Dev Biol* 353:344–353.

3

CRANIOFACIAL INTRAMEMBRANOUS BONE DEVELOPMENT AND REGENERATION

David P. Rice

Department of Orthodontics, Institute of Dentistry, University of Helsinki, and Oral and Maxillofacial Diseases, Helsinki University Central Hospital, Helsinki, Finland

Ritva Rice

Developmental Biology Program, Institute of Biotechnology, University of Helsinki, Helsinki, Finland

1 INTRODUCTION

Similar to bones elsewhere in the body, craniofacial bones have mechanical, supportive, protective, and homeostatic roles. They are integral to the processes of development and growth of the head and play a central role in eating, speaking (jaws), and in hearing (ear ossicles), and of course give us the foundation of our facial appearance.

The majority of the bones of the face and calvaria are formed by intramembranous ossification, and it is on these that we concentrate in this chapter. The bones of the skull base, the occiput, and some of the bones of the ear and face (inferior nasal concha) are formed by endochondral ossification. During endochondral ossification a cartilage scaffold is formed which gives the primitive bone mechanical strength and a capacity to grow prior to the eventual replacement of the cartilage by bone. Unlike the bones of the skull base, which grow in at the synchondroses by endochondral ossification, the majority of the facial and calvarial bones grow by sutural growth, surface apposition, and modeling and remodeling.

Although we focus on cellular differentiation and function during intramembranous osteogenesis, it should be remembered that the final product consists not only of cells but also canaliculi, vessels, nerves, and, primarily, mineralized extracellular matrix.

Stem Cells in Craniofacial Development and Regeneration, First Edition.
Edited by George T.-J. Huang, Irma Thesleff.

2 INITIAL STAGES OF CRANIOFACIAL BONE DEVELOPMENT

2.1 Cell Origin

Developmentally, the bones of the face, the anterior section of the calvaria, and the anterior cranial base are derived from neural crest cells. The remainder of the calvaria and posterior cranial base are derived from the paraxial and somatic mesoderm. The importance of whether one bone is of neural crest or mesodermal origin is unclear. What is clear is that both cranial neural crest and mesodermal cells have a high degree of plasticity (i.e., they can differentiate into many different cell types).

Functionally speaking, an osteoblast of cranial neural crest (CNC) or mesodermal origin appear to be identical. What is important is the environment in which cells are located. The local environment sends signals to the cells and informs them what to do. In the developing calvaria, what is also important is that CNC cells and mesodermal cells do not mix. The coronal suture forms at the boundary between the CNC and the mesoderm, and normally, cells of these two different origins stay segregated. Abnormal mixing of mesodermal and CNC cells at the coronal suture due to abnormal ephrin-Eph signaling can result in craniosynostosis. This is the case in the craniofrontonasal syndrome caused by mutations in *EFNB1* and also in *Twist1*$^{+/-}$ mice (Twigg et al., 2004; Merrill et al., (2006).

2.2 Patterning of Craniofacial Bones

In most of the vertebrate body, why one particular bone or group of bones forms at a specific location and is a specific shape is determined by the Hox code. This is not so in most of the skull. The Hox code contributes to the posterior section of the cranial base, the styloid process, and the stapes bone in the middle ear. But most of the craniofacial mesenchyme must stay Hox-negative for normal development to progress, and forced expression of *Hoxa2* in the craniofacial mesenchyme inhibits cranial bone development (Kanzler et al., 1998; Creuzet et al., 2002).

Distal-less homebox-containing (Dlx) transcription factors have a role in patterning the face with Dlx1 and 2 regulating upper jaw development and Dlx5 and 6 regulating lower jaw regionalization and identity (Depew et al., 2005; Jeong et al., 2010). Signaling through Endothelin 1, Hand2, and Jagged 1 also determine dorsoventral identity of the lower jaw (Clouthier et al., 2010; Zuniga et al., 2010, 2011). In the upper face, Shh, Fgf8, retinoic acid, and WNT signaling from the forebrain neuroepithelium and facial ectodermact in a coordinated network to pattern the nasal, orbital, premaxillary, and maxillary bones (Schneider et al., 2001; Wang et al., 2011). However, very little is known about why individual calvarial bones have unique shapes and why calvarial bones form at specific locations. Dlx5 and Dlx6 are likely to be involved, as *Dlx5*$^{-/-}$; *Dlx6*$^{-/-}$ compound mutants have no calvarial bone, but whether this is a patterning defect or is due to a defect in the osteogenesis process is not known (Depew et al., 2005). *Dlx5*$^{-/-}$ and *Dlx6*$^{-/-}$ single-mutant mice appear to have normal calvaria.

2.3 Skeletogenic Condensation

The first visible step in making a bone or a cartilage is the formation of a mesenchymal condensation. It forms in a certain location at a specific time and is responsible for the final shape and size of the skeletal element. Each skeletal element forms through a mesenchymal condensation. Some bones, such as the incus and malleus of the middle ear, form from a single condensation (Hall and Miyake, 2000). Multiple condensations can also form a single skeletal unit. For example, in mammalians, two paired chondrification centers, the trabecular and the polar hypophyseal, unite early and join with the basal plate to form a continuous plate of cartilage in the forming cranial base (Meikle, 2002).

Intramembranous bones arise from osteogenic condensations and cartilages from chondrogenic condensations. Cells within osteogenic condensations have already differentiated into osteoprogenitor cells, and formation of condensation amplifies their number, whereas chondrogenic condensations trigger chondrogenesis in undifferentiated mesenchymal cells (Dunlop and Hall, 1995). The basic cellular processes of condensation formation are the same. Initiation of mesenchymal condensation requires mesenchymal cell migration to and/or a failure to disperse a specific location where the cells form an aggregation. A boundary is set for the condensation that essentially makes it semi-independent from its surroundings. The cells within the established condensation will adhere together and have their own program for proliferation and growth. Once a critical size for the condensation has been reached, it will initiate differentiation, or switching to an organ-specific cell fate. Mesenchymal cell condensation is regulated by mitotic activity, cell–cell and cell–extracellular matrix contacts, and cytoskeletal tension affecting cell shape; and condensed mesenchyme is also less vascularized than loose, noncondensing mesenchyme (Hall and Miyake, 2000; Knothe Tate et al., 2008; Rice and Rice, 2008).

Mesenchymal condensations form in close proximity to epithelia. In chick embryos the osteogenic condensation contributing to maxilla and mandible forms adjacent to the facial ectoderm, and isolated cultures of mandibular arch mesenchyme do not form condensations. Recombining the mandibular mesenchyme with its epithelium rescues the ability of the mesenchyme to condense (Dunlop and Hall, 1995). During early tooth formation the dental epithelium has a developmental window at embryonic day E10–E11 to induce condensation of the adjacent mesenchyme (Mina and Kollar, 1987; Mammoto et al., 2011). The dental epithelium is a source of multiple signals that contribute to the mesenchymal cell migration, either by attraction or repulsion, or to mesenchymal cell adhesion through cell surface or extracellular matrix proteins (Hall and Miyake, 2000; Mammoto et al., 2011).

A lack of mesenchymal condensation leads to a lack of the corresponding skeletal element. For example, deletion of the transcription factor *Sox9* in the limb bud caused an absence of chondrogenic condensations in limb buds and a complete lack of limbs (Akiyama et al., 2002). Conditional deletion of *Sox9* in the cranial neural crest cells led to a lack of chondrogenic skeletal elements derived from this cell population (Mori-Akiyama et al., 2003). Interestingly, the authors showed that absence of *Sox9* in the cranial neural crest did not affect the intramembranous bone formation derived from the same cell population. Interestingly, haploinsufficiency of *Sox9* specifically in the cranial neural crest cells leads to patent posterior frontal suture in mice (Sahar et al., 2005). The mouse posterior frontal suture is a unique calvarial suture on account of both its morphology and function: It is comprised of exocranial and endocranial sutures between the opposing frontal bones, and only the endocranial posterior frontal suture fuses postnatally by endochondral ossification (Sahar et al., 2005). This is unusual, as most pathological or physiological suture fusion occurs intramembranously. The authors show that the endocranial posterior frontal suture fusion begins by mesenchymal cell

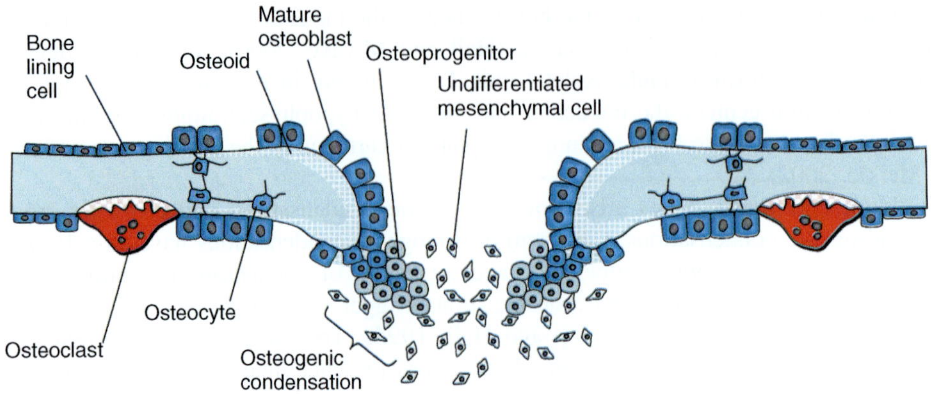

FIGURE 1 Schematic representation of a suture and intramembranous ossification highlighting: that all the stages of osteoblast differentiation can be seen at a single time point; that undifferentiated mesenchymal cells condense at the bone ends (osteogenic fronts) and differentiate directly into osteoblasts which lay down osteoid. The importance of osteoclastic differentiation and function early in intramembranous bone formation is also highlighted.

condensation between the opposing frontal bones, and this condensation coincides with increased expression of *Sox9* and *type II collagen* that are later followed by increased expression of *type X collagen* to indicate that the initial chondrogenesis was followed by hypertrophy, and that this was followed by osteogenesis and bone formation, as indicated by expression of *osteocalcin*.

Mesenchymal cells can be induced to condense and form chondrogenic condensations by growth factor TGFβ1. TGFβ1 upregulates the level of fibronectin, which regulates NCAM, a cell adhesion molecule that initiates the condensation (Hall and Miyake, 2000). When multipotent mesenchymal cells from postnatal sagittal and posterior frontal sutures were isolated and cultured in the presence of TGFβ1, proliferation decreased and osteogenesis was inhibited. In the mesenchymal cells derived from preclosure posterior frontal suture, TGFβ1 also increased the expression of cell adhesion molecules N-cadherin and fibronectin and induced mesenchymal cell condensation and the expression of chondrocyte-specific gene markers (Xu et al., 2008; James et al., 2009).

In many ways the osteogenic fronts that line the craniofacial intramembranous bones are analogous to the initial sites of skeletogenic condensation in the head. The same processes of mesenchymal cellular aggregation, compaction, differentiation, and function of osteoblasts occur at both locations (Fig. 1).

3 OSTEOBLAST DIFFERENTIATION, FUNCTION, AND FATE

Knowing how osteoblasts, osteocytes, and osteoclasts differentiate, function, and interact is central to understanding how intramembranous bones develop, grow, are remodeled, and function.

3.1 Osteoblast Differentiation

Osteoblasts and chondrocytes differentiate from common mesenchymal precursor cells, which given different transcriptional signals can differentiate into skeletal precursor cells and then progress down osteoblastic or chondroblastic lineages (Fig. 2). Sox9 is one protein that will drive cells into a chondroblastic fate. Skeletal precursor cells express both *Sox9* and the transcription factor *Runx2*. If Runx2 is suppressed, directly by Sox9 or indirectly via the homeobox containing transcription factor Nkx3.2, and cells express Sox9, 5, and 6, skeletal precursor cells will differentiate into chondroblasts. Sox9, 5, and 6 form homodimers which increase the efficiency of their cooperative binding to chondrocyte-specific enhancers (Akiyama and Lefebvre, 2011).

The initial stages of osteoblast differentiation from skeletal precursor cells into committed osteoprogenitors is controlled by Runx2, Osterix/Sp7 (Osx), and β-catenin (Kobayashi and Kronenberg, 2005; Hartmann, 2006). Runx2 lies at a nexus of control between skeletal precursor cells and osteoprogenitors, which still have the ability to differentiate into chondroblasts. Once osteoprogenitors express *Osx*, they are committed to an osteoblastic fate (Fig. 2). Runx2 is a master regulator of osteoblast differentiation and is under the transcriptional control of Msx2, Dlx5, and Sox8 and the posttranscriptional control of the transcription factor Twist1 and Histone deacetylases 3 (HDAC3) and 7 (Hartmann, 2009). The special importance of Runx2 and Osterix to bone development is demonstrated by Runx2- and Osterix-null allele mice, both of which have no osteoblasts (Komori et al., 1997; Otto et al., 1997; Nakashima et al., 2002). Interestingly, Sox9 and Runx2, which are essential for early stages of differentiation, are inhibitory for later stages and must be downregulated for osteoblast lineage progression.

During very early skeletogenic cell differentiation, β-catenin levels have to be maintained at a low level for skeletal precursor cell differentiation from mesenchymal stem cells. In contrast to this, for skeletal precursor cells to differentiate into osteoprogenitors, β-catenin levels must elevate. Continued low levels of β-catenin favor differentiation of the osteoprogenitors into the chondrocyte lineage (Fig. 2) (Hill et al., 2005). β-Catenin is an essential, obligatory, and nonredundant component of the canonical Wnt signaling pathway. Both canonical and noncanonical Wnt signaling are important at multiple stages of osteoblast differentiation and the control of pre- and postnatal osteoblast function. Similarly, signaling through the Hedgehog (HH), Notch, fibroblast growth factor (FGF) and bone morphogenetic protein (BMP) pathways are all important in regulating the stepwise progression of osteoblastic differentiation and function and thereby control intramembranous bone development. These families of developmental signals act by regulating the key transcription factors mentioned above (Long, 2012).

The importance of the genes and signaling pathways that regulate osteoblast development is highlighted by genetic mutations which result in human diseases involving intramembranous bones: namely, cleidocranial dysplasia, parietal foramina, Wormian (heterotopic) bones, and craniosynostosis (premature suture fusion) (Table 1).

3.2 Osteoblast Function

Once committed to an osteoblastic fate, cells progress through a preosteoblastic state before extracellular matrix production starts and cells differentiate into mature osteoblasts. Committed osteoprogenitors are still able to proliferate. The extracellular matrix that osteoblasts secrete is called *osteoid*. Osteoid is unmineralized and largely composed of type I collagen and is later mineralized through the formation of hydroxyapatite. The mineralization of osteoid is initiated in matrix vesicles which are

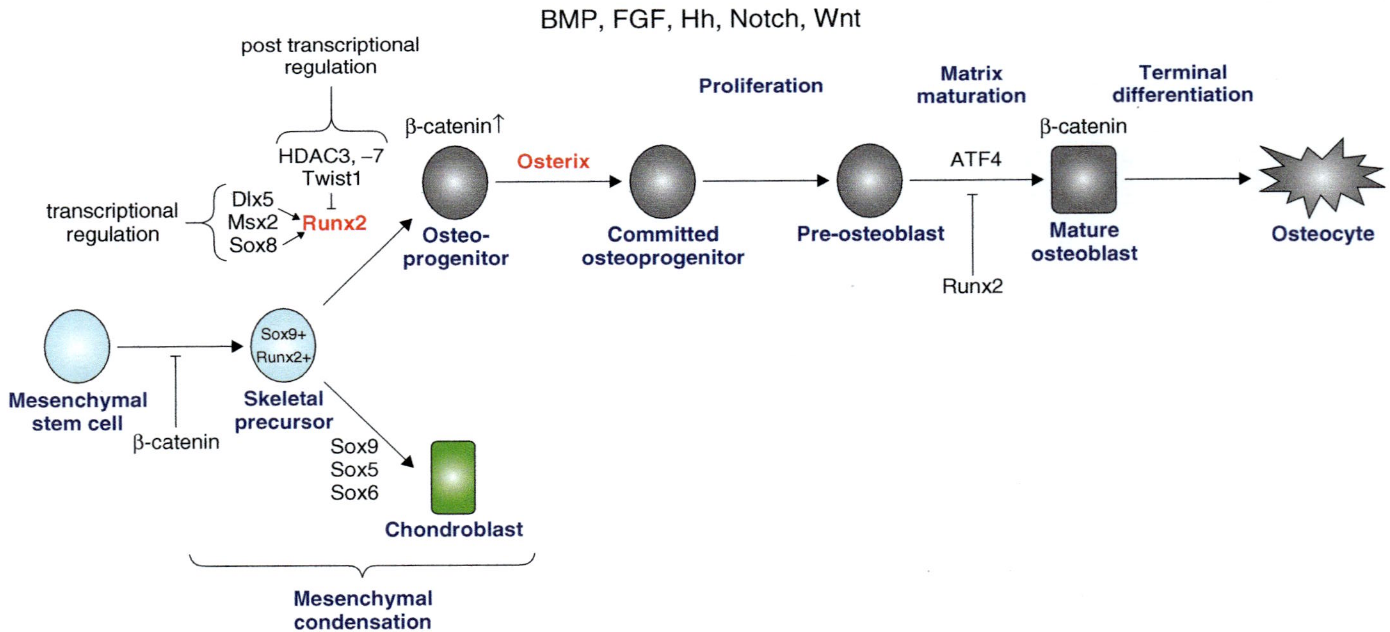

FIGURE 2 Osteoblast differentiation highlighting the mesenchymal stem cell origin of osteoblasts; the osteoblastic and chondroblastic bipotential of the Sox9-positive, Runx2-positive skeletal precursor cells; the key roles of Runx2 and Osterix; and the cells that can be found in skeletogenic condensations. BMP, FGF, Hh, Notch, and Wnt signaling play important roles at multiple steps of osteoblast differentiation and function. [Adapted from Hartmann (2009), with permission.]

TABLE 1 Human Gene Mutations Resulting in Calvarial (Intramembranous) Bone Phenotypes

Gene Mutated	Type of Protein	Action/ Function	Human Disease OMIM Number	Intramembranous Bone Phenotype	Other Features
RUNX2	Transcription factor	Master regulator of osteoblast differentiation	Cleidocranial dysplasia (119600)	Delayed intramembranous bone formation, Wormian bones, defective clavicle	Supernumerary teeth, delayed tooth eruption
TWIST1	Transcription factor	Inhibitor of *Runx2*	Saethre–Chotzen syndrome (101400)	Craniosynostosis	Syndactyly
FGFR1	Growth factor receptor		Pfeiffer syndrome (101600)	Craniosynostosis	Syndactyly, broad thumbs/toes
FGFR2	Growth factor receptor		Apert (101200)	Craniosynostosis	Syndactyly
			Crouzon (123500)	Craniosynostosis	
			Pfeiffer (101600)	Craniosynostosis	Syndactyly
			Antley–Bixler syndromes (201750, 207410)	Craniosynostosis	Radiohumeral synostosis
FGFR3	Growth factor receptor		Muenke syndrome (602849)	Craniosynostosis	Brachydactyly, carpal/tarsal fusion
			Thanatophoric dysplasia type II (187601)	Craniosynostosis	Short stature, neonatal mortality
MSX2	Transcription factor		Boston-type craniosynostosis (604757)	Craniosynostosis	
			Parietal foramina (168500), parietal foramina with cleidocranial dysplasia (168550)	Holes in the calvarial bones, delayed intramembranous bone formation	

(*Continued*)

TABLE 1 (*Continued*)

Gene Mutated	Type of Protein	Action/ Function	Human Disease OMIM Number	Intramembranous Bone Phenotype	Other Features
ALX4	Transcription factor		Parietal foramina 2 (609597)	Holes in the calvarial bones, delayed intramembranous bone formation	
IL11RA (interleukin 11 receptor, alpha)	Cytokine receptor	Regulates osteoblast and osteoclast function	Kreiborg–Pakistani syndrome (614188)	Craniosynostosis	Supernumerary teeth, delayed tooth eruption
EFNB1	Receptor protein—tyrosine kinase	Regulates cell–cell signaling and developmental boundary determination	Craniofrontonasal syndrome (304110)	Craniosynostosis	Females have frontonasal dysplasia, nail defects
GLI3	Transcription factor	Regulates Hh signaling	Greig cephalopolysyndactyly syndrome (175700)	Metopic suture synostosis	Polydactyly, syndactyly
RAB23	Small GTPase	Negative regulator of Hh signaling	Carpenter syndrome (201000)	Craniosynostosis	Polydactyly, syndactyly obesity, short stature

Source: OMIM Online, Mendelian inheritance of man.

extracellular membrane-invested vesicles produced by budding from the osteoblast membrane. As well as containing type I collagen, bone matrix vesicles also contain several noncollagenous matrix proteins, including osteocalcin and bone sialoprotein, and growth factors, including vascular endothelial growth factor and several BMPs (Xiao et al., 2007; Nahar et al., 2008).

A major determinant of osteoblastic function is activating transcription factor 4 (ATF4) (Karsenty, 2008). It interacts with Foxo2 to regulate protein synthesis in osteoblasts (Ambrogini et al., 2010). AFT4 regulates the transcription of osteocalcin directly and the synthesis of type I collagen indirectly, both major components of osteoid. AFT4 also regulates osteoclastic differentiation and function, and thus bone remodeling, by regulating the expression of RANKL (receptor activator of nuclear factor-κB ligand) (Yang et al., 2004). AFT4 activity is itself kept in check by FIAT (factor inhibiting AFT4-mediated transcription), which is expressed early in osteoblastogenesis (Yu et al., 2005). AFT4 activity can be modified by AP1 family proteins. Notably, Fra1 (FOS-related antigen 1) is likely to modify AFT4 by dimerizing with it. Fra1 also controls osteoid production by regulating the matrix genes osteocalcin, type I collagen, and matrix Gla protein (Eferl et al., 2004).

3.3 Osteoblast Fate

Once matrix production at a particular location is complete, osteogenesis is terminated and osteoblasts can either become quiescent bone-lining cells, undergo programmed cell death (apoptosis), or become surrounded by matrix and develop into osteocytes. Osteocytes are highly specialized cells that may act as mechanoreceptors, regulate osteoclast and osteoblast activity, and produce hormones (Bonewald, 2011).

4 OSTEOCLAST DIFFERENTIATION AND FUNCTION

Intramembranous bone development is not just about the formation of bone but also about bone degradation, modeling, and remodeling. These processes are tightly controlled, and the crosstalk between osteoblasts, osteoclasts, and osteocytes is central to this regulation. In the developing calvaria, osteoclasts appear relatively early during bone development, at E16.5 in the mouse, only a few days after the first osteoblasts have differentiated (Rice et al., 1997). Although some osteoclasts are located in the sutures, where they have a role in the maintenance of suture patency, most resorption occurs on the endocranial surface of the developing calvarial bones. This is balanced by apposition on the ectocranial surfaces so that the calvaria expands in unison with the developing brain. Thus, intramembranous bone development involves coordination of bone formation and degradation.

Osteoclasts are of myeloid origin. They are generally large polarized, multinucleated cells, although functioning small mononuclear osteoclasts have also been identified (Mellis et al., 2011). Osteoclasts are derived from mononuclear preosteoclasts, which are in turn derived from granulocyte–macrophage colony-forming units, which are derived from multipotent myeloid cells which are ultimately derived from hematopoietic stem cells. During early embryogenesis, osteoclast precursors are formed in the extraembryonic tissues. Once the circulatory system becomes established, precursor cells are generated in the liver and spleen, and once bone marrow formation occurs, they are then also formed in the bones themselves.

The transcription factor PU.1 is responsible for the early step in osteoclastogenesis of myeloid cells differentiating into the monocyte–macrophage lineage. Macrophage colony-stimulating factor (M-CSF) promotes macrophage survival and proliferation, and together with PU.1 it upregulates RANK. M-CSF and RANK ligand (RANKL) are necessary and sufficient for osteoclastogenesis from monocyte–macrophage precursors. Once activated, RANK, together with AP-1 transcription factor complex and microphthalmia-associated transcription factor (MITF) enter the nucleus to regulate the expression of multiple genes required for osteoclast differentiation (Mellis et al., 2011). M-CSF and RANKL are also involved in the osteoclast recruitment. Once recruited, osteoclasts adhere to bone and undergo cytodifferentiation and activation.

Osteoclasts destroy bone by the secretion of acid and hydrolase enzymes, including tartrate-resistant acid phosphatase and proteases such as cathepsin K, into the resorption lacuna. The acid dissolves the hydroxyapatite mineral, while the organic components of bone are degraded by the active enzymes. The lacuna is a sealed space, so that the active substances can reach the appropriate concentration levels and so that these potentially toxic substances are not released into the surrounding environment.

5 CELLULAR COORDINATION OF BONE DEPOSITION AND RESORPTION

Osteoblasts, osteoclasts, and osteocytes communicate with each other. This regulates both bone formation and destruction, presumably during both initial shaping and growth of bones (modeling) and subsequent reshaping of bones (remodeling).

Osteoblasts, bone-lining cells, and stromal cells induce osteoclast differentiation. This communication occurs through the growth factors M-CSF and RANKL, and the glycoprotein semaphorin 3B (Matsuo, 2009). The RANK–RANKL system seems to be of particular importance, as it requires physical contact of osteoclast precursors with osteoblasts, bone-lining cells, or stromal cells. This is because both RANKL and its receptor RANK are bound to the cell surface. Underlining the importance of this direct cell-to-cell communication, both $RANKL^{-/-}$ and $RANK^{-/-}$ mice have an identical osteopetrotic phenotype with a complete lack of osteoclasts. Osteoprotegerin (OPG) is a secreted receptor that lacks a transmembrane domain. OPG acts in a dominant negative manner to inhibit osteoclastogenesis by competing with RANK for RANKL-binding sites. Overexpression of *OPG* leads to osteopetrosis, whereas deletion of *OPG* causes enhanced osteoclastic activity, resulting in osteoporosis. In a regulatory feedback mechanism, OPG is produced by osteoblasts in response to estrogens, TGFβs, and BMPs and is thought to temper RANKL/RANK signaling (Fig. 3) (Boyle et al., 2003).

The activation of osteoclast differentiation and subsequent bone resorption through the RANK-RANKL system could, in theory, occur in two ways. First, osteoblasts or bone-lining cells could attract osteoclast precursors to the site where resorption is required, and osteoclastic induction and activation occur locally. Alternatively, osteoclast precursors may be induced at a remote location and the activated precursors are brought to the resorption site by the circulatory system.

Inhibition of osteoclast differentiation and stimulation of osteoblast differentiation can also occur through a direct cell–cell contact through ephrinB2-EphB4 signaling.

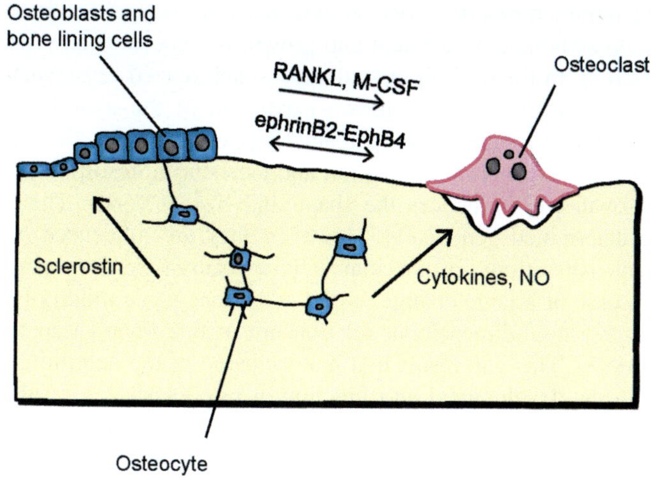

FIGURE 3 Network of communication between osteoblasts, osteocytes, and osteoclasts which controls bone formation and remodeling. [Modified from Matsuo (2009).]

Osteoclasts express the membrane-bound ligand ephrinB2, and osteoblasts express its cell surface receptor EphB4. As ephrin-Eph binding can result in bidirectional signal transduction, both osteoclastic and osteoblastic differentiation can be regulated simultaneously (Zhao et al., 2006).

Osteocytes have been proposed as orchestrators of bone remodeling through their role as mechanosensors. This is because of their location in bone, their complex dendritic network of canaliculi, and because their gene expression profile changes with bone loading and unloading (Bonewald, 2011). Osteocytes can regulate bone formation and destruction. In response to mechanical loading, osteocytes reduce the production of the Wnt signaling inhibitors sclerostin (SOST) and Dickkopf-1 (Dkk1). As both these factors are negative regulators of osteoblast function, bone formation is increased. Osteocytes can inhibit bone destruction through the effects of nitric oxide and TGFβ (Fig. 3) (Matsuo, 2009; Bonewald, 2011).

6 INTRAMEMBRANOUS BONE GROWTH, MODELING, REMODELING, SUTURE FORMATION, AND FUNCTION

Centers of ossification expand by appositional growth which occurs at the leading edges or osteogenic fronts of the immature intramembranous bones. When bone fronts converge, they either fuse or a suture or fontanel is formed some craniofacial bones, such as the human maxillae and zygomatic bones, are made from one ossification center, while others are made from two or more which coalesce to form a single bone. Examples of these include the human parietal bones, which form from two ossification sites, and the sphenoidal and temporal bones, which develop from multiple initiation sites. Some bones are formed entirely by intramembranous ossification: for example, the nasal bones. Others—for example, the temporal bones—are the result of fusion of both intramembranous and endochondral elements (Rice, 2008).

Craniofacial intramembranous bone growth not only occurs at sutures but also at other surfaces. Also, bone development and growth is a coordinated process of formation and destruction. In the developing calvaria, synchronized ectocranial appositional and endocranial resorptive activity allows expansion of the underlying brain while maintaining bone thickness and suture patency (Rice et al., 1997). Similarly in the face, the balance between osteoblastic apposition and osteoclastic resorption determines the rate of bone growth and influences the shape and size of bones. The dimensions of each bone are determined genetically. However, they are influenced by the soft tissue environment. Alterations in muscular activity, caused by, for example, muscular degenerative disease or a large change in diet consistency, can modify bone shape and size as well as the three-dimensional relationship of one bone to another (Kiliaridis and Katsaros, 1998). This can result in a malocclusion of the dentition.

Alterations in the development and function of osteoblasts can result in a delay of intramembranous bone growth, such as those seen in mice that have a heterozygotic loss of *Runx2* (Otto et al., 1997). However, little is known about the determinants of the particular direction and speed of intramembranous bone growth. We do know, however, that calvarial bone enlargement and the direction of growth are regulated by the transcription factor Foxc1. Mice lacking *Foxc1* havecalvarial bones that do not grow beyond a rudimentary size and remain at the sites of the initial skeletogenic condensations. Foxc1 regulates BMP-mediated osteoprogenitor proliferation specifically at the leading edge of the developing calvarial bones (Rice et al., 2003).

Sutures are major sites of osteogenesis and bone growth which modify calvarial as well as upper and midfacial growth. To ensure proper suture function, two adjacent bones must be held apart. When suture patency is lost, bones fuse and future growth at that location is not possible. The timing of suture fusion during skeletal growth is under delicate control. Suture fusion can occur with normal development and aging: for example, fusion of the midpalatal and interfrontal (metopic) sutures, or pathologically when it is termed craniosynostosis.

Embryonic sutures consist of two bone ends and the interposed mesenchymal tissue (Fig. 1). Later the bones become encapsulated by periosteum, with fibrous tissue in between. Contrary to theories put forth between the 1940s and 1970s that craniofacial sutures would possess an innate bone separating growth force (Moore, 1949; Azuma et al., 1976), sutures are now considered to be passive tension-adapted growth sites. Growth at the osteogenic fronts will occur in response to tension being exerted on the suture by, for example, the forces of mastication or the tension within the dura (Moss, 1957; Ryöppy, 1965). Tension can come, via the dura, from the expanding brain or raised intracranial pressure, or from mastication forces via muscles, ligaments, and the periosteum. A good example of the effects of tension of suture biogenesis comes from studies on rats that are fed soft diets. In these experiments, the tension on the facial bones and sutures is reduced, and this affects sutural apposition and bone growth in areas under the direct influence of masticatory muscles and their attachments. This, in turn, affects facial growth (Katsaros et al., 2006). The dura has been shown to have important roles in calvarial bone induction and in the maintenance of suture patency (Bradley et al., 1997).

At different stages of development, different areas of the craniofacial region grow at different velocities. The calvaria undergoes most of its growth during the embryonic and early postnatal periods, whereas the facial skeleton undergoes most of its rapid growth later. The rate of apposition or osteogenesis at each bone end in a suture may

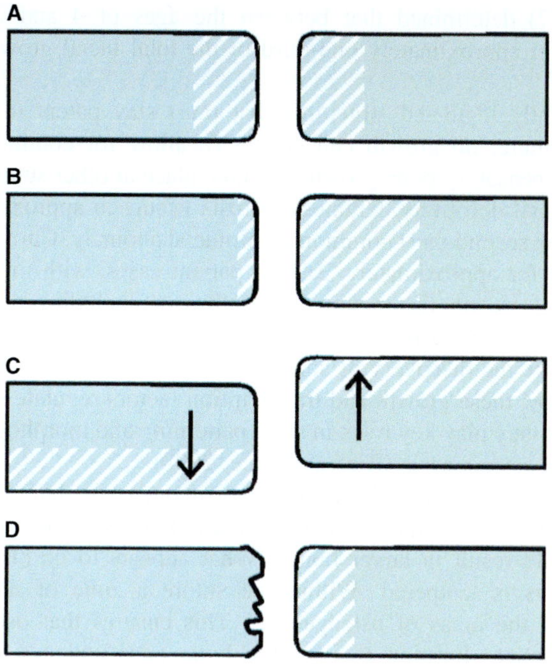

FIGURE 4 Sutural bone growth. (a) Equal bone apposition on both sides of the suture (e.g., the midpalatal suture). (b) Unequal bone apposition on the two sides of a suture (e.g., the frontonasal suture, where growth in the frontal bone can be five times that of the nasal bone). (c) Sliding of one osteogenic front in relation to its neighbor (arrows) (e.g., the nasopremaxillary suture). (d) Bone apposition on one side of a suture and resorption on the other (e.g., the transverse palatine suture). New bone growth is illustrated with diagonal shading. [Modified from Persson (1995).]

be equal and may be in opposing directions (Persson, 1995) (Fig. 4). However, this is not always the case. In the frontonasal suture, apposition on the frontal bone side is fivefold greater than that on the nasal bone side. Greater bone growth on one side of a suture compared to the other is sometimes seen in response to craniosynostosis. Compensatory growth in response to the premature fusion at one suture can occur at other sutures, probably caused by pressure from the growing brain, and this reaction may be asymmetric with apposition at one bone end greater than at its partner (Jane and Persing, 2001).

Two bones across a suture generally grow in opposite directions. But bones may slide across a suture line, as seen in the nasopremaxillary suture or rotate about a suture: for example, the median palatal suture. Here, there is differential growth in the posterior versus anterior sections of the suture, which results in the two maxillary bones rotating in relation to each other in the transverse plane (Bjork and Skieller, 1977).

The majority of the bones of the middle and upper parts of the face grow at the sutures, by surface remodeling, apposition, and resorption. The vertical growth in the maxilla and mandible is in large part due to growth of the alveolar bone in response to the development and eruption of the dentition. By following the displacement of metal implants with lateral cephalogram radiographs over time, Björk and Skieller (1977) determined that between the ages of 4 and 20 years, sutural growth accounts for approximately one-third of the total facial growth in the vertical dimension.

Sutures are tightly regulated structures that must stay patent to function. In the craniosynostotic suture, no growth can occur. To allow for continued calvarial and facial growth, compensatory bone growth can take place at other sutures, but this often results in craniofacial deformity. Craniosynostosis occurs in approximately 1 in 2500 live births and is the second most common craniofacial anomaly. Currently, known gene mutations account for approximately 20% of patient cases, with mutations occurring in genes in several growth factor signaling pathways, including the FGF (*FGFRs*), BMP (*MSX2*), HH (*GLI3, RAB23*), and ephrin-Eph (*EFNB1*) pathways (Johnson and Wilkie, 2011). Since the discovery of some of the gene mutations, we have learned a great deal about how these growth and transcription factors regulate suture biogenesis. We now know that they play key roles in skull patterning and morphogenesis, boundary determination, and osteogenesis (Rice and Rice, 2008).

Craniosynostosis is not merely excessive osteogenesis, but a combination of enhanced osteogenesis and loss of suture patency. Excessive osteogenesis and bone growth alone would result in larger bones. What appears to be of great importance is that osteogenesis is tempered. Within the suture a zone of inhibition must be established around the areas of osteogenesis. This ensures that osteogenesis occurs only at the appropriate location (i.e., at the bone ends and not in the midsutural region). Accordingly, the BMP antagonist Noggin, the Runx2 inhibitor Twist, and the Hedgehog repressor Gli3 have all been shown to have pivotal roles in maintaining suture patency (Rice et al., 2000, 2010; Warren et al., 2003; Bialek et al., 2004).

7 INTRAMEMBRANOUS BONE REGENERATION

The purpose of this section is to introduce the subject of craniofacial bone regeneration and emphasize the direct link between intramembranous bone development and bone regeneration. Here we set the scene for the topic of craniofacial bone regeneration, which is dealt with comprehensively in Chapters 19 to 21 and 23).

Most if not all multicellular organisms possess an innate ability to regenerate tissue. They can recreate or restore lost or damaged tissues, in some cases even organs and limbs. In amphibians this capability is dramatic, with regeneration of limbs, tails, jaws, eyes, and several internal structures. In higher vertebrates, including humans, tissue regeneration is more limited but still possible in many tissues that turn over during their life, including blood, skin, hairs, and in particular, bone.

There is great academic and clinical interest in harnessing this potential of bone in the field of craniofacial reconstruction. Reconstruction may be aimed at restoring defects due to developmental or growth deformity, destructive diseases such as tumors and periodontal disease, or trauma. Treatments may be dramatic, such as large-scale reconstruction of the facial skeleton following trauma, or more routine, such as reconstruction of the alveolar bone lost for various reasons, such as congenital bony clefting, tooth loss, or periodontal disease (Retzepi and Donos, 2010; Ward et al., 2010; Guo et al., 2011).

Most bone regeneration strategies rely on providing the ideal conditions for osteoinduction. In addition to bone induction, nonosteogenic cells can be mechanically excluded from entering the regeneration site by guided bone regeneration membranes. The conditions for osteoinduction are based largely on our knowledge of intramembranous bone development and remodeling. Bone is not only formed in the embryo, but our skeleton is turned over continually, which implies regenerative potential in the adult also. Most approaches used for bone tissue engineering mimic this development or turnover. Combinations of cells, biochemical signals, and materials are often used in concert (Ward et al., 2010). Material-based strategies provide support for cell recruitment, adhesion, survival, differentiation, and activation. They allow the correct temporally and spatially controlled release of the signals required for bone regeneration (Ji et al., 2012). They encourage correct vascularization and ensure that the new bone is the appropriate shape and size. Materials that have been used include natural and synthetic polymers, nanofibers, ceramics, and composites (Ward et al., 2010; King and Krebsbach, 2012).

For bone regeneration it is important to have cells capable of differentiation into functional osteoblasts. Multipotent mesenchymal cells and, by definition, mesenchymal stem cells have this ability. Stem cells are found in both embryonic and postnatal tissue and have the ability for self-renewal and asymmetric division potential (Friedenstein et al., 1966; Robey and Bianco, 2006). To do this, stem cells need to be able to initiate cell lineage–specific differentiation and commit to their chosen cell fate. Stem cells in culture have been shown to become osteoblasts by induction of changes in cell shape from spindle shaped to an elongated and spread shape. This osteoblastic differentiation requires an intact cytoskeleton and an appropriate cell density (Pittenger et al., 1999; McBeath et al., 2004).

Mesenchymal stem cells (MSCs) have been isolated from many different tissues, including bone marrow, adipose, umbilical cord, teeth, and the periodontal ligament. Umbilical cord–derived MSCs have been shown to produce bone, but as they can produce all three germ layer tissues (i.e., endoderm, mesoderm, and ectoderm), umbilical cord–derived stem cells have an enormous potential to produce a vast array of different tissues (Wang et al., 2009). Bone marrow–derived mesenchymal stem cells (BMSCs) have been used successfully in a number of different bone repair models (Ohgushi et al., 1989; Krebsbach et al., 1998). Of particular interest are adipose-derived stem cells (ADSCs). ADSCs have been shown to have good bone-healing ability, but unlike BMSCs, ADSCs are abundant and relatively easily sourced (Cowan et al., 2004; Panetta et al., 2010)

8 CONCLUSIONS AND FUTURE DIRECTIONS

Although intramembranous and endochondral bones form by different mechanisms, bones of both origins undergo appositional growth (increase in thickness of long bones), and they are modeled and remodeled via apposition and resorption. These are essentially mechanisms of intramembranous bone formation and are central to bone regeneration strategies of all bones. It is therefore important to understand the gene networks regulating osteoblast and osteoclast function, together with those regulating stem cell differentiation. A detailed understanding of the characteristics of MSCs and how they interact with various scaffolds will be key to the utilization of their potential. Rigorous in vitro and in vivo testing will be required before these types of bone regeneration therapies will be commonplace.

REFERENCES

Akiyama H, Lefebvre V. (2011). Unraveling the transcriptional regulatory machinery in chondrogenesis. *J Bone Miner Metab* 29(4):390–395.

Akiyama H, Chaboissier MC, Martin JF, Schedl A, et al. (2002). The transcription factor Sox9 has essential roles in successive steps of the chondrocyte differentiation pathway and is required for expression of *Sox5* and *Sox6*. *Genes Dev* 16(21):2813–2828.

Ambrogini E, Almeida M, Martin-Millan M, Paik JH, et al. (2010). FoxO-mediated defense against oxidative stress in osteoblasts is indispensable for skeletal homeostasis in mice. *Cell Metab* 11(2):136–146.

Azuma ME, Enlow DH, Fredrickson RG, Gaston LG. (1976). A myofibroblastic base for the physical forces that produce tooth drift and eruption, skeletal displacement at sutures, and periosteal migration. In: *Determinants of Mandibular Form and Growth*, McNamara, JA Jr, ed., The University of Michigan, Ann Arbor, MI, USA.

Bialek P, Kern B, Yang X, Schrock M, et al. (2004). A twist code determines the onset of osteoblast differentiation. *Dev Cell* 6(3):423–435.

Björk A, Skieller V. (1977). Growth of the maxilla in three dimensions as revealed radiographically by the implant method. *Br J Orthod* 4(2):53–64.

Bonewald LF. (2011). The amazing osteocyte. *J Bone Miner Res* 26(2):229–238.

Boyle WJ, Simonet WS, Lacey DL (2003). Osteoclast differentiation and activation. *Nature* 423(6937):337–342.

Bradley JP, Levine JP, McCarthy JG, Longaker MT (1997). Studies in cranial suture biology: regional dura mater determines in vitro cranial suture fusion. *Plast Reconstr Surg* 100(5):1091–1099; discussion 1100–1102.

Clouthier DE, Garcia E, Schilling TF (2010). Regulation of facial morphogenesis by endothelin signaling: insights from mice and fish. *Am J Med Genet A* 152A(12):2962–2973.

Cowan CM, Shi YY, Aalami OO, Chou YF, et al. (2004). Adipose-derived adult stromal cells heal critical-size mouse calvarial defects. *Nat Biotechnol* 22(5):560–567.

Creuzet S, Couly G, Vincent C, Le Douarin NM (2002). Negative effect of Hox gene expression on the development of the neural crest–derived facial skeleton. *Development* 129(18):4301–4313.

Depew MJ, Simpson CA, Morasso M, Rubenstein JL (2005). Reassessing the Dlx code: the genetic regulation of branchial arch skeletal pattern and development. *J Anat* 207(5):501–561.

Dunlop LL, Hall BK (1995). Relationships between cellular condensation, preosteoblast formation and epithelial–mesenchymal interactions in initiation of osteogenesis. *Int J Dev Biol* 39(2):357–371.

Eferl R, Hoebertz A, Schilling AF, Rath M, et al. (2004). The Fos-related antigen Fra-1 is an activator of bone matrix formation. *EMBO J* 23(14):2789–2799.

Friedenstein AJ, Piatetzky S, II, Petrakova KV (1966). Osteogenesis in transplants of bone marrow cells. *J Embryol Exp Morphol* 16(3):381–390.

Guo J, Li C, Zhang Q, Wu G, et al. (2011). Secondary bone grafting for alveolar cleft in children with cleft lip or cleft lip and palate. *Cochrane Database Syst Rev* 6:CD008050.

Hall BK, Miyake T (2000). All for one and one for all: condensations and the initiation of skeletal development. *Bioessays* 22(2):138–147.

Hartmann C. (2006). A Wnt canon orchestrating osteoblastogenesis. *Trends Cell Biol* 16(3):151–158.

Hartmann C (2009). Transcriptional networks controlling skeletal development. *Curr Opin Genet Dev* 19(5):437–443.

Hill TP, Spater D, Taketo MM, Birchmeier W, et al. (2005). Canonical Wnt/beta-catenin signaling prevents osteoblasts from differentiating into chondrocytes. *Dev Cell* 8(5):727–738.

James AW, Xu Y, Lee JK, Wang R, et al. (2009). Differential effects of TGF-beta1 and TGF-beta3 on chondrogenesis in posterofrontal cranial suture-derived mesenchymal cells in vitro. *Plast Reconstr Surg* 123(1):31–43.

Jane J, Persing J (2001). Neurosurgical treatment of craniosynostosis. In: *Craniosynostosis Diagnosis, Evaluation, and Management*, Cohen MJ, Maclean R, eds., Oxford University Press, Oxford, UK, Chap. 18.

Jeong J, Li X, McEvilly RJ, Rosenfeld MG, et al. (2008). *Dlx* genes pattern mammalian jaw primordium by regulating both lower jaw–specific and upper jaw–specific genetic programs. *Development* 135(17):2905–2916.

Ji W, Wang H, van den Beucken JJ, Yang F, et al. (2012). Local delivery of small and large biomolecules in craniomaxillofacial bone. *Adv Drug Deliv Rev* 12:1152–1164.

Johnson D Wilkie AO (2011). Craniosynostosis. *Eur J Hum Genet* 19(4):369–376.

Kanzler B, Kuschert SJ, Liu YH Mallo M (1998). Hoxa-2 restricts the chondrogenic domain and inhibits bone formation during development of the branchial area. *Development* 125(14):2587–2597.

Karsenty G (2008). Transcriptional control of skeletogenesis. *Annu Rev Genom Hum Genet* 9:183–196.

Katsaros C, Zissis A, Bresin A, Kiliaridis S (2006). Functional influence on sutural bone apposition in the growing rat. *Am J Orthod Dentofac Orthop* 129(3):352–357.

Kiliaridis S, Katsaros C (1998). The effects of myotonic dystrophy and Duchenne muscular dystrophy on the orofacial muscles and dentofacial morphology. *Acta Odontol Scand* 56(6):369–374.

King WJ, Krebsbach PH (2012). Growth factor delivery: how surface interactions modulate release in vitro and in vivo. *Adv Drug Deliv Rev*

Knothe Tate ML, Falls TD, McBride SH, Atit R, et al. (2008). Mechanical modulation of osteochondroprogenitor cell fate. *Int J Biochem Cell Biol* 40(12):2720–2738.

Kobayashi T, Kronenberg H (2005). Minireview: transcriptional regulation in development of bone. *Endocrinology* 146(3):1012–1017.

Komori T, Yagi H, Nomura S, Yamaguchi A, et al. (1997). Targeted disruption of Cbfa1 results in a complete lack of bone formation owing to maturational arrest of osteoblasts. *Cell* 89(5):755–764.

Krebsbach PH, Mankani MH, Satomura K, Kuznetsov SA, et al. (1998). Repair of craniotomy defects using bone marrow stromal cells. *Transplantation* 66(10):1272–1278.

Long F (2012). Building strong bones: molecular regulation of the osteoblast lineage. *Nat Rev Mol Cell Biol* 13(1):27–38.

Mammoto T, Mammoto A, Torisawa YS, Tat T, et al. (2011). Mechanochemical control of mesenchymal condensation and embryonic tooth organ formation. *Dev Cell* 21(4):758–769.

Matsuo K (2009). Cross-talk among bone cells. *Curr Opin Nephrol Hypertens* 18(4):292–297.

McBeath R, Pirone DM, Nelson CM, Bhadriraju K, et al. (2004). Cell shape, cytoskeletal tension, and RhoA regulate stem cell lineage commitment. *Dev Cell* 6(4):483–495.

Meikle MC (2002). Development of the face. In: Craniofacial Development, Growth and Evolution, Bateson Publishing, Bressingham, UK.

Mellis DJ, Itzstein C, Helfrich MH, Crockett JC (2011). The skeleton: a multi-functional complex organ: the role of key signalling pathways in osteoclast differentiation and in bone resorption. *J Endocrinol* 211(2):131–143.

Merrill AE, Bochukova EG, Brugger SM, Ishii M, et al. (2006). Cell mixing at a neural crest–mesoderm boundary and deficient ephrin-Eph signaling in the pathogenesis of craniosynostosis. *Hum Mol Genet* 15(8):1319–1328.

Mina M, Kollar EJ (1987). The induction of odontogenesis in non-dental mesenchyme combined with early murine mandibular arch epithelium. *Arch Oral Biol* 32(2):123–127.

Moore AW (1949). Head growth of the macaque monkey as revealed by vital staining, embedding, and undecalcified sectioning. *Am J Orthod* 35(9):654–671.

Mori-Akiyama Y, Akiyama H, Rowitch DH, de Crombrugghe B (2003). Sox9 is required for determination of the chondrogenic cell lineage in the cranial neural crest. *Proc Natl Acad Sci USA* 100(16):9360–9365.

Moss ML (1957). Experimental alteration of sutural area morphology. *Anat Rec* 127(3):569–589.

Nahar NN, Missana LR, Garimella R, Tague, SE, et al. (2008). Matrix vesicles are carriers of bone morphogenetic proteins (BMPs), vascular endothelial growth factor (VEGF), and noncollagenous matrix proteins. *J Bone Miner Metab* 26(5):514–519.

Nakashima K, Zhou X, Kunkel G, Zhang Z, et al. (2002). The novel zinc finger–containing transcription factor Osterix is required for osteoblast differentiation and bone formation. *Cell* 108(1):17–29.

Ohgushi H, Goldberg VM, Caplan AI (1989). Repair of bone defects with marrow cells and porous ceramic: experiments in rats. *Acta Orthop Scand* 60(3):334–339.

Otto F, Thornell AP, Crompton T, Denzel A, et al. (1997). *Cbfa1*, a candidate gene for cleidocranial dysplasia syndrome, is essential for osteoblast differentiation and bone development. *Cell* 89(5):765–771.

Panetta NJ, Gupta DM, Longaker MT (2010). Bone regeneration and repair. *Curr Stem Cell Res Ther* 5(2):122–128.

Persson M (1995). The role of sutures in normal and abnormal craniofacial growth. *Acta Odontol Scand* 53(3):152–161.

Pittenger MF, Mackay AM, Beck SC, Jaiswal RK, et al. (1999). Multilineage potential of adult human mesenchymal stem cells. *Science* 284(5411):143–147.

Retzepi M, Donos N (2010). Guided bone regeneration: biological principle and therapeutic applications. *Clin Oral Implants Res* 21(6):567–576.

Rice DP (2008). Developmental anatomy of craniofacial sutures. *Front Oral Biol* 12:1–21.

Rice DP, Rice R (2008). Locate, condense, differentiate, grow and confront: developmental mechanisms controlling intramembranous bone and suture formation and function. *Front Oral Biol* 12:22–40.

Rice DP, Kim HJ, Thesleff I (1997). Detection of gelatinase B expression reveals osteoclastic bone resorption as a feature of early calvarial bone development. *Bone* 21(6):479–486.

Rice DP, Aberg T, Chan Y, Tang Z, et al. (2000). Integration of FGF and Twist in calvarial bone and suture development. *Development* 127(9):1845–1855.

Rice DP, Connor EC, Veltmaat JM, Lana-Elola E, et al. (2010). *Gli3Xt-J/Xt-J* mice exhibit lambdoid suture craniosynostosis which results from altered osteoprogenitor proliferation and differentiation. *Hum Mol Genet* 19(17):3457–3467.

Rice R, Rice DP, Olsen BR, Thesleff I (2003). Progression of calvarial bone development requires Foxc1 regulation of Msx2 and Alx4. *Dev Biol* 262(1):75–87.

Robey PG, Bianco P (2006). The use of adult stem cells in rebuilding the human face. *J Am Dent Assoc* 137(7):961–972.

Ryöppy S (1965). Transplantation of epiphyseal cartilage and cranial suture: experimental studies on the preservation of the growth capacity in growing bone grafts. *Acta Orthop Scand Suppl* 82:1–106.

Sahar DE, Longaker MT, Quarto N (2005). *Sox9* neural crest determinant gene controls patterning and closure of the posterior frontal cranial suture. *Dev Biol* 280(2):344–361.

Schneider RA, Hu D, Rubenstein JL, Maden M, et al. (2001). Local retinoid signaling coordinates forebrain and facial morphogenesis by maintaining FGF8 and SHH. *Development* 128(14):2755–2767.

Twigg SR, Kan R, Babbs C, Bochukova EG, et al. (2004). Mutations of ephrin-B1 (EFNB1), a marker of tissue boundary formation, cause craniofrontonasal syndrome. *Proc Natl Acad Sci USA* 101(23):8652–8657.

Wang L, Singh M, Bonewald LF, Detamore MS (2009). Signalling strategies for osteogenic differentiation of human umbilical cord mesenchymal stromal cells for 3D bone tissue engineering. *J Tissue Eng Regen Med* 3(5):398–404.

Wang Y, Song L, Zhou CJ (2011). The canonical Wnt/beta-catenin signaling pathway regulates Fgf signaling for early facial development. *Dev Biol* 349(2):250–260.

Ward BB, Brown SE, Krebsbach PH (2010). Bioengineering strategies for regeneration of craniofacial bone: a review of emerging technologies. *Oral Dis* 16(8):709–716.

Warren SM, Brunet LJ, Harland RM, Economides AN, et al. (2003). The BMP antagonist Noggin regulates cranial suture fusion. *Nature* 422(6932):625–629.

Xiao Z, Camalier CE, Nagashima K, Chan KC, et al. (2007). Analysis of the extracellular matrix vesicle proteome in mineralizing osteoblasts. *J Cell Physiol* 210(2):325–335.

Xu Y, James AW, Longaker MT (2008). Transforming growth factor-beta1 stimulates chondrogenic differentiation of posterofrontal suture-derived mesenchymal cells in vitro. *Plast Reconstr Surg* 122(6):1649–1659.

Yang X, Matsuda K, Bialek P, Jacquot S, et al. (2004). ATF4 is a substrate of RSK2 and an essential regulator of osteoblast biology; implication for Coffin–Lowry syndrome. *Cell* 117(3):387–398.

Yu VW, Ambartsoumian G, Verlinden L, Moir JM, et al. (2005). FIAT represses ATF4-mediated transcription to regulate bone mass in transgenic mice. *J Cell Biol* 169(4):591–601.

Zhao C, Irie N, Takada Y, Shimoda K, et al. (2006). Bidirectional ephrinB2-EphB4 signaling controls bone homeostasis. *Cell Metab* 4(2):111–121.

Zuniga E, Stellabotte F, Crump JG (2010). Jagged-Notch signaling ensures dorsal skeletal identity in the vertebrate face. *Development* 137(11):1843–1852.

Zuniga E, Rippen M, Alexander C, Schilling TF, et al. (2011). Gremlin 2 regulates distinct roles of BMP and endothelin 1 signaling in dorsoventral patterning of the facial skeleton. *Development* 138(23):5147–5156.

4

TEMPOROMANDIBULAR JOINT DEVELOPMENT

Shuping Gu and YiPing Chen

Department of Cell and Molecular Biology, Tulane University, New Orleans, Louisiana

1 INTRODUCTION

The temporomandibular joint (TMJ) is a unique synovial joint found only in mammals. In mammalian head, there are two TMJs, one on each side, working in unison. In humans, the TMJs are involved not only in capturing, processing, and swallowing food, but also in speaking and maturing the facial contour. To accomplish these complicated roles, specific components from skull base and lower jaw have been integrated into the complex structure of TMJ during mammalian evolution.

The TMJ is classified as a synovial joint surrounded by a fibrous capsule and lubricated by synovial fluid. The TMJ is unique, with its sophisticated structure and flexible movement. The surface of the TMJ is covered with a fibrous layer and the joint cavity is divided into two by a specific articular disk made of fibrocartilage and positioned between the glenoid fossa and the condyle. This disk buffers compressing strokes from the joint and allows its rotational and translational movement (Herring, 2003; Sperber et al., 2010).

Evolutionally, the TMJ is a new creation in mammals and differs completely from the jaw joint in reptiles, which is a simple hinge junction between the quadrate of skull bone and the articular of the lower jaw, serving for grasping prey. The mammalian TMJ consists of five basic elements and functions synergistically with masticatory muscles, including temporals, masseters, and pterygoids. With its rotational and translational movement, the TMJ can perform grasping and shearing, as well as chewing (Kermack, 1972). However, there is considerable diversity in TMJ structure among mammals. In carnivores, the glenoid fossa of temporal bone develops into a cylindrical bony

Stem Cells in Craniofacial Development and Regeneration, First Edition.
Edited by George T.-J. Huang, Irma Thesleff.

flange and encircles the condyle of mandibular bone completely, such that the TMJ works as a strong hinge for secure grasping of prey during violent predatory action. In herbivores, extensive chewing is necessary for food intake, a process that requires the lower jaw to move horizontally rather than vertically. To accommodate such flat movement, the glenoid fossa and condyle of the TMJ develop a flat shape. In omnivores such as humans, both hinging and sliding movements are integrated in the TMJ, and the condyle is matched with the glenoid fossa, and the articular disk forms slopes, making anterior–posterior and lateral movement possible, as well as rotation and translation (Brodie, 1969).

In addition to being a key component of TMJ function, the condyle serves as a major site of lower jaw growth. It functions as a growth cartilage similarly as epiphyseal growth plates in long bones and contribute to growth of the ramus of the mandible in its length and height via cell division in the perichondrium and subsequent endochondrial ossification. The growth potential of the condyle persists until the age of maturity. Disturbances in this function of the condyle lead to defective development of the mandible, resulting in malocculusions and lateral mandibular displacement.

The TMJ is vulnerable to both internal and external conditions that attribute to its structural complexity and functional multiplicity. The TMJ is susceptible to factors affecting other joints in the human body, such as inflammation, trauma, tumors, overwork, and developmental anomalies, causing TMJ diseases, often referred to as TMJ disorder or TMJ syndrome. TMJ disorder usually involves more than one TMJ component and can be classified into three groups: congenital malformations of the TMJ, such as auriculo-condylar syndrome, primary growth disorders such as condylar hyperactivity, and acquired TMJ disorders such as juvenile idiopathic arthritis and the TMJ malformations associated with mandible trauma (Masotti et al., 2008; Pirttiniemi et al., 2009). The symptoms of TMJ disorder can be very complex, the most common ones being pain, sound, and restricted movement of the joints. Other signs of TMJ disorder include dizziness and hearing loss (Ramirez et al., 2005). Since the symptoms and etiology of TMJ disorder are complicated, there is no unambiguous correlation between certain signs and causes, making it difficult to give an accurate diagnosis and effective therapy.

2 DEVELOPMENT OF THE TMJ

The development of the mammalian TMJ is divided into three stages: the initiation stage, also called the blastema stage, the growth and cavitation stage, and the maturation or completion stage (Dixon, 1997; Merida-Velasco et al., 1999). Unlike the joints of long bones that are generated by cleavage or segmentation within a single skeletal condensation, the TMJ develops from two distinct and widely separated mesenchymal condensations: the glenoid fossa blastema, which derives from the otic capsule and undergoes intramembranous ossification, and the condyle primordium, which arises from the secondary cartilage of the mandible and forms bone through endochondral ossification. TMJ development begins with the appearance of mesenchymal cell condensation of the future condyle, which forms dorsal and superolateral to Meckel's cartilage and is independent of Meckel's cartilage. This is followed by the formation of the mesenchymal condensation of the glenoid fossa primordium located superior-laterally to the condylar condensation, which demarcates the initial development of the temporal component (glenoid fossa) of the TMJ. Between the condylar blastema and the glenoid fossa primordium is a wide area intervened with loose connective tissue that will later integrate into the articular disk. Subsequently, owing to rapid cell proliferation, the condylar blastema expands into a rectangular cell condensation situated lateral to and above Meckel's cartilage and attached by the lateral pterygoid muscle medially. At the same time, the primordium of glenoid fossa extends more anterior–medially above and caps the condylar blastema. A dense connective tissue preceding the future articular disk separates the two primordia of TMJ. As the condyle grows continuously upward, approaching the glenoid fossa, the wide space between them decreases, and the mesenchymal cell layer of the disk anlage condenses and centralizes, leading to the formation of TMJ cavity. Accompanied by the cell proliferation and increment in volume, the condyle anlage undergoes differentiation and is configured into a typical secondary cartilage covered superficially with a thick layer of flat fibrous cells. Corresponding to the differentiation of the condyle, intramembranous ossification occurs in the glenoid fossa. Along with the development of the skeletal elements of the TMJ, morphogenesis of the soft tissues that attach to the joint continues. After the completion of cavitation, TMJ development is characterized by dramatic growth and ossification of the condyle and glenoid fossa, functional remodeling of the articular disk through substantial condensation and avascular event, enclosure of the joint bone prominences and disk by the joint capsule, and development of the ligaments and muscles (Baume, 1962; Baume and Holz, 1970; Humphrey, 1971; McKay, 1992; Merida-Velasco et al., 1999, 2009; Ögütcen-Toller and Keskin, 2000).

Human TMJ development is initiated at around week 7 to 8 of gestation or 22 to 35-mm crown-rump length (CRL), with the formation of the condyle primordium and glenoid fossa blastema (Baume, 1962; Baume and Holz, 1970). Although there is disagreement concerning the exact time of the onset of human TMJ development, by the end of embryonic week 8, the five basic components of human TMJ—the glenoid fossa, the condyle, the articular disk, the capsule, and the ligament—have become discernible as mesenchymal condensations. Cavitation of the developing TMJ occurs at about weeks 9 to 11 of gestation. Toward week 12 of gestation, cavitation is completed by the formation of the upper (superior) and lower (inferior) cavities of the TMJ (Baume, 1962; Baume and Holz, 1970; Ramieri et al., 1996; Merida-Velasco et al., 1999, 2009).

The mouse has been used as a model to study TMJ organogenesis systematically because of genetic and developmental similarities to humans and amendable genetic manipulation. The mouse and human TMJ share similarities in structures (Fig. 1), as well as in the developmental processes.

In mice, the mesenchymal condensation of the condyle forms at embryonic day 13.5 (E13.5). Its location is depicted by the surrounding trigeminal ganglion, lateral pterygoid muscle, and Meckel's cartilage (Fig. 2A and B), and by the expression of *Sox9* (Fig. 2C and D). This condensation also undergoes rapid cell proliferation (Fig. 2E and F). On serial coronal sections, the anterior portion of the condylar primordium is closely connected to the primordium of the mandible (Fig. 2B), while its posterior portion is isolated from the mandible (Fig. 2A). These observations suggest that the mesenchymal cells at the dorsal end of the mandible proliferate superior–posteriorly to form the condylar blastema and are probably derived from the periosteum of the mandibular bone (Miyake et al., 1997). However, it has also been suggested that the mammalian condyle develops from a blastema that is separated from the mandibular

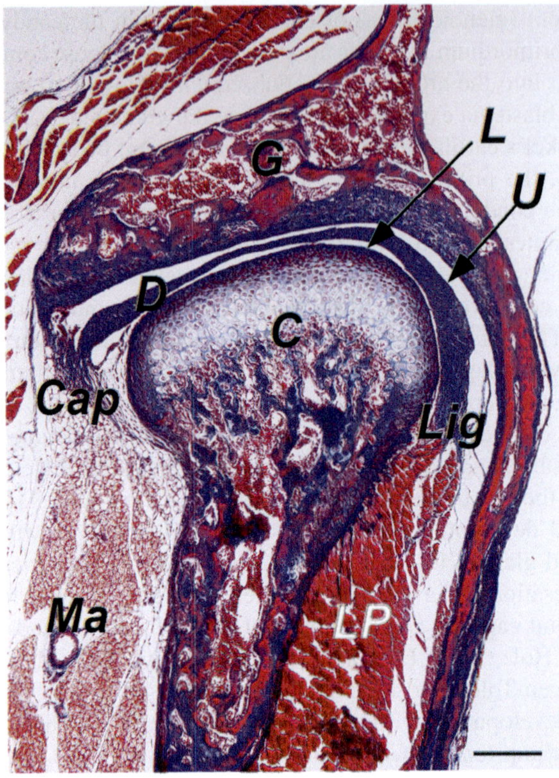

FIGURE 1 Coronal section through a P15 mouse TMJ shows the main structural components of a matured synovial joint. *C*, condyle; *Cap*, joint capsule; *D*, disk; *G*, glenoid fossa; *L*, lower cavity; *Lig*, ligament; *LP*, lateral pterygoid muscle; *Ma*, masseter muscle; *U*, upper cavity. Scale bar: 100 μm.

bone (Baume, 1962; Duterloo and Janson, 1970). At E14.5 the condylar primordium grows and condenses further as a result of rapid cell proliferation, and is rearranged and shaped into a reverse-positioned cone with its tip pointed toward the mandibular bone (Fig. 3A). Medially, the lateral pterygoid muscle runs into the cone-shaped condyle. Above the top of the condyle, the zygomatic process and squamosal plate join together to form a concave-shaped glenoid fossa primordium. At E15.5, cartilage development begins in the center of the condyle (Fig. 3B). Surrounding the cartilage is a thick layer of perichondrium, and the lower part of the condyle produces cartilaginous matrix such as type II collagen (Gu et al., 2008). At the apex of the developing condyle, several layers of cells that split from the surface of the condyle are tightened into a stripe, representing the anlage of the articular disk. The glenoid fossa also grows bigger, with ossification starting in the center of the fossa. However, the space between the developing condyle and glenoid fossa remains wide (Fig. 3B).

At E16.5, a cleft forms between the articulating surface of the glenoid fossa and the disk, marking the beginning of cavitation. This cleft is generated when loose mesenchymal cells in the articulating space are squeezed upward. Although there is no sign for lower cavity formation at this stage, the stripe of the disk anlage is separated completely from the surface of the condyle, with the cells of the medial

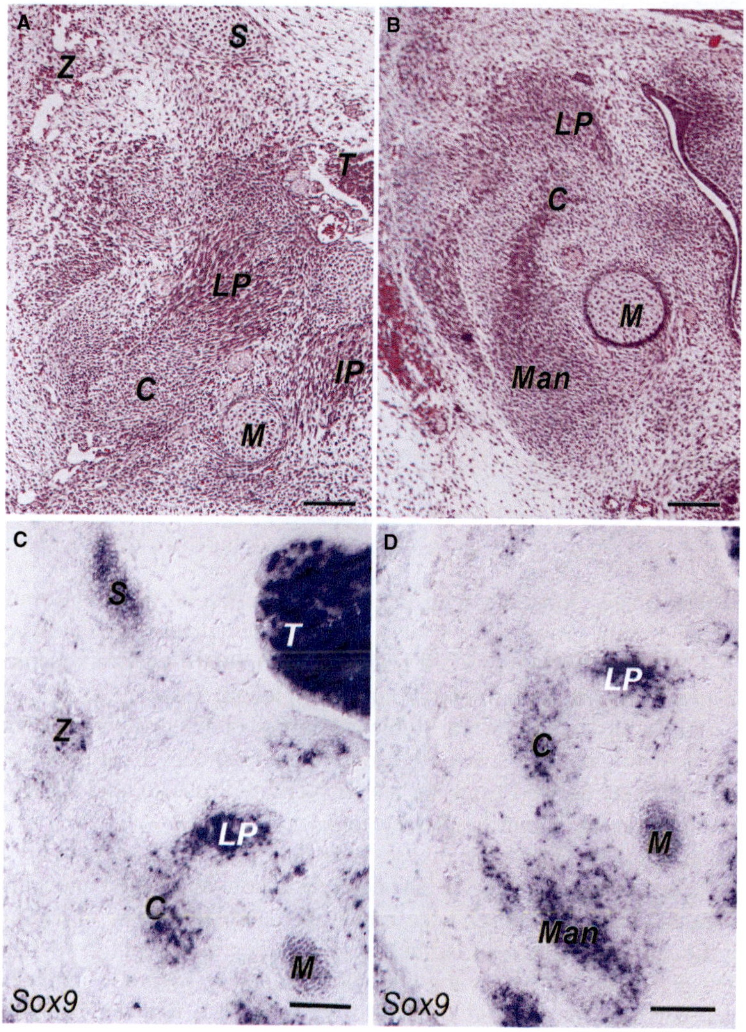

FIGURE 2 Coronal sections through an E13.5 mouse embryonic head show the position of the condylar primordium in relation to adjacent tissues. Sections in A, C, and E were made through the middle region of the trigeminal ganglion, and sections in B, D, and F, through the front edge of the trigeminal ganglion. (A and B) Condylar primodium appears as mesenchymal condensation positioned lateral-superiorly to Meckel's cartilage and lateral-inferiorly to the trigeminal ganglion. Its medial border is outlined by the lateral pterygoid muscle. (C and D) *Sox9* expression is detected in the primordium of the condyle, lateral pterygoid muscle, squamosal plate, and trigeminal ganglion. (E and F) BrdU labeling shows higher levels of cell proliferation in the primordia of condyle, lateral pterygoid muscle, squamosal plate, zygomatic process, and mandibular bone. *C*, condylar primordium; *IP*, internal pterygoid muscle; *LP*, lateral pterygoid muscle; *M*, Meckel's cartilage; *Man*, mandibular bone primordium; *S*, squmosal plate, *T*, trigeminal ganglion; *Z*, zygomatic process. Scale bar: 100 μm.

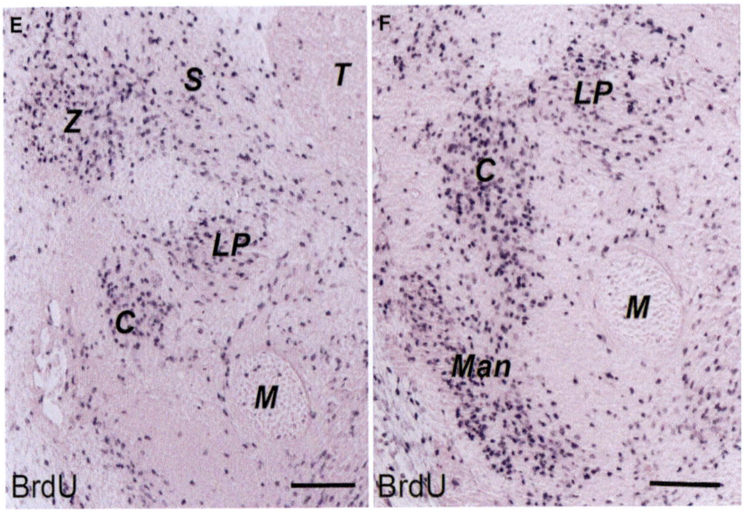

FIGURE 2 (*Continued*)

part of the disk anlage mingling with the cells of the lateral pterygoid muscle. The condyle itself not only grows fast but also differentiates, presenting the hypertrophic, flattened, polymorphic, and fibrous cell layers. The articulating space becomes narrow due to the rapid growth of the condyle and the glenoid fossa (Fig. 3C). At E17.5, a definite articular disk with several tightened layers of fibers forms, leading to the formation of a narrow cavity between the condyle and the disk. Its medial fibers blend with the tendon fibers derived from the lateral pterygoid muscle, and its lateral fibers tangle with that extended from the masseter muscle (Fig. 3D).

By postnatal day 15 (P15), the mouse TMJ is a completely developed and fully functional cartilaginous joint (Fig. 1). The condyle is now remodeled into a club shape, with its rounded head continuing to the ramus of mandibular bone through a narrow neck, with the surface of the rounded head covered by a thin, tight fibrous layer. The well-ossified glenoid fossa covers the condyle compatibly, with the joint disk sitting in the space between the glenoid fossa and condyle. The ligaments form from the peripheral fibers of the disk, which join with the tendon fibers from the surrounding muscles.

3 MAMMALIAN CONDYLE IS MADE OF SECONDARY CARTILAGE

The condylar cartilage is classified as secondary cartilage, differing from the primary skeletal cartilage (Beresford, 1981). Although the mandibular condyle undergoes endochondral ossification similar to the long bone formation, it is unique in its rapid differentiation from the progenitor cells to hypertrophic chondrocytes. This is evidenced by the fact that condylar cartilage cells express type II and type X collagens simultaneously (Shibata et al., 1997; Fukada et al., 1999; Gu et al., 2008), suggesting that the chondrocytes differentiate into hypertrophic chondrocytes very quickly or that the hypertrophic chondrocytes differentiate directly from the undifferentiated progenitor cells. In addition, the condylar cartilage cells also express type I collagen and are

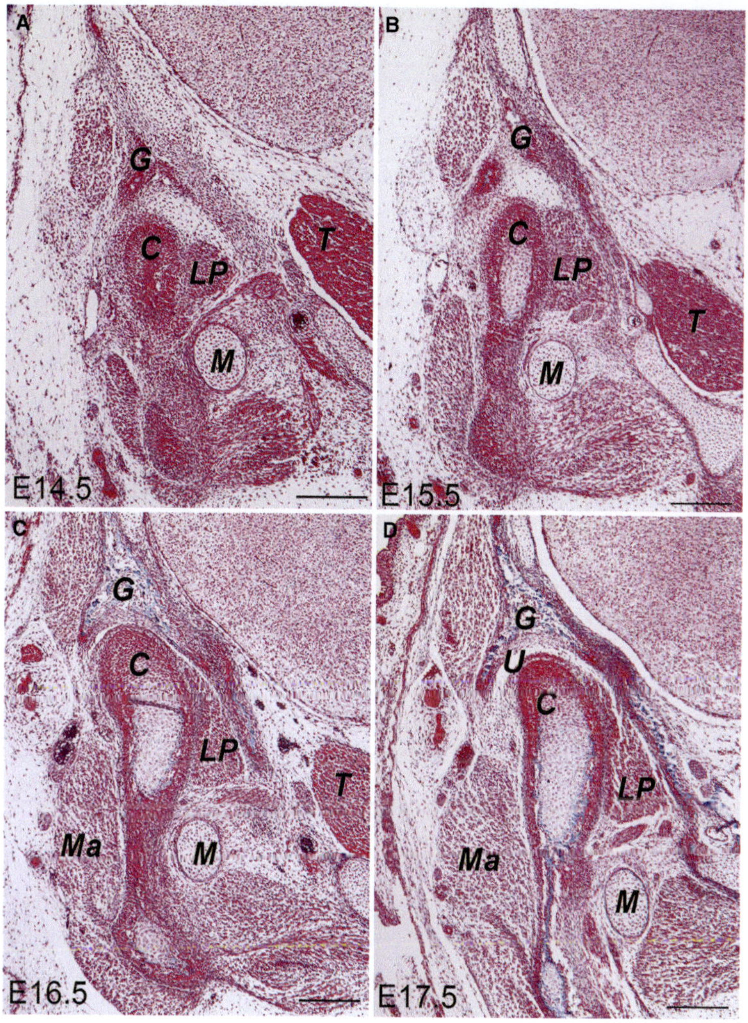

FIGURE 3 Early development of the mouse TMJ. Coronal sections were made through the middle area of the trigeminal ganglion. (A) At E14.5, a well-defined condylar primordium is observed under the cap-shaped glenoid fossa primordium. (B) The articular disk begins to form at E15.5. The developing condyle is still far from the glenoid fossa that is forming at this stage. (C) At E16.5, a differentiated condyle is covered by a well-ossified glenoid fossa. The upper cavity of the joint becomes discernible. (D) At E17.5, a cavitated TMJ is formed with the presence of all major tissue components. *C*, condyle; *G*, glenoid fossa; *LP*, lateral pterygoid muscle; *M*, Meckel's cartilage; *Ma*, masseter; *T*, trigeminal ganglion; *U*, upper cavity of the joint. Scale bar: 100 μm.

alkaline phosphatase positive (Silbermann et al., 1987; Shibata et al., 1997; Fukada et al., 1999; reviewed in Hall, 2005), indicating that these progenitor cells possess preosteoblast characteristics and have potential to differentiate into both chondrocytes and osteoblasts (Silbermann et al., 1987). The fact that the condylar primordial cells are closely associated with the periosteum of the mandible and are positive for alkaline phosphatase further supports the periosteal origin of the condyle (Miyake et al., 1997; Shibata et al., 1997).

4 MANY GENES ARE EXPRESSED SIMILARLY IN THE DEVELOPING CONDYLE AND PRIMARY CARTILAGES

Despite the well-documented structure and function of TMJ, the genetic control of TMJ development is poorly understood, in contrast to a wealth of information on the molecular regulation of synovial joint formation in the limbs. Gene expression has been examined during TMJ development, with a great emphasis on the developing condyle in the last two decades. This is likely due to the fact that (1) the condylar cartilage is an important growth site for the elongation of the mandibular ramus, and (2) condyle growth undergoes processes of chondrogenesis and endochondral ossification similar to those in long bone formation. These studies have documented in the developing rodent condyle the expression profiles of a number of genes that are known to be expressed and to play critical roles in the growth and differentiation of primary cartilage in long bone development, including *Bmp4, Fgf2, Ihh, Pthrp, Tgf*β2, *Vegf, Cbfa1, Osterix, Sox9, Aggrecan, Col2*, and *Col10* (Fukada et al., 1999; Rabie and Hägg, 2002; Kuboki et al., 2003; Ogawa et al., 2003; Watahiki et al., 2003; Tang et al., 2004; Shibata et al., 2006; Shibukawa et al., 2007). Some of these genes exhibit similar expression patterns in developing condylar and long bone cartilages, suggesting similar roles in the development of these two different types of cartilage.

5 TISSUE INTERACTIONS PLAY CRITICAL ROLES IN TMJ DEVELOPMENT

It is well documented that tissue interactions, mediated by diffusible signaling molecules, play essential roles in embryonic development and organ formation. They regulate multiple developmental processes during organogenesis, including determination, pattern formation, growth, and differentiation. While both the condyle and glenoid fossa, the skeletal elements of the TMJ, are derivatives of cranial neural crest cells (Gu et al., 2008), they develop from two distinct mesenchymal condensations and undergo different ossification processes. These two blastemas are initiated independently, and grow toward each other to form the TMJ. Using genetically modified mouse models, Wang et al., (2011) investigated tissue interactions between the developing condyle and glenoid fossa. It was shown that in the absence of the developing condyle, glenoid fossa development is arrested. In such a case, the glenoid fossa blastema forms initially, but regresses subsequently. A similar developmental defect was also observed in the glenoid fossa when the developing condyle is dislocated. These observations indicate that the presence and proximity of the developing condyle are essential for development of the glenoid fossa during TMJ formation. Interestingly, it was found that Meckel's cartilage is able to substitute for the condyle to support glenoid fossa development (Wang et al., 2011). These observations suggest that presently unknown signals from the developing condyle are required to sustain glenoid fossa development, and that Meckel's cartilage, which is a primary cartilage, appears to produce similar signaling molecules as the condyle. On the other hand, condyle development and differentiation appear to be independent of the presence of the glenoid fossa (Wang et al., 2011). This observation is consistent with previous reports that mesenchymal tissue isolated from the maxilla–mandibular junction region of rodent embryo developed into a differentiated condylar cartilage-like structure with secondary cartilage characteristics in organ culture in vitro (Glasstone, 1971; Vinkka-Puhakka and Thesleff, 1993). This conclusion is supported further by a recent study in which it was shown that glenoid fossa development is completely disrupted but condyle and disk formation are not affected in mice carrying null mutations in both *Spry1* and *Spry2* that encode antagonists of RTK signaling, including FGF signaling (Purcell et al., 2012).

6 REGULATION OF TMJ DEVELOPMENT BY TRANSCRIPTION FACTORS

Recent phenotypic analyses of genetically modified mouse models have revealed the functional importance of a number of genes in TMJ development, including transcription factors and growth factors. The transcription factor Sox9 is expressed in the mesenchymal condensations and proliferating chondrocytes of the condyle. Targeted inactivation of *Sox9* in the neural crest cells results in a complete ablation of the condyle due to a failed formation of the condylar blastema (Mori-Akiyama et al., 2003; Wang et al., 2011). A similar phenotype was observed in long bone development in mice carrying *Sox9* inactivation in the mesenchymal cells of the limb bud (Akiyama et al., 2002), indicating an identical function of *Sox9* in the development of the condyle and the long bone.

Unlike Sox9, the Runx2 transcription factor appears to play a distinct role in the development of the condylar cartilage and the primary cartilage. Similar to its expression during long bone formation, *Runx2* is initially expressed in the mesenchymal condensation of the condyle, then in the newly formed cartilage and the bone collar (Shibata et al., 2006; Gu et al., 2008; Wang et al., 2011). Although skeletal formation is completely disrupted in *Runx2*-deficient mice, the primary cartilage development is not affected largely (Ducy et al., 1997; Komori et al., 1997; Otto et al., 1997), but the mandibular condylar cartilage is completely absent (Shibata et al., 2004). This distinct function of *Runx2* in condylar cartilage development could be attributed to the periosteal origin of the condylar cartilage with preosteoblast characteristics.

The short-stature homeobox gene *Shox2* is expressed in the mesenchymal cells of the maxilla–mandibular junction as early as E10.5, and subsequently in the progenitor cells and undifferentiated chondrocytes of developing condyle and in the glenoid fossa (Gu et al., 2008). Conditional inactivation of *Shox2* in the cranial neural crest–derived cells leads to dysplasia of the condyle and glenoid fossa as well as TMJ ankylosis as manifested by the fusion of the articular disk with the fibrous layers of the condyle and glenoid fossa (Gu et al., 2008). Associated with these developmental abnormalities of the TMJ are a significantly reduced rate of cell proliferation and downregulation, but not completely, of several osteogenic genes, including *Sox9, Runx2, Osterix*, and *Ihh* in the developing condyle. Downregulation of these osteogenic genes apparently contributes to the significantly reduced bone formation in the *Shox2* mutant TMJ. Interestingly, *Ihh* is ectopically activated in the developing glenoid fossa, indicating opposite transcriptional activities of *Shox2* in these two skeletal elements of the TMJ. The downregulation of *Ihh* in the condyle and the ectopic activation of *Ihh* in the glenoid fossa could cause TMJ ankylosis in the mutant, since a similar phenotype is found in mice carrying *Ihh* ablation in cartilage cells at the neonatal stage (Ochiai et al., 2010).

7 GROWTH FACTOR–MEDIATED SIGNALING PATHWAYS PLAY MULTIPLE ROLES IN TMJ DEVELOPMENT

Many families of growth factors have been implicated in the development of primary cartilage and endochondral ossification. Among them is Ihh, a member of the Hedgehog (Hh) family that plays pivotal role in long bone development and digit joint formation (St-Jacques et al., 1999). Ihh regulates chondrocyte proliferation and rate of chondrocyte hypertrophy in cooperation with PthrP in the periarticular region. In mice, *Ihh* expression is initially detected in the condylar condensation and becomes strong in the condylar cartilage at E15.5, along with *PthrP* expression (Yamazaki et al., 1999; Shibukawa et al., 2007; Gu et al., 2008), suggesting that the Ihh-PthrP regulatory loop also operates in the developing condyle. Although *PthrP* is required for normal cell proliferation of the developing condylar cartilage, a TMJ with distinct structures forms in *PthrP*-deficient mice (Suda et al., 1999).

Accompanied by *Ihh* expression in the developing TMJ is the expression of Hh receptors and effector genes, including *Pct, Smo, Gli1, Gli2*, and *Gli3* (Shibukawa et al., 2007; Purcell et al., 2009). Consistent with *Ihh* expression and its role in the development of primary cartilage, TMJ development is severely compromised in *Ihh*-deficient mice, as manifested by inhibition of cell proliferation and growth of the condylar cartilage and by complete absence of the articular disk and joint cavities (Shibukawa et al., 2007). Although compounding the null allele of *Gli3*, the negative effector of Hh signaling, onto the *Ihh* mutant background partially rescues the defects, the $Ihh^{-/-}$; $Gli3^{-/-}$ double mutants still fail to form an articular disk primordium, suggesting that Ihh signaling is absolutely required for fate determination and subsequent development of the articular disk of the TMJ (Shibukawa et al., 2007). Similar TMJ defects were found in *Gli2*-deficient mice (Purcell et al., 2009). Interestingly, deletion of *Smo* from chondrocyte progenitors of the condyle allows disk formation, but the disk fails to separate from the apex of the condyle (Purcell et al., 2009). Thus, Ihh signaling appears to exert its function in two steps during TMJ development: (1) to initiate TMJ disk formation, and (2) to instruct the disk to undergo proper morphogenesis and to split from the condyle (Purcell et al., 2009). In addition to its role in TMJ development, Ihh may also function to maintain the proper structure and function of the TMJ after it forms. This is evidenced by the fact that ablation of *Ihh* in the cartilages of neonatal mice produces dysplastic TMJ associated with partial disk ankylosis (Ochiai et al., 2010).

The functional significance of TGFβ signaling in TMJ development has been investigated by conditional inactivation of type II TGFβ receptor in cranial neural crest cells in mice (Oka et al., 2007, 2008). In such mutant mice ($Wnt1\text{-}Cre;Tgfr2^{f/f}$), the condyle is absent. Detailed analysis revealed the presence of mesenchymal condensation of the condyle but failed formation of the condylar cartilage, suggesting that TGFβ signaling regulates the fate determination of secondary cartilage (Oka et al., 1997). Interestingly, the condylar cartilage appears to be replaced by bone in the mutant (Oka et al., 2008).

Thus, TGFβ signaling functions as a key inhibitor of osteoblast differentiation in the condylar precursor cells.

8 REGULATION OF TMJ DEVELOPMENT BY EXTRACELLULAR MATRIX

In addition to transcription factors and growth factors, the extracellular matrix appears to play critical roles in TMJ development. Heparan sulfate proteoglycans (HS-PGs), modified by deacetylase/N-sulfotransferases (Ndsts) after they are synthesized, are macromolecules of the cell surface and extracellular matrix. HS-PGs have been shown to interact with signaling molecules, including Hh and BMPs, in many developmental processes (Hacker et al., 2005). Mice lacking *Nsdt1* exhibit malformation of several organs, including the TMJ (Grobe et al., 2005; Yasuda et al., 2010) and they display different degrees of TMJ defects (Yasuda et al., 2010). In severe conditions the TMJ is completely absent. In mildly affected mice, although the TMJ is present, the condyle exhibits thickened superficial and polymorphic cell layers and is accompanied by increased cell proliferation, a much wider distribution of Ihh signaling, and ectopic ossification. These observations demonstrated that a controlled diffusion range of Ihh is essential for normal TMJ development. Apparently, the defective HS-PGs in the *Ndst1* mutant condyle could not bind Ihh well, allowing a much broader diffusion of Ihh to exert aberrant functions (Yasuda et al., 2010).

9 CONTRIBUTION OF EXTRINSIC FACTORS TO TMJ DEVELOPMENT

While the initial development and differentiation of the TMJ depend primarily on the intrinsic factors, as described above, it is evident that extrinsic factors such as biomechanical force also contribute to TMJ development, particularly in TMJ maturation and remodeling. In mice, the jaw movement, which generates endogenous biomechanical force, begins at E16.5. Restriction of lower jaw movement in E15.5 mouse embryos through exo uterus surgery disturbs TMJ development significantly, as demonstrated by the reduction of the condylar cartilage in size due to a decreased level of chondrocyte proliferation and disruption of endochondral bone formation in the condyle (Habib et al., 2005, 2007). Furthermore, development of the joint disk was interrupted. On the other hand, it was shown that small intermittent compressive forces stimulate proliferation of chondrocytes and deposition of extracellular matrix by hypertrophic chondrocytes, thus accelerating condyle growth (Copray et al., 1985a,b). In addition, loading of such force also increased the activities of alkaline phosphatase in the entire hypertrophic region of the condyle, which would induce the onset of ossification. The function of the biomechanical force, at least the stimulation of chondrocyte proliferation, may be mediated by Ihh (Tang et al., 2004).

10 CONCLUSIONS

In this chapter we summarized development of the human and mouse TMJ and reviewed recent progress in the studies of molecular regulation of TMJ development. Generally, TMJ development is an understudied field, and many questions remain to be answered. For example, how is the position and fate of TMJ determined? When is the fate of the articular disk established? What are the signaling molecules that mediate tissue interactions during TMJ development? Does the glenoid fossa play an inductive role in disk formation? With the availability of an increasing number of mutant mouse models and the development of new high-throughput technologies, combined with traditional experimental embryology, cell biology, and molecular biology approaches, we expect to see a major advance in elucidation of the morphogenetic and molecular mechanisms of TMJ development and disorders in the near future.

Acknowledgment

The studies cited from the authors' laboratory were supported by National Institutes of Health grant R01DE17792 (to Y.C.).

REFERENCES

Akiyama H, Chaboissier MC, Martin JF, Schedl A, de Crombrugghe B (2002). The transcription factor Sox9 has essential role in successive steps of the chondrocyte differentiation pathway and is required for expression of *Sox5* and *Sox5*. *Genes Dev* 16:2813–2828.

Baume LJ (1962). Ontogenesis of the human temporomandibular joint: 1. Development of the condyles. *J Dent Res* 41:1327–1339.

Baume LJ, Holz J (1970). Ontogenesis of the human temporomandibular joint: 2. Development of the temporal components. *J Dent Res* 49:864–875.

Beresford WA (1981). *Chondroid Bone, Secondary Cartilage and Metaplasia*, Urban & Schwarzenberg, Baltimore, MD.

Brodie AG (1969). The three arcs of mandibular movement as they affect the wear of teeth. *Angle Orthod* 39:217–229.

Copray JC, Jansen HW, Duterloo HS (1985a). Effect of compressive forces on phosphatase activity in mandibular condylar cartilage of the rat in vitro. *J Anat* 140:479–489.

Copray JC, Jansen HW, Duterloo HS (1985b). Effects of compressive forces on proliferation and matrix synthesis in mandibular condylar cartilage of the rat in vitro. *Arch Oral Biol* 30:299–304.

Dixon AD (1997). Formation of the cranial base and craniofacial joints. In: *Fundamentals of Craniofacial Growth*, Dixon AD, Hoyte DAN, Rönning O, eds., CRC Press, Boca Raton, FL, pp. 124–127.

Ducy P, Zhang R, Geoffroy V, Ridall A, Karsenty G (1997). Osf2/Cbfa1: a transcription activator of osteoblast differentiation. *Cell* 89:747–754.

Duterloo HS, Jansen HWB (1970). Chondrogenesis and osteogenesis in the mandibular condylar blasteme. *Trans Eur Orthod Soc* 1969:109–118.

Fukada K, Shibata S, Suzuki S, Ohya K, Kuroda T (1999). In situ hybridisation study of type I, II, X collagens and aggrecan mRNAs in the developing condylar cartilage of fetal mouse mandible. *J Anat* 195:321–329.

Glasstone S (1971). Differentiation of the mouse embryonic mandible and squamo-mandibular joint in organ culture. *Arch Oral Biol* 16:723–729.

Grobe K, Inatani M, Palleria SR, Castagnola J, Yamaguchi Y, Esko JD (2005). Cerebral hypoplasia and craniofacial defects in mice lacking heparin sulphate Ndst1 gene function. *Development* 132:3777–3786.

Gu S, Wei N, Yu L, Fei J, Chen Y (2008). Shox2-deficiency leads to dysplasia and ankylosis of the temporomandibular joint in mice. *Mech Dev* 125:729–742.

Habib H, Hatta T, Udagawa J, Zhang L, Yoshimura Y, Otani H (2005). Fetal jaw movement affects condylar cartilage development. *J Dent Res* 84:474–479.

Habib H, Hatta T, Rahman OI, Yoshimura Y, Otani H (2007). Fetal jaw movement affects development of articular disk in the temporomandibular joint. *Congenital Anom* 47:53–57.

Hacker U, Nybakken K, Perrimon N (2005). Heparan sulphate proteoglycans: the sweet side of development. *Nat Rev Mol Cell Biol* 6:530–541.

Hall BK (2005). *Bones and Cartilage: Developmental and Evolutionary Skeletal Biology*, Elsevier Academic Press, San Diego, CA.

Herring WH (2003). TMJ anatomy and animal models. *J Musculoskel Neuronal Interact* 3:391–396.

Humphrey T (1971). Development of oral and facial motor mechanisms in human fetuses and their relation to craniofacial growth. *J Dent Res* 50:1428–1441.

Kermack KA (1972). The origin of mammals and the evolution of the temporomandibular joint. *Proc R Soc Med* 65:389–392.

Komori T, Yagi H, Nomura S, Yamaguchi A, Sasaki K, Deguchi K (1997). Targeted disruption of Cbfa1 results in a complete lack of bone formation owing to maturational arrest of osteoblasts. *Cell* 89:755–764.

Kuboki T, Kanyama M, Nakanishi T, Akiyama K, Nawachi K, Yatani H, et al. (2003). *Cbfa1/Runx2* gene expression in articular chondrocytes of the mice temporomandibular and knee joints in vivo. *Arch Oral Biol* 48:519–525.

Masotti C, Oliveira KG, Poerner F, Splendore A, Souzza J, Freitas RS et al. (2008). Auriculo-condylar syndrome: mapping of a first locus and evidence for genetic heterogeneity. *Eur J Hum Genet* 16:145–152.

McKay GS (1992). The structure and function of the temporomandibular joint. *Br Dent J* 173:127–132.

Merida-Velasco JR, Rodriguez-Vazquez JF, Merida-Velasco JA, Sanchez-Montesinos I, Espin-Ferra J, Jimenez-Collado J (1999). Development of the human temporomandibular joint. *Anat Rec* 255:20–33.

Merida-Velasco JR, Rodriguez-Vazquez JF, Cuadra-Blanco C, Campos-Lopez R, Sanchez M, Merida-Velasco JA (2009). Development of the mandibular condylar cartilage in human specimens of 10–15 weeks' gestation. *J Anat* 214:56–64.

Miyake T, Cameron AM, Ha BK (1997). Stage-specific expression patterns of alkaline phosphatase during development of the first arch skeleton in inbred *C57BL/6* mouse embryos. *J Anat* 190:239–260.

Mori-Akiyama Y, Akiyama H, Rowitch DH, de Crombrugghe B (2003). Sox9 is required for determination of the chondrogenic cell lineage in the cranial neural crest. *Proc Natl Acad Sci USA* 100:9360–9365.

Ochiai T, Shibukawa Y, Nagayama M, Mundy C, Yasuda T, Okabe T, et al. (2010). Indian hedgehog roles in post-natal TMJ development and organization. *J Dent Res* 89:349–354.

Ogawa T, Shimokawa H, Fukada K, Suzuki S, Shibata S, Ohya K, et al. (2003). Localization and inhibitory effect of basic fibroblast growth factor on chondrogenesis in cultured mouse mandibular condyle. *J Bone Miner Metab* 21:145–153.

Oğütcen-Toller M, Keskin M (2000). Computerized 3-dimensional study of the embryologic development of the human masticatory muscles and temporomandibular joint. *J Oral Maxillofac Surg* 58:1381–1386.

Oka K, Oka S, Sasaki T, Ito Y, Bringas Jr P, Nonaka K, et al. (2007). The role of TGF-β signaling in regulating chondrogenesis and osteogenesis during mandibular development. *Dev Biol* 303:391–404.

Oka K, Oka S, Hosokawa R, Bringas Jr P, Brockhoff HC, Nonaka K, et al. (2008). TGF-β mediated Dlx5 signaling plays a crucial role in osteo-chondroprogenitor cell lineage determination during mandible development. *Dev Biol* 321:303–309.

Otto F, Thornell AP, Crompton T, Denzel A, Gilmour KC, Rosewell IR, et al. (1997). *Cbfa1*, a candidate gene for cleidocranial dysplasia syndrome, is essential for osteoblast differentiation and bone development. *Cell* 89:765–771.

Pirttiniemi P, Peltomäki T, Müller L, Luder HU (2009). Abnormal mandibular growth and the condylar cartilage. *Eur J Orthod* 31:1–11.

Purcell P, Joo BW, Hu JK, Tran PV, Calicchio ML, O'Connell DJ, et al. (2009). Temporomandibular joint formation requires two distinct Hedgehog-dependent steps. *Proc Natl Acad Sci USA* 106:18297–18302.

Purcell P, Jheon A, Vivero MP, Rahimi H, Joo A, Klein OD (2012) *Spry1* and *Spry2* are essential for development of the temporamandibular joint. *J Dent Res*. doi:10.117/0022034512438401.

Rabie AB, Hägg H (2002). Factors regulating mandibular condylar growth. *Am J Orthod Dentofac Orthop* 122:401–409.

Ramieri G, Bonardi G, Morani V, Panzica GC, Del Tetto F, Arisio R, et al. (1996). Development of nerve fibres in the temporomandibular joint of the human fetus. *Anat Embryol* 194:57–64.

Ramirez LM, Sandoval GP, Ballesteros LE (2005). Temporomandibular disorders: referred cranio-cervico-facial clinic. *Med Oral Pathol Oral Cir Bucal* 10:18–25.

Shibata S, Fukada K, Suzuki S, Yamashita Y (1997). Immunohistochemistry of collagen types II and X, and enzyme-histochemistry of alkaline phosphatase in the developing condylar cartilage of the fetal mouse mandible. *J Anat* 191:561–570.

Shibata S, Suda N, Yoda S, Fukuoka H, Ohyama K, Yamashita Y (2004). *Runx2*-deficient mice lack mandibular condylar cartilage and have deformed Meckel's cartilage. *Anat Embryol* 208:273–280.

Shibata S, Suda N, Suzuki S, Fukuoka H, Yamashita Y (2006). An in situ hybridization study of *Runx2, Osterix*, and *Sox9* at the onset of condylar cartilage formation in fetal mouse mandible. *J Anat* 208:169–177.

Shibukawa Y, Young B, Wu C, Yamada S, Long F, Pacifici M et al. (2007). Temporomandibular joint formation and condyle growth require Indian Hedgehog signaling. *Dev Dyn* 236:426–434.

Silbermann M, Reddi AH, Hand AR, Leapman RD, Von der Mark K, et al. (1987). Further characterisation of the extracellular matrix in the mandibular condyle in neonatal mice. *J Anat* 151:169–188.

Sperber HG, Sperber SM, Guttmann GD (2010). *Craniofacial Embryogenetics and Development*, PMPH-USA Press, Shelton, CT.

St-Jacques B, Hammerschmidt M, McMahon AP (1999). Indian hedgehog signaling regulates proliferation and differentiation of chondrocytes and is essential for bone formation. *Genes Dev* 13:2072–2086.

Suda N, Shibata S, Yamazaki K, Kuroda T, Senior PV, Beck F, et al. (1999). Parathyroid hormone-related protein regulates proliferation of condylar hypertrophic chondrocytes. *J Bone Miner Res* 14:1838–1847.

Tang GH, Rabie ABM, Hägg U (2004). Indian Hedgehog: A mechanotransduction mediator in condylar cartilage. *J Dent Res* 83:434–438.

Vinkka-Puhakka H, Thesleff I (1993). Initiation of secondary cartilage in the mandible of the Syrian hamster in the absence of muscle function. *Arch Oral Biol* 38:49–54.

Wang Y, Liu C, Rohr J, Liu H, He F, Yu J, et al. (2011). Tissue interaction is required for glenoid fossa development during temporomandibular joint formation. *Dev Dyn* 240:2466–2473.

Watahiki J, Yamaguchi T, Irie T, Nakano H, Marki K, Tachikawa T. 2003. Gene expression profiling of mouse condylar cartilage during mastication by means of laser microdissection and cDNA array. *J Dent Res* 83:245–249.

Yamazaki K, Suda N, Kuroda T (1999). Distribution of parathyroid hormone–related protein (PTHrP) and type I parathyroid hormone (PTH) PTHrP receptor in developing mouse mandibular condylar cartilage. *Arch Oral Biol* 44:853–860.

Yasuda T, Mundy C, Kinumatsu T, Shibukawa Y, Shibutani T, Koyama E, et al. (2010). Sulfotransferase Ndst1 is needed for mandibular and TMJ development. *J Dent Res* 89:1111–1116.

5

CRANIOFACIAL MUSCLE DEVELOPMENT

ROBERT G. KELLY

Developmental Biology Institute of Marseilles–Luminy, Aix–Marseille University, Marseille, France

1 INTRODUCTION

Skeletal muscle is essential not only for locomotion but also for bilateral vision, feeding, breathing, and vocal communication. The latter functions are controlled by craniofacial skeletal muscles. The convergent phenotype of head and trunk striated muscle with respect to histology and physiology belies the heterogeneous origins of muscle progenitor cells in the vertebrate embryo (Fig. 1A and B). All skeletal muscle is mesodermal in origin, yet different progenitor cell populations and regulatory programs drive myogenesis in different regions of the embryo. In contrast to the situation in the trunk, where all skeletal muscles, including appendicular muscles, originate from the myotomal compartment of the somites, muscles of the vertebrate head are derived from three progenitor cell sources. Anterior cranial mesoderm gives rise to extraocular muscles, required to move the eyeball; cranial mesoderm in the core of the pharyngeal or branchial arches gives rise to muscles of mastication, facial expression, and muscles involved in operating the larynx and pharynx; and anterior or occipital somites give rise to muscle progenitor cells that migrate secondarily into the head to form tongue and neck muscles. While skeletal muscle differentiation throughout the embryo is regulated by transcription factors of the MYOD family, fundamentally different intrinsic and extrinsic upstream pathways control the onset of myogenesis in cranial and somite-derived mesoderm. This is reflected in the distinct gene expression profiles and susceptibility to myopathies of head and trunk muscles. Similarly, resident skeletal muscle stem cells, or satellite cells, which lie under the basal lamina of muscle fibers and contribute to muscle homeostasis and repair, have different properties in muscles

Stem Cells in Craniofacial Development and Regeneration, First Edition.
Edited by George T.-J. Huang, Irma Thesleff.

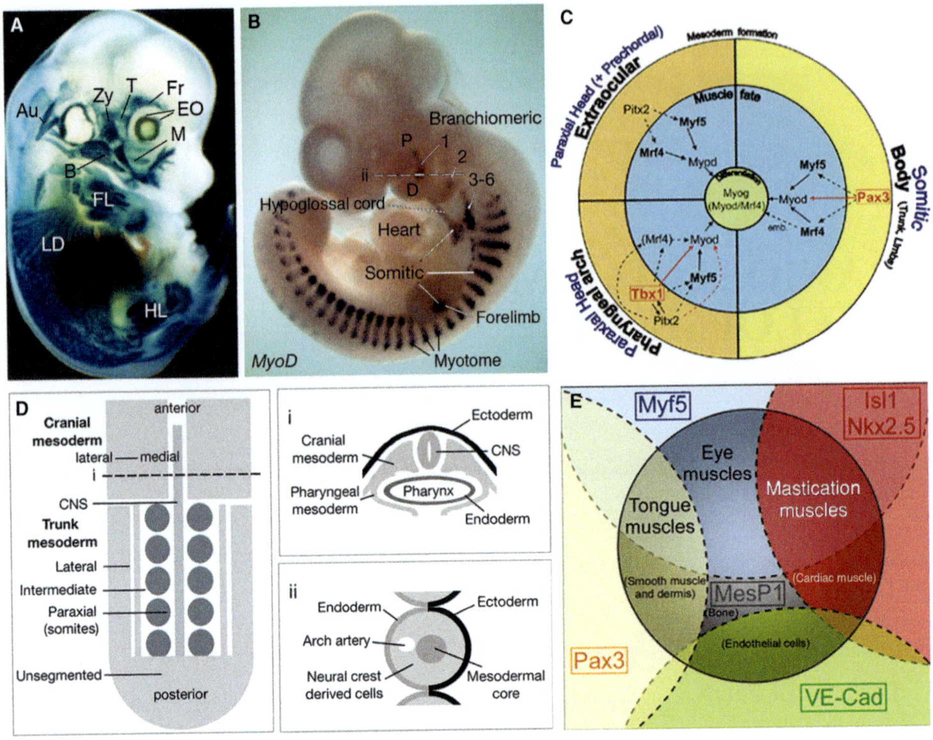

FIGURE 1 Skeletal muscle development in the mouse. (A) Expression of a transgene encoding β-galactosidase throughout the embryonic musculature, including branchiomeric, extraocular (EO), trunk and limb muscles, at embryonic day 13.5. (B) The distribution of *MyoD* transcripts at sites of branchiomeric (pharyngeal arches 1 to 6) and somite-derived myogenesis at day 10.5 of development. (C) Myogenic regulatory hierarchies at different sites of myogenesis. (D) Cartoon contrasting trunk and cranial mesoderm in the early embryo (left) and transverse sections at the level of cranial mesoderm and through a pharyngeal arch as indicated in (B). (E) Boundaries between craniofacial muscle groups revealed by the combinatorial use of different Cre lines. Au, auricularis; B, buccinator; Zy, zygomaticus; Fr, frontalis; T, temporalis; M, masseter; FL, forelimb; HL, hindlimb; LD, latissimus dorsi; P, proximal; D, distal [(B) and (D) reproduced from Grifone and Kelly (2007), with permission; (C) reproduced from Sambasivan et al. (2009), with permission; (E) reproduced from Harel et al. (2009), with permission.]

derived from cranial mesoderm or somites that are likely to contribute to disease mechanisms underlying muscle-restricted myopathies. Extensive embryological and anatomical studies have dissected the fate of cranial mesoderm and analyzed the development and differentiation of craniofacial muscles in different vertebrate models (reviewed in Noden and Francis-West, 2006; Bothe et al., 2007; Sambasivan et al., 2011). In this chapter we briefly discuss the control of skeletal myogenesis by the myogenic regulatory factor family of basic-helix loop helix transcription factors and then review the various intrinsic and extrinsic factors known to lie upstream of myogenic regulatory factor genes at different sites of muscle formation in the head, highlighting the modularity of craniofacial myogenesis. We then focus on the genetic regulation of branchiomeric myogenesis in core arch mesoderm. Recent findings using mouse genetics and experimental manipulation in the avian embryo have demonstrated that branchiomeric skeletal muscles share a common origin and overlapping regulatory program with cardiac progenitor cells in pharyngeal mesoderm. The evolutionary and clinical significance of this cardiocraniofacial developmental field is considered. Finally, the role of cranial neural crest cells in patterning skeletal muscles in the head is reviewed, focusing on the importance of mesoderm crest interactions during muscle development and evolution.

2 SOMITIC MYOGENESIS AND MYOGENIC REGULATORY HIERARCHIES

Skeletal muscle determination and differentiation is driven by members of the myogenic regulatory family (MRF) of basic-helix loop helix transcription factors: MYOD, MYF5, MRF4 (MYF6), and myogenin (reviewed by Bryson-Richardson and Currie, 2008). MRFs, interacting with other more broadly expressed transcription factors, including MADS box, homeobox, and chromatin remodeling proteins, drive myogenic determination and ultimately a convergent skeletal myogenic transcriptional program in all differentiated skeletal muscles. Superimposed on this program are differences in muscle fiber type and temporal and spatial differences associated with developmental stage and muscle identity. Before discussing cranial mesoderm and the development of extraocular and branchiomeric muscles, we briefly review myogenesis in the trunk and consider the secondary contribution of somite-derived progenitor cells to tongue and hypobranchial neck muscles. Epithelial somites in paraxial mesoderm flanking the neural tube are the source of all trunk and appendicular muscles. Somites differentiate rapidly into dermomyotome, myotome, and sclerotome; myogenesis is triggered by diverse intercellular signaling pathways, including promyogenic hedgehog and canonical WNT signals from the neural tube and overlying ectoderm and bone morphogenetic protein (BMP) signals that delay myogenesis in the lateral region of the somite (reviewed by Bothe et al., 2007; Bismuth and Relaix, 2010). Cells expressing the paired homeodomain transcription factors PAX3 and PAX7 continue to proliferate and give rise to the bulk of somite-derived skeletal muscle during fetal muscle growth as well as to skeletal muscle stem cells or satellite cells required for homeostasis and repair of adult muscles (Bismuth and Relaix, 2010). In the lateral region of limb-level somites, *Pax3*-expressing cells delaminate from the dermomyotome and migrate into the limb fields where MRF genes are activated (Fig. 1B). Dissection of the regulatory elements driving MRF transcription has identified multiple upstream enhancer elements responding to signal inputs in different regions of the somites and forming limb muscle masses, including target sites for PAX and SIX homeodomain transcription factors and signaling pathway target factors such as GLI (reviewed by Bismuth and Relaix, 2010). Somite-derived myogenic progenitor cells also contribute to the head musculature. In the most anterior, or occipital, somites, environmental cues trigger delamination of *Pax3*-expressing cells that converge ventrally in a structure termed the hypoglossal cord (Fig. 1B). Myogenic progenitor cells in the hypoglossal cord migrate anteriorly into the head, giving rise to intrinsic and extrinsic tongue muscles, including the hyoglossus and genioglossus, as well as hypobranchial ventral neck muscles such as the sternohyoid, thyrohyoid, and omohyoid muscles (Mackenzie et al., 1998; Huang et al., 1999).

The regulatory hierarchies driving myogenesis have been investigated using mouse genetics (reviewed by Bismuth and Relaix, 2010). MYF5, MRF4, or MYOD are required for myogenic determination, together with myogenin for differentiation. While the activation of *MyoD* rescues almost all sites of myogenesis in *Myf5 Mrf4*-null embryos, no myoblasts form in either the trunk or head in the absence of all three factors. PAX3 is required for migration of muscle progenitor cells to the limbs and hypoglossal cord (Franz et al., 1993; Tajbakhsh et al., 1997). In embryos lacking PAX3, MYF5, and MRF4, *MyoD* fails to be activated and all somite-derived muscles are absent (Tajbakhsh et al., 1997). In these embryos, branchiomeric myogenesis proceeds normally, suggesting that *MyoD* is activated by PAX3-independent pathways in the head (Tajbakhsh et al., 1997). The opposing phenotype is seen in embryos lacking only MYF5 and MYOD; trunk myogenesis proceeds normally while branchiomeric myogenesis fails, suggesting that MYF5 and MYOD, but not MRF4, are required for myogenic determination in pharyngeal arch mesoderm (Kassar-Duchossoy et al., 2004). Consistent with these conclusions, *Myf5* and *MyoD*, but not *Mrf4* or *Pax3*, are expressed at early stages of branchiomeric myogenesis; furthermore, the *Pax3*-related gene *Pax7* is activated after MRFs in the developing head (Hacker and Guthrie, 1998; Bothe and Dietrich, 2006; Tajbakhsh et al., 1997). As we will see below, in contrast to the situation in somitic or branchiomeric myogenesis, only MYF5 or MRF4 is required to initiate myogenesis in extraocular muscles. Myogenic regulatory hierarchies thus differ during the establishment of trunk, extraocular, and branchiomeric muscle development, reflecting the divergent sources of these muscle progenitor cells in the early embryo (Fig. 1C).

3 CRANIAL MESODERM

Myogenic progenitor cells giving rise to extraocular and branchiomeric skeletal muscles originate in cranial mesoderm (Fig. 1D). Cranial mesoderm migrates through the anterior primitive streak at gastrulation and gives rise to cardiac muscle and dorsal bones of the skull and vasculature, in addition to skeletal muscle. Cranial mesoderm becomes regionalized through external cues, prefiguring the localization of founder cell populations for various head muscles (Bothe et al., 2011). Reduced retinoic acid signaling in anterior cranial mesoderm defines anterior and posterior domains that are initially distinguished by expression of the genes encoding the transcription factors PITX2 and TBX1, respectively. This subdivision is subsequently refined by dynamic BMP and fibroblast growth factor (FGF) signaling, to establish extraocular and branchiomeric muscle progenitor cell territories (Bothe et al., 2011). Cranial mesoderm is also patterned along the medial lateral axis, although there is no morphological division into paraxial, intermediate, and lateral domains as seen at trunk levels (Fig. 1D). While the most lateral cranial mesoderm gives rise to the heart, the fate of lateral and medial domains of cranial mesoderm proximal to the neural tube (or paraxial cranial mesoderm) has been identified by labeling and transplantation experiments. The anterior domains give rise to extraocular muscles, and the lateral domains, overlying pharyngeal endoderm, contribute to the mesodermal core of pharyngeal arches and branchiomeric skeletal muscles (Noden, 1983; Couly et al., 1992; Schilling and Kimmel, 1994; Trainor et al., 1994; Hacker and Guthrie, 1998). Antagonistic signals from the neural tube and the pharyngeal arches specify extraocular versus branchiomeric muscle identity (von Scheven et al., 2006). Extraocular and branchiomeric muscles therefore develop from contiguous territories of cranial mesoderm. Indeed, retrospective clonal analysis in the mouse has demonstrated that all extraocular muscles share a common lineage relationship with branchiomeric muscles derived from the first arch (Lescroart et al., 2010). As will be seen below, this relationship extends to a requirement for PITX2 in myogenic progenitor cells.

4 GENETIC CONTROL OF EXTRAOCULAR MUSCLE DEVELOPMENT

Extraocular muscles develop from the most anterior (prechordal and anterior paraxial) cranial mesoderm that migrates to surround the developing eye. These six extrinsic eye muscles (superior, inferior, medial and lateral rectus muscles, and superior and inferior oblique muscles) orient the eyeball and are characterized by contractile speed, precision, and fatigue resistance (Spencer and Porter, 2006). Transcriptome analyses have revealed that extraocular myotubes and myoblasts express a distinct genetic program to other muscles (Fischer et al., 2002). This may result in part from their unique mode of myogenic regulation; genetic studies in the mouse have shown that MYF5 and MRF4 alone are required for determination of extraocular muscle fate within common *Myf5*-expressing bilateral primordia at midgestation (Fig. 1C; Sambasivan et al., 2009). Interestingly, experimental manipulation in avian embyros has shown that the initial development of extraocular muscles is independent of the eye itself (von Scheven et al., 2006). Subsequently, patterning of distinct extraocular muscles takes place during migration to the eye. In the absence of MYF5 and MRF4, extraocular muscle primordia undergo cell death, revealing that MRF activity is required for progenitor cell survival (Sambasivan et al., 2009). Unlike somite-derived muscles, extraocular muscles appear to lack an alternative pathway to activate *MyoD*; forced expression of *MyoD* under control of the *Myf5* locus rescues progenitor cell survival and extraocular muscle development in *Myf5 Mrf4* double-mutant embryos (Sambasivan et al., 2009).

The homeodomain transcription factor PITX2 is expressed early in cranial mesoderm and is a primary regulator of extraocular muscle specification, acting cell autonomously upstream of the MRFs (Gage et al., 1999; Kitamura et al., 1999; Zacharias et al., 2011). PITX2 can activate transcription from the *Myf5* and *MyoD* promoters, suggesting that it directly activates MRF gene expression in forming extraocular muscles (Zacharias et al., 2011). In addition, PITX2 is required for survival of extraocular muscle progenitor cells prior to MRF activation. Continued PITX2 expression is nevertheless insufficient to activate *MyoD* or ensure survival in the absence of MYF5 and MRF4, suggesting that multiple pathways control the survival of extraocular muscle primordia prior to myogenic differentiation (Sambasivan et al., 2009). Interestingly, extraocular myogenesis is sensitive to PITX2 dosage, since defects in extraocular muscle morphogenesis and MRF expression are also observed in *Pitx2* heterozygous-mutant mice (Diehl et al., 2006). Ocular anomalies are among the spectrum of defects found in human patients with Axenfeld–Rieger syndrome, carrying loss of function mutations in PITX2 (Semina et al., 1996). *Pitx2*, together with the related gene, *Pitx1*, continues to be expressed in satellite cells of definitive extraocular muscles (Sambasivan et al., 2009). Transplantation experiments into hindlimb muscles have, however, shown that *Pax7*-expressing extraocular muscle satellite cells differentiate according to the local myogenic program, demonstrating that the extraocular phenotype is not driven by intrinsic cues alone (Sambasivan et al., 2009).

5 BRANCHIOMERIC MYOGENESIS

In amniotes, the pharyngeal or branchial arches are transient structures that give rise to much of the face and neck. The five bilateral pairs of arches form in an anterior–posterior sequence, and each is surrounded by an epithelium comprised of an external ectodermal layer and an inner endodermal layer (Fig. 1D). Each arch comprises an arch artery and neural crest–derived mesenchyme surrounding a mesodermal core that contains the progenitor cells of branchiomeric skeletal muscles. Labeling and transplantation experiments in avian embryos have shown that the myogenic core of the arches is derived from cranial mesoderm overlying the foregut (Couly et al., 1992; Trainor et al., 1994; Hacker and Guthrie, 1998). Fluorescent labeling experiments have suggested that there is also a contribution to the distal mesodermal core from splanchnic mesoderm underlying the pharynx (Nathan et al., 2008). Trainor and colleagues observed a dorsoventral (or mediolateral) partition within murine cranial mesoderm that was transposed into a proximodistal distribution within the pharyngeal arches (Trainor et al., 1994). As cranial mesoderm is not morphologically divided into paraxial, intermediate, and lateral domains, the contribution of cranial mesoderm at different medial-to-lateral levels to core arch mesoderm arises along a continuum, and it is as yet unclear whether there is a boundary between cranial paraxial and lateral mesodermal contributions to the proximal and distal core (Noden and Trainor, 2005; Nathan et al., 2008). Outpocketing of pharyngeal endoderm and neural crest influx result in arch morphogenesis effectively segmenting cranial mesoderm in the arch core along the anterior–posterior embryonic axis. This segmental pattern is reflected in the anterior–posterior identity of crest-derived mesenchyme and motor innervation of the muscles derived from each arch (Kontges and Lumsden, 1996).

The onset of MRF expression and myogenic markers in forming branchiomeric muscles has revealed a delay between determination and differentiation that is significantly longer than that in somite-derived muscles, potentially due to the requirement for correct positioning of jaw-operating and facial expression muscles during head morphogenesis (Noden et al., 1999; Yamane, 2005). The pharyngeal arches give rise to the jaw and gill musculature in fish; over tetrapod evolution posterior pharyngeal arch–derived muscles have assumed a diversity of functions in the head and neck. In amniotes, the core of the first or mandibular arch gives rise to the muscles of mastication involved in opening and closing the jaw—the temporalis, masseter, pterygoids, mylohyoid, and anterior digastric—in addition to the tensor tympani and tensor veli palatini. These muscles are innervated by the maxillary and mandibular branches of the fifth or trigeminal cranial nerve. Muscle primordia from the second, or hyoid, arch undergo an extensive migration over the face to give rise to muscles of facial expression, including the orbicularis oculi, orbicularis oris, platysma, auricularis, frontalis, and buccinator muscles, in addition to the posterior digastric, stylohyoid, and stapedius muscles. These muscles are innervated by the seventh or facial nerve. The third arch, innervated by the ninth (glossopharyngeal) nerve, gives rise to the stylopharyngeal muscle, and the fourth and sixth arches give rise to muscles of the larynx and pharynx in addition to the trapezius and sternocleidomastoid muscle innervated by the tenth (vagus) and eleventh (accessory) cranial nerves.

Clonal analysis in the mouse has revealed that first- and second-arch-derived muscles correspond to two clonally distinct lineages that segregate early from lineages giving rise to somitic muscles (Lescroart et al., 2010). First-arch muscles share a clonal relationship with extraocular muscles, and segregation of first- and second-arch lineages precedes the bilateralization of mesoderm and thus occurs in the early embryo, prior to gastrulation (Lescroart et al., 2010). Similar clonal analyses remain to be applied to muscles derived from arches 3 to 6. Genetic lineage analysis using Cre recombinase under the regulatory control of different mesodermal regulatory elements has also provided important insights into branchiomeric muscle development (Fig. 1E; Harel et al., 2009). While *Pax3*-driven Cre recombinase marks the tongue musculature, derived from the hypoglossal chord, it does not label either branchiomeric or extraocular muscles. In contrast, Cre under transcriptional control of the gene encoding the LIM homeodomain transcription factor ISL1 reveals that all branchiomeric muscles have expressed *Isl1*, with a stronger labeling in distal arch-derived muscles, including the mylohyoid and anterior digastric. *Isl1-Cre* also labels neck muscles derived from the occipital lateral plate, including the trapezius and sternocleidomastoid, suggesting that these muscles share the branchiomeric transcriptional program and are also derived from pharyngeal mesoderm (Harel et al., 2009; Theis et al., 2010). The *Mesp1-Cre* line, defining a lineage of anterior mesoderm, including cranial and cardiac as well as anterior somitic mesoderm, labels all head muscles, including tongue and extraocular muscles (Harel et al., 2009). These experiments, summarized in Fig. 1E, show that the myogenic progenitor cell populations giving rise to different craniofacial muscle groups have distinct genetic signatures; all craniofacial muscles derive from *Mesp1*-expressing cells, branchiomeric muscles from Isl1-expressing cells, and tongue muscles from *Pax3*-expressing cells.

6 GENETIC CONTROL OF BRANCHIOMERIC MYOGENESIS

The signaling pathways and upstream transcriptional regulators of branchiomeric myogenesis differ from those operating in the somites. While BMP signaling inhibits myogenesis in both the somites and arches, WNT and Hedgehog (Hh) signals are potent inhibitors of cranial myogenesis, yet promote MRF expression in the somites (Tzahor et al., 2003). Crest influx into the arches may be permissive for cranial myogenesis, as BMP (Gremlin) and WNT (FRZB1) inhibitors are expressed in neural crest–derived arch mesenchyme, thus insulating the core from antimyogenic signals from pharyngeal epithelia (Tzahor et al., 2003). In addition, FGF signaling promotes myogenic progenitor cell proliferation and branchiomeric versus extraocular muscle identity in the first arch (von Scheven et al., 2006). Genetic studies in the mouse have identified a number of transcriptional regulators required for myogenic specification at sites of branchiomeric but not somite-derived muscle formation. These include the basic-helix loop helix transcription factors capsulin (TCF21) and MYOR, the homeodomain transcription factor PITX2, and the T-box containing transcriptional activator TBX1. These factors are expressed in highly dynamic and progressively regionalized domains of cranial mesoderm prior to MRF gene expression, at which stage they are coexpressed with ISL1 in the mesodermal core (Grifone et al., 2008; Nathan et al., 2008; Bothe et al., 2011). Mice lacking both Capsulin and MYOR, related transcriptional repressors, display an absence of jaw-closing muscles, including the masseter, temporalis, and pterygoid, derived from proximal first arch core mesoderm (Lu et al., 2002). Failure to activate MRF genes in Capsulin-expressing cells in the first arch of double-mutant embryos is associated with elevated cell death (Lu et al., 2002). In addition to its role in extraocular muscles, PITX2 is required cell autonomously to activate myogenesis in the first arch, and again, elevated cell death was observed in myogenic primordia of *Pitx2*-null embryos, suggesting a survival role (Dong et al., 2006; Shih et al., 2007). *TBX1* is a major candidate gene for the multigene haploinsufficiency del22q11.2 or DiGeorge syndrome, characterized by craniofacial and cardiovascular defects; in *Tbx1*-null mice premyogenic cells expressing *Capsulin* and *MyoR* are observed in the mandibular arch but fail to robustly activate transcription of *Myf5* and *MyoD* (Fig. 2; Kelly et al., 2004; Liao et al., 2004; Dastjerdi et al., 2007). However, low-level stochastic activation of MRF genes in *Tbx1*-null embryos results in the presence of sporadic and hypoplastic first arch–derived muscles at later stages of development. TBX1 thus ensures robust bilateral myogenic specification in the mandibular arch. Unlike PITX2, TBX1 does not appear to intervene in cell survival within core mesoderm, as no elevated cell death was observed (Grifone et al., 2008). All caudal arch–derived muscles are absent in *Tbx1*-null embryos, including largyngeal muscles and the trapezius and sternocleidomastoid muscle, while somite-derived tongue and hypobranchial muscles are unaffected (Fig. 2; Kelly et al., 2004; Theis et al., 2010). However, in the absence of TBX1, the caudal pharyngeal region is extremely hypoplastic and the loss of arch structures precludes direct investigation of the role of TBX1 in myogenesis in posterior arches (Kelly et al., 2004). *Tbx1* is also expressed in developing tongue and limb muscles, although after the activation

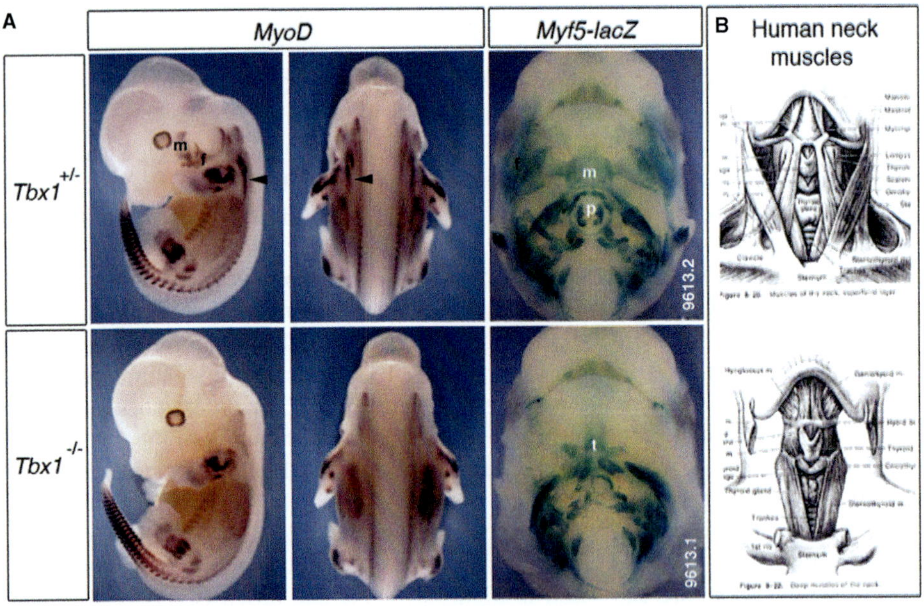

FIGURE 2 *Tbx1* regulates branchiomeric myogenesis. (A) *MyoD* transcript distribution (left and central panels) and expression of a *Myf5-lacZ* transgene (right panels) showing loss of muscles of facial expression (f), muscles underlying the mandible (m), muscles of the pharynx (p), and the cervical part of the trapezius (arrowhead) in *Tbx1*-null embryos at embryonic day 12.5; somite-derived muscles at the base of the tongue (t) are maintained. (B) Comparable sequential dissection of neck muscles in a human cadaver. [Reproduced from Jacob et al. (1982) and Kelly (2010), with permission.]

of MRFs (Dastjerdi et al., 2007; Grifone et al., 2008; Okano et al., 2008). No defects in the development of these muscles have been detected in *Tbx1*-null embryos, potentially due to overlapping functions of other members of the T-box transcription factor family (King et al., 2006; Grifone et al., 2008).

The sporadic branchiomeric muscles forming in *Tbx1* homozygous-mutant embryos develop as recognizable, although hypoplastic, muscles with normally distributed fiber-type groups and contain PAX7-positive myoblasts that proliferate and differentiate ex vivo with similar kinetics to cells from wild-type muscles (Grifone et al., 2008). These results suggest that the role of TBX1 in branchiomeric muscle development is restricted to ensuring robust bilateral specification at the earliest stages of myogenesis and that muscles that escape this requirement subsequently develop normally. The molecular mechanisms by which Capsulin, MYOR, PITX2, and TBX1 regulate myogenesis remain to be elucidated. Multiple enhancers upstream of the murine *Myf5* gene have been shown to be active in different pharyngeal arches, yet direct regulation by any of these factors has not been demonstrated in vivo (Carvajal et al., 2001). The sporadic myogenesis observed in *Tbx1*-null embryos is severely reduced in *Tbx1 Myf5* double-mutant embryos, positioning TBX1 upstream of *MyoD* in the regulatory hierarchy controlling myogenesis in the first arch, although it is likely that TBX1 also activates *Myf5* (Sambasivan et al., 2009). A direct role for TBX1 in activating MRF genes is supported by the identification of a T-box binding site mediating activation of the *Xenopus Myf5* gene in trunk muscles and the implication of ascidian T-box genes in early myogenic activation (Mitani et al., 2001; Lin et al., 2003). Most of our current understanding of the control of branchiomeric myogenesis concerns the mandibular arch; very little is known about myogenic regulatory circuits in the second and more posterior arches. Branchiomeric myogenesis is thus highly complex, and different regulatory circuits operate not only in different arches but also in different regions of an arch (proximal versus distal). Intriguingly, this diversity of regulatory programs in the pharyngeal region resembles the situation in skeletal muscle development in *Drosophila*, where different muscles develop from founder cells expressing specific transcription factors (Tixier et al., 2010). The *Drosophila* homolog of *Tbx1*, *Org1*, has recently been shown to be expressed in a small subset of skeletal muscle progenitor cells, where it directly activates muscle identity genes and controls muscle specification (Schaub et al., 2012).

Consistent with their distinct ontogeny, adult skeletal muscle stem cells, or satellite cells, from branchiomeric muscles have been shown to have a different gene expression profile to somite-derived muscles, including elevated *Capsulin* expression (Harel et al., 2009; Sambasivan et al., 2009). Satellite cells also share the genetic lineage of their associated muscles. Thus, satellite cells in somite-derived muscles, including tongue and hypobranchial muscles, are derived from *Pax3*-expressing progenitor cells, and branchiomeric muscle satellite cells are derived from *Isl1*-expressing progenitor cells, including the majority of satellite cells in the masseter (Harel et al., 2009). Furthermore, satellite cells from branchiomeric muscles retain cardiogenic potential and activate *Isl1* upon BMP exposure, a property that is not observed in somite-derived satellite cells (Harel et al., 2009). A recent comparison of satellite cells from the masseter versus somite-derived limb muscles has revealed that less satellite cells are associated with skeletal muscle fibers in the masseter and that they tend to be more proliferative and have delayed differentiation compared to those from limb muscles (Ono et al., 2010). These differences have important clinical consequences for myoblast therapy for muscle-restricted myopathies.

The zebrafish has provided important insights into the genetic regulation of craniofacial myogenesis. The homeodomain transcription factor SIX1 has been implicated in maintaining the branchiomeric myogenic program, playing a PAX3-like role upstream of *MyoD*, but not *Myf5*, independent of TBX1 (Lin et al., 2009). Branchiomeric myogenesis in zebrafish also requires TBX1, loss of which can be rescued by injection of *Myf5* cDNA but not *MyoD* or *Six1* cDNAs (Lin et al., 2009). In the mouse, severely hypoplastic branchiomeric skeletal muscles have been documented in embryos lacking both SIX1 and the SIX1 cofactor EYA1 (Guo et al., 2011). FGF signaling and TBX1 function are required independently to specify muscle identity in the zebrafish mandibular arch, as defined by expression of the homeodomain transcription factor EN2 (Knight et al., 2008). *En2* is also expressed in a subset of proximal mandibular arch myoblasts in the mouse, and subsequently in the masseter, temporalis, and pterygoid (Degenhardt et al., 2002). Of particular interest is the recent identification in the zebrafish of a specific genetic module required for the development of three muscles associated with the opercular bone, involved in respiration (Knight et al., 2011). Mutations in three different genes in the RET tyrosine kinase signaling pathway specifically affect opercular muscles, which are derived from two different arches. This suggests a mechanism by which muscles functioning together in a single module can coevolve during evolution, despite originating in different arches.

7 PHARYNGEAL MESODERM: CONNECTING HEAD AND HEART MUSCLE DEVELOPMENT

A specific pharyngeal mesodermal genetic program thus operates at sites of branchiomeric myogenesis that is fundamentally different from that elsewhere in the embryo. The genetic program of branchiomeric muscle progenitor cells, including expression of *Isl1, Tbx1, Fgf10, Capsulin*, and *MyoR* overlaps with that of a population of cardiac progenitor cells in the pharyngeal region, termed the second heart field (reviewed by Grifone and Kelly, 2007; Tzahor, 2009; Tzahor and Evans, 2011). These cardiac progenitor cells contribute cardiomyocytes to the growing poles of the heart tube during cardiac looping, ultimately giving rise to the right ventricle, ventricular outlets, and atrial myocardium of the definitive heart. Perturbation of second heart field development results in a shortened heart tube that fails to remodel correctly during cardiac septation, leading to congenital heart defects. The second heart field is characterized by continued proliferation and differentiation delay, and the progressive addition of these cells to the heart tube is regulated by a complex set of intercellular signals in the pharyngeal region (Dyer and Kirby, 2009; Rochais et al., 2009). In particular, antagonism between BMP signaling, promoting cardiac differentiation, and FGF signaling, promoting progenitor cell proliferation and delaying differentiation, has been shown to play a central role in orchestrating second heart field deployment (Tirosh-Finkel et al., 2010). *Isl1, Fgf10*, and *Tbx1* are expressed in mesoderm in the dorsal wall of the pericardial cavity, underlying pharyngeal endoderm, where TBX1 promotes proliferation and blocks differentiation of cardiac progenitor cells in the second heart field (Liao et al., 2008; Chen et al., 2009). *Pitx2* is also expressed in second heart field cells on the left side of the embryo, where it is required for cardiac outflow tract development (Ai et al., 2006). Six1 and Eya1 are also required for both cardiac and branchiomeric muscle development, operating downstream of

TBX1 (Guo et al., 2011). Most of these genes are downregulated on myogenic differentiation in both cardiomyocyte and branchiomeric skeletal muscle lineages. Isl1, for example, identifies multipotent progenitor cells that can give rise to myocardium endocardium, smooth muscle, and skeletal muscle (Laugwitz et al., 2008: Tzahor, 2009). Indeed, Isl1-expressing cells appear to define a progenitor cell pool that contributes secondarily to both the heart and head musculature (Fig. 3A; Tzahor, 2009). Consistent with a role in regulating progenitor cells, ISL1 has been shown to repress myogenic differentiation in avian experiments (Harel et al., 2009). As TBX1 and PITX2 are also known to regulate cell proliferation and MYOR and Capsulin have been shown to be transcriptional repressors, it has been proposed that these genes may delay the activation of the skeletal myogenic program in pharyngeal mesoderm, thus increasing the availability of progenitor cells for the heart (Bothe and Dietrich, 2006).

Pharyngeal mesoderm therefore has two myogenic derivatives: branchiomeric skeletal muscles and second heart field–derived myocardium (Fig. 3A). This shared origin is also reflected at the level of gene expression, the cardiac transcription factor NKX2-5 and cardiac α-myosin heavy chain being expressed in both the heart and branchiomeric muscles (Bredman et al., 1991; Kasahara et al., 1998). Retrospective lineage analysis in the mouse has recently shown that these divergent fates do not simply reflect shared gene expression patterns in neighboring progenitor cell populations but that common progenitor cells giving rise to both cardiac and skeletal muscle are present in pharyngeal mesoderm (Fig. 3B). A clonal relationship exists between cardiomyocytes in the right ventricle and muscles derived from the first pharyngeal arch and between myocytes of the cardiac outflow tract and second-arch-derived muscles (Lescroart et al., 2010). This apparently disparate relationship reflects the changing positions of the heart and head during pharyngeal morphogenesis (Fig. 3C). The early heart tube forms prior to arch morphogenesis, and the linear heart tube is attached only to the first-arch artery. At this stage, right ventricular progenitor cells are added to the arterial pole at the same time as splanchnic mesoderm contributes to core arch mesoderm. Subsequently, as the heart tube elongates, it is displaced caudally in the pharynx, becoming attached to the second arch. At this stage, progenitor cells giving rise to the cardiac outflow tract are added to the heart from splanchnic mesoderm, which also gives rise to the core of the second arch. Ultimately, the heart is attached to arches 3, 4, and 6, and although it has not yet been demonstrated, myocardium at the poles of the heart might be expected to share a clonal relationship with muscles derived from caudal pharyngeal mesoderm. The definitive configuration, when the heart lies in the thoracic cavity, thus masks the embryological proximity of the heart and face. These findings demonstrate that pharyngeal mesoderm makes progressive contributions to both the heart and branchiomeric skeletal muscle during early development. Perturbation of this cardiocraniofacial developmental field leads to defects in both craniofacial and cardiovascular morphogenesis, providing insights into the origins of human genetic syndromes, such as del22q11.2 syndrome.

8 VISCERAL ORIGINS OF BRANCHIOMERIC SKELETAL MUSCLES

Insights into the evolution of this cardiocraniofacial mesodermal developmental field have come from analysis of Isl-expressing cells in the protochordate *Ciona*

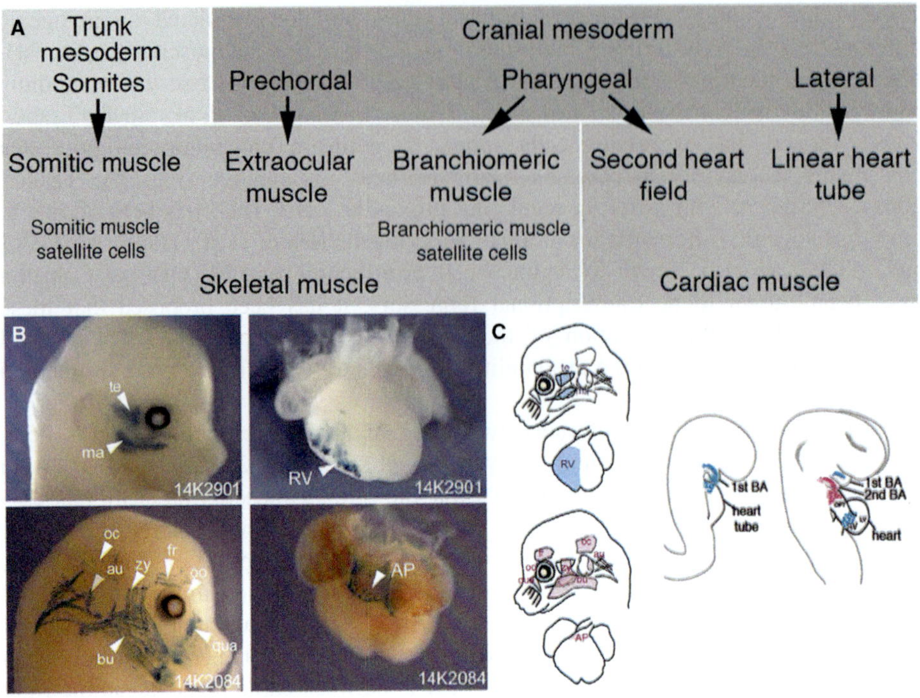

FIGURE 3 Connecting heart and head muscle development. (A) Cartoon showing the contribution of pharyngeal mesoderm to branchiomeric skeletal and cardiac muscle. Lateral cranial mesoderm gives rise to the early heart tube, medial cranial mesoderm to extraocular muscles, and all other skeletal muscles are derived from somites. (B) Embryos showing β-galactosidase labeling in clonally related skeletal muscle fibers in muscles derived from the first arch (te, temporalis; m, masseter) and right ventricular cardiomyocytes (top row) and in clonally related second arch–derived skeletal muscles and myocytes in the cardiac outflow tract (bottom row). (C) Cartoon showing the link between first-arch-derived muscles and the right ventricle and second-arch-derived muscles and the cardiac outflow tract (left) as the arterial pole of the elongating heart tube is displaced posteriorly during pharyngeal morphogenesis. [(A) reproduced from Kelly (2010), with permission; (B) and (C) reproduced from Lescroart et al. (2010), with permission.]

intestinalis. As *Ciona* undergoes metamorphosis from the larval form into the sessile adult sea squirt, ISL-positive cells detach from *Mesp1*-expressing cardiac cells and migrate to surround the atrial siphon, where they differentiate into skeletal muscles, expressing different myosin genes to those expressed into the heart, and also giving rise to *Tbx1*-positive longitudinal skeletal muscles (Stolfi et al., 2010). These results provide evolutionary support for a common origin of cardiac and skeletal muscle and suggest that over vertebrate evolution, ISL1-positive cells may have remained associated with the heart and contributed an increased number of cardiac progenitor cells in early vertebrates. The recent finding that the zebrafish also has a second heart field suggests that this model will be a powerful system to identify new genes and pathways regulating the contribution of pharyngeal mesodermal progenitor cells to the heart and head musculature (Zhou et al., 2011).

The common gene expression profile and lineage of branchiomeric muscles and cells giving rise to the heart supports a visceral, or splanchnic, mesodermal origin of craniofacial muscles, in contrast to the paraxial mesodermal origin of myogenic cells in the somites. Furthermore, branchiomeric muscles differ from all other skeletal muscles in that they are innervated by axons from PHOX2b-dependent visceral motor neuron columns positioned laterally in the CNS, as opposed to the somatic motor neurons that innervate extraocular and somite-derived muscles (Pattyn et al., 2000). Romer proposed that branchiomeric muscles and the smooth musculature of the gut arise in a similar fashion and are "the anterior and posterior parts of a single great visceral system of muscles whose primary locus is the walls of the digestive tract" (Romer and Parsons, 1977). Indeed, a putative branchial muscle precursor appears to be present in the pharyngeal floor of the nonvertebrate chordate amphioxus (Sambasivan et al., 2011). The requirement for food gathering and breathing at the anterior end of the digestive tract is likely to have driven the development of striated muscles in the pharyngeal region and to have been a key step in early vertebrate evolution (Gans and Northcutt, 1983). Although originally associated with gill musculature involved in feeding and breathing, over vertebrate evolution branchial muscles have become specialized for different functions, and moved away from the lining of the gut to form our craniofacial muscles. Visceral muscularization also gives rise to cardiac muscle (Gans and Northcutt, 1983). The new molecular genetic data discussed above reinforce such a model for head muscle evolution and suggest that myogenic potential has been acquired in cranial and trunk mesoderm through independent strategies. The acquisition of a muscular pharynx is likely to have predated chordate origins; analogy may be made between branchiomeric muscles and pharyngeal muscles, derived from myoepthelial cells in the coelomic lining, that are involved in regulating the feeding apparatus in sea urchins (Dolmatov et al., 2007). Even further removed in evolutionary terms, the nematode worm *Caenorhabditis elegans* has no heart but a contractile muscular pharynx required for filter feeding that shares electrical properties with the heart and requires the homologs of genes encoding the cardiac transcription factor NKX2-5 and the T-box factor TBX2 for normal development (Okkema and Fire, 1994; Smith and Mango, 2007).

9 CRANIAL NEURAL CREST PATTERNS DEVELOPING CRANIOFACIAL MUSCLE

The connective tissue and tendons of all head muscles, including tongue and hypobranchial muscles, are derived from cranial neural crest cells (Noden, 1988; Couly et al., 1992). Craniofacial muscle development involves extensive reciprocal interactions between mesoderm and neural crest cells (Fig. 4A; Noden and Trainor, 2005). Arch-derived muscles are associated with crest-derived tissue from the same arch and innervated by cranial nerves from the same axial level (Kontges and Lumsden, 1996). Craniofacial muscle development is distorted after neural crest ablation, and cranial neural crest cells thus identify a critical extrinsic regulatory mechanism during craniofacial myogenesis (reviewed by Noden and Trainor, 2005). In addition, arch mesoderm plays a role in maintaining crest fate and the establishment of regional identity in the arches (Trainor and Krumlauf, 2000). As neural crest cells migrate into the forming pharyngeal arches, they encircle premyogenic mesoderm in a central core. Neural crest cells also intermingle with mesoderm and a fraction of the crest cells remain

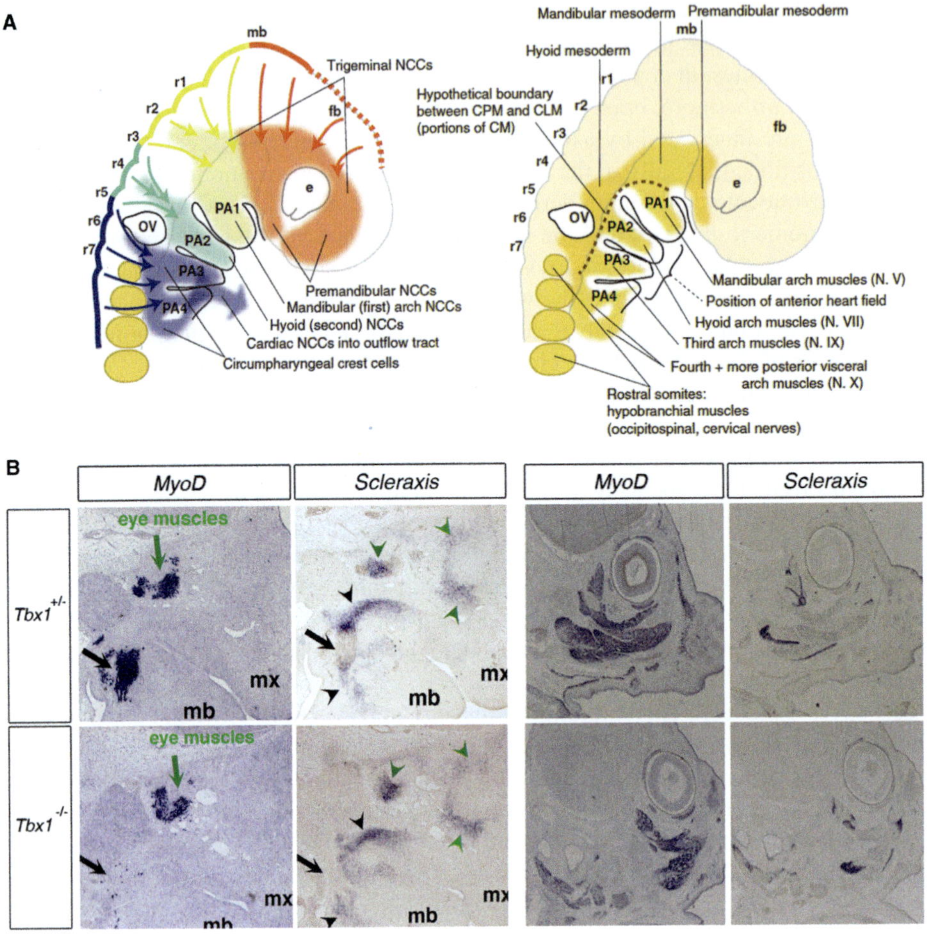

FIGURE 4 Cranial mesoderm and cranial neural crest. (A) Cartoons showing the movement of neural crest–derived cells from different axial levels of the dorsal neural tube into the pharyngeal arches (left) compared to the distribution of cranial mesoderm in the pharyngeal arches (right). (B) Initial differentiation of neural crest–derived mesenchyme, illustrated by Scleraxis expression, occurs independent of muscle development, as seen at sites where *MyoD* fails to be activated in *Tbx1*-null embryos (black arrow) at embryonic day 12.5 (left panels). Subsequent development is muscle dependent; at day 15.5 of development Scleraxis expression is observed associated only with hypoplastic skeletal muscles in *Tbx1*-null embryos (right panels). Mb, mandibular region; mx, maxillary region. [(A) reproduced from Sambasivan et al. (2011), with permission; (B) reproduced from Grenier et al., (2009) and Grifone et al. (2008), with permission.]

within the mesodermal core (Grenier et al., 2009). These neural crest cells are likely to be the precursors of connective tissue, including fascia and tendons, associated with branchiomeric muscles at later stages. However, despite this intermingling, the initial steps of cranial mesoderm and neural crest development are independent, and neural crest influx is not required for branchiomeric muscle specification, which occurs in the absence of neural crest cells in mouse and in chick embryos after cranial neural crest ablation (Rinon et al., 2007; von Scheven et al., 2006). In contrast, all subsequent steps of arch muscle development require interactions between mesoderm and crest, including migration, patterning, withdrawal from the cell cycle, and skeletal muscle differentiation (Rinon et al., 2007). This reflects reciprocal interactions between cranial mesoderm and neural crest and possibly a steric role of neural crest cells in separating the myogenic core of adjacent pharyngeal arches to ensure the axial registration of cranial mesoderm in the forming arches (Rinon et al., 2007). Crest cells also play a role in regulating FGF signaling to pharyngeal mesoderm, a process thought to be important in slowing proliferation in second heart field cells during heart tube elongation; elevated FGF8 expression after crest ablation may be associated with the elevated proliferation and differentiation delay observed in core mesoderm (Hutson et al., 2006; Rinon et al., 2007). Nor is activation of the myogenic program essential for the earliest steps of neural crest cell differentiation; initiation of *Scleraxis* expression, encoding a transcription factor expressed in differentiating tendon cells, occurs normally in the first arch of *Tbx1*-mutant embryos (Fig. 4B). In contrast, at later stages of development *Scleraxis* expression is downregulated in *Tbx1*-null embryos, except around sporadic hypoplastic muscles, suggesting that the maintenance of *Scleraxis* expression and tendon development are muscle dependent (Fig. 4B; Grenier et al., 2009; Grifone et al., 2008).

A number of genes regulating the dialogue between core arch mesoderm and neural crest cells during branchiomeric myogenesis have been identified. Although *Tbx1* is not itself expressed in neural crest cells, the mandible, entirely crest derived, is reduced in size and the coronoid process absent in *Tbx1*-null embryos (Jerome and Papaioannou, 2001). Altered gene expression is observed in the mandibular arch, including a proximal expansion of neural crest–derived mesenchymal expression domains normally restricted to the distal arch (Kelly et al., 2004). In addition, a medial shift in ectodermal expression domains was observed in the first arch of *Tbx1*-null embryos associated with an altered balance between BMP and FGF signaling within the first arch (Aggarwal et al., 2010). Altered patterning of gene expression in the mandibular arch has been observed in mesodermal-specific *Tbx1*-mutant embryos, showing that a TBX1-regulated signaling molecule in core mesoderm patterns the surrounding neural crest cells (Aggarwal et al., 2010). Signal exchange between core mesoderm and surrounding crest cells is thus important for mandibular development as well as for the patterning of mandibular arch muscles. This has important implications for human craniofacial disorders such as craniofacial microsomia, where arch muscles, as well as skeletal structures, are affected (Heude et al., 2011). In contrast to *Tbx1*, the homeodeomain containing transcription factors DLX5 and DLX6 are expressed in neural crest–derived cells in the first arch, where they are required to pattern the mandibular arch (Heude et al., 2010). In *Dlx5 Dlx6* double-mutant embryos first and second arch–derived muscles are absent and replaced by connective tissue; in addition, tongue muscles are absent. *Tbx1* and *Capsulin* are normally expressed in the premyogenic core of the first arch in *Dlx5 Dlx6* double-mutant embryos, as is *Pax3* in the hypoglossal cord. Reduced levels of *Myf5* and *MyoD* expression were observed at midgestation, and very few myosin-positive cells were found in the region of developing mandibular muscles in later *Dlx5 Dlx6* double-mutant embryos. These results suggest that DLX5 and DLX6 are required not for myogenic specification but rather to maintain the myogenic program in the mandibular arch and initiate muscle differentiation (Heude et al., 2010). Consistent with a failure of derepression of inhibitory signals, elevated *Bmp7* and *Wnt5a* expression was observed in the first arch of *Dlx5*

Dlx6 double-mutant embryos. The targets of TBX1, DLX5, and DLX6 that mediate cranial crest–mesoderm interactions remain to be identified.

Craniofacial muscles are extremely diverse in different vertebrate species and even within certain species, revealing extensive phenotypic plasticity. The key role of crest–mesodermal interactions in craniofacial muscle patterning suggests an involvement in the evolution of craniofacial muscle diversity. In support of such a hypothesis, exchange of neural crest from quail embryos to duck embryos has demonstrated that the presence of quail neural crest–derived connective tissues transforms duck host jaw muscles into quail-like muscle shapes (Tokita and Schneider, 2009). These alterations are associated with modified patterns of skeletal and connective tissue but not myogenic gene expression (Tokita and Schneider, 2009). Cranial neural crest thus appears to play a role not only in integrating muscle and skeletal elements during embryonic development but also in creating species-specific craniofacial muscle patterns over evolutionary time.

10 CONCLUSIONS AND PERSPECTIVES

Craniofacial muscles originate from progenitor cells in diverse regions of embryonic mesoderm: cranial mesoderm giving rise to extraocular and branchiomeric muscles and somatic mesoderm to tongue and hypobranchial muscles. The upstream events driving myogenesis in cranial mesoderm differ fundamentally from those at trunk levels, in particular at the level of myogenic specification—the signaling inputs and transcription factors activating myogenic determination genes of the MYOD family. These embryological differences are reflected in divergent transcriptional profiles in adult skeletal muscles and in the properties of adult skeletal muscle stem cells. A series of myogenic lineages segregating prior to gastrulation give rise to somitic, first arch and extraocular, and second-arch-derived muscles. Branchiomeric muscle progenitor cells are, in fact, more closely related to cardiac progenitor cells than to somite-derived skeletal muscles. Ongoing research aims to dissect the gene regulatory circuits operating in different modules of developing craniofacial musculature, define how divergent skeletal and cardiac myogenic fates arise in pharyngeal mesoderm, and identify the signaling events controlling the dialogue between cranial mesoderm and crest that regulate muscle identity and patterning. Better understanding of the regulation of muscle progenitor cells in the early embryo as well as that of adult skeletal muscle stem cells will in turn provide insights into the evolution of craniofacial muscle diversity and the origins of craniofacial congenital anomalies in humans.

Acknowledgments

This work was supported by the Fondation pour la Recherche Médicale, the EU FP7 contract CardioGeNet (Health-2007-B-223463), and the Association Française contre les Myopathies.

REFERENCES

Aggarwal VS, Carpenter C, Freyer L, Liao J, Petti M, Morrow BE (2010). Mesodermal Tbx1 is required for patterning the proximal mandible in mice. *Dev Biol* 344(2):669–681.

Ai D, Liu W, Ma L, Dong F, Lu MF, Wang D, et al. (2006). Pitx2 regulates cardiac left–right asymmetry by patterning second cardiac lineage–derived myocardium. *Dev Biol* 296(2):437–449.

Bismuth K, Relaix F (2010). Genetic regulation of skeletal muscle development. *Exp Cell Res* 316(18):3081–3086.

Bothe I, Dietrich S (2006). The molecular setup of the avian head mesoderm and its implication for craniofacial myogenesis. *Dev Dyn* 235(10):2845–2860.

Bothe I, Ahmed MU, Winterbottom FL, von Scheven G, Dietrich S (2007). Extrinsic versus intrinsic cues in avian paraxial mesoderm patterning and differentiation. *Dev Dyn* 236(9):2397–2409.

Bothe I, Tenin G, Oseni A, Dietrich S (2011). Dynamic control of head mesoderm patterning. *Development* 138(13):2807–2821.

Bredman JJ, Wessels A, Weijs WA, Korfage JA, Soffers CA, Moorman AF (1991). Demonstration of "cardiac-specific" myosin heavy chain in masticatory muscles of human and rabbit. *Histochem J* 23(4):160–170.

Bryson-Richardson RJ, Currie PD (2008). The genetics of vertebrate myogenesis. *Nat Rev Genet* 9(8):632–646.

Carvajal JJ, Cox D, Summerbell D, Rigby PW (2001). A BAC transgenic analysis of the Mrf4/Myf5 locus reveals interdigitated elements that control activation and maintenance of gene expression during muscle development. *Development* 128(10):1857–1868.

Chen L, Fulcoli FG, Tang S, Baldini A (2009). Tbx1 regulates proliferation and differentiation of multipotent heart progenitors. *Circ Res* 105(9):842–851.

Couly GF, Coltey PM, Le Douarin NM (1992). The developmental fate of the cephalic mesoderm in quail-chick chimeras. *Development* 114(1):1–15.

Dastjerdi A, Robson L, Walker R, Hadley J, Zhang Z, Rodriguez-Niedenfuhr M, et al. (2007). Tbx1 regulation of myogenic differentiation in the limb and cranial mesoderm. *Dev Dyn* 236(2):353–363.

Degenhardt K, Rentschler S, Fishman G, Sassoon DA (2002). Cellular and *cis*-regulation of En-2 expression in the mandibular arch. *Mech Dev* 111(1–2):125–136.

Diehl AG, Zareparsi S, Qian M, Khanna R, Angeles R, Gage PJ (2006). Extraocular muscle morphogenesis and gene expression are regulated by Pitx2 gene dose. *Invest Ophthal Vis Sci* 47:1785–1793.

Dolmatov IY, Mashanov VS, Zueva OR (2007). Derivation of muscles of the Aristotle's lantern from coelomic epithelia. *Cell Tissue Res* 327(2):371–384.

Dong F, Sun X, Liu W, Ai D, Klysik E, Lu MF, et al. (2006). Pitx2 promotes development of splanchnic mesoderm-derived branchiomeric muscle. *Development* 133(24):4891–4899.

Dyer LA, Kirby ML (2009). The role of secondary heart field in cardiac development. *Dev Biol* 336(2):137–144.

Fischer MD, Gorospe JR, Felder E, Bogdanovich S, Pedrosa-Domellof F, Ahima RS, et al. (2002). Expression profiling reveals metabolic and structural components of extraocular muscles. *Physiol Genom* 9(2):71–84.

Franz T, Kothary R, Surani MA, Halata Z, Grim M (1993). The Splotch mutation interferes with muscle development in the limbs. *Anat Embryol* 187(2):153–160.

Gage PJ, Suh H, Camper SA (1999). Dosage requirement of Pitx2 for development of multiple organs. *Development* 126(20):4643–4651.

Gans C, Northcutt RG (1983). Neural crest and the origin of vertebrates: a new head. *Science* 220(4594):268–273.

Grenier J, Teillet MA, Grifone R, Kelly RG, Duprez D (2009). Relationship between neural crest cells and cranial mesoderm during head muscle development. *PloS One* 4(2): e4381.

Grifone R, Kelly RG (2007). Heartening news for head muscle development. *Trends Genet* 3(8):365–369.

Grifone R, Jarry T, Dandonneau M, Grenier J, Duprez D, Kelly RG (2008). Properties of branchiomeric and somite-derived muscle development in *Tbx1* mutant embryos. *Dev Dyn* 237(10):3071–3078.

Guo C, Sun Y, Zhou B, Adam RM, Li X, Pu WT, et al. (2011). A Tbx1-Six1/Eya1-Fgf8 genetic pathway controls mammalian cardiovascular and craniofacial morphogenesis. *J Clin Investig* 121(4):1585–1595.

Hacker A, Guthrie S (1998). A distinct developmental programme for the cranial paraxial mesoderm in the chick embryo. *Development* 125(17):3461–3472.

Harel I, Nathan E, Tirosh-Finkel L, Zigdon H, Guimaraes-Camboa N, Evans SM, et al. (2009). Distinct origins and genetic programs of head muscle satellite cells. *Dev Cell* 16(6): 822–832.

Heude E, Bouhali K, Kurihara Y, Kurihara H, Couly G, Janvier P, et al. (2010). Jaw muscularization requires Dlx expression by cranial neural crest cells. *Proc Nat Acad Sci USA* 107(25):11441–11446.

Heude E, Rivals I, Couly G, Levi G (2011). Masticatory muscle defects in hemifacial microsomia: a new embryological concept. *Am J Med Genet A* 155A(8):1991–1995.

Huang R, Zhi Q, Izpisua-Belmonte JC, Christ B, Patel K (1999). Origin and development of the avian tongue muscles. *Anat Embryol* 200(2):137–152.

Hutson MR, Zhang P, Stadt HA, Sato AK, Li YX, Burch J, et al. (2006). Cardiac arterial pole alignment is sensitive to FGF8 signaling in the pharynx. *Dev Biol* 295(2):486–497.

Jacob SW, Francone CA, Lossow WL (1982). *Structure and Function in Man*, 5th ed., W.B. Saunders, Philadelphia, pp. 192–193.

Jerome LA, Papaioannou VE (2001). DiGeorge syndrome phenotype in mice mutant for the T-box gene, *Tbx1*. *Nat Genet* 27(3):286–291.

Kasahara H, Bartunkova S, Schinke M, Tanaka M, Izumo S (1998). Cardiac and extracardiac expression of Csx/Nkx2.5 homeodomain protein. *Circ Res* 82(9):936–946.

Kassar-Duchossoy L, Gayraud-Morel B, Gomes D, Rocancourt D, Buckingham M, Shinin V, et al. (2004). Mrf4 determines skeletal muscle identity in *Myf5:Myod* double-mutant mice. *Nature* 431(7007):466–471.

Kelly RG (2010). Core issues in craniofacial myogenesis. *Exp Cell Res* 316(18):3034–3041.

Kelly RG, Jerome-Majewska LA, Papaioannou VE (2004). The del22q11.2 candidate gene *Tbx1* regulates branchiomeric myogenesis. *Human Mol Genet* 13(22):2829–2840.

King M, Arnold JS, Shanske A, Morrow BE (2006). T-genes and limb bud development. *Am J Med Genet A* 140(13):1407–1413.

Kitamura K, Miura H, Miyagawa-Tomita S, Yanazawa M, Katoh-Fukui Y, Suzuki R, et al. (1999). Mouse Pitx2 deficiency leads to anomalies of the ventral body wall, heart, extra- and periocular mesoderm and right pulmonary isomerism. *Development* 126(24):5749–5758.

Knight RD, Mebus K, Roehl HH (2008). Mandibular arch muscle identity is regulated by a conserved molecular process during vertebrate development. *J Exp Zool B* 310(4): 355–369.

Knight RD, Mebus K, d'Angelo A, Yokoya K, Heanue T, Roehl H (2011). Ret signalling integrates a craniofacial muscle module during development. *Development* 138(10):2015–2024.

Kontges G, Lumsden A (1996). Rhombencephalic neural crest segmentation is preserved throughout craniofacial ontogeny. *Development* 122(10):3229–3242.

Laugwitz KL, Moretti A, Caron L, Nakano A, Chien KR (2008). Islet1 cardiovascular progenitors: a single source for heart lineages? *Development* 135(2):193–205.

Lescroart F, Kelly RG, Le Garrec JF, Nicolas JF, Meilhac SM, Buckingham M (2010). Clonal analysis reveals common lineage relationships between head muscles and second heart field derivatives in the mouse embryo. *Development* 137(19):3269–3279.

Liao J, Kochilas L, Nowotschin S, Arnold JS, Aggarwal VS, Epstein JA, et al. (2004). Full spectrum of malformations in velo-cardio-facial syndrome/DiGeorge syndrome mouse models by altering Tbx1 dosage. *Hum Mol Genet* 13(15):1577–1585.

Liao J, Aggarwal VS, Nowotschin S, Bondarev A, Lipner S, Morrow BE (2008). Identification of downstream genetic pathways of Tbx1 in the second heart field. *Dev Biol* 316(2):524–537.

Lin CY, Chen WT, Lee HC, Yang PH, Yang HJ, Tsai HJ (2009). The transcription factor Six1a plays an essential role in the craniofacial myogenesis of zebrafish. *Dev Biol* 331(2):152–166.

Lin GF, Geng X, Chen Y, Qu B, Wang F, Hu R, et al. (2003). T-box binding site mediates the dorsal activation of myf-5 in *Xenopus* gastrula embryos. *Dev Dyn* 226(1):51–58.

Lu JR, Bassel-Duby R, Hawkins A, Chang P, Valdez R, Wu H, et al. (2002). Control of facial muscle development by *MyoR* and *Capsulin*. *Science* 298(5602):2378–2381.

Mackenzie S, Walsh FS, Graham A (1998). Migration of hypoglossal myoblast precursors. *Dev Dyn* 213(4):349–358.

Mitani Y, Takahashi H, Satoh N (2001). Regulation of the muscle-specific expression and function of an ascidian T-box gene, *As-T2*. *Development* 128(19):3717–3728.

Nathan E, Monovich A, Tirosh-Finkel L, Harrelson Z, Rousso T, Rinon A, et al. (2008). The contribution of Islet1-expressing splanchnic mesoderm cells to distinct branchiomeric muscles reveals significant heterogeneity in head muscle development. *Development* 135(4):647–657.

Noden DM (1983). The embryonic origins of avian cephalic and cervical muscles and associated connective tissues. *Am J Anat* 168(3):257–276.

Noden DM (1988). Interactions and fates of avian craniofacial mesenchyme. *Development* 103 (Suppl):121–140.

Noden DM, Francis-West P (2006). The differentiation and morphogenesis of craniofacial muscles. *Dev Dyn* 235(5):1194–1218.

Noden DM, Trainor PA (2005). Relations and interactions between cranial mesoderm and neural crest populations. *J Anat* 207(5):575–601.

Noden DM, Marcucio R, Borycki AG, Emerson CP, Jr. (1999). Differentiation of avian craniofacial muscles: I. Patterns of early regulatory gene expression and myosin heavy chain synthesis. *Dev Dyn* 216(2):96–112.

Okano J, Sakai Y, Shiota K (2008). Retinoic acid down-regulates Tbx1 expression and induces abnormal differentiation of tongue muscles in fetal mice. *Dev Dyn* 237(10):3059–3070.

Okkema PG, Fire A (1994). The *Caenorhabditis elegans* NK-2 class homeoprotein CEH-22 is involved in combinatorial activation of gene expression in pharyngeal muscle. *Development* 120(8):2175–2186.

Ono Y, Boldrin L, Knopp P, Morgan JE, Zammit PS (2010). Muscle satellite cells are a functionally heterogeneous population in both somite-derived and branchiomeric muscles. *Dev Biol* 337(1):29–41.

Pattyn A, Hirsch M, Goridis C, Brunet JF (2000). Control of hindbrain motor neuron differentiation by the homeobox gene *Phox2b*. *Development* 127(7):1349–1358.

Rinon A, Lazar S, Marshall H, Buchmann-Moller S, Neufeld A, Elhanany-Tamir H, et al. (2007). Cranial neural crest cells regulate head muscle patterning and differentiation during vertebrate embryogenesis. *Development* 134(17):3065–3075.

Rochais F, Mesbah K, Kelly RG (2009). Signaling pathways controlling second heart field development. *Circ Res* 104(8):933–942.

Romer AS, Parsons TS (1977). *The Vertebrate Body*. 5th ed., W. B. Saunders, Philadelphia.

Sambasivan R, Gayraud-Morel B, Dumas G, Cimper C, Paisant S, Kelly RG, et al. (2009). Distinct regulatory cascades govern extraocular and pharyngeal arch muscle progenitor cell fates. *Dev Cell* 16(6):810–821.

Sambasivan R, Kuratani S, Tajbakhsh S (2011). An eye on the head: the development and evolution of craniofacial muscles. *Development* 138(12):2401–2415.

Schaub C, Nagaso H, Jin H, Frasch M (2012). Org-1, the *Drosophila* ortholog of Tbx1, is a direct activator of known identity genes during muscle specification. *Development* 139(5):1001–1012.

Schilling TF, Kimmel CB (1994). Segment and cell type lineage restrictions during pharyngeal arch development in the zebrafish embryo. *Development* 120(3):483–494.

Semina EV, Reiter R, Leysens NJ, Alward WL, Small KW, Datson NA, et al. (1996). Cloning and characterization of a novel bicoid-related homeobox transcription factor gene, *RIEG*, involved in Rieger syndrome. *Nat Genet* 14(4):392–399.

Shih HP, Gross MK, Kioussi C (2007). Cranial muscle defects of *Pitx2* mutants result from specification defects in the first branchial arch. *Proc Nat Acad Sci USA* 104(14):5907–5912.

Smith PA, Mango SE (2007). Role of T-box gene *Tbx-2* for anterior foregut muscle development in *C. elegans*. *Dev Biol* 302(1):25–39.

Spencer RF, Porter JD (2006). Biological organization of the extraocular muscles. *Prog Brain Res* 151:43–80.

Stanley EG, Biben C, Elefanty A, Barnett L, Koentgen F, Robb L, et al. (2002). Efficient Cre-mediated deletion in cardiac progenitor cells conferred by a 3′UTR-ires-Cre allele of the homeobox gene *Nkx2-5*. *Int J Dev Biol* 46(4):431–439.

Stolfi A, Gainous TB, Young JJ, Mori A, Levine M, Christiaen L (2010). Early chordate origins of the vertebrate second heart field. *Science* 329(5991):565–568.

Tajbakhsh S, Rocancourt D, Cossu G, Buckingham M (1997). Redefining the genetic hierarchies controlling skeletal myogenesis: Pax-3 and Myf-5 act upstream of MyoD. *Cell* 89(1):127–138.

Theis S, Patel K, Valasek P, Otto A, Pu Q, Harel I, et al. (2010). The occipital lateral plate mesoderm is a novel source for vertebrate neck musculature. *Development* 137(17):2961–2971.

Tirosh-Finkel L, Zeisel A, Brodt-Ivenshitz M, Shamai A, Yao Z, Seger R, et al. (2010). BMP-mediated inhibition of FGF signaling promotes cardiomyocyte differentiation of anterior heart field progenitors. *Development* 137(18):2989–3000.

Tixier V, Bataille L, Jagla K (2010). Diversification of muscle types: recent insights from *Drosophila*. *Exp Cell Res* 316(18):3019–3027.

Tokita M, Schneider RA (2009). Developmental origins of species-specific muscle pattern. *Dev Biol* 331(2):311–325.

Trainor P, Krumlauf R (2000). Plasticity in mouse neural crest cells reveals a new patterning role for cranial mesoderm. *Nat Cell Biol* 2(2):96–102.

Trainor PA, Tan SS, Tam PP (1994). Cranial paraxial mesoderm: regionalisation of cell fate and impact on craniofacial development in mouse embryos. *Development* 120(9):2397–2408.

Tzahor E (2009). Heart and craniofacial muscle development: a new developmental theme of distinct myogenic fields. *Dev Biol* 327(2):273–279.

Tzahor E, Evans SM (2011). Pharyngeal mesoderm development during embryogenesis: implications for both heart and head myogenesis. *Cardiovasc Res* 91(2):196–202.

Tzahor E, Kempf H, Mootoosamy RC, Poon AC, Abzhanov A, Tabin CJ, et al. (2003). Antagonists of Wnt and BMP signaling promote the formation of vertebrate head muscle. *Genes Dev* 17(24):3087–3099.

von Scheven G, Bothe I, Ahmed MU, Alvares LE, Dietrich S (2006). Protein and genomic organisation of vertebrate *MyoR* and *Capsulin* genes and their expression during avian development. *Gene Expr Patterns* 6(4):383–393.

Yamane A (2005). Embryonic and postnatal development of masticatory and tongue muscles. *Cell Tissue Res* 322(2):183–189.

Zacharias AL, Lewandoski M, Rudnicki MA, Gage PJ (2011). Pitx2 is an upstream activator of extraocular myogenesis and survival. *Dev Biol* 349(2):395–405.

Zhou Y, Cashman TJ, Nevis KR, Obregon P, Carney SA, Liu Y, et al. (2011). Latent TGF-beta binding protein 3 identifies a second heart field in zebrafish. *Nature* 474(7353):645–648.

6

TOOTH MORPHOGENESIS AND RENEWAL

MARIA JUSSILA, EMMA JUURI, AND IRMA THESLEFF

Developmental Biology Program, Institute of Biotechnology, University of Helsinki, Helsinki, Finland

1 INTRODUCTION

Teeth are an important innovation during vertebrate evolution, and there is extreme variation in tooth shape, size, and number between different species. Fish and reptiles have mostly simple-shaped conical teeth, which are replaced continuously during the life of the animal. Mammals have more complex tooth shapes, and their teeth are heterodont, meaning that they belong to different tooth classes: namely, incisors, canines, premolars, and molars. Mammalian teeth are replaced only once or not at all. Some species have developed specific ecological adaptations, such as rodents, which have continuously growing incisors. It has been shown that in mammals, tooth shape is functionally related to species-specific feeding habits (Evans et al., 2007).

The molecular regulation of tooth development has been studied primarily in the laboratory mouse, the model mammal in developmental biology, which has a highly specialized dentition. Mice have in each jaw quadrant one renewing incisor and three molars that are separated from the incisor by a toothless diastema, and their teeth are not replaced. Nevertheless, comparisons between mouse and other species have shown that the genes controlling tooth development are conserved between different tooth types and species (Fraser et al., 2009; Tummers and Thesleff, 2009; Richman and Handrigan, 2011).

The general signaling networks that regulate tooth development were present very early in evolution, when teeth first arose in the pharynx of jawless fish before the appearance of oral teeth and jaws (Fraser et al., 2009). The genes that regulate tooth

Stem Cells in Craniofacial Development and Regeneration, First Edition.
Edited by George T.-J. Huang, Irma Thesleff.

development are expressed in many other developing organs in the embryo and, in particular, in other ectodermal organs, such as hairs and different glands, which all develop as derivatives of the surface ectoderm. Nevertheless, the common toolkit of signal pathway molecules and transcription factors appears to regulate all steps of tooth morphogenesis and to specify the various dental cell lineages. In this chapter we describe the developmental anatomy and molecular regulation of tooth initiation, morphogenesis, and cell differentiation, as well as of tooth renewal and replacement.

2 DEVELOPMENTAL ANATOMY OF TOOTH FORMATION

Tooth morphogenesis is guided by interactions between epithelial and mesenchymal tissues and progresses through distinct stages defined by morphological features of the dental epithelium. In mammals, the dental epithelium is derived from the surface ectoderm, and the mesenchyme from the cranial neural crest cells that migrate into the first branchial arch and frontonasal process. Some fish (e.g., zebrafish) have pharyngeal teeth, where the epithelium is derived from the endoderm, and in some amphibians dental epithelium derives from both ectoderm and endoderm (Soukup et al., 2008).

The first morphological sign of tooth initiation in mammals is the formation of the dental lamina, a horseshoe-shaped epithelial thickening along the embryonic mandible and maxilla. The epithelium of the lamina thickens locally and forms dental placodes at the sites of tooth initiation. The underlying mesenchyme becomes condensed, and the proliferation of placodal cells leads to the formation of an epithelial bud, which grows into the mesenchyme (Fig. 1).

The formation of an epithelial cell aggregation called the primary enamel knot in the tip of the bud takes place during bud-to-cap stage transition. Enamel knots serve as signaling centers that orchestrate epithelial morphogenesis during the cap and bell stages and initiate the morphogenesis of the tooth crown. Cells in the enamel knot do not proliferate but they stimulate proliferation of the flanking epithelium. This results in the formation of cervical loops during cap stage. As the dental mesenchyme becomes surrounded by growing cervical loops, it forms the dental papilla. The peripheral part of condensed dental mesenchyme forms the dental follicle, which surrounds the tooth germ and eventually forms the periodontal tissues. The epithelium, referred to as an enamel organ, differentiates into distinct cell layers: a core of star-shaped cells with mesenchymal appearance called stellate reticulum (SR), and the inner and outer enamel epithelium (IEE and OEE) surrounding the SR. In mammals that replace their teeth, the bud of the permanent tooth protrudes from the lingual side of the enamel organ during the cap stage. The permanent tooth germ develops as its predecessor, the deciduous tooth, and stays enclosed in the alveolar bone beneath the deciduous tooth, waiting for its turn to erupt (Fig. 1).

During the bell stage, the tooth germ grows rapidly, the shape of the crown becomes evident, and cell differentiation as well as deposition and mineralization of the dental hard tissues begin. The secondary enamel knots are induced in the IEE, and they specify the sites where cusps form. Cusps are the occlusal tips of the tooth crown. The mesenchymal cells directly underlining the dental epithelium differentiate into odontoblasts, laying down the organic matrix of dentin. The epithelial cells, located next to the odontoblasts and separated from them by a basement membrane, differentiate into ameloblasts that deposit the enamel matrix. The mineralization

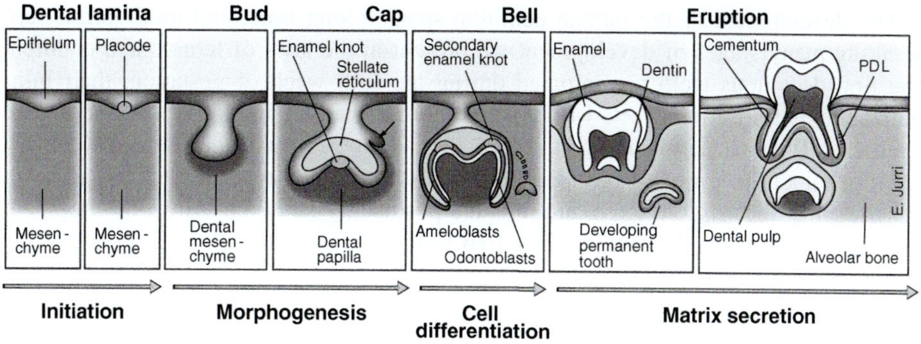

FIGURE 1 Schematic presentation of mammalian tooth development. Teeth develop from epithelial and mesenchymal tissues that interact, undergo complex morphogenesis, and differentiate into tooth-specific cell types. The signaling centers in the placode and subsequently in the enamel knots are key regulators of tooth development. Replacement teeth start to develop during embryogenesis from the successional dental lamina extending from the lingual side of the deciduous tooth (arrow at the cap stage). All images represent frontal planes of section through the deciduous premolar in the lower jaw. Lingual is on the right-hand side of each image.

starts from the cusp tips and moves toward the base of the tooth. Innervation and vascularization of the tooth are linked intimately with morphogenesis, and the first blood vessels and nerves invade the dental papilla mesenchyme during the early bell stage.

Root development begins from the cervical loops when ameloblast differentiation has reached the future crown–root junction. The cervical loops become depleted of SR cells as the IEE and OEE come into contact to form a bilayer. This bilayer is known as Hertwig's epithelial root sheath (HERS) and its growth directs the development of the root. The HERS cells also induce adjacent dental mesenchyme cells to differentiate into odontoblasts. Thus, dentin continues to form as the root grows, but because HERS cells do not differentiate into ameloblasts, no enamel forms in the root area. Instead, HERS loses its epithelial integrity, breaks down, and gives rise to the epithelial cell rests of Malassez (ERM), which eventually remains as a network covering the roots (Ten Cate, 1996; Tummers and Thesleff, 2008). The HERS continues to grow and moves away from the crown as an epithelial disk guiding root morphogenesis. The termination of HERS growth determines the length of the root. The dental follicle cells give rise to cementoblasts, osteoblasts, and periodontal ligament; cementoblasts deposit the cementum on the root surface on the top of the dentin, osteoblasts contribute to the formation of alveolar bone, and the periodontal ligament attaches the root to the alveolar bone. The formation of the root and periodontal ligament is described in more detail in Chapter 8.

The eruption of the tooth into the oral cavity begins after crown formation is completed and after the root has started to grow. Bone remodeling around the tooth is imperative for eruption and is regulated by the dental follicle (Marks and Cahill, 1987). The resorption of bone between the erupting tooth and the oral cavity depends on PTHrP, secreted by the reduced enamel epithelium covering the crown (Philbrick et al., 1998), and bone apposition at the base of the roots depends on the stimulation of osteoblast activity by the dental follicle. The enamel epithelium and dental follicle covering the tooth crown are degraded as the crown pushes its way into the oral cavity.

The development of the human dentition spans a long time, and tooth germs representing many stages of development are present in the jaws of fetuses and children. The first deciduous teeth are initiated during the fifth week of gestation, their mineralization starts during week 14, and they normally erupt in children at the age of 6 months. The first permanent teeth reach the bud stage around gestational week 14, begin to mineralize prior to birth, and erupt at the age of 5 to 6 years. The last teeth to be formed, the wisdom teeth, are initiated postnatally and their crown development is completed between the age of 12 and 16 years.

3 REGULATION OF TOOTH MORPHOGENESIS AND DENTAL SIGNALING CENTERS

Tooth morphogenesis is regulated by interactions between cells, in particular reciprocal and sequential interactions between the mesenchyme and epithelium (Kollar and Baird, 1969; Thesleff, 2003). The interactions are mediated by signals that belong to conserved signaling pathways, which include Hedgehog (Hh), Wnt, Fibroblast growth factor (Fgf), Tgfβ, Bone morphogenetic protein (Bmp), and Ectodysplasin (Eda). These pathways are integrated at many levels, and the gene regulatory networks composed of the signals, their target genes, and modulators have been studied actively over the past decades, leading to a detailed understanding of the roles of the gene regulatory networks in tooth development (Tummers and Thesleff, 2009; O'Connell et al., 2012; Chapter 9 in this volume). Expression patterns of most of the molecular factors discussed here have been elucidated in considerable detail in teeth (http://bite-it.helsinki.fi; Nieminen et al., 1998). How the odontogenic potential is established in tooth-forming tissues, and how distinct signaling centers guide tooth morphogenesis, are discussed below.

3.1 Induction of Odontogenic Potential in Epithelium and Mesenchyme

The question of how odontogenic tissues become specified was studied long before the signaling molecules regulating tooth development were identified. Shirley Glasstone, who pioneered the in vitro studies on tooth development, demonstrated that an early tooth bud can be dissected out and grown in tissue culture, where it will undergo normal tooth morphogenesis (Glasstone, 1936, 1967). Early tooth development is therefore not dependent on either the surrounding connective tissue and bone or on vascularisation or innervation. Similarly, when a single mouse molar tooth germ is placed into the anterior chamber of the eye of a recipient mouse, all three molars develop (Lumsden, 1979). These experiments indicate that the odontogenic potential becomes specified early during development, and it is an inherited property of the tissue. Once the odontogenic tissues have been induced, they can continue their development in an independent manner under suitable conditions.

Pioneering experiments, where different epithelial and mesenchymal tissues were recombined and cultured, demonstrated the key roles of epithelial mesenchymal interactions for tooth morphogenesis. They showed that the odontogenic competence resides first in the epithelium, and shifts later to the mesenchyme. In the mouse embryo this shift takes place between embryonic days E11 and E12 around the time of placode formation and condensation of the mesenchyme (Fig. 2A and 2B). Thus, when epithelium was separated from the first branchial arch before E12, recombined with neural crest–derived mesenchyme from the second arch, and placed in culture to the anterior chamber of the eye, a tooth developed (Mina and Kollar, 1987). If a similar experiment was performed after E12, no tooth germs formed, indicating that the epithelium had lost the tooth inductive potential. Mesenchyme from the first branchial arch prior to E12 does not induce tooth development when recombined with second branchial arch epithelium, but after this stage the dental mesenchyme has gained the odontogenic potential and forms a tooth when recombined with second-arch epithelium and even with limb epithelium (Kollar and Baird, 1970a; Mina and Kollar, 1987). All evidence so far indicates that at the initiation stage the odontogenic potential lies exclusively in the first-branchial-arch epithelium and that the interacting mesenchyme has to derive from the neural crest (Kollar and Baird, 1970a; Lumsden, 1988). Although the dental mesenchyme originates from the cranial neural crest, premigratory trunk neural crest cells can form a tooth when recombined with first-arch epithelium (Lumsden, 1988). By the time the dental mesenchyme has acquired the potential to induce tooth formation, it also contains the information on tooth identity. When epithelium and mesenchyme from incisors and molars were combined in heterologous recombinants, the resulting tooth shape formed according to the identity of the mesenchyme (Kollar and Baird, 1969).

It is not known how the odontogenic competence and the tooth identity become induced in the tissues at the molecular level. There are certain signaling molecules and transcription factors expressed in distinct patterns in the early branchial arch epithelium, such as *Bmp4* in the distal epithelium, *Fgf8* in more proximal regions, and *Shh* and *Pitx2* at the oral surface (Neubuser et al, 1997; Keränen et al., 1999; Fig. 2B). Wnt/β-catenin is so far the only signal pathway that has been shown to induce de novo tooth formation. It regulates *Fgf8* expression in the early epithelium, and tooth placodes do not form in mutant mice overexpressing Wnt inhibitor *Dkk1* in the epithelium (Andl et al., 2002; Wang et al., 2009). Activation of Wnt signaling in the epithelium induces supernumerary placodes throughout the surface epithelium of a mutant embryo, and these placodes give rise to various epithelial appendages, such as teeth and hair. Yet extra teeth form only in the region of dental arches, and mostly in connection with other teeth (Järvinen et al., 2006; Liu et al., 2008; Wang et al., 2009; Fig. 4C). This observation confirms that odontogenic competence is present only in the oral region.

The first morphological sign of tooth development is the formation of the dental lamina in the epithelium. It is not known how it becomes established, and currently few genes (e.g., *Shh, Pitx2*, *Foxi3*) are known to be expressed there (Keränen et al., 1999; Shirokova et al., 2013; Fig. 2B). In addition, there are no reports on mouse mutants where the dental lamina would not develop. All teeth develop within the dental lamina even in most mutant mice where extra teeth are formed. One exception is a mouse mutant lacking the mesenchymal transcription factor Osr2 that has an extra tooth developing lingual to the first molar (Zhang et al., 2009b). This phenotype was explained by a more broad *Bmp4* expression in the dental mesenchyme that leads to a broadening of the dental field. Tooth placodes form within the dental lamina at the sites of future teeth (Fig. 2B). The dental mesenchyme condenses around the placode at the time when the odontogenic competence shifts to the mesenchyme. Bmps in the epithelium induce expression of *Bmp4* in the mesenchyme at the placode stage, correlating with the shift of competence (Vainio et al., 1993). Wnt/β-catenin signaling promotes *Bmp4* expression in the early incisor mesenchyme, and this in turn regulates

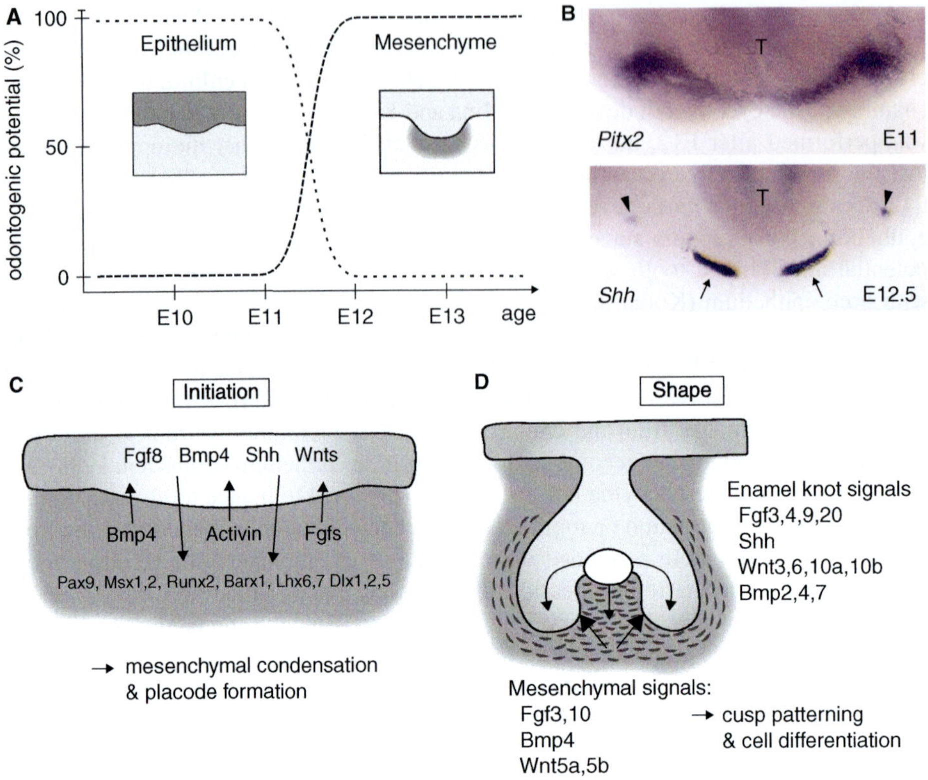

FIGURE 2 Odontogenic potential shifts from the epithelium to mesenchyme between the dental lamina and placode stages (A and B). Crosstalk between epithelium and mesenchyme is mediated by conserved signaling pathways, and it regulates tooth initiation and shape (C and D). Dental epithelium is capable of inducing tooth development when recombined with nondental mesenchyme before the E12 stage of mouse development. At E12 the odontogenic potential has shifted to the mesenchyme, which can now induce tooth development when recombined with nondental epithelium (Mina and Kollar, 1987) (A). *Pitx2* is expressed in the dental lamina of the mouse lower jaw at E11 (t = tongue), and *Shh* is expressed in the placodes of incisors (arrows) and molars (arrowheads) at E12.5 (B). When tooth development is initiated, signals from the epithelium activate a set of transcription factors in the mesenchyme, leading to condensation of the mesenchyme and formation of the epithelial placode and its subsequent budding (C). The primary enamel knot is a signaling center expressing multiple signaling molecules that regulate epithelial and mesenchymal proliferation and induce reciprocal signals from the mesenchyme. The enamel knot also determines the positions of the molar cusps and is thought to initiate the differentiation of odontoblasts (D).

Shh in the epithelium (Fujimori et al., 2010). Expression of transcription factor *Lef1*, a Wnt effector, shifts from the epithelium to the mesenchyme together with the shift in the odontogenic potential, and is regulated by Bmp4 (Kratochwil et al., 1996).

Expression of a number of other important transcription factors is induced in the mesenchyme when the odontogenic competence shifts. For example, epithelial Bmp4 induces the expression of *Msx1*, and Fgf8 induces *Pax9* (Vainio et al., 1993; Neubuser et al. 1997; Fig. 2C). It was shown recently that Fgf8 from the dental epithelium acts as an attractant stimulating the condensation of mesenchymal cells together with a repulsive factor, Semaphorin-3 (Mammoto et al., 2011). Interestingly, this condensation and packing of the mesenchymal cells induced changes in the cell shape that lead to upregulation of *Pax9* as well as *Msx1* and *Bmp4*. Thus, the effect of Fgf8 on the expression of some mesenchymal genes appears to be indirect. Other targets of Bmp and Fgf signaling in mesenchyme at this stage include *Lhx6,7, Barx1, Dlx1* and *2*, and *Runx2* (Bei, 2009b; Tummers and Thesleff, 2009). In addition, the Shh mediators *Gli2* and *Gli3* are expressed in the mesenchyme, and are required for tooth formation (Hardcastle et al., 1998). When the function of mesenchymal transcription factors is deleted either alone or together with another factor of the same family, tooth development is arrested at the placode or bud stage, emphasizing their importance for further development to proceed (Bei, 2009b). It is noteworthy that all of these transcription factors are also expressed in several other organs during development. It therefore appears that the odontogenic competence is not encoded by a tooth-specific gene but probably by a combination of mesenchymal transcription factors.

Insights into the key signaling events taking place during tooth initiation have been gained from model systems, where the developmental capacity that has been lost during evolution is reactivated experimentally. Modern birds do not develop teeth, but they do have a structure resembling the dental lamina, and certain genes (e.g. *Pitx2, Fgf8, Pax9*) have expression patterns similar to mouse jaws (Chen et al., 2000). Treating chicken mandibles with Bmp4 and Fgf4 proteins induces downgrowth of an epithelial structure that expresses *Shh* similar to the mouse enamel knot, and Bmp4-soaked beads are able to induce *Msx1* and *Msx2* expression in the chick mandibular mesenchyme (Chen et al., 2000). If early chicken neural crest is replaced by a mouse transplant, mouse cells that migrate to the chicken mandible induce chicken epithelium to form chimeric tooth germ–like structures (Mitsiadis et al., 2003). A *talpid2* chicken mutant develops epithelial buds that resemble morphologically and molecularly the tooth germs of crocodiles, the closest living relatives of birds, and forced expression of activated β-*catenin* in wild-type chicken epithelium results in the development of similar structures (Harris et al., 2006). Mice have vestigial tooth rudiments in their diastema region in front of molars that apparently represent the premolars lost during evolution. These rudiments develop into supernumerary teeth in several mutant mouse lines where the activity of different signaling pathways has been modulated, including the K14-Eda mouse line overexpressing *Ectodysplasin, Sprouty* mutants with enhanced Fgf signaling, *Polaris* mutants with enhanced Shh signaling, and mice lacking *Sostdc1*, a modulator of both Wnt and Bmp signaling pathways (Mustonen et al., 2003; Kassai et al., 2005; Klein et al., 2006; Ohazama et al., 2009; Ahn et al., 2010). These examples from chicken and mouse emphasize how early tooth induction is controlled by a network of important signaling pathways and their target genes.

3.2 Dental Placodes and Enamel Knots

There are three sets of transient signaling centers that play key roles in the regulation of patterning and morphogenesis of teeth: dental placodes and primary and secondary enamel knots. The reiterative appearance of epithelial signaling centers marks critical stages of tooth development; placodes initiate tooth formation, primary enamel knots initiate the morphogenesis of the tooth crown, and secondary enamel knots define the shape of the tooth crown in multicusped teeth. The signaling centers are characterized by convergence of several signal pathways. This is evidenced by the localized coexpression of multiple signaling molecules as well as their targets and mediators in the placodes and enamel knots. Largely the same set of genes is expressed repeatedly in all signaling centers of the tooth (Thesleff, 2003; Fig. 2C and D). Thus, their mutations potentially affect several stages of tooth development.

The dental placodes consist of thickened epithelium, which is associated with underlying neural crest–derived mesenchyme, and they are the first signaling centers of the tooth. The placodes initiate epithelial morphogenesis and induce condensation of the underlying mesenchyme (Fig. 2C). It is likely that each dental placode gives rise to an entire tooth class (incisor, canine, premolar/molar), with direct budding of only the first tooth of the class from the placode and subsequent formation of the successive teeth from the primed odontogenic epithelium and mesenchyme associated with the preceding tooth. According to this scenario, most mammalian teeth would be initiated from odontogenic tissue associated with previous teeth. It is also noteworthy that there may be differences in the molecular mechanisms of tooth initiation between the first teeth forming from the placodes and the successional teeth (see below), and that most current information on the genetic regulation of tooth initiation comes from work on mouse first molars, not from successionally forming teeth (see below).

The development of all ectodermal organs begins from placodes, and it is therefore conceivable that the same signaling pathways and molecular mechanisms that regulate the placode formation in teeth are shared with placodes of other ectodermal organs, such as hair, nails, and many exocrine glands (Mikkola, 2009). The similarities in the genetic programs in different ectodermal organs are illustrated by the more than 150 ectodermal dysplasia syndromes displaying congenital defects in two or more ectodermal appendages. The transcription factor *p63* is one of the necessary genes regulating the initiation of all ectodermal placodes. Mutations in the *p63* gene cause several human syndromes characterized by ectodermal dysplasias (Brunner et al., 2002). When *p63* function is deleted in mice, the placodes of teeth and other ectodermal appendages are completely absent, but the dental lamina forms (Laurikkala et al., 2006). In these mice the key signaling pathways, including Bmp, Fgf, Notch, and Eda, are impaired, and tooth development is arrested prior to epithelial budding.

The Ectodysplasin (Eda) signaling pathway is a specific regulator of ectodermal organ development, and it has a particular role in the formation of ectodermal placodes in all vertebrates (Mikkola, 2009). Eda belongs to the tumor necrosis factor signal family, and its receptor, *Edar*, is expressed specifically in placodes of all ectodermal appendages as well as in primary and secondary enamel knots. Mutations in the Eda pathway genes cause deficient formation of all ectodermal organs, which is manifested in the human hypohidrotic ectodermal dysplasia (HED) syndrome as missing teeth and defects in other ectodermal organs, particularly in hair and sweat glands (Mikkola, 2009). Mice with impaired Eda signaling often show absence of third molars or incisors and small first molars with defects in cusp patterning. The phenotype is due to the small size of placodes and primary enamel knots as well as fused secondary enamel knots, indicating a requirement of Eda in the functioning of all dental signaling centers (Pispa et al., 1999). In accordance with this, mice that overexpress *Eda* in epithelium have large placodes and enamel knots, aberrant molar cusp patters, and regularly develop an extra tooth anterior to the first molar (Pispa et al., 2004). The downstream targets of Eda include signals, mediators, and antagonists of all major signaling pathways, including the signal molecules Shh, Wnt10b, and Fgf20; the Bmp inhibitors Ctgf and

Follistatin; and the Wnt inhibitor Dkk4 (Fliniaux et al., 2008; Mikkola, 2009; Zhang et al. 2009a; Lefebvre et al., 2012). These data indicate the importance of Eda as a key modulator of the epithelial signaling centers.

The enamel knots are transient epithelial signaling centers composed of nondividing cells. The primary enamel knots appear in the IEE at the tips of the tooth buds during the transition from the bud to the cap stage. They initiate the morphogenesis of tooth crown and regulate cell proliferation in both the underlying mesenchyme forming the dental papilla and in the flanking epithelium forming the cervical loops (Fig. 2D). In addition, the enamel knots regulate formation of the secondary enamel knots appearing in the IEE during the early bell stage. The secondary enamel knots define the future positions of the cusp tips and thus the shape of the tooth crown.

The early markers of the primary enamel knot include *Msx2* and *p21*, which are induced by Bmp4 from the mesenchyme (Jernvall et al., 1998), and *Fgf4*, induced by epithelial Lef1, a mediator of Wnt signaling, whose absence leads to an arrest of tooth development from the bud to the cap stage (Kratochwil et al., 2002). The primary enamel knot expresses at least 13 different signaling molecules, including *Shh* and several *Fgfs, Wnts*, and *Bmps* (Fig. 2D). The primary enamel knot is a transient structure, and it is mostly removed by apoptosis during the transition from the cap to the bell stage. Secondary enamel knots express most of the same signaling molecules and transcription factors as those of primary enamel knots, including *Fgf4, Fgf9, Shh, Bmp4, Wnt10b, p21, Lef1*, and *Msx2* (Keränen et al., 1998; Kettunen and Thesleff, 1998).

Molar morphologies vary greatly between different mammalian species. This is due to the modulation and fine-tuning of enamel knot signaling, which affects the patterning of the secondary enamel knots and thereby the pattern of molar cusps. According to a morphodynamic mathematical model (Salazar-Ciudad and Jernvall, 2002), the balance and dynamics between the activators and inhibitors regulate the position of secondary enamel knot formation and cell differentiation in the tooth crown. Therefore, it is suggested that the modulation of the dynamics of activators and inhibitors leads to the differences in cusp patterns during evolution as it does in mathematical models (Salazar-Ciudad and Jernvall, 2002). Support for this is provided by the diverse tooth phenotypes of transgenic mice, in which signaling is modulated and the teeth resemble teeth of other species. For example, in the K14-Eda mouse line, where *Eda* is overexpressed in the epithelium, the teeth resemble those of kangaroo (Kangas et al., 2004), whereas the teeth of mice deficient in the Bmp and Wnt inhibitor Sostdc1 resemble rhino teeth (Kassai et al., 2005; Tummers and Thesleff, 2009). Both overexpression and deletion of *Follistatin*, an antagonist of Bmps and activin, results in aberrant cusp patterns in mice (Wang et al., 2004). Recently, Harjunmaa et al. (2012) showed that by adjusting multiple signaling pathways simultaneously, the number of tooth cusps can be increased in cultured mouse molars. Taken together, these findings indicate the importance of the balance of signaling in the regulation of enamel knot patterning.

4 DENTAL CELL LINEAGES AND THEIR DIFFERENTIATION

Differentiation of the distinct dental cell types is closely linked with morphogenesis. The mesenchyme of the dental papilla gives rise to odontoblasts and pulp fibroblasts, while the dental follicle mesenchyme generates cementoblasts, alveolar bone, and fibroblasts of the periodontal ligament. Dental epithelium gives rise to ameloblasts, which form the enamel, and the stellate reticulum and stratum intermedium cells, which support tooth morphogenesis and enamel formation, respectively. The epithelial cell rests of Malassez (ERM) covering the root surface also differentiate from the tooth epithelium. Little is known, however, of the functions of ERM and its differentiation. The major hard tissue producing cells in dental tissues are ameloblasts, odontoblasts, and cementoblasts, which form the extracellular matrices of enamel, dentin, and cementum, respectively. The differentiation of odontoblasts and ameloblasts has been studied in great detail and is addressed below, whereas little is known about cementoblast differentiation. It is generally thought that cementoblasts derive from dental follicle cells and that their differentiation is regulated by ERM cells (Hammarström et al., 1996). However, there is some evidence suggesting that cementoblasts may form via epithelial–mesenchymal transition from the ERM (Akimoto et al., 2011). See Chapter 8 for more details.

4.1 Odontoblasts and Dentin

Cytodifferentiation of the hard tissue–forming cells starts during the bell stage by the alignment of mesenchymal preodontoblasts under the basement membrane of the IEE at the cusp tips (Fig. 3). As the first odontoblasts differentiate next to the enamel knots, it has been suggested that signals produced by the enamel knot may be involved in the initiation of their differentiation (Thesleff et al., 2001). The dental epithelium is

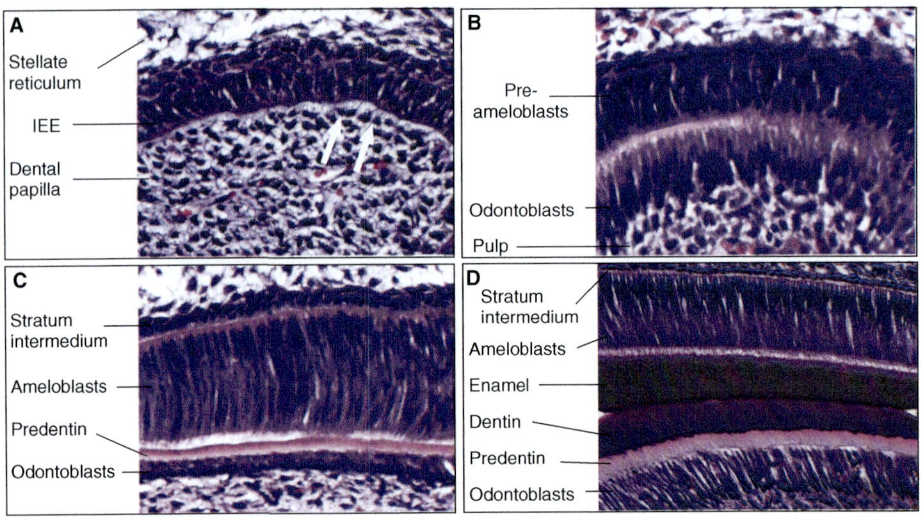

FIGURE 3 Successive steps of odontoblast and ameloblast differentiation, and deposition of dentin and enamel. Cell differentiation takes place at the interface between the inner enamel epithelium (IEE) and the dental papilla mesenchyme, and is regulated by reciprocal signaling between the tissues. At the early bell stage, dental papilla cells align under the basement membrane (arrows) in response to epithelial signals (A). The predontoblasts stimulate epithelial differentiation, and the preameloblasts and odontoblasts become tall polarized cells as the nuclei move away from the basement membrane (B). The odontoblasts secrete predentin, which starts to mineralize to dentin (C and D). The functional odontoblasts and predentin induce the further differentiation of ameloblasts, and enamel secretion begins (D).

required for odontoblast differentiation, and several signals and matrix molecules have been implicated in the process. Tgfβ family growth factors, in particular Bmps, have an essential role in odontoblast differentiation (Begue-Kirn et al., 1992; Ruch et al., 1995). As differentiation begins, cells first differentiate into cuboidal preodontoblasts expressing *Bmp4*. Later they elongate into large columnar and polarized odontoblasts, which no longer express *Bmp4* but begin to express *Bmp2* (Nakashima, 1994; Nakashima and Reddi, 2003; Yamashiro et al., 2003). Further support for the role of Bmps in odontoblast differentiation comes from studies where odontoblasts were induced to produce dentin in vitro by adding Bmp2, 4, and 7 (Nakashima, 1994; Rutherford et al., 1994). It was shown recently that the conditional loss of *Smad4*, a mediator of TGFβ/Bmp signaling, from the dental papilla prevents the terminal differentiation of odontoblasts and dentin deposition (Li et al., 2011).

In addition, Fgfs (Unda et al., 2000) and Wnt signals have been implicated in the regulation of odontoblast differentiation and dentin formation. *Wnt10a* is expressed in the enamel knots and later in the secretory odontoblasts together with *dentin sialophosphoprotein* (*Dspp*). It was suggested that Wnt10b regulates the expression of *Dspp* (Yamashiro et al., 2007). On the other hand, continuous stabilization of β-catenin, the mediator of Wnt signaling, in the mesenchyme leads to defects in odontoblasts and ameloblasts (Chen et al., 2009). Also, the localization of high Wnt reporter activity in odontoblasts is in accordance with the role of canonical Wnt signaling in odontoblast differentiation (Suomalainen and Thesleff, 2010). The basement membrane underlying the enamel epithelium is also important for polarization and differentiation of odontoblasts and may serve as a reservoir of signal molecules (Thesleff and Hurmerinta, 1981; Ruch et al., 1995).

Odontoblasts gradually move toward the center of the dental papilla as they secrete dentin matrix, and they leave behind a single cellular process that remains embedded in the dentinal tubule. After completion of tooth development, the odontoblasts remain as a continuous sheet lining the pulp chamber and root canals. Dentin matrix consists mainly of type I collagen, dentin phosphoprotein, and dentin sialophosphoprotein (Dspp), and mutations in these genes cause dentinogenesis imperfecta in humans (Shields et al., 1973).

The odontoblasts stay alive throughout life and continue to produce secondary dentin after tooth formation is complete. The production of secondary dentin is much slower than during tooth morphogenesis and leads to gradual narrowing of the pulp cavity. Tertiary, or reparative dentin, can be formed by odontoblasts that differentiate from dental pulp stem cells as a result of injury or bacterial attack. Tertiary dentin is produced to protect the pulp tissue from bacterial metabolites and to prevent pulpal necrosis. The pulp stem cells can be induced to differentiate into odontoblasts in vitro by Bmp2 recombinant protein (Iohara et al., 2004).

4.2 Ameloblasts and Enamel

Ameloblasts differentiate from the IEE adjacent to the basement membrane and the odontoblast layer. Ameloblasts are polarized columnar cells that synthesize and secrete enamel matrix. Their differentiation can be divided into the presecretory, secretory, and maturation stages (Fig. 3). At the secretory stage, ameloblasts deposit the enamel matrix, which becomes partially mineralized. Ameloblasts regulate mineralization by controlling the transfer of ions and other molecules, such as enzymes degrading the enamel proteins, to the matrix. Mineralization is completed during the maturation stage, concomitant with tooth eruption. All ameloblasts are lost after tooth eruption, hence enamel cannot renew.

The differentiation of the epithelial cells into ameloblasts is instructed by signals derived from the odontoblasts and predentin (Zeichner-David et al., 1995; Bei, 2009a). The morphology of ameloblasts changes from the early presecretory and cuboidal epithelial cells to elongated, polarized, secretory ameloblasts with a large cellular process from where the enamel matrix is secreted on the dentin surface (Nanci, 2008). Ameloblast differentiation begins from the cusp tips and is induced by signals secreted by preodontoblasts (Coin et al., 1999; Wang et al., 2004). Bmps have an important role in ameloblast differentiation. When *Follistatin*, a BMP inhibitor, was overexpressed in the epithelium of developing mouse incisors, the ameloblasts did not differentiate and no enamel formed, whereas in *Follistatin*-knockout mice, ameloblasts differentiated at the lingual surface of the tooth, where they are not normally present (Wang et al., 2004). Canonical Wnt signaling is present in the differentiating ameloblasts (Suomalainen and Thesleff, 2010), and epithelial overexpression of Wnt3 causes ameloblast loss from postnatal mouse incisors (Millar et al., 2003). Hedgehog signaling is required for ameloblast differentiation and maturation, and *Shh* is expressed by preameloblasts as well as the stratum intermedium cells (Dassule et al., 2000; Gritli-Linde et al., 2002). Recent studies have also proposed a role for miRNAs in the regulation of ameloblast differentiation. The conditional deletion of *Dicer-1* in the epithelium of developing teeth leads to impaired differentiation of ameloblasts, resulting in deficient enamel formation (Cao et al., 2010; Michon et al., 2010). During their differentiation process, ameloblasts express genes that code for various matrix proteins, enzymes, transcription factors, and signaling molecules of several pathways, such as Bmps, Tgfβ1, Shh, and Wnts (Bei, 2009a).

The integrity of the ameloblast cell layer and its tight contact with the stratum intermedium layer are essential for the formation of proper enamel matrix. The ameloblasts form an epithelial sheet where they are connected with desmosomes and tight junctions, and the ameloblast–stratum intermedium interface is sealed by desmosomes. It was shown that the cell–cell adhesion molecules Nectin-1 and Nectin-3 and the cell membrane protein PERP (P53 apoptosis effector related to PMP-22) are required for the formation of these cellular junctions and that their deletion in transgenic mice leads to irregularities in the ameloblast layer and splitting between the ameloblast and stratum intermedium layers, resulting in enamel defects (Barron et al., 2008; Yoshida et al., 2010; Jheon et al., 2011).

The enamel matrix is composed of unique proteins that regulate formation of the hardest mineralized tissue of the body. These proteins include amelogenin, enamelin, and ameloblastin. In addition, ameloblasts express specific proteinases, such as Mmp-20, which is involved in enamel matrix degradation. Mutations in several genes participating in amelogenesis lead to amelogenesis imperfecta, which is a heterogeneous group of inherited defects in enamel formation (Bei, 2009a).

5 TOOTH RENEWAL

Although the capacity for tooth regeneration and renewal has decreased significantly during vertebrate evolution, limited tooth renewal is still present in mammals.

Continuous growth of teeth has evolved in some mammals as a mechanism to compensate for reduction of tooth height resulting from chewing hard food. In addition to the continuously growing incisors, some rodent species, such as the sibling vole (*Microtus rossiaemeridionalis*), have continuously growing molars (Tummers and Thesleff, 2003). Stem cell–based tooth renewal has so far been demonstrated only in the mouse incisor, which grows continuously throughout the life of the animal. The growth is fueled by the action of both epithelial and mesenchymal stem cells. The epithelial stem cells are located in the labial cervical loop of the incisor, and they express the stem cell marker *Sox2*. Sox2-positive cells contribute to all epithelial cell lineages of the incisor (Juuri et al., 2012). The stem cell niche of the mouse incisor is discussed in more detail in Chapter 17. The major mechanism of tooth renewal in mammals is tooth replacement where either the entire deciduous dentition or part of it is replaced by a second set of functional teeth. The replacement teeth are formed successionally from their deciduous predecessors. Below we describe the morphological and molecular aspects of mammalian tooth replacement and compare it to the other form of successional tooth formation occurring during the serial addition of teeth in the primary dentition.

5.1 Tooth Replacement

Mammalian teeth, excluding molars, are replaced maximally once, whereas in reptiles and fish, tooth replacement may be continuous. So far, putative stem cells for tooth replacement have been identified in the dental lamina of the leopard gecko (a reptile, *Eublepharis macularius*). Reptilian tooth replacement is discussed in more detail in Chapter 6. Unlike reptiles and fish, which have mostly unicuspid and simple-shaped teeth, mammalian teeth have more complex shapes, and they belong to four tooth classes. Of the teeth in the primary dentition, the incisors, canines, and premolars are usually replaced once, but the molars are not replaced in any mammals. Tooth replacement is initiated from the successional dental lamina that detaches from the lingual side of each deciduous tooth and buds to form the permanent successor (Figs. 1 and 4A). The morphogenesis of the human replacement teeth starts before birth and is completed during early childhood.

The research done on the laboratory mouse has provided valuable information on the molecular regulation of primary tooth development and on the function of stem cells in the renewal of the ever-growing incisors. As mice do not replace their teeth, knowledge of the mechanisms and molecular regulation of tooth replacement has long been lacking. Other species have been studied to bridge the gap in our knowledge of the biology of tooth replacement. Reptiles and fish have proved to be very useful models because of their continuous replacement, and the presence of several stages of replacement tooth development at any point in time (Smith et al., 2009; Richman and Handrigan, 2011) (see Chapter 6). Nevertheless, understanding human tooth replacement requires the study of other mammalian species.

The histology of replacement tooth development in some mammalian species has been described as early as in the nineteenth-century literature (Leche, 1895; Luckett, 1993). The model animals used to study tooth replacement more recently have included the ferret (*Mustela putorius furo*), the minipig, and two species of shrews (*Sorex araneus* and *Suncus murinus*) (Järvinen et al., 2008, 2009; Stembirek et al., 2010; Yamanaka et al., 2010). The dental lamina epithelium is thought to play a key

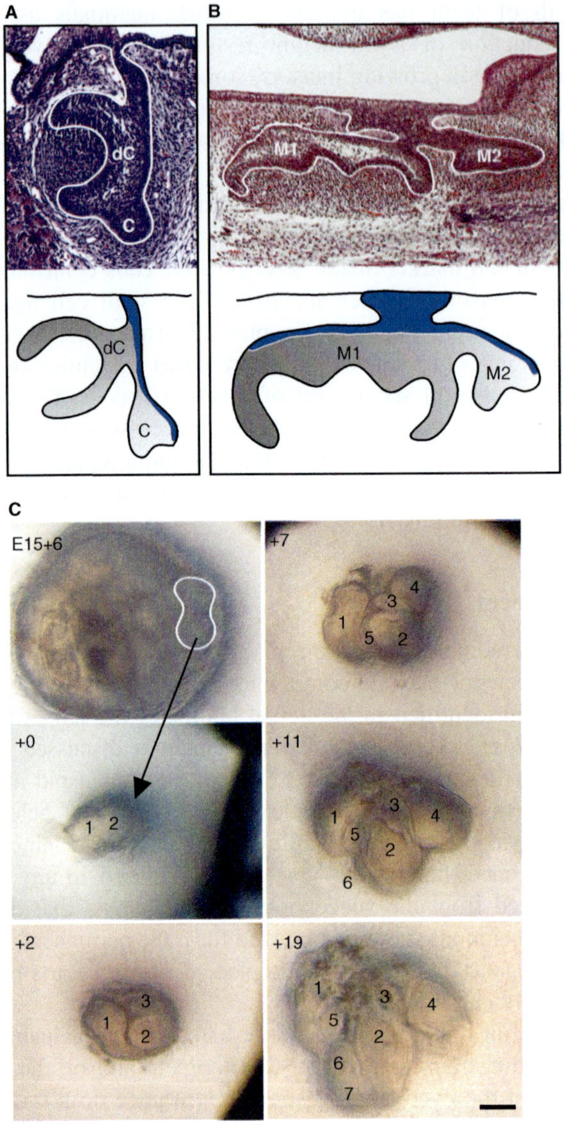

FIGURE 4 Successional tooth formation during ferret tooth replacement (A), during posterior addition of mouse molars (B), and in a mouse mutant with activated Wnt signaling (C). Permanent canine (C) develops as an extension of the successional lamina, which is embedded in the lingual side of the deciduous (primary) canine (dC) (A). Mouse second molar (M2) develops as an extension of the dental lamina embedded in the first molar (M1) (B). Forced activation of Wnt signaling by stabilization of β-catenin in the epithelium leads to successional tooth formation in mice (C). A single molar tooth from an E15 mutant embryo with stabilized β-catenin in the epithelium gives rise to multiple tooth germs during a 6-day in vitro culture (E15 + 6). When two tooth germs (+0) were dissected out from this explant, several new teeth formed successionally during a 19-day culture period (+2, +7, +11, and +19 days). Scale bar: 200 μm. A, frontal plane of section; B, sagittal plane of section. The white line indicates the border between epithelium and mesenchyme in A and B; dental lamina is represented in blue color in A, B and C, adapted from Järvinen et al. (2006).

role in tooth replacement in all vertebrates studied, including some fish, reptiles, and mammals (Järvinen et al., 2009; Smith et al., 2009; Richman and Handrigan, 2011). In reptiles, the successional dental lamina, which gives rise to the replacement teeth, expresses some of the same molecules as does the early dental lamina that forms when tooth development is initiated (Handrigan and Richman, 2010a). This suggests that the regulation of tooth initiation and replacement are largely similar, and that there may be a common origin of epithelial progenitor cells in teeth. It has also been proposed that there are stem cells in the early embryonic dental lamina and that these stem cells may give rise to all epithelial cells of teeth (Smith et al., 2009). In mouse the early dental lamina fragments at the time when the incisor and molar buds form, whereas in tooth-replacing mammals, such as humans and ferrets, the dental lamina grows down into the mesenchyme and contributes to successional tooth formation.

A detailed histological description of ferret tooth replacement has shown that all deciduous tooth germs are connected to each other like pearls in a string by the dental lamina spanning the entire jaw (Järvinen et al., 2009). This dental lamina is embedded to the lingual part of the enamel organ of each deciduous tooth (Luckett, 1993; Järvinen et al., 2009). When permanent tooth development is initiated in the ferret, a successional dental lamina forms by splitting off from the lingual side of the enamel organ of the deciduous tooth, grows down, and buds to form the permanent tooth enamel organ (Fig. 4A). In the ferret there are considerable differences in the timing of permanent tooth initiation between the canine and the premolars (Järvinen et al., 2009). The permanent canine development is initiated when the deciduous canine is in the early cap stage, whereas the permanent premolars start to develop only when their deciduous counterparts have advanced to the bell stage. Interestingly, in shrews the primary tooth begins to develop but becomes rudimentary once the replacement tooth is initiated, and never erupts (Järvinen et al., 2008; Yamanaka et al., 2010). It was suggested that the early-initiated permanent tooth inhibits further development of the deciduous tooth (Järvinen et al., 2008). There is a variation between shrew species, with only one tooth having a replacement tooth developing, or with all teeth except molars having rudimentary primary teeth (Järvinen et al., 2008; Yamanaka et al., 2010).

It is not known what restricts tooth replacement in mammals to one round only. Histological observations on human embryos and young children show the presence of a successional dental lamina in the developing permanent teeth (Ooë, 1981). Formation of a successional lamina was reported also in the bearded dragon (a reptile, *Pogona vitticeps*), which has only one set of functional teeth (Handrigan and Richman, 2010b). The lack of successional tooth formation was speculated to be due to increased apoptosis in the successional lamina and to decreased Wnt pathway signaling activity. In the minipig the dental lamina that connects the deciduous teeth to the oral epithelium seems to disappear through both epithelial-to-mesenchymal transition and apoptosis (Buchtová et al., 2012). Clearly, in mammals where tooth replacement ceases after permanent tooth development, there has to be a mechanism either to remove or to inhibit the cells that otherwise would be capable of further initiation of tooth development. It seems that this inhibitory mechanism may sometimes fail and that there actually is a capacity for continued tooth development, even in humans. There are human syndromes where the patients develop supernumerary teeth or even a partial tertiary dentition (as discussed below). Interestingly, excess dental epithelial tissue has been found in biopsies from jaws of patients with cleidocranial dysplasias with supernumerary teeth (Lukinmaa et al., 1995).

5.2 Successional Formation of Primary Teeth

In addition to the successionally developing replacement teeth, successional formation of teeth also takes place during development of the mammalian primary dentition, when incisors, premolars, and molars are added serially along the tooth row. These two processes of successional tooth formation share striking similarities. The replacement teeth bud from the epithelium at the lingual aspect of their primary tooth predecessors, while successional teeth bud from the anterior or posterior aspects of the primary incisor or premolar and later from the posterior aspect of the molar (Fig. 4B). Interestingly, there are rare examples of animals in which posterior addition of molars continues. One such species is the African mole rat (*Heliophobius argenteocinereus*). The new molars are added posteriorly, and the entire tooth row moves toward the front of the jaw as the most anterior molars are being resorbed (Rodrigues et al., 2011). This process thus has features of horizontal tooth replacement, although individual teeth are not replaced in this case. It has been shown in mouse molars that during the addition of the second and third molars there are inhibitory signals secreted from the previous molar that affect the size of the following tooth (Kavanagh et al., 2007). It is possible that this "inhibitory cascade" is responsible for the limited number of teeth in each tooth class.

5.3 Molecular Regulation of Tooth Replacement and Successional Formation of Primary Teeth

Most molecular data on tooth replacement come from reptiles and are discussed in Chapter 6. A general observation is that the signaling pathways operating during replacement tooth development seem to be the same as those regulating the development of the first set of teeth. As pointed out above, in mammals, most of the primary teeth form successionally in a manner similar to that of the replacement teeth within each tooth class, where only the first tooth forms from the placode. Therefore, the mechanisms regulating the development of all kinds of teeth may be quite similar. So far, the only signaling pathway regulating tooth placode initiation that appears not to be involved in replacement tooth initiation is the Shh pathway, as no Shh expression was detected during initiation of replacement teeth in either the ferret, gecko, or shrew (Järvinen et al., 2009; Handrigan and Richman, 2010b; Yamanaka et al., 2010). Wnt signaling has been shown to play a role in all types of tooth initiation. In the ferret, two Wnt signaling pathway modulators, *Sostdc1*, an inhibitor of Wnt and Bmp pathways, and *Axin2*, a Wnt feedback inhibitor, are expressed in association with the successional dental lamina (Järvinen et al., 2009). During snake tooth replacement, Wnt signaling is active in the tip of the successional lamina, and it is promoted by Bmp signaling from the mesenchyme and regulated negatively by mesenchymal Shh in surrounding cells (Richman and Handrigan, 2011).

Even though mice do not normally replace their teeth, phenotypes of some mutant and transgenic mouse lines have provided information on the genetic regulation of successional tooth formation, and these data may be applicable to replacement teeth as well. Mutant mice of the Wnt signaling pathway especially have confirmed the importance of this pathway on the regulation of tooth initiation and number. When β-catenin, a Wnt signaling mediator, is stabilized in the embryonic epithelium, successional tooth formation is activated (Järvinen et al., 2006; Liu et al., 2008). The mutant embryos develop epithelial buds with aberrant forms and multiple enamel knots, and a single tooth germ of these mutant embryos is able to give rise to tens of small teeth when grown under the kidney capsule (Järvinen et al., 2006). In vitro organ culture experiments showed that these teeth indeed developed sequentially through budding from previously initiated teeth (Fig. 4C). Similarly, epithelial deletion of *Apc*, an inhibitory component of the Wnt pathway, leads to supernumerary tooth development in mice (Wang et al., 2009). Extra teeth were initiated even when this deletion was induced in adult animals, but these teeth formed only around the continuously growing incisors, indicating that the incisor tissue has retained some odontogenic capacity to generate new teeth (Wang et al., 2009). $Sp6^{-/-}$ (*Epiprofin*) mutants have a similar tooth phenotype with multiple epithelial protrusions, but the association of this gene with Wnt signaling or other pathways is not clear (Nakamura et al., 2008).

Reduction in the number of molars (i.e., successional tooth formation) has been reported in many mouse models in which different signaling pathways are affected. For example, in Tabby mice that lack Eda signaling, missing third molars have been reported with variable frequency, from 17 to 55%, depending on the genetic background of the mouse lines (Grüneberg, 1966; Pispa et al., 1999; Kristenova-Cermankova et al., 2002). Overexpression of *Follistatin* or *Noggin* in oral epithelium causes the absence of all third molars (Wang et al., 2004; Plikus et al., 2005). *Noggin* overexpression further prevents formation of the second molars in some maxillary tooth rows, and first molars are also missing in the mandible. Additionally, reduction in the dose of transcription factor *Pax9* expression results in a graded loss of third and second molars (Kist et al., 2005). Pax9 is a target of Fgf signaling and regulates the expression of *Bmp4* in dental mesenchyme. Heterozygous loss of function of *PAX9* in humans results in severe tooth agenesis, affecting preferentially posterior teeth (Nieminen, 2009). In general, in human patients with hypodontia, the most commonly missing teeth are the ones that develop last in each tooth class (Nieminen, 2009).

In humans, mutations in the Wnt pathway lead to changes in tooth number. Mutations in human *AXIN2* lead to oligodontia, which affects specifically permanent teeth, suggesting that particularly tooth replacement is impaired (Lammi et al., 2004). Familial adenomatous polyposis (FAP) is caused by mutations in *APC*, and patients with FAP have both supernumerary teeth as well as odontomas, benign tumors composed of numerous small teeth (Wang and Fan, 2011). Patients with two different human syndromes, cleidocranial dysplasia (CCD) caused by mutations in the transcription factor *RUNX2* and a novel craniosynostosis syndrome caused by mutations in interleukin receptor *IL11RA*, develop supernumerary teeth that have been suggested to represent a third dentition (Jensen and Kreiborg, 1990; Nieminen et al., 2011). This indicates that the capacity for continuous tooth replacement may be unlocked in humans. Unfortunately, the mouse models for *Axin2, Runx*2, and *IL11Ra* mutations do not exhibit supernumerary teeth, probably because mouse teeth are not normally replaced (D'Souza et al., 1999; Nieminen et al., 2011; and unpublished observations from our laboratory). Based on the human phenotypes, it seems that Runx2 inhibits further replacement in humans and that the Wnt pathway plays both inhibitory and activatory roles. Runx2 has been associated with Fgf as well as Wnt signaling in tooth development since Runx2 is required for the induction of *Fgf3* and the Wnt inhibitor *Dkk1* in dental mesenchyme by epithelial Fgf4 (Åberg et al., 2004; James et al., 2006).

A human mutation associated with supernumerary tooth formation was recently identified in the *Sox2* gene (Numakura et al. 2010). Sox2 function has been linked to stem cell maintenance and iPS (induced pluripotent stem) cell induction, together with other stem cell factors (Takahashi and Yamanaka, 2006; Que et al., 2007; Kelberman et al., 2008). Recently we have localized *Sox2* expression in the lingual dental epithelium of deciduous teeth during tooth replacement in the ferret. This location correlates with the site of permanent tooth initiation, and Sox2 becomes downregulated in the permanent tooth (Juuri et al., 2013). *Sox2* expression is also present in human tooth replacement, in snakes with continuous tooth replacement, and interestingly, in mouse in a similar lingual location, even though in this species tooth replacement does not take place. Additionally, Sox2 localizes to the developing mouse and ferret molars, suggesting that it may serve as a marker of progenitor cells that give rise to successionally developing teeth in both serial primary tooth development and tooth replacement. *Sox2* is also expressed in the epithelial stem cells of the mouse incisor, where the *Sox2*-expressing cells contribute to all epithelial cell lineages of the tooth (Juuri et al., 2012). These results suggest that *Sox2* may serve as a common marker for epithelial cells with tooth-forming capacity.

6 DENTAL STEM CELLS, TOOTH RENEWAL, AND PROSPECTS FOR TOOTH REGENERATION

Tooth renewal and replacement require the action of stem cells that are capable of self-renewal and production of new progeny upon inductive signals. So far, the only putative stem cells identified during tooth replacement are the label-retaining cells in the dental lamina of the leopard gecko, a reptile with continuous tooth replacement (Handrigan et al., 2010). These label-retaining cells were shown to express a number of known stem cell marker genes. The presence of stem cell marker *Sox2* expression in the dental lamina of different tooth-replacing species suggests that the dental lamina cells could possess the odontogenic capacity in the epithelium (Juuri et al., 2013). Since *Sox2* is expressed in the dental lamina during successional molar development as well, there may be a common *Sox2*-positive progenitor cell pool set aside early during development, and maintained in the dental lamina. This may take place as early as at the initiation stage, since *Sox2* is expressed in the mouse dental lamina at E11 (Juuri et al. 2013). The dental lamina has in fact been proposed to be the source of stem cells in teeth (Smith et al., 2009).

The research on stem cells in tooth replacement has focused on the epithelium and the successional dental lamina. It is not known which cells in the underlying mesenchyme contribute to the dental mesenchyme of the replacement tooth, and whether the mesenchyme has been programmed to tooth fate. It is possible that there is a specific odontogenic mesenchymal cell population that is responsive to the stimulatory signals from the replacement tooth epithelium. Alternatively, the induction of replacement tooth mesenchyme may take place during primary tooth initiation, where nondental neural crest cell–derived mesenchyme responds to the inductive signals from the epithelium that has been specified by odontogenic fate.

After the tooth has completed development, there are still stem cells in the dental tissues. Human mesenchymal stem cells (MSCs) have been isolated from both deciduous and permanent teeth (Rodriguez-Lozano et al., 2011), and have been shown to be capable of regenerating the specific dental tissue from which they were isolated. However, there is so far no evidence that any of the tissue-specific stem cells would be capable of regenerating an entire tooth. Human dental MSCs were first isolated from the dental pulp of impacted third molars, and when transplanted into immunocompromised mice, they formed dentin (Gronthos et al., 2000). Similarly, stem cells from the periodontal ligament of human third molars were able to differentiate into cementoblast-like cells in culture and give rise to periodontal ligament-like structures when transplanted in mice (Seo et al., 2004). Epithelial stem cells have been isolated from the ERM cells of porcine teeth, and these cells could be induced into ameloblast-like cells (Shinmura et al., 2008). The various adult dental stem cells and their possible use in tissue regeneration is the subject of other chapters in this book.

Whole teeth have recently been bioengineered successfully in mice by modifying the classical tissue recombination experiments, which had demonstrated that complete teeth can develop when separated embryonic dental epithelium and mesenchyme are recombined and transplanted to adult mouse eye or kidney (Kollar and Baird, 1970b). The work from the Tsuji laboratory (see Chapter 25) has shown that the dental epithelium and mesenchyme from E14 mouse tooth germs can be disaggregated into single cells before reassociation and that these cells form a tooth with proper bone integration and periodontal ligament when grown in a three-dimensional device implanted into an adult jaw (Nakao et al., 2007; Oshima et al., 2011). If this approach would be used in human therapy, the embryonic dental cells should be replaced with adult cells with tooth-forming capacity.

7 CONCLUSIONS AND FUTURE DIRECTIONS

Most of our knowledge on the basic biology of tooth development has come from the laboratory mouse. We already know a great deal about the regulation of tooth initiation and morphogenesis and about the stem cells in the continuously growing mouse incisor. The study of tooth replacement using nonmodel animals such as reptiles and fish for continuous lifelong tooth replacement, and the ferret for mammalian replacement, is generating new information on the mechanisms of successional tooth formation and the characteristics of dental stem and progenitor cells. All these data are taking us closer to the long-cherished dream of biological replacement of lost and congenitally missing human teeth.

The various alternative methods of tooth bioengineering are discussed in other chapters. Perhaps the most realistic approach of whole-tooth bioengineering in the light of current knowledge is based on using cells with tooth-forming capacity that would be transplanted and left to develop in the jaw (Nakao et al., 2007; Oshima et al., 2011). Importantly, this method would allow the formation of a physiological root, which is a key feature required for the bioengineered tooth to be superior to titanium implants. However, for successful tooth regeneration through tissue engineering, more detailed understanding is required of the gene regulatory networks and cellular mechanisms guiding tooth development. Specific questions to be answered include the characterization of genes responsible for the odontogenic identity of the dental epithelium and mesenchyme, the genetic and cellular mechanisms controlling growth of the tooth crown and root, and the specification of important cell lineages, including ameloblasts, odontoblasts, and cementoblasts.

REFERENCES

Åberg T. Wang XP, Kim JH, Yamashiro T, Bei M, Rice R, et al. (2004). Runx2 mediates FGF signalling from epithelium to mesenchyme during tooth morphogenesis. *Dev Biol* 207:76–93.

Ahn Y, Sanderson BW, Klein OD, Krumlauf R (2010). Inhibition of Wnt signaling by Wise (Sostdc1) and negative feedback from Shh controls tooth number and patterning. *Development* 137:3221–3231.

Akimoto T, Fujiwara N, Kagiya T, Otsu K, Ishizeki K, Harada H (2011). Establishment of Hertwig's epithelial root sheath cell line from cells involved in epithelial–mesenchymal transition. *Biochem Biophys Res Commun* 404:308–312.

Andl T, Reddy ST, Gaddapara T, Millar SE (2002). WNT signals are required for the initiation of hair follicle development. *Dev Cell* 2:643–653.

Barron MJ, Brookes SJ, Draper CE, Garrod D, Kirkham J, Shore RC, et al. (2008). The cell adhesion molecule Nectin-1 is critical for normal enamel formation in mice. *Hum Mol Genet* 17:3509–3520.

Begue-Kirn C, Smith AJ, Ruch JV, Wozney JM, Purchio A, Hartmann D, et al. (1992). Effects of dentin proteins, transforming growth factor beta 1 (TGF beta 1) and bone morphogenetic protein 2 (BMP2) on the differentiation of odontoblast in vitro. *Int J Dev Biol* 36:491–503.

Bei M. (2009a). Molecular genetics of ameloblast cell lineage. *J Exp Zool B* 312B:437–444.

Bei M. (2009b). Molecular genetics of tooth development. *Curr Opin Genet Dev* 19:504–510.

Brunner HG, Hamel BC, Bokhoven H v H (2002). P63 gene mutations and human developmental syndromes. *Am J Med Genet* 112:284–290.

Buchtová M, Stembirek J, Glocova K, Matalova E, Tucker AS (2012). Early regression of the dental lamina underlies the development of diphyodont dentitions. *J Dent Res* 91:491–498.

Cao H, Wang J, Li X, Florez S, Huang Z, Venugopalan SR, et al. (2010). MicroRNAs play a critical role in tooth development. *J Dent Res* 89:779–784.

Chen J, Lan Y, Baek JA, Gao Y, Jiang R (2009). Wnt/beta-catenin signaling plays an essential role in activation of odontogenic mesenchyme during early tooth development. *Dev Biol* 334:174–185.

Chen Y, Zhang Y, Jiang TX, Barlow AJ, St Amand TR, Hu Y, et al. (2000). Conservation of early odontogenic signaling pathways in Aves. *Proc Natl Acad Sci USA* 97:10044–10049.

Coin R, Haikel Y, Ruch JV (1999). Effects of apatite, transforming growth factor beta-1, bone morphogenetic protein-2 and interleukin-7 on ameloblast differentiation in vitro. *Eur J Oral Sci* 107:487–495.

Dassule HR, Lewis P, Bei M, Maas R, McMahon AP (2000). Sonic hedgehog regulates growth and morphogenesis of the tooth. *Development* 127:4775–4785.

D'Souza RN, Aberg T, Gaikwad J, Cavender A, Owen M, Karsenty G, et al. (1999). Cbfa1 is required for epithelial–mesenchymal interactions regulating tooth development in mice. *Development* 126:2911–2920.

Evans AR, Wilson GP, Fortelius M, Jernvall J (2007). High-level similarity of dentitions in carnivorans and rodents. *Nature* 445:78–81.

Fliniaux I, Mikkola ML, Lefebvre S, Thesleff I (2008). Identification of dkk4 as a target of Eda-A1/Edar pathway reveals an unexpected role of ectodysplasin as inhibitor of Wnt signalling in ectodermal placodes. *Dev Biol* 320:60–71.

Fraser GJ, Hulsey CD, Bloomquist RF, Uyesugi K, Manley NR, Streelman JT (2009). An ancient gene network is co-opted for teeth on old and new jaws. *PLoS Biol* 7:e31.

Fujimori S, Novak H, Weissenbock M, Jussila M, Goncalves A, Zeller R, et al. (2010). Wnt/beta-catenin signaling in the dental mesenchyme regulates incisor development by regulating Bmp4. *Dev Biol* 348:97–106.

Glasstone S (1936). The development of tooth germs in vitro. *J Anat* 70:260–266.

Glasstone S (1967). Development of teeth in tissue culture. *J Dent Res* 46:858–861.

Gritli-Linde A, Bei M, Maas R, Zhang XM, Linde A, McMahon AP (2002). Shh signaling within the dental epithelium is necessary for cell proliferation, growth and polarization. *Development* 129:5323–5337.

Gronthos S, Mankani M, Brahim J, Robey PG, Shi S (2000). Postnatal human dental pulp stem cells (DPSCs) in vitro and in vivo. *Proc Natl Acad Sci USA* 97:13625–13630.

Grüneberg, H (1966). The molars of the Tabby mouse, and a test of the "single-active X-chromosome" hypothesis. *J Embryol Exp Morphol* 15:223–244.

Hammarström L, Alatli I, Fong CD (1996). Origins of cementum. *Oral Dis* 2:63–69.

Handrigan GR, Richman JM (2010a). A network of Wnt, Hedgehog and BMP signaling pathways regulates tooth replacement in snakes. *Dev Biol* 348:130–141.

Handrigan GR, Richman JM (2010b). Autocrine and paracrine Shh signaling are necessary for tooth morphogenesis, but not tooth replacement in snakes and lizards (Squamata). *Dev Biol* 337:171–186.

Handrigan GR, Leung KJ, Richman JM (2010). Identification of putative dental epithelial stem cells in a lizard with life-long tooth replacement. *Development* 137:3545–3549.

Hardcastle Z, Mo R, Hui C-, Sharpe PT (1998). The Shh signalling pathway in tooth development: defects in Gli2 and Gli3 mutants. *Development* 125:2803–2811.

Harjunmaa E, Kallonen A, Voutilainen M, Hämäläinen K, Mikkola ML, Jernvall J (2012). On the difficulty of increasing dental complexity. *Nature* 483:324–327.

Harris MP, Hasso SM, Ferguson MW, Fallon JF (2006). The development of archosaurian first-generation teeth in a chicken mutant. *Curr Biol* 16:371–377.

Iohara K, Nakashima M, Ito M, Ishikawa M, Nakasima A, Akamine A (2004). Dentin regeneration by dental pulp stem cell therapy with recombinant human bone morphogenetic protein 2. *J Dent Res* 83:590–595.

James MJ, Järvinen E, Wang XP, Thesleff I (2006). Different roles of Runx2 during early neural crest–derived bone and tooth development. *J Bone Miner Res* 21:1034–1044.

Järvinen E, Salazar-Ciudad I, Birchmeier W, Taketo MM, Jernvall J, Thesleff I (2006). Continuous tooth generation in mouse is induced by activated epithelial Wnt/beta-catenin signaling. *Proc Natl Acad Sci USA* 103:18627–18632.

Järvinen E, Välimäki K, Pummila M, Thesleff I, Jernvall J (2008). The taming of the shrew milk teeth. *Evol Dev* 10:477–486.

Järvinen E, Tummers M, Thesleff I (2009). The role of the dental lamina in mammalian tooth replacement. *J Exp Zool B* 312B:281–291.

Jensen BL, Kreiborg S (1990). Development of the dentition in cleidocranial dysplasia. *J Oral Pathol Med* 19:89–93.

Jernvall J, Åberg T, Kettunen P, Keränen S, Thesleff I (1998). The life history of an embryonic signalling center. BMP4 induces p21 and is associated with apoptosis in the mouse tooth enamel knot. *Development* 125:161–169.

Jheon AH, Mostowfi P, Snead ML, Ihrie RA, Sone E, Pramparo T, et al. (2011). PERP regulates enamel formation via effects on cell–cell adhesion and gene expression. *J Cell Sci* 124:745–754.

Juuri E, Jussila M, Seidel, K, Heikinheimo K, Richman J, Heikinheimo K, Chuong C-M, Arnold K, Hochedlinger K, Klein O, Michon F & Thesleff I (2013). Sox2 marks epithelial competence to generate teeth in mammals and reptiles. *Development*. In Press.

Juuri E, Saito K, Ahtiainen L, Seidel K, Tummers M, Hochedlinger K, Klein OD, Thesleff I & Michon F (2012). Sox2+ stem cells contribute to all epithelial lineages of the tooth via Sfrp5+ progenitors. *Developmental Cell* 14:317–328.

Kangas AT, Evans AR, Thesleff I, Jernvall J (2004). Nonindependence of mammalian dental characters. *Nature* 432:211–214.

Kassai Y, Munne P, Hotta Y, Penttilä E, Kavanagh K, Ohbayashi N, et al. (2005). Regulation of mammalian tooth cusp patterning by ectodin. *Science* 309:2067–2070.

Kavanagh KD, Evans AR, Jernvall J (2007). Predicting evolutionary patterns of mammalian teeth from development. *Nature* 449:427–432.

Kelberman D, de Castro SC, Huang S, Crolla JA, Palmer R, Gregory JW, et al. (2008). SOX2 plays a critical role in the pituitary, forebrain, and eye during human embryonic development. *J Clin Endocrinol Metab* 93:1865–1873.

Keränen SV, AAberg T, Kettunen P, Thesleff I, Jernvall J (1998). Association of developmental regulatory genes with the development of different molar tooth shapes in two species of rodents. *Dev Genes Evol* 208:477–486.

Keränen SV, Kettunen P, Åberg T, Thesleff I, Jernvall J (1999). Gene expression patterns associated with suppression of odontogenesis in mouse and vole diastema regions. *Dev Genes Evol* 209:495–506.

Kettunen P, Thesleff I (1998). Expression and function of FGFs-4, -8, and -9 suggest functional redundancy and repetitive use as epithelial signals during tooth morphogenesis. *Dev Dyn* 211:256–268.

Kist R, Watson M, Wang X, Cairns P, Miles C, Reid DJ, et al. (2005). Reduction of *Pax9* gene dosage in an allelic series of mouse mutants causes hypodontia and oligodontia. *Hum Mol Genet* 14:3605–3617.

Klein OD, Minowada G, Peterkova R, Kangas A, Yu BD, Lesot H, et al. (2006). Sprouty genes control diastema tooth development via bidirectional antagonism of epithelial–mesenchymal FGF signaling. *Dev Cell* 11:181–190.

Kollar EJ, Baird GR (1969). The influence of the dental papilla on the development of tooth shape in embryonic mouse tooth germs. *J Embryol Exp Morphol* 21:131–148.

Kollar EJ, Baird GR (1970a). Tissue interactions in embryonic mouse tooth germs: II. The inductive role of the dental papilla. *J Embryol Exp Morphol* 24:173–186.

Kollar EJ, Baird GR (1970b). Tissue interactions in embryonic mouse tooth germs: I. Reorganization of the dental epithelium during tooth-germ reconstruction. *J Embryol Exp Morphol* 24:159–171.

Kratochwil K, Dull M, Farinas I, Galceran J, Grosschedl R (1996). Lef1 expression is activated by BMP-4 and regulates inductive tissue interactions in tooth and hair development. *Genes Dev* 10:1382–1394.

Kratochwil K, Galceran J, Tontsch S, Roth W, Grosschedl R (2002). FGF4, a direct target of LEF1 and Wnt signaling, can rescue the arrest of tooth organogenesis in Lef1 (−/−) mice. *Genes Dev* 16:3173–3185.

Kristenova-Cermakova P, Peterka M, Lisi S, Lesot H, Peterkova R (2002). Postnatal lower jaw dentition in different phenotypes of Tabby mice. *Connect Tissue Res* 43:283–288.

Lammi L, Arte S, Somer M, Järvinen H, Lahermo P, Thesleff I, et al. (2004). Mutations in *AXIN2* cause familial tooth agenesis and predispose to colorectal cancer. *Am J Hum Genet* 74:1043–1050.

Laurikkala J, Mikkola ML, James M, Tummers M, Mills AA, Thesleff I (2006). P63 regulates multiple signalling pathways required for ectodermal organogenesis and differentiation. *Development* 133:1553–1563.

Leche W. (1895). Zur Entwicklungsgeschichte des Zahnsystems der Säugethiere zugleich ein Beitrag zur Stammesgeschichte dieser Thiergruppe. *Bibl Zool* 17.

Lefebvre S, Fliniaux I, Schneider P, Mikkola ML (2012). Identification of ectodysplasin target genes reveals the involvement of chemokines in hair development. *J Investig Dermatol* 132:1094–1102.

Li J, Huang X, Xu X, Mayo J, Bringas Jr P, Jiang R, Wang S, Chai Y (2011). SMAD4-mediated WNT signaling controls the fate of cranial neural crest cells during tooth morphogenesis. *Development* 138:1977–1989.

Liu F, Chu EY, Watt B, Zhang Y, Gallant NM, Andl T, et al. (2008). Wnt/beta-catenin signaling directs multiple stages of tooth morphogenesis. *Dev Biol* 313:210–224.

Luckett WP (1993). An ontogenetic assessment of dental homologies in therian mammals. In: *Mammal Phylogeny: Mesozoic Differentiation, Multituberculates, Monotremes, Early Therians, and Marsupials*, Szalay FS, Novacek MJ, McKenna MC eds., Springer-Verlag, New York, pp. 182–204.

Lukinmaa PL, Jensen BL, Thesleff I, Andreasen JO, Kreiborg S (1995). Histological observations of teeth and peridental tissues in cleidocranial dysplasia imply increased activity of odontogenic epithelium and abnormal bone remodeling. *J Craniofac Genet Dev Biol* 15:212–221.

Lumsden AG (1979). Pattern formation in the molar dentition of the mouse. *J Biol Buccale* 7:77–103.

Lumsden AG (1988). Spatial organization of the epithelium and the role of neural crest cells in the initiation of the mammalian tooth germ. *Development* 103(Suppl): 155–169.

Mammoto T, Mammoto A, Torisawa YS, Tat T, Gibbs A, Derda R, et al. (2011). Mechanochemical control of mesenchymal condensation and embryonic tooth organ formation. *Dev Cell* 21:758–769.

Marks SC, Jr, Cahill DR (1987). Regional control by the dental follicle of alterations in alveolar bone metabolism during tooth eruption. *J Oral Pathol* 16:164–169.

Michon F, Tummers M, Kyyrönen M, Frilander MJ, Thesleff I (2010). Tooth morphogenesis and ameloblast differentiation are regulated by micro-RNAs. *Dev Biol* 340:355–368.

Mikkola ML (2009). Molecular aspects of hypohidrotic ectodermal dysplasia. *Am J Med Genet A* 149A:2031–2036.

Millar S, Koyama E, Reddy ST, Andl, T, Gaddapara T, Piddington R, et al. (2003). Over- and ectopic expression of Wnt3 causes progressive loss of ameloblasts in postnatal mouse incisor teeth. *Connect Tissue Res* 44:124–129.

Mina M, Kollar EJ (1987). The induction of odontogenesis in non-dental mesenchyme combined with early murine mandibular arch epithelium. *Arch Oral Biol* 32:123–127.

Mitsiadis TA, Cheraud Y, Sharpe P, Fontaine-Perus J (2003). Development of teeth in chick embryos after mouse neural crest transplantations. *Proc Natl Acad Sci USA* 100:6541–6545.

Mustonen T, Pispa J, Mikkola ML, Pummila M, Kangas AT, Pakkasjärvi L, et al. (2003). Stimulation of ectodermal organ development by Ectodysplasin-A1. *Dev Biol* 259:123–136.

Nakamura T, de Vega S, Fukumoto S, Jimenez L, Unda F, Yamada Y (2008). Transcription factor epiprofin is essential for tooth morphogenesis by regulating epithelial cell fate and tooth number. *J Biol Chem* 283:4825–4833.

Nakao K, Morita R, Saji Y, Ishida K, Tomita Y, Ogawa M, et al. (2007). The development of a bioengineered organ germ method. *Nat Methods* 4:227–230.

Nakashima M. (1994) Induction of dentin formation on canine amputated pulp by recombinant human bone morphogenetic proteins (BMP)-2 and -4. *J Dent Res* 73:1515–1522.

Nakashima M, Reddi AH (2003) The application of bone morphogenetic proteins to dental tissue engineering. *Nat Biotechnol* 21:1025–1032.

Nanci A (2008). *Ten Cate's Oral Histology: Development, Structure and Function*, 7th ed., Mosby Elsevier, St. Louis, MO.

Neubuser A, Peters H, Balling R, Martin GR (1997). Antagonistic interactions between FGF and BMP signalling pathways: a mechanism for positioning the sites of tooth formation. *Cell* 90:247–255.

Nieminen P (2009). Genetic basis of tooth agenesis. *J Exp Zool B* 312B:320–342.

Nieminen P, Pekkanen M, Aberg T, Thesleff I (1998). A graphical WWW-database on gene expression in tooth. *Eur J Oral Sci* 106(Suppl 1): 7–11.

Nieminen P, Morgan NV, Fenwick AL, Parmanen S, Veistinen L, Mikkola ML, et al. (2011). Inactivation of IL11 signaling causes craniosynostosis, delayed tooth eruption, and supernumerary teeth. *Am J Hum Genet* 89:67–81.

Numakura C, Kitanaka S, Kato M, Ishikawa S, Hamamoto Y, Katsushima Y, et al. (2010). Supernumerary impacted teeth in a patient with SOX2 anophthalmia syndrome. *Am J Med Genet A* 152A:2355–2359.

O'Connell DJ, Ho JW, Mammoto T, Turbe-Doan A, O'Connell JT, Haseley PS, et al. (2012). A Wnt-Bmp feedback circuit controls intertissue signaling dynamics in tooth organogenesis. *Sci Signal* 5:ra4.

Ohazama A, Haycraft CJ, Seppälä M, Blackburn J, Ghafoor S, Cobourne M, et al. (2009). Primary cilia regulate Shh activity in the control of molar tooth number. *Development* 136:897–903.

Ooë T. (1981). *Human Tooth and Dental Arch Development*, Ishiyaku Publishers, Tokyo.

Oshima M, Mizuno M, Imamura A, Ogawa M, Yasukawa M, Yamazaki H, et al. (2011). Functional tooth regeneration using a bioengineered tooth unit as a mature organ replacement regenerative therapy. *PLoS One* 6:e21531.

Philbrick WM, Dreyer BE, Nakchbandi IA, Karaplis AC (1998). Parathyroid hormone–related protein is required for tooth eruption. *Proc Natl Acad Sci USA* 95:11846–11851.

Pispa J, Jung H, Jernvall J, Kettunen P, Mustonen T, Tabata MJ, et al. (1999). Cusp patterning defect in Tabby mouse teeth and its partial rescue by FGF. *Dev Biol* 216:521–534.

Pispa J, Mustonen T, Mikkola ML, Kangas AT, Koppinen P, Lukinmaa PL, et al. (2004). Tooth patterning and enamel formation can be manipulated by misexpression of TNF receptor Edar. *Dev Dyn* 231:432–440.

Plikus MV, Zeichner-David M, Mayer JA, Reyna J, Bringas P, Thewissen JG, et al. (2005). Morphoregulation of teeth: modulating the number, size, shape and differentiation by tuning Bmp activity. *Evol Dev* 7:440–457.

Que J, Okubo T, Goldenring JR, Nam KT, Kurotani R, Morrisey EE, et al. (2007). Multiple dose-dependent roles for Sox2 in the patterning and differentiation of anterior foregut endoderm. *Development* 134:2521–2531.

Richman JM, Handrigan GR (2011). Reptilian tooth development. *Genesis* 49:247–260.

Rodrigues HG, Marangoni P, Sumbera R, Tafforeau P, Wendelen W, Viriot L (2011). Continuous dental replacement in a hyper-chisel tooth digging rodent. *Proc Natl Acad Sci USA* 108:17355–17359.

Rodriguez-Lozano FJ, Bueno C, Insausti CL, Meseguer L, Ramirez MC, Blanquer M, et al. (2011). Mesenchymal stem cells derived from dental tissues. *Int Endod J* 44:800–806.

Ruch JV, Lesot H, Begue-Kirn C (1995). Odontoblast differentiation. *Int J Dev Biol* 39:51–68.

Rutherford RB, Spangberg L, Tucker M, Rueger D, Charette M (1994). The time-course of the induction of reparative dentine formation in monkeys by recombinant human osteogenic protein-1. *Arch Oral Biol* 39:833–838.

Salazar-Ciudad I, Jernvall J (2002). A gene network model accounting for development and evolution of mammalian teeth. *Proc Natl Acad Sci USA* 99:8116–8120.

Seo BM, Miura M, Gronthos S, Bartold PM, Batouli S, Brahim J, et al. (2004). Investigation of multipotent postnatal stem cells from human periodontal ligament. *Lancet* 364:149–155.

Shields ED, Bixler D, el-Kafrawy AM (1973). A proposed classification for heritable human dentine defects with a description of a new entity. *Arch Oral Biol* 18:543–553.

Shinmura Y, Tsuchiya S, Hata K, Honda MJ (2008). Quiescent epithelial cell rests of Malassez can differentiate into ameloblast-like cells. *J Cell Physiol* 217:728–738.

Shirokova V, Jussila M, Hytönen MK, Perälö N, Drögemüller C, Leeb T, Lohi H, Sainio K, Thesleff I, Mikkola MK (2013). Expression of *Foxi3* is regulated by Ectodysplasin in skin appendage placodes. *Developmental Dynamics*. In Press.

Smith MM, Fraser GJ, Mitsiadis TA (2009). Dental lamina as source of odontogenic stem cells: evolutionary origins and developmental control of tooth generation in gnathostomes. *J Exp Zool B* 312B:260–280.

Soukup V, Epperlein HH, Horacek I, Cerny R (2008). Dual epithelial origin of vertebrate oral teeth. *Nature* 455:795–798.

Stembirek J, Buchtova M, Kral T, Matalova E, Lozanoff S, Misek I (2010). Early morphogenesis of heterodont dentition in minipigs. *Eur J Oral Sci* 118:547–558.

Suomalainen M, Thesleff I, (2010). Patterns of Wnt pathway activity in the mouse incisor indicate absence of Wnt/beta-catenin signaling in the epithelial stem cells. *Dev Dyn* 239:364–372.

Takahashi K, Yamanaka S (2006). Induction of pluripotent stem cells from mouse embryonic and adult fibroblast cultures by defined factors. *Cell* 126:663–676.

Ten Cate AR (1996). The role of epithelium in the development, structure and function of the tissues of tooth support. *Oral Dis* 2:55–62.

Thesleff I (2003). Epithelial–mesenchymal signalling regulating tooth morphogenesis. *J Cell Sci* 116:1647–1648.

Thesleff I, Hurmerinta K (1981). Tissue interactions in tooth development. *Differentiation* 18:75–88.

Thesleff I, Keränen S, Jernvall J (2001). Enamel knots as signaling centers linking tooth morphogenesis and odontoblast differentiation. *Adv Dent Res* 15:14–18.

Tummers M, Thesleff I (2003). Root or crown: a developmental choice orchestrated by the differential regulation of the epithelial stem cell niche in the tooth of two rodent species. *Development* 130:1049–1057.

Tummers M, Thesleff I (2008). Observations on continuously growing roots of the sloth and the K14-Eda transgenic mice indicate that epithelial stem cells can give rise to both the ameloblast and root epithelium cell lineage creating distinct tooth patterns. *Evol Dev* 10:187–195.

Tummers M, Thesleff I (2009). The importance of signal pathway modulation in all aspects of tooth development. *J Exp Zool B* 312B:309–319.

Unda FJ, Martin A, Hilario E, Begue-Kirn C, Ruch JV, Arechaga J (2000). Dissection of the odontoblast differentiation process in vitro by a combination of FGF1, FGF2, and TGFbeta1. *Dev Dyn* 218:480–489.

Vainio S, Karavanova I, Jowett A, Thesleff I (1993). Identification of BMP-4 as a signal mediating secondary induction between epithelial and mesenchymal tissues during early tooth development. *Cell* 75:45–58.

Wang XP, Fan J (2011). Molecular genetics of supernumerary tooth formation. *Genesis* 49:261–277.

Wang XP, Suomalainen M, Jorgez CJ, Matzuk MM, Wankell M, Werner S, et al. (2004). Modulation of activin/bone morphogenetic protein signaling by follistatin is required for the morphogenesis of mouse molar teeth. *Dev Dyn* 231:98–108.

Wang XP, O'Connell DJ, Lund JJ, Saadi I, Kuraguchi M, Turbe-Doan A, et al. (2009). Apc inhibition of Wnt signaling regulates supernumerary tooth formation during embryogenesis and throughout adulthood. *Development* 136:1939–1949.

Yamanaka A, Yasui K, Sonomura T, Iwai H, Uemura M (2010). Development of deciduous and permanent dentitions in the upper jaw of the house shrew (*Suncus murinus*). *Arch Oral Biol* 55:279–287.

Yamashiro T, Tummers M, Thesleff I (2003). Expression of bone morphogenetic proteins and Msx genes during root formation. *J Dent Res* 82:172–176.

Yamashiro T, Zheng L, Shitaku Y, Saito M, Tsubakimoto T, Takada K, et al. (2007). Wnt10a regulates dentin sialophosphoprotein mRNA expression and possibly links odontoblast differentiation and tooth morphogenesis. *Differentiation* 75:452–462.

Yoshida T, Miyoshi J, Takai Y, Thesleff I (2010). Cooperation of Nectin-1 and Nectin-3 is required for normal ameloblast function and crown shape development in mouse teeth. *Dev Dyn* 239:2558–2569.

Zeichner-David M, Diekwisch T, Fincham A, Lau E, MacDougall M, Moradian-Oldak J, et al. (1995). Control of ameloblast differentiation. *Int J Dev Biol* 39:69–92.

Zhang Y, Tomann P, Andl T, Gallant NM, Huelsken J, Jerchow B, et al. (2009). Reciprocal requirements for EDA/EDAR/NF-kappaB and Wnt/beta-catenin signaling pathways in hair follicle induction. *Dev Cell* 17:49–61.

Zhang Z, Lan Y, Chai Y, Jiang R (2009). Antagonistic actions of Msx1 and Osr2 pattern mammalian teeth into a single row. *Science* 323:1232–1234.

7

REPTILIAN TOOTH REGENERATION

JOY M. RICHMAN, JOHN A. WHITLOCK, AND JOHN ABRAMYAN

Department of Oral Health Sciences, Life Sciences Institute, University of British Columbia, Vancouver, British Columbia, Canada

1 INTRODUCTION

As the closest living relatives of mammals to have retained continuous tooth replacement (polyphyodonty), reptiles are a potentially important resource for an understanding of the mechanisms controlling tooth regeneration. In addition, certain reptiles are tractable models for testing hypotheses about the capacity for tooth replacement. In this chapter we first review the evolution of tooth replacement or regeneration. Next we compare the capacity for tooth replacement in reptiles and mammals in order to determine whether there could be common ancestral ability to regenerate teeth. We then explore the cellular and molecular factors identified in reptiles that appear to play key roles in their enhanced ability to regenerate their teeth.

2 HOW ANCESTRAL IS TOOTH REGENERATION IN AMNIOTES?

The evolution of tooth regeneration forms an important foundation for this chapter. We first review the evolutionary relationships between living mammals and reptiles and focus on the evidence supporting indefinite tooth replacement not only in fossil reptiles but also in mammals.

The amniotes include all animals with extra embryonic membranes, consisting of an amnion, chorion, and allantois. There are two main branches: the reptilian (Sauropsida) and mammalian (Synapsida), and both are considered derived from an ill-defined extinct stem amniote. Therefore, reptiles evolved in parallel with mammals and are not more primitive, as is commonly thought. Teeth are present in both the mammalian and

Stem Cells in Craniofacial Development and Regeneration, First Edition.
Edited by George T.-J. Huang, Irma Thesleff.

reptilian branches of amniotes, but only reptiles have retained the ability to replace their teeth throughout life. Not all reptilia are dentate, however, so it is important to establish from the outset that in this chapter we will be focusing on the two branches with teeth, Squamata and Crocodilia (omitting Sphenodon, the rare lizard-like reptiles that are natives of New Zealand). The squamates include snakes and lizards, all of which have teeth. As we shall see, there is great variation in the type of teeth and the capacity for tooth replacement within squamates, just as there is in mammals. The difference is that in reptiles there are many examples of animals with lifelong tooth regeneration, whereas this is very rare in mammals and at best is limited to posterior regions of the dentition (Peyer, 1968).

3 BASAL AMNIOTES IN BOTH MAMMALIAN AND REPTILIAN LINEAGES COULD REPLACE THEIR TEETH

It may be surprising to learn that the extinct and most basal members of both "mammal-line amniota" (e.g., pelycosaurs) and "reptile"-line amniota (e.g., mesosaurs, captorhinids) displayed evidence of continuous tooth regeneration and replacement throughout their lifetimes (Edmund, 1960; Modesto and Reisz, 1992; Modesto, 1995; Modesto and Cisneros, 2009). Specifically, in the Carboniferous and early Permian, the fossils of early mammals showed evidence of resorption pits in adults as well as several generations of teeth (Hopson, 1964; Modesto and Reisz, 1992; Modesto, 1995). By the Late Permian, variation in rate of replacement between taxa has become evident (Modesto and Rybczynski, 2001). By the Triassic period, most extinct mammals had lost the ability to replace teeth in adulthood. Thus, mammals initially had retained the vertebrate capacity to form teeth that is characteristic of amphibians, fish, and sharks (Davit-Béal et al., 2007; Huysseune et al., 2009; Smith et al., 2009a). Teeth have also been lost entirely or drastically reduced in many different lineages of modern and ancient mammals (e.g., gomphodont cynodonts, *Sinoconodon*) (Hopson, 1964; Crompton and Jenkins, 1973, Davit-Béal et al. 2009). The evolutionary reasons behind this loss (and for the retention of the characteristic in reptiles) are uncertain; hypotheses vary from the determinate growth characteristics of mammals (Hopson, 1964) to the advent of lactation and parental care (Ewer, 1963; Hopson, 1973).

Whatever the reasons, the shared capacity for tooth replacement in long-extinct amniotes suggests that it might be possible to reactivate the programs controlling repeated tooth replacement in mammals. Indeed, mammals may have retained a vestigial post-permanent dentition as was described in human teeth (Ooë, 1981). In this case, the small rudiment that buds off the permanent tooth bud might be encouraged to survive and form another tooth. It is more likely, however, that the historical capacity to form teeth in mammals has been retained in the form of stem cells that are still resident in the dental tissues of modern animals. Evidence supporting the presence of such cells in mammals and humans is the focus of this book and is discussed in detail elsewhere.

Unfortunately, few diphyodont mammals are available for studies into stem cells and where they are located (Miyado et al., 2007; Yamanaka et al., 2007; Järvinen et al., 2006). The main mammalian model for dental development is the mouse, but tooth replacement does not occur in these animals. Instead, mice have a continuously erupting incisor, a diastema with vestigial dental epithelium and posterior molars that are not replaced (Tummers and Thesleff, 2009). Despite the lack of generational teeth in mice, there is evidence supporting evolutionary conserved stem cells in the cervical loop of the incisors (Harada et al., 1999; Wang et al., 2007; Suomalainen and Thesleff, 2010). As we shall see later, they have some similarities to reptilian dental epithelial stem cells. Several mouse knockouts may appear to have new tooth generations of teeth being formed in the diastema. However, these are usually not generational teeth (i.e., derived from molar tooth bud) but, instead, are derived from vestigial diastema teeth which survive rather than degenerate in certain experimental models (Peterkova et al., 2002; 2009, Ohazama et al., 2009). In other gain-of-function transgenic mice, there are numerous teeth forming, and these do appear to be generational (i.e., derived from one parent tooth bud) (Järvinen et al., 2006; Wang et al., 2009). These lines of evidence from the incisor and the gain of function in the Wingless signaling pathway (Wnt) support the idea that it is possible to reawaken the ancestral capacity to replace teeth in mammals.

4 GENERAL PROCESS OF TOOTH DEVELOPMENT AND REPLACEMENT IN SQUAMATE REPTILES

Just as in mammals, reptilian teeth form through epithelial and mesenchymal interactions. Initiation of teeth begins with an odontogenic band in the oral ectoderm (Smith et al., 2009b). The odontogenic band is characterized by a band of gene expression that marks the position of the future tooth rows (Richman and Handrigan, 2011). The odontogenic band gene which has been investigated in most detail in reptiles is *Shh* (Sonic hedgehog; Cobourne et al., 2004; Buchtová et al., 2008; Vonk et al., 2008). *Shh* is localized to a narrow band throughout all craniofacial processes where teeth will form, extending across the maxillary and premaxillary boundaries, as well as the pterygoid–palatine regions in snakes (Buchtová et al., 2008; Vonk et al., 2008). However, there are additional markers for the odontogenic band in mammals, such as *Pitx2* (paired-like homeodomain transcription factor 2, Mucchielli et al., 1997; Jernvall and Thesleff, 2000). Indeed, *Pitx2* expression is present in the chicken marginal oral epithelium, suggesting that this is a remnant of a vestigial odontogenic band that did not progress to form teeth (Chen et al., 2000; Mitsiadis et al., 2006). It is therefore likely that *Pitx2* will also have conserved expression in the dentate reptiles. In the mouse and presumably all mammals, the odontogenic band breaks down into placodes, marking the positions of future teeth. In diphyodont mammals, there are individual placodes for the primary dentition. However, in the reptiles that we have studied, individual placodes never form.

The next phase of tooth development is the invagination of the dental lamina into the adjacent jaw mesenchyme. Based on extensive work carried out in mice, epithelial–mesenchymal interactions are likely to be required for the formation of the dental lamina in reptiles. Once the primary palate has fused and the premaxilla is united with the maxillary regions, the dental lamina becomes continuous around the upper jaw. The mandibular prominence also has an uninterrupted dental lamina extending around the margin of the jaw. Experiments on snakes show that in all transverse sections, even those between teeth, there is a dental lamina, indicative of a continuous band throughout the jaws (Buchtová et al., 2007, 2008).

In squamates, the dental lamina is a tongue of epithelial cells that grows into the neural crest–derived mesenchyme at roughly a 60° angle slanting toward the lingual (Fig. 1A) (Buchtová et al., 2008; Handrigan and Richman, 2010a). Within the dental lamina, there are two external layers of cuboidal cells and an interstitial layer of cells. Signaling induced by the hedgehog pathway (Sonic hedgehog, Shh) has been shown to be required for the elongation and subsequent angulation of the dental lamina into the mesenchyme (Buchtová et al., 2008). Tooth formation is restricted to the labial or obtusely angled side. In snakes, the dental lamina bends at a 90° angle after the first generation of teeth are formed: a likely mechanism for offsetting the position of subsequent tooth generations in order to provide adequate spacing for carrying multiple tooth generations at one time (Buchtová et al., 2007, 2008).

Squamate teeth pass through stages comparable to those described for the mouse and human (Figs. 1 and 2). The first-generation teeth form close to the growing tip of the dental lamina. There is initially a bud stage, which is a periodic thickening of the labial side of the dental lamina. At sites where tooth formation occurs, there is a condensation of mesenchyme adjacent to this epithelial bud (Buchtová et al., 2007, 2008). The control of periodicity of tooth formation along the dental lamina is unknown at this point, but we speculate that mesenchymal signals play a role in selecting regions where first-generation teeth will form.

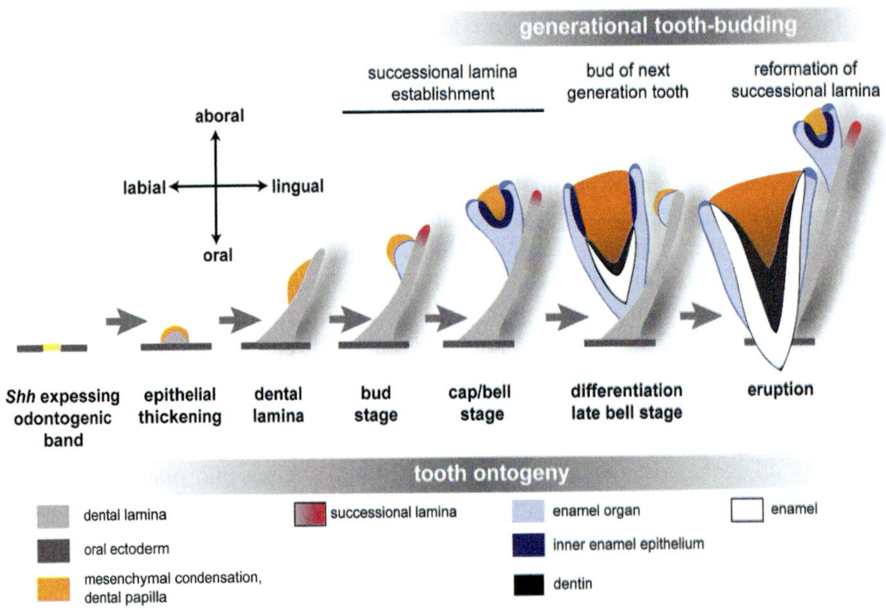

FIGURE 1 Summary of the stages of tooth development and regeneration in squamates. Schematic of tooth development beginning with specification of the number and position of the tooth rows by the odontogenic band. The successional lamina forms by the extension of the dental lamina into the jaw mesenchyme. There is a transient stage where the newest tooth bud is at the distal end of the dental lamina and the successional lamina has not yet re-formed; then when the newest tooth has reached the bell stage, the lamina is regenerated. Not shown here are the vestigial teeth which are present in most reptiles. They manifest as tooth buds that form close to or within the oral epithelium and arrest at the cap stage. Vestigial teeth do not become functional.

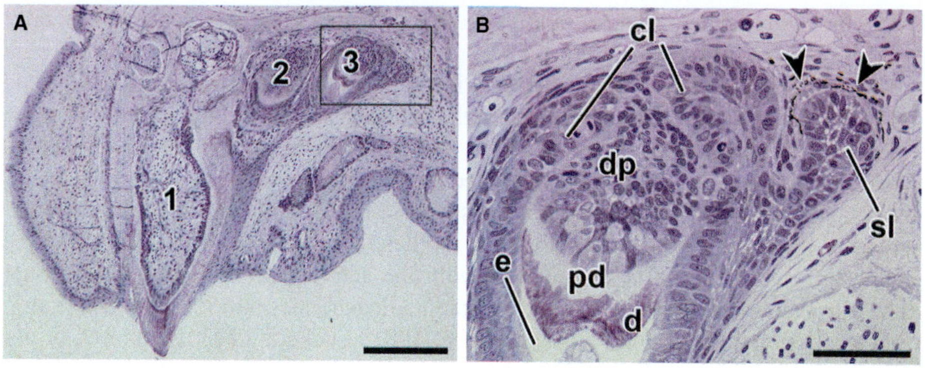

FIGURE 2 Histology of a tooth family in the maxilla of *Eublepharis macularius*. (A) This semithin resin section shows the relationship of the dental lamina to the three generations of teeth in this tooth family (numbers 1 to 3). The dental lamina is intact and runs along the lingual side of the erupted tooth, connecting each of the teeth. The enamel epithelial bulge is seen clearly in the third-generation tooth. (B) The cervical loops are forming near the apex of the tooth, and the successional lamina is close by. The successional lamina is ringed by pigmented cells in the mesenchyme that may be melanocytes (arrowheads). cl, cervical loop; d, dentin; dp, dental papilla; e, enamel; pd, predentin; sl, successional lamina. Scale bars: A = 200 μm, B = 100 μm.

After the bud stage, the cap and bell stages begin in a manner similar to that of the mammalian system, with a few differences that we will highlight. During these stages, the epithelial bud stage is transformed into a multilayered enamel organ which encloses the mesenchymal condensation or the dental papilla. The enamel organ is comprised of the inner and outer enamel epithelia, with a third layer of epithelial cells called the *stellate reticulum* sandwiched between them. The stellate reticulum is reduced significantly in reptiles (Westergaard and Ferguson, 1986, 1987; Delgado et al., 2005; Buchtová et al., 2007, 2008). In mammalian teeth, there is a fourth layer, called the *stratum intermedium*, which separates the inner enamel epithelium from the stellate reticulum; however, a comparable layer seems to be absent in all squamates and crocodilians examined to date (Westergaard and Ferguson, 1986, 1987, 1990; Delgado et al., 2005; Buchtová et al., 2008; Handrigan and Richman, 2010a, 2011). Additionally, the inner enamel epithelium of reptiles lacks the thickening called the *enamel knot*. Instead, the entire inner enamel epithelium shares features of the mammalian enamel knot, specifically the reduced proliferation and high levels of Shh expression (Buchtová et al., 2007, 2008; Zahradnicek et al., 2008; Handrigan and Richman, 2010a). Due to the simple cone shape of most squamate teeth, there is no requirement for secondary enamel knots, which are required for cusp formation in multicusped mammalian teeth (Thesleff et al., 2001; Wang et al., 2004). Geckos, which have bicuspid crown morphologies (Sumida and Murphy, 1987; Handrigan and Richman, 2011), have an enamel epithelial bulge (Fig. 2A) which synthesizes less enamel protein than do the adjacent cells. This leads to a central depression in the crown flanked by thicker areas of enamel. Focal Bmp (bone morphogenetic protein) expression in the bulge may help to encourage adjacent ameloblasts to secrete more enamel. Furthermore, since *Bmp2* is a marker of the primary enamel knot (Wang et al., 2004), the enamel epithelial bulge in the gecko may be a hybrid of a primary enamel knot, which induces odontoblasts, and a secondary enamel knot, which signals laterally to induce secretion of enamel.

The bell stage is when cytodifferentiation and matrix deposition begin. This is also when interspecific differences in tooth size and shape among reptiles become noticeable (Richman and Handrigan, 2011). This is also when secondary enamel knots appear in mammalian teeth, but as mentioned above, they do not form in reptilian teeth. Additionally, at this stage, initiation of the successional lamina begins from the lingual surface of the outer enamel epithelium. In reptiles, continuation of the dental lamina beyond the enamel organ is synonymous with the actively proliferating successional lamina. In mammals, however, the dental lamina degenerates and loses its connection to the oral epithelium at the bell stage (Ooë, 1981; Luckett, 1993; Stembirek et al., 2010), and the tooth together with its successional lamina are a separate unit.

When the successional lamina reaches a certain length, the next-generation tooth bud begins to form. Of course, in diphyodont mammals, there is only one successional lamina, and this forms the permanent dentition. In humans, successional laminae form only on the 20 primary tooth enamel organs. The permanent molars are derived from a distal extension of the dental lamina that originates from the lamina of the terminal primary molar (Ooë, 1981). In contrast to mammals, reptilian teeth re-form the successional lamina repeatedly throughout life. Thus, whole tooth "families" arise with one tooth being erupted and three or four teeth being unerupted and partially formed. The exact mechanism for determining the maximum number of replacement teeth forming at any one time is unknown, but work in dinosaurs suggests that diet has an influence on tooth family size. Fossil evidence tells us that the highest tooth replacement rates and the highest rates of dentine apposition are seen in the dinosaurs, specifically the herbivorous duck-billed hadrosaurs (Erickson, 1996) and the gigantic sauropods (Sereno et al., 2007; Whitlock et al., 2010). The sauropods in particular are known for their extraordinarily high rate of replacement, occasionally reaching rates as high as a new tooth per socket every 30 days, a rate that required the simultaneous development of up to seven replacement teeth per tooth position (Sereno et al., 2007). In contrast, the teeth of carnivorous theropod dinosaurs are replaced on annual to multiannual time scales (Erickson, 1996). Control of the replacement rate appears to be driven primarily by tooth wear, as rough diets (e.g., plants) are correlated strongly with rapid replacement, in contrast to the common axiom that "tooth replacement is not a direct response to wear" (Edmund, 1960). This relationship is not strictly a function of crown size, as large sauropod teeth (e.g., *Camarasaurus*) replace more rapidly than small theropod teeth (e.g., *Deinonychus*) (Erickson, 1996; Whitlock et al., 2010). In turn, this suggests that tooth family size in reptiles can be manipulated under experimental conditions. These approaches make it possible to determine factors that encourage or inhibit tooth regeneration.

5 THE RELATIONSHIP BETWEEN THE SUCCESSIONAL LAMINA AND TOOTH REGENERATION

The factors regulating successional lamina formation and tooth replacement are still poorly understood. We can learn about the role of the successional lamina from comparative work as well as functional studies in the lab. There are reptiles that lack indefinite tooth replacement capacity, the Agamids (Cooper et al., 1970). In these animals the first-generation teeth are rudimentary and are replaced by a functional set that remain throughout life (Handrigan and Richman, 2010a). Surprisingly, our examination of prehatching animals revealed that there is a successional lamina growing out from the functional teeth, although it is very small. This finding eliminates the possibility that the functional teeth are unable to form a successional lamina. Instead, the data suggest that failure of the successional lamina to survive is the reason for the cap on tooth family size. Consistent with this idea, we were able to detect increased apoptosis in this rudimentary lamina (Richman and Handrigan, 2011). In contrast, other squamates that we have examined do not have apoptosis in the successional lamina. Instead, programmed cell death is localized to the stellate reticulum (Buchtová et al., 2008). We have not yet examined posthatching agamids, but we predict that with this high degree of apoptosis the successional lamina will resorb. In humans there is usually just one successional lamina that forms on each of the 20 primary teeth; however, there is evidence of successional laminae forming on some of the permanent tooth germs (Ooë, 1981). This vestigial successional lamina may explain why some areas of the human mouth, such as the mandibular bicuspids, are more prone than others to forming supernumerary teeth (Solares and Romero, 2004). The ability to form and maintain a successional lamina is what distinguishes reptiles from mammals.

6 MOLECULAR CONTROLS OF THE REFORMATION OF THE SUCCESSIONAL LAMINA

We have carried out molecular studies on the bearded dragon, several types of snakes, and the leopard gecko to investigate the pathways that are most likely to be regulating tooth regeneration. The strategy was to examine expression of genes that are targets of the major signaling pathways get a readout of which pathways are likely to be active, to what tissues they are localized, and at what time during development they are most involved in signaling. The pathways we examined in most detail were Shh, Wnt, and Bmp. These three pathways have been shown to play roles in various aspects of mouse and human tooth development. We were particularly interested in the gene expression patterns surrounding and within the successional lamina, a tissue that is difficult to isolate and study in mammals.

6.1 Shh Pathway

As mentioned previously, Shh is an early marker of where teeth will form and then is expressed within the inner enamel epithelium. Careful examination of many squamates revealed that neither *Shh* nor its receptor and target gene, *Ptch1* (Patched), are expressed in or adjacent to the successional lamina (Buchtová et al., 2008; Handrigan and Richman, 2010a). We therefore do not believe that Shh plays a direct role in outgrowth of the successional lamina. As we shall see, it does, however, play a role in focusing the activity of other signaling molecules on the free end of the dental lamina (Handrigan and Richman, 2010b).

6.2 Wnt Pathway

In contrast, in reptiles that replace their teeth, such as geckos and snakes, there is heavy Wnt pathway activity in the successional lamina, as shown by strong *Lef1* and

Axin2 expression (Handrigan and Richman, 2010b; Handrigan et al., 2010). *Lef1* is a transcription factor that acts downstream of canonical Wnt signaling and is also a target of the pathway, whereas *Axin2* is an intracellular mediator of the pathway as well as a target gene (Filali et al., 2002; Jho et al., 2002). We also cloned several Wnt ligands, including *Wnt6, Wnt10b* (Handrigan and Richman, 2010b), and *Wnt7a* (Handrigan and Richman, 2010a). The overlap of *Lef1* with the highly proliferative successional lamina was very suggestive of Wnt activity being present in this region. There was also expression in the mesenchymal condensation from the bud stage onward. *Axin2* was expressed in a similar distribution but somewhat less specifically in the successional lamina and adjacent mesenchyme. Interestingly, all three Wnt ligands were restricted to dental epithelium, but none were expressed in the successional lamina, suggesting that we have not yet identified the specific Wnt that is involved in tooth replacement.

As a comparison, we also examined *Axin2* and another transcription factor in the Wnt pathway, *Tcf7* in the vestigial successional lamina of the bearded dragon. There was no expression of either gene in this small outgrowth of epithelium. This is an important finding and shows that one mechanism for the decreased size of tooth families is that successional lamina lacks certain molecular signaling pathways. Whether this is due to the absence of either the receptors or the ligands is unclear at this stage. Nonetheless, these data suggest that one approach that could be used to encourage new tooth formation is to supply factors that encourage cell survival and proliferation. Such treatments might promote the formation of a third generation of teeth.

6.3 Bmp Pathway

Bmps are also key factors in vertebrate tooth morphogenesis. In the reptile there is evidence of Bmp activity as shown by the presence of phosphorylated Smad proteins. Smads are nuclear transcription factors that are phosphorylated upon activating either the Bmp or transforming growth factor β pathways. We used an antibody that is specific for Bmp Smads 1, 5, and 8 and found strong epithelial expression at the tip of the successional lamina on the side where teeth will form (Handrigan and Richman, 2010b). However, the Bmps themselves were expressed in the adjacent mesenchyme. Thus, the evidence supports the idea that Bmps affect proliferation in the successional lamina in a noncell autonomous manner.

7 THE SUCCESSIONAL LAMINA RECAPITULATES GENE EXPRESSION PATTERNS OF THE EARLY DENTAL LAMINA

In our lab, The early dental lamina has been studied in the most detail, as this stage is easier to visualize in sections. However, it is important to point out that all of the genes we have been able to compare in the early lamina (as defined by the fact that first-generation teeth have not yet formed) and successional lamina are expressed in similar patterns in both tissues. We have proposed that this is not a coincidence but reflects the fact that the successional lamina is a miniature dental lamina and recapitulates early tooth formation (Handrigan and Richman, 2010b).

It is this ability of the successional lamina to regain its early embryonic character that is at the nub of regenerating teeth. Once the key factors that promote early dental lamina growth are identified, it is likely that these will be the same factors that will influence reformation of the successional lamina. As we shall see, gene expression patterns identical to the free end of the dental lamina appear in the successional lamina almost as soon as it forms. This suggests that the successional lamina is specified very early. One possibility is that the epithelium is instructive and induces the odontogenic mesenchyme; however, tissue interaction experiments will need to be studied to verify these ideas. Nevertheless, these data suggest that the dental epithelium holds the key to tooth replacement.

8 SIGNALING PATHWAY INTERACTIONS DURING FORMATION OF THE SUCCESSIONAL LAMINA

The formation of replacement teeth will probably involve negative and positive feedback loops between the various signaling pathways. We can gain some insights into which of these interactions might be working during tooth replacement from studies on the snake dental lamina. In the early reptilian dental lamina we have described complementary expression patterns between Wnt genes and Shh. In explant cultures, blocking of the Shh signal with the antagonist cyclopamine leads to an expansion of Wnt activity, whereas increasing Shh signaling represses expression of *Lef1*. This demonstrates that Shh is a negative regulator of Wnt signaling. We hypothesize that although *Shh* is not directly involved in formation of the successional lamina, there may be an earlier role in setting aside cells at the tip of the dental lamina that will later give rise to the successional lamina.

We have also investigated the interactions between mesenchymally expressed *Bmp*s and the focused Wnt activity at the tip of the dental lamina. *Bmp*s are expressed in the odontogenic mesenchyme directly adjacent to the successional lamina as well as the tooth-forming side of the early dental lamina (Handrigan and Richman, 2010b). Manipulating the level of Bmps using bead implants showed that there was an activation of the Wnt pathway by increasing the level of Bmps while blocking Bmp signaling with Noggin decreased *Lef1* expression. These data indicate that Bmps positively regulate Wnt activity, perhaps by inducing Wnt ligands. These ligands have not yet been identified. Finally, Bmps are relatively focused in their mesenchymal expression and we wanted to determine whether paracrine signaling from Shh would regulate expression of Bmp ligands. We did find clear induction of *Bmp2* when Shh was blocked; thus, Shh has a repressive effect on *Bmp2* expression similar to its role in negatively regulating Wnt activity. Thus, Bmps and Shh may act together to focus Wnt signaling to just the tip of the dental lamina. The setting aside of a small group of Wnt active dental epithelial cells is possibly one of the most important reasons that reptiles replace their teeth and mice do not (Handrigan and Richman, 2010b).

9 POTENTIAL INVOLVEMENT OF OTHER FACTORS IN SUCCESSIONAL LAMINA REFORMATION OR SURVIVAL: ECTODYSPLASIN AND FIBROBLAST GROWTH FACTORS

We have studied only three pathways in some detail, but there are others that could play important roles in tooth replacement. We discussed earlier that certain permanent teeth may have retained a vestigial successional lamina, and this may explain why some regions of the mouth are more likely than others to form supernumerary teeth. The investigation of signals that promote survival of vestigial dental structures is a worthwhile direction to follow in future studies.

In the mouse there are two major signaling pathways that promote survival of vestigial diastema teeth, ectodysplasin and fibroblast growth factors. The Eda pathway is a key component for the development of many ectodermal organs, including hair, scales, fins, and teeth. Disruption of the Eda pathway in humans either at the ligand, receptor, or adaptor protein levels results in the syndrome hypohidrotic ectodermal dysplasia (Mikkola, 2009). In this syndrome patients have far fewer permanent teeth than normal, even though the first-generation primary tooth may be present. This suggests that the successional lamina is susceptible to loss of the Eda signaling pathway.

The other evidence that supports a possible role for Eda activity during tooth replacement is the expression of the receptor *Edar* (ectodysplasin receptor) in the tooth-forming side of the early python dental lamina (Richman and Handrigan, 2011). This epithelial restriction of *Edar* expression is also seen in the mouse (Tucker et al., 2000; Laurikkala et al., 2001).

The phenotypes of the Tabby mouse and the Eda overexpression mouse also support a role for Eda in maintaining the successional lamina. The Tabby mouse is a loss of function of the *Eda* gene and the tooth number is reduced, mirroring the human condition (Pispa et al., 1999). Even more suggestive are the results from the overexpression of Eda in the ectoderm, where ectopic molars were induced in the mouse diastema (Mustonen et al., 2003). It will be worthwhile studying Eda-Edar signaling in the bearded dragon, where activation of this signaling pathway may promote development of the vestigial successional lamina.

Eda has also been shown to be regulated by the Wnt pathway in teeth. The induction of *Eda* expression in dental epithelium by Wnt6 was striking in molar organ cultures (Laurikkala et al., 2001). In addition, in the same study it was shown that *Eda* expression was dependent on *Lef1* expression. These data suggest that *Eda* may be mediating the effects that we have seen caused by activation of the Wnt pathway. More studies are needed to test this hypothesis. We predict that activation of Eda signaling might give similar effects to those of activation of the canonical Wnt pathway.

Of the 22 members of the fibroblast growth factor (FGF) family signals known to date, more than half are expressed during odontogenesis (Kettunen and Thesleff, 1998; Porntaveetus et al., 2011). One of the first signaling events in tooth development is the induction of the tooth mesenchyme by Shh and BMPs, followed by the progression from the bud to the bell stage, which is controlled by FGFs. Deletion of the antagonist *Spry2* and to a lesser extent, *Spry4*, promotes supernumerary tooth formation in the mouse diastema (Klein et al., 2006). In this situation, the rudimentary diastema tooth buds maintain cell proliferation and grow into full teeth as a result of increased FGF signaling (Peterkova et al., 2009). Future work will focus on the Fgf pathway to determine whether this group of growth factors works in concert with other signals to form generational teeth.

10 STEM CELLS IN SQUAMATE TOOTH REPLACEMENT

We and others (Huysseune and Thesleff, 2004; Smith et al., 2009b) believe that the dental epithelium contains stem cells that are capable of initiating new tooth formation.

We propose that during mammalian evolution, indefinite tooth replacement was lost due to an inability to re-form the successional lamina; however, the stem cells may still exist. These may be the same stem cells that we have recently identified in geckos (Handrigan et al., 2010). Since gecko teeth have no region akin to the hair follicle bulge or distended cervical loop of mouse incisors, there is no morphologically distinct tissue where stem cells might be predicted to reside. We therefore carried out a pulse-chase experiment in which a long exposure to BrdU (bromodeoxyuridine) is followed by a long chase period. This would label the majority of cells in the tooth after the long pulse, and only the most slowly dividing cells would retain the label after the long chase. Before carrying out this gecko study, we felt, as did many others, that the successional lamina would be the region where putative stem cells would be found. However, instead, we found that the label-retaining cells were located a distance away from the successional lamina, on the lingual side of the dental lamina. This makes sense since the lingual dental lamina does not form teeth and is a quiescent region with low proliferation. In contrast, the successional lamina is comprised of proliferative cells and not the slow-turnover cells that generally characterize stem cells (Fuchs, 2009).

The region of dental epithelium containing the label-retaining cells expressed several markers that were expressed in the cervical loop of incisors and the hair bulge. These genes include *Dkk3, Igfbp5*, and *Lgr5* (Morris et al., 2004; Tumbar et al., 2004).

The model we propose is that there is periodic stimulation of the stem cells within a niche in a temporally specific manner and that these lead to the formation of transit amplifying cells (Fig. 3A and B). We tested the responsiveness of different layers of the dental lamina to an increase in canonical Wnt signaling using a specific activator of the pathway, BIO. In these experiments, proliferation was stimulated on the lingual or label-retaining side of the dental lamina. This suggests that the area containing the stem cells is particularly sensitive to Wnt signaling, although these cells do not express Wnt ligands themselves. Thus, unidentified signals either coming from other regions of the dental epithelium or from the mesenchyme could activate stem cells.

There are many unanswered questions in the process of reptilian tooth regeneration. First, is there a stem cell niche, a collection of mesenchymal cells that maintain "stemness" in the adjacent niche in the dental lamina? Second, how are the stem cells replenished? Third, do stem cells give rise directly to the successional lamina and therefore the next-generation tooth? Fourth, what controls the activation of the stem cells and therefore regulates tooth family size and the spacing of teeth?

There are analogies between the dental epithelium and the mammalian hair. In hair, the stem cells are also epithelial and regenerate the hair follicle and the hair inside it. Elegant lineage tracing and pulse-chase experiments have determined the fates of the slowly dividing cells in the three phases of the hair cycle; resting, building, and destruction (telogen, anagen, and catagen) (Fuchs, 2009). Stem cells not only form the next hair germ but also contribute to the epidermis during wound healing (Blanpain and Fuchs, 2009). The outer root sheath contains stem cells that have exited the bulge. It was recently found that these cells can reenter the bulge to replenish the stem cells or can avoid destruction and give rise to new hair (Hsu et al., 2011). A similar cycling of cells between the lingual dental lamina and successional lamina may be operating in the tooth (Fig. 3A and B). In other words, we hypothesize that gecko stem cells in the lingual dental lamina are able (1) to replenish themselves, and (2) to form transit amplifying cells which then go on and re-form the successional lamina (Fig. 3A and B).

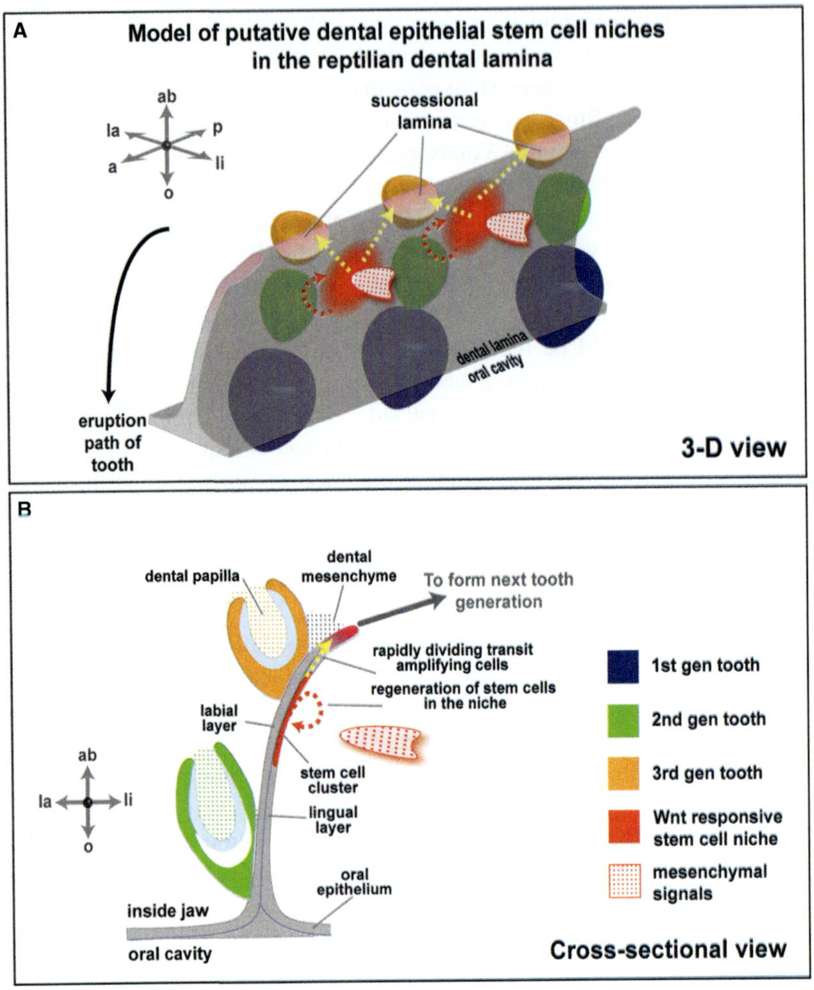

FIGURE 3 Model for the putative dental epithelial stem cell niches in reptilian teeth. A three-dimensional view along the dental arch (A) and a cross-sectional view through the putative stem cell niche (B). The location of the stem cell niches in the dental lamina is restricted to the lingual side, in between the newest-formed tooth and the one above it. This is based on the distribution of label-retaining cells in pulse-chase experiments. The hypothetical contribution of cells to both the successional lamina (yellow dashed arrow) and to replenishing the stem cell niche (red dashed arrow) is shown. It is possible that mesenchymal signals maintain the cluster of stem cells in this position. In addition, other instructive factors may stimulate the proliferation of the stem cells. The spacing of teeth within tooth families and between families may depend on concentration gradients of as yet unidentified mesenchymal signals.

The reptile dental laminae share some properties of the hair follicle, including the presence of slowly dividing cells and the expression of stem cell markers. However, in contrast to the hair, where a destructive stage must occur prior to formation of a new hair, there does not seem to a “catabolic” stage within the dental lamina in prehatching reptiles. This is something that needs to be examined more closely in adult animals.

11 CHALLENGES FOR THE FUTURE

Some of the signals that stimulate the growth of the early dental lamina have been worked out as well as the interactions between several signaling pathways. This has pointed us toward several candidate signals that might be important in stimulation tooth replacement in mammals. The area where stem cells are likely to be located has been determined and, surprisingly, these cells are not in the successional lamina. Next, we need to find ways to stimulate the label-retaining cells and show that they do contribute to next-generation teeth. It will be important to find ways to selectively stimulate the dental lamina itself and not the successional lamina.

The first challenge that must be met is to better label the potential stem cells. In our original study, approximately 40% of all the dental epithelial cells were labeled, thus omitting the majority of the slowly dividing stem cells. In hair stem cell research, the approach is to combine a pulse-chase experiment with other markers for "stemness". As long as the inducible promoter is turned on and then off prior to the first hair formation at postnatal day 21 in mice (Hsu et al., 2011), all stem cells will be labeled. It is therefore critical to find the means to label the dental stem cells of younger animals, preferably prior to hatching. This approach might mark the stem cells as they are being formed.

Once we have achieved more complete labeling, we would next determine whether the label-retaining cells contribute to the next generation of teeth. We can ask whether tooth replacement will be blocked if this population of stem cells is deleted selectively. Can stem cells be induced to divide by mechanical factors (injury)? In the hair bulge it is a simple matter to pluck hairs in order to initiate the process of hair replacement. Such a disruption of forming teeth in a reptile has never been attempted but could be very informative when combined with labeling techniques.

How can we best get access to the stem cell population while keeping the three-dimensional relationships of the developing teeth? It is very difficult to make enamel or teeth in culture, and even more challenging to culture ectothermic reptilian tissues. It is imperative to maintain the natural context as much as possible. Short-term organ cultures on squamates do allow manipulation of specific signals while studying the effects on short-term readouts such as proliferation, gene expression and apoptosis. However, morphogenesis cannot be followed in such a system. Other ideas we will pursue are the injection of small molecules into the eggs and postnatal injection into the jaw. These experiments would show within the natural context of the jaw whether we can increase the number of generations of teeth present at the time of hatching.

We can take inspiration for future molecular pathways to study from the continuously erupting mouse incisor (Chapter 17). Fgfs (Fgf10) are required for formation of the cervical loops in the continuously erupting incisor (Harada et al., 2002; Wang et al., 2007; Yokohama-Tamaki et al., 2006; Klein et al., 2008). The cervical loop has Fgf signaling activity in the epithelial component, as shown by the expression of downstream transcription factors *Etv4* and *Etv5* (Papagerakis et al., 2003) and the expression and function of the Sprouty genes inhibiting Fgf signaling (Klein et al., 2008). Several of the ligands are expressed in the mesenchyme adjacent to the cervical loop, suggesting that FGFs could maintain epithelial growth in the successional lamina. FGFs may also be key factors in promoting dental epithelial stem cell growth in reptiles.

At this time we believe that the leopard gecko is a promising in vivo system in which to manipulate tooth replacement. Thus, we are entering a new, exciting phase of tooth regeneration research, one in which the reptile offers great promise of an understanding of the molecular and cellular controls of polyphyodonty in an amniote.

REFERENCES

Blanpain C, Fuchs E (2009). Epidermal homeostasis: a balancing act of stem cells in the skin. *Nat Rev Mol Cell Biol* 10(3):207–217.

Buchtová M, Boughner JC, Fu K, Diewert VM, Richman JM (2007). Embryonic development of python sebae: II. Craniofacial microscopic anatomy, cell proliferation and apoptosis. *Zoology (Jena)* 110(3):231–251.

Buchtová M, Handrigan GR, Tucker AS, Lozanoff S, Town L, Fu K, et al. (2008). Initiation and patterning of the snake dentition are dependent on Sonic hedgehog signaling. *Dev Biol* 319(1):132–145.

Cabreira SF, Cisneros JC (2009). Tooth histology of the parareptile *Soturnia caliodon* from the Upper Triassic of Rio Grande do Sul, Brazil. *Acta Palaeontol Pol* 54:743–748.

Chen Y, Zhang Y, Jiang TX, Barlow AJ, St Amand TR, Hu Y, et al. (2000). Conservation of early odontogenic signaling pathways in Aves. *Proc Natl Acad Sci USA* 97(18):10044–10049.

Cobourne MT, Miletich I, Sharpe PT (2004). Restriction of Sonic hedgehog signalling during early tooth development. *Development* 131(12):2875–85.

Cooper JS, Poole D, Lawson R (1970). The dentition of agamid lizards with special reference to tooth replacement. *J Zool Lond* 162:85–98.

Crompton AW, Jenkins FA (1973). Mammals from reptiles: a review of mammalian origins. *Annu Rev Earth Planet Sci* 1:135–155.

Davit-Béal T, Chisaka H, Delgado S, Sire JY (2007). Amphibian teeth: current knowledge, unanswered questions, and some directions for future research. *Biol Rev* 82(1):49–81.

Davit-Béal T, Tucker AS, Sire JY (2009). Loss of teeth and enamel in tetrapods: fossil record, genetic data and morphological adaptations. *J Anat* 214:477–501.

Delgado S, Davit-Beal T, Allizard F, Sire JY (2005). Tooth development in a scincid lizard, *Chalcides viridanus* (Squamata), with particular attention to enamel formation. *Cell Tissue Res* 319(1):71–89.

Edmund AG (1960). Tooth replacement phenomena in the lower vertebrates. *Contrib Life Sci Div Royal Ontario Museum* 52:52–190.

Erickson GM (1996). Incremental lines of von Ebner in dinosaurs and the assessment of tooth replacement rates using growth line counts. *Proc Nat Acad Sci USA* 93:14623–14627.

Ewer RF (1963). Reptilian tooth replacement. *News Bull Zool Soc S Afr* 4:4–9.

Filali M, Cheng N, Abbott D, Leontiev V, Engelhardt JF (2002). Wnt-3A/beta-catenin signaling induces transcription from the LEF-1 promoter. *J Biol Chem* 277(36):33398–33410.

Fuchs E (2009). The tortoise and the hair: slow-cycling cells in the stem cell race. *Cell* 137(5):811–819.

Handrigan GR, Richman JM (2010a). Autocrine and paracrine Shh signaling are necessary for tooth morphogenesis, but not tooth replacement in snakes and lizards (Squamata). *Dev Biol* 337(1):171–186.

Handrigan GR, Richman JM (2010b). A network of Wnt, Hedgehog and BMP signaling pathways regulates tooth replacement in snakes. *Dev Biol* 348(1):130–141.

Handrigan GR, Richman JM (2011). Unicuspid and bicuspid tooth crown formation in squamates. *J Exp Zool B* 316(8):598–608.

Handrigan GR, Leung KJ, Richman JM (2010). Identification of putative dental epithelial stem cells in a lizard with life-long tooth replacement. *Development* 137(21):3545–3549.

Harada H, Kettunen P, Jung HS, Mustonen T, Wang YA, Thesleff I (1999). Localization of putative stem cells in dental epithelium and their association with Notch and FGF signaling. *J Cell Biol* 147(1):105–120.

Harada H, Toyono T, Toyoshima K, Yamasaki M, Itoh N, Kato S, et al. (2002). FGF10 maintains stem cell compartment in developing mouse incisors. *Development* 129(6):1533–1541.

Hopson JA (1964). Tooth replacement in cynodont, dicynodont, and therocephalian reptiles. *J Zool* 142(4):625–654.

Hopson JA (1973). Endothermy, small size and the origin of mammalian reproduction. *Am Nat* 107:446–452.

Hsu YC, Pasolli HA, Fuchs E (2011). Dynamics between stem cells, niche, and progeny in the hair follicle. *Cell* 144(1):92–105.

Huysseune A, Thesleff I (2004). Continuous tooth replacement: the possible involvement of epithelial stem cells. *Bioessays* 26(6):665–671.

Huysseune A, Sire J-Y, Witten PE (2009). Evolutionary and developmental origins of the vertebrate dentition. *J Anat* 214(4):465–476.

Järvinen E, Salazar-Ciudad I, Birchmeier W, Taketo MM, Jernvall J, Thesleff I (2006). Continuous tooth generation in mouse is induced by activated epithelial Wnt/beta-catenin signaling. *Proc Natl Acad Sci USA* 103(49):18627–18632.

Jernvall J, Thesleff I (2000). Reiterative signaling and patterning during mammalian tooth morphogenesis. *Mech Dev* 92(1):19–29.

Jho EH, Zhang T, Domon C, Joo CK, Freund JN, Costantini F (2002). Wnt/beta-catenin/Tcf signaling induces the transcription of Axin2, a negative regulator of the signaling pathway. *Mol Cell Biol* 22(4):1172–1183.

Kettunen P, Thesleff I (1998). Expression and function of FGFs-4, -8, and -9 suggest functional redundancy and repetitive use as epithelial signals during tooth morphogenesis. *Dev Dyn* 211(3):256–268.

Klein OD, Minowada G, Peterkova R, Kangas A, Yu BD, Lesot H, et al. (2006). Sprouty genes control diastema tooth development via bidirectional antagonism of epithelial–mesenchymal FGF signaling. *Dev Cell* 11(2):181–190.

Klein OD, Lyons DB, Balooch G, Marshall GW, Basson MA, Peterka M, et al. (2008). An FGF signaling loop sustains the generation of differentiated progeny from stem cells in mouse incisors. *Development* 135(2):377–385.

Laurikkala J, Mikkola M, Mustonen T, Aberg T, Koppinen P, Pispa J, et al. (2001). TNF signaling via the ligand–receptor pair ectodysplasin and Edar controls the function of epithelial signaling centers and is regulated by Wnt and activin during tooth organogenesis. *Dev Biol* 229(2):443–455.

Luckett WP (1993). Ontogenetic staging of the mammalian dentition, and its value for assessment of homology and heterochrony. *J Mamm Evol* 1:269–282.

Mikkola ML (2009). Molecular aspects of hypohidrotic ectodermal dysplasia. *Am J Med Genet A* 149A(9):2031–2036.

Mitsiadis TA, Caton J, Cobourne M (2006). Waking-up the sleeping beauty: recovery of the ancestral bird odontogenic program. *J Exp Zool B* 306(3):227–233.

Miyado M, Ogi H, Yamada G, Kitoh J, Jogahara T, Oda S, Sato I, Miyado K and Sunohara M (2007). Sonic hedgehog expression during early tooth development in Suncus murinus. *Biochem Biophys Res Commun* 363:269–275.

Modesto SP (1995). The skull of the herbivorous synapsid *Edaphosaurus boanerges* from the Lower Permian of Texas. *Palaeontology* 38(1):213–239.

Modesto SP, Reisz RR (1992). Restudy of Permo-Carboniferous synapsid *Edaphosaurus novomexicanus* Williston and Case, the oldest known herbivorous amniote. *Can J Earth Sci* 29:2653–2662.

Modesto SP, Rybczynski N (2001). The amniote faunas of the Russian Permian: implications for Late Permian terrestrial vertebrate biogeography. In: *The Age of Dinosaurs in Russia and Mongolia*, Benton MJ, Shishkin MA, Unwin DM, eds., Cambridge University Press, Cambridge, UK, pp. 17–34.

Modesto SP, Scott DM, Berman DS, Müller J, and Reisz RR (2007). The skull and paleoecological significance of *Labidosaurus hamatus*, a captorhinid reptile from the Lower Permian of Texas. *Zoo J Linn Soc* 149:237–262.

Morris RJ, Liu Y, Marles L, Yang Z, Trempus C, Li S, et al. (2004). Capturing and profiling adult hair follicle stem cells. *Nat Biotechnol* 22(4):411–417.

Mucchielli ML, Mitsiadis TA, Raffo S, Brunet JF, Proust JP, Goridis C (1997). Mouse *Otlx2/RIEG* expression in the odontogenic epithelium precedes tooth initiation and requires mesenchyme-derived signals for its maintenance. *Dev Biol* 189(2):275–284.

Mustonen T, Pispa J, Mikkola ML, Pummila M, Kangas AT, Pakkasjarvi L, et al. (2003). Stimulation of ectodermal organ development by Ectodysplasin-A1. *Dev Biol* 259(1):123–236.

Ohazama A, Haycraft CJ, Seppala M, Blackburn J, Ghafoor S, Cobourne M, et al. (2009). Primary cilia regulate Shh activity in the control of molar tooth number. *Development* 136(6):897–903.

Ooë T (1981). *Human Tooth and Dental Arch Development*, pp. 162, Ishiyaku Publishers, Tokyo.

Papagerakis P, MacDougall M, Hotton D, Bailleul-Forestier I, Oboeuf M, Berdal A (2003). Expression of amelogenin in odontoblasts. *Bone* 32(3):228–240.

Peterkova R, Peterka M, Viriot L, Lesot H (2002). Development of the vestigial tooth primordia as part of mouse odontogenesis. *Connect Tissue Res* 43(2–3):120–128.

Peterkova R, Churava S, Lesot H, Rothova M, Prochazka J, Peterka M, et al. (2009). Revitalization of a diastemal tooth primordium in *Spry2* null mice results from increased proliferation and decreased apoptosis. *J Exp Zool B* 312B(4):292–308.

Peyer B (1968). *Comparative Odontology*, University of Chicago Press, Chicago.

Pispa J, Jung HS, Jernvall J, Kettunen P, Mustonen T, Tabata MJ, et al. (1999). Cusp patterning defect in Tabby mouse teeth and its partial rescue by FGF. *Dev Biol* 216(2):521–534.

Porntaveetus T, Otsuka-Tanaka Y, Basson MA, Moon AM, Sharpe PT, Ohazama A (2011). Expression of fibroblast growth factors (Fgfs) in murine tooth development. *J Anat* 218(5):534–543.

Richman JM, Handrigan GR (2011). Reptilian tooth development. *Genesis* 49(4):247–260.

Sereno PC, Wilson JA, Witmer LM, Whitlock JA, Maga A, Ide O, et al. (2007). Structural extremes in a Cretaceous dinosaur. *PLoS One* 2(11):e1230.

Smith MM, Fraser GJ, Chaplin N, Hobbs C, Graham A (2009a). Reiterative pattern of sonic hedgehog expression in the catshark dentition reveals a phylogenetic template for jawed vertebrates. *Proc Biol Sci* 276(1660):1225–1233.

Smith MM, Fraser GJ, Mitsiadis TA (2009b). Dental lamina as source of odontogenic stem cells: evolutionary origins and developmental control of tooth generation in gnathostomes. *J Exp Zool B* 312B(4):260–280.

Solares R, Romero MI (2004). Supernumerary premolars: a literature review. *Pediatr Dent* 26(5):450–458.

Stembirek J, Buchtová M, Kral T, Matalova E, Lozanoff S, Misek I (2010). Early morphogenesis of heterodont dentition in minipigs. *Eur J Oral Sci* 118(6):547–558.

Sumida S, Murphy R (1987). Form and function of the tooth crown structure in gekkonid lizards (Reptilia, Squamata, Gekkonidae). *Can J Zool* 65:2886–2892.

Suomalainen M, Thesleff I (2010). Patterns of Wnt pathway activity in the mouse incisor indicate absence of Wnt/beta-catenin signaling in the epithelial stem cells. *Dev Dyn* 239(1):364–372.

Thesleff I, Keranen S, Jernvall J (2001). Enamel knots as signaling centers linking tooth morphogenesis and odontoblast differentiation. *Adv Dent Res* 15:14–18.

Tucker AS, Headon DJ, Schneider P, Ferguson BM, Overbeek P, Tschopp J, et al. (2000). Edar/Eda interactions regulate enamel knot formation in tooth morphogenesis. *Development* 127(21):4691–4700.

Tumbar T, Guasch G, Greco V, Blanpain C, Lowry WE, Rendl M, et al. (2004). Defining the epithelial stem cell niche in skin. *Science* 303(5656):359–363.

Tummers M, Thesleff I (2009). The importance of signal pathway modulation in all aspects of tooth development. *J Exp Zool B* 312B:309–319.

Vonk FJ, Admiraal JF, Jackson K, Reshef R, de Bakker MA, Vanderschoot K, et al. (2008). Evolutionary origin and development of snake fangs. *Nature* 454(7204):630–633.

Wang X-P, Suomalainen M, Jorgez CJ, Matzuk MM, Wankell M, Werner S, et al. (2004). Modulation of activin/bone morphogenetic protein signaling by Follistatin is required for the morphogenesis of mouse molar teeth. *Dev Dyn* 231(1):98–108.

Wang X-P, Suomalainen M, Felszeghy S, Zelarayan LC, Alonso MT, Plikus MV, et al. (2007). An integrated gene regulatory network controls stem cell proliferation in teeth. *PLoS Biol* 5(6):e159.

Wang X-P, O'Connell DJ, Lund JJ, Saadi I, Kuraguchi M, Turbe-Doan A, et al. (2009). Apc inhibition of Wnt signaling regulates supernumerary tooth formation during embryogenesis and throughout adulthood. *Development* 136(11):1939–1949.

Westergaard B, Ferguson MWJ (1986). Development of the dentition in *Alligator-mississippiensis*: early embryonic-development in the lower jaw. *J Zool* 210:575–597.

Westergaard B, Ferguson MWJ (1987). Development of the dentition in *Alligator-mississippiensis*: later development in the lower jaws of embryos, hatchlings and young juveniles. *J Zool* 212:191–222.

Westergaard B, Ferguson MWJ (1990). Development of the dentition in *Alligator-mississippiensis*: upper jaw dental and craniofacial development in embryos, hatchlings, and young juveniles, with a comparison to lower jaw development. *Am J Ana* 187(4):393–421.

Whitlock JA, D'Emic MD, Fisher DC, Smith K, Wilson JA (2010). Trends in sauropod tooth size, shape, and replacement rate. *J Vertebr Paleontol* 30(Suppl to 3):186A.

Yamanaka A, Yasui K, Sonomura T and Uemura M (2007). Development of heterodont dentition in house shrew (Suncus murinus). *Eur J Oral Sci* 115:433–40.

Yokohama-Tamaki T, Ohshima H, Fujiwara N, Takada Y, Ichimori Y, Wakisaka S, et al. (2006). Cessation of Fgf10 signaling, resulting in a defective dental epithelial stem cell compartment, leads to the transition from crown to root formation. *Development* 133(7):1359–1366.

Zahradnicek O, Horacek I, Tucker AS (2008). Viperous fangs: development and evolution of the venom canal. *Mech Dev* 125(9–10):786–796.

8

TOOTH ROOT DEVELOPMENT

BRIAN L. FOSTER

National Institute of Arthritis and Musculoskeletal and Skin Diseases, National Institutes of Health, Bethesda, Maryland

FRANCISCO H. NOCITI JR.

National Institute of Arthritis and Musculoskeletal and Skin Diseases, National Institutes of Health, Bethesda, Maryland; State University of Campinas School of Dentistry, Piracicaba, São Paulo, Brazil

MARTHA J. SOMERMAN

National Institute of Arthritis and Musculoskeletal and Skin Diseases, National Institutes of Health, Bethesda, Maryland

1 INTRODUCTION

Over the past decades, substantial insights have been gained in defining the mechanisms and factors controlling organogenesis in the body. These advances, based on new technologies and on the exponential growth in defining the genes and proteins regulating development, have propelled biomedical research into a period of dramatic discovery. In this context, the dental–oral–craniofacial region has received increased attention as an important aspect of overall health. Teeth and their supporting tissues are important not only for mastication, but also for speech and aesthetics. The visible portion of the tooth, the enamel-covered *clinical crown*, often preoccupies the common conception of the tooth. However, the root and supportive tissues are as critical for tooth function and longevity, providing mechanical support, innervation, and vascular supply.

The tooth root represents the only place in the body where three distinct mineralized tissues (dentin, cementum, and bone) are in close proximity and articulate via carefully maintained hard–soft tissue interfaces, in this case with the periodontal ligament

Stem Cells in Craniofacial Development and Regeneration, First Edition.
Edited by George T.-J. Huang, Irma Thesleff.

(PDL). This unique structure and composition gives the tooth root its multifunctional properties, important for providing attachment and support for the tooth, and absorbance and transmission of masticatory stresses and strains. Loss of integrity of the root structure as a result of inherited conditions, trauma, or disease, rapidly compromises the function of the tooth structure as a whole and can result in premature tooth loss. The current state of the art in treatment modalities for periodontal defects still leaves much room for improvement in terms of predictably restoring form and function to root tissues, and the use of stem cells for such enterprises has become an exciting prospect for future advances.

This chapter has been organized into three main sections, providing a broad overview of what is currently known about the structure and composition of tooth root tissues, the developmental processes in formation of these tissues, and the signals and influences directing root development. The chapter is intended not only as a primer for what is known about root development, but aims to identify areas requiring focused research in order to reach our ultimate goal: development of modalities for repair, regeneration, or engineering of tooth root tissues.

2 STRUCTURE AND COMPOSITION OF TOOTH ROOT TISSUES

Detailed evaluation of the composition of tooth tissues can be found in several histology-focused texts (see, e.g., Nanci and Somerman, 2008). Here, tissues of the root will be introduced briefly for the purpose of comparing and contrasting their structure and composition in order to understand how these are integral to functional properties and strategies for repair and regeneration. See Table 1 for an overview of information presented in this section.

2.1 Tissues of the Tooth Root

Dentin is the hard tissue forming the bulk of the tooth in both crown and root, and is mineralized matrix characterized secondarily by thin dentinal tubules that house odontoblast cell processes. Dentin provides strength and durability to the tooth, while its elasticity is essential for absorbing forces from the overlying hard but brittle enamel. Dentin bounds and protects the pulp chamber, a soft tissue that is highly vascular and innervated and home to mesenchymal stem cell populations among a heterogeneous population of other cell types.

Traditionally, cementum is classified into two major varieties, on the basis of whether cells are included and on the primary source of collagen fibers. Acellular cementum is a thin layer present on the cervical and middle portions of roots, characterized by lack of cell inclusions and by extrinsically produced collagen fibers that enter from the PDL. Sharpey's fibers from the PDL insert into an acellular cementum surface at a high density, emphasizing the importance of this cementum type for tooth attachment and distribution of forces generated by occlusal loading. The cellular type of cementum is distinguished by embedded cementocytes in a matrix of intrinsic collagen fibers, and this type is present on the apical portion of multirooted teeth, and sometimes in the furcation region, depending on species. Cellular cementum is sometimes referred to as *adaptive cementum* because it functions to maintain the tooth in

TABLE 1 Comparison of Key Characteristics of Tissues of the Tooth Root

Characteristic	Dentin	Acellular Cementum	Cellular Cementum	Periodontal Ligament	Alveolar Bone
Embryonic origin	Ectomesenchyme of the dental papilla descended from cranial neural crest (CNC)	Ectomesenchyme of the dental follicle descended from CNC, or transformed epithelium	Ectomesenchyme of the dental follicle descended from CNC or transformed epithelium	Ectomesenchyme of the dental follicle descended CNC	Ectomesenchyme of the dental follicle and perifollicular tissue descended from CNC
Associated cells	Odontoblasts, preodontoblasts, and heterogeneous pulp cells, including stem cells	Cementoblasts (cervical)	Cementoblasts (apical), cementocytes	Fibroblasts, stem, and progenitor cells; epithelial rests	Osteoblasts, osteocytes, lining cells, osteoclasts
Structural organization	Thick layer of mineralized matrix notable for dentin tubules and unmineralized predentin layer at the mineralization front	Thin layer of acellular mineralized matrix covering cervical portion of root dentin	Thick, bonelike layer of cellular mineralized matrix covering apical portion of root dentin and furcation areas	Collagen bundles extending from tooth to bone surfaces, providing continuity across ligament space	Compact and trabecular bone with complex geometry in relation to tooth roots
Mineral composition	70% Hydroxyapatite	45–50% Hydroxyapatite	45–50% Hydroxyapatite	Unmineralized	60–65% Hydroxyapatite
Mechanical properties	Less stiff than overlying enamel; resists propagation of cracks	Less mineralized and more compliant than underlying dentin	Less mineralized and more compliant than underlying dentin	Elastic and highly compliant	More compliant than dentin, less compliant than cementum

(*continued*)

TABLE 1 *(Continued)*

Characteristic	Dentin	Acellular Cementum	Cellular Cementum	Periodontal Ligament	Alveolar Bone
Extracellular matrix composition	Collagens (types I, III, VII), dentin sialoprotein (DSP), dentin phosphoprotein, dentin matrix protein 1, osteocalcin, proteoglycans	Collagens (types I, III, VII), bone sialoprotein, osteopontin, osteocalcin	Collagens (types I, III, VII), bone sialoprotein, osteopontin, osteocalcin, dentin matrix protein 1, proteoglycans	Collagens (types I, III, XII), elastic fibers, ground substance, periostin, proteoglycans	Collagens (types I, III, VII), bone sialoprotein, osteopontin, osteocalcin, dentin matrix protein 1, proteoglycans
Functions	Body of tooth; provides strength and durability; continuity from crown to root structures	Anchors tooth root to supporting bone by cementing embedded Sharpey's fibers	Maintains tooth in occlusal position; compensates for ongoing attrition	Supportive shock-absorbing ligament; invests periodontal tissues with vasculature and nervous tissues	Alveolar processes attach teeth to basal bone by Sharpey's fibers of PDL; a remodeling support structure

its proper occlusal position. Cellular cementum probably does not function in attachment because collagen fibers originate primarily from cells within the layer, run mostly parallel to the root surface, and do not protrude into the PDL space in large numbers.

The PDL is a specialized connective tissue stretching from the alveolar bone of the tooth socket to the cementum of the root surface, firmly attached to both sides by embedded Sharpey's fibers. That the PDL is highly elastic and remains unmineralized is critical for its functions in tooth attachment and as absorber and distributor of masticatory forces. However, the PDL is more than a tissue of attachment; it also supports a heterogeneous cell population (including stem and progenitor cells with osteo- and cementoprogenitor properties) and invests the periodontal region with vasculature and innervation.

Although alveolar bone is extradental by definition, it is useful to include in a consideration of tooth development and regeneration, for several reasons. First, the alveolar bone proper shares in common with other periodontal tissues an origin in the ectomesenchymal dental follicle (discussed in greater detail below). Second, alveolar bone functions as part of a unit with the cementum and PDL to provide attachment and support to the tooth. Third, alveolar bone formation is intimately associated with development of the tooth root, and this rapidly remodeling bone is highly responsive to changes in dental function (e.g., resorption under orthodontic tooth movement or loss of alveolar crest following edentulism). Fourth, alveolar bone loss is a common occurrence in periodontal disease, and the potential regeneration of this tissue through the use of stem cells is a topic relevant to the theme of this book.

2.2 Mineral Content

The hard tissues of the root, dentin, cementum, and alveolar bone have a similar mineral content and a large degree of overlap in their extracellular matrix compositions. More detailed evaluations of composition can be found in previous reviews on these tissues (Bosshardt, 2005; Foster et al., 2007), and key comparisons are presented in Table 1.

2.3 Extracellular Matrix Composition

Comparing and contrasting matrix composition reveals more similarities than differences among the mineralized tissues of the root. Type I collagen is the major matrix component, serving as a scaffold for hydroxyapatite mineral deposition in dentin, cementum, and bone, with lesser amounts of other collagens, including types III, V, VI, and XII. The remaining organic components of these matrices are made up of an assortment of noncollagenous proteins and proteoglycans which influence matrix properties and regulate hydroxyapatite crystal deposition.

Shared extracellular matrix proteins include small integrin-binding ligand *n*-linked glycoprotein (SIBLING) family members bone sialoprotein (BSP) and osteopontin (OPN), with dentin matrix protein 1 (DMP1) more enriched in dentin, cellular cementum, and bone, and dentin sialoprotein (DSP) and dentin phosphoprotein (DPP) expressed selectively in dentin. Osteocalcin (OCN), another mineral-regulating protein regarded as an osteoblast marker, is reported in all the hard tissues. Proteoglycans, including decorin, biglycan, versican, and lumican, also influence mineralization and have been localized to dentin, bone, and cellular cementum but not to acellular cementum.

Like hard tissues of the root, the PDL extracellular matrix is composed primarily of collagens, the predominant varieties being types I, III, and XII. The collagen fibrils are organized into directed fiber bundles which are remodeled to adapt to stresses placed on the tooth. PDL also includes oxytalan elastic fibers, which may be related to vascular elements. The PDL contains a diverse array of other extracellular components, including matrix proteins such as OPN and periostin, proteoglycans, and amorphous ground substance. Some extracellular matrix components may play a role in maintaining PDL in an unmineralized state, although this important property is not well understood at present.

3 DEVELOPMENTAL PROCESSES IN ROOT FORMATION

3.1 Overview

Root development is a late event in the scheme of odontogenesis, but like earlier stages of tooth formation described in previous chapters, is set in motion by cellular and molecular events taking place in the early embryo. It is useful to briefly overview major events leading up to the root formation; odontogenesis is a continuum of events, although we break it into discrete stages for purposes of study and for better understanding of the complex three-dimensional morphological changes that occur. The neural crest yields migratory cells with multipotent properties, giving rise to craniofacial mesenchymal cell populations as well as a diversity of other cell types, including neurons and melanocytes. Migrating cranial neural crest (CNC) cells stream down from the midbrain and first and second rhombomeres to populate the first branchial arch of the developing embryo (see Chapter 2).

Development of the tooth crown is driven by sequential, reciprocal, reiterative crosstalk between odontogenic epithelium and underlying dental mesenchyme, which is *ectomesenchyme* descended from the migratory CNC cells and is discussed in detail in Chapter 6. The product of the early stages of tooth development—the bud, cap, and bell stages—is a crown constructed of enamel and dentin. At this stage, the ectomesenchymal populations of the dental papilla and dental follicle (also called the dental sac), which were continuous with each other during earlier stages, have now been partitioned by the extended cervical loop of the folded enamel organ (Fig. 1A). The dental papilla within the enamel organ will give rise to the odontoblasts and pulp tissues, and the follicle and loosely arranged perifollicular tissues surrounding the tooth bud presage the periodontia, including cells of the cementum, PDL, and alveolar bone. With the culmination of the crown stage, and prior to tooth eruption, root formation is primed to begin.

3.2 Crown-to-Root Transition

After completion of crown morphogenesis, the inner and outer enamel epithelia (IEE and OEE, respectively) of the cervical loop reorganize to form a bilayered structure, Hertwig's epithelial root sheath (HERS), lacking stellate reticulum and stratum intermedium of the enamel organ (Figs. 1B and 2A). This transition is a key event in odontogenesis in teeth of limited eruption and represents a switch from crown to root

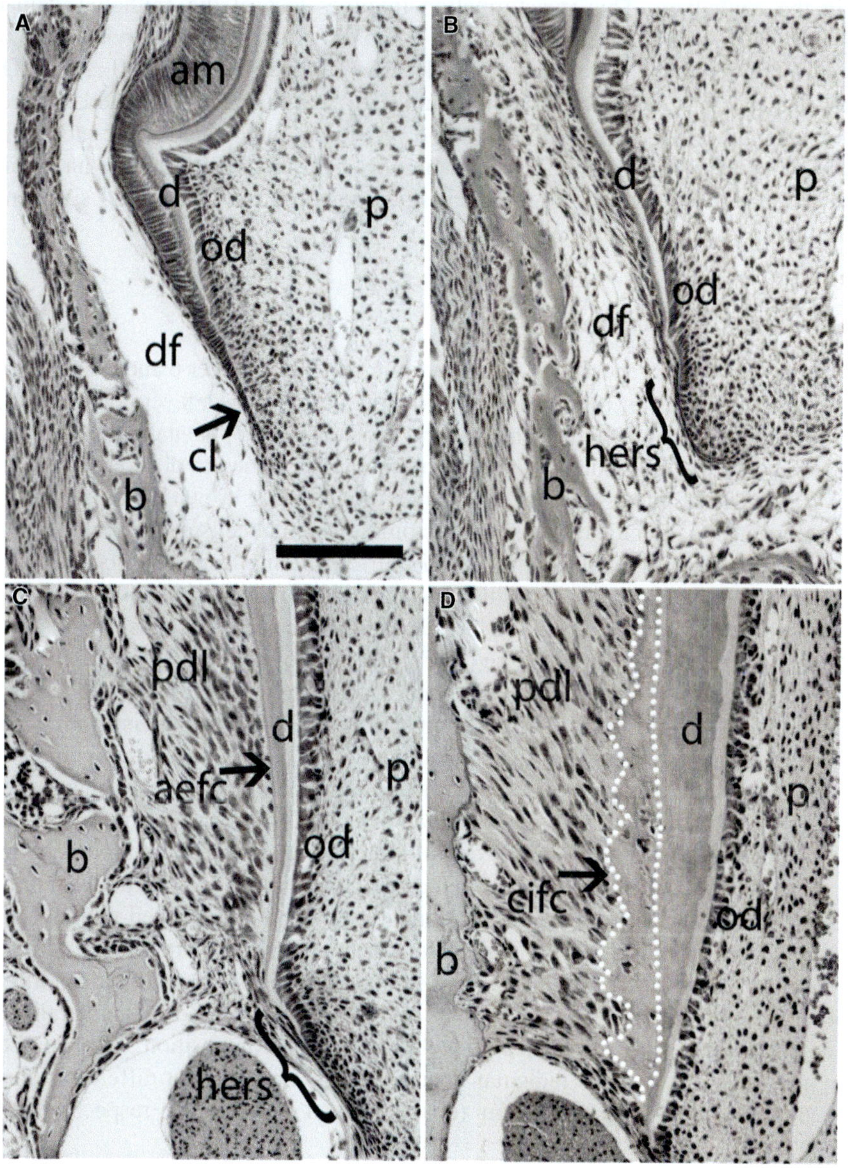

FIGURE 1 Tooth root development. Frontal sections of mouse mandibular first molar root over the course of development, 4 to 26 days postnatal (H&E preparation). (A) At the end of crown formation, the dental follicle (df) and dental pulp (p) are separated by the epithelial cervical loop (cl) at the apical portion of the crown. (B) A switch from crown to root formation involves transition of cervical loop to Hertwig's epithelial root sheath (hers). Dental papilla cells organize to differentiate to root odontoblasts (od). (C) Acellular extrinsic fiber cementum (aefc) formation occurs on the cervical portion of the root, in step with dentin (d) formation. Periodontal ligament (pdl) organization follows. (D) Cellular intrinsic fiber cementum (cifc) occurs on the apical portion of the root. d, dentin; am, ameloblasts; od, odontoblasts; df, dental follicle; p, dental pulp; cl, cervical loop; hers, Hertwig's epithelial root sheath; aefc, acellular extrinsic fiber cementum; cifc, cellular intrinsic fiber cementum; b, bone. Scale bar: 100 μm.

fate in the epithelium. This transformation involves loss of the epithelial stem cell niche associated with the stellate reticulum compartment within the cervical loop, and reorganization into the simpler structure of HERS (Tummers and Thesleff, 2003). Once created and segregated from crown epithelium, HERS cells undergo directed cell proliferation, migrating in the apical direction and providing a template for the morphology and length of the forming root(s). It is commonly held that HERS acts as an architect of the nascent root; however, mechanisms guiding tooth root shape and size remain poorly understood.

3.3 Root Dentin Formation

The process of dentinogenesis is covered in more detail in Chapter 6 and is summarized here as an integral part of root formation. Root dentin forms as a layer continuous with the crown dentin, employing the same general series of developmental stages, the major exception being that root odontoblasts are presumably induced to differentiate by cells of the IEE of HERS rather than by the enamel organ during crown dentinogenesis. As HERS elongates to define the root shape, ectomesenchymal cells of the peripheral dental papilla respond to signaling from the IEE, differentiating from small cells to larger, elongated preodontoblasts, and eventually to polarized and organized odontoblasts lining the pulp chamber (Fig. 1B and 1C). Odontoblasts are secretory cells responsible for synthesizing the organic matrix of the dentin and promoting its mineralization. Dentin is produced first as an unmineralized predentin layer into which odontoblasts extend long, branching cell processes. The first thin stratum of dentin formed is *mantle dentin*, and mineralization of this layer is dependent on matrix vesicles, small membrane-bound cell buds thought to initiate crystal precipitation. As dentinogenesis continues, mantle dentin gives way to the *circumpulpal dentin*, which ultimately will comprise the bulk of the dentin thickness. Circumpulpal dentin mineralization depends more on the influence of extracellular phosphoproteins rather than matrix vesicles, although how this shift occurs and what functional implications it may have are not well understood.

It should be mentioned that a number of differences in crown versus root dentin have been reported using various animal models. Some of these include differing odontoblast morphology, dissimilar rates of matrix production or mineralization, and distinctions in structure and extracellular matrix composition. The source of differences in coronal versus radicular dentin remains obscure, but different signaling between crown and root epithelium (enamel organ and HERS, respectively) and odontoblasts probably plays a role.

3.4 Acellular Cementum Formation

During root development, cementum forms in step with, and slightly after, the underlying dentin (Fig. 1C). The origin and nature of cementoblasts remains controversial. The *classical hypothesis* of cementum origins holds that cementoblasts are derived from the dental follicle which surrounds the developing tooth bud. This hypothesis is in line with the concept that cementum, PDL, and alveolar bone share a common ectomesenchymal origin. An *alternative hypothesis* for cementogenesis posits that epithelial cells from the outer layer of HERS take on a mesenchymal phenotype capable of secreting cementum matrix proteins and direct cementum mineralization on the dentin surface.

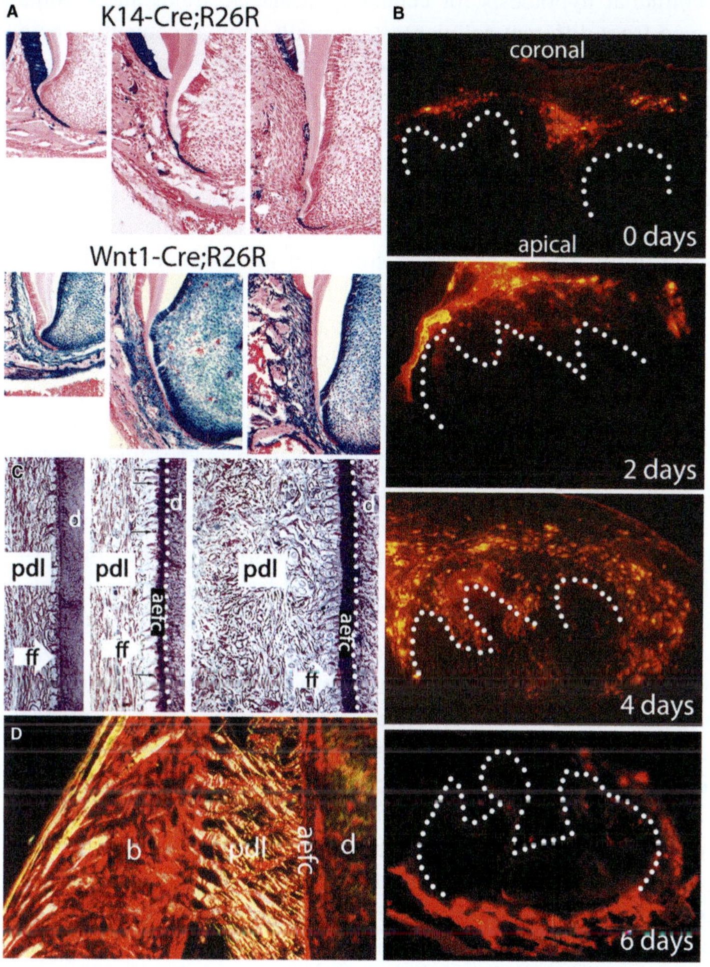

FIGURE 2 Origins and architecture of the periodontal attachment complex. (A) LacZ staining in keratin 14 (K14)-Cre; R26R mouse molars localizes to HERS (blue stain), the architect of the root that fenestrates as the root lengthens. Staining in Wnt1-Cre; R26R mouse molars demonstrates CNC-derived tissues (blue stain) giving rise to the periodontium. (Reprinted from Huang et al, (2010), with permission). (B) Fluorescent images demonstrating pathways of migratory cells in postnatal mouse molars following in vivo DiI labeling of coronal follicle. Follicle cells undergo a massive migration to populate the periodontal region, visualized by labeled cell displacement in the apical direction during postnatal days 6 to 12. The outline of the tooth cusps is indicated by a dotted white line. (Courtesy of T. G. H. Diekwisch, University of Chicago). (C) The acellular cementum (aefc) develops on the fringe fibers (ff) at the root surface, slowly undergoing apposition over time. The junction of the cementum and dentin (d) is demarcated by a dotted white line. [Reprinted from Bosshardt and Selvig (1997), with permission.] (D) The periodontium functions foremost as a complex for attachment and distribution of mechanical forces. The acellular extrinsic fiber cementum (aefc) attaches the tooth proper to the alveolar bone (b) via the nonmineralized periodontal ligament (pdl), by insertion of Sharpey's fibers into both cementum and bone. Mouse first molar section prepared by picrosirius red stain observed under polarized light. d, dentin; aefc, acellular cementum; p, periodontal ligament; b, alveolar bone; ff, fringe fibers.

Under the epithelial hypothesis for cementogenesis, acellular and cellular cementum may be derived from different cellular origins, with cellular cementum being generated by an osteoblast-like cell. Although some authors interpret recent in vitro and in vivo data to lend strength to the epithelial hypothesis, presently the larger body of work on cementogenesis supports the classical mesenchymal hypothesis; it is therefore this orthodoxy that is described in more detail below.

One of the first events contributing directly to cementogenesis is disruption of HERS to reveal the underlying dentin surface. Mechanisms for fenestration of HERS have been attributed to invasive migration of follicular mesenchyme and self-disruption by cells of HERS. Disruption of OEE and then IEE layers provides access for cementoblast precursors to extend cell processes or invade toward the exposed root dentin surface. Follicle cells have demonstrated substantial migratory capacity during root development in rodents as well as humans (Fig. 2B) (Diekwisch, 2001, 2002). Synthesis of collagen fibrils by these migrating follicle cells contributes to both establishment of Sharpey's fibers and to their linkage and continuity with dentin matrix fibers to make a strong attachment at the cementum–dentin junction (CDJ) (Bosshardt and Schroeder, 1996; Ho et al., 2009). Careful study of the structure and materials properties of the CDJ suggests that the graded nature of this region contributes to the ability of the root to accommodate occlusal loading, and the CDJ could be considered a fibrous joint akin to the PDL–cementum interface (Ho et al., 2007). In human teeth, tightly packed "fringe fibers" are important in the continuity of collagen fibers between PDL, cementum, and dentin in the region of acellular cementum (Fig. 2C) (Bosshardt and Schroeder, 1996). While cementum is being deposited and PDL fibers are reorganizing during root elongation, the alveolar bone surrounding the tooth also undergoes changes, increasing in height and incorporating Sharpey's fibers (Fig. 2D).

Cementoblasts surrounding the root dentin surface create a local environment conducive to cementum formation, specifically by directed apposition of hydroxyapatite mineral on the dentin surface to embed PDL fibers to the root. For acellular cementum, extrinsic collagen fibers from the PDL enter into and intermingle with dentin matrix. Cementoblasts produce characteristic noncollagenous ECM proteins, including BSP, OCN, and OPN, which are deposited into spaces between collagen fibers and probably influence hydroxyapatite crystal deposition and crystal growth. The enzyme, tissue-nonspecific alkaline phosphatase (TNAP), with a strong presence in the periodontal region, has been correlated positively with formation of acellular cementum and is a cementoblast and osteoblast marker. Cementoblasts responsible for acellular cementum have been described as fibroblastic–cuboidal in morphology and unipolar in their promotion of mineralization, accounting for the slow, controlled nature of acellular cementum synthesis. It has been remarked that acellular cementum seems to form by a process in which mineral is directed and ECM proteins are secreted concomitantly to the root dentin surface (Fig. 2C). The other mineralized tissues of the dentoalveolar complex—dentin, bone, and even cellular cementum—mineralize through a two-step process in which a nonmineralized matrix is first elaborated (i.e., predentin, osteoid, cementoid) and then undergoes gradual mineralization. Acellular cementum synthesis has been described as a progressive mineralization of the PDL fibers, or more specifically the fringe fibers, on the root surface (Bosshardt, 2005; Nanci and Somerman, 2008). However, the finer mechanisms of acellular cementogenesis await further resolution.

3.5 Cellular Cementum Formation

During eruption, and at about the time the tooth crown enters occlusion, the root is partially completed. The process of cementogenesis is altered to produce the thick, more rapidly forming cellular cementum (CIFC), which marks the apical root (Fig. 1D). Cells producing cellular cementum are cuboidal in morphology, resembling osteoblasts. A portion of these cells are embedded in the cementoid matrix, becoming cementocytes housed in lacunae, which feature a canalicular network of cell processes that reach toward the PDL interface. The stimulus for the change from acellular to cellular or mixed types of cementum remains unknown. It has been observed in rodents that cellular cementum forms around the time the tooth will enter occlusion, which is also a period when HERS proliferation is slowing. Some have hypothesized that cessation or slowing of HERS growth is a signal for cellular cementum, although how HERS is related to cementogenesis remains a matter of debate.

3.6 Tooth Eruption

The physiologic eruption of teeth into the oral cavity coincides with root formation and is necessary for the tooth to move from its developmental position in the bony crypt to its functional position in the occlusal plane. In addition to the ever-lengthening root, the period of eruption is marked by PDL organization and turnover, and bone remodeling, as described in the preceding section. Although not all mechanisms for eruption are understood, a series of clever experiments by multiple investigators have elucidated several aspects of this process [summarized by Wise and King (2008)]. Root lengthening coincides with eruption but is not causative. However, the ectomesenchymal dental follicle that gives rise to the periodontia plays a central role in regulating two aspects critical for eruption. The coronal portion of the follicle increases osteoclast formation by secretion of chemokines, including macrophage colony stimulating factor 1 (CSF1), receptor activator of nuclear factor κB ligand (RANKL), and monocyte chemotactic protein 1 (MCP1), and suppression of osteoprotegerin (OPG), creating an environment conducive to resorption of the bony crypt nearest the tooth crown. The mechanisms working to direct these actions by the coronal follicle are not well understood, but expression of parathyroid hormone–related protein (PTHrP) by the stellate reticulum of the enamel organ has been implicated in directing pro-osteoclastic factors by the follicle, in vivo and in vitro. Failure to activate osteoclasts results in delayed-eruption or nonerupting teeth, as has been observed in a variety of transgenic mice where bone resorption was impaired.

Conversely, the more basal portion of the follicle promotes bone formation by expression of osteogenic factors, including BMP2. Therefore, it is proposed that bone deposition on the crypt floor provides a motive force for eruption, while bone resorption creates the eruption pathway, allowing for emergence of the tooth. Thus, the dental follicle emerges not only as a formative tissue for the periodontia, but also as an orchestrator of the process of tooth eruption.

3.7 Development of Alveolar Bone and Periodontal Ligament

The alveolar bone proper consists of cortical and trabecular bone lining the tooth socket, or alveolus (Fig. 1). Alveolar bone is an extension of mandibular or maxillary basal bone, craniofacial bone that mineralizes by the process of intramembranous bone formation. Alveolar bone is part of the attachment complex of the periodontium, and like cells of the PDL and cementum, alveolar bone osteoblasts are derived from CNC cells, giving them ectomesenchymal origin [Fig. 2A; see also Chai et al. (2000)]. Alveolar bone forms in relation to teeth, dependent on interactions between the growing tooth root, the PDL in the midst of reorganization from the dental follicle, and outer regions of the dental follicle with osteogenic potential. As the erupting tooth enters the oral cavity and approaches its functional position, alveolar bone stabilizes the tooth to the basal bone by incorporating *Sharpey's fibers*, collagen fibers embedded from the PDL region (Fig. 2D). Like other bones of the skeleton, alveolar bone serves a biomechanical function in accommodating forces placed on teeth and transmitted by the PDL. The complex relationships between tooth roots, remodeling PDL and embedded Sharpey's fibers, and unique vector forces encountered result in bone with very complex organization lacking ordered arrangement of the Haversian systems found in long bones, and undergoing constant and rapid remodeling [see Sodek and McKee (2000) for more detailed discussion and citations].

The PDL arises from reorganization of the dental follicle during the period of root formation and tooth eruption. The unorganized and loosely arranged connective tissue of the follicle begins to organize by production of collagen fiber bundles that embed in the surfaces of cementum and alveolar bone (Sharpey's fibers; Fig. 2D) and provide continuity across the PDL space. Orientation of PDL fiber groups changes rapidly during eruptive tooth movement, the result of active remodeling. Remodeling of both PDL and bone is necessary as the tooth moves through the eruption pathway. During this process, the alveolar crest is remodeled to reach the level of the cemento–enamel junction (CEJ), and PDL fiber bundles are organized so as to withstand masticatory forces.

3.8 Remodeling and Homeostasis in the Tooth Root

The PDL and alveolar bone are dynamic tissues in perpetual turnover. These tissues form a specialized, adaptable unit which is mechanically sensitive and responds to the functional demands of the tooth proper during processes such as mastication and positional adaptation. The ability of the PDL to remain an elastic and unmineralized tissue, which is critical for its proper function, probably depends partially on the ability of PDL cells to influence cell behavior in the surrounding alveolar processes. For example, orthodontic tooth movement occurs by targeted remodeling of bone through resorption on the compression side and apposition on the tension side while PDL space is ultimately maintained (Wise and King, 2008). When the ability to modulate bone resorption and remodeling is lost, the PDL space can be compromised, and ankylosis may result.

Despite housing cells with osteogenic and cementogenic potential, the PDL displays a remarkable ability to resist mineralization and is an attribute critical for maintenance of its biomechanical function and indicative of the presence of factors that inhibit mineralization. These probably include putative negative regulators of mineral growth (e.g., OPN, osteocalcin, matrix gla protein, periostin, proteoglycans) as well as factors that indirectly prevent mineralization by influencing cell differentiation [e.g., bone morphogenetic antagonists BMP3, gremlin, and periodontal ligament-associated protein 1 (PLAP1, or asporin)]. Evidence to date suggests that the PDL relies on multiple redundant factors to maintain an unmineralized state.

The mineralized tissues of the tooth proper (enamel, dentin, and cementum) do not undergo tissue turnover as part of ongoing homeostasis. The primary dentin produced during tooth formation is succeeded by secondary dentin in an ongoing but much slower process of dentinogenesis that continues past tooth eruption and occlusion. Trauma, including severe caries, excessive attrition, or microbial infection, may spur formation of tertiary dentin. In cases where the insult injures local odontoblasts, the tertiary dentin produced is called *reparative* or *reactionary dentin*, and the requirement for newly differentiated odontoblasts was long held as evidence of a stem cell population residing in the pulp compartment. This hypothesis came to fruition when pluripotent cells were identified in pulp chambers of erupted teeth (Gronthos et al., 2000) as well as in root-associated apical papillae of developing teeth (Sonoyama et al., 2006).

The cellular type of cementum, often found as part of the healing process after removal of diseased cementum, is referred to as *reparative* or *regenerative cementum* (Bosshardt and Sculean, 2009). As with reparative dentin, the potential for cementum repair or regeneration supports the presence of multipotent progenitor cells in the periodontal region, a hypothesis proven with isolation and characterization of PDL stem cells capable of differentiating to osteo- and cementoblast-like cells (Seo et al., 2004). Like dentin, neither acellular nor cellular cementum participates in turnover, although cellular cementum especially may undergo small resorption and repair events. Some findings have indicated that maintenance of cellular cementum parallels bone in some respects. In a mouse model of hyperocclusion, cellular cementum apposition was disturbed and the tissue was susceptible to resorption, much like the adjacent alveolar bone proper (Walker et al., 2008). Orthodontic tooth movement sometimes results in root resorption, which can strike both types of cementum, with existing data suggestive of more frequent or more severe destruction of apical cellular cementum versus acellular cementum. As both types of cementum do not undergo regular turnover and osteoclasts more often target alveolar bone than cementum in pathological situations, it has been hypothesized that cementoblasts have a protective, antiresorptive role on the root surface. The role of cementoblasts and cementocytes in osteoclast biology remains to be defined.

4 SIGNALING AND INDUCTIVE INFLUENCES IN ROOT DEVELOPMENT

As we have described the major cellular and morphological events in tooth root formation, it is of value to summarize the current state of knowledge on the signaling pathways involved, as these will probably be important in stem cell–based strategies for regenerating root tissues or bioengineering roots. Some of the major signaling and inductive influences on tissue of the root are summarized in Table 2 and discussed in more detail below.

As highlighted in several chapters of this book, mouse models have enabled researchers to map genes, proteins, and signaling pathways required for proper tooth development. However, as root follows crown, transgenic manipulation arresting crown development at early stages necessarily precludes any root development and insight into the role of that particular gene in root formation. More recently, using a variety of newer techniques and strategies, including conditional knockout (tissue-targeted or temporal) tissue-specific overexpression of targeted genes and

TABLE 2 Signaling Pathways and Candidate Factors Associated with Tooth Root Formation

Pathway	Factor(s)[a]	Known Functions	Selected References
TGFβ superfamily	TGFβ1	TGFβ1 is expressed continuously in teeth during development. Overexpression of TGF β1 in odontoblasts interrupted differentiation and root formation. Knockout of TGFβ/BMP signal transduction protein Smad4 in odontoblasts also resulted in disrupted odontoblast differentiation and short roots.	Thyagarajan et al., 2001; Gao et al., 2009
	BMPs	BMPs 2, 4, and 7 have been localized to root tissues and may be involved in cell differentiation. BMP antagonists BMP3, noggin, and gremlin may be involved in resistance of PDL to mineralization.	Gao et al., 1998; Thomadakis et al., 1999; Yamashiro et al., 2003; Plikus et al., 2005; Nagatomo et al., 2008
	Msx2	BMP expression by root dental papilla and odontoblasts, and adjacent Msx2 expression in HERS suggests epithelial–mesenchymal interactions in root dentinogenesis. Msx2 ablation results in severe problems with root growth and morphogenesis, as well as alveolar bone modeling.	Yamashiro et al., 2003; Aiorb et al., 2007
	Runx family	Epithelial Runx1, 2, and 3 play a role in sustaining BMP-FGF signaling that supports maintenance and differentiation of the epithelial stem cell niche of the cervical loop, also intersecting with Shh signaling in this role. Modulation of this cascade may be related to crown-to-root transition.	Kurosaka et al., 2011
	Osterix	Cells of the tooth root express this osteogenic transcription factor. Conditional knockout of Osterix impaired cellular cementum formation, while overexpression increased cellular cementum.	Cao et al., 2012

FGF	FGF 10 and others	Decrease in mesenchymal FGF10 is associated with the depletion of the stem cell niche in the cervical loop and transition to HERS root epithelium.	Yokohama–Tamaki et al., 2006; Tummers and Thesleff, 2003
HH	SHH	SHH signaling within dental epithelium is involved in cell organization and tissue morphology in root development	Gritli-Linde et al., 2002
	Nfic	Transcription factor produced by pre-odontoblasts in response to Shh and indispensible for root dentin formation. Nfic induction lies downstream of TGFβ/BMP-mediated Smad4 signaling.	Steele-Perkins et al., 2003; Park et al., 2007; Lee et al., 2009; Huang et al., 2010; Gao et al., 2009
	Patched 1 Smoothened Gli1	SHH response elements expressed by dental papilla adjacent to HERS are probably involved in epithelial–mesenchymal interactions in root development.	Nakatomi et al., 2006
WNT	Wnts	Wnt signaling elements are expressed in HERS and developing pulp and periodontal tissues. Constitutive Wnt signaling and overexpression of Wnt signaling inhibitor DKK1 impaired both cell differentiation and root formation. Wnt and BMP crosstalk probably plays a role in differentiation.	Lohi et al., 2010; Kim et al., 2011; Rooker et al., 2010; Han et al., 2011
Other factors	IGF1	Drives growth of the HERS required for shaping and elongation of the tooth root.	Fujiwara et al., 2005
	Dlx genes	Homeobox gene *Dlx2* has been identified in cementoblasts. *Dlx3* mutations are associated with taurodontism.	Lezot et al., 2000; Choi et al., 2010; Wright, 2007

[a]Factors are organized by the relevant signal pathway, although in some cases, multiple signaling pathways are implicated for a particular factor. More details are available in the text and the selected references.

kidney capsules as an incubator for tooth root organogenesis, clues are emerging about genes modulating root development processes.

In this section we outline current knowledge regarding the role of classic signaling pathways in root development, discuss other potential inductive influences in root tissue formation, and touch on root pathology observation from humans and how these provide insight for understanding mechanisms modulating root development and, subsequently, strategies for root regeneration.

4.1 Signaling Pathways and Root Development

TGFβ signaling plays a large role in directing epithelium toward a crown (i.e., ameloblast) or root (i.e., HERS) fate, as the stem cells niche in the cervical loop is maintained via signaling interactions with the surrounding mesenchyme (Tummers and Thesleff, 2008), Mesenchymal FGF10 supports epithelial Notch signaling and continued stem cell proliferation, while diminution of FGF10 is linked to collapse of the cervical loop (to HERS) and transition to root fate (Yokohama-Tamaki et al., 2006). Interestingly, this pattern is maintained in continuously growing teeth with root analogs, including mouse incisors and vole molars (Harada et al., 2002; Tummers and Thesleff, 2003), where epithelial *Runx* genes are also implicated in maintaining BMP-FGF signaling loop critical for the epithelial stem cell niche (Kurosaka et al., 2011). After epithelial transition to HERS, cell proliferation is key to root length, demonstrated by studies showing a potent effect of insulin-like growth factor 1 (IGF1) on regulating HERS cell division (Fujiwara et al., 2005).

Recent data from several groups have highlighted a role for epithelial–mesenchymal signaling during early stages of root dentin development. In a seminal publication in 2003 (Steele-Perkins et al., 2003) it was reported that mice lacking transcription factor nuclear factor I/C (*Nfic*) lacked tooth roots, while crown formation was unperturbed (Fig. 3A). Further studies demonstrated that *Nfic* is essential for regulating the differentiation and cell function of odontoblasts, specifically during root development. Loss of *Nfic* resulted in dysmorphic odontoblasts lacking expression of markers such as dentin sialophosphoprotein (*Dspp*) (Park et al., 2007), as well as alterations in Smad phosphorylation and factors associated with cell cycle and cell death (Lee et al., 2009). Subsequent studies provided insight into the signaling pathway involved in *Nfic*-mediated control of root dentin development, implicating Smad4, a factor critical for elongation of HERS to guide root development. As a downstream target of Smad4-mediated TGFβ/BMP signaling in normal root development, Shh expression by HERS subsequently triggers expression of *Nfic* in CNC-derived mesenchymal cells (Huang et al., 2010). Shh acts through Gli1 to promote expression of *Nfic* in the dental mesenchyme. HERS induction of root odontoblast differentiation is in line with the epithelial–mesenchymal signaling that occurs to drive crown formation, although in this specific example of *Nfic*-associated signaling, crown and root are clearly operating under different developmental influences.

These studies on the function of *Nfic* have tapped into a parallel literature on signaling in root. Overexpression of TGFβ1 in odontoblasts resulted in a tooth phenotype similar to that seen in *Nfic*-null mice, both involving defects in Smad4 phosphorylation (Thyagarajan et al., 2001). Furthermore, Gao et al. (2009) reported that disruption of Smad4 in odontoblasts resulted in disruption in tooth root development, with associated keratocystic odontogenic tumors, confirming the importance of TGFβ family

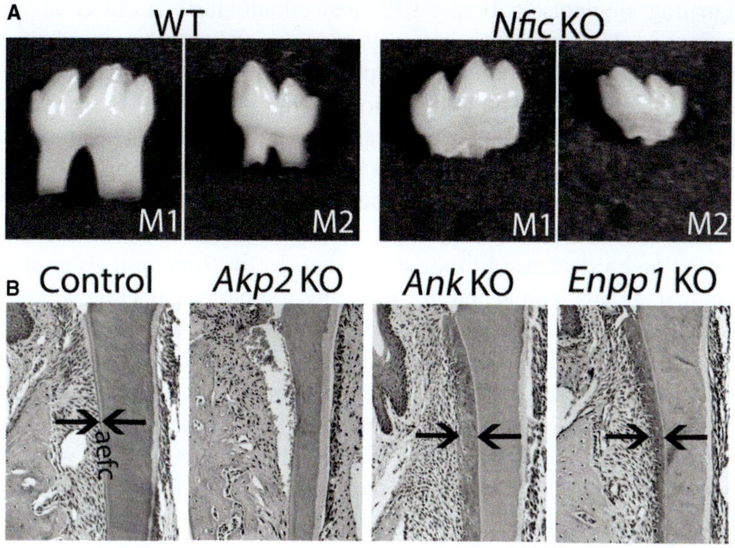

FIGURE 3 Mouse models for root development and associated pathologies. (A) Ablation of the transcription factor gene *Nfic* causes arrested development of tooth root formation in mice. The root defect is evident in both first (M1) and second molars (M2) at 15 days postnatal. (Courtesy of R. M. Gronostajski, State University of New York at Buffalo.) (B) Dysregulation in mineral ions such as phosphate and pyrophosphate has dramatic effects on mineralized tissues of the root. Increased pyrophosphate in the *Akp2*-knockout mice disrupts the mineralization of acellularcementum (aefc), while reduced pyrophosphate in *Ank*- or *Enpp1*-knockout mice drastically increases aefc formation (indicated by black arrows). Histology from 19 to 26 days postnatal.

signaling in root development. Recall that Smad4 is proposed to operate upstream from *Nfic*. Dental papilla cells adjacent to HERS express Shh target genes Patched1 (*Ptc1*), Smoothened (*Smo*), and *Gli1* (Nakatomi et al., 2006). Shh signaling is critical for tooth morphogenesis, as evidenced by the severe crown and root malformation following removal of *Smo* from the dental epithelium (Gritli-Linde et al., 2002). Shh controls expression of *Nfic* through the transcription factor *Gli1*. Thus, the upstream factors controlling *Nfic* were identified prior to findings of a defective root phenotype in *Nfic*-null mice. This appears to be an attractive pathway to explore to identify genes associated with defective tooth root development in humans.

Mapping of additional signaling molecules during root formation further supports a HERS–odontoblast communication. For example, coincident expression of Msx2 by HERS and Msx 1 and 2, and BMPs 2, 4, and 7 by early odontoblasts is suggestive of crosstalk parallel to that in crown (Yamashiro et al., 2003), and ablation of Msx2 resulted in severe tooth malformation, including dentin and periodontal defects (Aïoub et al., 2007).

Other candidate genes linked to abnormal root development in mice include agonists and antagonists modulating Wnt and BMP signaling, where there is strong evidence of crosstalk between these pathways Lohi et al., 2010; (Plikus et al., 2005; Liu et al., 2008; Nagatomo et al., 2008; Fujimori et al., 2010; Han et al., 2011). The Wnt signaling pathway has been implicated in controlling root dentin formation based on the presence of Wnt signaling elements in both HERS and odontoblasts (Lohi et al., 2010) and from alterations in dentinogenesis in transgenic mice. Wnt pathway activity has been demonstrated in developing periodontia, probably linked to cell processes such as proliferation and cell fate in the PDL region (Lohi et al., 2010; Rooker et al., 2010; Kim et al., 2011), although more studies are required to understand more fully how Wnt contributes to the development of root tissues. The overarching importance of the Wnt pathway in postnatal tooth development was underscored when overexpression of Wnt inhibitor (Dickkopf-related protein 1) DKK1 resulted in short roots, enlarged pulp chambers, and a decrease in the rate of root dentin formation (Han et al., 2011).

The proximal signals responsible for directing migration, attachment, and differentiation of cementoblasts remain elusive. Cemento-inductive roles have variously been ascribed to signals from HERS cells or odontoblasts, epithelial basement membrane components, or the surface of the mineralized dentin itself, although at present this question remains unanswered. The concept that HERS has an inductive role in initiating cementogenesis is appealing in the sense that it parallels the well-established epithelial–mesenchymal communication that drives crown formation (as well as the development of other tissues, e.g., limb, hair). It has been proposed that epithelium-derived enamel matrix proteins (EMPs) such as amelogenin or ameloblastin may play such a role in root formation. Numerous investigators over the last decades have focused on the localization and role of EMPs in HERS or cementum; however, the literature lacks consensus on this issue. Despite these lingering questions, a clinical product based on the concept of an epithelial role in cementogenesis was developed and is in clinical use with varied success (Giannobile and Somerman, 2003; Bosshardt and Sculean, 2009), although it has proven difficult to identify the bioactive component of this mixture that may be responsible.

Early studies reported widespread expression of TGFβ1 and BMPs in the developing periodontia (Gao et al., 1998; Thomadakis et al., 1999), and numerous in vitro experiments demonstrate that such bioactive factors can affect periodontal cell phenotypes. However, later reports call into question the presumptive roles for BMP in cementum formation. Dental follicle cells and cementoblasts were reported not to express BMP2, 4, and 7, though antagonist BMP3 was found (Yamashiro et al., 2003). BMP antagonist noggin has been localized to the developing periodontia (Kim et al., 2007), and noggin overexpression by epithelium resulted in severe disturbance of root morphogenesis, yet cementum formed (Plikus et al., 2005). Similarly, overexpression of BMP antagonist gremlin affected root dentin, whereas cementum was not overtly altered (Nagatomo et al., 2008).

Although little is known about the differentiation program in cementoblasts, their similarity to osteoblasts has suggested a parallel path. Currently, it remains unclear whether osteoblast master transcription factor Runx2 is essential for cementogenesis, but downstream transcription factor osterix was found to control cellular cementum in a manner parallel to its role in osteogenesis (Cao et al., 2012).

Signaling contributing to PDL establishment and organization is not well understood at present. As outlined above, PDL development occurs in tandem with root formation, cementogenesis, and alveolar bone remodeling; thus, there is a complex developmental process involving hard and soft tissues. Processes that are probably critical to proper PDL formation include those involved in collagen secretion, organization, and turnover, and those involved in maintenance of hard–soft tissue interfaces. For example, loss of membrane-type matrix metalloproteinase 1 (MT-MMP1), a collagen-degrading enzyme, results in severely inhibited root development and PDL establishment (Beertsen et al., 2002). As mentioned previously, signals to suppress osteo- and cementogenic differentiation and maintain progenitor and stem cells niches are probably important in PDL homeostasis, although these require further study for a more complete understanding of the dynamic nature of mineral regulation of this tissue. Alveolar bone is subject to developmental and postnatal signaling similar to that at other bone sites, and these are beyond the scope of this chapter. However, it is intriguing that differences in craniofacial versus axial and appendicular bone cells have begun to be recognized (Akintoye et al., 2006).

Thus, the major signaling pathways involved in crown formation (TGFβ, FGF, hedgehog/Shh, and Wnt) have been implicated in root formation, and some (e.g., the Shh-Nfic) have been well elucidated. An additional layer of signal regulation seems to lie in families of microRNAs, which have been shown to modulate signaling during crown formation (Michon et al., 2010) and are expressed differentially between crown and root epithelium (Jheon et al., 2011), although specific roles in root development remain unresolved at present.

4.2 Mineral Metabolism and Root Development

Classic hormonal regulators of mineral metabolism, vitamin D and parathyroid hormone, can affect all the mineralized tissues of the dentoalveolar complex when homeostasis is severely disrupted (Foster et al., 2008). More recent attention has focused on the phosphate-regulating hormone fibroblast growth factor 23 (FGF 23) and associated factors, with profound effects reported on the dentoalveolar complex when such factors are disrupted. For example, hypophosphatemic *Hyp* (mutation in the phosphate-regulating gene with homology to endopeptidases on the X chromosome, *Phex*) and *Dmp1*-knockout mice, and hyperphosphatemic *Fgf23*-knockout mice, feature severe defects in bone and dentin, with milder defects identified in acellular cementum (Foster et al., 2008; Ye et al., 2008; Fong et al., 2009; Chu et al., 2010). Effects of phosphate metabolism disorders on dentition are beginning to be described. For example, loss of function of GalNAc transferase 3 (GALNT3) decreases FGF23 function and has been identified as a cause of familial tumoral calcinosis (TC). Patients with TC have presented with extensive calcification of the pulp cavity (Fig. 4C).

The condition hypophosphatasia (HPP) is an inherited deficiency in tissue-nonspecific alkaline phosphatase (TNAP) which hydrolyzes the mineralization inhibitor pyrophosphate. HPP case reports going back decades suggest that compromised periodontal attachment resulted in exfoliation of teeth. Studies in mice lacking TNAP (Akp2 or Alp1KO) revealed that acellular cementum was almost completely abolished and Sharpey's fibers poorly attached to the root [Fig. 3B; Beertsen et al. (1999)]. The sensitivity of cementum to lack of TNAP is well demonstrated by a tooth-specific subtype of HPP, odontohypophosphatasia, wherein the dental phenotype is the sole manifestation (Mornet, 2007). The importance of mineral metabolism for acellular cementum formation has been underscored by mouse models wherein pyrophosphate levels are reduced. Ablation of either progressive ankylosis protein (ANK) or ectonucleotide pyrophosphatase/phosphodiesterase I (NPP1) causes several-fold expanded acellular cementum (Fig. 3B) (Nociti et al., 2002; Foster et al., 2011).

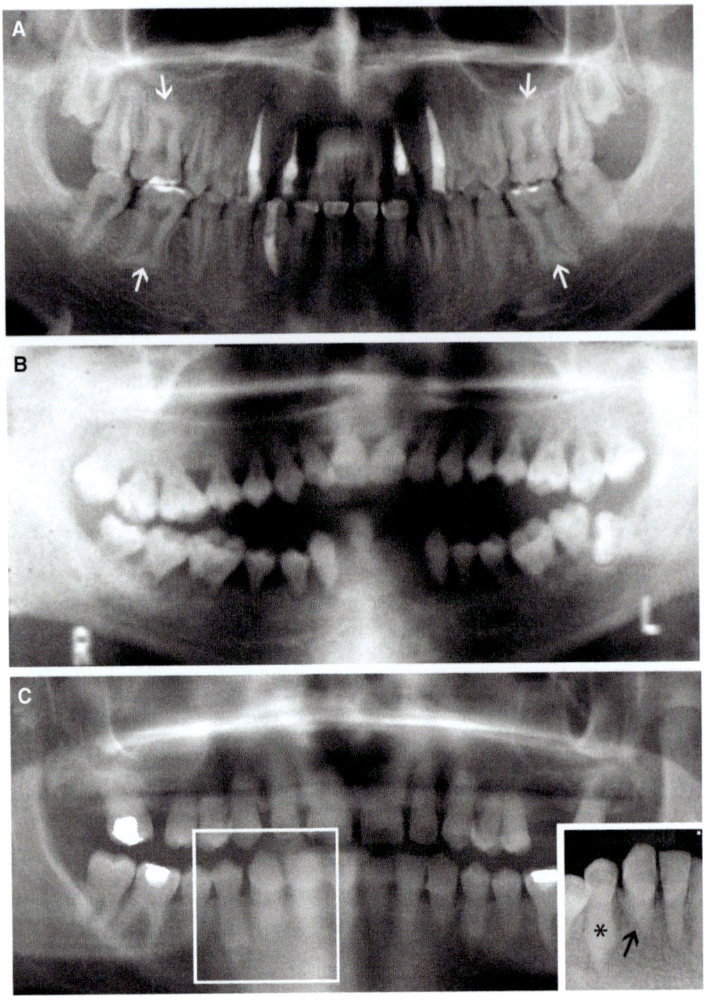

FIGURE 4 Root pathologies: taurodontism, short roots, and other variations. (A) Tricho-dento-osseous (TDO) syndrome is associated with taurodontism, as illustrated in this radiograph featuring severe hypertaurodontism of the first permanent molars (white arrows), with pyramid-shaped second molars. (Courtesy of J. T. Wright, University of North Carolina, Chapel Hill.) (B) Panoramic radiograph of Sponastrime dysplasia patient with multiple short tooth roots, which can lead to premature loss of the permanent dentition. [Reprinted from Gripp et al. (2008), with permission.] (C) Dysfunction of phosphate metabolism, as in hyperphosphatemia, can lead to dystrophic calcification of the pulp chamber, as shown in this radiograph taken from a patient with a Galnt3 mutation and with clinical findings consistent with familial tumoralcalcinosis. Further examination of mandibular incisors (the area outlined in the larger panoramic radiograph is highlighted in the inset, a periapical radiograph) suggested additional root abnormalities, including hypercementosis (star) and root resorption (black arrow). (Courtesy of M. T. Collins, National Institute of Dental and Craniofacial Research, Bethesda, MD.)

4.3 Root Pathology and Insights into Developmental Regulation

There are numerous syndromes, genetic mutations, and idiopathic conditions linked to variations in root development in humans. Many of these conditions affect both crowns and roots, including size, shape, and tissue composition. Some conditions are targeted to roots, either by some special expression or role for the associated factor in root versus crown formation, or because the severity of the defect is greater and thus more evident in the roots.

Numerous conditions are recognized to affect root morphology by impairing the function of HERS and odontoblasts. In multirooted teeth, invagination of the HERS defines the area of furcation where multiple roots separate during apical growth. The condition *taurodontism* ("bull-like teeth") results from delayed or abnormal HERS invagination, causing elongated root chambers, apical displacement of the pulp floor, and lack of constriction of the cementum–enamel junction [see reviews by Hu and Simmer (2007) and Wright (2007)]. Several syndromic conditions targeting epithelial tissues have been linked to taurodontism, including ectodermal dysplasias, amelogenesis imperfectas, and tricho-dento-osseous (TDO) syndrome [see Wright (2007) for more details].

Taurodontism, as well as more severe dysmorphic dental features, have been cataloged with ectodermal dysplasia, with a molecular mechanism probably related to interactions between EDA and BMP signaling pathways. Mutations in transcription factor distal-less homeobox gene (*Dlx3)* have been linked to enamel defects (e.g., amelogenesis imperfecta) as well as taurodontism in TDO syndrome (Fig. 4A) (Wright et al., 2008; Choi et al., 2010).

In addition to taurodontism, there is a growing list of inherited and idiopathic conditions linked to abnormally short or malformed roots. These include collagen disorders such as dentinogenesis imperfecta, dentin dysplasia type I, and Ehlers–Danlos syndrome, and conditions of varying etiology such as Rothmund–Thompson syndrome, Sponastrime dysplasia (Fig. 4B), Coffin–Lowry syndrome, and Hallermann–Streiff syndrome.

As described above, loss of *Nfic* results in arrested tooth root development in mice (Fig. 3A). This gene has been mapped to chromosome 19 in humans, although to date an association between this gene and tooth root defects in humans has not been reported. Several of the root defects noted above, such as short and thin types of root hypoplasia in the absence of other dental involvement, are candidates for *Nfic*-associated conditions.

These are but a subset of many examples of conditions linked to root hypoplasia, dysmorphism, or improper mineralization, and emphasize the point that there are many ways in which the process of root formation may be disrupted.

5 CONCLUSIONS AND FUTURE DIRECTIONS

Destruction of tooth root tissues, including dentin, acellular and cellular cementum, PDL, and bone, can result from trauma, periodontal disease, or hereditary conditions compromising tooth root tissue integrity. Predictable regeneration of root tissues and surrounding support tissues to normal structure and function is a desirable goal, with recognized challenges. As outlined in this chapter, much progress has been made in elucidating tooth root formation, although fundamental questions remain about the cells, signals, and events that contribute to this process. Human and animal studies show in principle that regeneration of the periodontium is possible; however, the complex nature of the soft–hard tissue interfaces has proven difficult to restore predictably. To date, there has been moderate success in periodontal regeneration with the use of a wide range of agents, including bone grafts, bone graft substitutes, antimicrobials, biologics, and biomimetics, with additional promising agents currently under investigation.

It has been recognized for decades that cells migrating to or present in the local periodontal region have the capacity to foster regeneration of tissues lost as a consequence of disease. As you will read in other chapters, more recent evidence has now corroborated that such cells have stem-like properties, and significant research efforts have now been focused on characterizing these stem cells, isolating and maintaining the cells, manipulating their multipotent properties, and identifying the ideal protocols and agents for use in regenerative therapies. With the great potential for use of these cells in regenerative therapies and tooth bioengineering, it is truly an exciting time to be involved in craniofacial research.

Acknowledgments

The authors would like to acknowledge and profoundly thank the pioneers in the fields of tooth development and biomineralization whose work is summarized here but who are too numerous to name because of space limitations.

REFERENCES

Aïoub M, Lézot F, Molla M, Castaneda B, Robert B, Goubin G, et al. (2007). Msx2$^{-/-}$ transgenic mice develop compound amelogenesis imperfecta, dentinogenesis imperfecta and periodental osteopetrosis. *Bone* 41(5):851–859.

Akintoye SO, Lam T, Shi S, Brahim J, Collins MT, Robey PG (2006). Skeletal site-specific characterization of orofacial and iliac crest human bone marrow stromal cells in same individuals. *Bone* 38(6):758–768.

Beertsen W, VandenBos T, Everts V (1999). Root development in mice lacking functional tissue non-specific alkaline phosphatase gene: inhibition of acellular cementum formation. *J Dent Res* 78(6):1221–1229.

Beertsen W, Holmbeck K, Niehof A, Bianco P, Chrysovergis K, Birkedal-Hansen H, et al. (2002). On the role of MT1-MMP, a matrix metalloproteinase essential to collagen remodeling, in murine molar eruption and root growth. *Eur J Oral Sci* 110(6):445–451.

Bosshardt DD (2005). Are cementoblasts a subpopulation of osteoblasts or a unique phenotype? *J Dent Res* 84(5):390–406.

Bosshardt DD, Schroeder HE (1996). Cementogenesis reviewed: a comparison between human premolars and rodent molars. *Anat Rec* 245(2):267–292.

Bosshardt DD, Sculean A (2009). Does periodontal tissue regeneration really work? *Periodontol 2000* 51:208–219.

Cao Z, Zhang H, Zhou X, Han X, Ren Y, Gao T, Xiao Y, de Crombrugghe B, Somerman MJ, Feng JQ (2012). Genetic evidence for the vital function of Osterix in cementogenesis. *J Bone Miner. Res* 27(5):1080–1092.

Chai Y, Jiang X, Ito Y, Bringas P, Jr., Han J, Rowitch DH, et al. (2000). Fate of the mammalian cranial neural crest during tooth and mandibular morphogenesis. *Development* 127(8):1671–1679.

Choi SJ, Song IS, Feng JQ, Gao T, Haruyama N, Gautam P, et al. (2010). Mutant *DLX 3* disrupts odontoblast polarization and dentin formation. *Dev Biol* 344(2):682–692.

Chu E, Fong H, Blethen F, Tompkins K, Foster B, Yeh K, et al. (2010). Ablation of systemic phosphate-regulating gene fibroblast growth factor 23 (Fgf23) compromises the dentoalveolar complex. *Anat Rec* (*Hoboken*) 293(7):1214–1226.

Diekwisch TG (2001). The developmental biology of cementum. *Int J Dev Biol* 45(5–6):695–706.

Diekwisch TG (2002). Pathways and fate of migratory cells during late tooth organogenesis. *Connect Tissue Res* 43(2–3):245–256.

Fong H, Chu EY, Tompkins KA, Foster BL, Sitara D, Lanske B, et al. (2009). Aberrant cementum phenotype associated with the hypophosphatemic *Hyp* mouse. *J Periodontol* 80(8):1348–1354.

Foster BL, Popowics TE, Fong HK, Somerman MJ (2007). Advances in defining regulators of cementum development and periodontal regeneration. *Curr Top Dev Biol* 78:47–126.

Foster BL, Tompkins KA, Rutherford RB, Zhang H, Chu EY, Fong H, et al. (2008). Phosphate: known and potential roles during development and regeneration of teeth and supporting structures. *Birth Defects Res C* 84(4):281–314.

Foster BL, Nagatomo KJ, Bamashmous SO, Tompkins KA, Fong H, Dunn D, et al. (2011). The progressive ankylosis protein regulates cementum apposition and extracellular matrix composition. *Cells Tissues Organs* 194(5):382–405.

Fujimori S, Novak H, Weissenböck M, Jussila M, Gonçalves A, Zeller R, et al. (2010). Wnt/β-catenin signaling in the dental mesenchyme regulates incisor development by regulating Bmp4. *Dev Biol* 348(1):97–106.

Fujiwara N, Tabata MJ, Endoh M, Ishizeki K, Nawa T (2005). Insulin-like growth factor-I stimulates cell proliferation in the outer layer of Hertwig's epithelial root sheath and elongation of the tooth root in mouse molars in vitro. *Cell Tissue Res* 320(1):69–75.

Gao J, Symons AL, Bartold PM (1998). Expression of transforming growth factor-beta 1 (TGF-beta1) in the developing periodontium of rats. *J Dent Res* 77(9):1708–1716.

Gao Y, Yang G, Weng T, Du J, Wang X, Zhou J, et al. (2009). Disruption of Smad4 in odontoblasts causes multiple keratocystic odontogenic tumors and tooth malformation in mice. *Mol Cell Biol* 29(21):5941–5951.

Giannobile WV, Somerman MJ (2003). Growth and amelogenin-like factors in periodontal wound healing: a systematic review. *Ann Periodontol* 8(1):193–204.

Gripp KW, Johnson C, Scott CI Jr, Nicholson L, Bober M, Butler MG, Shaw L, Gorlin RJ (2008). Expanding the phenotype of SPONASTRIME dysplasia to include short dental roots, hypogammaglobulinemia, and cataracts. *Am J Med Genet A* 146A(4):468–473.

Gritli-Linde A, Bei M, Maas R, Zhang XM, Linde A, McMahon AP (2002). Shh signaling within the dental epithelium is necessary for cell proliferation, growth and polarization. *Development* 129(23):5323–5337.

Gronthos S, Mankani M, Brahim J, Robey PG, Shi S (2000). Postnatal human dental pulp stem cells (DPSCs) in vitro and in vivo. *Proc Natl Acad Sci USA* 97(25):13625–13630.

Han XL, Liu M, Voisey A, Ren YS, Kurimoto P, Gao T, et al. (2011). Post-natal effect of overexpressed DKK1 on mandibular molar formation. *J Dent Res* 90(11):1312–1317.

Harada H, Toyono T, Toyoshima K, Yamasaki M, Itoh N, Kato S, et al. (2002). FGF10 maintains stem cell compartment in developing mouse incisors. *Development* 129(6):1533–1541.

Ho SP, Marshall SJ, Ryder MI, Marshall GW (2007). The tooth attachment mechanism defined by structure, chemical composition and mechanical properties of collagen fibers in the periodontium. *Biomaterials* 28(35):5238–5245.

Ho SP, Yu B, Yun W, Marshall GW, Ryder MI, Marshall SJ (2009). Structure, chemical composition and mechanical properties of human and rat cementum and its interface with root dentin. *Acta Biomater* 5(2):707–718.

Hu J, Simmer J (2007). Developmental biology and genetics of dental malformations. *Orthod Craniofac Res* 10(2):45–52.

Huang X, Xu X, Bringas P, Hung YP, Chai Y (2010). Smad4-Shh-Nfic signaling cascade-mediated epithelial-mesenchymal interaction is crucial in regulating tooth root development. *J Bone Miner Res* 25(5):1167–1178.

Jheon AH, Li CY, Wen T, Michon F, Klein OD (2011). Expression of microRNAs in the stem cell niche of the adult mouse incisor. *PLoS One* 6(9):e24536.

Kim JY, Cho SW, Hwang HJ, Lee MJ, Lee JM, Cai J, et al. (2007). Evidence for expansion-based temporal BMP4/Noggin interactions in specifying periodontium morphogenesis. *Cell Tissue Res* 330(1):123–132.

Kim TH, Lee JY, Baek JA, Lee JC, Yang X, Taketo MM, et al. (2011). Constitutive stabilization of β-catenin in the dental mesenchyme leads to excessive dentin and cementum formation. *Biochem Biophys Res Commun* 412(4):549–555.

Kurosaka H, Islam MN, Kuremoto K, Hayano S, Nakamura M, Kawanabe N, et al. (2011). Core binding factor beta functions in the maintenance of stem cells and orchestrates continuous proliferation and differentiation in mouse incisors. *Stem Cells* 29(11):1792–1803.

Lee DS, Park JT, Kim HM, Ko JS, Son HH, Gronostajski RM, et al. (2009). Nuclear factor I-C is essential for odontogenic cell proliferation and odontoblast differentiation during tooth root development. *J Biol Chem* 284(25):17293–17303.

Lézot F, Davideau JL, Thomas B, Sharpe P, Forest N, Berdal A (2000). Epithelial Dlx-2 homeogene expression and cementogenesis. *J Histochem Cytochem* 48(2):277–284.

Liu F, Chu EY, Watt B, Zhang Y, Gallant NM, Andl T, et al. (2008). Wnt/beta-catenin signaling directs multiple stages of tooth morphogenesis. *Dev Biol* 313(1):210–224.

Lohi M, Tucker AS, Sharpe PT (2010). Expression of Axin2 indicates a role for canonical Wnt signaling in development of the crown and root during pre- and postnatal tooth development. *Dev Dyn* 239(1):160–167.

Michon F, Tummers M, Kyyrönen M, Frilander MJ, Thesleff I (2010). Tooth morphogenesis and ameloblast differentiation are regulated by micro-RNAs. *Dev Biol* 340(2):355–368.

Mornet E (2007). Hypophosphatasia. *Orphanet J Rare Dis* 2:40.

Nagatomo KJ, Tompkins KA, Fong H, Zhang H, Foster BL, Chu EY, et al. (2008). Transgenic overexpression of gremlin results in developmental defects in enamel and dentin in mice. *Connect Tissue Res* 49(6):391–400.

Nakatomi M, Morita I, Eto K, Ota MS (2006). Sonic hedgehog signaling is important in tooth root development. *J Dent Res* 85(5):427–431.

Nanci A, Somerman M (2008). Periodontium. In: *Ten Cate's Oral Histology: Development, Structure, and Function*, Nanci A, ed., Elsevier St., Louis, MO, pp. 239–267.

Nociti FH, Jr., Berry JE, Foster BL, Gurley KA, Kingsley DM, Takata T, et al. (2002). Cementum: a phosphate-sensitive tissue. *J Dent Res* 81(12):817–821.

Park JC, Herr Y, Kim HJ, Gronostajski RM, Cho MI (2007). Nfic gene disruption inhibits differentiation of odontoblasts responsible for root formation and results in formation of short and abnormal roots in mice. *J Periodontol* 78(9):1795–1802.

Plikus MV, Zeichner-David M, Mayer JA, Reyna J, Bringas P, Thewissen JG, et al. (2005). Morphoregulation of teeth: modulating the number, size, shape and differentiation by tuning Bmp activity. *Evol Dev* 7(5):440–457.

Rooker SM, Liu B, Helms JA (2010). Role of Wnt signaling in the biology of the periodontium. *Dev Dyn* 239(1):140–147.

Seo B, Miura M, Gronthos S, Bartold P, Batouli S, Brahim J, et al. (2004). Investigation of multipotent postnatal stem cells from human periodontal ligament. *Lancet* 364(9429):149–155.

Sodek J, McKee MD (2000). Molecular and cellular biology of alveolar bone. *Periodontol 2000* 24:99–126.

Sonoyama W, Liu Y, Fang D, Yamaza T, Seo BM, Zhang C, et al. (2006). Mesenchymal stem cell–mediated functional tooth regeneration in swine. *PLoS One* 1:e79.

Steele-Perkins G, Butz KG, Lyons GE, Zeichner-David M, Kim HJ, Cho MI, et al. (2003). Essential role for NFI-C/CTF transcription-replication factor in tooth root development. *Mol Cell Biol* 23(3):1075–1084.

Thomadakis G, Ramoshebi LN, Crooks J, Rueger DC, Ripamonti U (1999). Immunolocalization of bone morphogenetic protein-2 and -3 and osteogenic protein-1 during murine tooth root morphogenesis and in other craniofacial structures. *Eur J Oral Sci* 107(5):368–377.

Thyagarajan T, Sreenath T, Cho A, Wright JT, Kulkarni AB (2001). Reduced expression of dentin sialophosphoprotein is associated with dysplastic dentin in mice overexpressing transforming growth factor-beta 1 in teeth. *J Biol Chem* 276(14):11016–11020.

Tummers M, Thesleff I (2003). Root or crown: a developmental choice orchestrated by the differential regulation of the epithelial stem cell niche in the tooth of two rodent species. *Development* 130(6):1049–1057.

Tummers M, Thesleff I (2008). Observations on continuously growing roots of the sloth and the K14-Eda transgenic mice indicate that epithelial stem cells can give rise to both the ameloblast and root epithelium cell lineage creating distinct tooth patterns. *Evol Dev* 10(2):187–195.

Tummers M, Thesleff I (2009). The importance of signal pathway modulation in all aspects of tooth development. *J Exp Zool B* 312B(4):309–319.

Walker CG, Ito Y, Dangaria S, Luan X, Diekwisch TG (2008). RANKL, osteopontin, and osteoclast homeostasis in a hyperocclusion mouse model. *Eur J Oral Sci* 116(4):312–318.

Wise GE, King GJ (2008). Mechanisms of tooth eruption and orthodontic tooth movement. *J Dent Res* 87(5):414–434.

Wright JT, Hong SP, Simmons D, Daly B, Uebelhart D, Luder HU (2008). DLX3 c.561_562delCT mutation causes attenuated phenotype of tricho-dento-osseous syndrome. *Am J Med Genet A* 146(3):343–349.

Wright T (2007). The molecular control of and clinical variations in root formation. *Cells Tissues Organs* 186(1):86–93.

Yamashiro T, Tummers M, Thesleff I (2003). Expression of bone morphogenetic proteins and Msx genes during root formation. *J Dent Res* 82(3):172–176.

Ye L, Zhang S, Ke H, Bonewald LF, Feng J (2008). Periodontal breakdown in the *dmp1* null mouse model of hypophosphatemic rickets. *J Dent Res* 87(7):624–629.

Yokohama-Tamaki T, Ohshima H, Fujiwara N, Takada Y, Ichimori Y, Wakisaka S, et al. (2006). Cessation of Fgf10 signaling, resulting in a defective dental epithelial stem cell compartment, leads to the transition from crown to root formation. *Development* 133(7):1359–1366.

9

SYSTEMS BIOLOGY OF EARLY TOOTH DEVELOPMENT

DANIEL J. O'CONNELL, JOSHUA W. K. HO, AND RICHARD L. MAAS
Division of Genetics, Department of Medicine, Brigham and Women's Hospital, Harvard Medical School, Boston, Massachusetts

1 INTRODUCTION

Mammalian organogenesis results in the acquisition of structure and function. However, even for organs as diverse as tooth, kidney, and lung, which originate from ectoderm, mesoderm, and endoderm, respectively, organogenesis is critically dependent on self-regulatory epithelial–mesenchymal (E-M) interactions. These E-M interactions typically involve temporally discrete events between the juxtaposed epithelium and mesenchyme to induce a stepwise program of differentiation. Furthermore, experimental embryologic and genetic analyses of organ development have established that sequential inductive interactions can be mediated by signals secreted between the juxtaposed epithelial and mesenchymal tissues (Grobstein, 1967; Saxén and Thesleff, 1992; Vainio et al., 1993). However, the integration of these molecular signaling events, which regulate organ differentiation and their reciprocal signaling dynamics, remains largely unexplored.

Experimental evidence from reductionist approaches has defined many of the key signaling and transcriptional regulatory events that control E-M interactions during formation of the murine first molar tooth. Traditional molecular genetics, in which a single gene or pathway is studied at a time, has proven highly effective at deciphering the molecular composition of genetic regulatory interactions. However, to begin to understand the dynamics of E-M interactions, where the integration of multiple signaling pathway inputs and transcriptional regulatory outputs results in a sequential and reciprocal exchange of instructions between the interacting tissues, it is necessary to obtain a global understanding of the system rather than just that of the individual

Stem Cells in Craniofacial Development and Regeneration, First Edition.
Edited by George T.-J. Huang, Irma Thesleff.

components. In the context of developmental biology, a systems approach explicitly recognizes that a process such as odontogenesis is a result of the spatiotemporally coordinated action of many genes and signaling pathways.

Systems biology is the study of *complex* biological organization and processes, such as organ development, in terms of the molecular constituents and the interactions among them (Kirschner, 2005). Recent advances in high-throughput genomic and proteomic technology have provided new tools to study complex gene regulation from a global perspective (Kitano, 2002; Kirschner, 2005). For example, it is now possible to investigate and model the collective behavior of many genes simultaneously, thereby revealing a variety of systems-level properties, such as self-sustaining feedback, controllability, and robustness (Kitano, 2002). A working premise that underlies our research is that knowledge of such system properties, in this case for odontogenesis, can powerfully assist efforts aimed at the bioengineering of tooth germs. The goal of this chapter, therefore, is to highlight the ability of systems biology approaches to provide insight into how a molar tooth is formed and into how that process might be recapitulated from a bioengineering standpoint.

To this end, we first provide a brief overview of relevant embryologic, genetic, and molecular knowledge about odontogenic E-M interactions. Then, to illustrate how gene expression profiling, literature-derived knowledge, integrative bioinformatics, and in vivo genetics can be combined to yield novel insights into the mechanisms that underlie E-M interactions, we present results from a recent systems biology study by our group on the early E-M interactions in molar odontogenesis (O'Connell et al., 2012). Finally, we discuss the potential applicability of a gene regulatory network (GRN) to tooth regeneration and stem cell–based bioengineering.

1.1 Sequential and Reciprocal E-M Signaling in Early Odontogenesis

In 1923, Hans Spemann and Hilde Mangold made the remarkable discovery that the transplantation of a small group of cells from one embryo to another resulted in a body axis duplication, thereby deriving the concepts of "the organizer" and of "embryonic induction" (Spemann and Mangold, 2001). *Induction* can therefore be defined as a tissue interaction in which one tissue initiates the differentiation of another. In this sense, embryogenesis is the culmination of a series of self-sustaining inductive tissue interactions.

Vertebrate organogenesis is often the product of inductive interactions between epithelial and mesenchymal tissues. A key feature of these inductive interactions is that an interaction leading to a subsequent change in both the epithelial and mesenchymal tissue components is frequently reciprocal (Grobstein, 1967; Saxén and Thesleff, 1992). The sequential and reciprocal inductive E-M interactions in odontogenesis were studied elegantly in 1987 by Mina and Kollar, who showed that the inductive potential for mouse tooth formation resides in the dental epithelium from embryonic day E9.5 to E11.5, and subsequently shifts to the dental mesenchyme at E12.5 (Mina and Kollar, 1987). Thus, in mice the potential to induce tooth formation shifts from dental epithelium to mesenchyme between E11.5 and E12.5.

The importance of diffusible signaling molecules in the mediation of inductive tissue interactions began with the discovery that it was possible to sustain developmental interactions between epithelial and mesenchymal tissues even in the presence of an intervening membrane (Grobstein, 1953, 1956; Koch, 1967). Despite the presence of cytoplasmic processes penetrating these transfilter experiments (Wartiovaara et al., 1974; Thesleff et al., 1977; Gurdon, 1987), the importance of diffusible factors mediating embryonic induction was confirmed further in the analysis of *Xenopus* mesoderm induction (Jessell and Melton, 1992)

In the tooth system the expression of *Bmp4*, the gene encoding the diffusible signaling molecule Bmp4 in the Tgfβ superfamily, shifts from dental epithelium at E11.5 to dental mesenchyme at E12.5, concomitant with the aforementioned shift in inductive tooth-forming potential (Mina and Kollar, 1987; Vainio et al., 1993). Furthermore, Bmp4 can partially substitute for the inductive function normally carried out by dental epithelium on dental mesenchyme (Vainio et al., 1993). Since then, the identities of and relationships between many genes and signaling pathways involved in E-M interactions have been discovered. Members of the Wnt, bone morphogenetic protein (Bmp), fibroblast growth factor (Fgf), and Hedgehog (Hh) families are among the most important signaling pathways used in odontogenesis, and throughout embryogenesis in general (Jernvall and Thesleff, 2000; Thesleff et al., 1995). Therefore, in-depth knowledge of the inductive E-M interactions in a model system such as the developing tooth has a direct bearing on our understanding of the general principles that govern organogenesis.

The initiation stage of mouse odontogenesis begins at E10.5 with formation of the primary dental lamina within a single cuboidal layer of epithelium in the first branchial arch (Dassule and McMahon, 1998; Jernvall and Thesleff, 2000; Thesleff, 2003). The primary dental lamina is an epithelial thickening in the oral ectoderm that defines the tooth-forming region. The primary dental lamina can also be recognized molecularly by the expression of several genes, including the transcription factors Pitx2 and Lef1 and the signaling molecules Bmp2, Wnt10b, and Shh (Dassule and McMahon, 1998; Jernvall and Thesleff, 2000; Thesleff, 2003).

During the initiation stage, the dental lamina resolves into molar and incisor placodes, and the expression of the transcription factors Pitx2 and Lef1, and the signaling molecules Bmp4, Bmp2, Wnt10b, and Shh become enriched in the placodes (Vainio et al., 1993; Dassule and McMahon, 1998; Jernvall and Thesleff, 2000; Zhang et al., 2000; Thesleff, 2003). E-M interactions are required for tooth progression from the E11.5 initiation stage to the bud stage at E12.5 (Thesleff, 2003). Furthermore, the signaling pathways that transduce Wnt (Andl et al., 2002, Fgf (Meyers et al., 1998), Shh (Hardcastle et al., 1998), and Bmp (Plikus et al., 2005) signals all play critical roles in the progression from initiation to the bud stage.

The bud stage begins at E12.5 as the epithelial dental placode invaginates into the underlying mesenchyme, and by E13.5 has progressed to a bud surrounded by condensed mesenchyme. This process is morphologically akin to early stages in the development of other organs (Thesleff et al., 1995). At E14.5, the tooth bud branches into a cap-stage tooth germ, a stage that marks the beginning of cytodifferentiation (Jernvall and Thesleff, 2000; Thesleff, 2003). E-M interactions are required for the progression from bud stage to cap stage, and mouse mutations that interrupt Bmp signaling (Chen et al., 1996; Peters et al., 1998; Andl et al., 2004; Xu et al., 2008), Wnt signaling (van Genderen et al., 1994; Kratochwil et al., 1996; Liu et al., 2008), and Fgf signaling (Wang et al., 2007) have been shown to cause an arrest in tooth development at the bud stage.

Also at the E13.5 bud stage, the enamel knot, a nonproliferative signaling center that expresses Wnt, Bmp, Fgf, and Shh, begins to form at the tip of the epithelial tooth bud

(Jernvall et al., 1994; Thesleff, 2003). Signaling by Bmp4 from the dental mesenchyme to epithelium between E12.5 and E13.5 is necessary for the bud-to-cap stage transition and for enamel knot formation (Chen et al., 1996; Jernvall et al., 1998; Bei et al., 2000; Thesleff, 2003; Andl et al., 2004). Wnt signal transduction within the epithelium has also been identified as critical to enamel knot formation (van Genderen et al., 1994; Kratochwil et al., 1996; 2002; Järvinen et al., 2006; Liu et al., 2008; Wang et al., 2009). Thereafter, molecular signals derived from the enamel knot direct subsequent cytodifferentiation and the acquisition of tooth shape (Jernvall et al., 1994). Hence, signal transduction mediated by diffusible signaling molecules in the Wnt, Bmp, Fgf, and Hh families is critical throughout the early phases of odontogenesis.

1.2 Wnt and Bmp Signaling in Early Odontogenesis

The Wnt and Bmp pathways are critical regulators of odontogenesis, and the evidence accumulated based on examining these two pathways during odontogenesis supports a stepwise interpretation of signaling requirements in E-M interactions. At the initiation stage, epithelial expression of *Bmp4* induces the expression of *Bmp4* in the dental mesenchyme through the dental mesenchymal transcription factor Msx1. The over-expression of the extracellular Wnt antagonist Dkk1 results in reduced mesenchymal *Bmp4* expression and an arrest of tooth formation at the initiation stage (Andl et al., 2002; Liu et al., 2008). Furthermore, compound mouse mutants that lack the canonical Wnt transcriptional effectors Tcf7 and Lef1 fail to induce mesenchymal *Bmp4* expression (Fujimori et al., 2010). Therefore, at the transition from the initiation to the bud stage, when inductive signaling is taking place from epithelium to mesenchyme, both Wnt and Bmp signals must be transduced into the mesenchyme.

After the initiation stage, when *Bmp4* expression has shifted from the dental epithelium to the dental mesenchyme (Vainio et al., 1993), there is a discrete order of Wnt and Bmp signaling requirements. For example, while canonical Wnt expression remains restricted to the dental epithelium from the initiation to the cap stage (Sarkar and Sharpe, 1999), there is a delay in the expression of some canonical Wnt target genes associated with the epithelial enamel knot. The expression of the direct Wnt target gene *Fgf4* (Kratochwil et al., 2002) first requires instructive signaling via mesenchymal Bmp4 to induce the enamel knot (Chen et al., 1996; Jernvall et al., 1998). Therefore, mesenchymal *Bmp4* expression precedes that of the direct Wnt target, *Fgf4*.

Furthermore, disruption of the mesenchymal transcription factors Pax9 or Msx1 results in a reduction of mesenchymal *Bmp4* expression and a bud-stage tooth arrest, prior to enamel knot formation. In the case of *Msx1* loss of function, it has been demonstrated that addition of exogenous Bmp4, but not Fgfs, is sufficient to rescue tooth formation in *Msx1*-null tissue (Bei et al., 2000). A recent study by Jia et al., has demonstrated that at least one other factor besides Bmp4 is required to transduce endogenous Msx1 function because the tooth developmental phenoytpe of Msx1 mutants is more severe than the mesenchymal deletion of *Bmp4* mediated by *Wnt1-Cre* (Jia et al., 2013). Disruption of the Wnt signal transduction gene *Lef1* phenocopies the bud-stage tooth arrest of *Pax9* and *Msx1* mouse mutants. In complete contrast to the *Msx1* mouse mutant, however, addition of exogenous Bmp4 fails to rescue *Lef1*-null tooth germs, whereas exogenous Fgf4 can.

These data establish a clear pattern of sequential and reciprocal signaling in the E-M interactions that underlie early odontogenesis. First, Wnt and Bmp epithelial expression at the initiation stage induces a paradigm of signaling to the dental mesenchyme, and of consequent gene expression, notably including that of Bmp4. After the dental mesenchyme has received the operative inductive signals from the dental epithelium, mesenchymal Bmp4 then acts reciprocally back on the dental epithelium to induce enamel knot formation and to activate the expression of direct Wnt target genes such as *Fgf4* in the enamel knot.

Particularly compelling evidence for the sequential and reciprocal nature of signaling between dental epithelium and mesenchyme derives from the ability to produce a fully differentiated tooth organ entirely from *Msx1*-null tissue in a double tissue recombination experiment (Bei et al., 2000). First, *Msx1*-null dental epithelium was recombined with wild-type dental mesenchyme during the bud and early cap stages (E13.5 to E14.5), when *Msx1* is required for mesenchymal signaling molecule expression. Next, after in vitro culture allowed transduction of the mesenchymal signal to the epithelium, the wild-type mesenchyme was replaced by *Msx1*-null mesenchyme, resulting in the generation of fully differentiated teeth entirely from *Msx1*-null tissues (Bei et al., 2000). Therefore, *Msx1* is required transiently only for dental mesenchymal programming of the dental epithelium. Once this signal is transduced, the function of *Msx1* becomes largely superfluous. Thus, in the E-M interactions in odontogenesis, tooth development is achieved through the transmission of discrete information, transduced in the form of signaling molecules, between the interacting tissue layers.

1.3 Systems Approach to Understand E-M Interactions

Although much is known about the individual pathways and transcription factors that participate in E-M interactions during early tooth development, it is less clear how the expression of key genes is temporally integrated between epithelium and mesenchyme. For example, what is the *mode of communication* between the developing epithelial and mesenchymal dental tissues? One proposed model is a *sequential* mode of communication, where the epithelium initially produces signals that act on the mesenchyme, which in turn acts back on the epithelium. This model is supported by embryonic tissue recombination experiments that indicate a discrete shift in instructive potential during the initiation-to-bud stage transition. In the alternative *simultaneous* mode of communication model, the epithelium and mesenchyme continuously and simultaneously exchange diffusible signaling molecules. Although the bulk of the biological evidence and common intuition support the sequential model, this question cannot be addressed fully without a detailed analysis of the genome-wide expression patterns of and the interactions between multiple signaling pathways that act during early odontogenesis. The study of E-M signaling interactions in odontogenesis provides an ideal test case for the application of systems biology to address this question.

Resolving the mechanistic differences between these two modes of communication will help inform strategies that are aimed at programming precursor cells for the in vitro or in vivo generation of multitissue organs. Below, we highlight key features from our recent study in which a systems approach was employed to address the nature of E-M signaling dynamics during early odontogenesis (O'Connell et al., 2012), including:

1. Changes in gene expression in developing dental epithelium and mesenchyme are tightly coupled.

2. Early odontogenic epithelial and mesenchymal gene regulatory networks (GRNs) are linked to each other through the action of diffusible signaling molecules.
3. Self-sustaining feedback systems are a characteristic feature of these GRNs, and reciprocal signaling dynamics may be a natural property of the GRN structure itself.
4. Knowledge of the GRN structure can inform effective organ regeneration strategies.

These conclusions are amplified in the following section.

2 SYSTEMS BIOLOGY OF E-M SIGNALING DYNAMICS DURING EARLY ODONTOGENESIS

2.1 Transcriptional Coupling of E-M Tissues During Organogenesis

To gain insight into the E-M signaling interactions in tooth development, we first created a spatiotemporal gene expression data set using microarrays. This data set allowed us to analyze the changes in gene expression as a dynamic time series from the morphologically distinct epithelial and mesenchymal compartments of the first lower molar in mouse during the reciprocal signaling phases spanning the initiation stage to the cap stage (E10 to E14.5; shown for E13.5 in Fig. 1A). To better understand the mode of communication during E-M interactions, we identified differentially regulated genes (DRGs) in each tissue that were up- or downregulated at the bud stage compared to the initiation stage. Surprisingly, of the genes that were differentially regulated from initiation to bud stage in both epithelium and mesenchyme, there was a highly significant proportion of genes with concordant changes in gene expression in both tissue compartments (i.e., upregulated in both epithelium and mesenchyme, or downregulated in both, compared to the proportion of genes with discordant changes) (Fig. 1B). This result indicates that the temporal shift in instructive tooth-forming potential from epithelium to mesenchyme between the initiation and bud stage (Mina and Kollar, 1987) is accompanied by a striking concordance in genome-wide gene expression changes across the E-M compartments. In turn, this result strongly suggested that gene expression changes in the two tissues were in some way intimately linked or coupled. Such coupling in gene expression occurred despite the fact that according to the classical sequential model, only one compartment at a time holds tooth-forming potential. This result supports a simultaneous mode of communication where the instructive signals secreted by one tissue compartment are transduced to both tissue compartments.

To reconcile the principle of reciprocal instructive potential (Mina and Kollar, 1987) and the observation of concordant gene expression changes at a genome-wide scale, we analyzed the expression of genes that encode signal transduction components. Despite the tight coupling of genome-wide expression changes between epithelium and mesenchyme, it was only the expression patterns of signaling molecule genes that temporally matches the shift in instructive potential from epithelium to mesenchyme at the initiation-to-bud stage transition. For example, whereas *Wnt3a* expression is dynamic and epithelial specific, and *Bmp4* expression alternates between epithelium and mesenchyme, the expression of their respective receptors, Bmpr1a and Lrp6, remains mostly constant in both epithelium and mesenchyme (Fig. 1C). Therefore, these data are

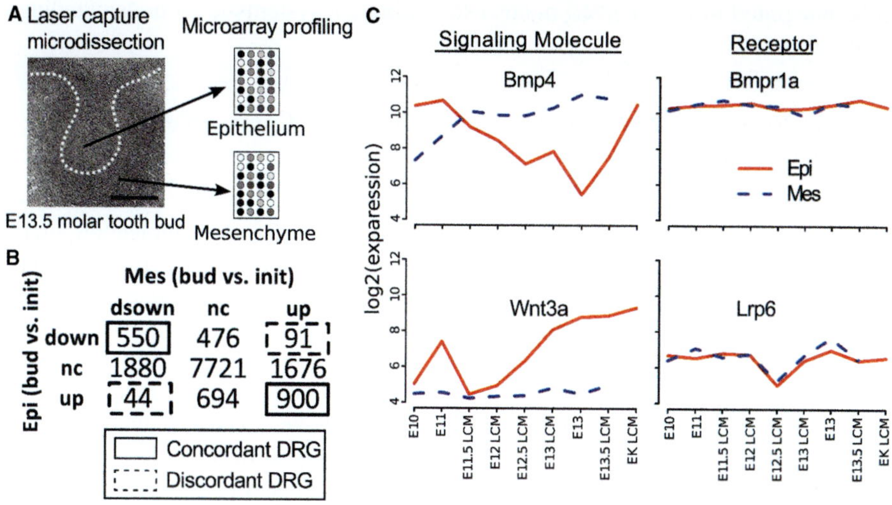

FIGURE 1 Genome-wide expression change in dental epithelium and mesenchyme are coupled temporally during the transition from initiation to bud stage. (A) Laser-capture microdissections (LCM) of E13.5 molar tooth epithelium and mesenchyme were subjected to microarray gene expression profiling. (B) A contingency table representing the intersection of differentially regulated genes (DRGs) between the initiation and bud stages for epithelium and mesenchyme. There is a statistically significant overlap of concordant DRGs: down–down in Epi and Mes, or up–up (solid boxes), compared to discordant DRGs (dashed boxes) ($p < 10^{-15}$, χ^2 test); bud: bud stage; init: initiation stage. (C) Microarray expression of representative signaling molecules and receptors in the Bmp and Wnt pathways. Signaling molecules show highly dynamic expression changes, whereas receptor expression appears relatively constant over time. [Adapted from Fig. 1, O'Connell et al. (2012).]

consistent with the idea that exposure to diffusible signaling molecules constitutes the primary mediator of sequential and reciprocal E-M interactions. Furthermore, exposure to diffusible signaling molecules, independent of their tissue site of synthesis (epithelium or mesenchyme), provides an attractive mechanism for how E-M compartments can be coupled simultaneously to yield concordant gene expression changes.

Although the expression patterns of many key genes in odontogenesis are already known and publicly accessible via the Helsinki BITE-IT database (http://bite-it.helsinki.fi/), our genome-wide spatiotemporal expression profiling provides additional information that could not be obtained by analyzing small sets of genes alone: (1) the expression changes in the developing dental epithelium and mesenchyme are tightly coupled, as indicated by the large number of genes with concordant gene expression changes; and (2) among all pathway components, only the expression of specific signaling molecules has a pattern that matches the sequential and reciprocal pattern of odontogenic instructive potential. The latter point is consistent with the hypothesis that extracellular signaling molecules themselves mediate E-M interactions. Although this view was already generally believed to be the case, these results demonstrate the value of an unbiased genome-wide analysis in the study of complex developmental processes and support the use of systems biology to understand organogenic E-M interactions. At the same time, these insights do not diminish the value of the large amounts of data produced from the many studies that have focused on individual genes and pathways. In fact, as we describe below, these published data can be integrated in a principled manner to facilitate a systems-level understanding of odontogenesis.

2.2 Gene Regulatory Network Reconstruction from a Compendium of Regulatory Data

An integrative analysis of signaling molecule expression and signaling pathway activity is essential to an understanding of how the expression of signaling molecules such as *Bmp4* is linked to the coupling of genome-wide expression changes across interacting epithelial and mesenchymal compartments. GRNs have become valuable tools to represent regulatory interactions between genes (McAdams and Shapiro, 1995; Bénazet et al., 2009; Davidson, 2010; Kirouac et al., 2010; Ho and Charleston, 2011; Peter and Davidson, 2011). A GRN is typically composed of nodes, representing genes, and edges, representing regulatory relationships between nodes. The key purpose of any GRN is to provide a framework for a systems-level analysis of the emergent behavior that arises due to interactions of multiple genes, proteins, and pathways.

There are many network modeling methodologies in the field, and these modeling paradigms differ in terms of the data required and the interpretation of the resulting network (Karlebach and Shamir, 2008; Ho and Charleston, 2011). To reconstruct a GRN for odontogenic E-M interactions, we chose to use a simple directed graph representation to model the first-order causal regulation between key genes and signaling pathways. This choice was motivated by the wealth of gene regulatory information related to tooth formation that has been derived from the study of tooth defects intransgenic mouse models (Thesleff, 2003; Bei, 2009).

Peer-reviewed studies on the molecular roles of genes involved in tooth formation have resulted in the accumulation of causal regulatory data during the key developmental steps in odontogenesis. We compiled the causal gene expression results from a manual survey of mouse tooth development publications since 1993. The analysis of over 100 publications resulted in over 1000 pieces of gene expression relationships derived from two types of experimental perturbation experiments: (1) treatment of explanted dental tissue with signaling molecules, or (2) knockout and knockdown mouse models of genetic perturbation (O'Connell et al., 2012) . Collectively, this experimental evidence provides causal regulatory information via experimental interpretations of positive, neutral, or negative relationships with respect to gene expression. When encoded in a tabular format, these data represent a high-quality source of literature-derived gene regulatory relationships. Currently, this valuable information cannot be achieved through computational text mining approaches. As a next step, this collection of regulatory evidence was augmented with our own microarray-based perturbation experimental data to provide a comprehensive collection of regulatory relationships for GRN reconstruction.

Given our interest in studying E-M signaling interactions, we generated a GRN in which nodes can represent either genes or signaling pathways, and edges represent causal regulatory relationships between two nodes. To infer the edges from experimental perturbation data, we used a probabilistic technique to assign the most likely mode of interaction between two nodes: activation, inhibition, or no effect (O'Connell et al., 2012). The nodes were then connected using the inferred edges to form a signaling-based GRN (Fig. 2A). This strategy generated one GRN for the bud-stage epithelium and one for the bud-stage mesenchyme (Fig. 2B).

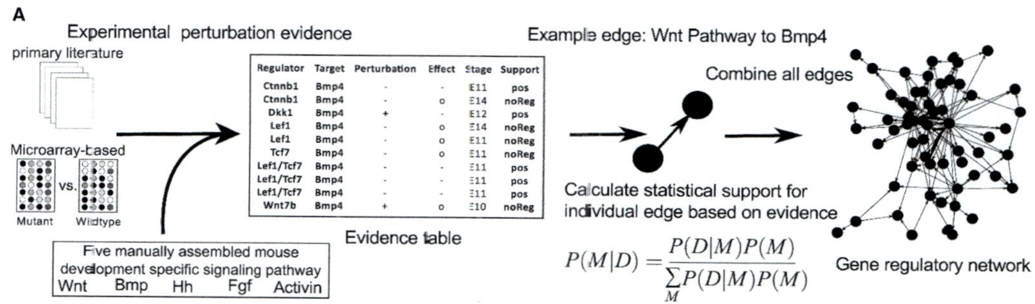

FIGURE 2 A signaling-based gene regulatory network (GRN) reveals a Wnt-Bmp feedback circuit that controls E-M signaling dynamics in tooth development. (A) Schematic of causal gene regulatory evidence data integration strategy. This example illustrates how the mode of interaction of one edge (e.g., Wnt pathway to *Bmp4*) is inferred based on perturbation data. All perturbation data describing the regulation of *Bmp4* by the Wnt pathway is summarized in an evidence table. Evidence is categorized as providing support for positive, negative, or no regulation. Bayes' rule was used to combine multiple pieces of evidence to estimate the probability of a particular mode of interaction between regulator and target. The mode of interaction with the highest probability (and above 0.6) is selected as the true mode of interaction, and an edge is assigned to connect two nodes that have an activating or inhibitory mode of interaction. This procedure was repeated for all potential interactions, and the results of all edges were then combined to create a GRN.

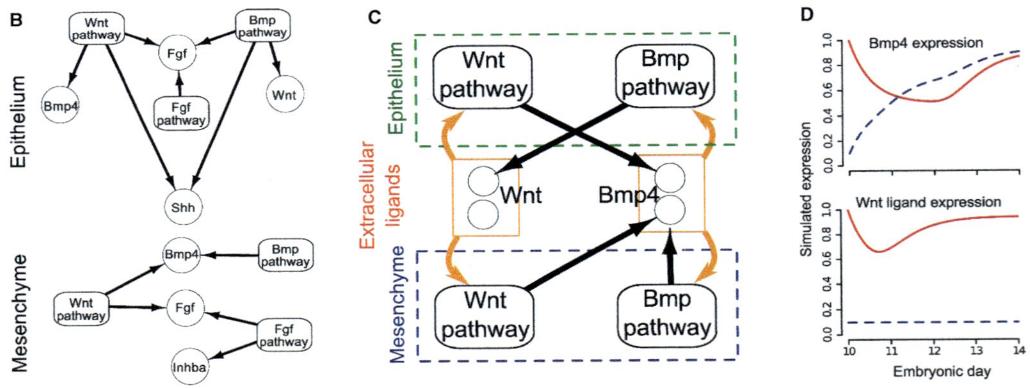

FIGURE 2 (*Continued*) (B) A small subnetwork of the full GRN for each tissue compartment is extracted to show the positive regulatory relationships between key signaling pathways and signaling molecules belonging to the Wnt, Bmp, Fgf, Shh, and Inhba pathways. (C) Epithelial and mesenchymal GRNs in (B) are coupled through the action of diffusible Wnt and Bmp4 signaling molecules under the assumption that they are equally effective at acting on the dental epithelium or mesenchyme regardless of the tissue of synthesis (i.e., dental epithelium or mesenchyme). This assumption affords a single integrated intertissue Wnt-Bmp feedback circuit. (D) Based on the Wnt-Bmp circuit structure, expression of *Bmp4* and *Wnt* ligands can be simulated using an ordinary differential equation model. The simulation shows that the circuit structure itself is sufficient to recapitulate the reciprocal expression dynamics of *Bmp4*, despite simultaneous transduction of Bmp and Wnt signaling to both epithelium and mesenchyme. [Adapted from Fig. 2, O'Connell et al. (2012).]

As mentioned earlier, the epithelial and mesenchymal GRNs in Fig. 2B represent just a subset of nodes of the complete GRN, in this case just those that correspond to extracellular signaling molecules Wnts, Bmp4, Fgfs, Shh, and Inhba (Activinβa), and to gene sets that represent the associated intracellular signal transduction pathway components. Other genes are not shown but, nonetheless, exist in the context of the complete GRN. The depiction of subnetworks within the complete GRN that were selected highlights the interactions among the various important signaling pathways. At a system-wide level, we find that the Wnt and Bmp pathways by themselves contribute to the positive regulation of most other signaling molecule expression in early odontogenesis. Stated differently, these GRNs, now selected to reveal only the regulatory relationships between the major operative signal transduction pathways, clearly indicate that the Wnt and Bmp pathways positively regulate the expression of *Bmp4, Shh, Fgf*s, and *Wnt*s in the epithelium, and of *Fgf*s and *Bmp4* in the mesenchyme.

Although Bmp4 and Wnt were already known to be important in odontogenesis from many studies, this result is nonetheless noteworthy in at least two ways. First, these results are derived from an unbiased approach, and second, the hierarchical relationship between the various signaling pathways depicted in Fig. 2B and 2C was previously not well understood. In addition, these GRNs reveal not only that the expression of some signaling molecules is tissue specific, but also that the positive regulatory structures controlling signaling molecule gene expression are also tissue specific. Thus, whereas expression of canonical *Wnt* ligands and *Shh* is detected only in dental epithelium (Thesleff, 2003; Bei, 2009) and *Inhba* is detected only in dental mesenchyme (Ferguson et al., 1998, 2001), *Bmp4*, whose expression shifts from dental epithelium to dental mesenchyme, is positively regulated in a tissue-specific manner. Taken together, these analyses reveal that in the epithelium, the *Bmp4* and *Wnt* signaling molecule genes are under *cross-regulatory control* by the Wnt and Bmp pathways, respectively. In contrast, the Wnt and Bmp pathways *jointly control* mesenchymal *Bmp4* expression, while mesenchymal *Wnt* is not under the control of the Wnt, Bmp, Hh, or Fgf signaling pathways.

To explain the result of an inexplicably high concordance between gene expression changes in dental epithelium and mesenchyme, we hypothesized that signaling molecules, regardless of their tissue site of synthesis, could act on both epithelium and mesenchyme. To model this postulated ability of the Wnt and Bmp signaling molecules to act on dental epithelium and mesenchyme simultaneously, we combined the E13.5 bud-stage epithelial GRN and the mesenchymal GRN to form one integrated intertissue GRN that is coupled through the activity of extracellular canonical Wnt and Bmp4 (Fig. 2C). This E-M GRN integration revealed the existence of a Wnt-Bmp intertissue feedback circuit within the GRN. Inspection of the circuit reveals *cross-regulation* of canonical *Wnt* and *Bmp4* expression by the complementary signaling pathways in the dental epithelium, and *joint regulation* by both the Wnt and Bmp4 pathways of *Bmp4* expression in the dental mesenchyme.

In light of this asymmetric Wnt-Bmp circuit structure, we wondered whether this feature could account for the reciprocal signaling dynamics that are typically observed in E-M interactions. This hypothesis was supported by an *in silico* simulation of an ordinary differential equation (ODE) model of the Wnt-Bmp feedback circuit. The simulation results largely confirm that the structure of the Wnt-Bmp circuit itself is sufficient to recapitulate key features of the sequential and reciprocal E-M signaling observed (Fig. 2D). The ODE simulation reveals that although Wnt and Bmp4 signaling molecules simultaneously activate their respective pathways in epithelial and mesenchymal compartments, the rate at which the epithelial GRN and the mesenchymal GRN respond through the expression of *Bmp4* and *Wnt* genes is different, thus accounting for the reciprocal pattern of signaling molecule gene expression observed (Fig. 1C). This in turn suggests that the reciprocal shift in the expression of signaling molecules between E-M tissues is an inherent property of the Wnt-Bmp feedback circuit itself. The systems-based construction of a GRN and the subsequent simulation of signaling dynamics provides a unifying framework to reconcile the sequential inductive tissue interactions and the tight coupling of transcriptional dynamics. Thus, simultaneous integration of signaling molecules by the epithelial and mesenchymal compartments can lead to reciprocal inductive phases of signaling through tissue-specific GRN structures.

2.3 ToothCODE: A Web Resource for Tooth Systems Biology

Public data access is an important principle to help ensure the utilization of systems approaches that involve the curation and analysis of large-scale data sets. This is well illustrated by the pioneering example of the Helsinki BITE-IT tooth gene expression database (http://bite-it.helsinki.fi/), which contains experimentally verified spatial and temporal expression data for more than 400 important genes in tooth development. In accord with the principle of open access, we recently developed an interactive Web resource for tooth systems biology, called ToothCODE (http://compbio.med.harvard.edu/ToothCODE). ToothCODE provides an intuitive user interface to browse and interrogate the compendium of gene expression patterns from (1) microarray profiling experiments, (2) the collection of literature-derived genetic regulatory evidence, (3) mouse-specific signaling network annotation, and (4) the Wnt-Bmp feedback circuit that underlies E-M interactions in early tooth development.

ToothCODE places a strong emphasis on linking every piece of evidence to its primary source, either from a microarray profiling study or from other forms of gene expression results published in the primary literature. The "Network View" page, a unique feature of ToothCODE, presents an expanded view of the Wnt-Bmp feedback circuit in an interactive network browser. Users can click on each edge in the circuit to interrogate the experimental evidence underlying the inference of that edge. We anticipate that public Web resources such as BITE-IT and ToothCODE will further facilitate the application of systems biology approaches to odontogenesis.

2.4 Systems Analysis of *Pax9*- and *Msx1*-Null Mutants

To investigate the role of transcription factors in the genetic regulation of inductive tissue interactions in odontogenesis from a systems perspective, we turned to two dental mesenchymal transcription factors that are required for odontogenesis, Pax9 and Msx1. Mutations in either gene are associated with nonsyndromic oligodontia in humans (Vastardis et al., 1996; Stockton et al., 2000; Jumlongras et al., 2001). Mice with null mutations in these transcription factors display complete tooth agenesis and a bud-stage arrest (Satokata and Maas, 1994; Peters et al., 1998). Furthermore, Pax9 and Msx1 have also been shown to interact physically in vitro (Ogawa et al., 2006), and $Msx1^{+/-}$; $Pax9^{+/-}$ compound heterozygotes in mice display incisor tooth abnormalities (Nakatomi et al., 2010).

Pax9 and *Msx1* are both necessary for mesenchymal *Bmp4* expression (Chen et al., 1996; Peters et al., 1998). Mesenchymally derived Bmp4 is a critical diffusible signaling molecule required to mediate the inductive signaling from mesenchyme to epithelium after the initiation stage. Epithelial reception of Bmps through Bmpr1a is required for the progression of tooth development and enamel knot formation (Jernvall et al., 1998; Andl et al., 2004).

Pax9 and Msx1 are thought to act within the same molecular pathway in odontogenesis: their loss-of-function mutations exhibit anodontia in mice and oligodontia in humans; their mutant alleles interact genetically, their gene products interact physically, and each is required for mesenchymal *Bmp4* expression (Chen et al., 1996; Peters et al., 1998; Ogawa et al., 2006; Nakatomi et al., 2010). Furthermore, it is possible to arrange *Pax9* and *Msx1* in a genetic hierarchy where *Pax9* acts upstream of *Msx1*, because *Pax9* expression persists in *Msx1* null tooth bud, whereas *Msx1* expression is reduced in *Pax9*-null tooth bud (Peters et al., 1998; Ogawa et al., 2006; Nakatomi et al., 2010). Interestingly, Pax9 is a putative transcriptional activator, whereas Msx1 is a putative repressor (Catron et al., 1995; Ogawa et al., 2006). Therefore, both positive and negative transcriptional inputs must presumably be integrated to achieve dental mesenchymal *Bmp4* expression.

A third transcription factor, Osr2, is also a critical player in the *Pax9-Msx1-Bmp4* GRN. Mice lacking *Osr2* display an expansion of mesenchymal *Bmp4* lingual to the first molar and develop accessory lingual molars (Zhang et al., 2009). Osr2 can interact physically with Pax9 and Msx1, and Pax9 is required for the expression of *Osr2* (Zhou et al., 2011). The regulatory relationships among *Pax9, Msx1, Osr2*, and *Bmp4* are summarized in a simplified GRN in Fig. 3A. This simplified GRN does not account for the spatial patterning of the genes within it. However, it is possible that the GRN structure is uniform throughout the dental mesenchyme and that expression levels of the transcription factors determines the phenotypic output of mesenchymal *Bmp4* expression.

To gain further insight into the dental mesenchymal functions of Pax9 and Msx1, we conducted a global gene expression analysis on the separated epithelial and mesenchymal compartments of bud-stage (E13.5) *Pax9*- and *Msx1*-null tooth germs using laser-capture microdissection (LCM) (data available via ToothCODE). Given the previously established molecular epistasis between *Pax9* and *Msx1*, whereby *Msx1* expression is reduced in *Pax9* mutants and *Pax9* expression persists in *Msx1* mutants, one might expect the set of Pax9 targets to encompass the set of Msx1 targets. However, what is clear from the global transcriptional analysis of these mutants is that many genes are regulated differentially in *Pax9*- and *Msx1*-mutant tissue independent of each other (Fig. 3B).

Interestingly, of the genes coregulated by *Pax9* and *Msx1*, a significant majority are regulated concordantly, including *Bmp4* (Fig. 3B). This observation fits with the view that Pax9 and Msx1 act in a common molecular pathway to regulate *Bmp4* expression and mesenchymal odontogenic potential. However, the simplest interpretation of the data may derive from the detection of a discordantly regulated gene, *Osr2* (Zhang et al., 2009; Zhou et al., 2011). In this scenario *Pax9* may be required for the transcriptional activation of *Bmp4* (Ogawa et al., 2006), and *Msx1* may be required to antagonize the repressive action of *Osr2* on *Bmp4* (Zhang et al., 2009; Zhou et al., 2011). The precise mechanism of *Pax9*- and *Msx1*-mediated *Bmp4* expression requires further investigation.

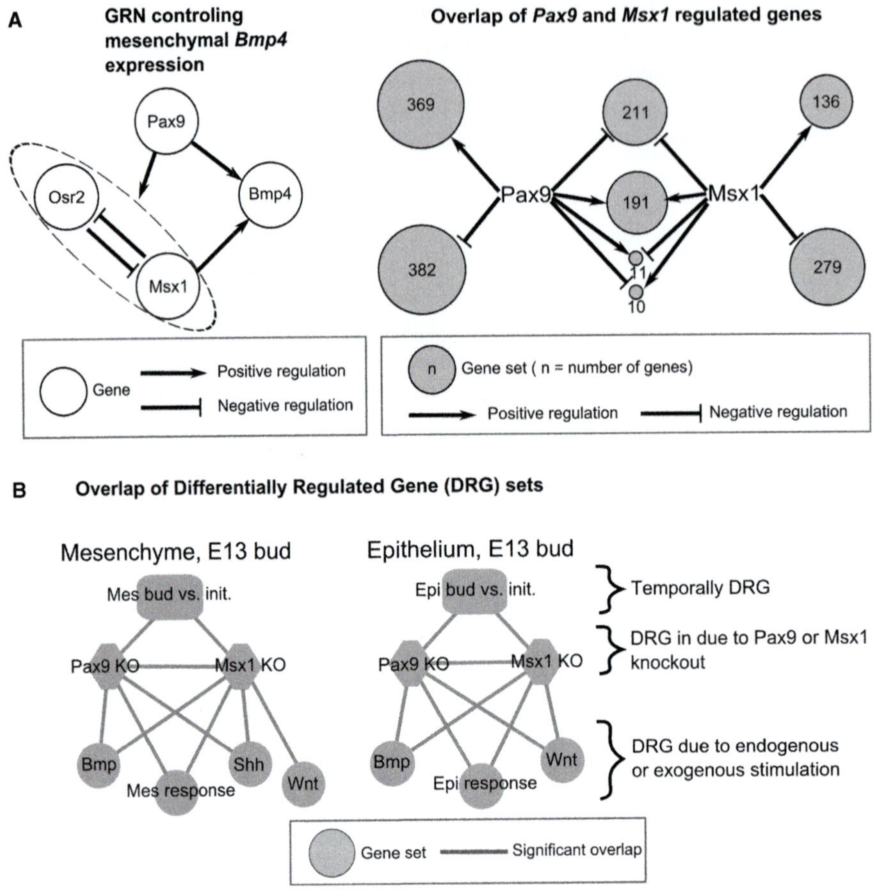

FIGURE 3 Systems analysis of *Pax9* and *Msx1* mutants. (A) A GRN that controls the expression of mesenchymal *Bmp4* expression. (B) Genome-wide microarray profiling of *Pax9*- and *Msx1*-null mutants reveals the extent of overlap of the differentially regulated gene (DRG) sets. (C) Molecular concept maps showing significant overlap between *Pax9* and *Msx1* target genes in dental epithelium and mesenchyme and their association with the endogenous tissue inductive response (Mes response; Epi response), differentially regulated genes (Mes bud stage vs. initiationstage; Epi bud stage vs. initiation stage), and tissue response to exogenous stimulation of Wnt, Bmp, Shh, and Fgf signaling (Wnt, Bmp, Fgf, Shh). Each edge demonstrates a statistically significant overlap of gene sets (FDR <0.05, χ^2 test; odds ratio >2). This analysis suggests that Wnt and Bmp signaling are perturbed in *Pax9*- and *Msx1*-null mutants.

Comparison of the differentially regulated genes (DRGs) in *Pax9* and *Msx1* mutants and the DRGs from morphogen addition experiments reveals that both Wnt and Bmp signaling targets are largely differentially regulated in *Pax9* and *Msx1* mutants. Furthermore, the DRGs are also enriched for genes that are differentially regulated in the bud stage compared to the initiation stage in both epithelium and mesenchyme (Fig. 3C). This finding agrees with the known roles of *Pax9* and *Msx1* in mesenchymal *Bmp4* expression and in the subsequent regulatory interdependence between Wnt and Bmp signaling identified via GRN reconstruction (Fig. 2C). This analysis may also help provide insight into the functional roles of *Pax9* and *Msx1* in odontogenesis and in the derivation of strategies for tooth regeneration.

2.5 Epithelial Wnt Activation Can Restore Dental Tissue Differentiation in *Pax9*- and *Msx1*-Mutant Mice

Pax9 and *Msx1* regulate *Bmp4* expression (Fig. 3A) and therefore affect the Wnt-Bmp feedback circuit. We therefore sought to validate the circuit by designing an experiment to restore dental differentiation in *Pax9* and *Msx1* mutants, which normally arrest at the bud stage. Based on the structure of the Wnt-Bmp circuit, we hypothesized that in *Pax9* or *Msx1* mouse mutants, constitutive activation of epithelial Wnt signaling via conditional inactivation of *Apc* would induce epithelial *Bmp4* expression and restore dental differentiation (Fig. 4A).

Indeed, this is the case: Constitutive activation of epithelial Wnt signaling bypasses the requirement for mesenchymal *Bmp4* expression and allows ameloblast and odontoblast differentiation in *Pax9* and *Msx1* mutants (Fig. 4B and 4C). For example, through ODE simulation of the Wnt-Bmp circuit in wild-type and *Pax9*-mutant genetic backgrounds, one can compute the expected expression dynamics of *Bmp4* in the dental epithelium and mesenchyme (Fig. 4D). To an initial first-order approximation, the spatiotemporal expression of *Bmp4* derived from ODE simulation can be evaluated experimentally in qualitative fashion using RNA in situ hybridization (Fig. 4D). Thus, as predicted by the structure of the Wnt-Bmp feedback circuit, constitutive activation of canonical Wnt signaling ($Apc^{f/f}$ genotypes in Fig. 4D) induces ectopic epithelial *Bmp4* expression and allows a bypass of the *Pax9*- and *Msx1*-dependent induction of *Bmp4* expression ($Pax9^{-/-}$ genotypes in Fig. 4D) in the dental mesenchyme, which is normally required for the bud-to-cap stage transition (Fig. 4D). This analysis corroborates the structure of the Wnt-Bmp feedback circuitin the early odontogenic GRN.

From an engineering perspective, the constitutive activation of canonical Wnt signaling in the dental epithelium is analogous to the condition known as an electrical "short-circuit." The *Apc* loss of function (or β-*catenin* gain of function) state effectively decouples epithelial Wnt pathway activity from reception of a Wnt signal and constitutively activates the intraepithelial Wnt-Bmp signaling circuit. This results in the unregulated or *ectopic* production of epithelial Bmp4. In the wild-type state, *Bmp4* expression shifts from the dental epithelium to the dental mesenchyme after the initiation stage (E10.5 to E11.5). Therefore, during the bud stage (E12.5 to E13.5), the intraepithelial Wnt-Bmp signaling circuit is dependent on mesenchymally synthesized Bmp4 to activate the Bmp signaling pathway in the dental epithelium (Fig. 2). Activation of the Bmp receptor is therefore required to produce *Wnt* expression effectively and to maintain Wnt signaling pathway activity in both the epithelium and the mesenchyme. The short-circuit arises because the constitutive activation of epithelial Wnt signaling results in the unregulated production of Bmp4, which can replace the mesenchymal source of Bmp4 and allow the restoration of dental differentiation in *Pax9* and *Msx1* mouse mutants. This analysis illustrates how one can begin to use knowledge derived from a GRN to design organ regeneration strategies and highlights the power of combining *in silico* computational analysis with genetic validation.

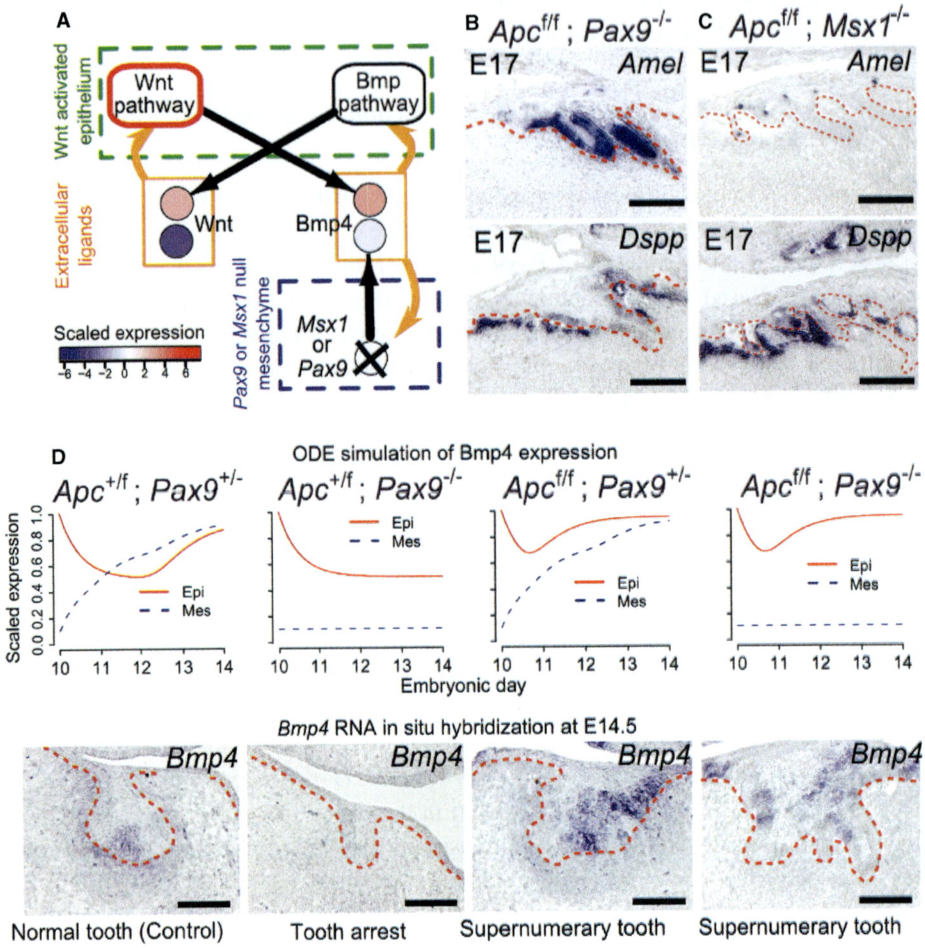

FIGURE 4 Constitutive epithelial Wnt activation induces ectopic *Bmp4* expression in dental epithelium, bypasses the mesenchymal *Bmp4* expression deficit in *Pax9*- and *Msx1*-null mutants, and restores dental differentiation. (A) Epithelial *Apc* loss of function (constitutive canonical Wnt pathway activity, outlined in red) short-circuits the intraepithelial Wnt-Bmp feedback circuit by stimulating ectopic epithelial *Bmp4* expression and by passing the requirement for mesenchymal *Bmp4* expression, which is decreased in *Pax9*- and *Msx1*-null mutants (X). This restores dental differentiation. (B and C) *Amelogenin* (*Amel*) and *Dentin sialophosphoprotein* (*Dspp*) expression confirms ameloblast and odontoblast differentiation, respectively, in E17.5 sagittal sections of compound epithelial *Apc* loss of function; *Pax9*-null mutants ($Apc^{f/f};Pax9^{-/-}$) and compound epithelial *Apc* loss of function; and *Msx1*-null mutants ($Apc^{f/f};Msx1^{-/-}$). The epithelium is denoted by dashed red lines. (D) (Upper) Ordinary differential equation (ODE) simulations of *Bmp4* expression dynamics in various *Apc;Pax9* genotypes. Mesenchymal *Bmp4* is decreased and flat in the $Pax9^{-/-}$ context, but in $Apc^{f/f};Pax9^{-/-}$ mutants the functional consequences of the decrease in mesenchymal *Bmp4* are supplanted by the ectopic expression of epithelial *Bmp4*. (Lower) Coronal sections at E14.0 showing that expression of epithelial *Bmp4* is increased and that of mesenchymal *Bmp4* expression is decreased in $Apc^{f/f};Pax9^{-/-}$ mutants compared to a control. [Adapted from Fig. 3, O'Connell et al. (2012).]

3 CONCLUSIONS AND FUTURE DIRECTIONS

3.1 Next-Generation GRNs for Organ Development

A central premise of the work described here is that fundamental principles for organ regeneration can be deduced from a detailed understanding of how organs form naturally. As illustrated here, multiple signaling pathways must be regulated coordinately in a tissue- and stage-specific manner for organogenesis and odontogenesis to occur. We posit that systems-based approaches can help provide the level of understanding of E-M interactions necessary to design efficient strategies for tooth and organ regeneration.

The Wnt-Bmp feedback circuit is a subnetwork contained within a larger odontogenic GRN. The systems analysis of Wnt and Bmp signaling in odontogenesis indicates that diffusible secreted signals can affect both tissue compartments simultaneously, and that the reciprocal signaling dynamics observed may arise as a natural consequence of tissue-specific differences in the GRN structure between the two compartments. Since the early stages of E-M interactions in many organs are functionally similar, intertissue self-sustaining feedback circuits of the type illustrated here may emerge as general features of organ formation that involve E-M interactions.

However, it is important to recognize the limitations imposed by the assumptions that underlie the odontogenic E-M GRN. For example, not all signaling pathways involved in tooth development involve diffusible signaling molecules. The Notch family of signaling molecules are important in tooth development and are not diffusible, but rather, mediate cell–cell communication directly between neighboring cells using a set of membrane receptors and ligands (Mitsiadis et al., 2010). In addition, it is unlikely that free diffusion models accurately reflect the action of signaling molecules during development. For example, although the effect of Wnt signaling molecules can be detected at a distance, they are hydrophobic due to covalent lipid modifications and have been reported to be membrane associated in many contexts (MacDonald et al., 2009). Similarly, other signaling molecules, such as Shh, Fgfs, and Bmps, are subject to postranslational modifications, ECM interactions, or other constraints that can modulate their activity.

In addition, alternative mRNA splicing of the Fgf receptor gene results in receptor variants Fgfr2b and Fgfr2c, which have different preferential binding affinity for Fgf ligands (Miki et al., 1992; Zhang et al., 2006). During tooth development, Fgfr2b is expressed predominantly in the dental epithelium, and Fgfr2c is expressed predominantly in the dental mesenchyme (Kettunen and Thesleff, 1998; Kettunen et al., 1998, 2000; Harada et al., 2002). Fgf10, which binds FGFR2b preferentially, is expressed in the dental mesenchyme and therefore seems to establish a unidirectional mode of communication from mesenchyme to epithelium. Furthermore, the ligands and receptors of the ectodysplasin (Eda) signaling pathway are important in tooth development and are expressed in distinct domains of the dental epithelium and therefore constitute an intraepithelial signaling network (Thesleff and Mikkola, 2002).

Beyond these limitations, the *in silico* simulation of Wnt-Bmp feedback circuit is largely qualitative and makes no attempt to quantify actual signaling molecule concentrations or effective pathway activities in the E-M compartments. The Wnt-Bmp feedback circuit also does not account for the heterogeneity in *Bmp4* expression in the lingual and buccal dental mesenchyme, or the intraepithelial patterning mechanism that enriches *Wnt* and *Bmp4* expression in the enamel knot. Furthermore, the identification of differentially regulated genes in Pax9 and Msx1 mouse mutants undoubtedly contains direct transcriptional targets as well as secondary or indirect targets. Hence, the Wnt-Bmp feedback circuit represents a substantially simplified model to help us understand the E-M signaling dynamics that regulate early tooth development. Nonetheless, just by making use of insights gleaned from simple odontogenic E-M GRNs, it was possible to devise a genetic strategy to restore dental differentiation in mouse mutants that have mutations in *Pax9* and *Msx1*.

Incorporation of additional high-throughput data will further refine the first-generation or "version 1.0" odontogenic GRN that we present here. For example, the application of advanced techniques in mass spectrometry will allow the quantification of protein signaling molecules and their inhibitors. Thus, incorporation of protein data will complement the extant transcriptional data in terms of quantifying the amount of bioactive ligands. Reporters of signaling pathway activity and signal transduction kinase assays can be used to quantify signaling pathway activity spatiotemporally. The application of RNA-seq will identify and quantify the expression of all RNA molecules, including microRNAs (miRNAs) and long noncoding RNAs (lncRNA). ChIP-seq can be used to identify binding sites of key transcription factors, such as Pax9 and Msx1, at a genome-wide scale. DNase-seq data can be used to detect dynamic changes in chromatin accessibility in dental epithelial and mesenchymal tissue during odontogenesis. In this way, through interdisciplinary collaborations between developmental biologists and computational and genome scientists, we envision the creation of new techniques in integrative genomics that will facilitate the continued incorporation of both systems-level and reductionist data into the odontogenic GRN. This approach, in turn, should allow scientists to refine our network understanding of early tooth development reiteratively and design effective strategies for tooth regeneration.

3.2 E-M Signaling Interaction and Strategies for Organ Generation

Systems biology has been applied effectively to the study of complex processes in stem cell biology, including self-renewal, differentiation, and reprogramming (Chang et al., 2008; Müller et al., 2008; MacArthur et al., 2009). Through the construction and analysis of a single GRN, these studies have yielded important insights into cell-fate decisions in embryonic and adult stem cells. Furthermore, protocols for the successful in vitro directed differentiation of ES cells to specific terminally differentiated cell types (e.g., pancreatic islet beta cells, neurons, cardiomyocytes) are frequently designed to approximate conditions that mimic endogenous embryonic development. To date, such protocols have been successful in the generation of several cell types (D'Amour et al., 2006) and tissues (Spence et al., 2011) from human embryonic and induced pluripotent stem cells (PSCs), in some cases with therapeutic efficacy in animal models (Kroon et al., 2008). However, in contrast to studies aimed at the generation of individual cell types, the challenge of how to properly assemble a multitude of cell types into a biologically complex multitissue organ for therapeutically effective organ function is even more difficult.

In strategies aimed at tooth regeneration, even the goal of biologically generating enamel covered dentin, or dental crown tissue, from stem cells raises fundamental questions. For example, does one take the approach of associating in vitro-derived ameloblast and odontoblast cell types, followed by engraftment, or does one aim to induce undifferentiated but odontogenically competent epithelial and mesenchymal precursors in a bioreactor and then guide this cell mixture along a path of odontogenic differentiation? Alternatively, what is the potential feasibility of harnessing endogenous progenitor populations in the jaw to form teeth in vivo?

One helpful and important unifying principle is that organ formation is the result of a series of autoregulatory, self-sustaining, and highly autonomous inductive tissue interactions. Therefore, in theory, if properly initiated and provided with the appropriate microenvironment, organogenesis can ensue largely independent of external intervention. This is illustrated by such striking biological results as the generation of ectopic eyes or eyelike structures in insects (Quiring et al., 1994; Halder et al., 1995), fish (Oliver et al., 1996), and amphibia (Chow et al., 1999) via the misexpression of components of a single regulatory pathway: that containing the transcription factor Pax6.

Odontogenesis also exemplifies this principle. For example, constitutive activation of canonical Wnt signaling in the oral epithelium leads to supernumerary tooth formation (Zhou et al., 1995; Järvinen et al., 2006; Liu et al., 2008; Wang et al., 2009). Moreover, when transplanted to a conducive environment in vivo, combinations of odontogenic embryonic epithelium as early as E9.5 and even premigratory neural crest has the intrinsic capacity to exchange signals with uninduced second arch mesenchyme and to form teeth (Mina and Kollar, 1987; Lumsden, 1988). In addition, the reassociation of single cells derived from the epithelium and mesenchyme of E14.5 cap-stage tooth germs allows for the reacquisition of positional identity and recommencement of morphogenesis in a mouse transplant model (Nakao et al., 2007; Ikeda et al., 2009; Takahashi et al., 2010). Therefore, a potentially efficient strategy for organ regeneration could involve assembly of mesenchyme encapsulated by a single layer of programmed epithelium, or vice versa. Such simple E-M combinations would be undifferentiated but poised for organogenic differentiation upon limited culture followed by engraftment into, for example, a tooth extraction site.

If tooth bioengineering strategies are to begin with assemblies of undifferentiated epithelial and mesenchymal cells, why does the theoretical mode of signaling tissue interaction really matter? Could one reasonably expect to achieve organogenesis simply by supplying these undifferentiated cell mixtures with an appropriate complement of signals and extracellular components? Some experimental support for this view comes from remarkable reconstructions of hearts (Ott et al., 2008) and lungs (Petersen et al., 2010) from progenitor heart and lung cells added to their respective decellularized extracellular matrices. However, the supply of available human hearts and lungs for decellularization and reseeding with progenitor cells is limited. The primary benefit of having an available GRN for organogenesis, although yet unproven, is that it provides a logical foundation from which to extract the minimal number of exogenous signals necessary to induce a cascade of self-sustaining E-M interactions that result in definitive multitissue organ formation.

It is from this approach, when attempting to induce a system that is composed of epithelium and mesenchyme into an odontogenic cascade, that a deep understanding of the mode of communication between the tissues is necessary. Our work on E-M GRNs indicates that the epithelial and mesenchymal GRNs are largely composed of different components with diverse response structures (Fig. 2B). In addition, the initiating signals are likely to affect both epithelial and mesenchymal compartments simultaneously and result in the differential production of subsequent signaling molecules that are necessary to sustain the developmental interaction. A systems-level understanding of gene regulation is important in this scenario because the manipulation of a signaling pathway in one tissue can affect multiple signaling pathways in both E-M tissue types. For example, epithelial Wnt pathway activation is able to restore dental differentiation in the mesenchymal defects of *Pax9* and *Msx1* mouse mutants. Understanding these signaling pathway responses and their connections from a network-based perspective will yield informed strategies for multitissue generation.

In summary, there is good reason to be optimistic about the future prospects for incorporating E-M GRNs into design strategies and tissue engineering protocols for the in vitro and in vivo synthesis of teeth and other organs. As stated earlier, it is our conviction that such work will require close interdisciplinary collaborations between stem cell and developmental biologists, bioengineers, and computational biologists. This chapter highlights the potential utility of systems biology to understand the fundamental principles of organ development and to assist in efforts to reconstitute organ formation de novo.

Acknowledgments

This work was supported by the National Institutes of Health Common Fund through National Institute of Dental and Craniofacial Research grants RL1DE019021 and UL1DE019581 and also by R37DE011697 (R.L.M.) and NIBIB grant RL9EB008539 (D.J.O., J.W.K.H.). We thank members of SysCODE and the Maas laboratory for insightful discussions.

REFERENCES

Andl T, Reddy ST, Gaddapara T, Millar SE (2002). WNT signals are required for the initiation of hair follicle development. *Dev Cell* 2:643–653.

Andl T, Ahn K, Kairo A, Chu EY, Wine-Lee L, Reddy ST, et al. (2004). Epithelial Bmpr1a regulates differentiation and proliferation in postnatal hair follicles and is essential for tooth development. *Development* 131:2257–2268.

Bei M (2009). Molecular genetics of tooth development. *Curr Opin Genet Dev* 19:504–510.

Bei M., Kratochwil K., Maas RL (2000). BMP4 rescues a non-cell-autonomous function of Msx1 in tooth development. *Development* 127:4711–4718.

Bénazet J-D, Bischofberger M, Tiecke E, calves AG, Martin JF, Zuniga A, et al. (2009). A self-regulatory system of interlinked signaling feedback loops controls mouse limb patterning. *Science* 323:1050–1053.

Catron KM, Zhang H, Marshall SC, Inostroza JA, Wilson JM, Abate C (1995). Transcriptional repression by Msx-1 does not require homeodomain DNA-binding sites. *Mol Cell Biol* 15:861–871.

Chang HH, Hemberg M, Barahona M, Ingber DE, Huang S (2008). Transcriptome-wide noise controls lineage choice in mammalian progenitor cells. *Nature* 453:544–547.

Chen Y, Bei M., Woo I, Satokata I, Maas R (1996). Msx1 controls inductive signaling in mammalian tooth morphogenesis. *Development* 122:3035–3044.

Chow RL, Altmann CR, Lang RA, Hemmati-Brivanlou A (1999). Pax6 induces ectopic eyes in a vertebrate. *Development* 126:4213–4222.

D'Amour KA, Bang AG, Eliazer S, Kelly OG, Agulnick AD, Smart NG, et al. (2006). Production of pancreatic hormone-expressing endocrine cells from human embryonic stem cells. *Nat Biotechnol* 24:1392–1401.

Dassule HR, McMahon AP (1998). Analysis of epithelial–mesenchymal interactions in the initial morphogenesis of the mammalian tooth. *Dev Biol* 202:215–227.

Davidson EH (2010). Emerging properties of animal gene regulatory networks. *Nature* 468:911–920.

Ferguson CA, Tucker AS, Christensen L, Lau AL, Matzuk MM, Sharpe PT (1998). Activin is an essential early mesenchymal signal in tooth development that is required for patterning of the murine dentition. *Genes Dev* 12:2636–2649.

Ferguson CA, Tucker AS, Heikinheimo K, Nomura M, Oh P, Li E, et al. (2001). The role of effectors of the activin signalling pathway, activin receptors IIA and IIB, and Smad2, in patterning of tooth development. *Development* 128:4605–4613.

Fujimori S, Novak H, Weissenböck M, Jussila M, calves AG, Zeller R, et al. (2010). Wnt/β-catenin signaling in the dental mesenchyme regulates incisor development by regulating Bmp4. *Dev Biol* 348:97–106.

Grobstein C (1953). Morphogenetic interaction between embryonic mouse tissues separated by a membrane filter. *Nature* 172:869–870.

Grobstein C (1956). Trans-filter induction of tubules in mouse metanephrogenic mesenchyme. *Exp Cell Res* 10:424–440.

Grobstein C (1967). Mechanisms of organogenetic tissue interaction. *Natl Cancer Inst Monogr* 26:279–299.

Gurdon JB (1987). Embryonic induction–molecular prospects. *Development* 99:285–306.

Halder G, Callaerts P, Gehring WJ (1995). Induction of ectopic eyes by targeted expression of the eyeless gene in *Drosophila*. *Science* 267:1788–1792.

Harada H, Toyono T, Toyoshima K, Yamasaki M, Itoh Nobuyuki, Kato S, et al. (2002). FGF10 maintains stem cell compartment in developing mouse incisors. *Development* 129:1533–1541.

Hardcastle Z, Mo R, Hui CC, Sharpe PT (1998). The Shh signalling pathway in tooth development: defects in *Gli2* and *Gli3* mutants. *Development* 125:2803–2811.

Ho JWK, Charleston MA (2011). Network modelling of gene regulation. *Biophys Rev* 3:1–13.

Ikeda E, Morita R, Nakao K, Ishida K, Nakamura T, Takano-Yamamoto T, et al. (2009). Fully functional bioengineered tooth replacement as an organ replacement therapy. *Proc Natl Acad Sci USA* 106:13475–13480.

Järvinen E, Salazar-Ciudad I, Birchmeier W, Taketo MM, Jernvall Jukka, Thesleff Irma (2006). Continuous tooth generation in mouse is induced by activated epithelial Wnt/beta-catenin signaling. *Proc Natl Acad Sci USA* 103:18627–18632.

Jernvall J, Thesleff I (2000). Reiterative signaling and patterning during mammalian tooth morphogenesis. *Mech Dev* 92:19–29.

Jernvall J., Kettunen P, Karavanova I, Martin LB, Thesleff I. (1994). Evidence for the role of the enamel knot as a control center in mammalian tooth cusp formation: non-dividing cells express growth stimulating *Fgf-4* gene. *Int J Dev Biol* 38:463–469.

Jernvall J, Aberg T, Kettunen P, Keränen S, Thesleff I (1998). The life history of an embryonic signaling center: BMP-4 induces p21 and is associated with apoptosis in the mouse tooth enamel knot. *Development* 125:161–169.

Jessell TM, Melton DA (1992). Diffusible factors in vertebrate embryonic induction. *Cell* 68:257–270.

Jia S, Zhou J, Gao Y, Baek J, Martin JF, Lan Y, Jiang R (2013). Roles of Bmp4 during tooth morphogenesis and sequential tooth formation. *Development* 140:423–432.

Jumlongras D, Bei M, Stimson JM, Wang WF, DePalma SR, Seidman CE, et al. (2001). A nonsense mutation in *MSX1* causes Witkop syndrome. *Am J Hum Genet* 69:67–74.

Karlebach G, Shamir R (2008). Modelling and analysis of gene regulatory networks. *Nat Rev Mol Cell Biol* 9:770–780.

Kettunen P, Thesleff I (1998). Expression and function of FGFs-4, -8, and -9 suggest functional redundancy and repetitive use as epithelial signals during tooth morphogenesis. *Dev Dyn* 211:256–268.

Kettunen P, Karavanova I, Thesleff I. (1998). Responsiveness of developing dental tissues to fibroblast growth factors: expression of splicing alternatives of FGFR1,-2,-3, and of FGFR4; and stimulation of cell proliferation by FGF-2, -4, -8, and -9. Dev Genet. *Dev Genet.* 22:374–85.

Kettunen P, Laurikkala J, Itäranta P, Vainio S, Itoh N., Thesleff I. (2000). Associations of FGF-3 and FGF-10 with signaling networks regulating tooth morphogenesis. *Dev Dyn* 219:322–332.

Kirouac DC, Ito C, Csaszar E, Roch A, Yu M, Sykes EA, et al. (2010). Dynamic interaction networks in a hierarchically organized tissue. *Mol Syst Biol* 6:1–16.

Kirschner MW (2005). The meaning of systems biology. *Cell* 121:503–504.

Kitano H (2002). Systems biology: a brief overview. *Science* 295:1662–1664.

Koch WE (1967). In vitro differentiation of tooth rudiments of embryonic mice: I. Transfilter interaction of embryonic incisor tissues. *J Exp Zool* 165:155–169.

Kratochwil K., Dull M, Farinas I, Galceran J., Grosschedl R. (1996). *Lef1* expression is activated by BMP-4 and regulates inductive tissue interactions in tooth and hair development. *Genes Dev* 10:1382–1394.

Kratochwil K, Galceran J, Tontsch S, Roth W, Grosschedl (2002). FGF4, a direct target of LEF1 and Wnt signaling, can rescue the arrest of tooth organogenesis in Lef1(−/−) mice. *Genes Dev* 16:3173–3185.

Kroon E, Martinson LA, Kadoya K, Bang AG, Kelly OG, Eliazer S, et al. (2008). Pancreatic endoderm derived from human embryonic stem cells generates glucose-responsive insulin-secreting cells in vivo. *Nat. Biotechnol* 26:443–452.

Liu F, Chu EY, Watt B, Zhang Yuhang, Gallant NM, Andl T, et al. (2008). Wnt/beta-catenin signaling directs multiple stages of tooth morphogenesis. *Dev Biol* 313:210–224.

Lumsden AG (1988). Spatial organization of the epithelium and the role of neural crest cells in the initiation of the mammalian tooth germ. *Development* 103(Suppl): 155–169.

MacArthur BD, Ma'ayan A, Lemischka IR (2009). Systems biology of stem cell fate and cellular reprogramming. *Nat Rev Mol Cell Biol* 10:672–681.

MacDonald BT, Tamai K, He X (2009). Wnt/β-catenin signaling: components, mechanisms, and diseases. *Dev Cell* 17:9–26.

McAdams HH, Shapiro L (1995). Circuit simulation of genetic networks. *Science* 269:650–656.

Meyers EN, Lewandoski M, Martin GR (1998). An *Fgf8* mutant allelic series generated by Cre- and Flp-mediated recombination. *Nat Genet* 18:136–141.

Miki T, Bottaro DP, Fleming TP, Smith CL, Burgess WH, Chan AM, et al. (1992). Determination of ligand-binding specificity by alternative splicing: two distinct growth factor receptors encoded by a single gene. *Proc Natl Acad Sci USA* 89:246–250.

Mina M, Kollar EJ (1987). The induction of odontogenesis in non-dental mesenchyme combined with early murine mandibular arch epithelium. *Arch Oral Biol* 32:123–127.

Mitsiadis TA, Graf D, Luder H, Gridley T, Bluteau G (2010). BMPs and FGFs target Notch signalling via Jagged 2 to regulate tooth morphogenesis and cytodifferentiation. *Development* 137:3025–3035.

Müller F-J, Laurent LC, Kostka D, Ulitsky I, Williams R, Lu C, et al. (2008). Regulatory networks define phenotypic classes of human stem cell lines. *Nature* 455:401–405.

Nakao K, Morita R, Saji Y, Ishida K, Tomita Y, Ogawa M, et al. (2007). The development of a bioengineered organ germ method. *Nat Methods* 4:227–230.

Nakatomi M, Wang X-P, Key D, Lund JJ, Turbe-Doan A, Kist R, et al. (2010). Genetic interactions between *Pax9* and *Msx1* regulate lip development and several stages of tooth morphogenesis. *Dev Biol* 340:438–449.

O'Connell DJ, Ho JWK, Mammoto T, Turbe-Doan A, O'Connell JT, Haseley PS, et al. (2012). A Wnt-Bmp feedback circuit controls intertissue signaling dynamics in tooth organogenesis. *Sci Signal* 5:ra4.

Ogawa T, Kapadia H, Feng JQ, Raghow R, Peters Heiko, D'Souza Rena N (2006). Functional consequences of interactions between *Pax9* and *Msx1* genes in normal and abnormal tooth development. *J Biol Chem* 281:18363–18369.

Oliver G, Loosli F, Köster R, Wittbrodt J, Gruss P (1996). Ectopic lens induction in fish in response to the murine homeobox gene *Six3*. *Mech Dev* 60:233–239.

Ott HC, Matthiesen TS, Goh S-K, Black LD, Kren SM, Netoff TI, et al. (2008). Perfusion-decellularized matrix: using nature's platform to engineer a bioartificial heart. *Nat Med* 14:213–221.

Peter IS, Davidson EH (2011). A gene regulatory network controlling the embryonic specification of endoderm. *Nature* 474:635–639.

Peters H., Neubüser A, Kratochwil K., Balling R (1998). Pax9-deficient mice lack pharyngeal pouch derivatives and teeth and exhibit craniofacial and limb abnormalities. *Genes Dev* 12:2735–2747.

Petersen TH, Calle EA, Zhao L, Lee EJ, Gui L, Raredon MB, et al. (2010). Tissue-engineered lungs for in vivo implantation. *Science* 329:538–541.

Plikus MV, Zeichner-David M, Mayer J-A, Reyna J, Bringas P, Thewissen JGM, et al. (2005). Morphoregulation of teeth: modulating the number, size, shape and differentiation by tuning Bmp activity. *Evol Dev* 7:440–457.

Quiring R, Walldorf U, Kloter U, Gehring WJ (1994). Homology of the eyeless gene of *Drosophila* to the Small eye gene in mice and Aniridia in humans. *Science* 265:785–789.

Sarkar L, Sharpe PT (1999). Expression of Wnt signalling pathway genes during tooth development. *Mech Dev* 85:197–200.

Satokata I, Maas R (1994). *Msx1* deficient mice exhibit cleft palate and abnormalities of craniofacial and tooth development. *Nat Genet* 6:348–356.

Saxén L, Thesleff I (1992). Epithelial–mesenchymal interactions in murine organogenesis. *Ciba Found Symp* 165:183–193; discussion 193–198.

Spemann H, Mangold H (2001). Induction of embryonic primordia by implantation of organizers from a different species 1923. *Int J Dev Biol* 45:13–38.

Spence JR, Mayhew CN, Rankin SA, Kuhar MF, Vallance JE, Tolle K, et al. (2011). Directed differentiation of human pluripotent stem cells into intestinal tissue in vitro. *Nature* 470:105–109.

Stockton DW, Das P, Goldenberg M, D'Souza R N, Patel PI (2000). Mutation of *PAX9* is associated with oligodontia. *Nat Genet* 24:18–19.

Takahashi C, Yoshida H, Komine A, Nakao K, Tsuji T, Tomooka Y (2010). Newly established cell lines from mouse oral epithelium regenerate teeth when combined with dental mesenchyme. In vitro *Cell Dev Biol Anim* 46:457–468.

Thesleff I (2003). Epithelial–mesenchymal signalling regulating tooth morphogenesis. *J Cell Sci* 116:1647–1648.

Thesleff I, Mikkola ML (2002). Death receptor signaling giving life to ectodermal organs. *Sci STKE* 2002:pe22.

Thesleff I., Vaahtokari A, Partanen AM (1995). Regulation of organogenesis.Common molecular mechanisms regulating the development of teeth and other organs. *Int J Dev Biol* 39:35–50.

Thesleff I., Lehtonen E, Wartiovaara J, Saxén L (1977). Interference of tooth differentiation with interposed filters. *Dev Biol* 58:197–203.

Vainio S, Karavanova I, Jowett A, Thesleff I (1993). Identification of BMP-4 as a signal mediating secondary induction between epithelial and mesenchymal tissues during early tooth development. *Cell* 75:45–58.

van Genderen C, Okamura RM, Fari nas I, Quo RG, Parslow TG, Bruhn L, et al. (1994). Development of several organs that require inductive epithelial–mesenchymal interactions is impaired in *LEF-1*-deficient mice. *Genes Dev* 8:2691–2703.

Vastardis H, Karimbux N, Guthua SW, Seidman JG, Seidman CE (1996). A human MSX1 homeodomain missense mutation causes selective tooth agenesis. *Nat Genet* 13:417–421.

Wang X-P, Suomalainen M, Felszeghy S, Zelarayan LC, Alonso MT, Plikus MV, et al. (2007). An integrated gene regulatory network controls stem cell proliferation in teeth. *PLoS Biol* 5:e159.

Wang X-P, O'Connell DJ, Lund JJ, Saadi I, Kuraguchi M, Turbe-Doan A, et al. (2009). Apc inhibition of Wnt signaling regulates supernumerary tooth formation during embryogenesis and throughout adulthood. *Development* 136:1939–1349.

Wartiovaara J, Nordling S, Lehtonen E, Saxén L (1974). Transfilter induction of kidney tubles: correlation with cytoplasmic penetration into nucleopore filters. *J Embryol Exp Morphol* 31:667–682.

Xu X, Han J, Ito Y, Bringas P, Deng C, Chai Y (2008). Ectodermal Smad4 and p38 MAPK are functionally redundant in mediating TGF-beta/BMP signaling during tooth and palate development. *Dev Cell* 15:322–329.

Zhang X, Ibrahimi OA, Olsen SK, Umemori H, Mohammadi M, Ornitz DM (2006). Receptor specificity of the fibroblast growth factor family: the complete mammalian FGF family. *J Biol Chem* 281:15694–15700.

Zhang Y., Zhang Z., Zhao X, Yu X, Hu Y, Geronimo B, et al. (2000). A new function of BMP4: dual role for BMP4 in regulation of Sonic hedgehog expression in the mouse tooth germ. *Development* 127:1431–1443.

Zhang Z, Lan Y, Chai Y, Jiang R (2009). Antagonistic actions of Msx1 and Osr2 pattern mammalian teeth into a single row. *Science* 323:1232–1234.

Zhou J, Gao Y, Zhang Zunyi, Zhang Yuan, Maltby KM, Liu Z, et al. (2011). Osr2 acts downstream of Pax9 and interacts with both Msx1 and Pax9 to pattern the tooth developmental field. *Dev Biol* 353:344–353.

Zhou P, Byrne C, Jacobs J, Fuchs E (1995). Lymphoid enhancer factor 1 directs hair follicle patterning and epithelial cell fate. *Genes Dev* 9:700–713.

PART II

STEM CELLS AND THEIR NICHES IN CRANIOFACIAL TISSUES

10

STEM CELLS, INDUCED PLURIPOTENT STEM CELLS, AND THEIR DIFFERENTIATION TO SPECIFIED LINEAGE FATES

GEORGE T.-J. HUANG

Department of Bioscience Research, College of Dentistry, The University of Tennessee Health Science Center, Memphis, Tennessee

XIAO-YING ZOU

Department of Endodontics, Boston University Henry M. Golden School of Dental Medicine, Boston, Massachusetts, and Department of Cariology, Endodontology, and Operative Dentistry, School and Hospital of Stomatology, Peking University, Beijing, China

XING YAN

Department of Endodontics, Boston University Henry M. Golden School of Dental Medicine, Boston, Massachusetts, and Department of Stomatology, Beijing Friendship Hospital, Capital Medical University, Beijing, China

KYLE J. HEWITT AND YULIA SHAMIS

Program in Cell, Molecular and Developmental Biology, Sackler School of Graduate Biomedical Sciences, Tufts University, Boston, Massachusetts

JONATHAN A. GARLICK

Division of Cancer Biology and Tissue Engineering, Department of Oral and Maxillofacial Pathology, School of Dental Medicine, Tufts University, Boston, Massachusetts

1 INTRODUCTION

Stem cell biology focuses on an understanding of a full spectrum of cell behaviors from the stem cell state to the fully differentiated condition. It examines the dynamics

Stem Cells in Craniofacial Development and Regeneration, First Edition.
Edited by George T.-J. Huang, Irma Thesleff.

of cell physiology as to how stem cells maintain their stemness and give rise to specialized cells. Furthermore, this field has become a fundamental underpinning for the establishment of regenerative medicine and cell-based therapies.

In the area of adult stem cells (also known as postnatal stem cells), studies of hematopoietic stem cells (HSCs) that defined their ability to undergo self-renewal and form clonogenic colonies have been applied to define mesenchymal stem cells (MSCs) (Becker et al., 1963; Wu et al., 1967). Bone marrow–derived mesenchymal stem cells (BMMSCs) are the best characterized MSCs, and they were described as colony-forming unit fibroblasts (CFU-Fs) in vitro (Cohnheim, 1867; Friedenstein et al., 1976; Caplan, 1991; Pittenger et al., 1999). The search for MSC-like cells in specific tissues has led to the discovery of a variety of stem cells in every organ and tissue in the body in the past decades, including adipose tissues, umbilical cord blood, and dental tissues (Mareschi et al., 2001; Zuk et al., 2001; Baksh et al., 2004; Porada et al., 2006; Kolf et al., 2007; Huang et al., 2009).

These adult stem cells are versatile in terms of their potential to differentiate into various lineages of cell types. However, this potency is not comparable to embryonic stem cells (ESCs), which can give rise to basically every cell type, a characteristic termed *pluripotency*. ESCs are being studied rigorously in the hope that one day they will be used in the clinic for tissue regeneration and curing diseases. The setback regarding ESCs is their involvement of embryos. Additionally, due to their potency in growth, tumor formation is a concern when transplanting ESCs into hosts for regeneration or therapeutic purposes. Nonetheless, ESCs have become a very importance cell type from the perspective of understanding the biology of stem cells.

Different approaches to manipulating cells to become ES-like cells have been tested. From a stem cell to a fully differentiated cell is not a unidirectional pathway but a reversible process. A differentiated somatic cell can be reprogrammed into a stem cell. Among various methods, a consistent and reliable technology discovered recently to reprogram somatic cells into ES-like cells termed *induced pluripotent stem cells* (iPSCs) opened a new page for understanding cell reprogramming (Takahashi and Yamanaka, 2006; Takahashi et al., 2007; Yu et al., 2007). iPSCs were first established by delivering the four factors *C-MYC/KLF4/OCT4/SOX2* or *LIN28/NANOG/OCT4/SOX2* into dermal fibroblasts (DFs). iPSCs provided insight into an understanding of cell differentiation and de-differentiation. From the perspective of clinical application, patient-specific iPSCs may be established for various medical purposes. There are, however, drawbacks in iPSCs, including mutations that occur during the process of reprogramming (Gore et al., 2011; Lister et al., 2011). Nonetheless, iPSCs may become a powerful cell source for clinical medicine in the future if the shortcomings can be overcome.

2 DEFINITION AND CONCEPTS OF STEM CELLS

2.1 Definitions

The general definition of stem cells is that they can self-renew and differentiate into different lineages of specialized cells. Depending on the type of stem cells and their ability and potency to become different tissues, the following categories of stem cells have been established: (1) totipotent stem cells: cells each of which is capable of developing into an entire organism; (2) pluripotent stem cells: cells from embryos—ESCs that when grown in the right environment in vivo are capable of forming all types

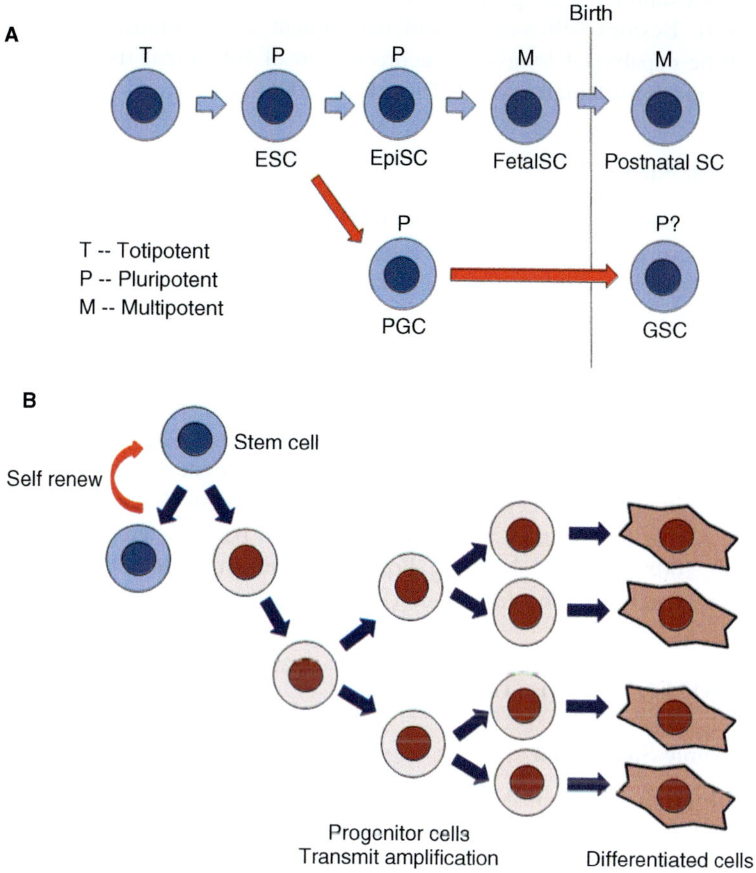

FIGURE 1 Stem cell potency at different developmental stages (A) and stem cell properties (B). ESC, embryonic stem cell; EpiSC, epiblast stem cell; PGC, primordial germ cell; GSC, germline stem cell.

of tissues; and (3) multipotent stem cells: fetal stem cells and postnatal stem cells (adult stem cells) that are capable of giving rise to multiple lineages of cells (Robey and Bianco, 2006). In mammalian systems, an organism begins as a fertilized egg that divides. During the first few divisions, each daughter cell is totipotent. In humans, cells at the four- to eight-cell stage are totipotent, after which the cells become pluripotent. At the blastocyst stage, cells in the inner cell mass can be isolated and grown in cultures and are termed *ESC lines*. The types of stem cells during early development are summarized in Fig. 1A.

Although the general definition of stem cells seems straightforward, the specific definition remains illusive, including the term *stemness*. Two theories of stemness have been considered: entity theory and state theory. Entity theory holds that stemness is a property of a cell: the self-renewal and multilineage differentiation capacity (Fig. 1B). State theory holds that stemness is a state of a cell: a state of high plasticity with either a less or more committed state (Leychkis et al., 2009). A more specific definition for stem cells examines their gene signatures, markers, niches, properties, and so on (Zipori, 2004). Besides self-renewal and multipotent differentiation properties, stem cells should be capable of long-term reconstitution of functional tissue as well as of exhibiting serial transplantability (Shafritz et al., 2006).

2.2 Stem Cell Niche

The term *niche* was introduced by Schofield (1978). The stem cell niche is based on the concept that there is a local microenvironment in the tissue where a stem cell resides and there are coordinated interactions between the stem cell and its local microenvironment. The players involved in these interactive activities include, but are not limited to, extracellular matrix, adjacent differentiated cells, secreted and cell surface molecules, mechanical signals, spatial arrangements, and certain metabolic conditions. Together they serve two purposes: (1) to maintain tissue homeostasis and (2) to regenerate damaged tissues (Moore and Lemischka, 2006; Roeder et al., 2011). Stem cell niches are considered to be tissue specific, such as hematopoietic stem cells (HSCs) in bone marrow and epithelial stem cells in the intestinal crypt with different local microenvironments. As for MSCs, they have been reported to be associated with vasculature (Crisan et al., 2008; Ergun et al., 2011). For specific information on the stem cell niche in dental tissues, refer to Chapters 15 & 18; and in muscle, refer to Chapter 13.

3 PLURIPOTENT STEM CELLS

3.1 Embryonic Stem Cells

Mouse (m) ESCs were established in 1981 (Evans and Kaufman, 1981; Martin, 1981) and in 1998 human (h) ESC isolation was reported (Thomson et al., 1998). hESCs are derived from embryos after in vitro fertilization. These extra embryos are not used clinically or are not suitable for implantation into a patient and therefore may be donated for research that would otherwise be discarded. The inner cell mass from the blastocyst stage of the embryo (5-day) is separated from the trophectoderm, which is removed by immunosurgery. Cells from the inner cell mass are plated onto a layer of feeder cells, which are usually mouse embryonic fibroblasts (MEFs). The hESC line is then established. These cells are maintained in an undifferentiated state and expanded on feeder cells or in a feeder-free culture system (Fig. 2A and 2B). This cell source, which is capable of giving rise to all types of cells, has served as a powerful tool in our understanding of stem cell stemness and differentiation. The ESC properties have been considered as a gold standard for pluripotent stem cell–based regenerative medicine. The major setback of hESCs for clinical use is the lack of identical genetics between the donor cells and the recipients. Nonetheless, the pluripotent nature of ESCs is unparalleled by any multipotent stem cells in terms of their versatility for tissue regeneration. Researchers have initiated clinical trials using ESCs for treatment of debilitating diseases. In 2009, the pharmaceutical company Geron was the first company approved by the U.S. Food and Drug Administration to carry out a clinical trial to treat spinal cord injury using hESCs, but for financial reasons that effort was abandoned in 2011. In 2012, a preliminary report of a clinical trial showed that two patients having Stargardt's macular dystrophy and dry age-related macular degeneration improved their vision significantly after receiving hESC-derived retinal pigment epithelium transplant.

The patients were under immunosuppression for 8 weeks. There were no signs of hyperproliferation, tumorigenicity, ectopic tissue formation, or apparent rejection after four months (Schwartz et al., 2012). This first clinical trial report demonstrates the promise of the clinical use of hESCs.

3.2 Verification of ESC Pluripotency

To examine ESC differentiation potential in vitro, the ESC colonies are grown in suspension to allow embryoid body (EB) formation (Fig. 2 D, F, and G). Within the EB, cells are randomly differentiated into various cell types. When EBs are allowed to attach to the culture dish, cells in the EBs will migrate out and differentiated into cell types representing all three germ layers. ESCs can also be guided to differentiate into specific cell types under determined chemical conditions. To verify their pluripotency in vivo, ESCs are injected into the subcutaneous space of immunocompromised mice. Teratomas will form and contain various primitive tissues representing all three germ layers (Fig. 2 C and E). Formation of teratoma in vivo is an important property for pluripotent cells such as ESCs.

In the mouse system, mESCs can form chimera animals after cells are injected into mouse blastocysts and transplanted into surrogate mothers. This ability to form chimeras is another important hallmark demonstrating cell pluripotency.

3.3 Cell Reprogramming

Besides ESCs, there are several approaches to acquiring pluripotent stem cells.

Somatic Cell Nuclear Transfer (SCNT). In mammals, an adult somatic cell nucleus can be reprogrammed to a state that confers totipotency, as evidenced in the creation of Dolly, the sheep (Wilmut et al., 1997). Although cloning of some mammalian species may be achieved (Miyashita et al., 2002; Gómez et al., 2006; Oh et al., 2009; Thuan et al., 2010), human SCNT so far has not been successful in giving rise to blastocyst embryos for ESC isolation.

Parthenogenesis. This is a phenomenon that does not occur in mammals but can be induced artificially. A diploid oocyte, obtained by blocking the second polar body extrusion, is stimulated to mimic fertilization. The oocyte divides and develops into a blastocyst, from which ESCs can be derived. Human parthenogenetic approaches that generate hESCs are potentially useful for clinical applications (Revazova et al., 2007, 2008).

Triploid Blastocyst. To generate human triploid blastocyst, the oocyte genome is not removed and the somatic cell genome is added. The resulting triploid cells develop to the blastocyst stage. Stem cell lines derived from these blastocysts differentiate into cell types of all three germ layers, and a pluripotent gene expression program is established on the genome derived from the somatic cell (Noggle et al., 2011).

In mouse systems, primordial germ cells or germline stem cells and bone marrow cells may become pluripotent in cultures (Matsui et al., 1992; Jiang et al., 2002; Kanatsu-Shinohara et al., 2004).

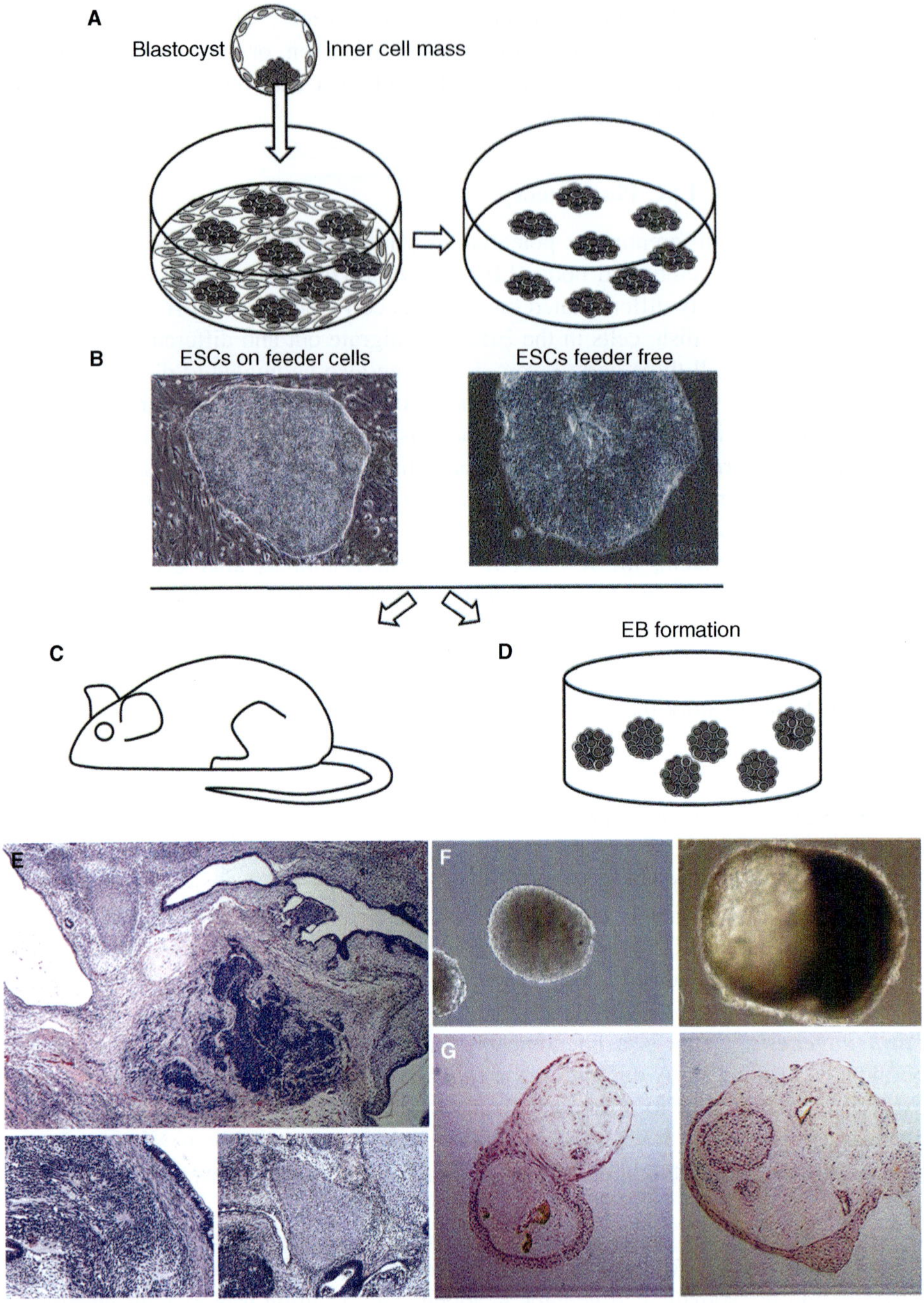

FIGURE 2 Derivation of hESCs: culturing and basic characteristics. (A) hESCs are isolated from the inner cell mass of a blastocyst and grown on feeder cells and may be maintained in feeder free cultures. (B) Bright-field microscopic images of hESCs grown on feeder cells or on feeder-free conditions. (C) The hESCs can be transplanted into immunocompromised mice to allow terotoma formation. (D) hESCs can form embryoid bodies (EBs) when detached from the culture dish and allowed to float in the culture medium. (E) Histological images of teratomas showing various primitive ectodermal (neural cells), mesodermal (cartilage), and endodermal (endothelium) tissues. (F) Microscopic images of EB in cultures. Left, after 10 days; Right, after 3 weeks. (G) Histological images of EB showing the formation of primitive tissues inside the mass.

Induced Pluripotent Stem Cells (iPSCs). Yamanaka and his team reported in 2006 that by delivering four factors—*c-Myc, Klf4, Oct4*, and *Sox2*—into mouse dermal fibroblasts (DFs) via viral vectors they were able to reprogram the DFs into ES-like cells termed iPSCs (Takahashi and Yamanaka, 2006). These miPSCs demonstrate the following features: (1) show ESC morphology in cultures, (2) have similar cell expansion rate to ESCs, (3) express key ESC genes, (4) have global gene profiles resembling those of ESCs, (5) have epigenetic profiles similar to those of genes involved in ESC maintenance and development, (6) form EBs in cultures, (7) can differentiate into cells of all germ layers in EBs in cultures, (8) form teratomas in vivo containing tissues of all germ layers, and (9) form chimeras after iPSCs injected into blastocysts.

Subsequently, Yamanaka's group demonstrated that the same approach can generate iPSCs in the human system (Takahashi et al., 2007). Independently, Thomson and his associates identified a core set of four genes—*OCT4, SOX2, NANOG*, and *LIN28*—that were able to reprogram human fetal and foreskin fibroblasts into iPSCs with ESC characteristics (Yu et al., 2007).

3.4 Vector- and Transgene-Free iPSCs

Exogenous vectors carrying transgenes were used to generate iPSCs. However, having integrated viral vectors with strong promoter driving the exogenous genes in the cells prevents their clinical applications. Tremendous efforts have been made to deliver the factors defined without having viral vector integration. The approaches include transient expression using adenoviral or nonviral vectors (Stadtfeld et al., 2008; Gonzalez et al., 2009) or nonintegrating episomal vectors (Yu et al., 2009), and removing the integrated vectors using piggyBac transposition or a loxP/Cre-recombinase excisable viral vector system (Gonzalez et al., 2009; Kaji et al., 2009; Soldner et al., 2009; Woltjen et al., 2009). Circumventing the use of vectors completely by delivery of the four recombinant protein-based factors to generate iPSCs in the mouse and human systems has been reported (Kim et al., 2009; Zhou et al., 2009).

3.5 Similarity of iPSCs to ESCs

Telomere and Telomerase Activity of iPSCs. Regaining telomere length is an important feature to indicate the completeness of reprogramming of iPSCs. High levels of telomerase activity were detected in hiPSCs (Takahashi et al., 2007). Regarding the telomere regaining length in iPSCs, it was addressed by using the disease state dyskeratosis congenita (DC), a disorder of telomere maintenance, as a study model. It was found that reprogrammed DC cells overcame a critical limitation in telomerase RNA component levels to restore telomere maintenance and self-renewal. Reprogramming restores telomere elongation in DC cells, despite genetic lesions affecting telomerase (Agarwal et al., 2010).

iPSCs Form Live Adult Animals via Tetraploid Complementation. An important hallmark of ES cells as pluripotent stem cells is the ability to form embryos and live animals via a tetraploid-complementation procedure. This process is commonly performed using a mouse system in which a diploid embryo at the two-cell stage is fused artificially into a tetraploid cell and allowed to form a blastocyst. ESCs are then injected into the blastocyst, which is then implanted into a surrogate mother mouse for pregnancy. Tetraploid cells will develop into extraembryonic tissues (e.g., placenta), while the injected ESCs will form the diploid embryo, which is born and develops into an adult mouse. Generation of live pups by iPSCs, some of whom lived to adulthood, was demonstrated by independent research groups (Boland et al., 2009; Kang et al., 2009; Zhao et al., 2009). However, hiPSCs cannot be tested via tetraploid complementation; therefore, further characterization at molecular levels must be carried out to understand the level of reprogramming in hiPSC clones.

Difference Between iPSCs and ESCs. Recent evidence showed that iPSCs may conserve some "memory" of the original cells. Since the iPSC derivation involved epigenetic reprogramming, it is considered that this memory may be determined by the epigenetic status. iPSCs derived from different cell types may also differ in their ability to undergo guided differentiation into specialized cells compared to ES cells (Yamanaka, 2009). Therefore, iPSCs should be generated from various easily accessible human tissues and characterized thoroughly. It has been noted that differences occur between the iPSCs generated from different cell types (e.g., mesenchymal vs. endodermal origin) in terms of the kinetics of reprogramming and the outcomes of the chimeric mice generated (Sridharan and Plath, 2008).

3.6 Current Issues on iPSCs for Therapeutic Use

Feasible Cell Types for iPSC Generation. Fibroblasts from oral mucosa can be reprogrammed into iPSCs (Miyoshi et al., 2010a). Because oral tissue biopsy does not leave a scar, oral mucosa fibroblasts appear to be the best cell source for iPSC derivation in terms of accessibility and feasibility. Blood cells would have been another easily obtainable cell type for iPSC generation. However, that requires isolation of a subpopulation for successful reprogramming (Loh et al., 2009; Seki et al., 2010). Discarded tissues or fluids regarded as biomedical waste can also serve as a source of iPSC reprogramming, such as amniotic fluid, umbilical cord, foreskin, adipose tissues, and shed or extracted teeth. Dental tissue–derived stem and progenitor cells, including dental pulp stem cells (DPSCs), stem cells from exfoliated deciduous teeth (SHED), and stem cells from apical papilla (SCAP), have been reprogrammed into iPSCs at a higher rate than that of fibroblasts (Yan et al., 2010).

Epigenetics and Reprogramming. Although the epigenetic profiles of iPSCs are similar to those of ESCs, the epigenomic reconfiguration of the iPSC genome remains incomplete. Large-scale methylation analysis has identified differences between ESCs and iPSCs as well as between individual iPSC clones. In general, levels of CpG methylation in iPSCs are higher than those in ESC cell lines (Deng et al., 2009).

Whole-genome screening of the methylome comparing hiPSC to hESC lines has revealed large differentially methylated regions (DMRs) within iPSCs that are resistant to reprogramming, which may lead to aberrant gene expression when cells are differentiated from them (Lister et al., 2011). The whole-genome profiles of DNA methylation at single-base resolution of hiPSC lines showed significant reprogramming variability, including somatic memory and aberrant reprogramming of DNA methylation. There is both incomplete reprogramming of non-CG methylation and differences in CG methylation and histone modifications. Furthermore, errors in reprogramming

CG methylation are transmitted at a high frequency, providing an iPSC reprogramming signature that is maintained after differentiation (Lister et al., 2011).

The DMRs found in iPSCs are thought to harbor a residual DNA methylation signature related to their cell of origin, termed *epigenetic memory*, that predisposes them toward differentiation along lineages related to the cell type that restricts differentiation to alternative cell fates (Kim et al., 2010; Polo et al., 2010). Epigenetic memory can also be correlated with a residual transcriptional profile in iPSCs that is related to the cell from which it was originally reprogrammed (Ghosh et al., 2010), which may contribute to differences in their phenotype compared to ESCs. Overall, most studies have shown a greater similarity in the methylation profile of cells derived from hESCs and hiPSCs than in the profile seen in similar cells derived from adult cells. The identification of specific gene promoters involved in the process of lineage specification through methylation analysis will help to clarify the role of DNA methylation during differentiation from pluripotent stem cell sources. Since the goal of ESC and iPSC differentiation is to recapitulate development and generate stable cell phenotypes, epigenetic analysis will provide a critical baseline for studying cellular changes occurring during controlled in vitro differentiation from these cell types, which will demonstrate the stability and utility of these cells for future therapies. Reprogramming-associated mutations also occur during or after reprogramming. It is suggested that extensive genetic screening should become a standard procedure to ensure hiPSC safety before clinical use (Gore et al., 2011).

4 DIFFERENTIATION OF STEM CELLS FOR REGENERATION AND THERAPY

4.1 Approaches for the Differentiation of Pluripotent Stem Cells

Two general approaches have been developed to differentiate ESCs and iPSCs into a broad spectrum of cell types. Each of these approaches has benefits and limitations that are dependent on the cell type undergoing differentiation and the stability of the cells derived. In the first method, differentiation can be accomplished by formation of EBs in which cells are aggregated in suspension and are subsequently differentiated toward all three germ layers in a pattern that is reminiscent of embryogenesis. Growth factors and media conditions can be altered within this environment to enhance differentiation toward specific cell types. EBs can then be further cultured and cells of interest can be selected based on growth conditions, morphological criteria, or sorted from this population based on a surface marker of interest. The second approach involves the differentiation of pluripotent cells upon plating them onto supportive protein substrate such as Matrigel, or onto a feeder layer substrate to enable the cell–cell or cell–matrix interactions that can direct differentiation responses. This differentiation approach provides a more controlled environment to direct cells to specific lineage fates and to monitor cell differentiation toward a desired germ layer or cell type. Both of these approaches can be fine-tuned based on cell morphology or protein expression profiles that meet criteria for cells of interest. Alternatively, cells can be sorted at any point in the differentiation protocol to isolate more uniform cell populations.

Many of the growth factor cues that direct pluripotent stem cell differentiation to specified lineage fates are mediated by a relatively small number of signals, including WNT, NODAL, and BMP, which are dynamically coordinated during development to specify the fate of specific cell types. Manipulation of regulatory pathways directed by these signals to differentiated cells from ESCs has revealed their dramatic effects on the specification of germ layer development. Such specification is seen upon activation of the NODAL pathway using activin A, which induces the derivation of endoderm and mesoderm in a concentration-dependent manner (Kubo et al., 2004). Outcomes of exposure of pluripotent stem cells to such signaling mediators are controlled temporally as well, as BMP signaling during the early stage of differentiation induces ectodermal specification, inhibits neuronal differentiation during later stages of differentiation, and promotes selection of definitive ectoderm or epidermal lineages (Aberdam et al., 2007). Thus, the development of well-defined protocols has established the capacity to differentiate specified cell types from pluripotent cell sources based on the presence of defined growth conditions that are directed by substrate and growth factor support.

4.2 Ectodermal Differentiation

The ectoderm gives rise to tissues of the nervous system, and epidermal tissues as well as craniofacial structures such as teeth. Differentiation of ectoderm from pluripotent stem cells can be achieved by directed differentiation upon seeding cells onto a stromal fibroblast feeder layer and supplementing the growth media with growth factors that have been found to be linked to the development of specified ectodermal lineages (Lee et al., 2010). Following the initial commitment of cells to ectodermal lineage, BMP4 supplementation allows for selection against neuronal differentiation when added to the growth milieu at specified times (Gambaro et al., 2006). Alternatively, inhibition of SMAD signaling mediated by Noggin has been shown to direct pluripotent stem cells to neural lineages (Chambers et al., 2009).

The timing of addition of either BMP4 or Noggin is critical in directing specific cell fates, as early addition of BMP4 to culture conditions can induce the differentiation of trophoblast cells (Xu et al., 2002), while the same growth conditions have been shown to induce differentiation of ectoderm when exposed to cells later in their differentiation (Chambers et al., 2009). Further refinements of this differentiation approach have allowed for the enrichment of mesenchymal, epithelial, or peripheral nerve cell differentiation (Lee et al., 2010). Together, these studies provide the framework to control and direct differentiation to cell types important to craniofacial development and provide experimental systems to study key regulatory signals that can select specific cell types.

5 APPLICATIONS USING CELLS DERIVED FROM PLURIPOTENT STEM CELLS

5.1 Tissue Engineering

Tissue engineering can be broadly defined as "the generation of biological substitutes that reproduce one or more functions of tissues or organs" (Langer and Vacanti, 1993). Successful tissue engineering approaches require the use of a supportive three-dimensional (3D) scaffold that can provide cells with a physical framework that will enable and optimize their spatial organization and function. Incorporation of stem cells or their differentiated progeny within an appropriate the 3D scaffold helps to organize these cells into the 3D architecture needed to achieve functional properties that can mimic the tissue or organ of interest. Since pluripotent stem cells can be differentiated and expanded to specified cell lineages, their incorporation into an optimal and engineered 3D scaffold microenvironment will allow these cells to self-organize into multicellular structures that mimic the morphologic and functional features of their in vivo counterparts (Zoldan and Levenberg, 2006).

Levenberg et al. (2003) established proof of the concept that ectodermal and mesenchymal cells differentiated from hESCs in vitro could assemble into tissues displaying in vivo–like epithelial tubular structures and neural tube–like rosettes that could integrate into the host vasculature. In addition, skin-like tissues that showed fully mature and functional skin upon engraftment to mice have been engineered from hESC-derived keratinocytes (Guenou et al., 2009). hESC-derived cells displaying characteristics of stromal fibroblasts have also been shown to support epithelial tissue development (Hewitt et al., 2009) and wound repair (Shamis et al., 2011), while iPSC-derived fibroblasts have also been shown to direct the development of engineered skin in an in vitro model of human skin that mimics many of the in vivo features of this tissue (Hewitt et al., 2011). These studies indicate that it is possible to construct tissues using ectodermal cells from pluripotent sources by providing them with the proper tissue microenvironment in vitro to acquire tissue organization and function that will lay the groundwork for tissue fabrication that may be used in future therapies (Fig. 3). In addition, such engineered tissues will provide an in vitro screening platform that can be used to analyze cell potency and function that can provide an in vivo–like tissue context to evaluate the safety of using these cells for human therapies that would not be provided by testing these differentiated cells in conventional two-dimensional monolayer cultures. Thus, construction of engineered hESC- or iPSC-derived tissues will facilitate the development of reliable methods to evaluate their functional properties before they can be translated for human therapy (Yamanaka, 2009).

5.2 Regenerative Therapy

Regenerative medicine is defined as "replacing or regenerating human cells, tissue, or organs, to restore or establish normal function" (Mason and Dunnill, 2008). The controlled isolation of specific functional cell types from the pluripotent stem cell sources and construction of in vitro tissues using tissue engineering approaches will be a promising step toward establishing stable sources of cells for the replacement of damaged organs and tissues. To achieve these goals it will be necessary to adapt and develop stem cell technologies in ways that will optimize the maintenance and expansion of therapeutically relevant cell populations. To be useful for regenerative therapies, cells derived from pluripotent stem cell sources must demonstrate functional features that are typically found in vivo while showing long-term stability of their differentiated phenotype that will ensure their safety following transplantation. Several recent examples illustrate the potential of using hESC- and iPSC-derived cell lines as an alternative or improved source of cells for regenerative therapies, thus providing evidence that this technology may be used in the future to treat human disease using patient-specific cells. Studies have highlighted the use of iPSC-based technologies to differentiate donor-specific cells into cell types that can be used for regenerative therapy (Zhang et al., 2011). For example, human iPSCs derived from amyotrophic lateral sclerosis (ALS) patient fibroblasts have subsequently been differentiated into motor neurons (Dimos et al., 2008), and murine iPSCs have been used to treat spinal

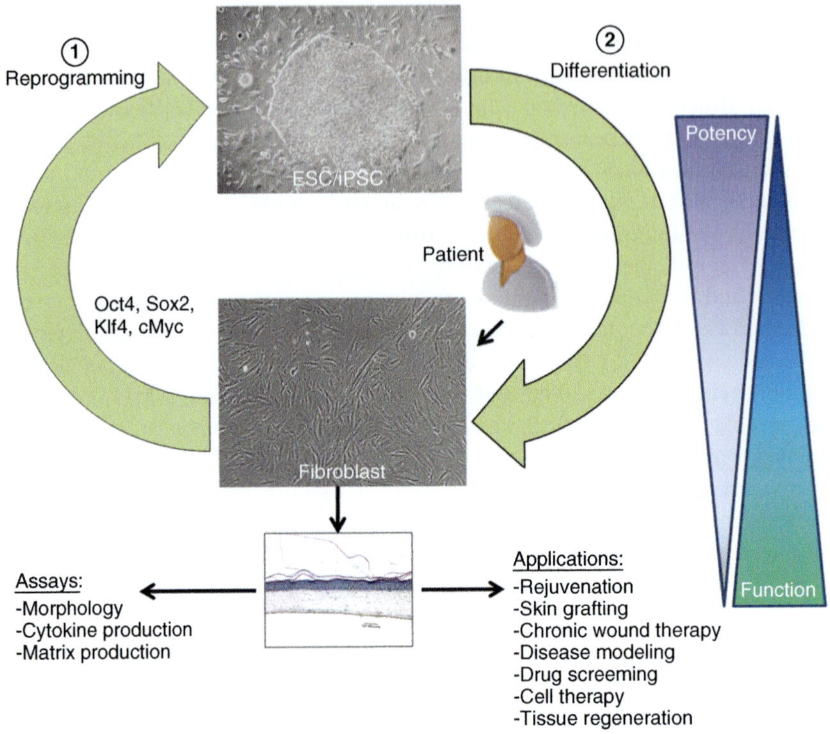

FIGURE 3 Experimental model and the potential therapeutic applications of iPSC-derived tissues. Fibroblasts grown from patients can be reprogrammed to a pluripotent state in culture. Following reprogramming, these iPSCs can be expanded indefinitely and differentiated into a number of more mature cell types that are also specific to a patient or disease. These cells can be used for modeling disease or screening drug compounds; engineering tissues such as skin rejuvenation, skin grafting, and chronic wound therapy; or used directly for cell-based patient transplantation.

cord injuries in mice (Tsuji et al., 2010). In addition, it has been shown in a rat model of Parkinson's disease that in vivo injection of iPSC-derived neuronal stem cells could migrate to appropriate brain regions and differentiate into dopamine-producing neurons (Wernig et al., 2008). Using activin A and BMP4 to differentiate hESCs to definitive endoderm, it has been shown that insulin-producing pancreatic β-like cells could be derived efficiently (Phillips et al., 2007) and that these cells could reverse the effects of hyperglycemia in mouse models of type 1 and type 2 diabetes (Alipio et al., 2010).

In addition to these disease-specific applications, there is intriguing evidence that iPSC reprogramming may serve to reset the biological clock and evade senescence normally associated with aged cells. It has recently been shown that the establishment of iPSCs that were reprogrammed from senescent fibroblasts from elderly patients, and subsequent differentiation of these cells to a fibroblast lineage, could restore functional features that were lost due to normal aging, including the elongation of telomeres and restoration of mitochondrial function (Suhr et al., 2010). Additional evidence for such cellular processes has been shown in the reprogramming of cancer cell lines to induced pluripotent cancer cells, which demonstrated that differentiated cells exhibited lower rates of proliferation and increased chemosensitivity, thus supporting the development of novel cancer treatments based on reprogramming approaches (Miyoshi et al., 2010b). Finally, mesenchymal stem cells derived from iPSCs have shown the improved rescue of limb ischemia compared to adult-derived MSCs (Lian et al., 2010), and fibroblasts derived from iPSCs produced greater amounts of extracellular matrix proteins than did the parental cells from which they were derived (Shamis et al., 2012). This suggests that the biological potential of iPSC-derived cells may be augmented following cellular reprogramming and subsequent differentiation to specific cell types.

5.3 Disease Modeling

The generation of iPSCs from patients with a spectrum of diseases can enable the differentiation of iPSC-derived cells that can recapitulate key stages of disease pathogenesis and be monitored in vitro. Such an application of this iPSC-based technology facilitates the development of in vitro disease models, laying the groundwork for large-scale screening of drug compounds that can be targeted to disrupt disease progression (Park et al., 2008). Disease-specific iPSCs have been generated from fibroblasts isolated from patients manifesting a wide range of inherited genetic conditions, including Down syndrome (DS), Becker muscular dystrophy (BMD), Duchenne muscular dystrophy (DMD), Parkinson's disease, and Huntington's disease (Park et al., 2008). Such disease modeling will allow us to compare the pathophysiology of cells differentiated from patient-specific iPSCs in ways that will offer an unprecedented opportunity to provide new platform technologies for drug screening. Beyond this, construction of 3D tissues harboring patient-specific iPSC-derived cells will move us from "disease in a dish" to "disease in a tissue."

6 LIMITATIONS AND PERSPECTIVE

Despite their potential, the application of cells derived from hESCs and iPSCs for human therapies faces many barriers that limit their implementation, due to concerns related to their safety, purity, and immunogenicity, as well as technical limitations in differentiating specific cell types (Belmonte et al., 2009). In addition, it has been demonstrated that these pluripotent cells can accumulate genetic mutations over time that may alter their phenotype and differentiation potential (Gore et al., 2011). Furthermore, while pluripotent cell types have been shown to be remarkably resistant to chromosomal abnormalities, both ESCs and iPSCs have been shown to acquire aberrant karyotypes following prolonged culture (Taapken et al., 2011). These genetic alterations, and epigenetic alterations such as aberrant DNA methylation found in iPSCs that are maintained in cells differentiated from iPSCs, may affect their ability to differentiate and may alter their stability as mature cell types (Lister et al., 2011). Another concern is that cells derived from pluripotent cells may be "contaminated" with small numbers of undifferentiated pluripotent iPSCs that can persist in long-term differentiated populations and lead to teratoma formation after transplantation in vivo (Fu et al., 2012). Despite extensive screening, this risk still seems to be a major complication of pluripotent-based therapies in animal transplantation models (Miura et al., 2009).

Acknowledgments

The work was supported in part by grants R01DE019156 (G.T.-J.H.) and R01DE017413 (J.A.G) from the National Institutes of Health/National Institute of Dental and Craniofacial Research.

REFERENCES

Aberdam D, Gambaro K, Medawar A, Aberdam E, Rostagno P, Forest Divone S, et al. (2007). Embryonic stem cells as a cellular model for neuroectodermal commitment and skin formation. *CR Biol* 330(6–7):479–484.

Agarwal S, Loh YH, McLoughlin EM, Huang J, Park IH, Miller JD, et al. (2010). Telomere elongation in induced pluripotent stem cells from dyskeratosis congenita patients. *Nature* 464(7286):292–296.

Alipio Z, Liao W, Roemer EJ, Waner M, Fink LM, Ward DC, et al. (2010). Reversal of hyperglycemia in diabetic mouse models using induced-pluripotent stem (iPS)-derived pancreatic beta-like cells. *Proc Natl Acad Sci USA* 107(30):13426–13431.

Baksh D, Song L, Tuan RS (2004). Adult mesenchymal stem cells: characterization, differentiation, and application in cell and gene therapy. *J Cell Mol Med* 8(3):301–316.

Becker AJ, McCulloch EA, Till JE. (1963). Cytological demonstration of the clonal nature of spleen colonies derived from transplanted mouse marrow cells.. *Nature* 2(197):452–4.

Belmonte JC, Ellis J, Hochedlinger K, Yamanaka S (2009). Induced pluripotent stem cells and reprogramming: seeing the science through the hype. *Nat Rev Genet* 10(12):878–883.

Boland MJ, Hazen JL, Nazor KL, Rodriguez AR, Gifford W, Martin G, et al. (2009). Adult mice generated from induced pluripotent stem cells. *Nature* 461(7260):91–94.

Caplan AI (1991). Mesenchymal stem cells. *J Orthop Res* 9(5):641–650.

Chambers SM, Fasano CA, Papapetrou EP, Tomishima M, Sadelain M, Studer L (2009). Highly efficient neural conversion of human ES and iPS cells by dual inhibition of SMAD signaling. *Nat Biotechnol* 27(3):275–280.

Cohnheim J (1867). Ueber Entzündung und Eiterung. *Arch Pathol Anat Physiol Klin Med* 40(1).

Crisan M, Yap S, Casteilla L, Chen C-W, Corselli M, Park TS, et al. (2008). A perivascular origin for mesenchymal stem cells in multiple human organs. *Cell Stem Cell* 3(3):301–313.

Deng J, Shoemaker R, Xie B, Gore A, Leproust EM, Antosiewicz-Bourget J, et al. (2009). Targeted bisulfite sequencing reveals changes in DNA methylation associated with nuclear reprogramming. *Nat Biotechnol* 27(4):353–360.

Dimos JT, Rodolfa KT, Niakan KK, Weisenthal LM, Mitsumoto H, Chung W, et al. (2008). Induced pluripotent stem cells generated from patients with ALS can be differentiated into motor neurons. *Science* 321(5893):1218–1221.

Ergun S, Tilki D, Klein D (2011). Vascular wall as a reservoir for different types of stem and progenitor cells. *Antioxid Redox Signal* 15(4):981–995.

Evans MJ, Kaufman MH (1981). Establishment in culture of pluripotential cells from mouse embryos. *Nature* 292(5819):154–156.

Friedenstein AJ, Gorskaja JF, Kulagina NN (1976). Fibroblast precursors in normal and irradiated mouse hematopoietic organs. *Exp Hematol* 4(5):267–274.

Fu W, Wang SJ, Zhou GD, Liu W, Cao Y, Zhang WJ (2012). Residual undifferentiated cells during differentiation of induced pluripotent stem cells in vitro and in vivo. *Stem Cells Dev* 21(4):521–529.

Gambaro K, Aberdam E, Virolle T, Aberdam D, Rouleau M (2006). BMP-4 induces a Smad-dependent apoptotic cell death of mouse embryonic stem cell-derived neural precursors. *Cell Death Differ* 13(7):1075–1087.

Ghosh Z, Wilson KD, Wu Y, Hu S, Quertermous T, Wu JC (2010). Persistent donor cell gene expression among human induced pluripotent stem cells contributes to differences with human embryonic stem cells. *PLoS One* 5(2):e8975.

Gómez MC, Pope CE, Dresser BL (2006). Nuclear transfer in cats and its application. *Theriogenology* 66(1):72–81.

Gonzalez F, Barragan Monasterio M, Tiscornia G, Montserrat Pulido N, Vassena R, Batlle Morera L, et al. (2009). Generation of mouse-induced pluripotent stem cells by transient expression of a single nonviral polycistronic vector. *Proc Nat Acad Sci USA* 106(22):8918–8922.

Gore A, Li Z, Fung HL, Young JE, Agarwal S, Antosiewicz-Bourget J, et al. (2011). Somatic coding mutations in human induced pluripotent stem cells. *Nature* 471(7336):63–67.

Guenou H, Nissan X, Larcher F, Feteira J, Lemaitre G, Saidani M, et al. (2009). Human embryonic stem-cell derivatives for full reconstruction of the pluristratified epidermis: a preclinical study. *Lancet* 374(9703):1745–1753.

Hewitt KJ, Shamis Y, Carlson MW, Aberdam E, Aberdam D, Garlick J (2009). Three-dimensional epithelial tissues generated from human embryonic stem cells. *Tissue Eng A*: 15(11):3417–3426.

Huang GT-J, Gronthos S, Shi S (2009). Mesenchymal stem cells derived from dental tissues vs.l those from other sources: their biology and role in regenerative medicine. *J Dent Res* 88(9):792–806.

Jiang Y, Jahagirdar BN, Reinhardt RL, Schwartz RE, Keene CD, Ortiz-Gonzalez XR, et al. (2002). Pluripotency of mesenchymal stem cells derived from adult marrow. *Nature* 418:41–49.

Kaji K, Norrby K, Paca A, Mileikovsky M, Mohseni P, Woltjen K (2009). Virus free induction of pluripotency and subsequent excision of reprogramming factors. *Nature* 458(7239):771–775.

Kanatsu-Shinohara M, Inoue K, Lee J, Yoshimoto M, Ogonuki N, Miki H, et al. (2004). Generation of pluripotent stem cells from neonatal mouse testis. *Cell* 119(7):1001–1012.

Kang L, Wang J, Zhang Y, Kou Z, Gao S (2009). iPS Cells can support full-term development of tetraploid blastocyst-complemented embryos. *Cell Stem Cell* 5(2):135–138.

Kim D, Kim C-H, Moon J-I, Chung Y-G, Chang M-Y, Han B-S, et al. (2009). Generation of human induced pluripotent stem cells by direct delivery of reprogramming proteins. *Cell Stem Cell* 4(6):472–476.

Kim K, Doi A, Wen B, Ng K, Zhao R, Cahan P, et al. (2010). Epigenetic memory in induced pluripotent stem cells. *Nature*: 467(7313):285–290.

Kolf C, Cho E, Tuan R (2007). Mesenchymal stromal cells. Biology of adult mesenchymal stem cells: regulation of niche, self-renewal and differentiation. *Arthritis Research & Therapy* 9(1):204.

Kubo A, Shinozaki K, Shannon JM, Kouskoff V, Kennedy M, Woo S, et al. (2004). Development of definitive endoderm from embryonic stem cells in culture. *Development* 131(7):1651–1662.

Langer R, Vacanti JP (1993). Tissue engineering. *Science* 260(5110):920–926.

Lee G, Chambers SM, Tomishima MJ, Studer L (2010). Derivation of neural crest cells from human pluripotent stem cells. *Nat Protoc* 5(4):688–701.

Levenberg S, Huang NF, Lavik E, Rogers AB, Itskovitz-Eldor J, Langer R (2003). Differentiation of human embryonic stem cells on three-dimensional polymer scaffolds. *Proc Natl Acad Sci USA* 100(22):12741–12746.

Leychkis Y, Munzer SR, Richardson JL (2009). What is stemness? *Stud Hist Philos Sci C* 40(4):312–320.

Lian Q, Zhang Y, Zhang J, Zhang HK, Wu X, Lam FF, et al. (2010). Functional mesenchymal stem cells derived from human induced pluripotent stem cells attenuate limb ischemia in mice. *Circulation* 121(9):1113–1123.

Lister R, Pelizzola M, Kida YS, Hawkins RD, Nery JR, Hon G, et al. (2011). Hotspots of aberrant epigenomic reprogramming in human induced pluripotent stem cells. *Nature* 471(7336):68–73.

Loh YH, Agarwal S, Park IH, Urbach A, Huo H, Heffner GC, et al. (2009). Generation of induced pluripotent stem cells from human blood. *Blood* 113(22):5476–5479.

Mareschi K, Biasin E, Piacibello W, Aglietta M, Madon E, Fagioli F (2001). Isolation of human mesenchymal stem cells: bone marrow versus umbilical cord blood. *Haematologica* 86(10):1099–1100.

Martin GR (1981). Isolation of a pluripotent cell line from early mouse embryos cultured in medium conditioned by teratocarcinoma stem cells. *Proc Natl Acad Sci USA* 78(12):7634–7638.

Mason C, Dunnill P (2008). A brief definition of regenerative medicine. *Regen Med* 3(1):1–5.

Matsui Y, Zsebo K, Hogan BL (1992). Derivation of pluripotential embryonic stem cells from murine primordial germ cells in culture. *Cell* 70(5):841–847.

Miura K, Okada Y, Aoi T, Okada A, Takahashi K, Okita K, et al. (2009). Variation in the safety of induced pluripotent stem cell lines. *Nat Biotechnol* 27(8):743–745.

Miyashita N, Shiga K, Yonai M, Kaneyama K, Kobayashi S, Kojima T, et al. (2002). Remarkable differences in telomere lengths among cloned cattle derived from different cell types. *Biol Reprod* 66(6):1649–1655.

Miyoshi K, Tsuji D, Kudoh K, Satomura K, Muto T, Itoh K, et al. (2010a). Generation of human induced pluripotent stem cells from oral mucosa. *J Biosci Bioeng*. 110(3):345–350.

Miyoshi N, Ishii H, Nagai K, Hoshino H, Mimori K, Tanaka F, et al. (2010b). Defined factors induce reprogramming of gastrointestinal cancer cells. *Proc Natl Acad Sci USA* 107(1):40–45.

Moore KA, Lemischka IR (2006). Stem cells and their niches. *Science* 311(5769):1880–1885.

Noggle S, Fung H-L, Gore A, Martinez H, Satriani KC, Prosser R, et al. (2011). Human oocytes reprogram somatic cells to a pluripotent state. *Nature* 478(7367):70–75.

Oh HJ, Hong SG, Park JE, Kang JT, Kim MJ, Kim MK, et al. (2009). Improved efficiency of canine nucleus transfer using roscovitine-treated canine fibroblasts. *Theriogenology* 72(4):461–470.

Park IH, Arora N, Huo H, Maherali N, Ahfeldt T, Shimamura A, et al. (2008). Disease-specific induced pluripotent stem cells. *Cell* 134(5):877–886.

Phillips BW, Hentze H, Rust WL, Chen QP, Chipperfield H, Tan EK, et al. (2007). Directed differentiation of human embryonic stem cells into the pancreatic endocrine lineage. *Stem Cells Dev* 16(4):561–578.

Pittenger MF, Mackay AM, Beck SC, Jaiswal RK, Douglas R, Mosca JD, et al. (1999). Multilineage potential of adult human mesenchymal stem cells. *Science* 284(5411):143–147.

Polo JM, Liu S, Figueroa ME, Kulalert W, Eminli S, Tan KY, et al. (2010). Cell type of origin influences the molecular and functional properties of mouse induced pluripotent stem cells. *Nat Biotechnol* 28(8):848–855.

Porada CD, Zanjani ED, Almeida-Porad G (2006). Adult mesenchymal stem cells: a pluripotent population with multiple applications. *Curr Stem Cell Res Ther* 1(3):365–369.

Revazova ES, Turovets NA, Kochetkova OD, Kindarova LB, Kuzmichev LN, Janus JD, et al. (2007). Patient-specific stem cell lines derived from human parthenogenetic blastocysts. *Cloning Stem Cells* 9(3):432–449.

Revazova ES, Turovets NA, Kochetkova OD, Agapova LS, Sebastian JL, Pryzhkova MV, et al. (2008). HLA homozygous stem cell lines derived from human parthenogenetic blastocysts. *Cloning and Cells* 10(1):11–24.

Robey PG, Bianco P (2006). The use of adult stem cells in rebuilding the human face. *J Am Dent Assoc* 137(7):961–972.

Roeder I, Loeffler M, Glauche I (2011). Towards a quantitative understanding of stem cell–niche interaction: experiments, models, and technologies. *Blood Cells Mol Dis* 46(4):308–317.

Schofield R (1978). The relationship between the spleen colony-forming cell and the haemopoietic stem cell. *Blood Cells* 4(1–2):7–25.

Schwartz SD, Hubschman J-P, Heilwell G, Franco-Cardenas V, Pan CK, Ostrick RM, et al. (2012). Embryonic stem cell trials for macular degeneration: a preliminary report. *Lancet.* 379(9817):713–720.

Seki T, Yuasa S, Oda M, Egashira T, Yae K, Kusumoto D, et al. (2010). Generation of induced pluripotent stem cells from human terminally differentiated circulating T cells. *Cell Stem Cell* 7(1):11–14.

Shafritz DA, Oertel M, Menthena A, Nierhoff D, Dabeva MD (2006). Liver stem cells and prospects for liver reconstitution by transplanted cells. *Hepatology* 43(S1):S89–S98.

Shamis Y, Hewitt KJ, Carlson MW, Margvelashvilli M, Dong S, Kuo CK, Daheron L, Egles C, Garlick A (2011). Fibroblasts derived from human embryonic stem cells direct development and repair of 3D human skin equivalents. *Stem Cell Res Ther.* 2(1):10.

Shamis Y, Hewitt KJ, Bear SE, Alt-Holland A, Qari H, Margvelashvilli M, Knight EB, Smith A, Garlick JA (2012). iPSC-derived fibroblasts demonstrate augmented production and assembly of extracellular matrix proteins. *In Vitro Cell Dev Biol Anim.* 48(2):112–22.

Soldner F, Hockemeyer D, Beard C, Gao Q, Bell GW, Cook EG, et al. (2009). Parkinson's disease patient-derived induced pluripotent stem cells free of viral reprogramming factors. *Cell* 136(5):964–977.

Sridharan R, Plath K (2008). Illuminating the black box of reprogramming. *Cell Stem Cell* 2(4):295–297.

Stadtfeld M, Nagaya M, Utikal J, Weir G, Hochedlinger K (2008). Induced pluripotent stem cells generated without viral integration. *Science* 322(5903):945–949.

Suhr ST, Chang EA, Tjong J, Alcasid N, Perkins GA, Goissis MD, et al. (2010). Mitochondrial rejuvenation after induced pluripotency. *PLoS One* 5(11):e14095.

Taapken SM, Nisler BS, Newton MA, Sampsell-Barron TL, Leonhard KA, McIntire EM, et al. (2011). Karotypic abnormalities in human induced pluripotent stem cells and embryonic stem cells. *Nat Biotechnol* 29(4):313–314.

Takahashi K, Yamanaka S (2006). Induction of pluripotent stem cells from mouse embryonic and adult fibroblast cultures by defined factors. *Cell* 126(4):663–676.

Takahashi K, Tanabe K, Ohnuki M, Narita M, Ichisaka T, Tomoda K, et al. (2007). Induction of pluripotent stem cells from adult human fibroblasts by defined factors. *Cell* 131(5):861–872.

Thomson JA, Itskovitz-Eldor J, Shapiro SS, Waknitz MA, Swiergiel JJ, Marshall VS, et al. (1998). Embryonic stem cell lines derived from human blastocysts. *Science* 282(5391):1145–1147.

Thuan NV, Kishigami S, Wakayama T (2010). How to improve the success rate of mouse cloning technology. *J Reprod Dev* 56(1):20–30.

Tsuji O, Miura K, Okada Y, Fujiyoshi K, Mukaino M, Nagoshi N, et al. (2010). Therapeutic potential of appropriately evaluated safe-induced pluripotent stem cells for spinal cord injury. *Proc Natl Acad Sci USA* 107(28):12704–12709.

Wernig M, Zhao JP, Pruszak J, Hedlund E, Fu D, Soldner F, et al. (2008). Neurons derived from reprogrammed fibroblasts functionally integrate into the fetal brain and improve symptoms of rats with Parkinson's disease. *Proc Natl Acad Sci USA* 105(15):5856–5861.

Wilmut I, Schnieke AE, McWhir J, Kind AJ, Campbell KH (1997). Viable offspring derived from fetal and adult mammalian cells. *Nature* 385(6619):810–813.

Woltjen K, Michael IP, Mohseni P, Desai R, Mileikovsky M, Hamalainen R, et al. (2009). piggyBac transposition reprograms fibroblasts to induced pluripotent stem cells. *Nature* 458(7239):766–770.

Wu AM, Till JE, Siminovitch L, McCulloch EA, (1967). A cytological study of the capacity for differentiation of normal hemopoietic colony-forming cells. *J Cell Physiol* 69:177–184.

Xu RH, Chen X, Li DS, Li R, Addicks GC, Glennon C, et al. (2002). BMP4 initiates human embryonic stem cell differentiation to trophoblast. *Nat Biotechnol* 20(12):1261–1264.

Yamanaka S (2009). A fresh look at iPS cells. *Cell* 137(1):13–17.

Yan X, Qin H, Qu C, Tuan RS, Shi S, Huang GT-J (2010). iPS cells reprogrammed from human mesenchymal-like stem/progenitor cells of dental tissue origin. *Stem Cells Dev* 19(4):469–480.

Yu J, Vodyanik MA, Smuga-Otto K, Antosiewicz-Bourget J, Frane JL, Tian S et al. (2007). Induced pluripotent stem cell lines derived from human somatic cells. *Science* 318(5858):1917–1920.

Yu J, Hu K, Smuga-Otto K, Tian S, Stewart R, Slukvin II, et al. (2009). Human induced pluripotent stem cells free of vector and transgene sequences. *Science* 324(5928):797–801.

Zhang F, Citra F, Wang DA (2011). Prospects of induced pluripotent stem cell technology in regenerative medicine. *Tissue Eng B* 17(2):115–124.

Zhao X-y, Li W, Lv Z, Liu L, Tong M, Hai T, et al. (2009). iPS cells produce viable mice through tetraploid complementation. *Nature* 461(7260):86–90.

Zhou H, Wu S, Joo JY, Zhu S, Han DW, Lin T, et al. (2009). Generation of induced pluripotent stem cells using recombinant proteins. *Cell Stem Cell* 4(5):381–384.

Zipori D (2004). The nature of stem cells: state rather than entity. *Nat Rev Genet* 5(11):873–878.

Zoldan J, Levenberg S (2006). Engineering three-dimensional tissue structures using stem cells. *Methods Enzymol* 420:381–391.

Zuk PA, Zhu M, Mizuno H, Huang J, Futrell JW, Katz AJ et al. (2001). Multilineage cells from human adipose tissue: implications for cell-based therapies. *Tissue Eng* 7(2):211–228.

11

BONE MARROW MESENCHYMAL STEM CELLS

SONGTAO SHI

Center for Craniofacial Molecular Biology, Herman Ostrow School of Dentistry, University of Southern California, Los Angeles, Califonia

STAN GRONTHOS

Mesenchymal Stem Cell Laboratory, Department of Haematology, University of Adelaide, Adelaide, South Australia, Australia

1 INTRODUCTION

An appreciable amount of evidence supports the notion that skeletal and bone marrow stromal tissues are interrelated both functionally, in their regulation of hematopoiesis, and through a common ontogeny. Postnatal bone marrow tissue is known to contain multipotential bone marrow mesenchymal stem cells (MSCs) with the capacity to develop into various mature marrow stromal elements and associated skeletal tissues. Over the last few decades, substantial efforts have been made to unlock the potential of MSCs with a view to utilizing these cells for various tissue engineering and gene therapy applications. However, the progress of these studies has been restrained, due to a lack of understanding of the basic nature and properties of MSCs. In this chapter we review the properties of clonogenic stromal cell types within postnatal bone marrow tissue and address a poorly understood aspect of their cell biology: namely, the developmental relationships between the different bone marrow stromal cell lineages. In particular, evidence for the existence of a multipotential bone marrow MSC is discussed, including an examination of the cellular and molecular characteristics of MSC and potential clinical applications.

Stem Cells in Craniofacial Development and Regeneration, First Edition.
Edited by George T.-J. Huang, Irma Thesleff.

2 BONE MARROW STROMAL SYSTEM

The repair and slow turnover of postnatal skeletal and marrow stromal connective tissue is attributed to the proliferation and differentiation of MSCs with the capacity to give rise to committed progenitor cells that develop into the various stromal components present in the bone marrow microenvironment and skeletal tissues. Although it is tempting to draw analogies with the highly regenerative hematopoietic system, characterized by a well-defined hierarchy of cellular differentiation, further research is required to determine the exact nature of the molecular pathways leading to the turnover and maintenance of the postnatal bone marrow stromal and skeletal systems. Moreover, much work is still needed to elucidate the precise relationships between these different skeletal precursor populations with comparable MSC-like populations identified in other tissues (Shi et al., 2005; Gronthos et al., 2006).

Bone marrow MSCs were first identified in aspirates of rodent marrow by their ability to form small cell clusters of greater than 50 adherent cells morphologically resembling fibroblasts (colony-forming units–fibroblastic; CFU-F) when grown at low cell densities in short-term liquid serum–replete culture conditions (Friedenstein et al., 1970, 1976). Subsequent studies identified the presence of CFU-F in a number of other species, including humans (Castro-Malaspina et al., 1980). Limiting dilution and chromosomal mixing experiments showed that each colony was ultimately derived from the clonal expansion of a single adherent progenitor cell (Friedenstein et al., 1970, 1976; Castro-Malaspina et al., 1980; Owen et al., 1987; Perkins and Fleischman, 1990; Gronthos et al., 2003). Moreover, the majority of CFU-F contained within marrow aspirates was found to be noncycling prior to culture, based on their lack of [^{3}H]thymidine labeling, resistance to 5-fluoracil treatment, and their negativity to the proliferation antigen, Ki67 (Friedenstein et al., 1976; Castro-Malaspina et al., 1980; Van Vlasselaer et al., 1994; Gronthos et al., 2003).

Ex vivo–expanded CFU-F, derived from a number of mammalian species, demonstrates great heterogeneity in respect to morphology, proliferation, and differentiation potential. Various studies have shown that a proportion of individual CFU-F clonal cell lines (10 to 20%) display an extensive potential to proliferate beyond 20 population doublings (PDs), with an upper growth limit ranging from 40 to 50 PDs, following continuous subculture (Friedenstein, 1976, 1980; Owen and Friedenstein, 1988; Gronthos et al., 2003). Conversely, the majority of CFU-F clones appear to exhaust their growth potential after only 10 or 15 PDs in vitro. Other studies have shown that ex vivo–expanded bone marrow MSCs lose telomerase activity rapidly (Shi et al., 2002; Simonsen et al., 2002; Zimmermann et al., 2003), a marker highly expressed by different stem cell populations and most cancer cell lines. The telomerase enzyme complex is responsible for maintaining telomere lengths at the end of chromosomes, providing genetic stability during cell division and preventing the onset of cellular senescence. It has been reported that CFU-F clonal cell lines exhibiting high proliferation potential in vitro are correlated with a longer-than-average telomere length (Banfi et al., 2002). These observations support the proposal that the CFU-F compartment contains a mixed population of progenitor cells at various stages of development, as proposed by Owen and Friedenstein (Owen, 1988; Owen and Friedenstein, 1988) (Fig. 1).

The MSC hypothesis was supported initially by a series of elegant experiments devised to assess the developmental capacity of individual CFU-F clonal cell lines in vivo. Rodent bone marrow–derived CFU-F clones were first isolated and expanded

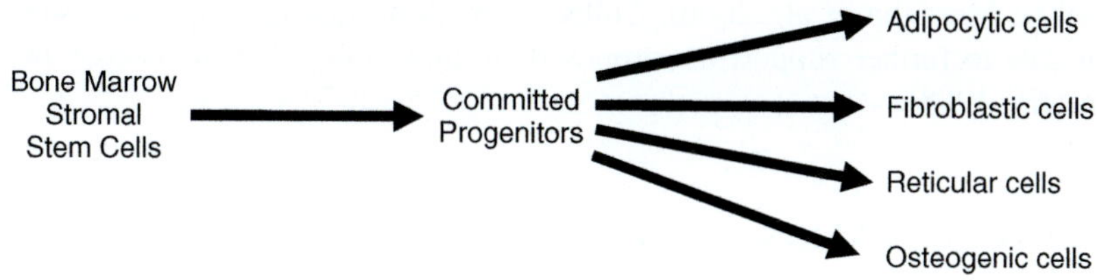

FIGURE 1 Bone marrow stromal system proposed by Owen and Friedenstein (1988).

in vitro, then transplanted beneath the renal capsules of littermates (Friedenstein, 1980). Analysis of the ectopic transplants after several weeks posttransplantation revealed that about 15% of the CFU-F clonal cell lines had formed a vascularized bone marrow connective tissue with associated calcified bone spicules. Moreover, the ectopic stromal organs were also shown to have the capacity to support local hematopoiesis. A further 15% of the bone marrow CFU-F clones produced a calcified bone matrix only, while the remainder formed either a soft connective tissue or failed to give rise to any recognizable tissue (Friedenstein, 1980). The hematopoietic cells contained within the stromal organs were identified as being derived from the host animal, while the fibroblast-like cells were determined to be of donor origin, using immunological and chromosome analyses. Similar studies have showed that a proportion of rabbit-derived CFU-F clonal cell lines could be induced to differentiate into adipocytes, in the presence of the glu cocorticoid hydrocortisone (Bennett et al., 1991). Furthermore, approximately 40% of CFU-F clones comprising either fibroblastic or adipocytic cells were also induced to form a calcified bone matrix when transplanted in vivo using diffusion chambers implanted into the peritoneal cavity of Nude mice (Bennett et al., 1991). The minor population of CFU-F clones that formed bone marrow organs were hypothesized to be derived from multipotential MSC, while those CFU-F clones that gave rise only to bone or soft connective tissue were proposed to be committed progenitors with restricted developmental potential (Owen and Friedenstein, 1988).

Other reports have subsequently confirmed the potential of ex vivo expanded bone marrow CFU-F, derived from various mammalian species, to develop into multiple tissue types in vivo (Ashton et al., 1980; Friedenstein, 1980; Goshima et al., 1991; Huss et al., 1995; Krebsbach et al., 1997; McCarty et al., 2009). Comparable experiments utilizing ex vivo expanded adult human CFU-F reported the formation of bone, cartilage, and hematopoietic supportive stroma when transplanted into immunocompromised rodents using a variety of carrier vehicles (Bab et al., 1988; Dennis et al., 1992; Haynesworth et al., 1992; Gundle et al., 1995; Krebsbach et al., 1997; Kuznetsov et al., 1997; Gronthos et al., 2000, 2003; At a clonal level, only a proportion of individually expanded human bone marrow CFU-F (50 to 60%) was capable of forming bone in vivo, following cotransplantation with hydroxyapatite/tricalcium ceramic particles into immunocompromised mice (Kuznetsov et al., 1997; Gronthos et al., 2003. In analogy with the rodent studies of Owen and Friedenstein (1988), only a minor subpopulation (approximately 30%) of these CFU-F clones also demonstrated the capacity to support local hematopoiesis. Parallel in vitro experiments showed that different human-derived CFU-F clones displayed a differential capacity for multidifferentiation into myelosupportive stroma, osteoblasts, chondrocytes, smooth muscle cells, and adipocytes (Gronthos et al., 1994, 2003; Menicanin et al., 2010; Pittenger et al., 1999; Muraglia et al., 2000). Collectively, these findings provide a substantial body of data to further support the proposed stromal stem cell hypothesis (Owen and Friedenstein, 1988).

3 CHARACTERIZATION OF BONE MARROW MSC

The Mesenchymal and Tissue Stem Cell Committee of the ISCT has designated a minimal criterion for defining multipotent mesenchymal stem cells: adherence to plastic during culture, a specific cell-surface antigen expression profile, and a multipotent differentiation potential (Dominici et al., 2006). Specifically, MSCs must be plastic-adherent when cultured in standard culture conditions in standard culture flasks. At least 95% of a MSC population must also express CD105, CD73, and CD90. Furthermore, less than 2% of the population should express markers: CD45, CD34, CD14 or CD11b, CD79α or CD19, and HLA class II. Finally, MSC populations must be able to undergo trilineage mesenchymal differentiation to form osteoblasts, adipocytes, and chondroblasts under standard in vitro induction conditions (Dominici et al., 2006). However, the notion that all cells present in cultures derived from human CFU-F are multipotential bone marrow MSCs is a gross misinterpretation, despite the uniform expression of a few selected markers (Pittenger et al., 1999) that are coincidently expressed by virtually all cultured fibroblast populations examined. More rigorous criteria are clearly required when describing what is essentially a heterogeneous cell mix, with only a minor subpopulation within any given culture flask being able to satisfy the criteria that define MSCs.

Difficulties have arisen over the last three decades concerning the characterization of bone marrow MSCs, due to the extreme low incidence (1 : 10,000 to 100,000) of CFU-F in bone marrow aspirates and the absence of markers specific to long-lived multipotential MSCs, which constitute only a minor proportion of the total CFU-F population. Therefore, it is not surprising that most studies have relied on traditional plastic adherence to generate crude preparations of primary bone marrow stromal cultures (Prockop, 1997). As mentioned above, cultures prepared in this fashion are not homogeneous but, rather, are an assortment of mesenchymal and stromal precursor cells, differentiated mesenchymal and stromal cells, endothelial cells, hematopoietic progenitors, and macrophage, which can persist for several cell passages (Psaltis et al., 2011). Head-to-head comparisons of prospectively isolated versus plastic adherence–selected MSCs derived from the same bone marrow aspirates demonstrated differences in the growth and functional properties of these populations in favor of immunoselection (Psaltis et al., 2011).

The ability to distinguish hematopoietic and stromal cells based on their cell-surface antigen expression patterns was a major step forward in designing bone marrow MSC purification strategies. Early studies examining murine bone marrow identified a population of 5-fluoracil (5-FU)–resistant stromal progenitor cells exhibiting an immunophenotype lacking lymphoid and myeloid cell lineage markers (Lin^-) but expressing stem cell antigen 1 ($Sca1^+$) and reacting strongly with wheat germ agglutinin (WGA^{bright}) (Falla et al., 1993; Van Vlasselaer et al., 1994). The $Lin^-/Sca1^+/WGA^{bright}$ cell population was shown to possess the potential to support hematopoiesis and develop a calcified bone matrix in vitro in the presence of ascorbate and β-glycerolphosphate.

Consistent with the observations in the murine system, differences in the cell surface expression of hematopoietic and stromal cells have been reported in human studies. For example, cell preparations containing a higher incidence of CFU-F can be obtained from human bone marrow aspirates due to their binding of the lectin, soybean agglutinin, which binds poorly to HSCs (Sutherland et al., 1989). Bone marrow CFU-F can also be partially enriched from human bone marrow aspirates based on their lack of expression of cell-surface antigens characteristic of myeloid, erythroid, and lymphoid cells, including CD3, CD14, CD19, CD38, CD45, CD50, glycophorin-A, and c-kit (Van Vlasselaer et al., 1994; Waller et al., 1995; Rickard et al., 1996; Zannettino et al., 2007).

Alternatively, other strategies have pursued positive selection methodologies using markers highly expressed by CFU-F, such as nerve growth factor receptor (CD271), PDGF-R, EGF-R, IGF1-R, various CD29 integrin family members, CD10, CD13, heat-shock protein-90β (STRO4), and alkaline phosphatase (Simmons et al., 1994; Gronthos and Simmons, 1995; Gronthos et al., 2001, 2007, 2009). However, these antigens are also present on marrow reticular cells along with mature osteogenic populations and some hematopoietic lineages and therefore can only serve to partially enrich for CFU-F. The generation of antibody reagents with restricted specificity for CFU-F has yet to be developed where none have been reported to distinguish between primitive and more committed CFU-F populations. For these reasons, the isolation of CFU-F based on the expression of any combination of the antigens described above represents an insufficient strategy for obtaining pure populations of adult human bone marrow CFU-F.

The murine monoclonal antibody STRO1, developed by Simmons and Torok-Storb (1991), was one of the first reagents with the unique property of binding to all detectable human bone marrow CFU-F. The antibody identifies less than 10% of adult human bone marrow mononuclear cells, where over 90% of the $STRO1^{+}$ cells were found to be nucleated glycophorin A^{+} erythroblasts, and some $CD19^{+}$ B-lymphocytes. Importantly, CFU-F were found to be restricted to the minor population of $STRO1^{bright}/CD34^{-}/CD45^{-}/CD14^{-}/CD19^{-}$/glycophorin-$A^{-}$ cells (Gronthos et al., 2003; Zannettino et al., 2007). Moreover, hematopoietic stem and progenitor cell populations appeared to lack any expression of the STRO1 antigen.

Adult bone marrow mononuclear cells sorted on the basis of STRO-1 expression have been shown to develop into adherent stromal layers when grown under "Dexter-type" long-term bone marrow culture conditions and exhibited a greater capacity to support hematopoiesis than that in stroma derived from unfractionated marrow (Simmons and Torok-Storb, 1991). More recently, STRO1 selected cells have been shown to increase the support and engraftment of hematopoietic stem and progenitor cells in vitro and in vivo (Bensidhoum et al., 2004; Goncalves et al., 2006). Cultured bone marrow MSCs selected on the basis of STRO1 expression also exhibit a higher growth capacity, a greater potency for inducing angiogenesis in vivo, increased immunomodulatory properties, and a higher rate of bone formation than that of MSCs isolated by plastic adherence or based on their lack of STRO1 expression (Martens et al., 2006; Nasef et al., 2009; Psaltis et al., 2011). The adherent layers derived from $STRO1^{+}$ cells consisted of a number of phenotypically distinct stromal cell types, including fibroblasts, smooth muscle cells, and adipocytes, but lacked the presence of macrophage and endothelial cells, which are readily detected in primary Dexter-type long-term bone marrow cultures derived from unfractionated bone marrow (Simmons and Torok-Storb, 1991). Furthermore, studies have demonstrated that adult bone marrow

$STRO1^+$-derived stromal cultures also have the capacity for differentiation into other cell lineages, including osteoblasts and chondrocytes (Dennis et al., 2002; Gronthos et al., 2003; Gronthos, 2004). However, the use of STRO1 antibody alone is not sufficient to obtain high purity CFU-F, due to the presence of contaminating populations of glycophorin-A^+ nucleated red cells and a minor subset of $CD19^+$ B-lymphocytes.

The development of further purification steps to increase the incidence of CFU-F was accomplished by a systematic analysis of a wide range of cell-surface molecules coexpressed on $STRO1^+$ bone marrow cells by means of two-color FACS (Simmons et al., 1994). From these studies, the expression of two immunoglobulin superfamily members, CD106 and CD146, were found to be restricted to a minor fraction of a $STRO1^{bright}$ bone marrow mononuclear cell population that contained all assayable CFU-F and were not readily detectable on bone marrow reticular cells, adipocytes, osteoblasts, chondrocytes, hematopoietic stem cells, and their progeny (Gronthos et al., 2003; Shi and Gronthos, 2003). The CD106 or vascular cell adhesion molecule 1 (VCAM1) antigen is a transmembrane glycoprotein present on the membranes of reticular and endothelial cells (Osborn et al., 1989; Simmons et al., 1992; Jacobsen et al., 1996). It binds the integrin receptor $\alpha_4\beta_1$ (VLA4) present on hematopoietic stem/progenitor cells and is involved in the recruitment of lymphocytes and monocytes expressing $\alpha_4\beta_1$ to sites of infection and inflammation (Elices et al., 1990; Simmons et al., 1992). The CD146 molecule (MUC18/MEL-CAM) antigen known to be expressed by metastatic melanoma cells, Schwann cells, endothelial cells, and smooth muscle cells (Kuzu et al., 1993; Bardin et al., 1996) has also been shown to be highly expressed by clonogenic bone marrow CFU-F (Shi and Gronthos, 2003). While the precise function of CD146 is not known, it has been linked to various cellular processes, including cell adhesion, cytoskeletal reorganization, cell migration, and proliferation. Overall, both CD106 and CD146 are considered as appropriate candidate antigens for developing additional enrichment strategies for purifying CFU-F utilizing a combination of MACS isolated $STRO1^+$ bone marrow mononuclear cells followed by two-color FACS based on $STRO1^{bright}/CD106^+$ or $STRO1^{bright}/CD146^+$ expression (Gronthos et al., 2003; Shi and Gronthos, 2003).

4 PERIVASCULAR PROPERTIES OF HUMAN BONE MARROW MSC

Until recently, the identification and exact locality of multipotential MSCs in bone marrow had not been determined, due to an inability to discriminate between bone marrow MSCs and more differentiated stromal cell types residing in the bone marrow microenvironment. Electron microscopic examination and cytochemical staining of marrow tissue sections first identified reticular fibroblast cells as the predominant stromal cell type present in the bone marrow spaces. The reticular cells have characteristic large irregular bodies, with long cytoplasmic processes, and are responsible for secreting the extracellular matrix fibers that support the hematopoietic tissue (Weiss, 1965; Lichtman, 1981; Bianco, 1998). The extracellular matrix is composed mainly of glycosaminoglycans attached to proteoglycans, fibronectin, and collagen types I, III, V, and VI (Bentley et al., 1981; Bentley, 1982; Andersen et al. 2011). In situ, reticular cells express nerve growth factor receptor (CD271), endopeptidase (CD13), and alkaline phosphatase, but appear to lack expression of more specific markers associated with endothelial (factor VIII and CD34), neural (CD56 and neurofilament), and hemopoietic cell lineages (CD45 and CD68) (Westen and Bainton, 1979; Cattoretti et al., 1993). The expression of the osteoblast-related marker alkaline phosphatase suggests an association with bone-forming cells, or osteoblasts. Indeed, it has been proposed that alkaline phosphatase–positive reticular cells in the vicinity of the endosteal bone surface or adventitial reticular cells lining the marrow sinuses may serve as reservoirs of osteogenic, adipogenic, and myelosupportive stromal precursor cells (Westen and Bainton, 1979; Bianco and Boyde, 1993). It is now thought that ALP is expressed by stromal precursor populations and is later downregulated during osteogenic differentiation until it is reexpressed following the formation of mature functional osteoblasts (Gronthos et al., 1999, 2007).

For the first time, the development of strategies to purify the CFU-F population allowed examination of the physical properties of human CFU-F residing in bone marrow aspirates. In analogy to murine studies (Falla et al., 1993), $STRO1^{bright}/CD106^{+}$ or $STRO1^{bright}/CD146^{+}$ human CFU-F exhibited high forward and perpendicular light-scattering properties compared to small hematopoietic stem cells (Gronthos et al., 2003; Shi and Gronthos, 2003). Preparations of freshly isolated $STRO1^{bright}/CD106^{+}$ displayed a morphology characteristic of large cells (15 to 20 μm) with long cytoplasmic processes and an open chromatin structure, using electron and light microscopy (Gronthos et al., 2003; Shi and Gronthos, 2003).

Although the precise location of CFU-F in bone marrow is largely unknown, mounting evidence suggests that CFU-F may reside in a perivascular niche within human bone marrow tissues (Bianco et al., 2001; Gronthos et al., 2003; Shi and Gronthos, 2003). Importantly, freshly isolated $STRO1^{bright}/CD106^{+}$ or $STRO1^{bright}/CD146^{+}$ does not express characteristic endothelial Weibel–Palade bodies and lacks expression of the endothelial-specific markers von Willebrand factor and CD31 (Bianco et al., 2001; Gronthos et al., 2003; Shi and Gronthos, 2003). However, a high proportion of $STRO1^{bright}/CD106^{+}$ or $STRO1^{bright}/CD146^{+}$ CFU-F were found to express α smooth muscle actin, an antigen restricted to pericytes and vascular smooth muscle cells in the bone marrow. The possible perivascular origin of CFU-F was substantiated further using a combination of immunohistochemical and immunoselection methods to isolate different purified populations of CFU-F based on their expression of STRO-1, the smooth muscle and perivascular markers, CD146, α-smooth muscle actin, and the pericyte-associated markers 3G5 and PDGF receptor (Gronthos and Simmons, 1995; Filshie et al., 1998; Shi and Gronthos, 2003). This work was also supported by studies depicting colocalization of STRO1/CD146 antigens in perivascular cells in situ, surrounding large blood vessels and arterioles in the bone marrow. More recently, Sacchetti and colleagues (2007) demonstrated that clonal populations of CD146-selected multipotential MSCs could regenerate a vascularized bone marrow microenvironment containing bone trabeculae and active hematopoiesis in vivo, where some donor cells remained undifferentiated as CD146-expressing perivascular cells with the potential to regenerate a bone marrow organ in secondary transplants, demonstrating their capacity to undergo self-renewal.

Collectively, these data imply that the majority of human bone marrow CFU-F exhibits a phenotype consistent with smooth muscle cells while a minor population is characteristic of pericytes. It is still unknown, however, whether highly proliferating multipotential bone marrow MSCs are contained within the 3G5-positive or 3G5-negative cell fractions, or even if the two cell populations share a common developmental origin. Nevertheless, these observations are in accord with the findings of

Charbord and Dennis, which describe human bone marrow MSCs as having a smooth muscle or perivascular-like phenotype in vitro (Charbord et al., 1985; Dennis and Charbord, 2002). Furthermore, previous studies have eluded to the possibility that 3G5- and STRO1-positive pericytes, obtained from bovine retina, may themselves be a source of multipotential cells (Doherty et al., 1998). Following on from this work, the perivascular niche has been proposed as a potential location for MSC-like populations identified in various tissues, such as adipose, skeletal muscle, umbilical cord, placenta, dental pulp, and periodontal ligament, using the perivascular markers CD146, 3G5, and STRO1 (MUC18) (Castrechini et al., 2010; Shi and Gronthos, 2003; Covas et al., 2008; Crisan et al., 2008; Zannettino et al., 2008). Continued efforts toward identification of the elusive stem cell niche may eventually help identify the appropriate conditions necessary for selective maintenance and expansion of multipotential stromal stem cells ex vivo.

5 IMMUNOMODULATORY PROPERTIES

Previously, studies have reported that cultured bone marrow MSCs display an immunoprivileged quality due to a lack of immune helper antigens such as HLA-DR, CD40, CD54, CD80, and CD86 (Wada et al., 2009). Moreover, third-party bone marrow MSCs were found to inhibit allogeneic mixed-lymphocyte reactions, to suppress T-lymphocyte proliferation following mitogenic stimulation in vitro, and to reduce graft-versus-host responses in vivo, despite mismatching of the class I major histocompatibly complex antigen (Bartholomew et al., 2002; Di Nicola et al., 2002; Le Blanc et al., 2003). The various mechanisms by which different MSC-like populations mediate immunomodulatory effects include the secretion of soluble factors such as tumor growth factor β1, hepatocyte growth factor (Di Nicola et al., 2002; Le Blanc et al., 2003; Meisel et al., 2004; Ryan et al., 2007) and prostaglandin E2 (Agarwal and Pittenger, 2005). Furthermore, the intracellular enzyme indolen-2,3-dioxygenase (IDO) has been shown to be one of the critical immunosuppressive factors produced by human BMSC-like populations (Meisel et al., 2004; Ryan et al., 2007). IDO is known to be induced by interferon-γ and other proinflammatory cytokines and catalyzes and degrades tryptophan, an essential amino acid for cell growth (King and Thomas, 2007), which is involved with inhibition of T-cell responses to autoantigens and alloantigens in vivo (Munn et al., 1998, 1999; Hwu et al., 2000). Nitric oxide (NO) has also been reported to be one of the major mediators of T-cell suppression by bone marrow MSCs but even more so by murine cells (Mazzoni et al., 2002; Sato et al., 2007). NO produced by inducible NO synthase (iNOS) in bone marrow MSCs suppresses phosphorylation of Stat5, which plays an important role in the T-cell proliferation signaling pathway (Mazzoni et al., 2002). In addition to the inhibitory effects of MSCs on activated immune cells, MSCs have been shown to work by stimulating Foxp3^{+} regulatory T-cells, which play a role in dampening immune responses under pathological conditions (Kikuiri et al., 2010; Liu et al., 2011; Tatara et al., 2009; Zhang et al., 2009). Furthermore, other reports have proposed that cell–cell contact is the major mechanism responsible for the immunomodulatory effects of MSCs highlighting the complex nature of interactions between MSCs and various activated immune cells (Krampera et al., 2003). These studies support the proposal that MSCs may be ideal candidates for allogeneic tissue regeneration applications and possible therapeutic vehicles for various inflammatory and autoimmune-based diseases without the complications of immunorejection and the requirement for immunosuppressive therapy. However, the view that this is a unique property of MSCs needs to be reconsidered, based on the findings of various published reports describing similar immunomodulatory properties of cultured fibroblast populations derived from different tissues expressing a MSC-like immunophenotype but exhibiting little or no capacity for multidifferentiation (Jones et al., 2007; Wada et al., 2009, 2011).

6 BONE MARROW MSC-BASED THERAPIES FOR TREATING CRANIOFACIAL DEFECTS

Bone autografts are the mainstream treatment for craniofacial repair that can be sourced from secondary sites such as the scapula, ribs, fibula, or iliac crest. However, major complications using this approach relate to the limited tissue available for large defects and the risk of secondary-site morbidity. Alternative approaches have included the use of allografts and artificial bone substitutes plus or minus autologous bone marrow with varying success.

More recently, the regenerative capacity of bone marrow MSCs for skeletal tissue repair has been demonstrated in preclinical large-animal studies and more recently in phase I and II human trials for a range of orthopedic-related indications (Miura et al., 2006; Arthur et al., 2009). It is proposed that the use of MSCs in combination with biocompatible scaffolds may result in comparable or better outcomes than conventional treatment options without the issue of large bone harvests and secondary-site morbidity (Miura et al., 2006; Arthur et al., 2009). Interestingly, studies using various preclinical large animal bone defect models have reported that MSCs are a reliable alternative for bone repair of cranial bone defects compared to other stromal cell populations (Krebsbach et al., 1998).

Human clinical investigations have been conducted using either autologous MSCs seeded into either collagen-coated scaffolds, polylactic polymer scaffolds, hydroxyapatite (HA) ceramics, tricalcium phosphate (TCP), or HA/TCP composites for treating alveolar cleft defects and general cranial defects (Velardi et al., 2006; Gimbel et al., 2007; Behnia et al., 2009; Shayesteh et al., 2008; Mendonca and Juiz-Lopez, 2010) Rickert et al., 2011. However, it has been observed that different biomaterial scaffolds have a differential effect on the bone-forming response of bone marrow MSCs derived from various species, where scaffold composites containing a percentage of HA (Fig. 2) exhibit superior osteoconductive properties when seeded with human bone marrow MSCs above other biomaterial composites in vivo (Krebsbach et al., 1997; Zannettino et al. 2011). Nevertheless, promising outcomes were reported from the majority of pilot clinical trials, where patients generally displayed increased bone formation and a significant reduction in defect size at the transplant sites (Velardi et al., 2006; Gimbel et al., 2007; Mendonca and Juiz-Lopez, 2010; Shayesteh et al., 2008; Behnia et al., 2009; Rickert et al., 2011) In a long-term follow-up study, 12 patients undergoing maxillary sinus augmentation showed regenerative bone formation and integration with surrounding tissue with no adverse effects, up to two years posttransplantation (Yamada et al., 2008). These preliminary clinical studies suggest that MSC therapy is potentially a viable treatment for craniofacial regeneration compared to more conventional allografts, autografts, and artificial bone substitutes. It is anticipated that future

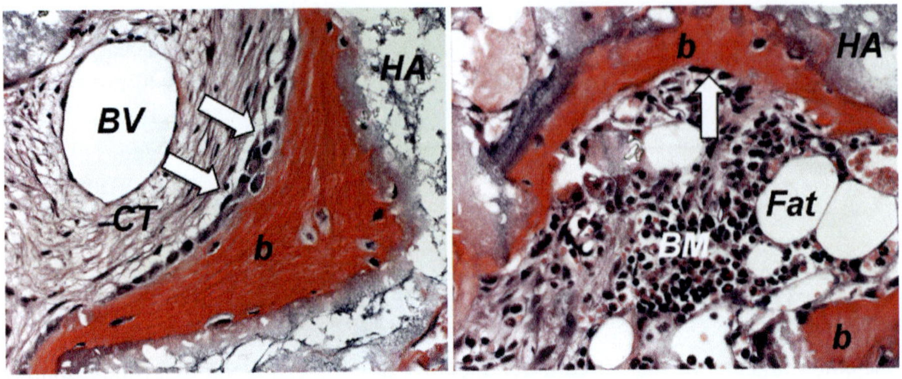

FIGURE 2 Differential osteoconductive properties of HA biomaterial composites. Cross sections of 8-week-old transplants of human bone marrow MSCs seeded onto HA/TCP (HA) ceramic particles implanted into NOD/SCID mice. Tissue sections stained with H&E showing connective fibrous tissue (CT), differentiated osteoblasts (arrows) new bone formation (b), adipose tissue (Fat), blood vessels (BV), and active hematopoietic bone marrow (BM).

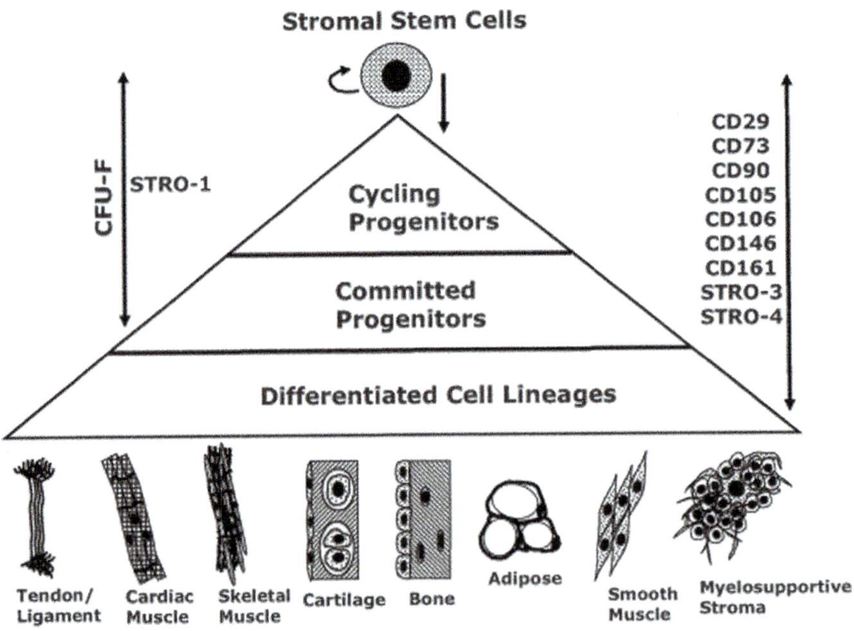

FIGURE 3 Expanded hierarchy of the bone marrow stromal system.

studies will assess the efficacy of allogeneic preparations of BMSCs with or without the presence of exogenous bone promoting growth factors.

7 CONCLUDING REMARKS

Preliminary studies have indicated the efficacy of employing ex vivo–expanded bone marrow MSCs for different tissue engineering orthopedic applications, particularly with respect to bone regeneration. What is clear is that a tremendous amount of work is still required to identify, isolate, and maintain multipotential bone marrow MSCs beyond the basic established criteria (Fig. 3), to complement recent advances in tissue engineering and gene manipulation technologies. Some of the main cell biology issues that remain to be addressed properly include the identification of various stromal precursor and committed progenitor cell populations following ex vivo expansion, establishment of growth and differentiation conditions that induce lineage specific differentiation, and the development of suitable carriers able to help implants integrate into the surrounding environment for accurate reconstruction of functional bone. This multilevel approach should help develop practical and workable MSC-mediated therapeutic alternatives in the future.

REFERENCES

Agarwal S, Pittenger MF (2005). Human mesenchymal stem cells modulate allogeneic immune cell responses. *Blood* 105(4):1815–1822.

Andersen DC, Kortesidis A, Zannettino AC, Kratchmarova I, Chen L, Jensen ON, et al. (2011). Development of novel monoclonal antibodies that define differentiation stages of human stromal (mesenchymal) stem cells. *Mol Cells* 32(2):133–142.

Arthur A, Zannettino A, Gronthos S (2009). The therapeutic applications of multipotential mesenchymal/stromal stem cells in skeletal tissue repair. *J Cell Physiol* 218(2):237–245.

Ashton BA, Allen TD, Howlett CR, Eaglesom CC, Hattori A, Owen M (1980). Formation of bone and cartilage by marrow stromal cells in diffusion chambers in vivo. *Clin Orthop* 151:294–307.

Bab I, Passi-Even L, Gazit D, Sekeles E, Ashton BA, Peylan-Ramu N, et al. (1988). Osteogenesis in in vivo diffusion chamber cultures of human marrow cells. *Bone Miner* 4(4):373–386.

Banfi A, Bianchi G, Notaro R, Luzzatto L, Cancedda R, Quarto R (2002). Replicative aging and gene expression in long-term cultures of human bone marrow stromal cells. *Tissue Eng* 8(6):901–910.

Bardin N, Frances V, Lesaule G, Horschowski N, George F, Sampol J (1996). Identification of the S-Endo 1 endothelial-associated antigen. *Biochem Biophys Res Commun* 218(1):210–216.

Bartholomew A, Sturgeon C, Siatskas M, Ferrer K, McIntosh K, Patil S, et al. (2002). Mesenchymal stem cells suppress lymphocyte proliferation in vitro and prolong skin graft survival in vivo. *Exp Hematol* 30(1):42–48.

Behnia H, Khojasteh A, Soleimani M, Tehranchi A, Khoshzaban A, Keshel SH, et al. (2009). Secondary repair of alveolar clefts using human mesenchymal stem cells. *Oral Surg Oral Med Oral Pathol Oral Radiol Endod* 108(2):e1–e6.

Bennett JH, Joyner CJ, Triffitt JT, Owen ME (1991). Adipocytic cells cultured from marrow have osteogenic potential. *J Cell Sci* 99(Pt 1):131–139.

Bensidhoum M, Chapel A, Francois S, Demarquay C, Mazurier C, Fouillard L, et al. (2004). Homing of in vitro expanded Stro-1^- or Stro-1^+ human mesenchymal stem cells into the NOD/SCID mouse and their role in supporting human CD34 cell engraftment. *Blood* 103(9):3313–3319.

Bentley SA (1982). Bone marrow connective tissue and the haemopoietic microenvironment. *Br J Haematol* 50(1):1–6.

Bentley SA, Alabaster O, Foidart JM (1981). Collagen heterogeneity in normal human bone marrow. *Br J Haematol* 48(2):287–291.

Bianco P, Boyde A (1993). Confocal images of marrow stromal (Westen–Bainton) cells. *Histochemistry* 100(2):93–99.

Bianco P, Riminucci M, Gronthos S, Robey PG (2001). Bone marrow stromal stem cells: nature, biology, and potential applications. *Stem Cells* 19(3): 180–192.

Bianco P, Riminucci M (1998). The bone marrow stroma in vivo: ontogeny, structure, cellular composition and changes in disease. In: *Marrow Stromal Cell Culture*, Beresford JN, ed., Cambridge University Press, Cambridge, UK, pp. 10–25.

Castrechini NM, Murthi P, Gude NM, Erwich JJ, Gronthos S, Zannettino A, et al. (2010). Mesenchymal stem cells in human placental chorionic villi reside in a vascular Niche. *Placenta* 31(3):203–212.

Castro-Malaspina H, Gay RE, Resnick G, Kapoor N, Meyers P, Chiarieri D, et al. (1980). Characterization of human bone marrow fibroblast colony-forming cells (CFU-F) and their progeny. *Blood* 56(2):289–301.

Cattoretti G, Schiro R, Orazi A, Soligo D, Colombo MP (1993). Bone marrow stroma in humans: anti-nerve growth factor receptor antibodies selectively stain reticular cells in vivo and in vitro. *Blood* 81(7):1726–1738.

Charbord P, Gown AM, Keating A, Singer JW (1985). CGA-7 and HHF, two monoclonal antibodies that recognize muscle actin and react with adherent cells in human long-term bone marrow cultures. *Blood* 66(5):1138–1142.

Covas DT, Panepucci RA, Fontes AM, Silva WA, Jr., Orellana MD, Freitas MC, et al. (2008). Multipotent mesenchymal stromal cells obtained from diverse human tissues share functional properties and gene-expression profile with $CD146^+$ perivascular cells and fibroblasts. *Exp Hematol* 36(5):642–654.

Crisan M, Yap S, Casteilla L, Chen CW, Corselli M, Park TS, et al. (2008). A perivascular origin for mesenchymal stem cells in multiple human organs. *Cell Stem Cell* 3(3):301–313.

Dennis JE, Charbord P (2002). Origin and differentiation of human and murine stroma. *Stem Cells* 20(3):205–214.

Dennis JE, Haynesworth SE, Young RG, Caplan AI (1992). Osteogenesis in marrow-derived mesenchymal cell porous ceramic composites transplanted subcutaneously: effect of fibronectin and laminin on cell retention and rate of osteogenic expression. *Cell Transplant* 1(1):23–32.

Dennis JE, Carbillet JP, Caplan AI, Charbord P (2002). The STRO−1+ marrow cell population is multipotential. *Cells Tissues Organs* 170(2–3):73–82.

Di Nicola M, Carlo-Stella C, Magni M, Milanesi M, Longoni PD, Matteucci P, et al. (2002). Human bone marrow stromal cells suppress T-lymphocyte proliferation induced by cellular or nonspecific mitogenic stimuli. *Blood* 99(10):3838–3843.

Doherty MJ, Ashton BA, Walsh S, Beresford JN, Grant ME, Canfield AE (1998). Vascular pericytes express osteogenic potential in vitro and in vivo. *J Bone Miner Res* 13(5):828–838.

Dominici M, Le Blanc K, Mueller I, Slaper-Cortenbach I, Marini F, Krause D, et al. (2006). Minimal criteria for defining multipotent mesenchymal stromal cells: The International Society for Cellular Therapy position statement. *Cytotherapy* 8(4):315–317.

Elices MJ, Osborn L, Takada Y, Crouse C, Luhowskyj S, Hemler ME, et al. (1990). VCAM-1 on activated endothelium interacts with the leukocyte integrin VLA-4 at a site distinct from the VLA-4/fibronectin binding site. *Cell* 60(4):577–584.

Falla N, Van V, Bierkens J, Borremans B, Schoeters G, Van Gorp U (1993). Characterization of a 5-fluorouracil-enriched osteoprogenitor population of the murine bone marrow. *Blood* 82(12):3580–3591.

Filshie RJ, Zannettino AC, Makrynikola V, Gronthos S, Henniker AJ, Bendall LJ, et al. (1998). MUC18, a member of the immunoglobulin superfamily, is expressed on bone marrow fibroblasts and a subset of hematological malignancies. *Leukemia* 12(3):414–421.

Friedenstein AJ (1976). Precursor cells of mechanocytes. *Int Rev Cytol* 47(5):327–359.

Friedenstein AJ (1980). Stromal mechanisms of bone marrow: cloning in vitro and retransplantation in vivo. *Haematol Bluttransfus* 25: 19–29.

Friedenstein AJ, Chailakhjan RK, Lalykina KS (1970). The development of fibroblast colonies in monolayer cultures of guinea-pig bone marrow and spleen cells. *Cell Tissue Kinet* 3(4):393–403.

Friedenstein AJ, Gorskaja JF, Kulagina NN (1976). Fibroblast precursors in normal and irradiated mouse hematopoietic organs. *Exp Hematol* 4(5):267–274.

Gimbel M, Ashley RK, Sisodia M, Gabbay JS, Wasson KL, Heller J, et al. (2007). Repair of alveolar cleft defects: reduced morbidity with bone marrow stem cells in a resorbable matrix. *J Craniofac Surg* 18(4):895–901.

Goncalves R, Lobato da Silva C, Cabral JM, Zanjani ED, Almeida-Porada G (2006). A Stro-1(+) human universal stromal feeder layer to expand/maintain human bone marrow hematopoietic stem/progenitor cells in a serum-free culture system. *Exp Hematol* 34(10):1353–1359.

Goshima J, Goldberg VM, Caplan AI (1991). The osteogenic potential of culture-expanded rat marrow mesenchymal cells assayed in vivo in calcium phosphate ceramic blocks. *Clin Orthop* 262: 298–311.

Gronthos S, Simmons PJ (1995). The growth factor requirements of STRO-1-positive human bone marrow stromal precursors under serum-deprived conditions in vitro. *Blood* 85(4):929–940.

Gronthos S (2004). Reconstruction of human mandible by tissue engineering. *Lancet* 364(9436):735–736.

Gronthos S, Graves SE, Ohta S, Simmons PJ (1994). The STRO-1^+ fraction of adult human bone marrow contains the osteogenic precursors. *Blood* 84(12):4164–4173.

Gronthos S, Zannettino AC, Graves SE, Ohta S, Hay SJ, Simmons PJ (1999). Differential cell surface expression of the STRO-1 and alkaline phosphatase antigens on discrete developmental stages in primary cultures of human bone cells. *J Bone Miner Res* 14(1):47–56.

Gronthos S, Mankani M, Brahim J, et al. (2000). Postnatal human dental pulp stem cells (DPSCs) in vitro and in vivo. *Proc Natl Acad Sci USA* 97(25):13625–13630.

Gronthos S, Simmons PJ, Graves SE, Robey PG (2001). Integrin-mediated interactions between human bone marrow stromal precursor cells and the extracellular matrix. *Bone* 28(2):174–181.

Gronthos S, Zannettino AC, Hay SJ, Shi S, Graves SE, Kortesidis A, et al. (2003). Molecular and cellular characterisation of highly purified stromal stem cells derived from human bone marrow. *J Cell Sci* 116(Pt 9):1827–1835.

Gronthos S, Akintoye SO, Wang CY, Shi S (2006). Bone marrow stromal stem cells for tissue engineering. *Periodontol 2000* 41:188–195.

Gronthos S, Fitter S, Diamond P, Simmons PJ, Itescu S, Zannettino AC (2007). A novel monoclonal antibody (STRO-3) identifies an isoform of tissue nonspecific alkaline phosphatase expressed by multipotent bone marrow stromal stem cells. *Stem Cells Dev* 16(6):953–963.

Gronthos S, McCarty R, Mrozik K, Fitter S, Paton S, Menicanin D, et al. (2009). Heat shock protein-90 beta (Hsp90ss) is expressed at the surface of multipotential mesenchymal precursor cells (MPC): generation of a novel monoclonal antibody, STRO-4, with specificity for MPC from human and ovine tissues. *Stem Cells Dev* 18(9):1253–1262.

Gundle R, Joyner CJ, Triffitt JT (1995). Human bone tissue formation in diffusion chamber culture in vivo by bone-derived cells and marrow stromal fibroblastic cells. *Bone* 16(6):597–601.

Haynesworth SE, Goshima J, Goldberg VM, Caplan AI (1992). Characterization of cells with osteogenic potential from human marrow. *Bone* 13(1):81–88.

Huss R, Hong DS, McSweeney PA, Hoy CA, Deeg HJ (1995). Differentiation of canine bone marrow cells with hemopoietic characteristics from an adherent stromal cell precursor. *Proc Natl Acad Sci USA* 92(3):748–752.

Hwu P, Du MX, Lapointe R, Do M, Taylor MW, Young HA (2000). Indoleamine 2,3-dioxygenase production by human dendritic cells results in the inhibition of T cell proliferation. *J Immunol* 164(7):3596–3599.

Jacobsen K, Kravitz J, Kincade PW, Osmond DG (1996). Adhesion receptors on bone marrow stromal cells: in vivo expression of vascular cell adhesion molecule-1 by reticular cells and sinusoidal endothelium in normal and gamma-irradiated mice. *Blood* 87(1):73–82.

Jones S, Horwood N, Cope A, Dazzi F (2007). The antiproliferative effect of mesenchymal stem cells is a fundamental property shared by all stromal cells. *J Immunol* 179(5):2824–2831.

Kikuiri T, Kim I, Yamaza T, Akiyama K, Zhang Q, Li Y, et al. (2010). Cell-based immunotherapy with mesenchymal stem cells cures bisphosphonate-related osteonecrosis of the jaw-like disease in mice. *J Bone Miner Res* 25(7):1668–1679.

King NJ, Thomas SR (2007). Molecules in focus: indoleamine 2,3-dioxygenase. *Int J Biochem Cell Biol* 39(12):2167–2172.

Krampera M, Glennie S, Dyson J, Scott D, Laylor R, Simpson E, et al. (2003). Bone marrow mesenchymal stem cells inhibit the response of naive and memory antigen-specific T cells to their cognate peptide. *Blood* 101(9):3722–3729.

Krebsbach PH, Kuznetsov SA, Satomura K, Emmons RV, Rowe DW, Robey PG (1997). Bone formation in vivo: comparison of osteogenesis by transplanted mouse and human marrow stromal fibroblasts. *Transplantation* 63(8):1059–1069.

Krebsbach PH, Mankani MH, Satomura K, Kuznetsov SA, Robey PG (1998). Repair of craniotomy defects using bone marrow stromal cells. *Transplantation* 66(10):1272–1278.

Kuznetsov SA, Krebsbach PH, Satomura K, Kerr J, Riminucci M, Benayahu D, et al. (1997). Single-colony derived strains of human marrow stromal fibroblasts form bone after transplantation in vivo. *J Bone Miner Res* 12(9):1335–1347.

Kuzu I, Bicknell R, Fletcher CD, Gatter KC (1993). Expression of adhesion molecules on the endothelium of normal tissue vessels and vascular tumors. *Lab Investig* 69(3):322–328.

Le Blanc K, Tammik L, Sundberg B, Haynesworth SE, Ringden O (2003). Mesenchymal stem cells inhibit and stimulate mixed lymphocyte cultures and mitogenic responses independently of the major histocompatibility complex. *Scand J Immunol* 57(1):11–20.

Lichtman MA (1981). The ultrastructure of the hemopoietic environment of the marrow: a review. *Exp Hematol* 9(4):391–410.

Liu Y, Wang L, Kikuiri T, Akiyama K, Chen C, Xu X, et al. (2011). Mesenchymal stem cell–based tissue regeneration is governed by recipient T lymphocytes via IFN-gamma and TNF-alpha. *Nat Med* 17(12):1594–1601.

Martens TP, See F, Schuster MD, Sondermeijer HP, Hefti MM, Zannettino A, et al. (2006). Mesenchymal lineage precursor cells induce vascular network formation in ischemic myocardium. *Nat Clin Pract Cardiovasc Med* 3(Suppl 1):S18–S22.

Mazzoni A, Bronte V, Visintin A, Spitzer JH, Apolloni E, Serafini P, et al. (2002). Myeloid suppressor lines inhibit T cell responses by an NO-dependent mechanism. *J Immunol* 168(2):689–695.

McCarty RC, Gronthos S, Zannettino AC, Foster BK, Xian CJ (2009). Characterisation and developmental potential of ovine bone marrow derived mesenchymal stem cells. *J Cell Physiol* 219(2):324–333.

Meisel R, Zibert A, Laryea M, Gobel U, Daubener W, Dilloo D (2004). Human bone marrow stromal cells inhibit allogeneic T-cell responses by indoleamine 2,3-dioxygenase-mediated tryptophan degradation. *Blood* 103(12):4619–4621.

Mendonca JJ, Juiz-Lopez P (2010). Regenerative facial reconstruction of terminal stage osteoradionecrosis and other advanced craniofacial diseases with adult cultured stem and progenitor cells. *Plast Reconstr Surg* 126(5):1699–1709.

Menicanin D, Bartold PM, Zannettino AC, Gronthos S (2010). Identification of a common gene expression signature associated with immature clonal mesenchymal cell populations derived from bone marrow and dental tissues. *Stem Cells Dev* 19(10):1501–1510.

Miura M, Miura Y, Sonoyama W, Yamaza T, Gronthos S, Shi S (2006). Bone marrow–derived mesenchymal stem cells for regenerative medicine in craniofacial region. *Oral Dis* 12(6):514–522.

Munn DH, Zhou M, Attwood JT, Bondarev I, Conway SJ, Marshall B, et al. (1998). Prevention of allogeneic fetal rejection by tryptophan catabolism. *Science* 281(5380):1191–1193.

Munn DH, Shafizadeh E, Attwood JT, Bondarev I, Pashine A, Mellor AL (1999). Inhibition of T cell proliferation by macrophage tryptophan catabolism. *J Exp Med* 189(9):1363–1372.

Muraglia A, Cancedda R, Quarto R (2000). Clonal mesenchymal progenitors from human bone marrow differentiate in vitro according to a hierarchical model. *J Cell Sci* 113(Pt 7):1161–1166.

Nasef A, Zhang YZ, Mazurier C, Bouchet S, Bensidhoum M, Francois S, et al. (2009). Selected Stro-1-enriched bone marrow stromal cells display a major suppressive effect on lymphocyte proliferation. *Int J Lab Hematol* 31(1):9 19.

Osborn L, Hession C, Tizard R, Vassallo C, Luhowskyj S, Chi-Rosso G, et al. (1989). Direct expression cloning of vascular cell adhesion molecule 1, a cytokine-induced endothelial protein that binds to lymphocytes. *Cell* 59(6):1203–1211.

Owen M (1988). Marrow stromal stem cells. *J Cell Sci Suppl* 10: 63–76.

Owen M, Friedenstein AJ (1988). Stromal stem cells: marrow-derived osteogenic precursors. *Ciba Found Symp* 136(29):42–60.

Owen ME, Cave J, Joyner CJ (1987). Clonal analysis in vitro of osteogenic differentiation of marrow CFU-F. *J Cell Sci* 87(Pt 5):731–738.

Perkins S, Fleischman RA (1990). Stromal cell progeny of murine bone marrow fibroblast colony-forming units are clonal endothelial-like cells that express collagen IV and laminin. *Blood* 75(3):620–625.

Pittenger MF, Mackay AM, Beck SC, Jaiswal RK, Douglas R, Mosca JD, et al. (1999). Multilineage potential of adult human mesenchymal stem cells. *Science* 284(5411):143–147.

Prockop DJ (1997). Marrow stromal cells as stem cells for nonhematopoietic tissues. *Science* 276(5309):71–74.

Psaltis PJ, Paton S, See F, Arthur A, Martin S, Itescu S, et al (2011). Enrichment for STRO-1 expression enhances the cardiovascular paracrine activity of human bone marrow–derived mesenchymal cell populations. *J Cell Physiol* 223(2):530–540.

Rickard DJ, Kassem M, Hefferan TE, Sarkar G, Spelsberg TC, Riggs BL (1996). Isolation and characterization of osteoblast precursor cells from human bone marrow. *J Bone Miner Res* 11(3):312–324.

Rickert D, Sauerbier S, Nagursky H, Menne D, Vissink A, Raghoebar GM (2011). Maxillary sinus floor elevation with bovine bone mineral combined with either autogenous bone or autogenous stem cells: a prospective randomized clinical trial. *Clin Oral Implants Res* 22(3):251–258.

Ryan JM, Barry F, Murphy JM, Mahon BP (2007). Interferon-gamma does not break, but promotes the immunosuppressive capacity of adult human mesenchymal stem cells. *Clin Exp Immunol* 149(2):353–363.

Sacchetti B, Funari A, Michienzi S, Di Cesare S, Piersanti S, Saggio I, et al. (2007). Self-renewing osteoprogenitors in bone marrow sinusoids can organize a hematopoietic microenvironment. *Cell* 131(2):324–336.

Sato K, Ozaki K, Oh I, Meguro A, Hatanaka K, Nagai T, et al. (2007). Nitric oxide plays a critical role in suppression of T-cell proliferation by mesenchymal stem cells. *Blood* 109(1):228–234.

Shayesteh YS, Khojasteh A, Soleimani M, Alikhasi M, Khoshzaban A, Ahmadbeigi N (2008). Sinus augmentation using human mesenchymal stem cells loaded into a beta-tricalcium phosphate/hydroxyapatite scaffold. *Oral Surg Oral Med Oral Pathol Oral Radiol Endod* 106(2):203 209.

Shi S, Gronthos S (2003). Perivascular niche of postnatal mesenchymal stem cells in human bone marrow and dental pulp. *J Bone Miner Res* 18(4):696–704.

Shi S, Gronthos S, Chen S, Reddi A, Counter CM, Robey PG, et al. (2002). Bone formation by human postnatal bone marrow stromal stem cells is enhanced by telomerase expression. *Nat Biotechnol* 20(6):587–591.

Shi S, Bartold PM, Miura M, Seo BM, Robey PG, Gronthos S (2005). The efficacy of mesenchymal stem cells to regenerate and repair dental structures. *Orthod Craniofac Res* 8(3):191–199.

Simmons PJ, Torok-Storb B (1991). Identification of stromal cell precursors in human bone marrow by a novel monoclonal antibody, STRO-1. *Blood* 78(1):55–62.

Simmons PJ, Masinovsky B, Longenecker BM, Berenson R, Torok-Storb B, Gallatin WM (1992). Vascular cell adhesion molecule-1 expressed by bone marrow stromal cells mediates the binding of hematopoietic progenitor cells. *Blood* 80(2):388–395.

Simmons PJ, Gronthos S, Zannettino A, Ohta S, Graves S (1994). Isolation, characterization and functional activity of human marrow stromal progenitors in hemopoiesis. *Prog Clin Biol Res* 389: 271–280.

Simonsen JL, Rosada C, Serakinci N, Justesen J, Stenderup K, Rattan SI, et al. (2002). Telomerase expression extends the proliferative life-span and maintains the osteogenic potential of human bone marrow stromal cells. *Nat Biotechnol* 20(6):592–596.

Sutherland HJ, Eaves CJ, Eaves AC, Dragowska W, Lansdorp PM (1989). Characterization and partial purification of human marrow cells capable of initiating long-term hematopoiesis in vitro. *Blood* 74(5):1563–1570.

Tatara R, Ozaki K, Kikuchi Y, Hatanaka K, Oh I, Meguro A, et al. (2011). Mesenchymal stromal cells inhibit Th17 but not regulatory T-cell differentiation. *Cytotherapy* 13(6):686–694.

Van Vlasselaer P, Falla N, Snoeck H, Mathieu E (1994). Characterization and purification of osteogenic cells from murine bone marrow by two-color cell sorting using anti-Sca-1 monoclonal antibody and wheat germ agglutinin. *Blood* 84(3):753–563.

Velardi F, Amante PR, Caniglia M, De Rossi G, Gaglini P, Isacchi G, et al. (2006). Osteogenesis induced by autologous bone marrow cells transplant in the pediatric skull. *Child's Nerv Syst* 22(9):1158–1166.

Wada N, Menicanin D, Shi S, Bartold PM, Gronthos S (2009). Immunomodulatory properties of human periodontal ligament stem cells. *J Cell Physiol* 219(3):667–676.

Wada N, Bartold PM, Gronthos S (2011) Human foreskin fibroblasts exert immunomodulatory properties by a different mechanism to bone marrow stromal/stem cells. *Stem Cells Dev* 20(4):647–659.

Waller EK, Olweus J, Lund-Johansen F, Huang S, Nguyen M, Guo GR, et al. (1995). The "common stem cell" hypothesis reevaluated: human fetal bone marrow contains separate populations of hematopoietic and stromal progenitors. *Blood* 85(9):2422–2435.

Weiss L (1965). The structure of bone marrow: functional interrelationships of vascular and hematopoietic compartments in experimental hemolytic anemia: an electron microscopic study. *J Morphol* 117(3):467–537.

Westen H, Bainton DF (1979). Association of alkaline-phosphatase-positive reticulum cells in bone marrow with granulocytic precursors. *J Exp Med* 150(4):919–937.

Yamada Y, Nakamura S, Ito K, Kohgo T, Hibi H, Nagasaka T, et al. (2008). Injectable tissue-engineered bone using autogenous bone marrow–derived stromal cells for maxillary sinus augmentation: clinical application report from a 2–6-year follow-up. *Tissue Eng A* 14(10):1699–1707.

Zannettino AC, Paton S, Kortesidis A, Khor F, Itescu S, Gronthos S (2007). Human mulipotential mesenchymal/stromal stem cells are derived from a discrete subpopulation of STRO-1(bright)/CD34(-)/CD45(-)/glycophorin-A(-) bone marrow cells. *Haematologica* 92(12):1707–1708.

Zannettino AC, Paton S, Arthur A, Khor F, Itescu S, Gimble JM, et al. (2008). Multipotential human adipose-derived stromal stem cells exhibit a perivascular phenotype in vitro and in vivo. *J Cell Physiol* 214(2):413–421.

Zhang Q, Shi S, Liu Y, Uyanne J, Shi Y, Le AD (2009). Mesenchymal stem cells derived from human gingiva are capable of immunomodulatory functions and ameliorate inflammation-related tissue destruction in experimental colitis. *J Immunol* 183(12):7787–7798.

Zannettino AC, Paton S, Itescu S, Gronthos S (2011). Comparative assessment of the osteoconductive properties of different biomaterials in vivo seeded with human or ovine mesenchymal stem/stromal cells. *Tissue Eng A* 16(12):3579–3587.

Zimmermann S, Voss M, Kaiser S, Kapp U, Waller CF, Martens UM (2003). Lack of telomerase activity in human mesenchymal stem cells. *Leukemia* 17(6):1146–1149.

12

ADIPOSE TISSUE–DERIVED STEM CELLS AND THEIR REGENERATION POTENTIAL

JEFFREY GIMBLE

Stem Cell Biology Laboratory, Pennington Biomedical Research Center, Louisiana State University System, Baton Rouge, Louisiana

MARYAM REZAI RAD AND SHAOMIAN YAO

Department of Comparative Biomedical Sciences, Louisiana State University School of Veterinary Medicine, Baton Rouge, Louisiana

1 INTRODUCTION

During the past decade, adipose tissue has drawn significant attention from the fields of regenerative medicine and tissue engineering for its use as a source of stem cells for craniofacial repair (Halvorsen et al., 2000; Zuk et al., 2001, 2002; Gimble and Guilak, 2003). The number of publications referring to "adipose stem cells" has increased exponentially. Adipose tissue constitutes from 5 to over 50% of a person's body mass, and surgeons can access subcutaneous depots using minimally invasive liposuction techniques. Thus, it is feasible to harvest adipose-derived cells from the majority of patients requiring a regenerative procedure. In this chapter we review the in vitro and in vivo characterization of adipose derived cells and recent advances of craniofacial repair and regeneration using adipose tissue–derived stromal and stem cells (ASCs) in both preclinical and human models. The potential of ASCs for dental tissue regeneration is also discussed.

Stem Cells in Craniofacial Development and Regeneration, First Edition.
Edited by George T.-J. Huang, Irma Thesleff.

2 CHARACTERIZATION OF STROMAL VASCULAR FRACTION CELLS AND ADIPOSE TISSUE–DERIVED STROMAL AND STEM CELLS

2.1 Isolation Methods

Plastic surgeons harvest adipose tissue through tumescent liposuction techniques or direct tissue resection (abdominoplasty or mammoplasty reduction surgeries). Each method yields tissue that contains viable cells, although the quantity is reduced significantly if liposuction is performed using ultrasound assistance (Oedayrajsingh-Varma et al., 2006). Routinely, the adipose tissue is transferred to the laboratory for cell isolation immediately; however, independent studies have documented that tissue can be stored up to 24 hours at 4°C or room temperature before processing without significant reduction in cell recovery (Matsumoto et al., 2007; Carvalho et al., 2011). In lipoaspirates, cells can be isolated directly by centrifugation of the infranatant bloody tumescent fluid phase and subsequent cell culture methods (Yoshimura et al., 2006); however, the vast majority of protocols rely on an enzymatic digestion of the supernatant floating adipose tissue phase of the lipoaspirate (Zuk et al., 2001; Yu et al., 2011) (Fig. 1). Collagenase digestion releases the floating mature adipocytes from a heterogeneous population of higher-density stromal vascular fraction (SVF) cells that can be separated by differential centrifugation. The SVF cells include pre-adipocyte or adherent adipose tissue–derived stromal and stem cells (ASCs) as well as endothelial cells, fibroblasts, lymphocytes, myeloid cells, and pericytes based on flow cytometric and immunohistochemical analyses (Sengenes et al., 2005; McIntosh et al., 2006; Mitchell, 2006; Khan et al., 2008; Traktuev et al., 2008; Zannettino, 2008; Yang et al., 2010; Zimmerlin et al., 2010). When seeded in tissue culture flasks, the ASC subpopulation of the SVF cells adheres to the plastic surface and expands with culture doubling times of about 48 to 72 hours for humans and more rapidly in canine, equine,

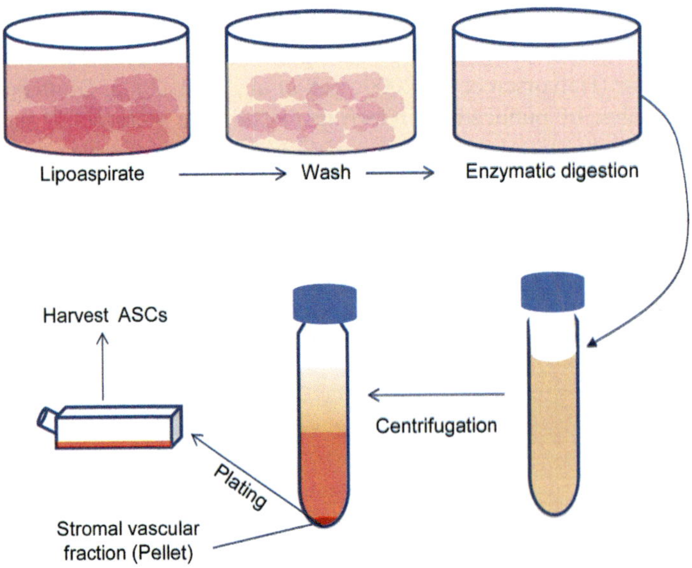

FIGURE 1 Isolation of adipose-derived stromal and stem cells.

and other species (Mitchell, 2006; Vidal et al., 2007; Spencer et al., 2011) (Fig. 1). The ASCs maintain their immunophenotype with extended passage in vitro and continue to proliferate, consistent with their expression of telomerase (Estes et al., 2006; Izadpanah et al., 2006); however, these features may vary depending on the species of origin (Danoviz et al., 2011). The genomic stability of late-passage ASCs has been brought into question based on genetic analyses and reports of tumorigenic transformation (Rubio et al., 2005; Danoviz et al., 2011). Although the publication reporting malignant transformation of human ASCs has been retracted, further studies will be needed to document the safety of late-passage ASCs for regenerative medicine and tissue engineering (de la Fuente et al., 2010; Garcia et al., 2010).

2.2 Differentiation Potential

The ASCs were first characterized based on their mesenchymal differentiation characteristics. Independent studies documented the adipogenic, chondrogenic, myogenic, and osteogenic capabilities of human ASCs in vitro in response to chemical and hormonal factors, and this has been confirmed by multiple independent laboratories (Halvorsen et al., 2001a,b; Zuk et al., 2001, 2002; Erickson et al., 2002; Gimble et al., 2007, 2008). In related studies, the ASCs were found to undergo cardiomyocyte differentiation based on electrophysiological behavior and biochemical markers (Planat-Benard et al., 2004a). Subsequent investigations demonstrated the ability of ASCs to exhibit endothelial cell characteristics when cultured in Matrigel with the addition of angiogenic growth factors (Miranville et al., 2004; Planat-Benard et al., 2004b; Rehman et al., 2004). The ASC differentiation potential is not limited to the mesodermal and mesenchymal lineages, however. Independent studies have shown that their ability to differentiate along epithelial, hepatocyte, and neuronal-like cell pathways, consistent with both ectodermal and endodermal lineages (Safford et al., 2002, 2004; Zuk et al., 2002; Ashjian et al., 2003; Brzoska et al., 2005; Guilak et al., 2006; Banas et al., 2007, 2008; Ruiz et al., 2010; Baer et al., 2011). Despite these findings, the majority of the literature refers to ASCs as "multipotent" rather than "pluripotent."

Questions have been raised concerning the ability of ASCs to undergo osteogenic differentiation relative to bone marrow mesenchymal stromal and stem cells (BMSCs) and other sources of adult stem cells (De Ugarte et al., 2003; Kern et al., 2006; Liu et al., 2007; Rebelatto et al., 2008). While some studies indicate that the osteogenic properties of ASCs are equivalent to those of BMSCs, others suggest that BMSCs display a superior differentiation capacity (De Ugarte et al., 2003; Kern et al., 2006; Liu et al., 2007; Rebelatto et al., 2008). Furthermore, recent work found that the osteogenic capacity of ASCs is impaired subsequent to cryopreservation, although this could be recovered with insulin-like growth factor 1 or bone morphogenetic protein 4 (BMP4) supplementation (James et al., 2011). Nevertheless, there is a substantial body of literature supporting the potential use of ASCs for bone and craniofacial regeneration (Cowan et al., 2004, 2005; Hicok et al., 2004; Justesen et al., 2004)

2.3 Immunomodulatory Regulation and Immunogenicity

Human SVF cells are surface positive for both HLA-ABC and HLA-DR and, as a result, elicit a robust proliferative response from peripheral blood mononuclear cells (PBMNs) in a mixed-lymphocyte reaction (MLR) (Gronthos et al., 2001; Zuk et al., 2002; McIntosh et al., 2006). In contrast, culture-expanded ASCs, like BMSCs, have a reduced HLA-DR expression and fail to stimulate PBMN proliferation in vitro (Bartholomew et al., 2002; Le Blanc et al., 2003, 2004; Puissant et al., 2005; McIntosh et al., 2006). The addition of ASCs suppresses the proliferative response of alloreactive lymphocytes in an active MLR in vitro and in graft-versus-host disease models in vivo (Puissant et al., 2005; McIntosh et al., 2006; Yanez et al., 2006). This immunosuppressive activity has been correlated with the ASC expression of prostaglandin E2, indoleamine-2, 3-dioxygenase (IDO), interleukin 10 (IL-10), Jagged-1, and transforming growth factor β (Cui et al., 2007b; Kang et al., 2008; DelaRosa et al., 2009; Crop et al., 2010a,b; Yanez et al., 2010; Shi et al., 2011). The adherence- and passage-dependent change in the surface immunophenotype implies that it will be possible to transplant allogeneic ASCs but not SVF cells across traditional immune barriers. The ASC immunosuppressive properties have potential benefits and drawbacks with respect to regenerative medicine. While immunosuppression may reduce inflammatory responses secondary to engraftment, foreign body reaction, or acute or chronic disease, it may also protect and promote the growth of residual (postmastectomy) or occult tumor cells in the vicinity of the ASC implant. Substantial evidence indicates that ASCs are recruited to tumor sites and can enhance their growth and proliferation in vitro and in vivo (Zhang et al., 2009, 2010; Donnenberg et al., 2010; Lin et al., 2010; Prantl et al., 2010; Zimmerlin et al., 2011).

2.4 Paracrine Actions

Like BMSCs, ASCs secrete a diverse range of adipokines, cytokines, and signal transducers with both anabolic and catabolic consequences that modulate cellular activity within their local microenvironment (Rehman et al., 2004; Puissant et al., 2005; Trayhurn and Wood, 2005; Kilroy, 2007; Zvonic et al., 2007a). It is now accepted that a substantial portion of the regenerative capacity of ASCs is secondary to the paracrine action of these secreted proteins. Both transplanted SVF cells and ASCs release cytokines such as stromal derived factor 1 (SDF1), which can recruit local, recipient-derived stem cells to a site of injury (Kondo et al., 2009). A growing body of data indicates that SVF cells, ASCs, and adipose tissue itself promote angiogenesis and vascularization in vitro and in vivo (Fukumura et al., 2003; Miranville et al., 2004; Planat-Benard et al., 2004b; Rehman et al., 2004; Cao et al., 2005; Moon et al., 2006; Kim et al., 2007; Sumi et al., 2007; Traktuev et al., 2008; Blanton et al., 2009; Cai et al., 2009; Kondo et al., 2009; Kang et al., 2010; Shoji et al., 2010; Bhang et al., 2011; Eto et al., 2011; Levi et al., 2011c). This is mediated, in part, through their release of hepatocyte growth factor (HGF), vascular endothelial growth factor (VEGF), and related factors (Rehman et al., 2004; Kilroy, 2007). By manipulating the physical environment, it is possible to modulate the level of angiogenic cytokine expression in ASCs. When exposed to oxygen tensions of 5% or less, the ASCs alter their secretion of HGF, VEGF, and other angiogenic factors (Rehman et al., 2004; Thangarajah et al., 2009; Wei, 2009; Amos et al., 2011; Rasmussen et al., 2011). By acting as a localized cytokine factory, transplanted SVF cells and ASCs improve recovery from hindlimb ischemia, myocardial infarction, and related ischemic insults (Miranville et al., 2004; Planat-Benard et al., 2004b; Rehman et al., 2004; Miyahara et al., 2006; Sumi et al., 2007; Bel et al., 2010; Kang et al., 2010). This same paracrine-mediated vasculogenic mechanism is postulated to underlie the ability of ASCs to promote osteogenesis and bone regeneration (Shoji et al., 2010; Behr et al., 2011).

3 FINDINGS FROM CRANIOFACIAL PRECLINICAL ANIMAL MODELS

3.1 ASCs for Repairing Critical-Sized Cranial Defects

Multiple groups have used canine, leporid, or rodent models to evaluate the utility of adipose tissue and ASCs to repair critical-sized cranial defects (Cowan et al., 2004, 2005; Dudas et al., 2006; Cui et al., 2007a; Yoon et al., 2007; Betz et al., 2010; Levi et al., 2010, 2011b,d; Kwan et al., 2011). In the presence of an osteoinductive scaffold, autologous ASC implants promote repair of critical-sized calvarial defects in dogs, mice, rats, and rabbits within a 4- to 8-week period (Cowan et al., 2004; Dudas et al., 2006; Cui et al., 2007a; Yoon et al., 2007) (Fig. 2). The addition of fibroblast growth factor 2 (FGF2) or BMP2 can augment or accelerate this process (Betz et al., 2010; Kwan et al., 2011). When human-derived ASCs with scaffolds were examined in immunocompromised mice, they exhibited osteogenic potential and supported regeneration of acute calvarial defects (Levi et al., 2010). The mechanism underlying this repair relies, in part, on the paracrine activation of the Sonic hedgehog pathway in adjacent murine osteoblasts by the ASCs (Levi et al., 2011a). Additionally, the underlying dura mater contributes to the process by its production of BMP2 (Levi et al., 2011d).

The timing of the ASC delivery to the injury site has important consequences. If the introduction of the ASCs was delayed for a period of 8 weeks after the original injury in mice, they failed to support repair of the chronic defect, possibly due to a lack of endogenous BMP production (Levi et al., 2011b). More subtle aspects of timing could also influence regenerative outcomes. Transcriptomic analyses have documented that circadian mechanisms influence murine calvarial tissue metabolism in vivo (Zvonic et al., 2007b). Similarly, in vitro studies have demonstrated the expression of the circadian regulatory apparatus in BMSCs (Wu et al., 2008b). Consistent with these findings, a luciferase transgene driven by the osteocalcin promoter displayed a circadian oscillatory expression profile in mice (Gafni et al., 2009). Since similar circadian mechanisms have been implicated in both adipose and bone biology, these data suggest that circadian regulation may prove to be an important component of ASC function in regenerative medicine (Zvonic et al., 2006; Wu et al., 2007; Kawai and Rosen, 2010).

3.2 Tooth Regeneration Studies Using ASCs

Recent advances in stem cell and regenerative medical research have made possible the regeneration of a functional and living tooth to replace a missing, diseased, or damaged tooth (Ohazama et al., 2004; Shi et al., 2005). In that vein, tooth regeneration using various adult stem cells has been attempted. Stem cells showing the greatest potential for tooth regeneration are non-dental-derived mesenchymal stem cells (Ohazama et al., 2004) and dental tissue–derived stem cells, such as stem cells from human exfoliated deciduous teeth (SHEDs), adult dental pulp stem cells (DPSCs), stem cells from the apical part of the papilla (SCAPs), stem cells from the dental follicle (DFSCs), and periodontal ligament stem cells (PDLSCs).

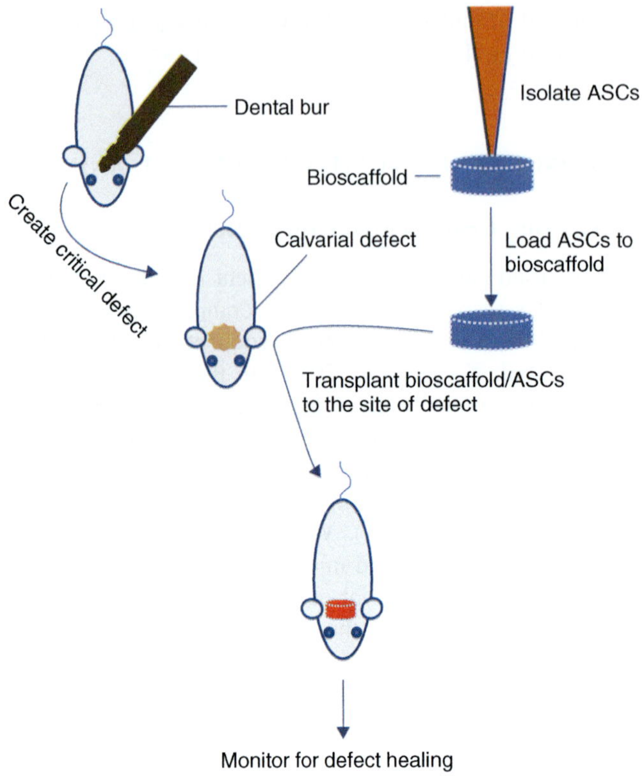

FIGURE 2 Animal model for cranial defect repair.

For tooth regeneration experiments, it is logical to begin by mimicking the natural tooth development process (Chapters 24 and 25). During tooth development, reciprocal interactions between ectomesenchymal cells and epithelium allow the oral epithelium to provide instructive information and signals for tooth initiation (Ferguson et al., 2000; Thesleff, 2003). Mimicking this natural tooth developmental process, Ohazama et al. (2004) found that recombinations of non-dental-derived mesenchyme cell aggregations and embryonic oral epithelium stimulate the stem cells to express odontogenic genes. When the stem cell recombinations were transplanted into adult renal capsules or tooth sockets, the cells formed tooth structures and associated bone.

Whereas bone marrow stem cells (BMSCs) may be relatively easy to obtain, BMSCs lose their differentiation capability in aged patients, which is especially a concern as tooth loss most often occurs in elderly patients (Stenderup et al., 2003) Other autologous stem cell sources, including stem cells from skin dermis, hair follicles, and adipose tissue, have been considered as alternatives for tooth regeneration (Jing et al., 2008; Wu et al., 2009). In this regard, ASCs would be especially appealing for the following reasons: (1) a large quantity of adipose tissue is usually available for ASC isolation; (2) less invasive procedures are required to obtain adipose tissues; (3) ASCs possess a high growth rate and differentiation potential (Schaffler and Buchler, 2007), as discussed earlier in the chapter; and (4) ASCs do not decrease differentiation capability with the increasing age of donors (Cowan et al., 2004; Shi et al., 2005).

Because BMSCs have been shown to have the capability of odontogenic differentiation (Li et al., 2007; Liu et al., 2007), and ASCs and BMSCs displayed similar gene expression profiles and multidifferentiation abilities (Izadpanah et al., 2006; Kern et al., 2006), it is speculated that ASCs would possess odontogenic differentiation ability (Jing et al., 2008). When ASCs were enforced to express dentin sialophosphoprotein (DSPP), a protein regulating dentin mineralization and tooth development, enhanced expression of genes related to mineralization, such as *Cbfa1*, *Osx, BSP*, *OCN*, and *DMP1* in ASCs, was observed. More important, the DSPP-overexpressed ASCs could express the early odontogenic marker genes *Msx1*, *Msx2*, *Lhx7*, and *Pax9* (Lu et al., 2008). Collectively, these results, together with the osteogenic capability of ASCs as discussed in this chapter, suggest that ASCs could be an alternative stem cell source for tooth regeneration.

In an in vitro cell culture experiment, human ASCs were mixed with agarose to form a three-dimensional structure and cultured in differentiation medium (BJG medium with the addition of BMP2, BMP4, BMP5, 1,25-hydroxy-D_3 vitamin, fibronectin, calcitonin, insulin-like growth factor 2, TGFβ1, VEGF, and other components). After one month of incubation, the ASC aggregates were sectioned and H&E staining of the sections revealed that the ASCs formed three distinct layers expressing ameloblastic and odontoblastic differentiation markers, such as ameloblastin, enamelin, amelogenin, and MMP20 (Ferro et al., 2011) Moreover, oriented hydroxyapatite crystals as displayed by enamel and dentin were also detected in this three–layer structure. This experiment demonstrated that ASCs could be induced in vitro to undergo dental development by supplementing proper molecules, but without reciprocal interactions to epithelium signals (Ferro et al., 2011).

Recently, an in vivo study was conducted to test the tooth regeneration potential of ASCs compared with DPSCs in a preclinical model animal (Hung et al., 2011). Rabbit ASCs and DPSCs were implanted into incisor sockets in combination with type I collagen and BMP2. In the presence (but not absence) of BMP2, the ASCs and DPSC implants formed mineralized toothlike structures at high frequency after 15 weeks (Hung et al., 2011). Further histological studies of the regenerated tooth structures showed well-developed vascular and nervous system and dentinal structures (Hung et al., 2011). This body of evidence suggests that ASCs have potential applications in tooth regeneration.

3.3 Soft Tissue Reconstruction

Biomaterial scaffolds and ASCs have been used for soft tissue regeneration in vitro and in vivo [reviewed by Choi et al. (2010) and Marra et al. (2011)]. Adipogenic differentiation by ASCs is well established and can be augmented by the presence of FGF2, glucocorticoids, thiazolidinediones, and related growth factors (Hauner et al., 1989; Kawaguchi et al., 1998; Halvorsen et al., 2001a; Zuk et al., 2001; Marra et al., 2008). While ASCs maintained in two-dimensional cultures undergo adipogenesis, the process can be promoted in three-dimensional scaffolds derived from silk or adipose tissue extracellular matrix itself (Flynn et al., 2007, 2008; Mauney et al., 2007; Brown et al., 2011). These three-dimensional constructs exhibit metabolic activity comparable to that of intact adipose tissue fragments based on lipolysis and insulin sensitivity (Kang et al., 2009; Choi et al., 2011). The combination of ASCs and extracellular matrix scaffolds has potential utility for the regeneration of facial and cranial soft tissue defects secondary to trauma, cancer, or genetic disorders.

3.4 Clinical Applications

Maxillofacial surgeons have begun to use adipose-derived cells and tissues for craniofacial repair. The first case report in 2004 described a pediatric patient in Germany who presented with a poorly healing traumatic cranial wound complicated by infection (Lendeckel et al., 2004). In a single operation, the patient was treated with a combination of autologous bone graft from the iliac crest, fibrin glue, and SVF cells harvested from subcutaneous adipose tissue intraoperatively (Lendeckel et al., 2004). The patient showed signs of new bone formation and healing based on computerized tomography (CT) scans performed three months later (Lendeckel et al., 2004). Following this report, another pediatric case report from the United States described repair of hereditary orbitozygomatic defects in a patient with Treacher–Collins syndrome (Taylor, 2010). A Finnish group expanded these approaches to adults, with a case report describing an elderly hemimaxillectomy patient who had a hard palate defect (Mesimaki et al., 2009). The patient's subcutaneous adipose tissue was harvested and autologous ASCs were manufactured according to current good manufacturing practice protocols (Mesimaki et al., 2009). In the meantime, a β-tricalcium phosphate (β-TCP) matrix was fabricated based on a CT scan of the craniofacial defect (Mesimaki et al., 2009). This scaffold was seeded with the expanded autologous ASCs together with BMP2, implanted into the patient's rectus abdominis muscle in close proximity to the inferior epigastric artery, and followed radiographically for evidence of bone formation (Mesimaki et al., 2009). After eight months, the tissue engineered bone was transplanted to the defect site, where vascularization was achieved by anastomosis of the inferior epigastric artery to the facial artery (Fig. 3). Ultimately, the graft integrated successfully, with the formation of oral mucosa, dental implants were placed, and the patient recovered full function (Mesimaki et al., 2009) (Fig. 3). Since then, this same research

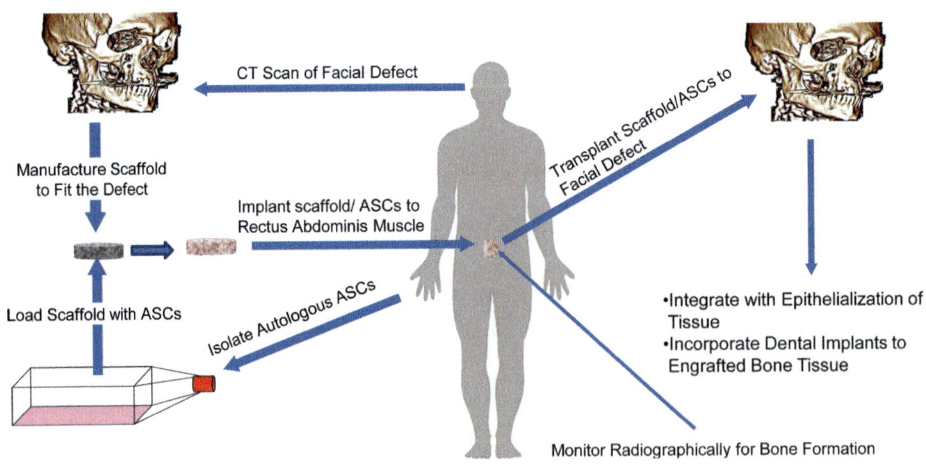

FIGURE 3 Human model for maxillofacial repair.

team has reported successful repair of calvarial defects in four patients using β-TCP and autologous ASCs manufactured under cGMP protocols (Thesleff et al., 2011).

Craniofacial defect repair with adipose tissue and cells is not limited to bone. Plastic surgeons in Japan have coined the term *cell-assisted lipotransfer* to describe the use of autologous adipose tissue combined with SVF cells for facial soft tissue reconstruction (Yoshimura et al., 2008). These investigators performed lipotransfer with or without SVF cells in three patients with facial lipoatrophy due to Parry–Romberg syndrome or related defects (Yoshimura et al., 2008). Although the cohort size was too small to achieve statistically significant outcomes, there was a trend toward enhanced cosmetic outcomes when SVF cells were used in the procedure (Yoshimura et al., 2008). A clinical trial using ASCs to treat Romberg's disease has been registered and completed in South Korea; however, the outcomes of this work have not been reported to date (www.clinicaltrials.gov). Independent clinical trials from Italy have explored the use of platelet-rich plasma (PRP), alone or in combination with adipose-derived cells, for craniofacial soft and hard tissue repair as well as the treatment of trauma-related lower-extremity ulcers (Cervelli et al., 2009, 2011; Gentile et al., 2010). The presence of PRP significantly improved fat grafting outcomes for the treatment of Romberg's syndrome and facial atrophy, with 70% retention of volume as compared to 31% in controls one year postoperatively (Cervelli et al., 2009). Although these studies remain preliminary, together they demonstrate the feasibility of autologous SVF cells and ASCs for craniofacial bone and soft tissue reconstruction in chronic or genetic defects. Further studies are necessary documenting the safety and efficacy of these procedures. Future work should explore the utility of allogeneic as well as autologous ASCs for craniofacial reconstruction.

4 FUTURE DIRECTIONS

The use of adipose-derived cells (SVF cells and ASCs) and matrices for craniofacial regeneration is still in its infancy from a clinical perspective. The use of adipose tissue as an autologous or allogeneic resource holds substantial promise. One exciting possibility that has been suggested by a leader in the field is their use to repair cleft palate defects (Zuk, 2008). Nevertheless, there is a need to develop the underlying basic and clinical science supporting these applications. It will be important to define the mechanisms responsible for the contribution of SVF cells and ASCs to regeneration at the signal transduction and/or differentiation levels. This information may lead to improved combinations of growth factors and/or osteo-inductive and osteo—conductive matrices to accelerate repair. Clinically, it is necessary to monitor the safety and efficacy of these procedures over time with appropriate patient follow-up (Gimble et al., 2011). Ultimately, the field must be prepared to present an evidence-based medical approach, including meta-analyses collating data from multiple centers, justifying to patients, physicians, and health care regulators the continued use of ASC and SVF cell procedures (Gimble et al., 2011).

Acknowledgments

This work was supported in part with funding from the Pennington Biomedical Research Foundation. The authors acknowledge the administrative and research assistance of Ms. Laura Dallam.

REFERENCES

Amos PJ, Mulvey CL, Seaman SA, Walpole J, Degen KE, Shang H, et al. (2011). Hypoxic culture and in vivo inflammatory environments affect the assumption of pericyte characteristics by human adipose and bone marrow progenitor cells. *Am J Physiol Cell Physiol* 301(6):C1378–C1388.

Ashjian PH, Elbarbary AS, Edmonds B, DeUgarte D, Zhu M, Zuk PA, et al. (2003). In vitro differentiation of human processed lipoaspirate cells into early neural progenitors. *Plast Reconstr Surg* 111(6):1922–1931.

Baer PC, Brzoska M, Geiger H (2011). Epithelial differentiation of human adipose-derived stem cells. *Methods Mol Biol* 702:289–298.

Banas A, Teratani T, Yamamoto Y, Tokuhara M, Takeshita F, Quinn G, et al. (2007). Adipose tissue–derived mesenchymal stem cells as a source of human hepatocytes. *Hepatology* 46(1):219–228.

Banas A, Teratani T, Yamamoto Y, Tokuhara M, Takeshita F, Osaki M, et al. (2008). IFATS collection: in vivo therapeutic potential of human adipose tissue mesenchymal stem cells after transplantation into mice with liver injury. *Stem Cells* 26(10):2705–2712.

Bartholomew A, Sturgeon C, Siatskas M, Ferrer K, McIntosh K, Patil S, et al. (2002). Mesenchymal stem cells suppress lymphocyte proliferation in vitro and prolong skin graft survival in vivo. *Exp Hematol* 30(1):42–48.

Behr B, Tang C, Germann G, Longaker MT, Quarto N (2011). Locally applied vascular endothelial growth factor A increases the osteogenic healing capacity of human adipose-derived stem cells by promoting osteogenic and endothelial differentiation. *Stem Cells* 29(2): 286–296.

Bel A, Planat-Bernard V, Saito A, Bonnevie L, Bellamy V, Sabbah L, et al. (2010). Composite cell sheets: a further step toward safe and effective myocardial regeneration by cardiac progenitors derived from embryonic stem cells. *Circulation* 122(11 Suppl): S118–S123.

Betz OB, Betz VM, Abdulazim A, Penzkofer R, Schmitt B, Schroder C, et al. (2010). The Repair of critical size bone defects using expedited, autologous BMP-2 gene activated fat implants. *Tissue Eng A* 16:1093–1101.

Bhang SH, Cho SW, La WG, Lee TJ, Yang HS, Sun AY, et al. (2011). Angiogenesis in ischemic tissue produced by spheroid grafting of human adipose–derived stromal cells. *Biomaterials* 32(11):2734–2747.

Blanton MW, Hadad I, Johnstone BH, Mund JA, Rogers PI, Eppley BL, et al. (2009). Adipose stromal cells and platelet-rich plasma therapies synergistically increase revascularization during wound healing. *Plast Reconstr Surg* 123(2 Suppl):56S–64S.

Brown BN, Freund JM, Han L, Rubin JP, Reing JE, Jeffries EM, et al. (2011). Comparison of three methods for the derivation of a biologic scaffold composed of adipose tissue extracellular matrix. *Tissue Eng C* 17(4):411–421.

Brzoska M, Geiger H, Gauer S, Baer P (2005). Epithelial differentiation of human adipose tissue–derived adult stem cells. *Biochem Biophys Res Commun* 330(1):142–150.

Cai L, Johnstone BH, Cook TG, Tan J, Fishbein MC, Chen PS, et al. (2009). IFATS collection: human adipose tissue–derived stem cells induce angiogenesis and nerve sprouting following myocardial infarction, in conjunction with potent preservation of cardiac function. *Stem Cells* 27(1):230–237.

Cao Y, Sun Z, Liao L, Meng Y, Han Q, Zhao RC (2005). Human adipose tissue–derived stem cells differentiate into endothelial cells in vitro and improve postnatal neovascularization in vivo. *Biochem Biophys Res Commun* 332(2):370–379.

Carvalho PP, Wu X, Yu G, Dias IR, Gomes ME, Reis RL, et al. (2011). The effect of storage time on adipose-derived stem cell recovery from human lipoaspirates. *Cells Tissues Organs* 194(6):494–500.

Cervelli V, Gentile P, Scioli MG, Grimaldi M, Casciani CU, Spagnoli LG, et al. (2009). Application of platelet-rich plasma in plastic surgery: clinical and in vitro evaluation. *Tissue Eng C* 15(4):625–634.

Cervelli V, Gentile P, De Angelis B, Calabrese C, Di Stefani A, Scioli MG, et al. (2011). Application of enhanced stromal vascular fraction and fat grafting mixed with PRP in post-traumatic lower extremity ulcers. *Stem Cell Res* 6(2):103–111.

Choi JH, Gimble JM, Lee K, Marra KG, Rubin JP, Yoo JJ, et al. (2010). Adipose tissue engineering for soft tissue regeneration. *Tissue Eng B* 16(4):413–426.

Choi JH, Bellas E, Gimble J, Vunjak-Novakovic G, Kaplan DL (2011). Lipolytic Function of Adipocyte/Endothelial Co-Cultures. *Tissue Eng A* 17:1437–1444.

Cowan CM, Shi YY, Aalami OO, Chou YF, Mari C, Thomas R, et al. (2004). Adipose-derived adult stromal cells heal critical-size mouse calvarial defects. *Nat Biotechnol* 22(5):560–567.

Cowan CM, Aalami OO, Shi YY, Chou YF, Mari C, Thomas R, et al. (2005). Bone morphogenetic protein 2 and retinoic acid accelerate in vivo bone formation, osteoclast recruitment, and bone turnover. *Tissue Eng* 11(3–4):645–658.

Crop MJ, Baan CC, Korevaar SS, Ijzermans JN, Pescatori M, Stubbs AP, et al. (2010a). Inflammatory conditions affect gene expression and function of human adipose tissue–derived mesenchymal stem cells. *Clin Exp Immunol* 162(3):474–486.

Crop MJ, Baan CC, Korevaar SS, Ijzermans JN, Weimar W, Hoogduijn MJ (2010b). Human adipose tissue–derived mesenchymal stem cells induce explosive T-cell proliferation. *Stem Cells Dev* 19(12):1843–1853.

Cui L, Liu B, Liu G, Zhang W, Cen L, Sun J, et al. (2007a). Repair of cranial bone defects with adipose derived stem cells and coral scaffold in a canine model. *Biomaterials* 28(36):5477–5486.

Cui L, Yin S, Liu W, Li N, Zhang W, Cao Y (2007b). Expanded adipose-derived stem cells suppress mixed lymphocyte reaction by secretion of prostaglandin E2. *Tissue Eng* 13:1185–1195.

Danoviz ME, Bassaneze V, Nakamuta JS, dos Santos-Junior GR, Saint-Clair D, Bajgelman MC, et al. (2011). Adipose tissue–derived stem cells from humans and mice differ in proliferative capacity and genome stability in long-term cultures. *Stem Cells Dev* 20(4):661–670.

de la Fuente R, Bernad A, Garcia-Castro J, Martin MC, Cigudosa JC (2010). Retraction: spontaneous human adult stem cell transformation. *Cancer Res* 70(16):6682.

De Ugarte DA, Morizono K, Elbarbary A, Alfonso Z, Zuk PA, Zhu M, et al. (2003). Comparison of multi-lineage cells from human adipose tissue and bone marrow. *Cells Tissues Organs* 174(3):101–109.

DelaRosa O, Lombardo E, Beraza A, Mancheno-Corvo P, Ramirez C, Menta R, et al. (2009). Requirement of IFN-gamma-mediated indoleamine 2,3-dioxygenase expression in the modulation of lymphocyte proliferation by human adipose-derived stem cells. *Tissue Eng A* 15(10):2795–2806.

Donnenberg VS, Zimmerlin L, Rubin JP, Donnenberg AD (2010). Regenerative therapy after cancer: What are the risks? *Tissue Eng B* 16(6):567–575.

Dudas JR, Marra KG, Cooper GM, Penascino VM, Mooney MP, Jiang S, et al. (2006). The osteogenic potential of adipose-derived stem cells for the repair of rabbit calvarial defects. *Ann Plast Surg* 56(5):543–548.

Erickson GR, Gimble JM, Franklin DM, Rice HE, Awad H, Guilak F (2002). Chondrogenic potential of adipose tissue–derived stromal cells in vitro and in vivo. *Biochem Biophys Res Commun* 290(2):763–769.

Estes BT, Wu AW, Storms RW, Guilak F (2006). Extended passaging, but not aldehyde dehydrogenase activity, increases the chondrogenic potential of human adipose-derived adult stem cells. *J Cell Physiol* 209(3):987–995.

Eto H, Suga H, Inoue K, Aoi N, Kato H, Araki J, et al. (2011). Adipose injury-associated factors mitigate hypoxia in ischemic tissues through activation of adipose-derived stem/progenitor/stromal cells and induction of angiogenesis. *Am J Pathol* 178(5):2322–2332.

Ferguson CA, Tucker AS, Sharpe PT (2000). Temporospatial cell interactions regulating mandibular and maxillary arch patterning. *Development* 127(2):403–412.

Ferro F, Spelat R, Falini G, Gallelli A, D'Aurizio F, Puppato E, et al. (2011). Adipose tissue–derived stem cell in vitro differentiation in a three-dimensional dental bud structure. *Am J Pathol* 178(5):2299–2310.

Flynn L, Prestwich GD, Semple JL, Woodhouse KA (2007). Adipose tissue engineering with naturally derived scaffolds and adipose-derived stem cells. *Biomaterials* 28(26):3834–3842.

Flynn LE, Prestwich GD, Semple JL, Woodhouse KA (2008). Proliferation and differentiation of adipose-derived stem cells on naturally derived scaffolds. *Biomaterials* 29(12):1862–1871.

Fukumura D, Ushiyama A, Duda DG, Xu L, Tam J, Krishna V, et al. (2003). Paracrine regulation of angiogenesis and adipocyte differentiation during in vivo adipogenesis. *Circ Res* 93(9):e88–e97.

Gafni Y, Ptitsyn AA, Zilberman Y, Pelled G, Gimble JM, Gazit D (2009). Circadian rhythm of osteocalcin in the maxillomandibular complex. *J Dent Res* 88(1):45–50.

Garcia S, Bernad A, Martin MC, Cigudosa JC, Garcia-Castro J, de la Fuente R (2010). Pitfalls in spontaneous in vitro transformation of human mesenchymal stem cells. *Exp Cell Res* 316(9):1648–1650.

Gentile P, Bottini DJ, Spallone D, Curcio BC, Cervelli V (2010). Application of platelet-rich plasma in maxillofacial surgery: clinical evaluation. *J Craniofac Surg* 21(3):900–904.

Gimble J, Guilak F (2003). Adipose-derived adult stem cells: isolation, characterization, and differentiation potential. *Cytotherapy* 5(5):362–369.

Gimble JM, Katz AJ, Bunnell BA (2007). Adipose-derived stem cells for regenerative medicine. *Circ Res* 100(9):1249–1260.

Gimble JM, Bunnell BA, Chiu ES, Guilak F (2011). Concise review: adipose-derived stromal vascular fraction cells and stem cells: Let's not get lost in translation. *Stem Cells* 29(5):749–754.

Gimble JM, Guilak F, Nutall ME, Sathishkumar S, Vidal MA, Bunnell BA (2008). In vitro differentiation potential of mesenchymal stem cells. *Transfus Medi Hemother* 35:228–238.

Gronthos S, Franklin DM, Leddy HA, Robey PG, Storms RW, Gimble JM (2001). Surface protein characterization of human adipose tissue-derived stromal cells. *J Cell Physiol* 189(1):54–63.

Guilak F, Lott KE, Awad HA, Cao Q, Hicok KC, Fermor B, et al. (2006). Clonal analysis of the differentiation potential of human adipose-derived adult stem cells. *J Cell Physiol* 206(1):229–237.

Halvorsen YC, Wilkison WO, Gimble JM (2000). Adipose-derived stromal cells: their utility and potential in bone formation. *Int J Obes Relat Metab Disord* 24 (Suppl 4):S41–S44.

Halvorsen YD, Bond A, Sen A, Franklin DM, Lea-Currie YR, Sujkowski D, et al. (2001a). Thiazolidinediones and glucocorticoids synergistically induce differentiation of human adipose tissue stromal cells: biochemical, cellular, and molecular analysis. *Metabolism* 50(4):407–413.

Halvorsen YD, Franklin D, Bond AL, Hitt DC, Auchter C, Boskey AL, et al. (2001b). Extracellular matrix mineralization and osteoblast gene expression by human adipose tissue–derived stromal cells. *Tissue Eng* 7(6):729–741.

Hauner H, Entenmann G, Wabitsch M, Gaillard D, Ailhaud G, Negrel R, et al. (1989). Promoting effect of glucocorticoids on the differentiation of human adipocyte precursor cells cultured in a chemically defined medium. *J Clin Investig* 84(5):1663–1670.

Hicok KC, Du Laney TV, Zhou YS, Halvorsen YD, Hitt DC, Cooper LF, et al. (2004). Human adipose-derived adult stem cells produce osteoid in vivo. *Tissue Eng* 10(3–4):371–380.

Hung CN, Mar K, Chang HC, Chiang YL, Hu HY, Lai CC, et al. (2011). A comparison between adipose tissue and dental pulp as sources of MSCs for tooth regeneration. *Biomaterials* 32(29):6995–7005.

Izadpanah R, Trygg C, Patel B, Kriedt C, Dufour J, Gimble JM, et al. (2006). Biologic properties of mesenchymal stem cells derived from bone marrow and adipose tissue. *J Cell Biochem* 99:1286–1297.

James AW, Levi B, Nelson ER, Peng M, Commons GW, Lee M, et al. (2011). Deleterious effects of freezing on osteogenic differentiation of human adipose-derived stromal cells in vitro and in vivo. *Stem Cells Dev* 20(3):427–439.

Jing W, Wu L, Lin Y, Liu L, Tang W, Tian W (2008). Odontogenic differentiation of adipose-derived stem cells for tooth regeneration: necessity, possibility, and strategy. *Med Hypotheses* 70(3):540–542.

Justesen J, Pedersen SB, Stenderup K, Kassem M (2004). Subcutaneous adipocytes can differentiate into bone-forming cells in vitro and in vivo. *Tissue Eng* 10(3–4):381–391.

Kang JH, Gimble JM, Kaplan DL (2009). In vitro 3D model for human vascularized adipose tissue. *Tissue Eng A* 15(8):2227–2236.

Kang JW, Kang KS, Koo HC, Park JR, Choi EW, Park YH (2008). Soluble factors–mediated immunomodulatory effects of canine adipose tissue–derived mesenchymal stem cells. *Stem Cells Dev* 17(4):681–693.

Kang Y, Park C, Kim D, Seong CM, Kwon K, Choi C (2010). Unsorted human adipose tissue–derived stem cells promote angiogenesis and myogenesis in murine ischemic hindlimb model. *Microvasc Res* 80(3):310–316.

Kawaguchi N, Toriyama K, Nicodemou-Lena E, Inou K, Torii S, Kitagawa Y (1998). De novo adipogenesis in mice at the site of injection of basement membrane and basic fibroblast growth factor. *Proc Natl Acad Sci USA* 95(3):1062–1066.

Kawai M, Rosen CJ (2010). PPARgamma: a circadian transcription factor in adipogenesis and osteogenesis. *Nat Rev Endocrinol* 6(11):629–636.

Kern S, Eichler H, Stoeve J, Kluter H, Bieback K (2006). Comparative analysis of mesenchymal stem cells from bone marrow, umbilical cord blood, or adipose tissue. *Stem Cells* 24(5):1294–1301.

Khan WS, Tew SR, Adesida AB, Hardingham TE (2008). Human infrapatellar fat pad–derived stem cells express the pericyte marker 3G5 and show enhanced chondrogenesis after expansion in fibroblast growth factor–2. *Arthritis Res Ther* 10(4):R74.

Kilroy GE, Foster S, Wu X, Ruiz J, Sherwood S, Heifetz A, Ludlow JW, Stricker DM, Potiny S, Green P, Halvorsen YDC, Cheatham B, Storms RW, Gimble JM (2007). Cytokine profile of human adipose-derived stem cells: expression of angiogenic, hematopoietic, and pro-inflammatory factors. *J Cell Physiol* 212(3):702–709.

Kim Y, Kim H, Cho H, Bae Y, Suh K, Jung J (2007). Direct comparison of human mesenchymal stem cells derived from adipose tissues and bone marrow in mediating neovascularization in response to vascular ischemia. *Cell Physiol Biochem* 20(6):867–876.

Kondo K, Shintani S, Shibata R, Murakami H, Murakami R, Imaizumi M, et al. (2009). Implantation of adipose-derived regenerative cells enhances ischemia-induced angiogenesis. *Arterioscler Thromb Vasc Biol* 29(1):61–66.

Kwan MD, Sellmyer MA, Quarto N, Ho AM, Wandless TJ, Longaker MT (2011). Chemical control of FGF-2 release for promoting calvarial healing with adipose stem cells. *J Biol Chem* 286(13):11307–11313.

Le Blanc K, Tammik C, Rosendahl K, Zetterberg E, Ringden O (2003). HLA expression and immunologic properties of differentiated and undifferentiated mesenchymal stem cells. *Exp Hematol* 31(10):890–896.

Le Blanc K, Rasmusson I, Gotherstrom C, Seidel C, Sundberg B, Sundin M, et al. (2004). Mesenchymal stem cells inhibit the expression of CD25 (interleukin-2 receptor) and CD38 on phytohaemagglutinin-activated lymphocytes. *Scand J Immunol* 60(3):307–315.

Lendeckel S, Jodicke A, Christophis P, Heidinger K, Wolff J, Fraser JK, et al. (2004). Autologous stem cells (adipose) and fibrin glue used to treat widespread traumatic calvarial defects: case report. *J Craniomaxillofac Surg* 32(6):370–373.

Levi B, James AW, Nelson ER, Vistnes D, Wu B, Lee M, et al. (2010). Human adipose derived stromal cells heal critical size mouse calvarial defects. *PLoS One* 5(6):e11177.

Levi B, James AW, Nelson ER, Li S, Peng M, Commons GW, et al. (2011a). Human adipose-derived stromal cells stimulate autogenous skeletal repair via paracrine Hedgehog signaling with calvarial osteoblasts. *Stem Cells Dev* 20(2):243–257.

Levi B, James AW, Nelson ER, Peng M, Wan DC, Commons GW, et al. (2011b). Acute skeletal injury is necessary for human adipose-derived stromal cell-mediated calvarial regeneration. *Plast Reconstr Surg* 127(3):1118–1129.

Levi B, Nelson ER, Hyun JS, Glotzbach JP, Li S, Nauta A, et al. (2011c). Enhancement of human adipose-derived stromal cell angiogenesis through knockdown of a BMP-2 inhibitor. *Plast Reconstr Surg* 9(1):53–66.

Levi B, Nelson ER, Li S, James AW, Hyun JS, Montoro DT, et al. (2011d). Dura mater stimulates human adipose-derived stromal cells to undergo bone formation in mouse calvarial defects. *Stem Cells* 29(8):1241–1255.

Li ZY, Chen L, Liu L, Lin YF, Li SW, Tian WD (2007). Odontogenic potential of bone marrow mesenchymal stem cells. *J Oral Maxillofac Surg* 65(3):494–500.

Lin G, Yang R, Banie L, Wang G, Ning H, Li LC, et al. (2010). Effects of transplantation of adipose tissue–derived stem cells on prostate tumor. *Prostate* 70(10):1066–1073.

Liu TM, Martina M, Hutmacher DW, Hoi J, Hui P, Lee EH, et al. (2007). Identification of common pathways mediating differentiation of bone marrow and adipose tissues derived human mesenchymal stem cells (MSCs) into three mesenchymal lineages. *Stem Cells* 25(3):750–760.

Marra KG, Defail AJ, Clavijo-Alvarez JA, Badylak SF, Taieb A, Schipper B, et al. (2008). FGF-2 enhances vascularization for adipose tissue engineering. *Plast Reconstr Surg* 121(4):1153–1164.

Marra KG, Brayfield CA, Rubin JP (2011). Adipose stem cell differentiation into smooth muscle cells. *Methods Mol Biol* 702:261–268.

Matsumoto D, Shigeura T, Sato K, Inoue K, Suga H, Kato H, et al. (2007). Influences of preservation at various temperatures on liposuction aspirates. *Plast Reconstr Surg* 120(6):1510–1517.

Mauney JR, Nguyen T, Gillen K, Kirker-Head C, Gimble JM, Kaplan DL (2007). Engineering adipose-like tissue in vitro and in vivo utilizing human bone marrow and adipose-derived mesenchymal stem cells with silk fibroin 3D scaffolds. *Biomaterials* 28(35):5280–5290.

McIntosh K, Zvonic S, Garrett S, Mitchell JB, Floyd ZE, Hammill L, et al. (2006). The immunogenicity of human adipose derived cells: temporal changes in vitro. *Stem Cells* 24:1245–1253.

Mesimaki K, Lindroos B, Tornwall J, Mauno J, Lindqvist C, Kontio R, et al. (2009). Novel maxillary reconstruction with ectopic bone formation by GMP adipose stem cells. *Int J Oral Maxillofac Surg* 38(3):201–209.

Miranville A, Heeschen C, Sengenes C, Curat CA, Busse R, Bouloumie A (2004). Improvement of postnatal neovascularization by human adipose tissue–derived stem cells. *Circulation* 110(3):349–355.

Mitchell JB, Mcintosh K, Zvonic S, Garrett S, Floyd ZE, Kloster A et al. (2006). The immunophenotype of human adipose derived cells: temporal changes in stromal- and stem cell–associated markers. *Stem Cells* 24(2):376–385.

Miyahara Y, Nagaya N, Kataoka M, Yanagawa B, Tanaka K, Hao H, et al. (2006). Monolayered mesenchymal stem cells repair scarred myocardium after myocardial infarction. *Nat Med* 12(4):459–465.

Moon MH, Kim SY, Kim YJ, Kim SJ, Lee JB, Bae YC, et al. (2006). Human adipose tissue–derived mesenchymal stem cells improve postnatal neovascularization in a mouse model of hindlimb ischemia. *Cell Physiol Biochem* 17(5–6):279–290.

Oedayrajsingh-Varma MJ, van Ham SM, Knippenberg M, Helder MN, Klein-Nulend J, Schouten TE, et al. (2006). Adipose tissue–derived mesenchymal stem cell yield and growth characteristics are affected by the tissue-harvesting procedure. *Cytotherapy* 8(2):166–177.

Ohazama A, Modino SA, Miletich I, Sharpe PT (2004). Stem-cell-based tissue engineering of murine teeth. *J Dent Res* 83(7):518–522.

Planat-Benard V, Menard C, Andre M, Puceat M, Perez A, Garcia-Verdugo JM, et al. (2004a). Spontaneous cardiomyocyte differentiation from adipose tissue stroma cells. *Circ Res* 94(2):223–229.

Planat-Benard V, Silvestre JS, Cousin B, Andre M, Nibbelink M, Tamarat R, et al. (2004b). Plasticity of human adipose lineage cells toward endothelial cells: physiological and therapeutic perspectives. *Circulation* 109(5):656–63.

Prantl L, Muehlberg F, Navone NM, Song YH, Vykoukal J, Logothetis CJ, et al. (2010). Adipose tissue–derived stem cells promote prostate tumor growth. *Prostate* 70(15):1709–1715.

Puissant B, Barreau C, Bourin P, Clavel C, Corre J, Bousquet C, et al. (2005). Immunomodulatory effect of human adipose tissue–derived adult stem cells: comparison with bone marrow mesenchymal stem cells. *Br J Haematol* 129(1):118–129.

Rasmussen JG, Frobert O, Pilgaard L, Kastrup J, Simonsen U, Zachar V, et al. (2011). Prolonged hypoxic culture and trypsinization increase the pro-angiogenic potential of human adipose tissue–derived stem cells. *Cytotherapy* 13(3):318–328.

Rebelatto CK, Aguiar AM, Moretao MP, Senegaglia AC, Hansen P, Barchiki F, et al. (2008). Dissimilar differentiation of mesenchymal stem cells from bone marrow, umbilical cord blood, and adipose tissue. *Exp Biol Med (Maywood)* 233(7):901–913.

Rehman J, Traktuev D, Li J, Merfeld-Clauss S, Temm-Grove CJ, Bovenkerk JE, et al. (2004). Secretion of angiogenic and antiapoptotic factors by human adipose stromal cells. *Circulation* 109(10):1292–1298.

Rubio D, Garcia-Castro J, Martin MC, de la Fuente R, Cigudosa JC, Lloyd AC, et al. (2005). Spontaneous human adult stem cell transformation. *Cancer Res* 65(8):3035–3039.

Ruiz JC, Ludlow JW, Sherwood S, Yu G, Wu X, Gimble JM (2010). Differentiated human adipose-derived stem cells exhibit hepatogenic capability in vitro and in vivo. *J Cell Physiol* 225(2):429–436.

Safford KM, Hicok KC, Safford SD, Halvorsen YD, Wilkison WO, Gimble JM, et al. (2002). Neurogenic differentiation of murine and human adipose-derived stromal cells. *Biochem Biophys Res Commun* 294(2):371–379.

Safford KM, Safford SD, Gimble JM, Shetty AK, Rice HE (2004). Characterization of neuronal/glial differentiation of murine adipose-derived adult stromal cells. *Exp Neurol* 187(2):319–328.

Schaffler A, Buchler C (2007). Concise review: adipose tissue–derived stromal cells: basic and clinical implications for novel cell-based therapies. *Stem Cells* 25(4):818–827.

Sengenes C, Lolmede K, Zakaroff-Girard A, Busse R, Bouloumie A (2005). Preadipocytes in the human subcutaneous adipose tissue display distinct features from the adult mesenchymal and hematopoietic stem cells. *J Cell Physiol* 205(1):114–122.

Shi D, Liao L, Zhang B, Liu R, Dou X, Li J, et al. (2011). Human adipose tissue–derived mesenchymal stem cells facilitate the immunosuppressive effect of cyclosporin A on T lymphocytes through Jagged-1-mediated inhibition of NF-kappaB signaling. *Exp Hematol* 39(2):214–224 e1.

Shi S, Bartold PM, Miura M, Seo BM, Robey PG, Gronthos S (2005). The efficacy of mesenchymal stem cells to regenerate and repair dental structures. *Orthod Craniofac Res* 8(3):191–199.

Shoji T, Ii M, Mifune Y, Matsumoto T, Kawamoto A, Kwon SM, et al. (2010). Local transplantation of human multipotent adipose-derived stem cells accelerates fracture healing via enhanced osteogenesis and angiogenesis. *Lab Invest* 90(4):637–649.

Spencer ND, Chun R, Vidal MA, Gimble JM, Lopez MJ (2011). In vitro expansion and differentiation of fresh and revitalized adult canine bone marrow–derived and adipose tissue–derived stromal cells. *Vet J* 191(2):231–239.

Stenderup K, Justesen J, Clausen C, Kassem M (2003). Aging is associated with decreased maximal life span and accelerated senescence of bone marrow stromal cells. *Bone* 33(6):919–926.

Sumi M, Sata M, Toya N, Yanaga K, Ohki T, Nagai R (2007). Transplantation of adipose stromal cells, but not mature adipocytes, augments ischemia-induced angiogenesis. *Life Sci* 80(6):559–565.

Taylor JA (2010). Bilateral orbitozygomatic reconstruction with tissue-engineered bone. *J Craniofac Surg* 21(5):1612–1614.

Thangarajah H, Vial IN, Chang E, El-Ftesi S, Januszyk M, Chang EI, et al. (2009). IFATS collection: adipose stromal cells adopt a proangiogenic phenotype under the influence of hypoxia. *Stem Cells* 27(1):266–274.

Thesleff I (2003). Developmental biology and building a tooth. *Quintessence Int* 34(8):613–620.

Thesleff T, Lehtimaki K, Niskakangas T, Mannerstrom B, Miettinen S, Suuronen R, et al. (2011). Cranioplasty with adipose-derived stem cells and biomaterial: a novel method for cranial reconstruction. *Neurosurgery* 68(6):1535–1540.

Traktuev DO, Merfeld-Clauss S, Li J, Kolonin M, Arap W, Pasqualini R, et al. (2008). A population of multipotent CD34-positive adipose stromal cells share pericyte and mesenchymal surface markers, reside in a periendothelial location, and stabilize endothelial networks. *Circ Res* 102(1):77–85.

Trayhurn P, Wood IS (2005). Signalling role of adipose tissue: adipokines and inflammation in obesity. *Biochem Soc Trans* 33(Pt 5):1078–1081.

Vidal MA, Kilroy GE, Lopez MJ, Johnson JR, Moore RM, Gimble JM (2007). Characterization of equine adipose tissue–derived stromal cells: adipogenic and osteogenic capacity and comparison with bone marrow–derived mesenchymal stromal cells. *Vet Surg* 36(7):613–622.

Wei X DZ, Zhao L, Feng D, Wei G, He Y, Tan J, Lee WH, Hampel H, Dodel R, Johnstone BH, March KL, Farlow MR, Du Y. (2009). IFATS collection: the conditioned media of adipose stromal cells protect against hypoxia-ischemia–induced brain damage in neonatal rats. *Stem Cells* 27(2):478–488.

Wu G, Deng ZH, Fan XJ, Ma ZF, Sun YJ, Ma DD, et al. (2009). Odontogenic potential of mesenchymal cells from hair follicle dermal papilla. *Stem Cells Dev* 18(4):583–589.

Wu L, Zhu F, Wu Y, Lin Y, Nie X, Jing W, et al. (2008). Dentin sialophosphoprotein-promoted mineralization and expression of odontogenic genes in adipose-derived stromal cells. *Cells Tissues Organs* 187(2):103–112.

Wu X, Zvonic S, Floyd ZE, Kilroy G, Goh BC, Hernandez TL, et al. (2007). Induction of circadian gene expression in human subcutaneous adipose-derived stem cells. *Obesity (Silver Spring)* 15(11):2560–2570.

Wu X, Yu G, Parks H, Hebert T, Goh BC, Dietrich MA, et al. (2008). Circadian mechanisms in murine and human bone marrow mesenchymal stem cells following dexamethasone exposure. *Bone* 42(5):861–870.

Yanez R, Lamana ML, Garcia-Castro J, Colmenero I, Ramirez M, Bueren JA (2006). Adipose tissue–derived mesenchymal stem cells have in vivo immunosuppressive properties applicable for the control of the graft-versus-host disease. *Stem Cells* 24(11): 2582–2591.

Yanez R, Oviedo A, Aldea M, Bueren JA, Lamana ML (2010). Prostaglandin E2 plays a key role in the immunosuppressive properties of adipose and bone marrow tissue–derived mesenchymal stromal cells. *Exp Cell Res* 316(19):3109–3123.

Yang H, Youm YH, Vandanmagsar B, Ravussin A, Gimble JM, Greenway F, et al. (2010). Obesity increases the production of proinflammatory mediators from adipose tissue T cells and compromises TCR repertoire diversity: implications for systemic inflammation and insulin resistance. *J Immunol* 185(3):1836–1845.

Yoon E, Dhar S, Chun DE, Gharibjanian NA, Evans GR (2007). In Vivo osteogenic potential of human adipose-derived stem cells/poly lactide-*co*-glycolic acid constructs for bone regeneration in a rat critical-sized calvarial defect model. *Tissue Eng* 13(3):619–627.

Yoshimura K, Shigeura T, Matsumoto D, Sato T, Takaki Y, Aiba-Kojima E, et al. (2006). Characterization of freshly isolated and cultured cells derived from the fatty and fluid portions of liposuction aspirates. *J Cell Physiol* 208(1):64–76.

Yoshimura K, Sato K, Aoi N, Kurita M, Inoue K, Suga H, Eto H, Kato H, Hirohi T, Harii K (2008). Cell-assisted lipotransfer for facial lipoatrophy: efficacy of clinical use of adipose-derived stem cells. *Dermatol Surg* 34:1178–1185.

Yu G, Floyd ZE, Wu X, Halvorsen YD, Gimble JM (2011). Isolation of human adipose-derived stem cells from lipoaspirates. *Methods Mol Biol* 702:17–27.

Zannettino ACW PS, Khor F, Itescu S, Gimble JM, Gronthos S (2008). Multipotential human adipose-derived stromal stem cells exhibit a perivascular phenotype in vitro and in vivo. *J Cell Physiol* 214:413–421.

Zhang Y, Daquinag A, Traktuev DO, Amaya-Manzanares F, Simmons PJ, March KL, et al. (2009). White adipose tissue cells are recruited by experimental tumors and promote cancer progression in mouse models. *Cancer Res* 69(12):5259–5266.

Zhang Y, Bellows CF, Kolonin MG (2010). Adipose tissue–derived progenitor cells and cancer. *World J Stem Cells* 2(5):103–113.

Zimmerlin L, Donnenberg VS, Pfeifer ME, Meyer EM, Peault B, Rubin JP, et al. (2010). Stromal vascular progenitors in adult human adipose tissue. *Cytometry A* 77(1):22–30.

Zimmerlin L, Donnenberg AD, Rubin JP, Basse P, Landreneau RJ, Donnenberg VS (2011). Regenerative therapy and cancer: in vitro and in vivo studies of the interaction between adipose-derived stem cells and breast cancer cells from clinical isolates. *Tissue Eng A* 17(1–2):93–106.

Zuk PA (2008). Tissue engineering craniofacial defects with adult stem cells? Are we ready yet? *Pediatr Res* 63(5):478–486.

Zuk PA, Zhu M, Mizuno H, Huang J, Futrell JW, Katz AJ, et al. (2001). Multilineage cells from human adipose tissue: implications for cell-based therapies. *Tissue Eng* 7(2):211–228.

Zuk PA, Zhu M, Ashjian P, De Ugarte DA, Huang JI, Mizuno H, et al. (2002). Human adipose tissue is a source of multipotent stem cells. *Mol Biol Cell* 13(12):4279–4295.

Zvonic S, Ptitsyn AA, Conrad SA, Scott LK, Floyd ZE, Kilroy G, et al. (2006). Characterization of peripheral circadian clocks in adipose tissues. *Diabetes* 55(4):962–970.

Zvonic S, Lefevre M, Kilroy G, Floyd ZE, Delany JP, Kheterpal I, et al. (2007a). Secretome of primary cultures of human adipose-derived stem cells (ASCs): modulation of serpins by adipogenesis. *Mol Cell Proteomi* 6:18–28.

Zvonic S, Ptitsyn AA, Kilroy G, Wu X, Conrad SA, Scott LK, et al. (2007b). Circadian oscillation of gene expression in murine calvarial bone. *J Bone Miner Res* 22:357–365.

13

SKELETAL MUSCLE STEM CELLS: THEIR ORIGIN AND NICHE FACTORS

JOHANNES W. VON DEN HOFF

Department of Orthodontics and Craniofacial Biology, Radboud University Nijmegen Medical Centre, Nijmegen, The Netherlands

SANDER GREFTE

Department of Biochemistry, Radboud University Nijmegen Medical Centre, Nijmegen, The Netherlands

1 INTRODUCTION

Skeletal muscle represents nearly half of the total human body mass and thus is the most abundant tissue of the human body. Skeletal muscles are generally able to self-regenerate after injury. Crucial cells in this process are the satellite cells (SCs), which are located between the sarcolemma and the basal lamina of the myofiber (Mauro, 1961; Muir et al., 1965). These cells are normally quiescent but are activated after myotrauma; they then proliferate, self-renew, and finally, differentiate into multinucleated myofibers (Chargé and Rudnicki, 2004; Shi and Garry, 2006). Since the original identification of SCs in 1961 (Mauro, 1961), it has been hypothesized that SCs are remaining embryonic myoblasts from the developing somites. Several studies (Kassar-Duchossoy et al., 2005; Relaix et al., 2005) have demonstrated that progenitor cells from the dermomyotome give rise to SCs, but it remains unclear whether these cells are the only SC precursors (Kuang and Rudnicki, 2008). After birth, SCs proliferate extensively and play a major role in skeletal muscle growth and regeneration (Relaix et al., 2005). The specific microenvironment of the SCs, the niche, and many growth factors regulate SC physiology (Ten Broek et al., 2010). Upon injury, the SCs are activated and play a key role in muscle healing (Shi and Garry, 2006). The formation of scar tissue (fibrosis) during healing may cause persistent problems in muscle function

Stem Cells in Craniofacial Development and Regeneration, First Edition.
Edited by George T.-J. Huang, Irma Thesleff.

(Li and Huard, 2002). Thus, the identification of factors that influence the regeneration of injured muscle is of great interest. In this chapter we discuss the embryonic origin of craniofacial and trunk and limb muscles and the niche factors that regulate SC activity.

2 EMBRYONIC MYOGENESIS

2.1 Somite Development

In the early stages of embryonic development the major function of gastrulation is to create a mesodermal layer between the ectoderm and the endoderm. The mesoderm forms the blood, blood vessels, bones, cartilage, connective tissue, and the muscles of the trunk and limbs (Fig. 1). On either side of the neural tube this mesoderm is divided into the axial mesoderm (notochord), intermediate mesoderm, paraxial mesoderm, and the lateral plate mesoderm (Christ and Ordahl, 1995). With the exception of the craniofacial muscles, nearly all embryonic skeletal muscles are derived from the paraxial mesoderm (Harel et al., 2009). The paraxial mesoderm, which generates the trunk and limb muscles, separates into cell clusters, the somites, starting at the head region and progressing sequentially caudally. Cells of the ventral part of the somites undergo an epithelial-to-mesenchymal transition, thereby forming the sclerotome, which eventually forms the vertebrae and ribs. In the chick this is characterized by the downregulation of Pax3 and Pax7, two members of the family of paired/homeodomain transcription factors. Members of this family play an essential role in embryonic organogenesis (Tremblay and Gruss, 1994). Cells of the dorsal part of the somites maintain Pax3 and Pax7 expression and form the dermomyotome. This dermomyotome is responsible for the musculature and the dermis and is divided into an epaxial and a hypaxial part, which form the deep back muscles and the intercostal, abdominal, and limb muscles (Christ and Ordahl, 1995).

2.2 Myotome Development

A crucial step in the formation of skeletal muscle is the appearance of the myotome. First, muscle progenitor cells delaminate from the four edges of the dermomyotome (Gros et al., 2004). In addition, muscle progenitor cells migrate into the limb buds (Fig. 1). c-Met, a tyrosine kinase receptor that binds hepatocyte growth factor (HGF) (Dietrich et al., 1999), and Pax3 are major contributors to delamination and migration, since mouse embryos lacking functional c-Met and Pax3 do not form skeletal muscle in the limbs. At the edges of the dermomyotome Pax3 ensures survival of these cells (Goulding et al., 1994). These delaminating progenitor cells downregulate Pax3 and become myoblasts through the action of the myogenic regulatory factors (MRFs), a family of basic helix–loop–helix transcription factors that regulate myogenesis. The myoblasts differentiate into myocytes through the action of myogenin, Mrf4, and MyoD (Tajbakhsh and Buckingham, 2000). The myocytes eventually fuse and mature into multinucleated muscle fibers, forming a continuous muscle layer, the myotome. These processes are stimulated by signals from adjacent structures such as Sonic hedgehog (Shh) and Wnt proteins, two families of secreted signaling molecules. These proteins are released from the neural tube, notochord, and surface ectoderm. Bone morphogenic proteins (BMPs) are released from the neural tube and the lateral plate mesoderm, and inhibit myogenesis (Hawke and Garry, 2001).

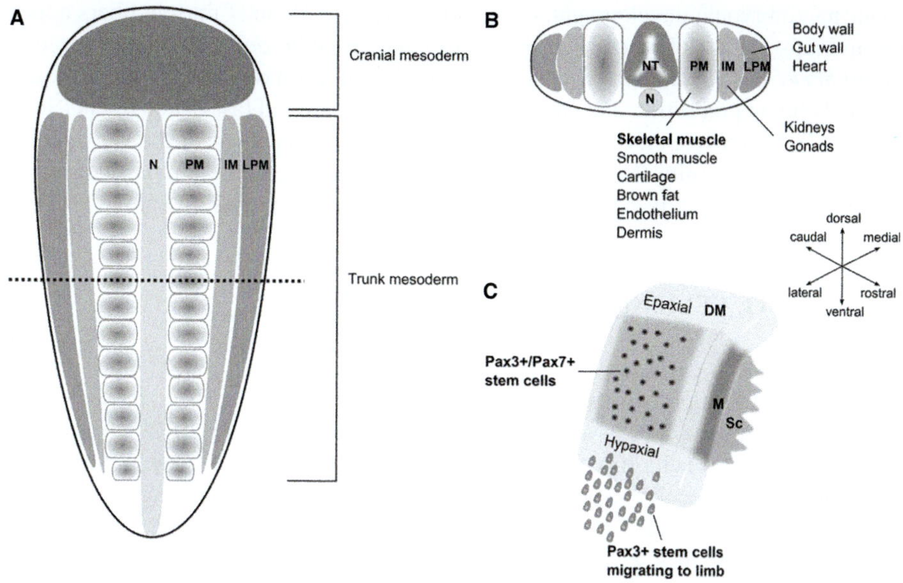

FIGURE 1 Model of embryonic myogenesis. The mesoderm is divided into the cranial and trunk mesoderm (A). (B) On either side of the neural tube (NT), the mesoderm is divided into the axial mesoderm or notochord (N), paraxial mesoderm (PM), intermediate mesoderm (IM), and the lateral plate mesoderm (LPM). The paraxial mesoderm (PM) segments into somites. (C) The dorsal part of the somites, the dermomyotome (DM), harbors the muscle progenitor cells. From the dorsomedial lip (epaxial) of the DM, these progenitors are the first cells in forming the myotome (M). Pax3-positive cells from the ventrolateral lip (hypaxial) of the DM migrate to form limb muscles. Pax3- and Pax7-positive cells from the central part of the DM migrate into the M and eventually give rise to the postnatal satellite cells. Sc, sclerotome.

2.3 Embryonic Muscle and Satellite Cells

Given that the dermomyotome disintegrates progressively (Christ and Ordahl, 1995), and the myotome is already postmitotic, these two structures cannot account for the massive muscle development in the embryo. It appears that cells expressing Pax3 and Pax7, but no myogenic markers, migrate from the central dermomyotome directly into the myotome. These cells contribute to muscle growth and are maintained within the muscle mass. As skeletal muscle develops in the limb buds, these precursor cells probably proliferate extensively to form the tissue (Gros et al., 2005). As mentioned earlier, c-Met and Pax3 regulate the migration of cells from the somites into the limb buds, and only upon arrival in the limb start to express MyoD and Myf5 (Tajbakhsh and Buckingham, 1994). Furthermore, the Pax3- and Pax7-positive cells derived from the central dermomyotome also give rise to most if not all of the SCs that emerge later (Relaix et al., 2005). Although the paraxial mesoderm in the embryonic body is completely segmented into somites, the head contains only seven somitomeres. Most head muscles, such as the masticatory, jaw-opening, and eye muscles, are derived from the paraxial head mesoderm, but the tongue muscles are derived from the somites (Christ and Ordahl, 1995; Yamane, 2005). In addition, differences have been described in the regulation of myogenesis in the head, which is reviewed elsewhere (Yamane, 2005).

During embryonic development, two distinct types of skeletal muscle fibers appear. The first muscle fibers that emerge are the *primary* or *embryonic fibers*; the *secondary* or *fetal fibers* arise later. The primary and secondary fibers have distinct morphological and biochemical properties, and can be classified into slow and fast twitch fibers (Stockdale, 1997). Moreover, this commitment seems to be independent of the surroundings and occurs in the somite (Nikovits et al., 2001). Toward the end of embryogenesis the SCs, the major players in postnatal muscle growth and regeneration, appear.

3 SATELLITE CELL NICHE

A *stem cell niche* is defined as a specific tissue microenvironment that influences stem cell behavior and supports their maintenance. The SC niche differs from most stem cell niches, as it keeps the SCs in a quiescent state. Only upon damage, SCs are activated to produce daughter cells (myoblasts) and to self-renew. SCs are heterogeneous with respect to their molecular signature and their ability to differentiate and self-renew. Many cell surface markers, such as CD34, integrin-α7, integrin-β1, and CXCR4, have been used to isolate SCs, but none of these are expressed uniformly by all SCs (Sherwood et al., 2004). More important, with Myf5-reporter mice, two populations of SCs were identified. The larger population (about 90%) expresses both Pax7 and Myf5, whereas the smaller population (about 10%) expresses Pax7 only (Kuang et al., 2007). The latter population of SCs shows a larger potential for regeneration and self-renewal. One of the main questions is how the niche determines the specific properties of these populations of SCs.

In the following sections we discuss the niche elements that are in direct contact with or near the SCs and the main soluble factors that influence their function.

3.1 Inside the Structural Niche

In skeletal muscle, SCs are flanked by the adjacent myofiber and the basal lamina that surrounds the myofiber (Fig. 2). Therefore, the SCs are exposed to an asymmetrical or bipolar niche. This might be critical for the asymmetric self-renewal of the SCs. At the basal side of the SCs, exposed to the basal lamina, integrin α7β1 is expressed. The basal lamina itself is composed primarily of laminin-2, type IV collagen, and heparan sulfate proteoglycans (HSPGs) (Boonen and Post, 2008). Integrin α7β1 binds to laminin-2 and thereby anchors the SC to the basal lamina (Burkin and Kaufman, 1999; Boonen and Post, 2008). Integrins generally regulate cell migration, cell shape, and cell–cell interactions, and thus α7β1 might play a major role in SC physiology. In contrast, at the apical side of the SC and on the myofiber, M-cadherin is expressed, which allows mutual binding (Cornelison and Wold, 1997; Burkin and Kaufman, 1999; Boonen and Post, 2008).

Asymmetrical division of SCs results in two daughter cells, of which one remains in contact with the myofiber and the other with the basal lamina. The one in contact with the basal lamina remains quiescent, while the other one fuses with the myofiber to maintain tissue homeostasis. This suggests that binding to the basal lamina is essential for self-renewal (Kuang et al., 2007). Moreover, knocking out laminin-2 in mice results in a decreased SC number (Girgenrath et al., 2005). However, it remains unknown which molecular signals regulate this process.

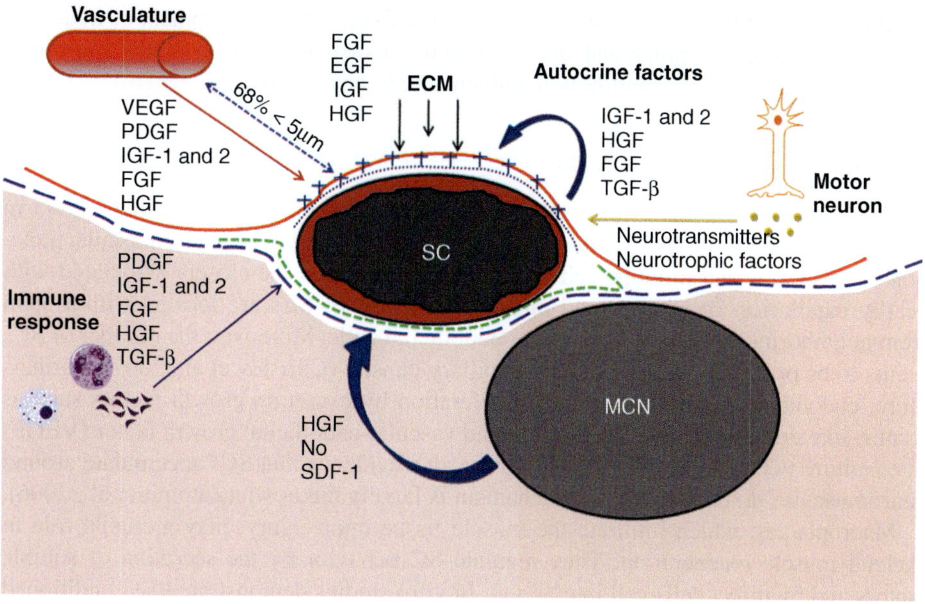

FIGURE 2 Satellite cell niche. SCs are located in a specific niche between the sarcolemma (long dashed blue line) and the basal lamina (solid red line). The basal side of the SC expresses integrin α7β1 (purple dotted line), which links the SC with the laminin (blue crosses) in the basal membrane. The apical side expresses M-cadherin (green dashed line), which attaches the SC to the adjacent myofiber. These areas are essential for signal transduction between the SCs and adjacent structures. SCs are normally quiescent and are activated after myotrauma, through the action of many regulatory factors. The proximity of the microvasculature suggests a reciprocal interaction between SCs and these vessels. ECM, extracellular matrix; MCN, myocyte nucleus; SC, satellite cell. Reproduced with permission from John Wiley and Sons (Ten Broek et al., 2010).

The basal lamina and the myofiber might also regulate SC physiology in other ways. For example, the basal lamina and other ECM components might preserve SC quiescence by capturing activating growth factors such as HGF. In contrast, by capturing growth factors, the basal lamina might also present these to the SCs in a high local concentration. The myofiber also secretes SDF1α, which binds to the CXCR4 receptor on SCs and thus induces migration (Sherwood et al., 2004; Grefte et al., 2010). Additionally, mechanical and electrical signals from the myofiber could also influence SC function. The biomechanical properties of the tissue seem to be a crucial component of the SC niche. Skeletal muscle tissue has a stiffness of about 12 kPa. In vitro, C2C12 mouse myoblasts differentiate most efficiently at this stiffness, but less on softer and stiffer substrates (Engler et al., 2004). Moreover, SCs cultured on substrates with a stiffness of 12 kPa self-renew and improve muscle regeneration significantly after transplantation (Gilbert et al., 2010). In contrast, the proliferation of primary myoblasts is highest on substrates with a stiffness of 21 kPa (Boonen et al., 2009). As the biomechanical properties of muscle change during aging and after muscle injury or disease, the activation, proliferation, and differentiation of SCs might be affected. This might be mediated by changes in the deposition of ECM by fibroblasts. Overall, the main niche elements that regulate SC activity are the basal lamina and the myofiber, and the biomechanical properties of the tissue. The development of treatment strategies for muscle injury and disease requires exact deciphering of the molecular and biomechanical niche, and how it influences SC physiology.

3.2 Outside the Structural Niche

Next to the niche, SC behavior is also regulated by components or cells that are not in direct contact with the SCs. These include the microvasculature, neuromuscular junctions, fibroblasts, and immune cells (Fig. 2). SCs appear to be closely associated with nearby capillaries. In humans and mice the majority of SCs are located within 5μm from neighboring capillaries or vascular endothelial cells. Moreover, the number of SCs seems to be positively correlated with capillary density (Christov et al., 2007). Furthermore, endothelial cells stimulate SC proliferation by excreting growth factors such as insulin-like growth factor (IGF) I, HGF, and vascular endothelial growth factor (VEGF) in coculture (Christov et al., 2007). There is also evidence that SCs accumulate around neuromuscular junctions, but the mechanism is largely unknown (Zammit et al., 2006).

Macrophages, which infiltrate the muscle tissue upon injury, play a crucial role in skeletal muscle regeneration. They regulate SC behavior by the secretion of soluble factors and by direct cell–cell interaction. In vitro studies demonstrate that conditioned medium of proinflammatory (M1) macrophages induces SC proliferation, while that of anti-inflammatory (M2) macrophages induces differentiation (Chazaud et al., 2009). This dual role of macrophages has also been demonstrated in vivo. Depletion of all infiltrating macrophages completely prevents muscle regeneration. However, selective depletion of anti-inflammatory macrophages only reduces the diameter of regenerating myofibers (Arnold et al., 2007). Direct cell–cell contact between macrophages and SCs may protect an SC from apoptosis (Chazaud et al., 2003)

Fibroblasts are abundantly present within the connective tissue of skeletal muscle and probably affect the function of SCs. These fibroblasts specifically express the transcription factor (Tcf)-4. Coculture experiments show that muscle fibroblasts regulate the type and maturation of myofibers (Mathew et al., 2011). Additionally, partial ablation of these fibroblasts in vivo reduces the SC pool, induces premature differentiation, and leads to smaller myofibers (Murphy et al., 2011). In short, factors produced by endothelial cells, fibroblasts, and macrophages adjacent to the SC niche may also regulate SC physiology.

4 GROWTH FACTORS

A plethora of growth factors is involved in the regulation of SCs. These growth factors are produced after injury by many cell types near the SC niche (Fig. 3). Next to that, growth factors are "stored" in the ECM and released upon tissue damage. In addition, SCs may upregulate matrix metalloproteinases 2 and 9, which degrade the ECM and liberate the growth factors (Koskinen et al., 2002). Specific growth factors either stimulate or inhibit the proliferation and differentiation of SCs.

4.1 Stimulatory Growth Factors

Several growth factors stimulate the proliferation and differentiation of SCs, including IGF1, HGF, fibroblast growth factor (FGF) 2 and 6, SDF1α, VEGF,

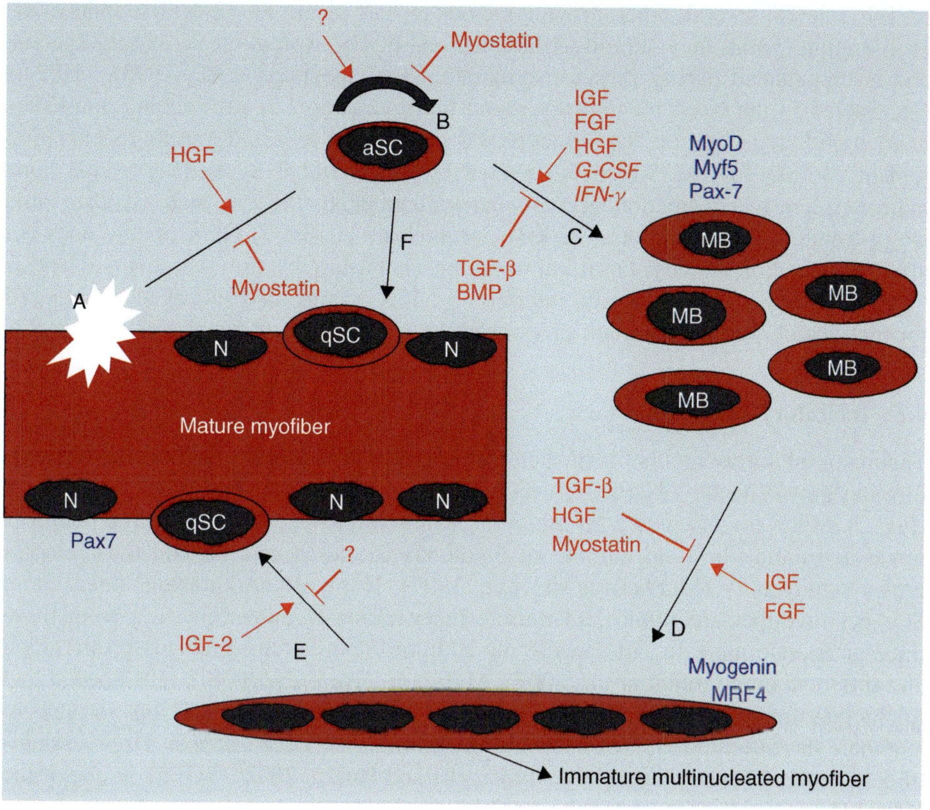

FIGURE 3 Regulatory factors in muscle regeneration. Following myotrauma (A), quiescent satellite cells are activated (by HGF) to enter the cell cycle for self-renewal (B) and proliferation (C). Activated satellite cells are characterized by a high expression of Pax7, MyoD, and/or Myf5. Subsequent differentiation is marked by the downregulation of Pax7 and the upregulation of Mrf4 and myogenin. The differentiated myoblasts form new immature multinucleated myofibers (D) or fuse to damaged myofibers (not shown). Finally, the central SC nuclei migrate to a subsarcolemmal position in mature myofibers (E). After SC activation, a subset of activated SCs reenters the quiescent state to replenish the satellite cell pool (F). aSC, activated satellite cell; qSC, quiescent satellite cell; MB, myoblast; N, nucleus. Reproduced with permission from John Wiley and Sons (Ten Broek et al., 2010).

and platelet-derived growth factor (PDGF) AA and BB (Fig. 3). The most potent stimulatory factor seems to be IGF1. Systemic administration of IGF1 results in increased DNA and protein in muscle tissue. Mice overexpressing human IGF1 show muscle hypertrophy (Adams and McCue, 1998), while the injection of IGF1 into muscle wounds of wild-type mice improves muscle regeneration (Sato et al., 2003).

HGF binds to the c-met receptor on the SCs and subsequently induces proliferation and migration. However, differentiation is inhibited by HGF. Correlating with this, HGF expression is increased during the early proliferation phase of muscle regeneration, but injection of HGF in later stages does not improve regeneration (Tatsumi et al., 1998). Furthermore, injection of HGF during the early stage of muscle regeneration also increases the number of SCs. In contrast, HGF inhibits the formation of multinuclear myotubes at later stages (Hayashi et al., 2004).

The role of several other growth factors is less clear. FGF2 and 6 induce SC proliferation, while they inhibit differentiation. FGF6 expression is muscle specific and is upregulated during muscle regeneration (deLapeyriere et al., 1993). FGF2 is sequestered in the basal membrane around the myotubes, but neutralizing antibodies delay muscle regeneration. The exact effects of FGFs on SCs and muscle regeneration remain unclear. SDF1α, which is secreted by the myofiber as well as by the bone marrow, seem to be functioning as a chemoattractant for SCs as well as other muscle progenitors (Ratajczak et al., 2003; Grefte et al., 2010). VEGF probably does not affect SC function directly, but it improves muscle healing by stimulating angiogenesis (Springer et al., 1998). Recently, interferon-γ (IFN-γ) was also shown to be required for muscle cell proliferation and proper muscle regeneration (Cheng et al., 2008).

4.2 Inhibitory Growth Factors

The major inhibitory regulators of skeletal muscle regeneration are myostatin and transforming growth factor (TGF) β1, which are both members of the TGFβ superfamily (Fig. 3). Myostatin is a powerful inhibitor of muscle growth. In cattle with a nonfunctional myostatin, the total number of myofibers increases and skeletal muscle mass grows significantly (McPherron and Lee, 1997). In myostatin-knockout mice, more SCs per myofiber are found and muscle regeneration is enhanced. SCs from these mice proliferate more in vitro, while the addition of myostatin prevents proliferation and differentiation (Joulia et al., 2003). Myostatin also prevents SC self-renewal and might thus maintain their quiescence (McCroskery et al., 2003). TGFβ1 is able to stimulate the differentiation of SCs into myofibroblasts, which deposit large amounts of collagen and hence may induce fibrosis (Li and Huard, 2002). TGFβ1 is one of the main inducers of muscle fibrosis and might also lead to depletion of the SC pool. The inhibition of TGFβ1 activity might reduce fibrosis and enhance muscle regeneration after injury.

In summary, many factors in the SC niche influence the behavior of SCs. The basement membrane and the myofiber are in direct contact with the SCs and influence their functioning by cell–cell or cell–ECM interactions. More distinct players, such as the microvasculature, neuromuscular junctions, fibroblasts, and immune cells, produce factors that also influence SCs. Their regulatory role has been identified primarily by studies on muscle regeneration. Recently, the biomechanical properties of the muscle tissue have also gained interest in SC physiology. Although it begins to become clear how SCs are activated, it remains largely unknown how quiescence is maintained.

5 DIFFERENCES IN CRANIOFACIAL AND LIMB MUSCLES

In the last few years it has become clear that SCs and muscles from the craniofacial region differ from the trunk and limb muscles. *Pax3*-knockout mice do not develop trunk and limb muscles, whereas their craniofacial muscles are normal (Buckingham, 2001). Developmental research has shown that the branchiomeric muscles from the head are derived from the cranial paraxial mesoderm and the lateral splanchnic mesoderm, whereas the trunk and limb muscles are derived from the somites (McLoon et al., 2007). Furthermore, lineage-tracing studies demonstrate that craniofacial SCs originate from the head mesoderm and have a phenotype overlapping with that of cardiogenic cells (Harel et al., 2009). Craniofacial SCs also follow a different genetic program, although a core set of factors, such as Pax7, Myf5, and MyoD, are found in all SCs (Sambasivan et al., 2009). These differences might explain the unequal occurrence of muscle myopathies in craniofacial and limb muscles (Muller et al., 2001) and the lower regeneration capacity of craniofacial muscles (Pavlath et al., 1998). However, the SC phenotype is also strongly influenced by the local environment. Isolated SCs from extraocular muscles lose their specific myosin expression in culture or after transplantation into a limb muscle (Sambasivan et al., 2009). Furthermore, isolated craniofacial and limb SCs show similar proliferation and differentiation capacities under standard culture conditions (Grefte et al., 2012). Also, after implantation into limb muscle, craniofacial SCs regenerate the host muscle with the same efficiency as that of limb SCs (Sambasivan et al., 2011). Thus, the phenotype of SCs from a certain muscle group is not dictated by genetic factors alone, but the microenvironment plays a significant role as well.

6 CONCLUSIONS AND FUTURE PERSPECTIVE

Satellite cells (SCs) are the tissue stem cells of skeletal muscle and as such are responsible for muscle growth, maintenance, and repair. The embryonic origin of craniofacial skeletal muscles and their associated SCs differs from that of skeletal muscle in the trunk and limbs. Also in the adult, these muscles and SCs still have a distinct genetic signature, which may partially explain their different susceptibility to congenital myopathies such as the muscular dystrophies. However, the tissue microenvironment in the craniofacial and trunk and limb muscles might also be different and induce tissue-specific behavior of SCs.

The activity of SCs is determined primarily by the local microenvironment, the niche, but also by soluble factors from nearby cells and by the stiffness of the tissue. These factors determine whether the SC remains quiescent or is activated and then able to proliferate, differentiate, and self-renew. Crucial components of the niche are the basal lamina and myofiber that form the bipolar niche, which allows asymmetrical self-renewal. Furthermore, several other cell types, such as endothelial cells, macrophages, and fibroblasts, influence SC function. Through direct cell–cell contact or the secretion of growth factors, these cells can influence the activation, proliferation, and differentiation of SCs. The most important stimulatory growth factors are HGF, IGF, and FGF; TGFβ1 and myostatin are the major inhibitory growth factors.

Future studies should identify the specific niche and environmental factors involved and how they regulate SC functioning. Furthermore, the effects of substrate stiffness and mechanical stimuli should be studied in more detail, since these influence myofiber formation and orientation. The knowledge gained will significantly improve skeletal muscle engineering to treat a wide variety of muscle defects and diseases (see also Chapter 22).

Acknowledgments

We are grateful to Ruurd Torensma and Roel ten Broek for valuable discussions. This work was supported by a grant from the Radboud University Nijmegen Medical Centre.

REFERENCES

Adams GR, McCue SA (1998). Localized infusion of IGF-I results in skeletal muscle hypertrophy in rats. *J Appl Physiol* 84:1716–1722.

Arnold L, Henry A, Poron F, Baba-Amer Y, van Rooijen N, Plonquet A, et al. (2007). Inflammatory monocytes recruited after skeletal muscle injury switch into antiinflammatory macrophages to support myogenesis. *J Exp Med* 204:1057–1069.

Boonen KJ, Post MJ (2008). The muscle stem cell niche: regulation of satellite cells during regeneration. *Tissue Eng B* 14:419–431.

Boonen KJ, Rosaria-Chak KY, Baaijens FP, van der Schaft DW, Post MJ (2009). Essential environmental cues from the satellite cell niche: optimizing proliferation and differentiation. *Am J Physiol Cell Physiol* 296:C1338–C1345.

Buckingham M (2001). Skeletal muscle formation in vertebrates. *Curr Opin Genet Dev* 11:440–448.

Burkin DJ, Kaufman SJ (1999). The alpha7beta1 integrin in muscle development and disease. *Cell Tissue Res* 296:183–190.

Chargé SB, Rudnicki MA (2004). Cellular and molecular regulation of muscle regeneration. *Physiol Rev* 84:209–238.

Chazaud B, Sonnet C, Lafuste P, Bassez G, Rimaniol AC, Poron F, et al. (2003). Satellite cells attract monocytes and use macrophages as a support to escape apoptosis and enhance muscle growth. *J Cell Biol* 163:1133–1143.

Chazaud B, Brigitte M, Yacoub-Youssef H, Arnold L, Gherardi R, Sonnet C, et al. (2009). Dual and beneficial roles of macrophages during skeletal muscle regeneration. *Exerc Sport Sci Rev* 37:18–22.

Cheng M, Nguyen MH, Fantuzzi G, Koh TJ (2008). Endogenous interferon-gamma is required for efficient skeletal muscle regeneration. *Am J Physiol Cell Physiol* 294:C1183–C1191.

Christ B, Ordahl CP (1995). Early stages of chick somite development. *Anat Embryol (Berl)* 191:381–396.

Christov C, Chretien F, Abou-Khalil R, Bassez G, Vallet G, Authier FJ, et al. (2007). Muscle satellite cells and endothelial cells: close neighbors and privileged partners. *Mol Biol Cell* 18:1397–1409.

Cornelison DD, Wold BJ (1997). Single-cell analysis of regulatory gene expression in quiescent and activated mouse skeletal muscle satellite cells. *Dev Biol* 191:270–283.

deLapeyriere O, Ollendorff V, Planche J, Ott MO, Pizette S, Coulier F, et al. (1993). Expression of the *Fgf6* gene is restricted to developing skeletal muscle in the mouse embryo. *Development* 118:601–611.

Dietrich S, Abou-Rebyeh F, Brohmann H, Bladt F, Sonnenberg-Riethmacher E, Yamaai T, et al. (1999). The role of SF/HGF and c-Met in the development of skeletal muscle. *Development* 126:1621–1629.

Engler AJ, Griffin MA, Sen S, Bonnemann CG, Sweeney HL, Discher DE (2004). Myotubes differentiate optimally on substrates with tissue-like stiffness: pathological implications for soft or stiff microenvironments. *J Cell Biol* 166:877–887.

Gilbert PM, Havenstrite KL, Magnusson KE, Sacco A, Leonardi NA, Kraft P, et al. (2010). Substrate elasticity regulates skeletal muscle stem cell self-renewal in culture. *Science* 329:1078–1081.

Girgenrath M, Kostek CA, Miller JB (2005). Diseased muscles that lack dystrophin or laminin-alpha2 have altered compositions and proliferation of mononuclear cell populations. *BMC Neurol* 5:7.

Goulding M, Lumsden A, Paquette AJ (1994). Regulation of *Pax-3* expression in the dermomyotome and its role in muscle development. *Development* 120:957–971.

Grefte S, Kuijpers-Jagtman AM, Torensma R and Von den Hoff JW (2010). Skeletal muscle fibrosis: the effect of stromal-derived factor-1alpha-loaded collagen scaffolds. *Regen Med* 5:737–747.

Grefte S, Kuijpers MA, Kuijpers-Jagtman AM, Torensma R, Von den Hoff JW (2012). Myogenic capacity of muscle progenitor cells from head and limb muscles. *Eur J Oral Sci* 120:38–45.

Gros J, Scaal M, Marcelle C (2004). A two-step mechanism for myotome formation in chick. *Dev Cell* 6:875–882.

Gros J, Manceau M, Thome V, Marcelle C (2005). A common somitic origin for embryonic muscle progenitors and satellite cells. *Nature* 435:954–958.

Harel I, Nathan E, Tirosh-Finkel L, Zigdon H, Guimaraes-Camboa N, Evans SM, et al. (2009). Distinct origins and genetic programs of head muscle satellite cells. *Dev Cell* 16:822–832.

Hawke TJ, Garry DJ (2001). Myogenic satellite cells: physiology to molecular biology. *J Appl Physiol* 91:534–551.

Hayashi S, Aso H, Watanabe K, Nara H, Rose MT, Ohwada S, et al. (2004). Sequence of IGF-I, IGF-II, and HGF expression in regenerating skeletal muscle. *Histochem Cell Biol* 122:427–434.

Joulia D, Bernardi H, Garandel V, Rabenoelina F, Vernus B, Cabello G (2003). Mechanisms involved in the inhibition of myoblast proliferation and differentiation by myostatin. *Exp Cell Res* 286:263–275.

Kassar-Duchossoy L, Giacone E, Gayraud-Morel B, Jory A, Gomes D, Tajbakhsh S (2005). Pax3/Pax7 mark a novel population of primitive myogenic cells during development. *Genes Dev* 19:1426–1431.

Koskinen SO, Ahtikoski AM, Komulainen J, Hesselink MK, Drost MR, Takala TE (2002). Short-term effects of forced eccentric contractions on collagen synthesis and degradation in rat skeletal muscle. *Pfluegers Arch* 444:59–72.

Kuang S, Rudnicki MA (2008). The emerging biology of satellite cells and their therapeutic potential. *Trends Mol Med* 14:82–91.

Kuang S, Kuroda K, Le Grand F, Rudnicki MA (2007). Asymmetric self-renewal and commitment of satellite stem cells in muscle. *Cell* 129:999–1010.

Li Y, Huard J (2002). Differentiation of muscle-derived cells into myofibroblasts in injured skeletal muscle. *Am J Pathol* 161:895–907.

Mathew SJ, Hansen JM, Merrell AJ, Murphy MM, Lawson JA, Hutcheson DA, et al. (2011). Connective tissue fibroblasts and Tcf4 regulate myogenesis. *Development* 138:371–384.

Mauro A (1961). Satellite cell of skeletal muscle fibers. *J Biophys Biochem Cytol* 9:493–495.

McCroskery S, Thomas M, Maxwell L, Sharma M, Kambadur R (2003). Myostatin negatively regulates satellite cell activation and self-renewal. *J Cell Biol* 162:1135–1147.

McLoon LK, Thorstenson KM, Solomon A and Lewis MP (2007). Myogenic precursor cells in craniofacial muscles. *Oral Dis* 13:134–140.

McPherron AC, Lee SJ (1997). Double muscling in cattle due to mutations in the myostatin gene. *Proc Natl Acad Sci USA* 94:12457–12461.

Muir AR, Kanji AH, Allbrook D (1965). The structure of the satellite cells in skeletal muscle. *J Anat* 99:435–444.

Muller J, Vayssiere N, Royuela M, Leger ME, Muller A, Bacou F, et al. (2001). Comparative evolution of muscular dystrophy in diaphragm, gastrocnemius and masseter muscles from old male *mdx* mice. *J Muscle Res Cell Motil* 22:133–139.

Murphy MM, Lawson JA, Mathew SJ, Hutcheson DA, Kardon G (2011). Satellite cells, connective tissue fibroblasts and their interactions are crucial for muscle regeneration. *Development* 138:3625–3637.

Nikovits W, Jr., Cann GM, Huang R, Christ B, Stockdale FE (2001). Patterning of fast and slow fibers within embryonic muscles is established independently of signals from the surrounding mesenchyme. *Development* 128:2537–2544.

Pavlath GK, Thaloor D, Rando TA, Cheong M, English AW, Zheng B (1998). Heterogeneity among muscle precursor cells in adult skeletal muscles with differing regenerative capacities. *Dev Dyn* 212:495–508.

Ratajczak MZ, Majka M, Kucia M, Drukala J, Pietrzkowski Z, Peiper S, et al. (2003). Expression of functional CXCR4 by muscle satellite cells and secretion of SDF-1 by muscle-derived fibroblasts is associated with the presence of both muscle progenitors in bone marrow and hematopoietic stem/progenitor cells in muscles. *Stem Cells* 21:363–371.

Relaix F, Rocancourt D, Mansouri A, Buckingham M (2005). A Pax3/Pax7-dependent population of skeletal muscle progenitor cells. *Nature* 435:948–953.

Sambasivan R, Gayraud-Morel B, Dumas G, Cimper C, Paisant S, Kelly RG, et al. (2009). Distinct regulatory cascades govern extraocular and pharyngeal arch muscle progenitor cell fates. *Dev Cell* 16:810–821.

Sambasivan R, Yao R, Kissenpfennig A, Van Wittenberghe L, Paldi A, Gayraud-Morel B, et al. (2011). Pax7-expressing satellite cells are indispensable for adult skeletal muscle regeneration. *Development* 138:3647–3656.

Sato K, Li Y, Foster W, Fukushima K, Badlani N, Adachi N, et al. (2003). Improvement of muscle healing through enhancement of muscle regeneration and prevention of fibrosis. *Muscle Nerve* 28:365–372.

Sherwood RI, Christensen JL, Conboy IM, Conboy MJ, Rando TA, Weissman IL, et al. (2004). Isolation of adult mouse myogenic progenitors: functional heterogeneity of cells within and engrafting skeletal muscle. *Cell* 119:543–554.

Shi X, Garry DJ (2006). Muscle stem cells in development, regeneration, and disease. *Genes Dev* 20:1692–1708.

Springer ML, Chen AS, Kraft PE, Bednarski M, Blau HM (1998). VEGF gene delivery to muscle: potential role for vasculogenesis in adults. *Mol Cell* 2:549–558.

Stockdale FE (1997). Mechanisms of formation of muscle fiber types. *Cell Struct Funct* 22:37–43.

Tajbakhsh S, Buckingham ME (1994). Mouse limb muscle is determined in the absence of the earliest myogenic factor myf-5. *Proc Natl Acad Sci USA* 91:747–751.

Tajbakhsh S, Buckingham M (2000). The birth of muscle progenitor cells in the mouse: spatiotemporal considerations. *Curr Top Dev Biol* 48:225–268.

Tatsumi R, Anderson JE, Nevoret CJ, Halevy O, Allen RE (1998). HGF/SF is present in normal adult skeletal muscle and is capable of activating satellite cells. *Dev Biol* 194:114–128.

Ten Broek RW, Grefte S, Von den Hoff JW (2010). Regulatory factors and cell populations involved in skeletal muscle regeneration. *J Cell Physiol* 224:7–16.

Tremblay P, Gruss P (1994). Pax: genes for mice and men. *Pharmacol Ther* 61:205–226.

Yamane A (2005). Embryonic and postnatal development of masticatory and tongue muscles. *Cell Tissue Res* 322:183–189.

Zammit PS, Partridge TA, Yablonka-Reuveni Z (2006). The skeletal muscle satellite cell: the stem cell that came in from the cold. *J Histochem Cytochem* 54:1177–1191.

14

STEM CELLS IN SALIVARY GLAND DEVELOPMENT AND REGENERATION

ISABELLE M. A. LOMBAERT AND MATTHEW P. HOFFMAN
National Institute of Dental and Craniofacial Research, National Institutes of Health, Bethesda, Maryland

1 INTRODUCTION

Salivary glands are a component of the craniofacial complex that are amenable to a stem cell–based approach to regeneration after injury. The clinical scenario driving much of this research is irradiation treatment of head and neck cancer in which the salivary glands are damaged. This often results in a permanent loss of saliva production. Chronic hyposalivation is caused by the massive cell death of saliva-producing acinar cells, which are no longer replenished by the epithelial salivary gland stem cells. Recent studies also indicate that irradiation alters the local microenvironment, or niche, that surrounds the epithelial stem cells. Since the niche is essential to direct stem cell function, irradiation-surviving stem cells are no longer stimulated to form acinar cells. As such, researchers are investigating strategies to regenerate irradiated glands by both stem cell replacement and/or repair of the damaged niche. However, identifying the salivary gland stem cells and understanding the molecular mechanisms that drive stem cell maintenance and differentiation remain major challenges in the field. The emerging concepts from recent studies suggest that multiple stem cell populations are present in the salivary gland. Furthermore, understanding how the stem cell niche functions to instruct stem cells will be essential for regeneration purposes. Here we review literature that identifies and characterizes salivary gland stem cells and investigates their function in the glandular environment during development, homeostasis, and regeneration. We conclude with current efforts to translate these findings into clinically relevant therapies for treatment of irradiated salivary glands.

Stem Cells in Craniofacial Development and Regeneration, First Edition.
Edited by George T.-J. Huang, Irma Thesleff.

2 SALIVARY GLAND DEVELOPMENT

The major salivary glands are comprised of three pairs of organs: the submandibular (SMG), the sublingual, and the parotid glands. This review focuses on mouse submandibular glands (SMGs), which initiate at embryonic day E11.5 with the formation of an initial placode from the oral epithelium. The placode enlarges and invaginates into a condensing mesenchyme that contains nerves and blood vessels, and forms a primary end bud distal to the duct. Branching morphogenesis is defined by continuous rounds of proliferation and clefting within the end bud combined with the formation of secondary ducts. The epithelial patterning continues in the SMG until about E16, when acinar cell differentiation begins. This is followed by further ductal lumenization, cell polarization, and differentiation into functional acinar cells, so that at birth the SMG can secrete saliva. The reader is directed to several recent reviews on salivary gland development for more detailed descriptions (Patel et al., 2006; Tucker, 2007; Knosp et al., 2012).

Adult glands are comprised primarily of epithelial acini, ducts, and myoepithelial cells. The ductal compartment is comprised of intercalated ducts (IDs), striated ducts (SDs), granular convoluted tubules (GCTs), and excretory ducts (EDs) (Fig. 1). Proximal to the acini lie the smallest ducts, the IDs, which collect the saliva from the acinar cells. Saliva then flows through larger SDs and GCTs, which modify the electrolytes in the saliva. The EDs join the Wharton's duct, from which the saliva enters the oral cavity. The GCTs and SDs are formed postnatally under hormonal control (Gresik, 1980), and their distribution in the gland may vary (Gresik, 1994). Glands are also comprised of nerves, blood vessels, and interstitial cells, which are located in the surrounding microenvironment.

A reservoir of epithelial stem cells needs to be preserved during gland development for adult homeostasis and regeneration. Thus, a stem cell must undergo the process of "self-renewal." Within this process, a stem cell can either make a symmetric division to expand the stem cell pool or an asymmetric division to maintain the stem cell pool while giving rise to a differentiated cell type. In contrast, stem cells are depleted when they form two differentiated cell types along a specific cell lineage. Direct progeny of stem cells are the progenitors. The only difference between these cell types lies in the reduced potency of progenitors to self-renew and differentiate into multiple cell types. Since it is difficult to distinguish between a stem cell and a progenitor, and most reports in the literature cannot discriminate between them, in this review we refer to them both as progenitors.

3 THE CLINICAL NEED FOR PROGENITOR CELLS

There is a major clinical need to identify salivary progenitors and to understand how they function, due to their potential use in regenerating damaged adult salivary glands. Hyposalivation affects more than 300,000 people annually and can be associated with xerostomia, the perception of a dry mouth. Head and neck cancer patients undergoing radiation therapy are prone to hyposalivation because their salivary glands lie in the field of radiation during treatment [reviewed by Vissink et al. (2003)]. Irradiated glands have lost the majority of acinar cells and are often comprised of fibrotic tissue that

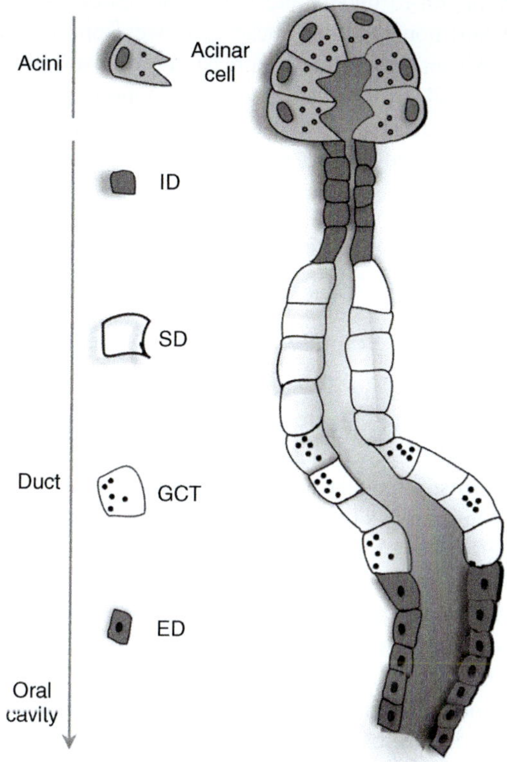

FIGURE 1 Model of adult mouse salivary gland epithelia. Saliva is produced by the acinar cells and transported via the duct system into the oral cavity. Different duct cell types modify the saliva by exchanging electrolytes. Intercalated duct (ID) cells collect the saliva that flows into the striated duct (SD), granular convoluted tubule (GCT), excretory duct (ED), and Wharton's duct, which connects to the oral cavity.

surrounds the remaining ducts. The inability of the irradiated gland to form new acinar cells is associated with a loss of progenitors (Lombaert et al., 2008c). However, recent reports suggest that the irradiated environment, or niche, also prevents regeneration. For example, irradiation-surviving progenitors in vivo are no longer stimulated to form acinar cells (Tatsuishi et al., 2009). In addition, loss of parasympathetic function and innervation prevent proper regeneration (Knox et al., 2013). Although loss of saliva production is not life threatening, it seriously impairs the quality of life of these patients. It affects everyday activities, including speaking, swallowing, taste perception and food processing and enjoyment. Furthermore, it increases the risk of mucosal ulcerations, dental caries, and oral infections. Thus, permanent salivary gland hypofunction impairs the oral health of patients and broadly affects their quality of life. Current treatments for hyposalivation after radiotherapy include administration of saliva substitutes, saliva-stimulating drugs, or gene therapy [reviewed by Vissink et al. (2010)]. However, future approaches for glandular regeneration may include transplanting progenitor cells into irradiated glands, increasing the number and function of surviving progenitor cells, or preserving the damaged niche.

We have recently begun to understand the dynamics of salivary gland progenitors during development and homeostasis. These cells were characterized based on either their topographical position in the gland, expression of transcription factors and proteins that are found on stem cells in other systems, or by unique progenitor properties, such as the exclusion of toxic dyes or detoxifying aldehydes. An emerging concept in the field is that multiple gland progenitors exist and may have overlapping functions.

4 MULTIPLE PROGENITORS PARTICIPATE IN DEVELOPMENT AND REGENERATION

4.1 Intercalated Progenitors Rapidly Repopulate Acinar Cells

A commonly used technique for localizing progenitors is the label-retaining assay (Potten et al., 2002). This technique uses the incorporation of a nucleotide analog such as bromodeoxyuridine (BrdU) or tritiated thymidine (3HTdR) into DNA to label cells within a defined administration period. Subsequent rounds of cell divisions result in dilution of the dye, which marks highly proliferative cells as negative and slow-cycling cells as positive. The slow-cycling cells, termed quiescent cells, are hypothesized to be salivary progenitors. These quiescent progenitors are localized within the IDs and EDs (Zajicek et al., 1985; Man et al., 2001). It was hypothesized that ED progenitors are more likely to replace SDs, whereas acinar cells and GCTs derive from the ID progenitors. This is, at least in part, confirmed by reversible duct ligation studies. This model involves mechanical obstruction of the main salivary gland duct, which causes acinar cells to undergo apoptosis. Subsequent removal of the ligation results in the new formation of acinar cells by ID progenitors [reviewed by Carpenter and Cotroneo (2010)].

Based on detailed anatomical study of IDs, we know that the ID progenitors are a heterogeneous population of cells (Denny et al., 1997). This was confirmed by the observation that the IDs consist of $BrdU^+$ quiescent progenitors and a separate side population (SP) cell (Kim et al., 2008). The SP progenitor is characterized by its exclusion of toxic dyes via the ABC cell membrane transporter. In adult SMGs, the SP comprises only a small percentage of cells (0.2 to 1%) (Lombaert, 2008; Purwanti et al., 2011), and increases threefold in cell number after duct ligation. It will be of interest in the future to evaluate how these progenitors differ from each other and how potent they must be to form the acinar cell pool. Furthermore, it is not clear whether acinar cells have some progenitor potency since all salivary epithelial cells retained a basal level of proliferative capacity (Denny et al., 1997).

4.2 There Are Multiple Progenitors Within the Salivary Gland That Can Form Specific Lineages and Can Compensate for Each Other

BrdU pulse labeling has provided essential information about progenitors but may have limitations in terms of defining all salivary progenitors. Since salivary glands have a slow turnover rate of 2 to 4 months (Zajicek et al., 1985), slowly dividing progenitors are easily missed during the BrdU or 3HTdR labeling period. A complementary approach is to use genetic cell tracing to identify progenitors. Genetic lineage tracing marks all the progeny cells that derive from a specific cell type and from a given period in organ development. As such, the potency of a progenitor can be distinguished based on the number of cell types that it forms. A seminal report in the salivary field used

speculate that the regeneration initiated by Fgf7 might involve the expansion of Kit^+ cells.

Another way to increase progenitor cell numbers was demonstrated by using an activator of aldhehyde 3 (Aldh3), which is an isozyme that detoxifies aldehydes. This detoxification is a characteristic of progenitors since SMG Kit^+ cells contain significantly higher levels of *Aldh3* versus Kit^- cells (Banh et al., 2011). Although the role of Aldh isozymes in progenitors is still unclear, the use of an Aldh3-specific activator doubled the number of Kit^+Thy1^+ progenitors in vivo. Further data will reveal whether this also positively influences the functional repair of irradiated glands.

Expansion of a specific progenitor such as the adult SMG $K5^+$ progenitor was shown by genetic activation of Wnt signaling (Hai et al., 2010). This $K5^+$ expansion leads to improved regeneration of irradiated glands (Hai et al., 2012), indicating that Wnt signaling could be a potential strategy for regenerating irradiated glands. Furthermore, Wnt not only stimulated proliferation in $K5^+$ cells, but also inhibited apoptosis. In a similar manner, other growth factors, such as EGF, IGF, and FGF2, can inhibit epithelial apoptosis induced by irradiation [reviewed by Coppes and Stokman (2011)].

Aside from protecting and stimulating the epithelial compartment, maintenance of the progenitor niche is just as critical for regeneration of the injured gland. For example, the density of blood vessels already decreases by half in 4 hours postirradation [reviewed by Coppes and Stokman (2011)], and ultimately results in blood vessel dilation and decreased function (Lombaert et al., 2008b). However, preirradiation treatment with FGF2-containing virus vector (Cotrim et al., 2007) or G-CSF–induced mobilization of bone marrow–derived cells (Lombaert et al., 2008b) can help protect the blood vessels from degeneration.

Our recent study also showed that protection of the parasympathetic nerves can improve SMG regeneration after irradiation (Knox et al., 2013). Fetal SMGs were damaged with irradiation, resulting in reduced epithelial morphogenesis and a reduction of innervation due to apoptosis within the PSG. Treatment with neurturin, a neurotrophic factor, reduced radiation-induced apoptosis in the PSG and improved PSG function, which increased epithelial growth (Fig. 3). Irradiation also induced extensive apoptosis in the end buds, which produce neurturin (Knox et al., 2013). It is known that neurturin is also important for normal organ development. Multiple organs show reduced innervation and have epithelial defects in mice when neurturin or its receptors, Gfra2 and Ret, are deleted (Rossi et al., 1999, 2003; Enomoto et al., 2000). Importantly, PSG function and innervation were also significantly reduced in irradiated human adult SMGs (Knox et al., 2013). Taken together, our data suggest that protection of radiation-induced damage to the nerves could be an alternative strategy to improve salivary gland regeneration. We speculate that treatment of SMGs with neurturin, possibly by gene therapy before irradiation, may protect the PSG from apoptosis and improve epithelial regeneration.

7 FUTURE DIRECTIONS IN THE SALIVARY GLAND FIELD

Recent developments in the salivary gland field suggest that progenitor cell therapy has the potential to cure hyposalivation. However, many challenges must be addressed before progenitor cell therapy can be used in clinical settings. With the characterization

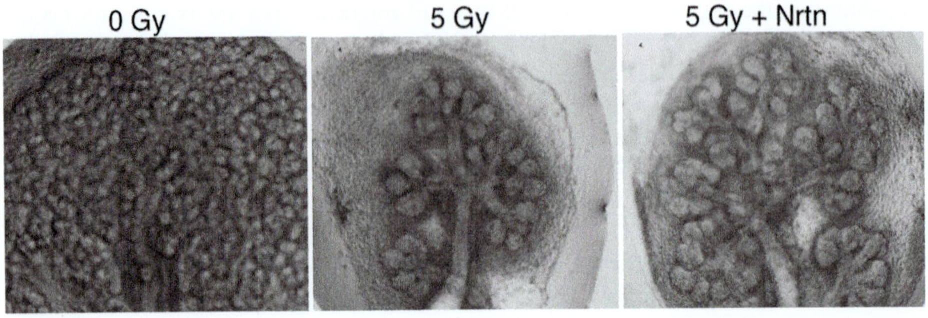

FIGURE 3 Irradiation of fetal SMGs reduces morphogenesis, but the addition of neurturin (Nrtn, 1 ng/mL) protects the parasympathetic ganglia and improves regeneration. Bright-field pictures of E13 SMGs cultured for 3 days. SMGs were radiated with 0 or 5 Gy and treated with neurturin.

and discovery of multiple progenitors in the salivary gland, the relative importance of the individual progenitor populations is not well understood. If we hypothesize that every progenitor can compensate for the loss of another one, gland regeneration should be achieved with any progenitor. However, if different types of progenitors have different self-renewing capacities, identification of the optimal type is a requisite for efficient regeneration.

Although several cell types are proposed to be human SMG progenitors, most information has been gathered on human Kit^+ cells. Similar to the rodent model, Kit^+ progenitors can be isolated (Feng et al., 2009; Banh et al., 2011) and cultured in vitro as floating “salispheres” or in three-dimensional matrixes to demonstrate their differentiation potential. However, human Kit^+ cells are also a heterogeneous cell population since they can coexpress other markers, such as CD34, THY1 (CD90), or CD44 (Banh et al., 2011). It is not clear whether one of these Kit^+ cell types is more potent as a progenitor. Furthermore, the challenge with regard to Kit^+ cells lies in the number of cells that we can obtain from a gland biopsy before the patient undergoes radiotherapy treatment. Although the number of Kit^+ cells in a human SMG is not known, Kit^+ cells in the mouse SMG are rare (0.3 to 3%) (Lombaert et al., 2008a). This suggests that induction of proliferation in vitro after biopsy or in vivo before biopsy may be an alternative for expanding the progenitor pool. Potentially, Kit^+ progenitors can be expanded by ALDH activators, as high levels of *ALDH3* and *ALDH1* were detected in the human Kit^+ cell population. Another study showed that human SMG progenitors resemble pancreatic progenitors but do not express Kit. These isolated human SMG cells were $ITGA6^+$, $THY1^+$, $CD44^+$, ENG $(CD105)^+$ $NGFR^+$, and could differentiate into pancreatic cell types (Gorjup et al., 2009), suggesting that human SMG progenitors can be multipotent and that multiple progenitors could be present in the human gland.

As suggested above, an additional research direction would be to stimulate or enhance the progenitor niche to optimize organ regeneration after injury. Since nerves and blood vessels are both essential parts of the maintenance and differentiation of epithelial progenitors, it will be important to investigate the crosstalk between different cell types within the SMG. We anticipate that understanding the cellular mechanisms of progenitor self-renewal and identifying the progenitor lineages will help us to direct the expansion of specific epithelial progenitors in vitro in order to use them to regenerate injured salivary glands.

REFERENCES

Arany S, Catalan MA, Roztocil E, Ovitt CE (2011). *Ascl3* knockout and cell ablation models reveal complexity of salivary gland maintenance and regeneration. *Dev Biol*. doi:10.1016/j.ydbio.2011.02.025.

Arnold K, Sarkar A, Yram MA, Polo JM, Bronson R, Sengupta S, Seandel M, Geijsen N, Hochedlinger K (2011). Sox2(+) adult stem and progenitor cells are important for tissue regeneration and survival of mice. *Cell Stem Cell* 9:317–329.

Ball WD (1974). Development of the rat salivary glands: 3. Mesenchymal specificity in the morphogenesis of the embryonic submaxillary and sublingual glands of the rat. *J Exp Zool* 188:277–288.

Banh A, Xiao N, Cao H, Chen CH, Kuo P, Krakow T, Bavan B, Khong B, Yao M, Ha C, et al. (2011). A novel aldehyde dehydrogenase-3 activator leads to adult salivary stem cell enrichment in vivo. *Clin Cancer Res* 17:7265–7272.

Bullard T, Koek L, Roztocil E, Kingsley PD, Mirels L, Ovitt CE (2008). *Ascl3* expression marks a progenitor population of both acinar and ductal cells in mouse salivary glands. *Dev Biol* 320:72–78.

Carpenter GH, Cotroneo E (2010). Salivary gland regeneration. *Front Oral Biol* 14:107–128.

Carpenter GH, Khosravani N, Ekstrom J, Osailan SM, Paterson KP, Proctor GB (2009). Altered plasticity of the parasympathetic innervation in the recovering rat submandibular gland following extensive atrophy. *Exp Physiol* 94:213–219.

Coppes RP, Stokman MA (2011). Stem cells and the repair of radiation-induced salivary gland damage. *Oral Dis* 17:143–153.

Cotrim AP, Sowers A, Mitchell JB, Baum BJ (2007). Prevention of irradiation-induced salivary hypofunction by microvessel protection in mouse salivary glands. *Mol Ther* 15:2101–2106.

Cotroneo E, Proctor GB, Paterson KL, Carpenter GH (2008). Early markers of regeneration following ductal ligation in rat submandibular gland. *Cell Tissue Res* 332:227–235.

Cutler LS (1980). The dependent and independent relationships between cytodifferentiation and morphogenesis in developing salivary gland secretory cells. *Anat Rec* 196:341–347.

David R, Shai E, Aframian DJ, Palmon A (2008). Isolation and cultivation of integrin alpha(6)beta(1)-expressing salivary gland graft cells: a model for use with an artificial salivary gland. *Tissue Eng A* 14:331–337.

De Moerlooze L, Spencer-Dene B, Revest J, Hajihosseini M, Rosewell I, Dickson C (2000). An important role for the IIIb isoform of fibroblast growth factor receptor 2 (FGFR2) in mesenchymal–epithelial signalling during mouse organogenesis. *Development (Camb)* 127:483–492.

Denny PC, Ball WD, Redman RS (1997). Salivary glands: a paradigm for diversity of gland development. *Crit Rev Oral Biol Med* 8:51–75.

Enomoto H, Heuckeroth RO, Golden JP, Johnson EM, Milbrandt J (2000). Development of cranial parasympathetic ganglia requires sequential actions of GDNF and neurturin. *Development (Camb)* 127:4877–4889.

Feng J, van der Zwaag M, Stokman MA, van Os R, Coppes RP (2009). Isolation and characterization of human salivary gland cells for stem cell transplantation to reduce radiation-induced hyposalivation. *Radiother Oncol* 92:466–471.

Gorjup E, Danner S, Rotter N, Habermann J, Brassat U, Brummendorf TH, Wien S, Meyerhans A, Wollenberg B, Kruse C, et al. (2009). Glandular tissue from human pancreas and salivary gland yields similar stem cell populations. *Eur J Cell Biol* 88:409–421.

Gouon-Evans V, Rothenberg ME, Pollard JW (2000). Postnatal mammary gland development requires macrophages and eosinophils. *Development (Camb)* 127:2269–2282.

Gresik EW (1980). Postnatal developmental changes in submandibular glands of rats and mice. *J Histochem Cytochem* 28:860–870.

Gresik EW (1994). The granular convoluted tubule (GCT) cell of rodent submandibular glands. *Microsc Res Tech* 27:1–24.

Grobstein C (1953). Inductive epitheliomesenchymal interaction in cultured organ rudiments of the mouse. *Science* 118:52–55.

Gyorki DE, Asselin-Labat ML, van Rooijen N, Lindeman GJ, Visvader JE (2009). Resident macrophages influence stem cell activity in the mammary gland. *Breast Cancer Res* 11:R62.

Hai B, Yang Z, Millar SE, Choi YS, Taketo MM, Nagy A, Liu F (2010). Wnt/beta-catenin signaling regulates postnatal development and regeneration of the salivary gland. *Stem Cells Dev* 19:1793–1801.

Hai B, Yang Z, Shangguan L, Zhao Y, Boyer A, Liu F (2012). Concurrent transient activation of Wnt/beta-catenin pathway prevents radiation damage to salivary glands. *Int J Radiat Oncol, Biol Phys*.

Hick AC, van Eyll JM, Cordi S, Forez C, Passante L, Kohara H, Nagasawa T, Vanderhaeghen P, Courtoy PJ, Rousseau GG, et al. (2009). Mechanism of primitive duct formation in the pancreas and submandibular glands: a role for SDF-1. *BMC Dev Biol* 9:66.

Hisatomi Y, Okumura K, Nakamura K, Matsumoto S, Satoh A, Nagano K, Yamamoto T, Endo F (2004). Flow cytometric isolation of endodermal progenitors from mouse salivary gland differentiate into hepatic and pancreatic lineages. *Hepatology* 39:667–675.

Kim YJ, Kwon HJ, Shinozaki N, Hashimoto S, Shimono M, Cho SW, Jung HS (2008). Comparative analysis of ABCG2-expressing and label-retaining cells in mouse submandibular gland. *Cell Tissue Res* 334:47–53.

Kishi T, Takao T, Fujita K, Taniguchi H (2006). Clonal proliferation of multipotent stem/progenitor cells in the neonatal and adult salivary glands. *Biochem Biophys Res Commun* 340:544–552.

Knosp WM, Knox SM, Hoffman MP (2012). Salivary gland organogenesis. *Dev Biol* 1:69–82.

Knox SM, Lombaert IM, Reed X, Vitale-Cross L, Gutkind JS, Hoffman MP (2010). Parasympathetic innervation maintains epithelial progenitor cells during salivary organogenesis. *Science* 329:1645–1647.

Knox SM, Lombaert IM, Haddox CL, Abrams SR, Cotrim A, Wilson AJ, Hoffman MP (2013). Parasympathetic stimulation improves epithelial organ regeneration. *Nat Commun*, in Press.

Lammert E, Cleaver O, Melton D (2001). Induction of pancreatic differentiation by signals from blood vessels. *Science* 294:564–567.

Lombaert IMA (2008). *Regeneration of Irradiated Salivary Glands by Stem Cell Therapy*, Van Denderen, Groningen, The Netherlands.

Lombaert IM, Wierenga PK, Kok T, Kampinga HH, de Haan G, Coppes RP (2006). Mobilization of bone marrow stem cells by granulocyte colony-stimulating factor ameliorates radiation-induced damage to salivary glands. *Clin Cancer Res* 12:1804–1812.

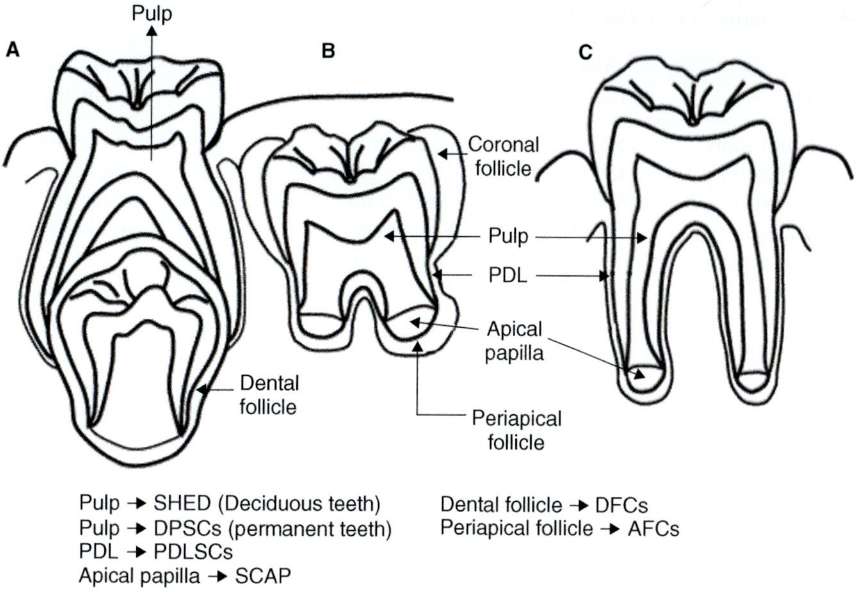

FIGURE 1 Sources of stem cells from dental tissues. (A) A shedding deciduous molar with an emerging permanent premolar underneath. (B) An erupting permanent molar emerging through the alveolar crest bone underneath the gingival tissue. (C) A fully erupted permanent molar with immature apices. DPSCs, dental pulp stem cells; SHED, stem cells from exfoliated deciduous teeth; PDL, periodontal ligament; PDLSCs, periodontal ligament stem cells.

2 DENTAL PULP

2.1 DPSCs

Undifferentiated dental mesenchymal cells are still present after tooth eruption, and it was possible to locate stem cells in the dental mesenchyme (Shi and Gronthos, 2003). For the first time human dental stem cells were isolated from the dental pulp of extracted wisdom teeth (Gronthos et al., 2000). DPSCs can be isolated from postnatal teeth (Gronthos et al., 2000) as well as from the very rare natal teeth (Karaöz et al., 2010). DPSCs display features similar to those of bone marrow–derived MSCs. For example, both cell types adhere to cell culture dishes and form clonogenic colonies. The capacity to form clonogenic colonies is indicative of their ability to undergo self-renewal, which is a characteristic of stem cells. DPSCs are heterogeneous with different cell morphologies and sizes. However, only 20% of selected single-colony-derived strains of DPSCs were capable of proliferating over 20 population doublings and generated enough cells for studies of their developmental potential (Gronthos et al., 2002). Interestingly, the number of clonogenic colonies from single-cell suspensions of the dental pulp increased up to 100% after selection with immunomagentic bead-based protocols with specific antibodies for STRO-1, CD146, and 3G5 (Shi and Gronthos, 2003). The self-renewal capability of proliferating DPSCs was also demonstrated after the isolation of human dentinogenic cells from DPSC transplants in immunocompromised mice (Gronthos et al., 2002). DPSCs are located perivascularly and express typical markers of smooth muscle cells and pericytes, such as smooth muscle actin, CD146, and the

TABLE 1 Summary of Dental Stem Cell Types

Cell Type	Markers (Selected)	Multipotentiality	Ectopic Tissue Formation
Dental pulp stem cells (including stem cells from exfoliated deciduous teeth)	STRO-1 CD105 CD44 CD146	Osteogenic, Dentinogenic Adipogenic Chondrogenic Neurogenic Myogenic Bone induction	Dentin–pulp-like complex (tissues) Bone-like tissues
Stem cell apical papilla (including apical pad–derived dental neural crest progenitor cells)	CD105 CD106 CD90 CD146	Osteogenic, Dentinogenic Adipogenic Chondrogenic Neurogenic	Dentin–pulp-like complex (tissues) Bone-like tissues
Periodontal ligament stem cells	STRO-1 CD105 CD44 CD146	Osteogenic, Cementogenic Adipogenic Chondrogenic Neurogenic	Periodontal tissues
Dental follicle cells (or dental follicle precursor cells)	STRO-1 CD105 CD44 Notch1	Cementogenic Osteogenic Adipogenic Chondrogenic Neurogenic	Periodontal tissues

pericyte-associated antigen (3G5) (Shi and Gronthos, 2003). Moreover, they shared characteristics of osteoblast-like cells and express a number of bone marrow–derived MSC markers, such as CD105, CD146, CD44, and STRO-1 (5 to 10%) (Huang et al., 2009).

In contrast to bone marrow–derived MSCs, DPSCs were found to differentiate into odontoblast-like cells. This dentinogenic phenotype appeared especially when DPSCs were seeded onto dentin. Here, DPSCs convert into cells with a polarized cell body and a cell process extending into the existing dentinal tubules (Huang et al., 2006). Previous studies have shown that the generation of a dentin-pulp-like complex is possible after transplantation of DPSCs (which were seeded, for example, on dentin disks or cylinders) into immunocompromised mice (Gronthos et al., 2000; Sakai et al., 2011). Fig. 2A shows DPSC-generated dentin-pulp-like complex. The differentiation into a pulp–dentin complex provides new opportunities for the regeneration of lost or degenerated dentin and/or dental pulp due to caries and pulpitis. Recently, complete pulp regeneration after pulpectomy was demonstrated with autologous DPSCs in dogs (Iohara et al., 2011). More information about pulp and dentin regeneration may be obtained in Chapter 26.

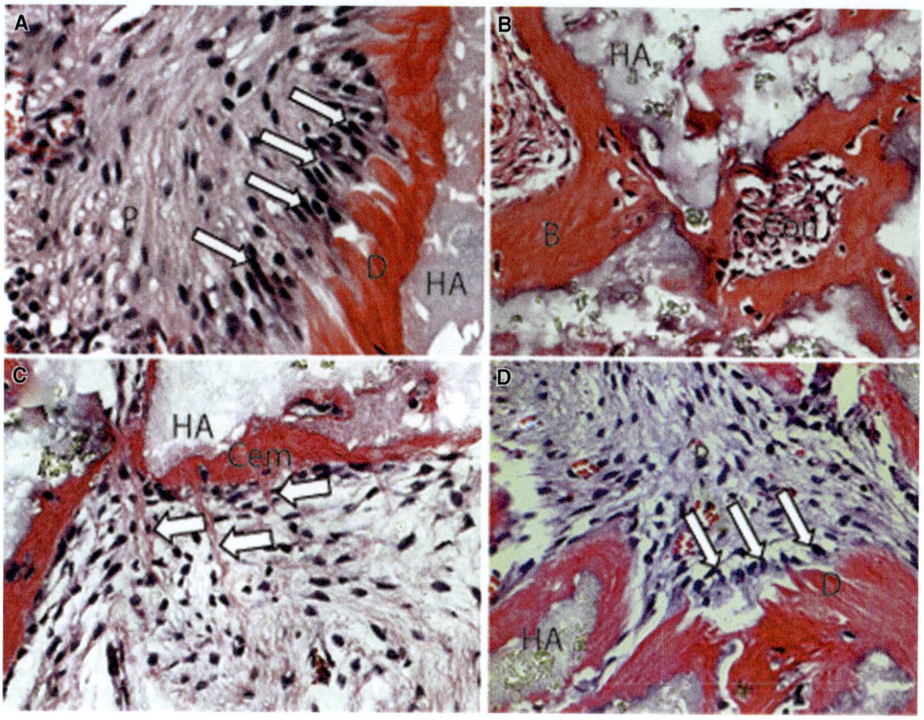

FIGURE 2 Dental stem cells were implanted into immunocompromised mice subcutaneously for 8 weeks using tricalcium phosphate hydroxyapatite (*HA*) as a carrier. (A) DPSCs differentiate into odontobalst (arrows) to generate dentin (*D*) and pulplike tissue (*P*). (B) SHED regenerate bonelike tissue (*B*) and connective tissue (*Con*). (C) PDLSCs generate cementum (*Cem*) and Sharpey's fibers (arrows). (D) SCAP regenerate dentin (*D*) and pulplike tissue (*P*).

DPSCs are multipotent stem cells (Gronthos et al., 2002) that undergo myogenic, chondrogenic, and osteogenic differentiation in vitro (Gronthos et al., 2002; Laino et al., 2005; d'Aquino et al., 2007). They are also an exciting stem cell source for the regeneration of nondental tissues. One important target for cellular therapy in modern medicine is the regeneration of neural tissues. The regeneration of degenerated neural tissues by endogeneous neural stem or progenitor cells is very limited, and many people suffer severely or die, because a stem cell–based therapy is not available. Unfortunately, the isolation of human neural stem cells is difficult, and protocols for the generation of neural stem cells from embryonic stem cells or induced pluripotent stem cells (iPSCs) are still under development. Although dental stem cells are a good source of iPSCs (Tamaoki et al., 2010; Yan et al., 2010), DPSCs were successfully differentiated into neural–like cells. Recently, Arthur et al. (2008) have shown that DPSCs can differentiate under in vitro conditions toward functionally active neurons. Moreover, DPSCs can also induce endogenous axon guidance (Arthur et al., 2009), which makes these cells not only attractive for the regeneration of neurons of the central nervous system but also for the guidance of disconnected nerve fibers.

Another example for the extraordinary potential of dental stem cells was shown by a subpopulation of DPSCs. Papaccio and colleagues identified a c-Kit^{+}/CD34^{+}/CD45^{-} cell population of stromal bone producing DPSCs (SBP/DPSCs) (Laino et al., 2005) and isolated a clonogenic DPSC population by fluorescence-activated cell sorting. In contrast to initially isolated DPSCs, which were isolated as plastic-adherent and clonogenic cells (Gronthos et al., 2000), these cells express the glycoprotein CD34. CD34 is a typical marker of hematopoetic precursor cells, and it is also expressed by murine epithelial stem cells in the bulge of the hair follicle, which is the niche of hair follicle stem cells (Morrison and Spradling, 2008). At present, little is known about the relation between different DPSC populations. Papaccio and colleagues describe not only the isolation of a special population of DPSCs but also the differentiation into living bone tissue and the production of bone for clinical applications, which makes this DPSC population very special (Laino et al., 2005; Carinci et al., 2008).

DPSCs can also be isolated from inflamed pulp, termed DPSCs-IP (Alongi et al., 2010). Some of these DPSCs-IP exhibit properties identical or similar to those from normal pulp. They express similar CD markers, such as CD73 and CD146, and form a pulp–dentin complex in immunocompromised mice. The qualities of the DPSCs-IP appear to be lower than those of DPSCs from normal pulp, presumably due to the result of inflammation.

2.2 SHED

Stem cells were also isolated from human primary teeth; for example, SHED (stem cells of human exfoliated deciduous teeth) were isolated from the dental pulp of exfoliated primary teeth (Miura et al., 2003). These dental stem cells can be cultivated either as plastic-adherent cells, or like neural stem cells as sphere-like cell clusters, called *neurospheres*. SHED are multipotent and capable of differentiation into odontoblast, adipocytes, and neural cells (Miura et al., 2003). For example, SHED were capable of biomineralized tissue formation under in vivo conditions in immunocompromised animals; and they were able to survive and migrate in murine brain after transplantation (Miura et al., 2003). Transplantation of SHED to the severed site of spinal cord in rats promotes locomotor recovery (Sakai et al, 2012). Interestingly, unlike DPSCs, SHED are unable to regenerate a complete dentin–pulp complex in vivo (Miura et al., 2003). This property of the differentiation potential may reflect that SHED were isolated from “dying”exfoliated deciduous teeth. However, SHED appeared to induce new bone formation by forming an osteo-inductive template to recruit host osteoblasts after transplantation into mice (Miura et al., 2003). Moreover, the osteo-inductive effect of SHED was sufficient to repair critical-sized calvarial defects in mice (Seo et al., 2004). Figure 2B shows the regenerate bonelike tissue (B) and connective tissue (Con).

Similar to SHED, another group of dental pulp–derived stem cells from exfoliated deciduous teeth was isolated. Kerkis and colleagues (2006) isolated a $CD34^-$ population of immature dental pulp stem cells (IDPSCs), which express embryonic stem cell–associated markers such as OCT4, NANOG, SSEA3, SSEA4, TRA-1-60, and TRA-1-81. However, in contrast to DPSCs or SHED, IDPSCs were isolated from dental pulp tissue explants. IDPSCs are highly potent and undergo uniform differentiation into smooth and skeletal muscle cells, neurons, cartilage cells, and bone cells under chemically defined culture conditions. Interestingly, the human DNA of IDPSCs was detectable in lungs, liver, kidney, and brain after engraftment of these cells following intraperitoneal injection into nude mice. This experiment supports the fact that stem cells from the dental pulp are an interesting source for cellular therapies.

3 PERIODONTAL LIGAMENT

The periodontal ligament (PDL) connects cementum and alveolar bone and it is the connective tissue of the tooth attachment apparatus. All three tissues of the periodontium are under a strong mechanical load. For this reason it was suggested that cells of the PDL are involved in the homeostasis and regeneration of the periodontal apparatus long before stem cells were discovered in the PDL (McCulloch and Bordin, 1991).

3.1 PDLSCs

For the isolation of PDL stem cells (PDLSCs), single cells were obtained after the PDL was dissolved enzymatically. A small subpopulation of these liberated PDL cells were plastic-adherent and clonogenic cells. The identification of PDLSCs was shown by the expression of mesenchymal stem cell markers such as STRO-1 and CD146/MUC18. Although they are DPSC-like, PDLSCs are a novel type of dental ectomesenchymal stem cells (Seo et al., 2004). In contrast to DPSCs, PDLSCs express scleraxis, which is a tendon cell–specific transcription factor. As multipotent stem cells, PDLSCs differentiate into adipogenic- and chondrogenic-like cells under in vitro conditions, but their osteogenic differentiation potential was weak compared to that of DPSCs (Seo et al., 2004). However, PDLSCs differentiate into periodontal cell types under in vitro conditions, and they generate a periodontium-like tissue under in vivo conditions after transplantation into immunocompromised mice. The structure of this periodontium-like tissue is totally different from that of bone marrow or dental pulp–dentin structures, which are generated by MSCs or DPSCs, respectively. These PDL-like structures of differentiated PDLSCs contain dense collagenous fibers with sparse cells. Collagen fibers mimicked the attachment of Sharpey's fibers and were able to connect with PDLSC-generated mineralized tissues (cementum) (Seo et al., 2004). Figure 2C shows PDLSC-generated cementum with Sharpey's fibers. In the same study, human PDLSCs were able to regenerate periodontal defects in immunocompromised mice. Here human cells were also identified to be closely associated with regeneration of the trabecular bone, which was next to the PDLSC-regenerated PDL.

It is important to mention that PDLSCs could also be isolated from inflamed PDL tissues (ihPDLSCs) (Park et al., 2011). PDLSCs and ihPDLSCs expressed the same stem cell markers (e.g., CD146 and STRO-1) and they had the same multipotent cell differentiation potential. Autologous PDLSCs are therefore a good source for a stem cell–based therapy of periodontitis. In recent years several studies have investigated the potential of undifferentiated PDL cells for periodontitis. For example, a PDLSC treatment of periodontitis was successful in conjugation with a standard surgical procedure in a swine model (Liu et al., 2008). Moreover, a pilot study with three patients was published about the regeneration of lost or diseased periodontium after cell therapy with PDL progenitors (PDLPs) (Feng et al., 2010). PDLPs were isolated from periodontal tissue explants, and they are an undifferentiated PDL cell population with typical stem cell–like properties. They express characteristic surface marker molecules for MSCs, although typical PDLSC markers such as scleraxis, CD146, and STRO-1 were expressed lower than were PDLSCs. Moreover, PDLPs showed decreased capacities to differentiate into osteo- and adipogenic cells under in vitro condititions, but they still have the ability to regenerate cementum–collagen–Sharpey's fibers under in vivo conditions. PDLPs represent a later lineage cell population of the PDL, when compared with PDLSCs. This PDLP pilot study was begun in the 1990s, some years before PDLSCs were isolated. Although PDLPs are not typical dental stem cells, it is important to mention that this study showed for the first time the safety of autologous PDL cells in the treatment of periodontitis. The treatment of periodontitis will be one of the first targets for dental stem cell–based therapies. An overview of periodontal bioengineering strategies is provided in Chapter 28.

One problem with PDL tissue engineering or regeneration of the periodontium with PDLSCs could be the limited number of somatic stem cells present in the PDL. An alternative source for stem cell–based therapies of the PDL could be induced pluripotent stem cells (iPSCs). Although iPSCs can be produced from any somatic cell of the body, it has been proposed that reprogramming could be more efficient from populations of undifferentiated stem cells. The PDL is therefore an excellent source of undifferentiated cells, and very recently, iPSCs were easily produced from a mixture of PDL cells (Wada et al., 2011). However, currently we can only speculate about the success and safety of iPSCs for cellular therapies. Moreover, we do not have appropriate protocols that make it possible to use PDL-derived iPSCs to produce large numbers of PDLSCs for regeneration of the periodontium, although this might be only a technical problem.

4 APICAL PAPILLA

The tooth development begins with the condensation of dental epithelium and continues with interplay between tooth germ tissue cells, which are derived from the dental epithelium and dental mesenchyme. However, most tissues of the human tooth germ and their undifferentiated cells are not available for regenerative dentistry, because they are lost during tooth development before tooth eruption. One example is the human dental organ with dental epithelial stem cells and ameloblast precursors mentioned earlier. These tooth germ tissues are not available in an impacted third molar tooth or are available only in a very small quantity. For the generation of dental enamel or for whole-tooth engineering with autologous human stem cells, these dental epithelial stem cells are missing (Ikeda et al., 2009). In contrast to the dental organ, dental mesenchymal tissues of the tooth germ are available from impacted third molars. A good example is the dental papilla, also known as apical pad-like tissue, which is attached to the apex of a developing tooth. However, it is only possible to isolate this tissue from impacted human third molars with immature roots, which are, for example, extracted routinely for orthodontic reasons from adolescent patients. In this section we present stem cells and progenitor cells from the dental papilla, which is also called the dental apical pad-like tissue (Degistirici et al., 2008; Sonoyama et al., 2008). The dental papilla represents the progenitor of the dental pulp–dentin complex. However, it is also probably involved in the development of alveolar bone (Degistirici et al., 2008; Huang et al., 2008).

To date, very little is known about the status of dental papilla from extracted human third molars. The dental papilla is adjacent to the apex of the developing tooth and can be obtained only from teeth of an early stage of development, before tooth eruption. This pad-like tissue is apically of the developing dental pulp–dentin complex of the root and demarcated from the tooth root by the apical cell–rich zone and the epithelial root diaphragm (Degisterici et al., 2008; Sonoyama et al., 2008). It is not clear whether the dental papilla can be divided into different parts: one part adjacent to the developing dental pulp, with specialized progenitors for the root pulp–dentin complex, and an "apical" part, which is located close to the alveolar bone, with osteogenic progenitors for the alveolar bone. However, the dental pad-like papilla can be clearly demarcated from the dental pulp in histological stainings with Alcian Blue (Degistirici et al., 2008). In contrast to the dental pulp, the dental papilla contains an abundance of proteoglycans (Degistirici et al., 2008). Furthermore, histological analysis showed that cells are sparsely distributed within the matrix of this tissue, in contrast to the dental pulp, which consists of a dense cellular and fibrous tissue with more vessels and nerve fibers (Huang et al., 2008; Sonoyama et al., 2008).

4.1 SCAP

Like DPSCs, stem cells from the human dental apical papilla (SCAP) were isolated from the dental pulp as plastic-adherent and clonogenic cells. In comparison to DPSCs, SCAP express lower levels of odontoblast markers, such as dentin sialoprotein and matrix extracellular phosphoglycoprotein. However, SCAP but not DPSCs express the heat-stable antigen CD24, which is a sialoglycoprotein also found on mature granulocytes. Nonetheless, SCAP have an excellent odontogenic cell differentiation potential, which could be demonstrated by the expression of typical odontoblast markers after differentiation into mineralizing cells (Sonoyama et al., 2006, 2008). Moreover, after transplantation into immunocompromised mice, SCAP on a carrier of hydroxylapatite produce a typical dental pulp/dentin-like complex (Sonoyama et al., 2006). Figure 2D shows that SCAP regenerate dentin and pulp-like tissue. In the same study, SCAP, in combination with PDLSCs, generated a biological tooth root under in vivo conditions. This result also shows that SCAP are involved in development of the dental pulp–dentin complex. Moreover, SCAP are multipotent and differentiate into adipogenic cells and neural cells under appropriate in vitro conditions. After cultivation in a serum-free neural differentiation medium, SCAP express obviously neuronal markers such as neuron specific enolase, neurofilament M, and neuronal nuclear antigen. SCAP are multipotent cells and dental pulp progenitors, but they are probably more undifferentiated than are DPSCs (Sonoyama et al., 2008).

5 DENTAL NEURAL CREST–DERIVED PROGENITOR CELLS

In addition to SCAP, another type of undifferentiated cell was identified in the dental papilla. In this approach, migratory undifferentiated cells called dental neural crest–derived progenitor cells (dNC-PCs) were isolated from explants of dental pad-like tissues. This strategy was the same as that used for the isolation of PDLPs and IDPSCs (Kerkis et al., 2006; Feng et al., 2010). dNC-PCs are plastic-adherent proliferating cells, and have a typical fibroblast-like morphology. In flow cytometry analysis a large number of dNC-PCs are positive for CD49d (α4 integrin), CD56 (NCAM), and PDGFRa, but negative for CD31, CD34, CD45, and STRO-1. Based on the expression pattern of somatic stem cell markers, dNC-PCs are different from most other dental stem cells, such as SCAP. However, dNC-PCs are typical ectomesenchymal progenitor cells, because they express neural crest marker genes such as *Sox9, Snail1, Snail2, Twist1, Msx2*, and *Dlx6* and could be differentiated into neurogenic, chondrogenic, and osteogenic lineages (Degistirici et al., 2008).

Moreover, it was possible to induce bone formation after transplantation in immunocompromised mice: for example, by combining dNC-PCs with a scaffold of bovine bone-derived granulates (Degistirici et al., 2008). dNC-PCs generated thick biomineralized structures under in vivo conditions that are comparable to those produced by bone marrow–derived MSCs (Degistirici et al., 2008). Under cell culture conditions dNC-PCs responded to osteogenic stimulation by progressing through a series of structural and biochemical changes that terminated in the formation of a simple structured bone-like matrix (Degistirici et al., 2010). In contrast to $CD34^{+}$ DPSCs (Laino et al., 2005) osteogenic-stimulated dNC-PCs did not generate a fully organized and anatomically structured bone (Degistirici et al., 2008, 2010). The lack of formation of capillary-like structures, for example, may be due to a number of factors, one of them being the limited differentiation potential of dNC-PCs. At the moment, however, we do not know much about the dentinogenic differentiation potential of dNC-PCs.

In an unpublished study, dNC-PCs expressed typical markers for odontoblasts such as DMP1 and DSPP after cultivation in an osteogenic differentiation medium. Moreover, after differentiation under in vivo conditions, histological examinations revealed obvious differences among dNC-PCs and bone marrow–derived MSCs (Degistirici et al., 2008), although tubular dentin was not observed for dNC-PCs. We can only speculate on the direct relation between dNC-PCs and SCAP. Although dNC-PCs express typical markers for stem cells and have a multipotent differentiation potential, but they do not express the mesenchymal stem cell marker STRO-1. We suppose that the relation between SCAP and dNC-PCs is simply the relation between stem cells and stem cell–derived migratory progenitor cells. Figure 3 illustrates different strategies for the isolation of SCAP and dNC-PCs. A similar relation is also described for PDLSCs and PDLPs (see above).

The relation between DPSCs and SCAP/dNC-PCs is also not clear, but SCAP are probably the more undifferentiated stem cells, with a faster cell proliferation rate and more population doublings (Sonoyama et al., 2006). A current model for SCAP and DPSCs may reveal an explanation for their biological relation during tooth development and regeneration of the dental pulp–dentin complex. The dental papilla–derived SCAP (but not DPSCs) are the cell source for primary odontoblasts and for the production of the dental pulp–dentin complex of the tooth root. These differentiated primary odontoblasts produce primary and secondary dentin (root dentin). In contrast to SCAP, DPSCs and probably also SHED are progenitor cells for odontoblast-like cells, which are also known as *replacement odontoblasts*. These cells replace degenerated primary odontoblast and produce tertiary (regenerative) dentin, for example, after severe mechanical stress and subsequent degeneration of the dentin. If this is true, the isolation of stem cells and progenitor cells from the dental papilla and the dental pulp will help us to understand the development and regeneration of the dental pulp–dentin complex in more detail.

6 DENTAL FOLLICLE

It is widely known that the dental follicle plays a crucial role in tooth eruption and during the development of a tooth root. The dental follicle is separated from the

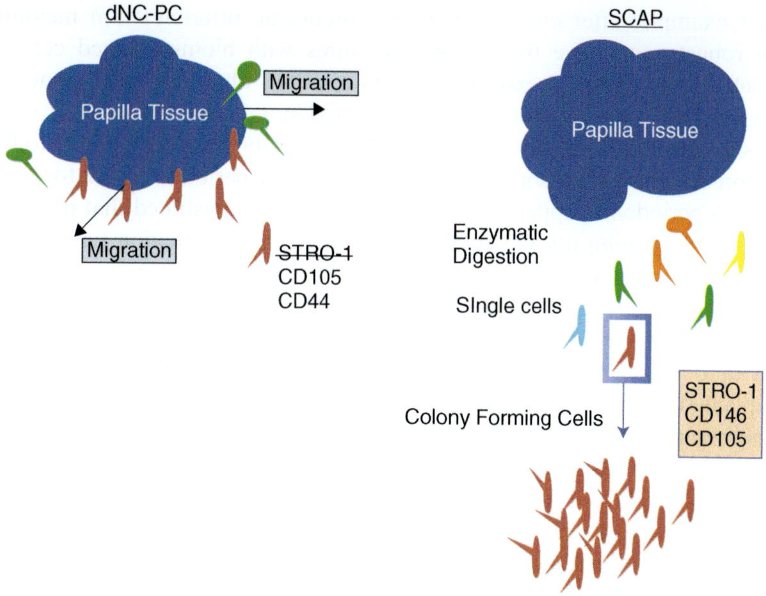

FIGURE 3 Strategies for the isolation of dNC-PCs and SCAP from the apical papilla of impacted wisdom teeth.

developing dentin by an epithelial cell layer (Hertwig's epithelial root sheath) during early stages of periodontal development. Similar to the dental papilla, the dental follicle can be separated from an impacted human wisdom tooth with an immature tooth root. The follicle contains progenitors for the periodontal apparatus, consisting of the alveolar bone, the PDL, and cementum (Diekwisch, 2001).

6.1 DFCs

The dental follicle contains multipotent ectomesenchymal cells. These neural crest–derived precursor cells, also known as dental follicle precursor cells or dental follicle cells (DFCs), can be isolated as plastic-adherent and clonogenic cells (Morsczeck et al. 2005). DFCs have a fibroblastic DPSC-like morphology and also express typical markers of progenitor or stem cells, such as nestin, notch-1, CD44, CD105, and STRO-1 (Morsczeck et al., 2005; Kémoun et al., 2007). Like other dental stem cell types, DFCs have a high proliferation rate. It has been shown that the number of proliferating DFCs is higher than that of bone marrow–derived MSCs. Similar to SHED, they form nonadherent cell clusters like neural stem cells after cultivation in serum-replacement medium and also differentiate into neural cells (Völlner et al., 2009). DFCs can be cultivated under serum-free conditions for a longer period and behave like neural progenitor cells. The proteome of these cell clusters is comparable to that of de-differentiated retinal Müller progenitor cells, which form similar cell clusters after cultivation in serum-replacement medium (Beck et al., 2011).

Under in vitro conditions, DFCs differentiate into different cell types, especially into cells of the periodontium (Morsczeck et al., 2005, Kémoun et al., 2007; Völlner et al.,

2009). For example, after cultivation in an osteogenic differentiation medium, they can form robust connective tissue–like structures with biomineralized cell clusters. Differentiated cells typically express markers for PDL fibroblasts and cementoblasts. Interestingly, DFCs differentiate into this periodontium-like tissue with occasional blood vessel–like structures (Morsczeck et al., 2005). DFCs are comparable to PDLSCs and therefore an alternative source for the treatment of periodontitis or for the reconstruction of a periodontal apparatus. Recently, it was demonstrated that porcine DFCs were capable of forming a tooth root–like structure in combination with DPSCs and enamel stem cells under in vivo conditions in the omentum of immunocompromised rats (Honda et al., 2010). In another study, single cell–derived rat DFCs generate a whole tooth root without PDLSCs and epithelial stem cells (Guo et al., 2011). DFCs produced all tooth root tissue when seeded on scaffolds of treated dentin matrix (TDM) and transplanted in an alveolar fossa microenvironment. However, this special environment is necessary for the generation of a tooth root, because they did not differentiate into tooth root cells after transplantation in skull and omental pockets (Guo et al., 2011). Although DFCs do not express odontoblast markers before and after cultivation in a standard medium for osteogenic differentiation, TDM induces the differentiation of DFCs into odontoblast-like cells (Guo et al., 2009). This study suggests that DFCs are also an appropriate source for the regeneration of the dental pulp–dentin complex.

Although DFCs differentiate into biomineralizing cells with cementoblast-like characteristics under in vitro conditions (Morsczeck et al., 2005; Kémoun et al., 2007), these cells were also evaluated for bone regeneration. In a recent study, single cell–derived DFCs were obtained from porcine dental follicles at an early stage of tooth crown formation (Honda et al., 2010). The osteogenic differentiation potential of these DFCs under in vivo conditions was characterized and compared with PDLSCs and bone marrow–derived MSCs. All stem cell types had similar capacities for bone regeneration of critical size defects in the calvaria of immuno compromised rats (Honda et al., 2010). Compared to untreated animals, stem cells improved the process of bone regeneration. In another study, three clonal-derived human DFC lines, which exhibited different colony morphologies and gene expression patterns, also promoted bone formation when transplanted into surgically created critical-sized defects in immunodeficient rat calvaria (Honda et al., 2010). These results show that DFCs have an excellent potency for differentiation into osteoblast-like cells and for the regeneration of alveolar bone. DFCs are the genuine progenitors of alveolar osteoblasts and an attractive source for bone tissue engineering.

We do not know much about molecular processes of DFCs, but as progenitor cells of the periodontium they may use variable mechanisms of cell differentiation (Saugspier et al., 2010). The initiation of differentiation in the stem cell niche is modulated by a number of different extracellular factors, such as growth factors, cell–cell contacts, extracellular matrix, and mechanical loading (Morrison and Spradling, 2008). Although each factor may not be sufficient to induce the differentiation process directly, together they may coordinately influence or modulate this process or facilitate the process of differentiation into a particular functional body cell type. The direction of differentiation, for example, also depends on the stiffness of the extracellular environment. The multipotent differentiation of naive MSCs is extremely sensitive to tissue-level stiffness (Engler et al., 2006). By changing the stiffness of the substrate, human MSCs could be directed toward neuronal (soft matrices that mimic brain) or bone lineages (rigid matrices that mimic collagenous bone). Interestingly, this is not the case for DFCs.

The differentiation of DFCs into biomineralizing cells is supported on soft connective tissue–like stiffness but is inhibited on rigid stiffness (Viale-Bouroncle et al., 2011). This result may reflect the biological function of DFCs during development of the periodontium, which comprises three sandwich-like tissues: two rigid matrices with a soft matrix in between. A balanced distribution of hard tissue (cementum/bone) and connective tissue (PDL) is an important feature of a functional periodontium. DFCs have to control the proportional amount of the PDL, cementum, and alveolar bone during the periodontal development. Therefore, more PDL and increased softness in the periodontium, for example, supports the differentiation of DFCs into cementoblasts and osteoblasts. This could explain why a soft extracellular matrix supports the osteogenic differentiation of DFCs. However, stiffness modulates but does not induce or completely inhibit the process of differentiation. In contrast, soluble factors such as BMP2 induce the differentiation of DFCs into biomineralizing cells (Kémoun et al., 2007; Saugspier et al., 2010).

6.2 Periapical Follicle

The dental follicle surrounds the tooth germ during early stages of tooth development (Fig. 1A). However, during tooth root development the dental follicle begins to disappear, and the remainder can be separated into a coronal and a periapical part (Fig. 1B). In histology the coronal part of the follicle is close to the developing periodontium and tooth crown before tooth eruption. During root development, Hertwig's epithelial root sheath dissociates and cells of the attached follicle differentiate into functional cells of the periodontium. In separated dental follicles of an extracted wisdom tooth the disintegrating epithelial cell layer and islands of epithelial cell clusters can be found in the coronal part of the dental follicle (Morsczeck et al., 2005). DFCs as periodontal precursor cells can be isolated from this part of the follicle (Morsczeck et al., 2005). The other main portion of the disintegrating follicle (Fig. 1B) has been reported to be attached to the apical pad or papilla of the tooth germ (Sonoyama et al., 2006; Jo et al., 2007; Han et al., 2010). This tissue is called the periapical follicle, and undifferentiated cells from this part of the dental follicle can be isolated as periapical dental follicle stem cells (PAFSCs). PAFSCs express typical MSC markers such as CD44 and STRO-1 and have an excellent cell differentiation potential for all cell types of the periodontium (Jo et al., 2007; Han et al., 2010). Therefore, they are also potential candidates for the regeneration of periodontal tissues (Han et al., 2010). Presently, the relation between DFCs and PAFSCs is unclear, but both cell types are probably closely related in terms of their stem cell–like properties.

6.3 Follicle-Derived Embryonic Neural Crest Stem Cells

The dental follicle also contains a population of embryonic-like stem cells. This special cell type, called follicle-derived embryonic neural crest (FENC) stem cells, can be isolated from DFC cultures after five cell passages and cell sorting with the embryonic stem cell marker SSEA4 (d'Aquino et al., 2011). FENC stem cells express typical markers of embryonic stem cells, and they are multipotent. In particular, these cells share the following features with embryonic stem cells: (1) high levels of embryonic stem cell markers (TRA1-60, TRA1-81, OCT4); (2) mRNA transcripts for *NANOG* and *REX1*; (3) pluripotency (FENC stem cells differentiate into the cell types of all three germ layers, including smooth and skeletal muscle, osteoblasts, neurons, glial cells, and adipocytes); (4) high levels of telomerase activity; (5) the ability to form embryoid bodies; (6) the ability, after injection in murine blastocysts, to be localized within the inner cell mass; (7) no teratoma formation after injection; and (8) in vivo tissue formation after transplantation (d'Aquino et al., 2011). The relation of FENC cells and DFCs is unclear. Nevertheless, it is obvious that the dental follicle is like the dental papilla (Sonoyama et al., 2006), as tooth germ tissues contain a number of undifferentiated embryonic-like cell types. FENC stem cells were isolated from DFC cultures at the fifth passage, and therefore it is very likely that FENC cells constitute a typical subpopulation of all DFC cultures. If this is true, it opens new questions about the biology of subpopulations in primary dental cell lines.

7 GINGIVAL TISSUE–DERIVED MSCS

Oral mucoal tissue, including gingiva, is rich in blood supply and has great healing potential. This highly regenerative capacity prompted researchers to investigate the presence of MSCs in oral mucosa. It was reported that human gingival and alveolar mucosa harbor robust stem cells termed gingival tissue–derived MSCs (GMSCs), located in lamina propria (Marynka-Kalmani et al., 2010). Regardless of donor age, of the 95% of GMSCs that express MSC markers, 40 to 60% were positive for OCT4, SOX2, and NANOG, and 60 to 80% expressed constitutively neural and neural crest SC markers. GMSCs differentiated in culture into osteoblastic, chondroblastic, adipocytic, and neuronal lineages. Furthermore, GMSCs treated with dexamethasone formed benign tumors consisting of ectodermal and mesodermalgerm layer–derived tissues when transplanted in scid mice. Human GMSCs appear to be as osteogenic in vivo as BMMSCs, but with a higher proliferation rate, displaying a stable phenotype with normal karyotype and telomerase activity in long-term cultures (Tomar et al., 2010).

In vivo orthotopic bone tissue regeneration using GMSCs was also demonstrated by repairing mandiblular and critical-sized calvarial defects in Sprague–Dawley rats (Wang et al., 2011). Zhang et al. (2009) reported that these GMSCs are capable of immunomodulatory functions; specifically, they suppress peripheral blood lymphocyte proliferation and induce expression of a wide panel of immunosuppressive factors, including IL10, IDO, inducible NO synthase, and cyclooxygenase 2, in response to the inflammatory cytokine IFN-γ. This research team also tested cell-based therapy using systemic infusion of GMSCs in experimental colitis and found that this approach significantly ameliorated both the clinical and histopathological severity of colonic inflammation in mice.

8 CONCLUSIONS

Stem and progenitor cells from different locations in teeth and different stages during their development exhibit subtle differences. Their natural functions are to form tooth structures and to maintain the physiological homeostasis of the structures. They represent a specialized subpopulation of mesenchymal type of stem and progenitor cells. Although most postnatal stem cells are multipotent, dental tissue–derived stem and progenitor cells possess versatile features in regard to their potential applications in regenerative medicine. Besides their natural ability to differentiate into odontogenic cell types under the right type of environmental cues, they can also be guided to differentiate into various cell lineages, especially neurogenic and osteogenic cell types. Additionally, FENC stem cells demonstrate pluripotent capacities, indicating their tremendous potential for various clinical applications. GMSCs are also a promising cell source for various medical applications and are much easier to access than most other MSCs. Like all types of stem cells, many of their cellular and molecular aspects are still not well understood. To utilize them predictably and safely for various medical applications, further research will be required to understand their behavior and fate in vivo as well as how they are maintained as stem cells and their differentiation mechanisms in vitro.

Acknowledgments

This work was supported in part by grants from Deutsche Forschungsgemeinschaft (DFG MO1875/2-2)(to C.M.) and from the U.S. National Institutes of Health/Institute of Dental and Craniofacial Research: R01DE019156 (to G.T.-J.H.) and R01DE017449 (to S.S.).

REFERENCES

Alongi DJ, Yamaza T, Song Y, Fouad AF, Romberg EE, Shi S, et al. (2010). Stem/progenitor cells from inflamed human dental pulp retain tissue regeneration potential. *Regen Med* 5(4):617–631.

Arthur A, Rychkov G, Shi S, Koblar SA, Gronthos S (2008). Adult human dental pulp stem cells differentiate toward functionally active neurons under appropriate environmental cues. *Stem Cells* 26(7):1787–1795.

Arthur A, Shi S, Zannettino AC, Fujii N, Gronthos S, Koblar SA (2009). Implanted adult human dental pulp stem cells induce endogenous axon guidance. *Stem Cells* 27(9):2229–2237.

Beck HC, Petersen J, Felthaus O, Schmalz G, Morsczeck C (2011). Comparison of neurosphere-like cell clusters derived from dental follicle precursor cells andretinal Muller cells. *Neurochem Res* 36:2002–2007.

Carinci F, Papaccio G, Laino G, Palmieri A, Brunelli G, D'Aquino R, et al. (2008). Comparison between genetic portraits of osteoblasts derived from primary cultures and osteoblasts obtained from human pulpar stem cells. *J Craniofac Surg* 19(3):616–625.

Chai Y, Jiang X, Ito Y, Bringas P Jr, Han J, Rowitch DH, et al. (2000). Fate of the mammalian cranial neural crest during tooth and mandibular morphogenesis. *Development* 127(8):1671–1679.

d'Aquino R, Graziano A, Sampaolesi M, Laino G, Pirozzi G, De Rosa A, et al. (2007). Human postnatal dental pulp cells codifferentiate into osteoblasts and endotheliocytes: a pivotal synergy leading to adult bone tissue formation. *Cell Death Differ* 14(6):1162–1171.

d'Aquino R, Tirino V, Desiderio V, Studer M, De Angelis GC, Laino L, et al. (2011). Human neural crest–derived postnatal cells exhibit remarkable embryonic attributes either in vitro or in vivo. *Eur Cell Mater* 21:304–316.

Degistirici O, Jaquiery C, Schönebeck B, Siemonsmeier J, Götz W, Martin I, et al. (2008). Defining properties of neural crest–derived progenitor cells from the apex of human developing tooth. *Tissue Eng A* 14(2):317–330.

Degistirici O, Grabellus F, Irsen S, Schmid KW, Thie M (2010). Using human neural crest–derived progenitor cells to investigate osteogenesis: an in vitro study. *Matrix Biol* 29(3):219–227.

Diekwisch TG (2001). The developmental biology of cementum. *Int J Dev Biol* 45(5–6):695–706.

Engler AJ, Sen S, Sweeney HL, Discher DE (2006). Matrix elasticity directs stem cell lineage specification. *Cell* 126(4):677–689.

Feng F, Akiyama K, Liu Y, Yamaza T, Wang TM, Chen JH, et al. (2010). Utility of PDL progenitors for in vivo tissue regeneration: a report of 3 cases. *Oral Dis* 16(1):20–28.

Friedenstein A, Kuralesova AI (1971). Osteogenic precursor cells of bone marrow in radiation chimeras. *Transplantation* 12(2):99–108.

Gronthos S, Mankani M, Brahim J, Robey PG, Shi S (2000). Postnatal human dental pulp stem cells (DPSCs) in vitro and in vivo. *Proc Natl Acad Sci USA* 97(25):13625–13630.

Gronthos S, Brahim J, Li W, Fisher LW, Cherman N, Boyde A, et al. (2002). Stem cell properties of human dental pulp stem cells. *J Dent Res* 81(8):531–535.

Guo W, He Y, Zhang X, Lu W, Wang C, Yu H, et al. (2009). The use of dentin matrix scaffold and dental follicle cells for dentin regeneration. *Biomaterials* 30(35):6708–6723.

Guo W, Gong K, Shi H, Zhu G, He Y, Ding B, et al. (2011). Dental follicle cells and treated dentin matrix scaffold for tissue engineering the tooth root. *Biomaterials* PMID:22088889.

Han C, Yang Z, Zhou W, Jin F, Song Y, Wang Y, et al. (2010) Periapical follicle stem cell: a promising candidate for cementum/periodontal ligament regeneration and bio-root engineering. *Stem Cells Dev* 19(9):1405–1415.

Harada H, Kettunen P, Jung HS, Mustonen T, Wang YA, Thesleff I (1999). Localization of putative stem cells in dental epithelium and their association with Notch and FGF signaling. *J Cell Biol* 147(1):105–120.

Harada H, Toyono T, Toyoshima K, Yamasaki M, Itoh N, Kato S, et al. (2002). FGF10 maintains stem cell compartment in developing mouse incisors. *Development* 129(6):1533–1541.

Honda MJ, Imaizumi M, Tsuchiya S, Morsczeck C (2010). Dental follicle stem cells and tissue engineering. *J Oral Sci* 52(4):541–552.

Huang GT, Shagramanova K, Chan SW (2006). Formation of odontoblast-like cells from cultured human dental pulp cells on dentin in vitro. *J Endod* 32(11):1066–1073.

Huang GT, Sonoyama W, Liu Y, Liu H, Wang S, Shi S (2008). The hidden treasure in apical papilla: the potential role in pulp/dentin regeneration and bioroot engineering. *J Endod* 34(6):645–51. Review. PubMed PMID: 18498881; PubMed Central PMCID: PMC2653220.

Huang GT, Gronthos S, Shi S (2009). Mesenchymal stem cells derived from dental tissues vs. those from other sources: their biology and role in regenerative medicine. *J Dent Res* 88(9):792–806.

Ikeda E, Morita R, Nakao K, Ishida K, Nakamura T, Takano-Yamamoto T, et al. (2009). Fully functional bioengineered tooth replacement as an organ replacement therapy. *Proc Natl Acad Sci USA* 106(32):13475–13480.

Iohara K, Imabayashi K, Ishizaka R, Watanabe A, Nabekura J, Ito M, et al. (2011). Complete pulp regeneration after pulpectomy by transplantation of $CD105^+$ stem cells with stromal cell–derived factor-1. *Tissue Eng A* 17(15–16):1911–1920.

Jo YY, Lee HJ, Kook SY, Choung HW, Park JY, Chung JH, et al. (2007). Isolation and characterization of postnatal stem cells from human dental tissues. *Tissue Eng* 13(4):767–773.

Karaöz E, Doğan BN, Aksoy A, Gacar G, Akyüz S, Ayhan S, et al. (2010). Isolation and in vitro characterisation of dental pulp stem cells from natal teeth. *Histochem Cell Biol* 133(1):95–112.

Kémoun P, Laurencin-Dalicieux S, Rue J, Farges JC, Gennero I, Conte-Auriol F, et al. (2007). Human dental follicle cells acquire cementoblast features under stimulation by BMP-2/-7 and enamel matrix derivatives (EMD) in vitro. *Cell Tissue Res* 329(2):283–294.

Kerkis I, Kerkis A, Dozortsev D, Stukart-Parsons GC, Gomes Massironi SM, Pereira LV, et al. (2006). Isolation and characterization of a population of immature dental pulp stem cells expressing OCT-4 and other embryonic stem cell markers. *Cells Tissues Organs* 184(3–4):105–116.

Laino G, d'Aquino R, Graziano A, Lanza V, Carinci F, Naro F, et al. (2005). A new population of human adult dental pulp stem cells: a useful source of living autologous fibrous bone tissue (LAB). *J Bone Miner Res* 20(8):1394–1402.

Liu Y, Zheng Y, Ding G, Fang D, Zhang C, Bartold PM, et al. (2008). Periodontal ligament stem cell–mediated treatment for periodontitis in miniature swine. *Stem Cells* 26(4):1065–1073.

Marynka-Kalmani K, Treves S, Yafee M, Rachima H, Gafni Y, Cohen MA, et al. (2010). The lamina propria of adult human oral mucosa harbors a novel stem cell population. *Stem Cells* 28(5):984–995.

McCulloch CA, Bordin S (1991). Role of fibroblast subpopulations in periodontal physiology and pathology. *J Periodont Res* 26:144–154.

Mitsiadis TA, Barrandon O, Rochat A, Barrandon Y, De Bari C (2007). Stem cell niches in mammals. *Exp Cell Res* 313(16):3377–3385.

Miura M, Gronthos S, Zhao M, Lu B, Fisher LW, Robey PG, et al. (2003). SHED: stem cells from human exfoliated deciduous teeth. *Proc Natl Acad Sci USA* 100(10):5807–5812.

Morrison SJ, Spradling AC (2008). Stem cells and niches: mechanisms that promote stem cell maintenance throughout life. *Cell* 132(4):598–611.

Morsczeck C, Götz W, Schierholz J, Zeilhofer F, Kühn U, Möhl C, et al. (2005). Isolation of precursor cells (PCs) from human dental follicle of wisdom teeth. *Matrix Biol* 24(2):155–165.

Ohazama A, Modino SA, Miletich I, Sharpe PT (2004) Stem-cellbased tissue engineering of murine teeth. *J Dent Res* 83:518–522.

Park JC, Kim JM, Jung IH, Kim JC, Choi SH, Cho KS, et al. (2011). Isolation and characterization of human periodontal ligament (PDL) stem cells (PDLSCs) from the inflamed PDL tissue: in vitro and in vivo evaluations. *J Clin Periodontol* 38(8):721–731.

Sakai K, Yamamoto A, Matsubara K, Nakamura S, Naruse M, Yamagata M et al. (2012). Human dental pulp-derived stem cells promote locomotor recovery after complete transection of the rat spinal cord by multiple neuro-regenerative mechanisms. *J Clin Invest* 122(1):80–90. doi: 10.1172/JCI59251. Epub 2011 Dec 1.

Sakai VT, Cordeiro MM, Dong Z, Zhang Z, Zeitlin BD, Nör JE (2011). Tooth slice/scaffold model of dental pulp tissue engineering. *Adv Dent Res* 23(3):325–332.

Saugspier M, Felthaus O, Viale-Bouroncle S, Driemel O, Reichert TE, Schmalz G, et al. (2010). The differentiation and gene expression profile of human dentalfollicle cells. *Stem Cells Dev* 19(5):707–717.

Seo BM, Miura M, Gronthos S, Bartold PM, Batouli S, Brahim J, et al. (2004). Investigation of multipotent postnatal stem cells from human periodontal ligament. *Lancet* 364(9429):149–155.

Shi S, Gronthos S (2003). Perivascular niche of postnatal mesenchymal stem cells in human bone marrow and dental pulp. *J Bone Miner Res* 18(4):696–704.

Sonoyama W, Liu Y, Fang D, Yamaza T, Seo BM, Zhang C, et al. (2006). Mesenchymal stem cell–mediated functional tooth regeneration in swine. *PLoS One* 20(1):e79.

Sonoyama W, Liu Y, Yamaza T, Tuan RS, Wang S, Shi S, et al. (2008). Characterization of the apical papilla and its residing stem cells from human immature permanent teeth: a pilot study. *J Endod* 34(2):166–171.

Tamaoki N, Takahashi K, Tanaka T, Ichisaka T, Aoki H, Takeda-Kawaguchi T, et al. (2010). Dental pulp cells for induced pluripotent stem cell banking. *J Dent Res* 89(8):773–778.

Tomar GB, Srivastava RK, Gupta N, Barhanpurkar AP, Pote ST, Jhaveri HM, et al. (2010). Human gingiva-derived mesenchymal stem cells are superior to bone marrow–derived mesenchymal stem cells for cell therapy in regenerative medicine. *Biochem Biophys Res Commun* 393(3):377–383.

Tomita M, Mori T, Maruyama K, Zahir T, Ward M, Umezawa A, et al. (2006). A comparison of neural differentiation and retinal transplantation with bone marrow–derived cells and retinal progenitor cells. *Stem Cells* 24(10):2270–2278.

Viale-Bouroncle S, Völlner F, Möhl C, Küpper K, Brockhoff G, Reichert TE, et al. (2011). Soft matrix supports osteogenic differentiation of human dental follicle cells. *Biochem Biophys Res Commun* 410(3):587–592.

Völlner F, Ernst W, Driemel O, Morsczeck C (2009). A two-step strategy for neuronal differentiation in vitro of human dental follicle cells. *Differentiation* 77(5):433–441.

Wada N, Wang B, Lin NH, Laslett AL, Gronthos S, Bartold PM (2011). Induced pluripotent stem cell lines derived from human gingival fibroblasts and periodontal ligament fibroblasts. *J Periodont Res* 46(4):438–447.

Wang F, Yu M, Yan X, Wen Y, Zeng Q, Yue W, et al. (2011). Gingiva-derived mesenchymal stem cell–mediated therapeutic approach for bone tissue regeneration. *Stem Cells Dev* 20(12):2093–2102.

Yan X, Qin H, Qu C, Tuan RS, Shi S, Huang GT-J (2010). iPS cells reprogrammed from human mesenchymal-like stem/progenitor cells of dental tissue origin. *Stem Cells Dev* 19(4):469–480.

Yokohama-Tamaki T, Ohshima H, Fujiwara N, Takada Y, Ichimori Y, Wakisaka S, et al. (2006). Cessation of Fgf10 signaling, resulting in a defective dental epithelial stem cell compartment, leads to the transition from crown to root formation. *Development* 133(7):1359–1366.

Zhang Q, Shi S, Liu Y, Uyanne J, Shi Y, Le AD (2009). Mesenchymal stem cells derived from human gingiva are capable of immunomodulatory functions and ameliorate inflammation-related tissue destruction in experimental colitis. *J Immunol* 183(12):7787–7798.

16

REGULATION AND DIFFERENTIATION POTENTIAL OF DENTAL MESENCHYMAL STEM CELLS

LEI WANG

Center for Craniofacial Molecular Biology, Herman Ostrow School of Dentistry, University of Southern California, Los Angeles, California; and Department of Oral and Maxillofacial Surgery, School of Stomatology, Fourth Military Medical University, Shaanxi, China

CHRISTIAN MORSCZECK

Department of Cranio- and Maxillofacial Surgery, University Hospital Regensburg, Regensburg, Germany

STAN GRONTHOS

Mesenchymal Stem Cell Laboratory, Department of Haematology, University of Adelaide, Adelaide, South Australia, Australia

SONGTAO SHI

Center for Craniofacial Molecular Biology, Herman Ostrow School of Dentistry, University of Southern California, Los Angeles, California

1 INTRODUCTION

Mesenchymal stem cells (MSCs) and hematopoietic stem cells (HSCs) both contribute to tissue homeostasis and repair through differentiation. Control over MSC differentiation potential is therefore a crucial regulatory and clinical requirement for the retention of MSC functions (Augello and De Bari, 2010; Medema and Vermeulen, 2011; Wang and Wagers, 2011). Accumulating evidence has shown that the MSC differentiation potential can be regulated at the cellular, molecular, and transcriptional levels (Nishimura et al., 2008; Augello and De Bari, 2010; Liu et al., 2011a,b). Reports have also shown that physical factors, including bioscaffolds and extracellular matrix, as well as newly emerged epigenetic regulation, contribute

Stem Cells in Craniofacial Development and Regeneration, First Edition.
Edited by George T.-J. Huang, Irma Thesleff.

to MSC differentiation (Engler et al., 2006; Lakshmipathy and Hart, 2008; Fan et al., 2009). In this chapter we summarize the regulatory factors that influence the differentiation potential of dental MSCs, including dental pulp stem cells (DPSCs), stem cells from human exfoliated deciduous teeth (SHEDs), periodontal ligament stem cells (PDLSCs), dental follicle cells (DFCs), stem cells from root apical papilla (SCAPs), and gingival-derived mesenchymal stem cells (GMSCs).

2 DIFFERENTIATION POTENTIAL OF DENTAL MESENCHYMAL STEM CELLS

Dental MSCs can be differentiated into multiple lineages, as summarized in Table 1. To facilitate the selection of dental MSCs for clinical applications, a few studies compared the differentiation potential among different types of dental MSCs. For example, Yu et al. (2007) used in vitro culture assays to compare the odontogenic capability of dental and nondental stem cell populations, and they found that DPSCs presented more striking odontogenic capability than bone marrow stromal stem cells (BMSSCs) under the induction of postnatal apical bud cells (ABCs). Lei et al. (2011) found further that immature root papilla stem cells (iRPSCs) presented stronger dentinogenesis, but weaker osteogenesis, than those of mature root pulp stem cells (mRPSCs), suggesting that the dentinogenic competence of root MSCs decreases, whereas their osteogenic potential increases following maturation of the tooth root.

Since dental MSCs are potentially derived from neural crest cells, their neurogenic potential has attracted attention. For example, SHEDs are able to improve the apomorphine-evoked rotation of behavioral disorders in parkinsonian rat models by differentiating into dopaminergic neurons(Wang et al., 2010). Although SHEDs and DFCs can both differentiate into neurogenic cells under cell culture conditions, they had different neural cell marker expression patterns. For example, gene expression of the late neural cell marker microtubule-associated protein 2 was upregulated in DFCs but downregulated in SHEDs. In contrast, SHEDs formed neurosphere-like cell clusters and expressed the glial cell marker glial fibrillary acidic protein, which, in contrast, was expressed weakly in DFCs or not at all (Morsczeck et al., 2010).

TABLE 1 Differentiation Potential of Dental Mesenchymal Stem Cells

	Differentiation Potential					
Cell Type	Osteoblast	Cartilage	Adipocyte	Neuron	Dentin	Periodontium–Cementum
DPSC	+	+	+	+	+	−
SHED	+	+	+	+	−	−
PDLS	+	+	+	+	−	+
SCAF	+	+	+	+	+	−
DFC	+	+	+	+	+	+
OMSC	+	+	+	+	−	−
GMSC	+	N/A	+	+	−	−

3 REGULATION OF STEM CELL DIFFERENTIATION BY T-CELLS

It is well established that MSCs maintain their stemness with the ability to self-renew and differentiation, thereby having the ability to replace damaged and diseased tissue. In addition, MSCs, including dental MSCs, either alone or in combination with biomaterial scaffolds (Seo et al., 2008; Bueno and Glowacki, 2009; Tasso et al., 2010), are able to inhibit proinflammatory cytokines by targeting several subsets of immune cells, including T-cells, B-cells, dendritic cells (DCs), and natural killer (NK) cells (Ren et al., 2008; Zhao et al., 2010). This immunoregulatory property of MSCs provides the clinical foundation for treating a variety of autoimmune diseases, such as acute graft-versus-host disease and systemic lupus erythematosus (SLE), as well as promoting engraftment (Le Blanc et al., 2008; Sun et al., 2009; English et al., 2010). Several types of dental mesenchymal stem cells have been demonstrated to present immunoregulatory properties. SHEDs can inhibit secretion of IL17 in vitro, and they are capable of effectively reversing SLE-associated disorders in MRL/lpr mice by elevating the ratio of regulatory T-cells (Tregs) to Th17 cells (Yamaza et al., 2010). Ding et al. (2010) demonstrated that swine SCAPs could suppress T-cell proliferation in vitro through an apoptosis-independent mechanism. PDLSCs and DPSCs also possess immunosuppressive properties mediated, in part, by soluble factors such as transforming growth factor β1 (TGFβ1), hepatocyte growth factor (HGF), and indoleamine 2,3-dioxygenase (IDO) produced by activated peripheral blood mononuclear cells (PBMNCs) (Wada et al., 2009). Additionally, GMSCs are capable of suppressing peripheral blood lymphocyte proliferation by inducing expression of immunosuppressive factors, including interleukin-10 (IL10), IDO, inducible NO synthase, and cyclooxygenase 2 (COX-2) (Zhang et al., 2009). Conversely, immune components can also regulate the differentiation potential of MSCs. Suzawa et al. indicated that TNFα inhibits adipogenesis and osteoblastogenesis in MSCs (Suzawa et al., 2003; Lu et al., 2011). Interleukin-2–activated NK cells and T-cells activated by CD3 and CD28 can induce bone marrow MSC apoptosis through the Fas/Fas ligand (FasL) pathway (Spaggiari et al., 2006; Yamaza et al., 2008). Moreover, activated T-lymphocytes also impair orofacial bone/bone marrow–derived MSCs (OMSCs) via the Fas/FasL pathway, suggesting that OMSCs play an important role in systemic immunity (Yamaza et al., 2011). However, the specific function of the recipient cells, especially immune cells, in regulating stem cell differentiation remains largely unknown.

Most recently, Liu et al. (2011b) reported that host lymphocytes secrete interferon-γ (IFN-γ) and tumor necrosis factor-α (TNFα) to initiate apoptosis and/or inhibit osteogenic potential of transplanted MSCs, and that Tregs and aspirin can alleviate these effects to improve bone repair, respectively. By grafting MSCs together with carrier hydroxyapatite tricalcium phosphate (HA-TCP) particles subcutaneously, the authors found that no bone formation occurred when MSCs were implanted into immune-competent mice in which IFN-γ and TNF-α produced by distinct T-cell subsets played a critical role in determining the outcome for the implanted MSCs, whereas the implants into immune-incompetent nude mice showed significant bone growth. When MSCs were treated in vitro with IFN-γ alone, or with a combination of IFN-γ and TNFα, the authors found that IFN-γ alone blocked osteogenic differentiation by inducing upregulation of smad6, thereby inhibiting Runt-related transcription factor 2 (Runx2), a key transcription factor associated with osteoblast differentiation. In contrast, the combination of IFN-γ and TNFα induced bone marrow MSC apoptosis

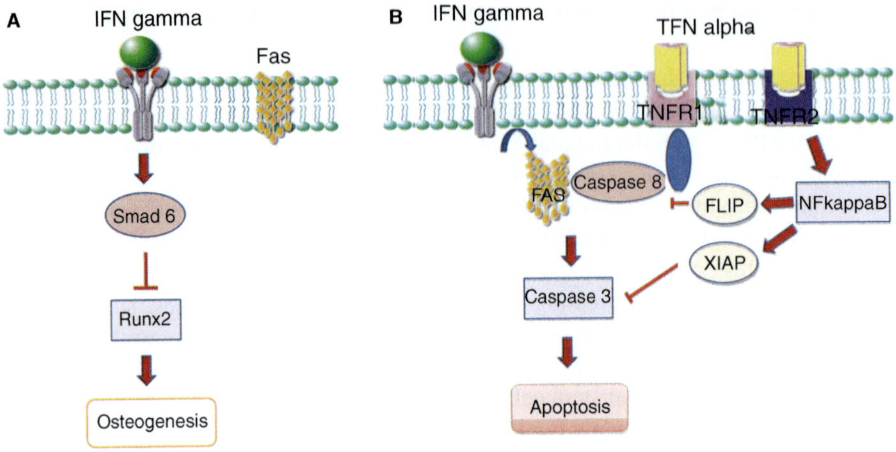

FIGURE 1 Schematic diagram of regulatory effects of proinflammatory cytokines on MSCs. (A) IFNγ blocks osteogenic differentiation by inducing upregulation of smad6, which inhibits Runx2 expression. (B) Combination of IFNγ and TNFα induces MSC apoptosis through internalization of Fas, which is enhanced when there is a reduction of the antiapoptotic factors NF-κB, XIAP, and FLIP.

through internalization of Fas receptor, which is a death receptor also known as tumor necrosis factor receptor superfamily member 6, with reduction of the antiapoptotic factors nuclear factor kappa-light-chain enhancer of activated B-cells (NF-κB), X-linked inhibitor of apoptosis protein (XIAP), and FLICE-like inhibitory protein (FLIP) (Fig. 1).

4 MOLECULAR REGULATION OF DENTAL MESENCHYMAL STEM CELL DIFFERENTIATION

Studies have shown that several molecules are involved in the regulation of dental MSC differentiation. In this chapter we focus on the major signaling pathways involved in the modulation of differentiation potential in dental MSCs (Fig. 2).

Both canonical and noncanonical Wnt signaling pathways play a critical role in tooth development and stem cell self-renewal through β-catenin (Gong et al., 2001; Chang et al., 2007; Ling et al., 2009). For example, Wnt4 enhances in vitro osteogenic differentiation of PDLSCs via activated p38 mitogen-activated protein kinase (MAPK) in a novel noncanonical signaling pathway (Chang et al., 2007). Surprisingly, canonical Wnt signaling negatively regulates the odontoblast-like differentiation of DPSCs, inhibits cementoblast differentiation, and promotes cell proliferation. Scheller et al. (2008) found that Wnt1 inhibited alkaline phosphatase (ALP) activity and the formation of mineralized nodules in DPSCs and that overexpression of β-catenin was also sufficient to suppress the mineralization of DPSCs. This conclusion was supported by the fact that Wnt signaling inhibits cementogenesis and promotes cell proliferation in cementoblasts (Nemoto et al., 2009).

The TGFβ superfamily consists of many growth factors, including bone morphogenetic proteins (BMPs), growth and differentiation factors (GDFs), Activin, Nodal and

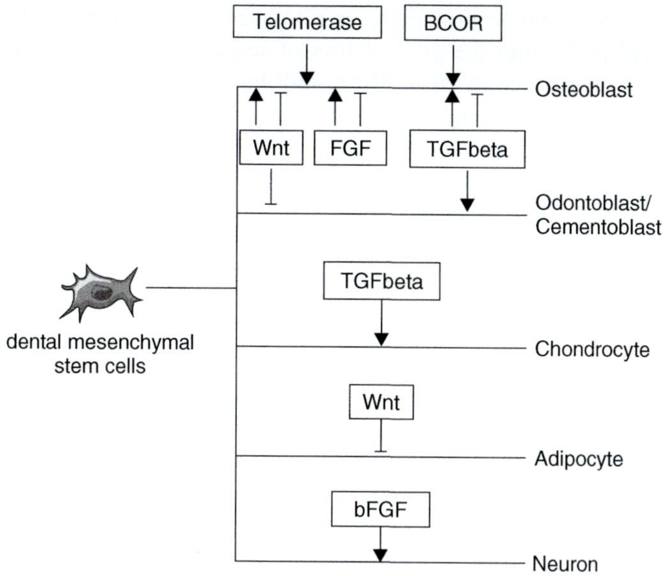

FIGURE 2 Signaling regulation of dental stem cell differentiation. Various signaling molecules are involved in either positive or negative regulation of dental stem cell differentiation into multiple cell lineages.

TGFβ, all of which play roles in developmental skeletogenesis and postnatal skeletal homeostasis (Piek et al., 1999). Here we concentrate on the regulatory effects of TGFβ and BMPs on dental MSC differentiation.

The regulatory effects of TGFβ are diverse. On the one hand, Miura et al. (2004) found delayed ossification and decreased bone mineral density in caspase-3–deficient mice caused by attenuated osteogenic differentiation of bone marrow MSCs. They further demonstrated that caspase-3 is crucial for the differentiation of bone marrow MSCs by influencing the TGFβ/Smad2 pathway and cell cycle progression. On the other hand, TGFβ3 has been proven able to induce ectopic mineralization through upregulation of osteocalcin and type I collagen expression in DPSCs, and it may also regulate the differentiation of DPSCs into odontoblasts (Huojia et al., 2005). As members of TGFβ superfamily, BMPs show positive effects on MSC differentiation. For example, BMP2 could enhance osteogenic and chondrogenic differentiation of MSCs, partly modulated by N-cadherin, a Ca^{2+}-dependent adhesion molecule (Rickard et al., 1994; Sekiya et al., 2005).

Fibroblast growth factors (FGFs) participate in both hematopoiesis and osteogenesis by maintaining osteogenic precursors in human bone marrow (Martin et al., 1997). FGF2 could enhance ALP activity, osteogenic marker expression and osteogenic differentiation in human dental pulp cells; thus, it plays a role as a differentiation-inducing factor in the injury repair processes of pulpal tissue (Kim et al., 2010b). Basic fibroblast growth factors (bFGFs) could regulate such DPSC behavior as maintaining stemness and directing differentiation. Specifically, bFGFs could inhibit osteogenesis, whereas they could enhance neurogenic differentiation through a FGFR and phospholipase C-γ signaling pathway in DPSCs (Osathanon et al., 2011).

Telomerase is a ribonucleoprotein reverse transcriptase responsible for elongating the telomere and preventing the gradual loss of sequence from chromosome ends to avoid replicative senescence in vitro. Most adult human somatic cells lack significant telomerase activity, whereas highly proliferative germline cells, hematopoietic stem cells, and many cancer cells retain detectable levels of telomerase activity (Nakamura et al., 1997). It has been shown that ectopic expression of human telomerase reverse transcriptase in bone marrow MSCs could extend their lifespan and maintain their osteogenic potential, thus providing a useful strategy for bone regeneration and repair (Shi et al., 2002). It was further demonstrated that telomerase-expressing bone marrow MSCs could yield more bone formation both in vitro and in vivo and that the underlying mechanism might be acceleration of cell cycle progression from the G1 to the S phase, as well as an upregulation of important osteogenic genes, such as core-binding factor (CBFA-1, also called Runx2), osterix, and osteocalcin (Gronthos et al., 2003).

The BCL-6 co-repressor (BCOR)/activating protein 2α (AP−2α) pathway also contributes to differentiation of dental MSCs. By studying the oculo-facio-cardio-dental (OFCD) syndrome, a rare human genetic disease with BCOR mutation characterized by canine teeth with extremely long roots, Fan et al. (2009) found that BCOR mutation increased the osteo-dentinogenic potential of SCAPs through abnormal activation of AP−2α, a key transcription factor for osteo-dentinogenic differentiation.

Other signaling molecules are also involved in the differentiation pathways of dental MSCs. Statin, 3-hydroxy-3-methylglutaryl coenzyme A reductase inhibitor, could markedly upregulate osteocalcin and dentin sialophosphoprotein expression in DPSCs and might be an ideal active ingredient to accelerate the differentiation of DPSCs (Okamoto et al., 2009). Odontoblastic differentiation and growth of human DPSCs are positively regulated by heme oxygenase1 (HO1) activity induction(Kim et al., 2010a), vascular endothelial growth factor (VEGF) (I et al., 2011) and twist-related protein 1 (Twist1) (Li et al., 2011). In contrast, CD271 was demonstrated to inhibit the differentiation of deciduous dental pulp stem cells (DDPSCs) (Mikami et al., 2011).

5 TRANSCRIPTIONAL REGULATION OF DENTAL MESENCHYMAL STEM CELL DIFFERENTIATION

Herein, we summarize the transcription factors that influence the differentiation of dental MSCs. In particular, the principal transcription factors involved in osteogenic differentiation are CBFA-1/Runx2 and Osterix (Osx) (Komori, 2010), whereas Sox9 is a transcription factor required for cartilage formation (Akiyama et al., 2004), and CCAAT-enhancer-binding proteins (C/EBPs) and peroxisome proliferator-activated receptor γ (PPARγ) are considered to control adipogenic differentiation (Nishimura et al., 2008).

A large body of evidence indicates an essential role of CBFA-1/Runx2 and Osx in bone formation and osteoblastogenesis (Komori, 2010). TNFα inhibits osteogenesis by activation of the paired related homeobox protein (Prx1), which, in turn, inhibits transcription of Osx and Runx2 (Lu et al., 2011). Recently, transcription factor DLX3 was proven able to influence cell viability and stimulate the osteogenic differentiation of DFCs via the BMP2/Smad1 pathway (Viale-Bouroncle et al., 2012). Moreover, transcription factors SP1 and TP53 were found to influence cell proliferation and differentiation in both SHEDs and DFCs (Felthaus et al., 2012).

Sox9, which contains an SRY-related high-mobility group box, plays a critical role in the commitment of mesenchymal cells into chondrocytes and the expression of cartilage-specific extracellular matrix genes, including type II collagen, type XI collagen, and aggrecan (de Crombrugghe et al., 2001). The functional and physical crosstalk between Sox9 and Wnt signaling during chondrogenesis differentiation has also been demonstrated (Akiyama et al., 2004).

The C/EBP and PPAR family members play crucial roles in the regulation of adipogenesis. C/EBPβ and C/EBPσ regulate PPARγ, thereby promoting the adipogenic differentiation of MSCs (Wu et al., 1995). TNFα inhibits adipogenesis through inhibition of PPAR-γ (Suzawa et al., 2003). It has been demonstrated that the switch between adipogenic and osteogenic differentiation may be attributed to some transcription factors. For example, a naturally occurring isoform of C/EBP, C/EBPbeta-liver inhibitory protein (LIP), functions as a strong dominant-negative mutant for C/EBP family members and inhibits adipogenesis (Hata et al., 2005). In addition, Msx2 has reciprocal roles in osteogenesis and adipogenesis (Ichida et al., 2004).

6 EPIGENETIC REGULATION OF DENTAL MESENCHYMAL STEM CELL DIFFERENTIATION BY MICRORNAs

Epigenetic regulation, such as covalent modification of histones, DNA methylation, or regulation of gene expression by microRNAs, acts in chromatin remodeling that controls structural changes of DNA required for the binding of transcription factors to promoter regions (Benayahu et al., 2009). Mutation of the BCOR pathway results in epigenetic mechanisms of histone H3K4 and H3K36 methylation in SCAPs, thereby enhancing the osteodentinogenic potential of SCAPs through reactivating transcription of silenced target genes such as AP−2α (Fan et al., 2009).

MicroRNAs (miRNAs) are single-stranded RNAs of 19 to 23 nucleotides that derive from nucleotide precursors and are found in a wide variety of organisms, from plants to humans. After transcription, microRNA precursors are cleaved in the nucleus, exported to cytoplasm, and inserted into an RNA-induced silencing complex after additional cleavage (Lakshmipathy and Hart, 2008). They are crucial for self-renewal and the behavior of embryonic stem cells (Wang et al., 2007). Furthermore, accumulating evidence has demonstrated the requirement of miRNAs in controlling key steps of MSC differentiation into certain cell lineages (Lakshmipathy and Hart, 2008).

Quite a number of studies have identified the roles of miRNAs in MSC differentiation or targeting pathways. For example, modulation of miR-143 and microRNA-369-5p could promote the formation of adipocytes from MSCs (Esau et al., 2004; Bork et al., 2011). Interestingly, Michon et al. (2010) found that conditional deletion of Dicer-1 in the epithelium showed impaired ameloblast differentiation, suggesting that miRNAs modulate tooth morphogenesis largely by fine-tuning conserved signaling networks. Very recently, Liu et al. (2011a) isolated PDLSCs from periodontitis-affected periodontal ligament tissue and found that excessive inflammatory cytokine levels, miR-17, and Smad ubiquitin regulatory factor 1 (Smurf1), an important negative regulator of MSC osteogenic differentiation, were all involved in a coherent feedforward loop. Inflammatory cytokines led to downregulation of miR-17 and direct activation of Smurf1, thereby downregulating the osteogenic differentiation of PDLSCs.

7 PHYSICAL FACTORS IN DENTAL MESENCHYMAL STEM CELL DIFFERENTIATION

Since MSCs interact physically with the surrounding environment, physical parameters such as bioscaffolds and extracellular matrix also contribute to the regulation of dental MSCs differentiation. Many studies have reported on the regulatory effects of bioscaffolds on stem cell differentiation. For example, nanofibrous poly(L-lactic acid) (PLLA) scaffolds are advantageous in promoting odontogenic differentiation and hard tissue formation over solid-walled PLLA scaffolds (Wang et al., 2011). Other dental materials, including 2-hydroxyethyl methacrylate, triethylene glycol dimethacrylate, three-dimensional calcium phosphate porous granules, and zinc chloride, were also proved to facilitate odontogenic differentiation of DPSCs (Bakopoulou et al., 2011; Lin et al., 2011; Nam et al., 2011).

In addition, extracellular matrix provides microenvironments that are important for stem cell lineage specification. It has been shown that MSCs committed to phenotypes are extremely sensitive to tissue-level elasticity, suggesting that the physical effects of in vivo microenvironments are important for the therapeutic use of stem cells (Engler et al., 2006). Dentin matrix protein 1 (DMP1) is a dentin noncollagenous extracellular matrix protein involved in the regulation of mineralization. Almushayt et al. (2006) have demonstrated that DMP1 induces differentiation of DPSCs into odontoblast-like cells and formation of reparative dentin in a rat model. In another example, collagen type I matrix could enhance proliferation and osteogenic differentiation of both DFCs and PDLSCs (Tsuchiya et al., 2008). Futhermore, controlling the membrane potential of MSCs can influence their differentiation (Sundelacruz et al., 2008).

8 SUMMARY

In conclusion, the dental MSC differentiation is regulated intensively at the cellular, molecular, and transcriptional levels. A deeper understanding of the mechanisms underlying dental MSC differentiation and its regulation will benefit the successful development of dental stem cell–based therapeutic approaches.

Acknowledgments

This work was supported by grants from the U.S. National Institute of Dental and Craniofacial Research, National Institutes of Health (R01DE017449, R01DE019932, and R01DE019413 to S.S.), a grant from the California Institute for Regenerative Medicine (RN1-00572 to S.S.), and by the Intramural Program of the National Institute of Dental and Craniofacial Research.

REFERENCES

Akiyama H, Lyons JP, Mori-Akiyama Y, Yang X, Zhang R, Zhang Z, et al. (2004). Interactions between Sox9 and beta-catenin control chondrocyte differentiation. *Genes Dev* 18(9):1072–1087.

Almushayt A, Narayanan K, Zaki AE, George A (2006). Dentin matrix protein 1 induces cytodifferentiation of dental pulp stem cells into odontoblasts. *Gene Ther* 13(7):611–620.

Augello A, De Bari C (2010). The regulation of differentiation in mesenchymal stem cells. *Hum Gene Ther* 21(10):1226–1238.

Bakopoulou A, Leyhausen G, Volk J, Tsiftsoglou A, Garefis P, Koidis P, et al. (2011). Effects of HEMA and TEDGMA on the in vitro odontogenic differentiation potential of human pulp stem/progenitor cells derived from deciduous teeth. *Dent Mater* 27(6):608–617.

Benayahu D, Shefer G, Shur I (2009). Insights into the transcriptional and chromatin regulation of mesenchymal stem cells in musculo-skeletal tissues. *Ann Anat* 191(1):2–12.

Bork S, Horn P, Castoldi M, Hellwig I, Ho AD, Wagner W (2011). Adipogenic differentiation of human mesenchymal stromal cells is down-regulated by microRNA-369-5p and up-regulated by microRNA-371. *J Cell Physiol* 226(9):2226–2234.

Bueno EM, Glowacki J (2009). Cell-free and cell-based approaches for bone regeneration. *Nat Rev Rheumatol* 5(12):685–697.

Chang J, Sonoyama W, Wang Z, Jin Q, Zhang C, Krebsbach PH, et al. (2007). Noncanonical Wnt-4 signaling enhances bone regeneration of mesenchymal stem cells in craniofacial defects through activation of p38 MAPK. *J Biol Chem* 282(42):30938–30948.

de Crombrugghe B, Lefebvre V, Nakashima K (2001). Regulatory mechanisms in the pathways of cartilage and bone formation. *Curr Opin Cell Biol* 13(6):721–727.

Ding G, Liu Y, An Y, Zhang C, Shi S, Wang W, et al. (2010). Suppression of T cell proliferation by root apical papilla stem cells in vitro. *Cells Tissues Organs* 191(5):357–364.

Engler AJ, Sen S, Sweeney HL, Discher DE (2006). Matrix elasticity directs stem cell lineage specification. *Cell* 126(4):677–689.

English K, French A, Wood KJ (2010). Mesenchymal stromal cells: Facilitators of successful transplantation? *Cell Stem Cell* 7(4):431–442.

Esau C, Kang X, Peralta E, Hanson E, Marcusson EG, Ravichandran LV, et al. (2004). MicroRNA-143 regulates adipocyte differentiation. *J Biol Chem* 279(50):52361–52365.

Fan Z, Yamaza T, Lee JS, Yu J, Wang S, Fan G, et al. (2009). BCOR regulates mesenchymal stem cell function by epigenetic mechanisms. *Nat Cell Biol* 11(8):1002–1009.

Felthaus O, Viale-Bouroncle S, Driemel O, Reichert TE, Schmalz G, Morsczeck C (2012). Transcription factors TP53 and SP1 and the osteogenic differentiation of dental stem cells. *Differentiation* 83(1):10–16.

Gong Y, Slee RB, Fukai N, Rawadi G, Roman-Roman S, Reginato AM, et al. (2001). LDL receptor–related protein 5 (LRP5) affects bone accrual and eye development. *Cell* 107(4):513–523.

Gronthos S, Chen S, Wang CY, Robey PG, Shi S (2003). Telomerase accelerates osteogenesis of bone marrow stromal stem cells by upregulation of CBFA1, osterix, and osteocalcin. *J Bone Miner Res* 18(4):716–722.

Hata K, Nishimura R, Ueda M, Ikeda F, Matsubara T, Ichida F, et al. (2005). A CCAAT/enhancer binding protein beta isoform, liver-enriched inhibitory protein, regulates commitment of osteoblasts and adipocytes. *Mol Cell Biol* 25(5):1971–1979.

Huojia M, Muraoka N, Yoshizaki K, Fukumoto S, Nakashima M, Akamine A, et al. (2005). TGF-beta3 induces ectopic mineralization in fetal mouse dental pulp during tooth germ development. *Dev Growth Differ* 47(3):141–152.

I DA, Nargi E, Mastrangelo F, Falco G, Lanuti P, Marchisio M, et al. (2011). Vascular endothelial growth factor enhances in vitro proliferation and osteogenic differentiation of human dental pulp stem cells. *J Biol Regul Homeost Agents* 25(1):57–69.

Ichida F, Nishimura R, Hata K, Matsubara T, Ikeda F, Hisada K, et al. (2004). Reciprocal roles of MSX2 in regulation of osteoblast and adipocyte differentiation. *J Biol Chem* 279(32):34015–34022.

Kim SJ, Min KS, Ryu HW, Lee HJ, Kim EC (2010a). The role of heme oxygenase-1 in the proliferation and odontoblastic differentiation of human dental pulp cells. *J Endod* 36(8):1326–1331.

Kim YS, Min KS, Jeong DH, Jang JH, Kim HW, Kim EC (2010b). Effects of fibroblast growth factor-2 on the expression and regulation of chemokines in human dental pulp cells. *J Endod* 36(11):1824–1830.

Komori T (2010). Regulation of bone development and extracellular matrix protein genes by RUNX2. *Cell Tissue Res* 339(1):189–195.

Lakshmipathy U, Hart RP (2008). Concise review: microRNA expression in multipotent mesenchymal stromal cells. *Stem Cells* 26(2):356–363.

Le Blanc K, Frassoni F, Ball L, Locatelli F, Roelofs H, Lewis I, et al. (2008). Mesenchymal stem cells for treatment of steroid-resistant, severe, acute graft-versus-host disease: a phase II study. *Lancet* 371(9624):1579–1586.

Lei G, Yan M, Wang Z, Yu Y, Tang C, Yu J, et al. (2011). Dentinogenic capacity: immature root papilla stem cells versus mature root pulp stem cells. *Biol Cell* 103(4):185–196.

Li Y, Lu Y, Maciejewska I, Galler KM, Cavender A, D'Souza RN (2011). TWIST1 promotes the odontoblast-like differentiation of dental stem cells. *Adv Dent Res* 23(3):280–284.

Lin CY, Lin HH, Tsai MH, Lin SP, Chen MH (2011). Zinc chloride for odontogenesis of dental pulp stem cells via metallothionein up-regulation. *J Endod* 37(2):211–216.

Ling L, Nurcombe V, Cool SM (2009). Wnt signaling controls the fate of mesenchymal stem cells. *Gene* 433(1–2):1–7.

Liu Y, Liu W, Hu C, Xue Z, Wang G, Ding B, et al. (2011a). MiR-17 modulates osteogenic differentiation through a coherent feed-forward loop in mesenchymal stem cells isolated from periodontal ligaments of patients with periodontitis. *Stem Cells* 29(11):1804–1816.

Liu Y, Wang L, Kikuiri T, Akiyama K, Chen C, Xu X, et al. (2011b). Mesenchymal stem cell–based tissue regeneration is governed by recipient T lymphocytes via IFN-gamma and TNF-alpha. *Nat Med* 17(12):1594–1601.

Lu X, Beck GR, Jr., Gilbert LC, Camalier CE, Bateman NW, Hood BL, et al. (2011). Identification of the homeobox protein Prx1 (MHox, Prrx-1) as a regulator of osterix expression and mediator of tumor necrosis factor alpha action in osteoblast differentiation. *J Bone Miner Res* 26(1):209–219.

Martin I, Muraglia A, Campanile G, Cancedda R, Quarto R (1997). Fibroblast growth factor-2 supports ex vivo expansion and maintenance of osteogenic precursors from human bone marrow. *Endocrinology* 138(10):4456–4462.

Medema JP, Vermeulen L (2011). Microenvironmental regulation of stem cells in intestinal homeostasis and cancer. *Nature* 474(7351):318–326.

Michon F, Tummers M, Kyyronen M, Frilander MJ, Thesleff I (2010). Tooth morphogenesis and ameloblast differentiation are regulated by micro-RNAs. *Dev Biol* 340(2):355–368.

Mikami Y, Ishii Y, Watanabe N, Shirakawa T, Suzuki S, Irie S, et al. (2011). CD271/p75(NTR) inhibits the differentiation of mesenchymal stem cells into osteogenic, adipogenic, chondrogenic, and myogenic lineages. *Stem Cells Dev* 20(5):901–913.

Miura M, Chen XD, Allen MR, Bi Y, Gronthos S, Seo BM, et al. (2004). A crucial role of caspase-3 in osteogenic differentiation of bone marrow stromal stem cells. *J Clin Investig* 114(12):1704–1713.

Morsczeck C, Vollner F, Saugspier M, Brandl C, Reichert TE, Driemel O, et al. (2010). Comparison of human dental follicle cells (DFCs) and stem cells from human exfoliated deciduous teeth (SHED) after neural differentiation in vitro. *Clin Oral Investig* 14(4):433–440.

Nakamura TM, Morin GB, Chapman KB, Weinrich SL, Andrews WH, Lingner J, et al. (1997). Telomerase catalytic subunit homologs from fission yeast and human. *Science* 277(5328):955–959.

Nam S, Won JE, Kim CH, Kim HW (2011). Odontogenic differentiation of human dental pulp stem cells stimulated by the calcium phosphate porous granules. *J Tissue Eng* 2011: 812547.

Nemoto E, Koshikawa Y, Kanaya S, Tsuchiya M, Tamura M, Somerman MJ, et al. (2009). Wnt signaling inhibits cementoblast differentiation and promotes proliferation. *Bone* 44(5):805–812.

Nishimura R, Hata K, Ikeda F, Ichida F, Shimoyama A, Matsubara T, et al. (2008). Signal transduction and transcriptional regulation during mesenchymal cell differentiation. *J Bone Miner Metab* 26(3):203–212.

Okamoto Y, Sonoyama W, Ono M, Akiyama K, Fujisawa T, Oshima M, et al. (2009). Simvastatin induces the odontogenic differentiation of human dental pulp stem cells in vitro and in vivo. *J Endod* 35(3):367–372.

Osathanon T, Nowwarote N, Pavasant P (2011). Basic fibroblast growth factor inhibits mineralization but induces neuronal differentiation by human dental pulp stem cells through a FGFR and PLCgamma signaling pathway. *J Cell Biochem* 112(7):1807–1816.

Piek E, Heldin CH, Ten Dijke P (1999). Specificity, diversity, and regulation in TGF-beta superfamily signaling. *FASEB J* 13(15):2105–2124.

Ren G, Zhang L, Zhao X, Xu G, Zhang Y, Roberts AI, et al. (2008). Mesenchymal stem cell–mediated immunosuppression occurs via concerted action of chemokines and nitric oxide. *Cell Stem Cell* 2(2):141–150.

Rickard DJ, Sullivan TA, Shenker BJ, Leboy PS, Kazhdan I (1994). Induction of rapid osteoblast differentiation in rat bone marrow stromal cell cultures by dexamethasone and BMP-2. *Dev Biol* 161(1):218–228.

Scheller EL, Chang J, Wang CY (2008). Wnt/beta catenin inhibits dental pulp stem cell differentiation. *J Dent Res* 87(2):126–130.

Sekiya I, Larson BL, Vuoristo JT, Reger RL, Prockop DJ (2005). Comparison of effect of BMP-2, -4, and -6 on in vitro cartilage formation of human adult stem cells from bone marrow stroma. *Cell Tissue Res* 320(2):269–276.

Seo BM, Sonoyama W, Yamaza T, Coppe C, Kikuiri T, Akiyama K, et al. (2008). SHED repair critical-size calvarial defects in mice. *Oral Dis* 14(5):428–434.

Shi S, Gronthos S, Chen S, Reddi A, Counter CM, Robey PG, et al. (2002). Bone formation by human postnatal bone marrow stromal stem cells is enhanced by telomerase expression. *Nat Biotechnol* 20(6):587–591.

Spaggiari GM, Capobianco A, Becchetti S, Mingari MC, Moretta L (2006). Mesenchymal stem cell–natural killer cell interactions: evidence that activated NK cells are capable of killing MSCs, whereas MSCs can inhibit IL-2–induced NK-cell proliferation. *Blood* 107(4):1484–1490.

Sun L, Akiyama K, Zhang H, Yamaza T, Hou Y, Zhao S, et al. (2009). Mesenchymal stem cell transplantation reverses multiorgan dysfunction in systemic lupus erythematosus mice and humans. *Stem Cells* 27(6):1421–1432.

Sundelacruz S, Levin M, Kaplan DL (2008). Membrane potential controls adipogenic and osteogenic differentiation of mesenchymal stem cells. *PLoS One* 3(11): e3737.

Suzawa M, Takada I, Yanagisawa J, Ohtake F, Ogawa S, Yamauchi T, et al. (2003). Cytokines suppress adipogenesis and PPAR-gamma function through the TAK1/TAB1/NIK cascade. *Nat Cell Biol* 5(3):224–230.

Tasso R, Fais F, Reverberi D, Tortelli F, Cancedda R (2010). The recruitment of two consecutive and different waves of host stem/progenitor cells during the development of tissue-engineered bone in a murine model. *Biomaterials* 31(8):2121–2129.

Tsuchiya S, Honda MJ, Shinohara Y, Saito M, Ueda M (2008). Collagen type I matrix affects molecular and cellular behavior of purified porcine dental follicle cells. *Cell Tissue Res* 331(2):447–459.

Viale-Bouroncle S, Felthaus O, Schmalz G, Brockhoff G, Reichert T, Morsczeck C (2012). The transcription factor DLX3 regulates the osteogenic differentiation of human dental follicle precursor cells. *Stem Cells Dev* 21(11):1936–1947.

Wada N, Menicanin D, Shi S, Bartold PM, Gronthos S (2009). Immunomodulatory properties of human periodontal ligament stem cells. *J Cell Physiol* 219(3):667–676.

Wang J, Wang X, Sun Z, Yang H, Shi S, Wang S (2010). Stem cells from human-exfoliated deciduous teeth can differentiate into dopaminergic neuron-like cells. *Stem Cells Dev* 19(9):1375–1383.

Wang J, Ma H, Jin X, Hu J, Liu X, Ni L, et al. (2011). The effect of scaffold architecture on odontogenic differentiation of human dental pulp stem cells. *Biomaterials* 32(31):7822–7830.

Wang LD, Wagers AJ (2011). Dynamic niches in the origination and differentiation of haematopoietic stem cells. *Nat Rev Mol Cell Biol* 12(10):643–655.

Wang Y, Medvid R, Melton C, Jaenisch R, Blelloch R (2007). DGCR8 is essential for microRNA biogenesis and silencing of embryonic stem cell self-renewal. *Nat Genet* 39(3):380–385.

Wu Z, Xie Y, Bucher NL, Farmer SR (1995). Conditional ectopic expression of C/EBP beta in NIH-3T3 cells induces PPAR gamma and stimulates adipogenesis. *Genes Dev* 9(19):2350–2363.

Yamaza T, Miura Y, Bi Y, Liu Y, Akiyama K, Sonoyama W, et al. (2008). Pharmacologic stem cell based intervention as a new approach to osteoporosis treatment in rodents. *PLoS One* 3(7): e2615.

Yamaza T, Kentaro A, Chen C, Liu Y, Shi Y, Gronthos S, et al. (2010). Immunomodulatory properties of stem cells from human exfoliated deciduous teeth. *Stem Cell Res Ther* 1(1): 5.

Yamaza T, Ren G, Akiyama K, Chen C, Shi Y, Shi S (2011). Mouse mandible contains distinctive mesenchymal stem cells. *J Dent Res* 90(3):317–324.

Yu J, Wang Y, Deng Z, Tang L, Li Y, Shi J, et al. (2007). Odontogenic capability: bone marrow stromal stem cells versus dental pulp stem cells. *Biol Cell* 99(8):465–474.

Zhang Q, Shi S, Liu Y, Uyanne J, Shi Y, Le AD (2009). Mesenchymal stem cells derived from human gingiva are capable of immunomodulatory functions and ameliorate inflammation-related tissue destruction in experimental colitis. *J Immunol* 183(12):7787–7798.

Zhao S, Wehner R, Bornhauser M, Wassmuth R, Bachmann M, Schmitz M (2010). Immunomodulatory properties of mesenchymal stromal cells and their therapeutic consequences for immune-mediated disorders. *Stem Cells Dev* 19(5):607–614.

17

AN INCISIVE LOOK AT STEM CELLS: THE MOUSE INCISOR AS AN EMERGING MODEL FOR TOOTH RENEWAL

FREDERIC MICHON

Institute of Biotechnology, Developmental Biology Program, University of Helsinki, Helsinki, Finland

ANDREW H. JHEON AND KERSTIN SEIDEL

Department of Orofacial Sciences and Program in Craniofacial and Mesenchymal Biology, University of California–San Francisco, San Francisco, California

OPHIR D. KLEIN

Departments of Orofacial Sciences and Pediatrics, Program in Craniofacial and Mesenchymal Biology, and Institutes for Human Genetics and Regeneration Medicine, University of California–San Francisco, San Francisco, California

1 INTRODUCTION

The homeostasis of adult tissues requires continuous generation of differentiated progeny from stem cells. Many organs, such as the skin, blood, and intestine, renew throughout the life of an animal. Similarly, teeth that grow continuously are found in a number of mammalian species. In mice, the ever-growing incisors undergo constant renewal, fueled by specialized populations of adult stem cells, which is counterbalanced by the abrasion of teeth by hard food in the diet and occlusal forces. The study of this stem cell–based growth is of interest to scientists at the molecular, cellular, and evolutionary levels. Whereas developmental biologists are exploiting the incisor to elucidate reciprocal epithelial–mesenchymal interactions during organogenesis, evolutionary biologists are comparing the fossil record to

Stem Cells in Craniofacial Development and Regeneration, First Edition.
Edited by George T.-J. Huang, Irma Thesleff.

specific mouse mutants, and stem cell biologists are evaluating the possibilities of using dental stem cells in regenerative medicine. Here we describe the recent advances in the field of incisor stem cell regulation, including new findings in developmental genetics and in regulation by miRNAs.

2 FROM REPLACEMENT TO EVER-GROWING TEETH

2.1 Loss of Tooth Rows and Replacement in Mammals

Comparison of teeth in fish and mammals reveals interesting adaptations to the environment and provides information about feeding habits. Fish are principally *homodonts*, meaning that within any given species, all teeth have the same shape (mainly conical), and they are polyphyodont, meaning that continuous replacement generates new teeth throughout their lives. Newly forming teeth appear in rows behind the main tooth row, such that each exfoliated tooth is replaced quickly to ensure continuous hunting and feeding throughout the fish's lifespan (Verstraeten et al., 2010).

Both humans and rodents evolved from a common mammalian ancestor that is thought to have had a full complement of teeth, comprising three incisors, one canine, four premolars, and three molars in each dental quadrant, with the dentition replaced a single time (Line, 2003). Because mammals possess teeth of different shapes in the same species, they are referred to as *heterodonts*. During mammalian evolution, the number of teeth was decreased in the lineages that gave rise to both rodents and humans. Humans have all four of the major classes of teeth but have lost one incisor and two premolars per jaw quadrant. Rodent ancestors underwent a further reduction of teeth, such that mice possess only one incisor and three molars per quadrant, and have no replacement teeth. Rodent teeth are considered to be deciduous teeth that do not undergo replacement (Moss-Salentijn, 1975). However, some rodents, such as the African rat-mole (*Heliophobius argenteocinereus*), are known to continuously generate molars in the back of the jaw that secondarily move toward the front of the jaw, where the teeth are resorbed, resulting in an horizontal escalator-like tooth replacement mechanism (Gomes-Rodrigues et al., 2011).

As discussed in the remainder of this chapter, the loss of tooth rows, the decrease in tooth number, and the diminution in tooth replacement capacity in mammals as compared to fish and reptiles have been compensated for by the evolution of continuously growing teeth. One remarkable example of such teeth is the mouse incisor, which has been gaining traction as a model of stem cell–fueled renewal (Mitsiadis et al., 2007; Wang et al., 2007; Parsa et al., 2010; Seidel et al., 2010; Jheon et al., 2011).

2.2 Evolution of Ever-Growing Teeth

The mouse incisor is one type of continuously growing tooth (Fig. 1A); such teeth have evolved independently in a number of mammalian species. In all cases of continuously growing, or *hypselodont*, teeth, an evolutionary intermediate is thought to be a high-crowned, or *hypsodont*, tooth. Hypsodonty results from an extended crown growth period to compensate for extensive crown abrasion and is probably fueled by a transitory maintenance of stem cells located at the base of the crown. This results in a delayed crown-to-root transition and a higher crown compared to the *brachydont*, or short-crowned, molars that are present in mice. An increase in the maintenance of

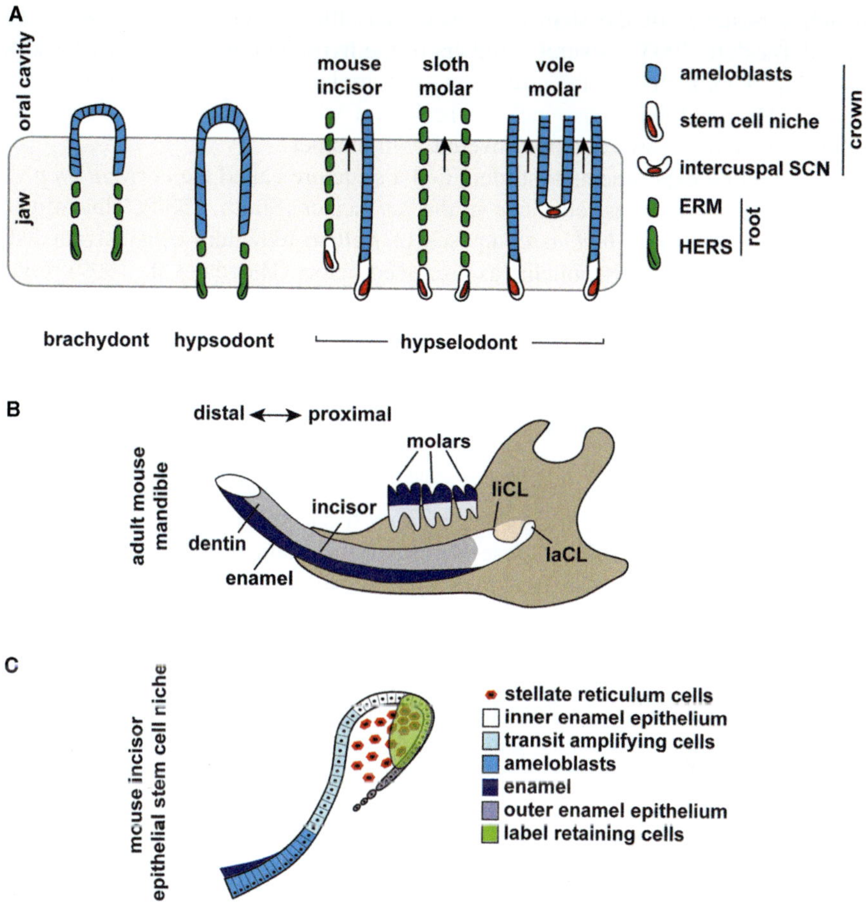

FIGURE 1 Evolution of continuously growing teeth and schematic of the epithelial stem cell niche of the mouse incisor. (A) The variation in timing of stem cell maintenance during tooth development gives rise to different tooth patterns. As long as the epithelial stem cells persist, continuous growth of the crown and/or root is observed. The crown-to-root transition is associated with disappearance of the stem cells and the formation of Hertwig's epithelial root sheath (HERS). Most mammalian teeth are brachyodont (i.e., short crown relative to root). When the disappearance of the stem cells is delayed, a longer crown develops (relative to root), forming a hypsodont tooth. Lifelong maintenance of the stem cell niche, sustaining continuous production of ameloblasts and ERM cells giving rise to the crown and root domain, respectively, is the feature of hypselodont (i.e., ever-growing) teeth. The continuous growth of teeth (arrows) can give rise to a crownless tooth (sloth molar), to a rootless tooth (vole molar), or to a tooth composed of a crown and root domain (mouse incisor). Arrows indicate the direction of growth (adapted from Tummers and Thesleff, 2008). (B) Schematic illustration of the mandibular mouse incisor. The sagittal view demonstrates the proximal location of the labial (laCL) and lingual (liCL) cervical loop, and the asymmetry of the enamel deposition on the labial side of the incisor. (C) The continuous growth of the mouse incisor is supported by the labial CL housing the epithelial stem cells. Before differentiation into enamel-forming ameloblasts, the stem cells become progenitors that actively proliferate as transit-amplifying cells.

stem cells has been suggested as the cause for the appearance of hypselodont teeth, in which persistence of the stem cell niche fuels lifelong continuous growth (Tummers and Thesleff, 2003). Several members of the hypselodont tooth family have been studied (Fig. 1A), including the crownless sloth molar, the vole molar (Tummers and Thesleff, 2003), and the mouse incisor, the last of which contains both a root analog (enamel-free) and a crown analog (covered with enamel).

Label-retaining experiments first identified a structure called the *cervical loop* (CL) as the location of the stem cell niche in the rat incisor (Smith, 1980). This structure, also known as the *apical bud*, is composed of stellate reticulum cells surrounded by the inner and outer enamel epithelia, as discussed below (Harada et al., 1999) (Fig. 1B and C).

3 THE EPITHELIAL STEM CELL NICHE IN THE CONTINUOUSLY GROWING MOUSE INCISOR

Similar to other organs, such as hair (Nowak et al., 2008) and intestine (Snippert et al., 2010), renewal of the mouse incisor depends on tissue-specific stem cell populations (Thesleff and Tummers, 2009). These stem cells maintain homeostasis of the organ by replenishing the tissue lost by abrasion and attrition. We prefer the term *renewal* to *regeneration*, as the latter implies a more complete reformation of the damaged part of an organ and suggests active morphogenesis, which does not appear to be the case in mouse incisor renewal. Although the study of incisor stem cells is a relatively new field, significant advances have been made in the identification of these cells, the understanding of their functions, and the characterization of molecular mechanisms regulating their behavior.

3.1 Localization of Epithelial Stem Cells in the Incisor

Whereas some rodents, such as certain species of voles, have continuously growing molars and incisors, mice renew only their incisors. This renewal capacity depends on the presence of epithelial and mesenchymal stem cells that give rise to the different cell lineages of the tooth, including ameloblasts (Am), odontoblasts (Od), stellate reticulum (SR), stratum intermedium (SI), inner enamel epithelium (IEE), outer enamel epithelium (OEE), transit-amplifying (T-A) cells, Hertwig's epithelial root sheath (HERS), and epithelial cell rests of Malassez (ERMs) (Figs. 1C and 2A).

Early labeling experiments examining the rate of ameloblast and odontoblast displacement in mice and rats gave the first clues that turnover of these specialized cells was rapid, underlining the need for progenitor pools to resupply differentiated cell populations (Hwang and Tonna, 1965; Smith and Warshawsky, 1975; Smith, 1980) (Fig. 2B). Moreover, initial studies using explant cultures of the CL region from 2-day-old mice showed that new epithelial structures could be generated in culture, indicating that the labial CL houses the dental stem cells that give rise to ameloblasts and the SI (Harada et al., 1999).

The incisor stem cells are located in two structures called the labial and lingual CL (Figs. 1B and 2A). In the labial CL, the early stem cell progeny contribute to a population of T-A cells. Those cells undergo several rounds of cell division before differentiating into ameloblasts (Thesleff and Tummers, 2009). As in the intestine, the early progeny are thought to push the differentiated cells distally, similar to a conveyor

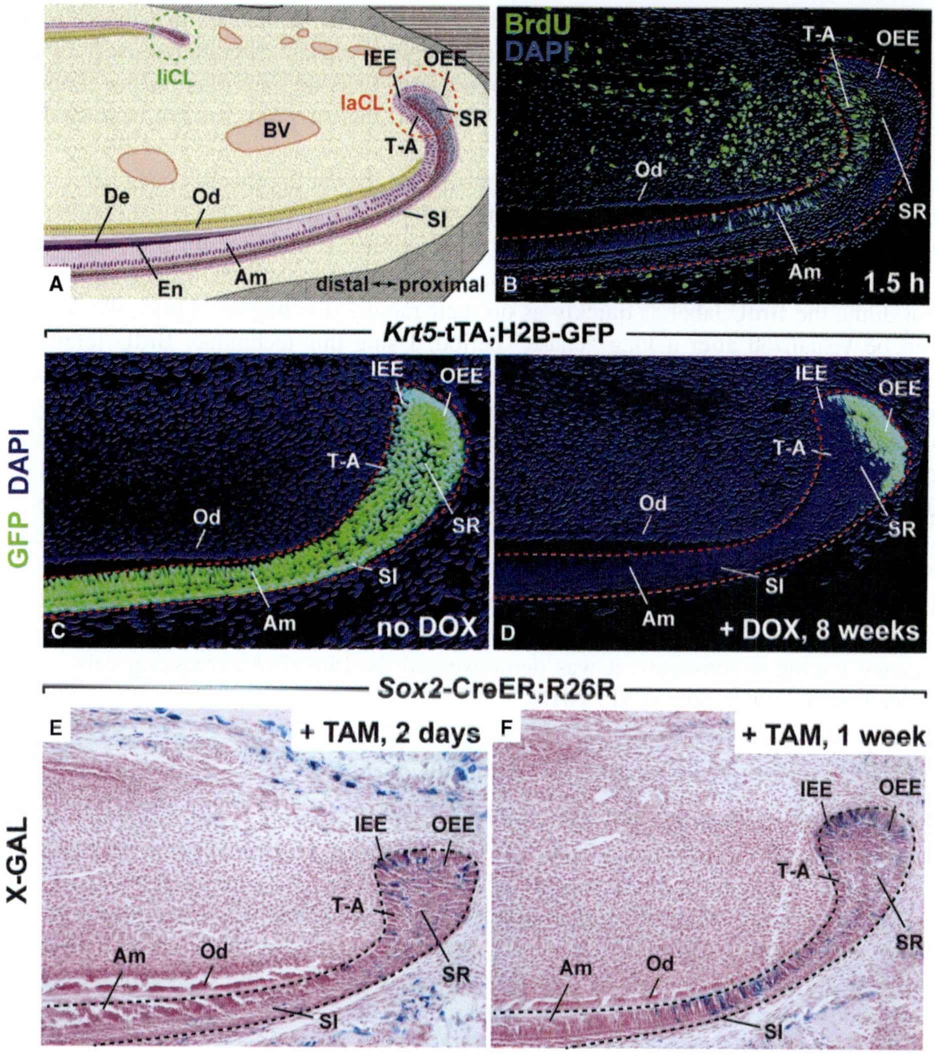

FIGURE 2 Stem cell niche in the adult mouse incisor. (A to D) The lower incisor is shown in sagittal view. (A) Encircled are the two epithelial stem cell compartments in the lingual (liCL) and labial (laCL) cervical loops. Also shown are the inner enamel epithelium (IEE), from which the transit-amplifying (T-A) cells and ameloblasts (Am) arise, the outer enamel epithelium (OEE), stellate reticulum (SR), stratum intermedium (SI), odontoblasts (Od), dentin (De), enamel (En), and blood vessels (BV). (B) Adult mice were injected with BrdU for 1.5 h. BrdU-positive cells indicate rapidly proliferating cells in the T-A region. (C and D) Images from incisors of *Krt5*-tTA; H2B-GFP mice. In the absence of doxycycline (no DOX), GFP is present in all cells expressing *Krt5*, which includes the OEE, IEE, SR, SI, and Am. In the presence of doxycycline (+ DOX) for 8 weeks, H2B-GFP expression was turned off, leading to the retention of GFP in the slowly proliferating label-retaining cells (LRCs) of the OEE. The LRCs are putative dental epithelial stem cells. (E and F) The use of *Sox2*-CreER;R26R mice enabled the postnatal labeling of the *Sox2*-positive cell progeny during incisor renewal. The tamoxifen (TAM) pulse at P2, followed by a 48-h and 1-week chase, demonstrated that *Sox2*-positive cells are stem cells during incisor renewal, as progeny of these cells were detected in the OEE, SR, SI, and T-A cells, and Am. The dashed lines outline the epithelium.

belt. Although this has not yet been established formally, it appears that the cells do not actively migrate after reaching the enamel epithelium but migrate passively or are displaced. The lingual CL presumably houses stem cells that are required for renewal of the root analog surface of the incisor. However, this structure has not been studied in as much detail as the labial CL.

Since the initial labeling studies were performed, newer approaches have been used to identify incisor stem cells. Slow-cycling stem cell populations have been identified in several systems through label retention experiments, traditionally utilizing BrdU incorporation (Harada et al., 1999; Fuchs, 2009). Because cells that divide slowly do not dilute the BrdU label as quickly as do their rapidly dividing neighbors, the former can be visualized after a long "chase" period. Using this technique, BrdU retaining cells were identified in the labial CL of cultured perinatal incisors and in adult incisors in situ. Another approach to label retention is the use of transgenic mice harboring a tetracycline-sensitive histone H2B-GFP cassette under the control of a tissue-specific trans-activator (Tumbar et al., 2004). Expression of H2B-GFP is initially activated in all cells of the tissue of interest, followed by a chase period when the transgene is repressed by exposure of the animal to doxycycline, such that rapidly dividing cells dilute the label (Fig. 2C and D). This technique was used to identify label-retaining cells (LRCs) in the OEE and SR of the adult labial CL (Seidel et al., 2010). In addition, the LRCs of the dental epithelium expressed *Gli1*, a target of Shh signaling, and through lineage tracing experiments, it was demonstrated that the *Gli1*-expressing cells were indeed stem cells (Seidel et al., 2010).

3.2 Segregation of the Stem Cells Within the Niche

Despite the increase in studies related to incisor epithelial stem cells, the identification of specific markers for these cells has been challenging. Such markers will be critical for more in-depth analysis of the stem cell niche and of incisor renewal.

Recently, a subset of cells in the mouse incisor CL was identified that expressed *Sox2* (Juuri et al., 2012). The analysis of this expression during tooth formation indicated the similarity of this expression with *Sox9* expression during the segregation of stem cells in the hair bulge (Nowak et al., 2008). Early in tooth development, a few cells of the suprabasal layers of the dental placode begin to express a Sox family member, in this case *Sox2*. In both hair and tooth, the Sox-positive cells gather together during organ morphogenesis and eventually end up in the stem cell niche.

Through genetic-fate mapping experiments, the *Sox2*-positive cells were confirmed to be the ancestors of all dental epithelial lineages during incisor renewal (Juuri et al., 2012). However, it remains unclear whether the adult stem cells are derived from cells that are prevented from commitment and are set aside to become adult stem cells early in development, or whether the stem cells arise from a pool of cells that undergo dedifferentiation within the incisor stem cell niche. Also, questions regarding the timing of stem cell fate determination remain to be answered.

3.3 Regulation of Stem Cell Maintenance and Differentiation and Tissue Homeostasis by Signaling Molecules

The identification of organ-specific adult stem cells will be an important step toward the development of strategies for regenerative medicine. Subsequently, characterization of the factors involved in the maintenance and differentiation of these stem cells, as well as an understanding of how the stem cell niche is established during development, will be important goals of research.

Over the past 25 years, the roles of the principal morphogenetic signaling pathways in tooth development have been studied, leading to a deep understanding of the functions of these pathways. Notably, the same pathways that are involved in morphogenesis are also involved in tooth renewal. Discoveries over the past decade of multiple activators and inhibitors of conserved signaling pathways have increased our understanding of the complexities of the genetic networks regulating tooth formation and renewal (Tummers and Thesleff, 2009). Expression data of most of the molecular factors discussed here can be found on the publicly available database bite-it (http://bite-it.helsinki.fi) (Nieminen et al., 1998).

The principal conserved signaling pathways, such as Notch, Bmp, Wnt, Shh, and Fgf, were found to be involved in incisor development, morphogenesis and homeostasis. Over the past several years, the Fgf signaling pathway has become a particular focus of interest with regard to stem cell maintenance and differentiation (Harada et al., 1999, 2002; Parsa et al., 2010). Specifically, *Fgf3* and *Fgf10* are expressed in the mesenchyme immediately adjacent to and surrounding the labial CL, whereas *Fgfr1b* and *Fgfr2b* are enriched in the epithelium. Incisors of *Fgf3* null mice at P0 are grossly indistinguishable from those in wild-type mice, yet $Fgf3^{-/-}$; $Fgf10^{+/-}$compound mutants reveal a severely hypoplastic labial CL, indicating that precise levels of Fgf signaling are responsible for regulating the size and shape of the stem cell niche (Wang et al., 2007). Consistent with this result, attenuation of signaling via Fgfr2b in the embryo using a tetracycline-inducible dominant negative system led to a similar phenotype (Parsa et al., 2010). The Sprouty genes are targets of Fgf signaling and encode proteins that function in a negative regulatory feedback loop. Sprouty genes are expressed in both the lingual and labial epithelia as well as in the mesenchyme adjacent to the labial CL. Loss of Sprouty function in the incisor resulted in higher sensitivity to Fgf signaling in the lingual epithelium and mesenchyme and the presence of ameloblasts on the lingual side (Klein et al., 2008). This study highlighted the importance of balanced Fgf signaling in the incisor to maintain asymmetric production of ameloblasts on the labial side while preventing ameloblast formation on the lingual side.

Moreover, several other pathways directly or indirectly converge on Fgf signaling via regulation of *Fgf3* to control proliferation in developing incisors. Ectopic expression of follistatin (*Fst*) throughout the dental epithelium under the control of the *Krt14* promoter resulted in complete downregulation of mesenchymal *Fgf3* expression with reduction in epithelial proliferation and severely hypoplastic labial CLs (Wang et al., 2007). Conversely, epithelial loss of *Fst* in the incisors led to upregulated *Fgf3* expression in mesenchyme adjacent to the lingual CL, causing increased proliferation and expansion of the lingual CL (Wang et al., 2007). Inactivation of the *Alk5/Tgfbr1* receptor in the mesenchyme led to downregulation of *Fgf3, Fgf9*, and *Fgf10*, and reduced the size of the labial CL, probably due to proliferation defects (Zhao et al., 2011). Notably, fewer LRCs survived in this mutant, and the defect could be reversed by the addition of exogenous FGF10. Thus, several lines of evidence indicate that Fgf signaling is involved in the maintenance of incisor stem cell number.

In addition to Fgf signaling, incisor stem cells require the activity of the Notch pathway to ensure development and survival. Three Notch receptors are expressed in the CL regions in the developing incisor. *Notch1* and *Notch2* are expressed in the epithelium and mesenchyme, whereas *Ntch3* is restricted to the mesenchyme (Harada et al., 1999). The Notch ligand encoded by *Jag2* is strongly expressed in the epithelium, and *Jag2*-null mice show incisors with defects in cellular morphology (Mitsiadis et al., 2010). Inhibition of Notch signaling with DAPT led to a reduction in the size of the labial CL in explant experiments (Felszeghy et al., 2010).

Surprisingly, and in contrast to other organs, such as skin (Huelsken et al., 2001), intestine (He et al., 2004), and cornea (McGowan et al., 2007), inhibition rather than activation of Wnt signaling occurs during incisor renewal. Although several Wnt ligands are expressed at E16 and E18 in the incisor epithelial and mesenchymal compartments, no canonical Wnt activity can be detected in the dental epithelium with TOPGAL or BATGAL reporter mice, and only faint *Axin2* expression was detected in the pre-ameloblasts at those stages (Suomalainen and Thesleff, 2010). Moreover, the over- and ectopic expression of *Wnt3* in the dental epithelium led to a progressive loss of ameloblast differentiation during incisor renewal (Millar et al., 2003).

Continuous growth of the mouse incisor compensates for the constant abrasion and attrition from masticatory and occlusion forces. As in other renewing organs, the stem cells require instructions from specific factors to constantly replenish the loss of epithelial cells. Recently, it was demonstrated that *Shh* is expressed by differentiating progeny and signals back to the stem cells to promote production of more progeny (Seidel et al., 2010). The use of a specific Hedgehog (Hh) pathway inhibitor led to the arrest of ameloblast production by the stem cells, pointing to a positive feedback loop between the stem cells and their progeny.

The effect of the Hh pathway is not limited to the epithelial compartment. Lineage tracing experiments revealed that Hh-responsive cells are found as well in the mesenchymal compartment, constantly producing new odontoblasts. Interestingly, there appear to be two sources of dental mesenchymal stem cells, one involved in normal renewal and another in repair after injury (Feng et al., 2011).

3.4 Regulation of Incisor Stem Cells by MicroRNAs

MicroRNAs (miRNAs) are endogenously expressed, short (about 21 nucleotides), noncoding RNA molecules that affect protein synthesis by posttranscriptional mechanisms. miRNAs are processed by several enzymes, including DICER1. To date, only two studies have deleted *Dicer1* in order to analyze miRNA function during tooth development and renewal. The first reported the deletion of *Dicer1* in the epithelium by *Pitx2*-Cre and in the mesenchyme by *Wnt1*-Cre (Cao et al., 2010). The craniofacial defects after *Dicer1* deletion by *Wnt1*-Cre, which is expressed in the neural crest cells, were too severe to analyze any tooth phenotype. However, the epithelial deletion of *Dicer1* using *Pitx2*-Cre led to multiple-branched enamel-free incisors and cusp-less molars. These observations indicated an involvement of miRNAs in tooth morphogenesis and patterning, as well as in terminal cell differentiation and tissue homeostasis. The second study reported a milder epithelial phenotype after deletion using *Krt14*-Cre (Michon et al., 2010). The expression of *Krt14* occurs later in development, and no obvious phenotype was observed during tooth morphogenesis at embryonic stages. However, soon after birth (P5), the epithelium formed invaginated grooves in the mesenchymal compartment due to an excess of cell proliferation. Later (P25), those grooves extended toward the incisor tip, due to the constant growth of the incisor. At this stage, the erupted molars exhibited a modified cusp pattern. An enamel defect, due to impaired differentiation of ameloblasts, was present in both incisors and molars. Another recent study has proposed a role for specific miRNAs in ameloblast differentiation during mouse incisor renewal. Numerous differentially expressed miRNAs were identified in ameloblasts and the lingual and labial CLs, reflecting the requirement for specific fine-tuning of signaling pathways in various incisor cell types (Jheon et al., 2011). Taken together, these results demonstrated the dynamic involvement of miRNA-mediated regulation during both tooth development and renewal.

3.5 Crown-to-Root Transition and the Loss of Stem Cells

Important observations have been made in the incisor not only by studying the enamel-producing cell layer on the labial side (crown analog), but also by examining the root-like structure of the lingual side (root analog) (Amar et al., 1986). The analysis of mouse mutants that exhibit either no enamel or abnormalities in the asymmetry between the labial and lingual sides of the incisor has provided clues regarding the mechanisms of crown or root development. Several mouse mutants are known to develop such defects in incisor asymmetry, including *Fst* and Sprouty nulls (Wang et al., 2004; Klein et al., 2008), as well as *Krt14*-driven overexpression of *Fst* and *Eda* (Mustonen et al., 2003; Wang et al., 2004).

It appears that depletion of SR cells of the CL in rooted teeth has led to the transition from crown to root. Ever-growing molars in voles and sloths do not undergo such a transition, and enamel is continuously generated throughout the life of the animal (Tummers and Thesleff, 2008) (Fig. 1A). Interestingly, brachydont mouse and hypselodont vole molars are practically identical in morphology and in the expression pattern of key genes until E17, when the molars are in the late bell stage. Given the roles of Fgf and Notch signaling in maintenance of the stem cell niche in the mouse incisor (Harada et al., 1999, 2002), these pathways were compared at stages during which mouse molars initiate root formation. The development of roots in the mouse coincides with loss of epithelial Notch and mesenchymal Fgf expression, whereas vole molars continue to express key signaling components and partially bypass the root fate (Tummers and Thesleff, 2003). This idea was substantiated by evidence showing that modulation of *Fgf10* led to maintenance of the CLs and continuous growth of the mouse molar crown *in vitro* (Harada et al., 2002; Yokohama-Tamaki et al., 2008). Although correlative, these studies suggest that vole molars have evolved mechanisms that maintain the necessary morphology for continuous growth fueled by molar CLs, which may be analogous to the stem cell niche of the mouse incisor.

4 CONCLUDING REMARKS AND OPEN QUESTIONS

Since the early descriptions of continuously growing incisors of rodents, such as by Waterhouse (1848), their evolution and development in several taxa have been the focus of interest, gradually advancing us toward potential applications in regenerative medicine. However, much work remains to be done, and the use of cellular, molecular, and genetic approaches, as well as paleontological techniques, will enable continued progress.

One unknown aspect of incisor stem cell biology concerns the mechanism of stem cell segregation and homing in the stem cell niche. Another area that requires further investigation is the possible molecular differences between stem cells involved in organ renewal, such as those in the CL, and stem cells able to generate a new organ, such as those found in the dental lamina (Li et al., 2011). So far, only a few types of adult stem cells, such as those found in mammary gland (Shackleton et al., 2006) or intestinal crypt (Sato et al., 2009), were shown to be able to regenerate an entire organ *in vitro*. Teeth provide an interesting opportunity for the study of regeneration because two types of dental stem cells are found in two different niches (CL and dental lamina) and are able, respectively, to renew or regenerate an organ. The molecular study of these two kinds of dental stem cells will lead to vital information regarding the ability of different stem cells to recapitulate embryonic development, as opposed to the capacity to contribute to renewal.

A recent study has pointed out that early stem cell progeny can revert to a stem cell state in case of destruction of the stem cell niche (Arnold et al., 2011). Thus, using early stem cell progeny as starting material in regenerative protocols instead of the generally less abundant stem cells themselves could be a promising alternative route for regenerative studies.

Another question of interest is the involvement of a possible negative feedback loop that prevents the overproduction of progeny during incisor renewal; this concept has already been proposed in other tissues (Lander et al., 2009; Lo et al., 2009; Kirouac et al., 2010). A part of this loop may be the level of mesenchymal *Fgf10* expression, which is in turn regulated by FGF9 produced by the T-A cells (Klein et al., 2008; Yokohama-Tamaki et al., 2008). It is likely that the stem cells, through yet unknown mechanisms, are able to sense the level of progeny in order to adapt to the pace of renewal, as suggested by the Hh data described above (Seidel et al., 2010).

Finally, a more philosophical question concerns the nature of the epithelial stem cells. It is still unclear whether these cells reside in the SR, in the dental epithelium at the proximal tip of the CL (OEE), or in both regions (Ohshima et al., 2005; Parsa et al., 2010; Seidel et al., 2010, Juuri et al., 2012). The architecture of the cells in these two regions is drastically different, exhibiting a mesenchymal morphology in the SR and an epithelial morphology in the OEE. In epithelial mammary stem cells, it was shown that an increase in stem cell activity is acquired after undergoing a process reminiscent of epithelial–mesenchymal transition (EMT) (Mani et al., 2008). This was not a true EMT, as the cells did not become mesenchymal, but they did reach a level of plasticity allowing for better self-renewal and mammosphere formation. In the tooth context, some HERS cells are known to undergo EMT-like behaviors (Akimoto et al., 2011). It will be interesting to determine if mechanisms involving plasticity exist that maintain a pool of multipotent stem cells in the developing tooth.

These open questions, together with the increasing numbers of molecular tools available, make this an exciting time to study the development and evolution of stem cells in teeth.

REFERENCES

Akimoto T, Fujiwara N, Kagiya T, Otsu K, Ishizeki K, Harada H (2011). Establishment of Hertwig's epithelial root sheath cell line from cells involved in epithelial–mesenchymal transition. *Biochem Biophys Res Commun* 404:308–312.

Amar S, Karcher-Djuricic V, Meyer JM, Ruch JV (1986). The lingual (root analogue) and the labial (crown analogue) mouse incisor dentin promotes ameloblast differentiation. *Arch Anat Microsc Morphol Exp* 75:229–239.

Arnold K, Sarkar A, Yram MA, Polo JM, Bronson R, Sengupta S, Seandel M, Geijsen N, Hochedlinger K (2011). Sox2(+) adult stem and progenitor cells are important for tissue regeneration and survival of mice. *Cell Stem Cell* 9:317–329.

Cao H, Wang J, Li X, Florez S, Huang Z, Venugopalan SR, Elangovan S, Skobe Z, Margolis HC, Martin JF, Amendt BA (2010). MicroRNAs play a critical role in tooth development. *J Dent Res* 89:779–784.

Felszeghy S, Suomalainen M, Thesleff I (2010). Notch signalling is required for the survival of epithelial stem cells in the continuously growing mouse incisor. *Differentiation* 80:241–248.

Feng J, Mantesso A, De Bari C, Nishiyama A, Sharpe PT (2011). Dual origin of mesenchymal stem cells contributing to organ growth and repair. *Proc Natl Acad Sci USA* 108:6503–6508.

Fuchs E (2009). The tortoise and the hair: slow-cycling cells in the stem cell race. *Cell* 137:811–819.

Gomes-Rodrigues H, Marangoni P, Sumbera R, Tafforeau P, Wendelen W, Viriot L (2011). Continuous dental replacement in a hyper-chisel tooth digging rodent. *Proc Natl Acad Sci USA* 108:17355–17359.

Harada H, Kettunen P, Jung HS, Mustonen T, Wang YA, Thesleff I (1999). Localization of putative stem cells in dental epithelium and their association with Notch and FGF signaling. *J Cell Biol* 147:105–120.

Harada H, Toyono T, Toyoshima K, Ohuchi H (2002). FGF10 maintains stem cell population during mouse incisor development. *Connect Tissue Res* 43:201–204.

He XC, Zhang J, Tong WG, Tawfik O, Ross J, Scoville DH, Tian Q, Zeng X, He X, Wiedemann LM, Mishina Y, Li L (2004). BMP signaling inhibits intestinal stem cell self-renewal through suppression of Wnt-beta-catenin signaling. *Nat Genet* 36:1117–1121.

Huelsken J, Vogel R, Erdmann B, Cotsarelis G, Birchmeier W (2001). beta-Catenin controls hair follicle morphogenesis and stem cell differentiation in the skin. *Cell* 105:533–545.

Hwang WSS, Tonna EA (1965). Autoradiographic analysis of labeling indices and migration rates of cellular component of mouse incisors using tritiated thymidine (H3TDR). *J Dent Res* 44:42.

Jheon AH, Li CY, Wen T, Michon F, Klein OD (2011). Expression of MicroRNAs in the stem cell niche of the adult mouse incisor. *PLoS One* 6:e24536.

Juuri E, Saito K, Ahtiainen L, Seidel K, Tummers M, Hochedlinger K, Klein OD, Thesleff I, Michon F (2012). $Sox2^+$ stem cells contribute to all epithelial lineages of the tooth via $Sfrp5^+$ progenitors. *Dev Cell* 23:317–28.

Kirouac DC, Ito C, Csaszar E, Roch A, Yu M, Sykes EA, Bader GD, Zandstra PW (2010). Dynamic interaction networks in a hierarchically organized tissue. *Mol Syst Biol* 6:417.

Klein OD, Lyons DB, Balooch G, Marshall GW, Basson MA, Peterka M, Boran T, Peterkova R, Martin GR (2008). An FGF signaling loop sustains the generation of differentiated progeny from stem cells in mouse incisors. *Development* 135:377–385.

Lander AD, Gokoffski KK, Wan FY, Nie Q, Calof AL (2009). Cell lineages and the logic of proliferative control. *PLoS Biol* 7:e15.

Li L, Kwon HJ, Harada H, Ohshima H, Cho SW, Jung HS (2011). Expression patterns of ABCG2, Bmi-1, Oct-3/4, and Yap in the developing mouse incisor. *Gene Expr Patterns* 11:163–170.

Line SR (2003). Variation of tooth number in mammalian dentition: connecting genetics, development, and evolution. *Evol Dev* 5:295–304.

Lo WC, Chou CS, Gokoffski KK, Wan FYM, Lander AD, Calof AL, Nie Q (2009). Feedback regulation in multistage cell lineages *Math Biosci Eng* 6:59.

Mani SA, Guo W, Liao MJ, Eaton EN, Ayyanan A, Zhou AY, Brooks M, Reinhard F, Zhang CC, Shipitsin M, Campbell LL, Polyak K, Brisken C, Yang J, Weinberg RA (2008). The epithelial–mesenchymal transition generates cells with properties of stem cells. *Cell* 133:704–715.

McGowan SL, Edelhauser HF, Pfister RR, Whikehart DR (2007). Stem cell markers in the human posterior limbus and corneal endothelium of unwounded and wounded corneas. *Mol Vis* 13:1984–2000.

Michon F, Tummers M, Kyyrönen M, Frilander MJ, Thesleff I (2010). Tooth morphogenesis and ameloblast differentiation are regulated by micro-RNAs. *Dev Biol* 340:355–368.

Millar SE, Koyama E, Reddy ST, Andl T, Gaddapara T, Piddington R, Gibson CW (2003). Over- and ectopic expression of Wnt3 causes progressive loss of ameloblasts in postnatal mouse incisor teeth. *Connect Tissue Res* 44 (Suppl 1): 124–129.

Mitsiadis TA, Barrandon O, Rochat A, Barrandon Y, De Bari C (2007). Stem cell niches in mammals. *Exp Cell Res* 313:3377–3385.

Mitsiadis TA, Graf D, Luder H, Gridley T, Bluteau G (2010). BMPs and FGFs target Notch signalling via Jagged 2 to regulate tooth morphogenesis and cytodifferentiation. *Development* 137:3025–3035.

Moss-Salentijn L (1975). Studies on dentin: 2. Vestigial lacteal incisor teeth of the rat. *Acta Anat (Basel)* 92:329–350.

Mustonen T, Pispa J, Mikkola ML, Pummila M, Kangas AT, Pakkasjärvi L, Jaatinen R, Thesleff I (2003). Stimulation of ectodermal organ development by ectodysplasin A1. *Dev Biol* 259:123–136.

Nieminen P, Pekkanen M, Aberg T, Thesleff I (1998). A graphical WWW-database on gene expression in tooth. *Eur J Oral Sci* 106 (Suppl 1): 7–11.

Nowak JA, Polak L, Pasolli HA, Fuchs E (2008). Hair follicle stem cells are specified and function in early skin morphogenesis. *Cell Stem Cell* 3:33–43.

Ohshima H, Nakasone N, Hashimoto E, Sakai H, Nakakura-Ohshima K, Harada H (2005). The eternal tooth germ is formed at the apical end of continuously growing teeth. *Arch Oral Biol* 50:153–157.

Parsa S, Kuremoto K, Seidel K, Tabatabai R, Mackenzie B, Yamaza T, Akiyama K, Branch J, Koh CJ, Al Alam D, Klein OD, Bellusci S (2010). Signaling by FGFR2b controls the regenerative capacity of adult mouse incisors. *Development* 137:3743–3752.

Sato T, Vries RG, Snippert HJ, van de Wetering M, Barker N, Stange DE, van Es JH, Abo A, Kujala P, Peters PJ, Clevers H (2009). Single Lgr5 stem cells build crypt-villus structures in vitro without a mesenchymal niche. *Nature* 459:262–265.

Seidel K, Ahn CP, Lyons D, Nee A, Ting K, Brownell I, Cao T, Carano RA, Curran T, Schober M, Fuchs E, Joyner A, Martin GR, de Sauvage FJ, Klein OD (2010). Hedgehog signaling regulates the generation of ameloblast progenitors in the continuously growing mouse incisor. *Development* 137:3753–3761.

Shackleton M, Vaillant F, Simpson KJ, Stingl J, Smyth GK, Asselin-Labat ML, Wu L, Lindeman GJ, Visvader JE (2006). Generation of a functional mammary gland from a single stem cell. *Nature* 439:84–88.

Smith CE, Warshawsky H (1975). Cellular renewal in the enamel organ and the odontoblast layer of the rat incisor as followed by radioautography using ^{3}H thymidine. *Anat Rec* 183:523–561.

Smith CE (1980). Cell turnover in the odontogenic organ of the rat incisor as visualized by graphic reconstructions following a single injection of ^{3}H-thymidine. *Am J Anat* 158:321–343.

Snippert HJ, van der Flier LG, Sato T, van Es JH, van den Born M, Kroon-Veenboer C, Barker N, Klein AM, van Rheenen J, Simons BD, Clevers H (2010). Intestinal crypt homeostasis results from neutral competition between symmetrically dividing Lgr5 stem cells. *Cell* 143:134–144.

Suomalainen M, Thesleff I (2010). Patterns of Wnt pathway activity in the mouse incisor indicate absence of Wnt/beta-catenin signaling in the epithelial stem cells. *Dev Dyn* 239:364–372.

Thesleff I, Tummers M (2009). Tooth organogenesis and regeneration. Cambridge (MA): *Harvard Stem Cell Institute*; 2008–2009 Jan 31.

Tumbar T, Guasch G, Greco V, Blanpain C, Lowry WE, Rendl M, Fuchs E (2004). Defining the epithelial stem cell niche in skin. *Science* 303:359–363.

Tummers M, Thesleff I (2003). Root or crown: a developmental choice orchestrated by the differential regulation of the epithelial stem cell niche in the tooth of two rodent species *Development* 130:1049.

Tummers M, Thesleff I (2008). Observations on continuously growing roots of the sloth and the K14-Eda transgenic mice indicate that epithelial stem cells can give rise to both the ameloblast and root epithelium cell lineage creating distinct tooth patterns. *Evol Dev* 10:187–195.

Tummers M, Thesleff I (2009). The importance of signal pathway modulation in all aspects of tooth development. *J Exp Zool B* 312B:309–319.

Verstraeten B, Sanders E, van Hengel J, Huysseune A (2010). Zebrafish teeth as a model for repetitive epithelial morphogenesis: dynamics of E-cadherin expression. *BMC Dev Biol* 10:58.

Wang XP, Suomalainen M, Jorgez CJ, Matzuk MM, Werner S, Thesleff I (2004). Follistatin regulates enamel patterning in mouse incisors by asymmetrically inhibiting BMP signaling and ameloblast differentiation. *Dev Cell* 7:719–730.

Wang XP, Suomalainen M, Felszeghy S, Zelarayan LC, Alonso MT, Plikus MV, Maas RL, Chuong CM, Schimmang T, Thesleff I (2007). An integrated gene regulatory network controls stem cell proliferation in teeth. *PLoS Biol* 5:e159.

Waterhouse GR (1848). *A Natural History of the Mammalia*. H. Baillière, London.

Yokohama-Tamaki T, Fujiwara N, Shibata S, Wakisaka S, Harada H (2008). The epithelial–mesenchymal interaction plays a role in the maintenance of the stem cell niche of mouse incisors via Fgf10 and Fgf9 signaling. *Open Biotechnol J* 2:111–115.

Zhao H, Li S, Han D, Kaartinen V, Chai Y (2011). Alk5-mediated transforming growth factor β signaling acts upstream of fibroblast growth factor 10 to regulate the proliferation and maintenance of dental epithelial stem cells. *Mol Cell Biol* 31:2079–2089.

18

MESENCHYMAL STEM CELL NICHES IN RODENT TOOTH PULP

Jifan Feng and Paul T. Sharpe

Craniofacial Development and Stem Cell Biology, Biomedical Research Centre and MRC Centre for Transplantation, Dental Institute, King's College London, UK

1 INTRODUCTION

Mesenchymal stem cells (MSCs) are found in many adult tissues and organs. In tissues that grow continuously or exhibit high rates of cell turnover, MSCs provide a renewable source of progenitor cells to differentiate and replace those lost. In the majority of tissues, MSCs are found in small numbers and remain quiescent until mobilized in response to tissue damage. The rodent incisor is a continuously growing organ where MSCs provide both growth and repair functions. The locations of these cells and aspects of their function are beginning to be explored.

2 RODENT INCISOR MESENCHYMAL STEM CELL NICHE(S)

Rodent incisors are highly specialized teeth, adapted for cutting hard food. This is achieved in three ways. Enamel is present on only one side (labial) of the tooth, so that as the upper and lower incisors occlude during feeding, the lingual surface wears away much more quickly than does the harder labial surface. This leads to the tips of the incisors forming very sharp points that enable the animal to cut hard food. As a consequence of this asymmetric wear, the incisors lose tissue from their tips whenever the animal feeds, and thus the incisors get shorter. If this shortening continued, the incisors would not occlude and the animal would not be able to feed properly. The second specialized adaptation is thus continuous growth of the incisors

Stem Cells in Craniofacial Development and Regeneration, First Edition.
Edited by George T.-J. Huang, Irma Thesleff.

from new tissue added at the apical end that is achieved by mesenchymal and epithelial stem cell populations located in specific areas of the apical region of the incisors. The third adaptation arises from the need to quickly repair the incisal tips as they wear. The highly vascularized incisor pulp chamber occurs along the entire length of the incisor, and thus the asymmetric wear during feeding exposes the pulp to the external environment. As the animal feeds, blood loss could occur from the incisor tips and exposed pulp could become infected. This is prevented by a coagulation-like process of the blood and rapid formation of mineralized tissue to cover the exposed pulp.

During the evolution of rodents the acquisition of self-sharpening incisors provided them with a significant selective advantage. The multiple "changes" that occurred to result in continuously growing incisors were unlikely to have occurred from de novo mechanisms but more likely from adaptations of existing mechanisms used in non-continuously growing teeth. Thus, what seem like specialized adaptations unique to rodent incisors may have parallels in other teeth.

2.1 Apical Niche

Since new odontoblasts are formed continuously from the apical-most region of the incisor, the source of MSCs must be located in this area. In addition, the location of the epithelial stem cell niches in the cervical loops (Chapter 17) also supports the notion that the MSCs are located close to the epithelial stem cells to offer some form of coordination of growth of the tooth.

One of the main issues with identification of MSCs *in vivo* is the lack of suitably definitive markers. Most markers of MSCs also occur on perivascular cells, and the abundance of small blood vessels in the tooth pulp thus makes use of these markers problematic (see below). Alternative approaches involve (1) identification of better markers for incisor MSCs, (2) use of functional assays to detect the MSCs, and (3) identification of gene mutations that affect incisor growth and MSC function. To identify genes whose expression may correlate with cells in the MSC niche, microarrays have been used to compare transcripts isolated from different regions of incisor pulp or between incisor and molar pulp (on the basis that molars, being nongrowing, do not have an equivalent specific MSC niche) (Fig. 1).

The use of functional approaches has consisted primarily of genetic lineage tracing, cell labeling, and cell proliferation assays. Genetic lineage tracing involves the application of Cre-mediated recombination of specific reporter genes to label cells indelibly. In the incisor, if labeled mesenchymal cells are found continuously in all mesenchymal derivatives, this means that MSCs have been labeled. If mesenchymal cell derivatives are labeled but disappear with time (growth), this would suggest that MSCs have not been labeled.

Stem cells are often (but not always) considered to be cells that divide very slowly (slow-cycling), whereas their immediate progeny, often called transit-amplifying cells (TACs), usually proliferate very rapidly. By injecting mice with a nucleoside analog such as 5-Bromo-2′-deoxyuridine (BrdU) that is incorporated into DNA during DNA synthesis (the S-phase), cells in the cell cycle can be labeled and detected with antibodies against BrdU. If BrdU is administered over a long period of time (e.g., several continuous days) to label all cells and then animals left for a long period (e.g., several months), so that dividing cells progressively lose all label, only cells that cycle very slowly (stem cells) will retain the label.

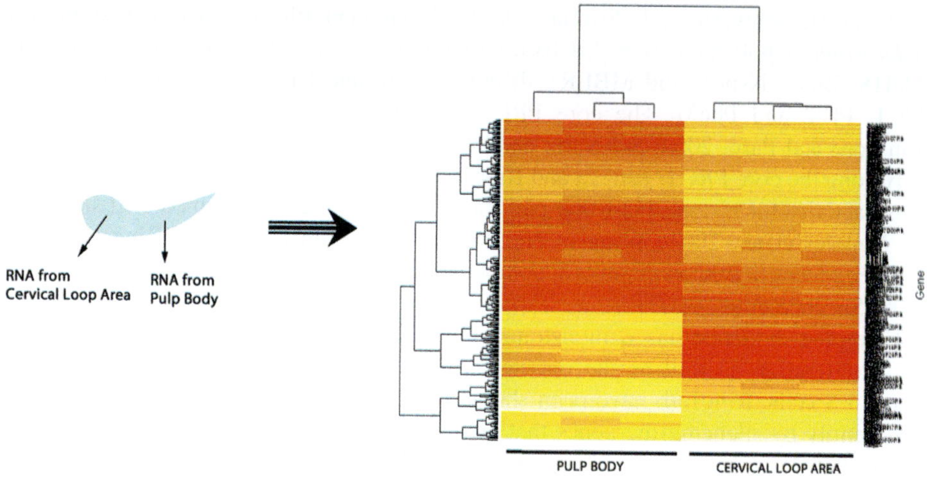

FIGURE 1 Heat map of entire genome microarrays of transcripts from pulp of the incisor body and cervical loop area.

If a single dose of BrdU is administered and the label chased after a very short period (2 hours), only those cells that are proliferating rapidly will be detected. These can represent transit-amplifying populations. When incisors are labeled in these two different ways in different studies, two adjacent mesenchymal cell populations are labeled at the apical end of the tooth (Fig. 2). Slowly cycling cells are located in a small area at the apex between the ends of the epithelial cervical loops. Rapidly dividing cells (TACs) are located adjacent to both cervical loops and extend in width across the pulp (Fig. 2) (Seidel et al., 2011; Lapthanasupkul et al., 2012). These spatial distributions thus define the incisor apical MSC niche where mesenchymal progenitors are generated.

Cell populations within the MSC niche are slowly beginning to be defined molecularly. The location of TACs correlates with expression of members of the polycomb repressor complex (PRC1) group of proteins. These include the proteins Ring1a and Ring1b, Fbxl10, Nspc1, Skp1, and Bcor (Lapthanasupkul et al, 2012). In mammals, two distinct PcG complexes have been studied extensively: polycomb repressive complex (PRC) 1 and PRC2. While PRC1 is essential for stable maintenance of gene repression by preventing nucleosome remodeling, PRC2 is involved in the initiation of gene repression and functions as a histone methyltransferase that specifically methylates lysine 27 of histone H3 (H3K27) in nucleosomes (Cao et al., 2002; Valk-Lingbeek

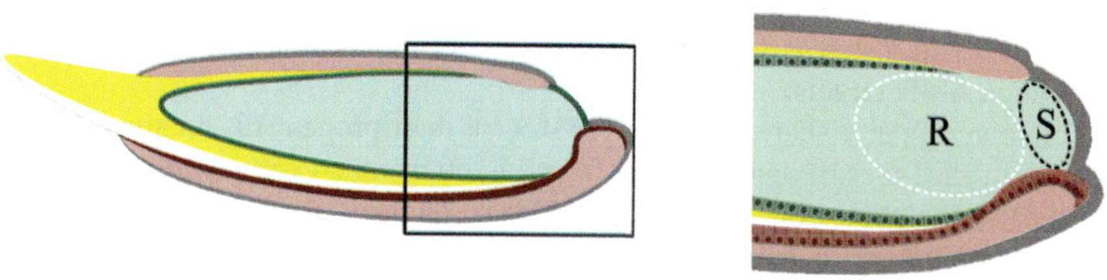

FIGURE 2 Schematic drawing of rapid (R) and slow (S) cycling cell populations in incisor mesenchyme.

et al., 2004; Schwartz and Pirrotta, 2007). Mammalian PRC1 consists of orthologs of *Drosophila* polycomb (Cbx2, Cbx4, Cbx6, Cbx7, and Cbx8), posterior sex combs (Mel18, Bmi1, Nspc1, and MBLR), dRing (Ring1a and Ring1b), and polyhomeotics (Phc1, Phc2 and Phc3). The core PRC2 consists of Suz12, Eed, Ezh1, and Ezh2 (Ringrose and Paro, 2004; Schwartz and Pirrotta, 2007).

Among the core PRC1, Ring1a and Ring1b have been shown to possess E3 ubiquitin ligase activity for histone H2A that plays an important role in PcG-mediated silencing (de Napoles et al., 2004; Wang et al., 2004). Apart from the PRC1, Ring1a/b have also been identified in other protein complexes with BCor and Fbxl10 (Gearhart et al., 2006; Sanchez et al., 2007).

Emerging evidence has demonstrated possible roles of Ring1b in the control of cell proliferation as well as the maintenance of ES cells. Using *Ring1a/b*-knockout mouse ES cells, Ring1a/b have been shown to be required for the maintenance of ES cell identity by silencing the genes that govern differentiation of ES cells. It has been demonstrated that transcriptional repression mediated by Ring1a/b is Oct3/4-dependent. Moreover, in the presence of enforced expression of the differentiation inducer *Gata6*, Ring1a/b target genes become derepressed and the Ring1a/b binding is also reduced significantly. These results indicate that Ring1a/b act downstream of the core transcriptional regulatory circuit to regulate ES cell self-renewal (Endoh et al., 2008). Subsequent studies further support the essential role of Ring1b in stable maintenance of mouse ES cells. To maintain undifferentiated ES cells, Ring1b is needed to silence a particular subset of genes, which are cooccupied by ES cell regulators, including Oct4 and Nanog. These Ring1b target genes also possess bivalent histone marks with CpG-rich promoters and include developmental transcriptional factors, morphogens, and cell-surface markers (van der Stoop et al., 2008).

Ring1a-null mutant mice survive and are relatively normal whereas *Ring1b*-null mice die in early gestation (Voncken et al., 2003). Using an inducible conditional strategy, $Ring1a^{-/-}$;$Ring1b^{fl/fl}$ mice were crossed with tamoxifen-inducible ubiquitous Cre-expressing mice, ERT2-Cre mice, and tamoxifen was administered to postnatal animals to generate $Ring1a^{-/-}$; $Ring1b^{cko/cko}$ mice. These mice thus lost the function of both Ring proteins during postnatal incisor growth, and the result was arrested incisor growth and severely compromised ameloblast and odontoblast differentiation (Lapthanasupkul et al., 2012). Molecular analysis revealed that expression of Fgf10 and Fgf3 in the pulp mesenchyme was lost in the mutants, and consequently, Fgf activity in the cervical loops was downregulated (Lapthanasupkul et al., 2012). This explains the failure of ameloblast development since Fgf signaling is required to maintain the epithelial stem cell niche (Harada et al., 1999, 2002). Failure of odontoblast development is probably a result of loss of TACs, since proliferation was reduced in the mutant mesenchyme (Fig 3). The PRC1 complex thus appears to play a key role directly in maintaining TACs and indirectly in the cervical loops by regulating mesenchymal FGF signaling. The next step is to understand the molecular details of PRC1 function by identification of its targets.

A common property shared by many MSCs (or their progeny) is an apparent ability to migrate toward sites of damage. This function, often called *cell homing*, is an important aspect of the ability of MSCs to repair tissue damage. Based on the observations that the natural repair of severe tooth damage is believed to involve homing of pulp MSCs, cells close to the apical end of mouse incisors were labeled with the lipophilic

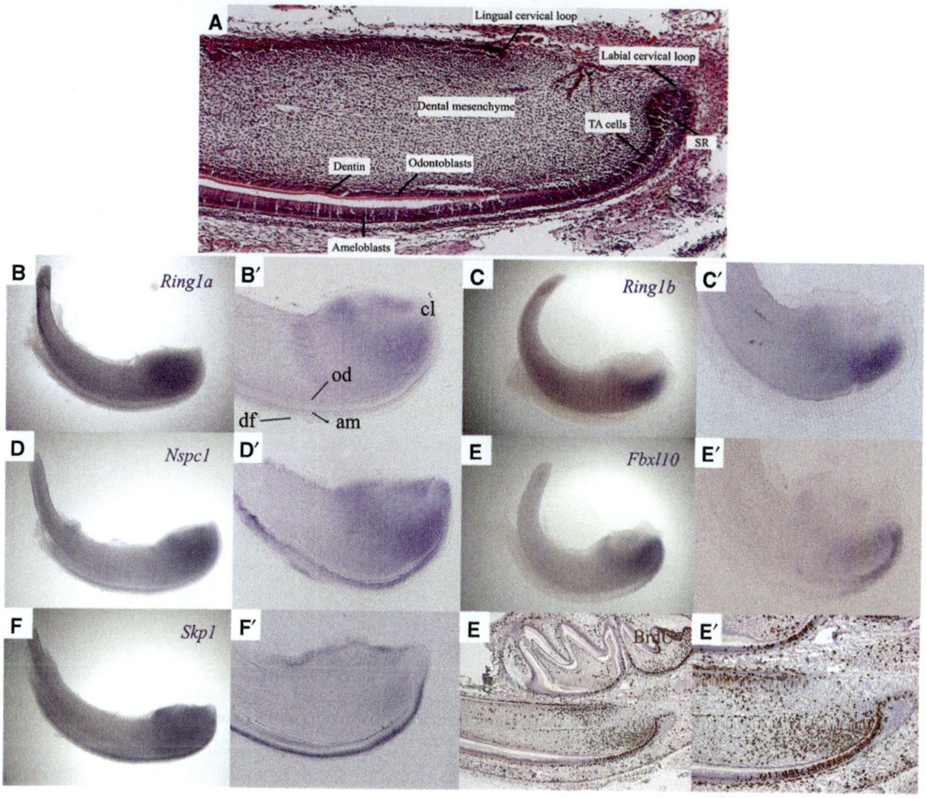

FIGURE 3 Coexpression of genes encoding proteins of the PRC1 complex and cell proliferation marker BrdU in 2-day-old mouse mandibular incisor apical mesenchyme.(A) The apical end of incisor consists of cervical loops at both the lingual and labial sides surrounding the dental mesenchyme. The labial cervical loop contains a stellate reticulum (SR) core surrounded by dental epithelium. TA, transit amplifying. (B, B′, C, and C′) *Ring1a* and *Ring1b* are strongly expressed in the dental mesenchyme close to both the labial and lingual cervical loops of the incisor and in the area of cells with high rates of proliferation (TA cells). (D and D′) *Nspc1* is expressed in the dental mesenchyme near labial and lingual cervical loops, and in the transit-ampifying cells of the dental epithelium. (E and E′) *Fbxl10* expression is notable in the labial mesenchyme and the transit-ampifying cells of the dental epithelium. (F and F′) *Skp1* is weakly expressed in the mesenchyme and more highly expressed in cells of the dental follicle (df) of the incisor germ and the ameloblasts (Am). (G and G′) BrdU expression indicates that highly proliferative cells were located predominantly in the dental mesenchyme near labial and lingual cervical loops, and also in the transit-ampifying cells of the dental epithelium. [Reproduced from Lapthanasupkul et al. (2012)].

dye DiI and their response to tooth damage followed (Feng et al., 2011). When cells at the most apical end of the incisor corresponding to the location of slowly cycling cells were labeled, there was no indication of any migration toward the site of damage (Fig. 4). However, when cells corresponding to those in the area of the highest rates of cell proliferation (see Fig. 2) were labeled, cells could be observed migrating toward the site of damage (Fig. 4).

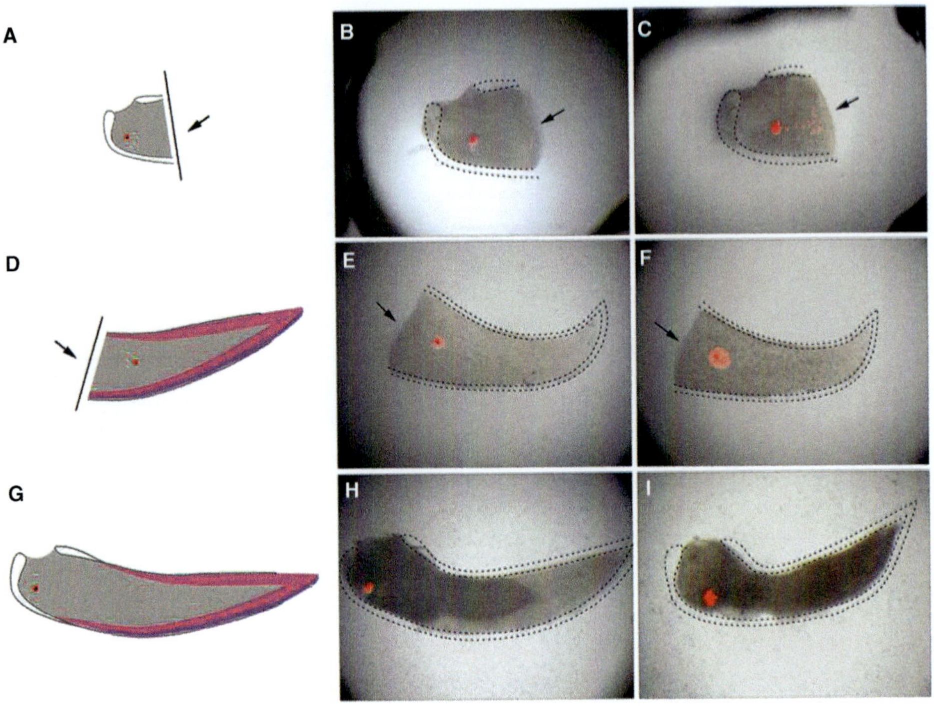

FIGURE 4 Directed cell migration toward tissue damage. Cervical loop area DiI labeling of incisor mesenchyme damaged pulp cells area (A to C), damaged pulp body (D to F), and control teeth with no damage (G to I). A lesion separating the cervical loop and the pulp body was provoked in incisors and 48 h later, mesenchymal cells were capable of migrating toward the site of the damage (A to C), whereas the mesenchymal cells from the pulp body showed no migration (D to F). In the absence of the lesion, no directed migration of cervical loop mesenchymal cells was observed (G to I). [Reproduced from Feng et al. (2011)].

2.2 Perivascular Niche

Perivascular cells, pericytes, have long been associated with MSCs. Most "markers" used to detect and isolate MSCs are expressed by perictyes. Pericytes are widely distributed in the body, and their location on capillaries means that processes such as inflammation in response to tissue damage will naturally lead to recruitment of capillary pericytes to the damage site. Tooth pulp is highly vascular and provides the main nutrient reservoir for the tooth. Pulp cells suggested to have MSC properties are indistinguishable from pericytes by molecular markers. As in all tissues, this begs the question of how we distinguish a pericyte from an MSC.

Confirmation that pericytes are a source of MSCs can, realistically, only come from genetic lineage tracing experiments. Using expression of the pericyte marker NG2 gene, the inducible NG2-Cre-expressing mice crossed with R26R reporter mice results in permanent labeling of NG2-expressing perictyes and their derivatives with β-galactosidase ($LacZ^+$). In these animals the contribution of pericytes to incisor mesenchymal cell differentiation during growth and in response to damage can be followed (Feng et al., 2011). $LacZ^+$ (pericyte-derived) odontoblasts were observed in very small numbers in

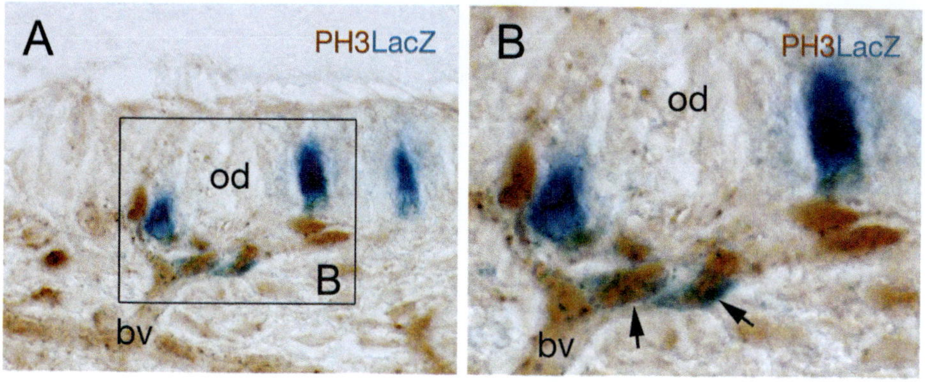

FIGURE 5 Coexpression of proliferation marker PH3 and LacZ$^+$ (NG2-Cre-Labelled) pericytes following *in vivo* damage. Higher magnification of region B showed proliferating pericytes (arrows) are located on blood vessels (bv) extending towards odontoblasts (od) in the damaged area. [Reproduced from Feng et al. (2011)].

postnatal incisors. This suggests that continuous renewal of odontoblasts during normal incisor growth is from a source of MSCs that are not perivascular. The rare occurrence of a perivascular-derived odontoblast may be indicative of a stochastic contribution to the MSC precursor pool of a pericyte MSC.

Following incisor damage, a major contribution of pericyte-derived cells to odontoblast formation was observed in *Ng2-Cre/Rosa26R* animals. Up to 15% of newly formed odontoblasts identified at the site of damage were pericyte-derived. Phosphohistone 3 staining revealed proliferation of the normally quiescent pericytes at the site of damage (Fig 5). Thus, in response to damage, pericytes are stimulated to proliferate and differentiate into odontoblasts that produce reparative dentine. During such response, therefore, two different MSC niches in the incisor are mobilized: the perivascular niche and the apical MSC niche. This dual origin of MSCs involved in incisor repair is indicative of the special properties of rodent incisors and suggests that the perivascular response is generic for all teeth and the additional apical MSC response occurs only in continuously growing teeth. Consistent with this are observations that when molars and incisors are damaged to a similar extent in the same animals, incisors repair more quickly than molars (Sharpe lab, unpublished data).

2.3 Incisal Tip Niche

The wearing of incisor tips during feeding exposes the pulp to the environment, and to prevent bleeding and infection, the cells at the tip undergo a rapid mineralization process that seals off the pulp when the animal stops feeding (Fig. 6). At present nothing is known about the processes that mediate this rapid "repair" may involve pericyte-derived cells, or it is possible that pulp cells at the incisor tip represent a specialized niche devoted to very rapid mineralization. Migration of cells from the apical MSC niche is unlikely since migration is not rapid enough for cells to reach the tip (Fig. 6).

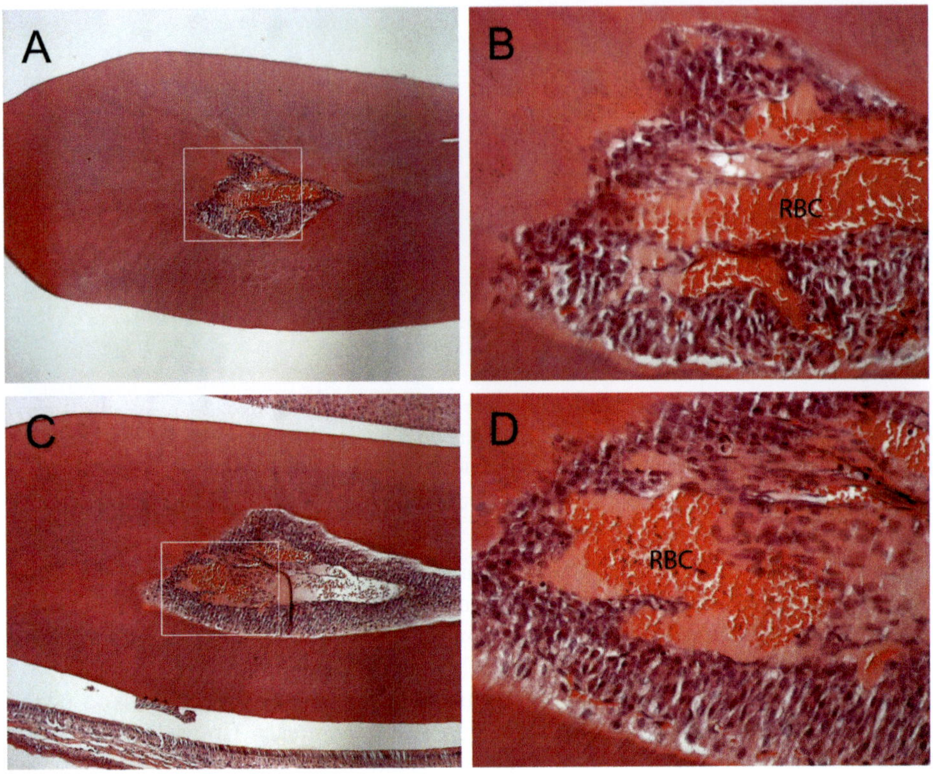

FIGURE 6 H&E staining of sections of the tip of an adult incisor, showing red blood cell (RBC) coagulation and mineralization.

3 MOLAR MESENCHYMAL STEM CELL NICHE

Much of the published research on dental pulp MSCs comes from studies using in vitro culture of cells from human teeth. Since human teeth do not grow continuously, there is no obvious location to search for an MSC niche in the pulp. MSCs resident in the pulp of a human tooth such as a mouse molar are there for tissue repair purposes only. Obvious candidates for the source of MSCs in human pulp are thus pericytes that can readily be visualized on vessels in mouse molar pulp (Fig. 7). In addition, individual pericytes, not localized on vessels can also be seen in molar pulp.

4 FUTURE PROSPECTS

MSCs have great therapeutic potential in tissue regeneration and repair, and for immune suppression. MSC populations that reside in tooth pulp provide an easily accessible cell source. Understanding dental pulp MSC characteristics and behavior is thus becoming increasingly important. The mouse incisor is an excellent model to study pulp MSCs since it uses MSCs for both growth and repair. This, together with the ability to label cells genetically and the use of genetically modified mouse strains, provides significant opportunities to study these cells in ways that are more difficult in other tissues,

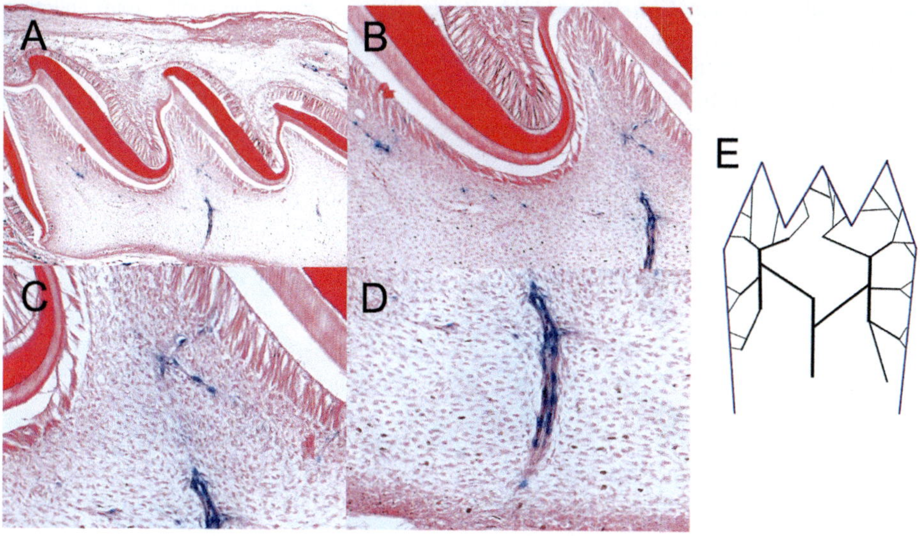

FIGURE 7 Sections of a postnatal day 5 (P5) molar from pericyte LacZ reporter mice (*XLacZ4*) showing localization of pericytes on blood vessels. The diagram in E shows is a representation of blood vessels in the molar pulp.

and impossible in humans. The mouse incisor can thus provide not only invaluable information regarding the processes of tooth growth and repair but also inform our understanding of MSC behavior in general.

Acknowledgments

Research in the Sharpe laboratory is supported by the MRC and Comprehensive BRC. We would like to thank Dr. Andrea Mantesso for contributing to the *XLacZ4* data and Dr. Puangwan Lapthanasupkul to the incisor pulp microarray.

REFERENCES

Cao R, Wang L, Wang H, Xia L, Erdjument-Bromage H, Tempst P, Jones RS, Zhang Y (2002). Role of histone H3 lysine 27 methylation in Polycomb-group silencing. *Science* 298:1039–1043.

de Napoles M, Mermoud JE, Wakao R, Tang YA, Endoh M, Appanah R, Nesterova TB, Silva J, Otte AP, Vidal M, Koseki H, Brockdorff N (2004). Polycomb group proteins Ring1A/B link ubiquitylation of histone H2A to heritable gene silencing and X inactivation. *Dev Cell* 7:663–676.

Endoh M, Endo TA, Endoh T, Fujimura Y, Ohara O, Toyoda T, Otte AP, Okano M, Brockdorff N, Vidal M, Koseki H (2008). Polycomb group proteins Ring1A/B are functionally linked to the core transcriptional regulatory circuitry to maintain ES cell identity. *Development* 135:1513–1524.

Feng J, Mantesso A, de Bari C, Nishiyama A, Sharpe PT (2011). Dual origin of mesenchymal stem cells contributing to organ growth and repair. *Proc Natl Acad Sci USA* 108:6503–6508.

Gearhart MD, Corcoran CM, Wamstad JA, Bardwell VJ (2006). Polycomb group and SCF ubiquitin ligases are found in a novel BCOR complex that is recruited to BCL6 targets. *Mol Cell Biol* 26:6880–6889.

Harada H, Kettunen P, Jung HS, Mustonen T, Wang YA, Thesleff I (1999). Localization of putative stem cells in dental epithelium and their association with Notch and FGF signaling. *J Cell Biol* 147:105–120.

Harada H, Toyono T, Toyoshima K, Yamasaki M, Itoh N, Kato S, Sekine K, Ohuchi H (2002). FGF10 maintains stem cell compartment in developing mouse incisors. *Development* 129:1533–1541.

Lapthanasupkul P, Feng J, Mantesso A, Takada-Horisawa Y, Vidal M, Koseki H, Miletich I, Sharpe PT (2012). Ring1a/b polycomb proteins regulate the mesenchymal stem cell niche in continuously growing incisors. *Dev Biol* 367:140–153.

Ringrose L, Paro R (2004). Epigenetic regulation of cellular memory by the Polycomb and Trithorax group proteins. *Annu Rev Genet* 38:413–443.

Sanchez C, Sanchez I, Demmers JA, Rodriguez P, Strouboulis J, Vidal M (2007). Proteomics analysis of Ring1B/Rnf2 interactors identifies a novel complex with the Fbxl10/Jhdm1B histone demethylase and the Bcl6 interacting corepressor. *Mol Cell Proteom* 6:820–834.

Schwartz YB, Pirrotta V (2007). Polycomb silencing mechanisms and the management of genomic programmes. *Nat Rev Genet* 8:9–22.

Seidel K, Ahn CP, Lyons D, Nee A, Ting K, Brownell I, Cao T, Carano RAD, Curran T, Schober M, Fuchs E, Joyner A, Martin GR, de Sauvage FJ, Klein OD (2010). Hedgehog signalling regulates the generation of ameloblast progenitors in the continuously growing mouse incisor. *Development* 137:3753–3761.

Valk-Lingbeek ME, Bruggeman SW, van Lohuizen M (2004). Stem cells and cancer; the Polycomb connection. *Cell* 118:409–418.

van der Stoop P, Boutsma EA, Hulsman D, Noback S, Heimerikx M, Kerkhoven RM, Voncken JW, Wessels LF, van Lohuizen M (2008). Ubiquitin E3 ligase Ring1b/Rnf2 of Polycomb repressive complex 1 contributes to stable maintenance of mouse embryonic stem cells. *PLoS One* 3: e2235.

Voncken JW, Roelen BA, Roefs M, de Vries S, Verhoeven E, Marino S, Deschamps J, van Lohuizen M (2003). Rnf2 (Ring1b) deficiency causes gastrulation arrest and cell cycle inhibition. *Proc Natl Acad Sci USA* 100:2468–2473.

Wang H, Wang L, Erdjument-Bromage H, Vidal M, Tempst P, Jones RS, Zhang Y (2004). Role of histone H2A ubiquitination in Polycomb silencing. *Nature* 431:873–878.

PART III

STEM CELL–MEDIATED CRANIOFACIAL TISSUE BIOENGINEERING

19

BONE BIOENGINEERING: SCAFFOLDS, GROWTH FACTORS, AND STEM CELLS

Christopher S. D. Lee, Christopher D. Hermann, Rolando Gittens, Rene Olivares-Navarrete, Zvi Schwartz, and Barbara D. Boyan

Wallace H. Coulter Department of Biomedical Engineering, Georgia Institute of Technology, Atlanta, Georgia

1 INTRODUCTION

Craniofacial bone reconstruction is typically required because of trauma, skeletal disease, congenital malformations, periodontal disease, and following cancer surgery. These reconstruction procedures aim to restore the anatomical shape, mechanical stability, and biological properties of craniofacial bone. Despite recent advances in autologous reconstruction, even the most sophisticated techniques are often inadequate to replace normal form and function of craniofacial bone. Additionally, these procedures require extended hospitalization and bone extraction from a secondary site that may lead to morbidity and further complications. Although allo-transplantation can address many of the shortcomings associated with autologous reconstruction, the side effects of lifelong immunosuppression and disease transmission hinder the clinical uptake of this technique. Therefore, craniofacial bone bioengineering is of particular interest.

The combination of scaffolds, growth factors, and stem cells is currently being investigated in preclinical and clinical studies. These different components can be delivered directly to stimulate bone regeneration or as an engineered graft. In the latter method, stem cells and progenitor cells are seeded on a scaffold, differentiated into osteoblasts, and cultured in vitro until mineralization of the construct occurs. In this chapter we describe the various components, different enabling technologies to enhance these components, and their application in craniofacial bone repair.

Stem Cells in Craniofacial Development and Regeneration, First Edition.
Edited by George T.-J. Huang, Irma Thesleff.

2 SCAFFOLDS

In craniofacial bone bioengineering, the scaffold can be used as an osteogenicmaterial to deliver growth factors or progenitor cells or as an osteoconductive matrix to facilitate host cell recruitment. When used to deliver bone morphogenetic protein (BMP), scaffolds may also be osteoinductive. The major classes of materials used as scaffolds are natural polymers, synthetic polymers, ceramics, and composite materials. Each of these materials has unique properties that are advantageous for different bone bioengineering applications (Table 1). The bulk (i.e., tissue level) properties of these scaffolds are important for imparting functionality to bioengineered bones, whereas micro- and nanoscale (i.e., cell and protein level, respectively) properties are important for eliciting a desired biological response (Fig. 1). Therefore, selecting both the appropriate material and the proper design characteristics are important for craniofacial bone regeneration.

2.1 Anatomical Shape and Architecture

Reconstructing anatomical shape and architecture is especially challenging for craniofacial bone bioengineering because restoring natural appearance is of utmost importance. Three-dimensional (3D) imaging modalities such as magnetic resonance imaging (MRI) and computed tomography (CT) are used to fabricate customized bone constructs at the macroscopic and microscopic levels. Microscopic biomaterial architecture parameters such as pore size, density, and interconnectivity are important since these have subsequent effects on local mechanical properties, mass transport, blood vessel invasion, osteoblast proliferation, and osteogenic behavior. Traditional methods of achieving appropriate shape and architecture have included carving, molding, and salt leaching. Scaffolds such as collagen (Chen et al., 2006), poly(caprolactone) (Rohner et al., 2002), hydroxyapatite (HA) (Furst et al., 2003), and β-tricalcium phosphate (TCP) (Zerbo et al., 2005) and hydrogels such as alginate (Weng et al., 2001) and poly(ethylene glycol) diacrylate (Burdick and Anseth, 2002) can be modified using these methods for bone bioengineering.

2.2 Mechanical Properties

Scaffold mechanical properties that mimic those of bone are important not only to provide stability and protection but also to elicit a desired osteogenic response. Stiffer matrices facilitate cell differentiating down the osteogenic lineage, whereas softer matrices that mimic the mechanics of other tissues do not (Engler et al., 2006; Smith et al., 2010). Stiff matrices used for craniofacial bone bioengineering include ceramics such as HA and TCP or composites based on these materials (Shayesteh et al., 2008). Scaffolds with high toughness levels, or the ability to absorb large amounts of energy at failure are also commonly used for bone bioengineering. These materials include natural polymers such as collagen and fibronectin. Crystalline polymers such as poly(methyl methacrylate) (PMMA), have increased stiffness and fatigue strength for craniofacial bone bioengineering (Bruens et al., 2003). Furthermore, scaffold mechanical properties can be varied by adjusting the local architecture or using different biomaterial compositions (Hollister, 2005; Stevens, 2008).

TABLE 1 Characteristics of Selected Materials Used as Scaffolds for Craniofacial Bone Bioengineering

Tissue/ Scaffold	Tensile/Compressive Strength (MPa)	Tensile/Compressive Modulus (GPa)	T_g (°C)	Characteristics for Craniofacial Application	References
Cranial bone	3.9–10.9	1.0–3.9	—	Anisotropic composite material; composition: 60–70% hydroxyapatite, 30–40% organic/H_2O; mechanical properties depend on the direction of the force applied (longitudinal, transverse)	McElhaney et al., 1970
Mandibular bone (cancellous)	3.5–125.6	0.02–0.24	—	Complex geometries, anisotropic mechanical properties	Misch et al., 1999
Femoral bone (cancellous)	0.6–9.3	0.01–0.9	—	—	Hing et al., 1998
Femoral bone (cortical)	133–193	17	—	—	Hing et al., 1998
Alginate (hydrogel)	0.01–0.1	1×10^{-5}–2×10^{-4}	—	Natural hydrogel, moldable, biomimetic modifications	Weng et al., 2001
Collagen (scaffold)	50–100	1	10	Extracellular matrix, enhances toughness, cell adhesion; ceramic, enhances stiffness	Chen et al., 2006
Hydroxyapatite (HA)	1–11	0.2–3.1	—	Biological HA contains impurities such as Mg, Na, and CO_2^{3-}; mechanical properties are dependent on impurities, porosity	Hing et al., 1999a,b
Poly (ε-caprolactone)	16	0.4	−62	Synthetic polymer	Yang et al., 2009; Yeo et al., 2012

(continued)

TABLE 1 ***(Continued)***

Tissue/ Scaffold	Tensile/Compressive Strength (MPa)	Tensile/Compressive Modulus (GPa)	T_g (°C)	Characteristics for Craniofacial Application	References
Poly (ethylene glycol) diacrylaate (hydrogel)	0.01–0.1	1×10^{-6} –5×10^{-3}	—	Synthetic hydrogel, moldable, biomimetic modifications	Burdick and Anseth, 2002
Poly(glycolic acid) (PGA)	647	6.5	36	Synthetic polymer, crystalline, controllable degradation as copolymer with L-PLA	Athanasiou et al., 1995; Oest et al., 2007
Poly(L-lactic acid) L-PLA	28–48	1.2–3.0	54–59	Synthetic polymer, controllable degradation as copolymer with PGA	Pitt et al., 1979a
Poly(methyl methacrylate)	38–80	3–4.8	106–115	Synthetic polymer	Bruens et al., 2003
β-Tricalcium Phosphate (TCP)	2.5–13.5	0.05–0.45	—	Ceramic, enhances stiffness; mechanical data collected for TCP/PLGA construct	Sherwood et al., 2002; Yu et al., 2008

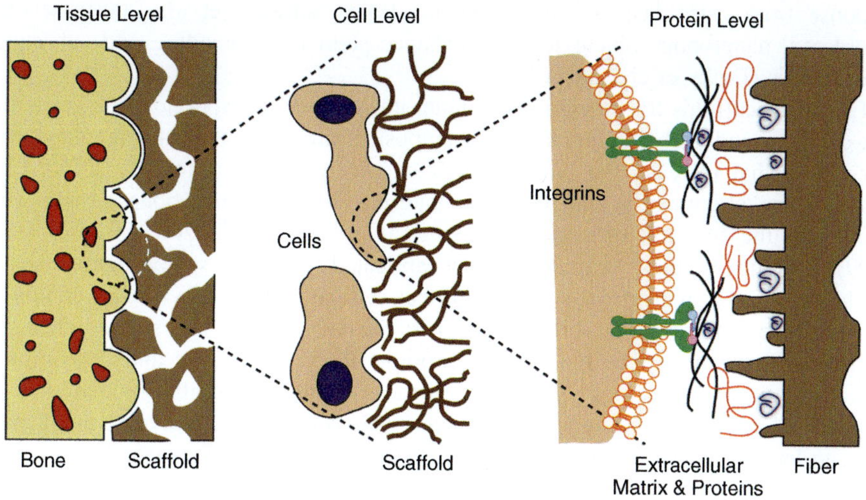

FIGURE 1 Scaffold considerations at the tissue, cell, and protein levels for craniofacial bone bioengineering. These considerations include anatomical shape, architecture, mechanical properties, degradation, surface properties, topographical cues, and cellular adhesion.

2.3 Degradation

Scaffold degradation refers to the process in which covalent bonds are cleaved either chemically (e.g., hydrolysis) or biologically (e.g., enzymatic cleavage). The degradation rate of bioengineered bone scaffolds must be fast enough to accommodate extracellular matrix deposition, and tissue ingrowth while be slow enough to maintain mechanical stability. The degradation products must also be biocompatible while eliciting minimal inflammation. Tissue-derived polymers such as collagen scaffolds can have favorable degradation properties.

Several synthetic, degradable polymers have gained approval by the U.S. Food and Drug Administration for a range of clinical applications. Poly(glycolic acid) (PGA), poly(lactic acid) (PLA), and their copolymers are widely investigated synthetic biomaterials for tissue bioengineering. PGA is highly crystalline, but this property is rapidly lost in copolymers of glycolic acid and lactic acid (PGLA). This change leads to an increase in hydration and hydrolysis, which makes the 50:50 PLGA copolymer degrade faster than either PGA or PLA. PCL is a semicrystalline polymer that degrades at a slower rate than PLA. In preclinical studies, the degradation of bioengineered bone scaffolds can also be used to control the release of stimulatory signaling molecules to promote bone regeneration (Athanasiou et al., 1995). Therefore, PCL is a scaffold biomaterial that can also serve as a drug delivery device that can remain active for over one year (Pitt et al., 1979b).

2.4 Surface Properties and Topographical Cues

Surface properties such as wettability and topography have well-defined influences on osteogenic behavior and endow these scaffolds with their osteoconductive properties. Osteoblasts and stem cells respond to groves, ridges, wells, and other features at the micro- and nanoscale (Healy and Guldberg, 2007; Gittens et al., 2011). On biomaterial surfaces, rougher hydrophilic interfaces elicit a stronger osteogenic response from osteoblasts (Schwartz et al., 1997; Schwarz et al., 2009), whereas disordered, nanotopographical features enhance stem cell spreading and subsequent osteogenesis (Dalby et al., 2007).

Effective scaffolds for bioengineered craniofacial bone incorporate these properties. HA-collagen and carbon polymer nanocomposites are examples of nanophase reinforcements that increase the bioactivity of engineered grafts (Stevens, 2008). Fiber diameter and alignment within these biomaterials also influence bone growth while mimicking the function of periosteal tissues (Guldberg et al., 2007; Bashur et al., 2009). Specifically, these scaffolds can be fabricated with high spatial interconnectivity, high porosity, and oriented fibers to direct cellular morphology and migration for in vivo bone regeneration (Shin et al., 2004; Stevens, 2008). The surface chemistry of these fibers can also be modified to promote direct nucleation and deposition of hydroxyapatite (Hartgerink et al., 2001). Surfaces of fibers and other biomaterials can be modified with other biological components such as glycosaminoglycans and heparin that can localize growth factors important for bone formation (Stevens, 2008).

2.5 Cellular Interaction

The ability of a scaffold not only to support cellular adhesion, but also to direct it, is critical for successful regeneration and osseointegration. Cellular adhesion influences osteoblast migration and spreading and also provides cellular signaling to influence cell behavior and differentiation. Natural scaffolds such as collagen and fibronectin have binding domains for specific heterodimeric transmembrane glycoproteins, or integrin receptors, found on osteoblasts. The integrin receptors consist of two types of subunits: α and β. For osteoblasts, $\alpha_1\beta_1$, $\alpha_2\beta_1$, and $\alpha_3\beta_1$ mediate osteoblast binding and responses to type I collagen, whereas $\alpha_3\beta_1$, $\alpha_4\beta_1$, $\alpha_5\beta_1$, $\alpha_v\beta_1$, and $\alpha_v\beta_3$ mediate osteoblast binding to fibronectin (Anselme, 2000). Integrins on osteoblasts also mediate their responses to microscale surface structure and chemistry on biomaterial surfaces (Olivares-Navarrete et al., 2008). This response is initiated by integrin clustering, which leads to focal adhesions that are important for stimulating osteogenesis and directing bone growth. To enhance the osteogenic properties of synthetic scaffolds, both collagen and fibronectin are used to coat or be incorporated into these structures to engage osteoblastic integrins.

3 GROWTH FACTORS

Prenatal craniofacial bone development and postnatal craniofacial bone regeneration are orchestrated by an array of growth factors. These soluble signaling molecules are responsible for initiating and transitioning between the various stages of intramembranous bone formation and endochondral ossification. In both of these processes, growth factors are required to stimulate and transition between angiogenesis and ossification (Fig. 2).

3.1 Angiogenic

Angiogenic factors stimulate angiogenesis, the growth and sprouting of new blood vessels from preexisting vessels. Angiogenesis is an essential process for bone formation since the newly formed blood vessels facilitate nutrient transport and osteoprogenitor cell migration. Vascular endothelial growth factor (VEGF) promotes endothelial cell growth and migration, and this growth factor is also essential for

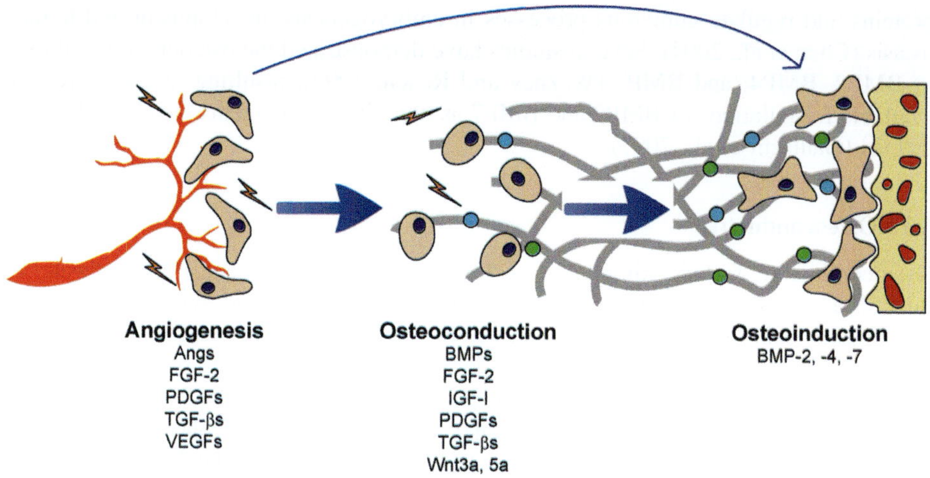

FIGURE 2 Growth factors involved in bone regeneration. Angiogenic factors stimulate the growth and sprouting of new blood vessels from preexisting ones. Osteoinductive or osteoconductive growth factors then stimulate bone formation. Osteoinductive factors act upon undifferentiated stem cells and differentiate them into cells of the osteochondral lineage. These factors can stimulate bone formation de novo. Osteoconductive factors stimulate osteoprogenitor cells in adjacent bone tissue to migrate, proliferate, and synthesize and mineralize osteoid. Osteoconduction and osteoinduction appear to be interlinked, and the order in which osteoconductive or osteoinductive factors act on surrounding cells during bone regeneration may vary or occur simultaneously.

initiating angiogenesis (Gerhardt et al., 2003). Other growth factors known to be mitogenic for endothelial cells include transforming growth factor beta (TGFβ) and fibroblast growth factor 2 (FGF2)(Iruela-Arispe and Sage, 1993; Pepper et al., 1993). Platelet-derived growth factor (PDGF) is responsible for recruiting pericytes to form walls around newly formed vessels, whereas angiopoietins (Angs) are also responsible for stabilizing new capillary sprouting (Lindahl et al., 1997a,b; Suri et al., 1996).

3.2 Osteoinductive

Osteoinductive growth factors are molecules that stimulate the differentiation of stem or progenitor cells toward a boneforming cell lineage. These molecules can induce de novo bone formation in unrelated bone areas where progenitor or stem cells are present. Osteoinduction was first described in the 1930s through the 1950s as the induction of heterotopic bone formation (De Bruyn, 1947; Hartley et al., 1949). However, those initial studies did not describe specific proteins or factors responsible for the osteoinductive capacity. In 1965, Marshal Urist et al., published the seminal experiment with demineralized bone matrix (DBM), which indicated that factors present in bone matrix were responsible for the osteoinduction (Van de Putte and Urist, 1965). The muscle pouch model described by Urist is still the most accepted method to demonstrate osteoinduction. Urist isolated a soluble glycoprotein he termed *bone morphogenetic protein* (BMP) as being responsible for the osteoinductive capacity.

Multiple BMPs have now been reported, several of which are associated with the development of musculoskeletal tissues. BMPs belong to the TGFβ superfamily of proteins and regulate numerous processes in embryogenesis, development and homeostasis (Chen et al., 2004). Several studies have demonstrated the osteoinductive ability of BMP2, BMP4, and BMP7 (Wozney and Rosen, 1998), resulting in commercially available formulations of BMP2 and BMP7 as bone inductive agents for clinical applications (Gautschi et al., 2007).

3.3 Osteoconductive

Osteoconductive factors stimulate osteoprogenitor cells in adjacent bone tissue to migrate and proliferate. These actions lead to subsequent osteoid synthesis and mineralization, resulting in new bone formation that is contiguous with existing bone. Current craniofacial reconstruction procedures may combine materials or grafts with platelet-rich plasma (PRP) to help stimulate bone regeneration. PRP is defined as a sample of plasma with a twofold or greater increase in platelet concentration above baseline levels (Hall et al., 2009). Similar to clot formation, the platelets and fibrin in PRP can form a scaffold when delivered. Within these platelets, over 1500 proteins are present, and many are believed to help facilitate bone regeneration (Nurden et al., 2008). In preclinical trials, PRP treatments enhanced osteoprogenitor cell proliferation, fracture healing, and bone regeneration (Marx, 2001; Choi et al., 2005). However, other preclinical and clinical studies suggest that PRP does not enhance bone healing in oral and maxillofacial applications and may even inhibit the osteoinductivity of DBM (Furst et al., 2003; Jakse et al., 2003; Raghoebar et al., 2005; Ranly et al., 2007). PRP contains high concentrations of PDGFs, and although these growth factors are potent in stimulating osteoblast migration and proliferation, they inhibit osteoinductive agents such as BMPs and DBM (Goodkin and Pierce, 1993; Ranly et al., 2005). Regardless of the therapeutic benefits of PRP, it contains several other osteogenic growth factors, including VEGFs, FGF2, and TGFβs, and insulin-like growth factors (IGFs) (Ranly et al., 2007; Qureshi et al., 2009; Senzel et al., 2009).

FGF signaling via multiple receptors is especially important for skeletal development, and multiple FGFs are involved at different stages of bone formation (Naski and Ornitz, 1998; Miraoui and Marie, 2010). Wnts are also associated with cartilage and bone differentiation, morphogenesis, and development (Tamura et al., 2010; Franz-Odendaal, 2012). It is not completely clear if these factors alone are osteoinductive or if they regulate or interact with BMPs to result in bone formation (Augello and De Bari, 2010; Miraoui and Marie, 2010). However, it is clear that these factors are needed and contribute to bone formation, remodeling, and homeostasis.

Wnt proteins are secreted glycoproteins that control development through regulation of cell proliferation, differentiation, adhesion, and migration in many organs and tissues of mammals (Cadigan, 2012). Specifically, Wnt proteins are important regulators of osteoblast and chondrocyte differentiation and maturation (Boland et al., 2004). These proteins are divided into two classes, based on their biological activities and the terminal fate of β-catenin, which is responsible for the activation of specific transcription factors that regulate cell growth and differentiation. The first class is known as Wnt-β-catenin dependent, or the canonical pathway (Buechling and Boutros, 2011). In this class, activators of the Wnt-canonical pathway are responsible for the stabilization of β-catenin in the cytoplasm. This stabilization leads to nuclear translocation of β-catenin and resulting activation of transcription factors. The second class is known as Wnt-β-catenin independent or Wnt-noncanonical pathways. This class includes at least another two established signaling pathways: the Wnt-Ca-dependent pathway (De, 2011) and the Wnt-planar cell polarity pathway (Wansleeben and Meijlink, 2011). Several

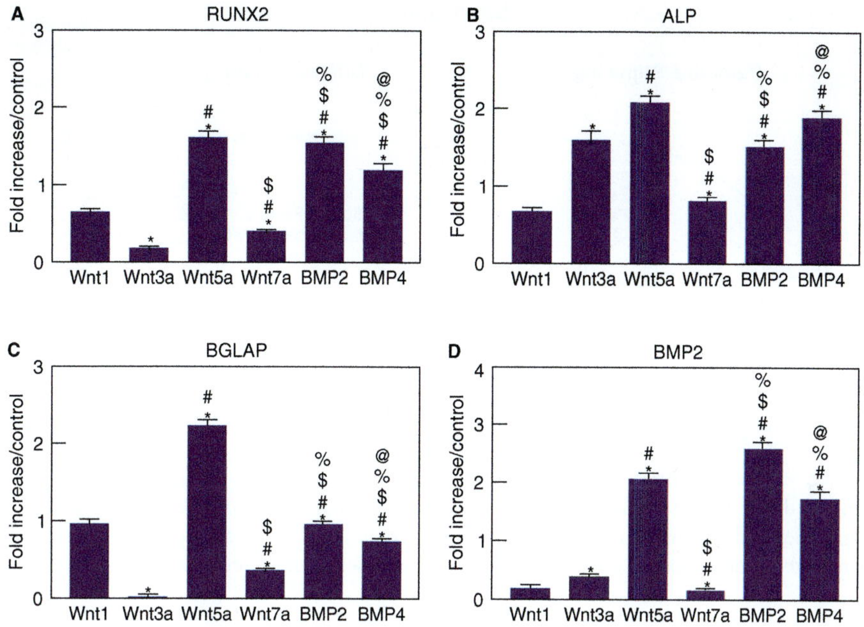

FIGURE 3 Osteogenic response of human bone marrow–derived mesenchymal stem cells (MSCs) to Wnt and BMP ligands. MSCs treated with canonical Wnts (Wnt1, Wnt3a), non-canonical Wnts(Wnt5a, Wnt7a), and BMPs (BMP2, BMP4) regulate mRNA levels of osteogenic markers RUNX2 (A), alkaline phosphatase (ALP, B), osteocalcin (BGLAP, C), and BMP2 (D). $^{*}p < 0.05$, versus Wnt1; $^{\#}p < 0.05$, versus Wnt3a; $^{\$}p < 0.05$, versus Wnt5a; $^{\%}p < 0.05$, versus Wnt7a; $^{@}p < 0.05$, versus BMP2.

studies have shown the contribution of the Wnt-canonical pathway on osteoblastogenesis and bone formation (Marsell et al., 2012). In addition, high- and low-bone-mass phenotypes have been associated with mutation in the low-density lipoprotein receptor 5 (LRP5), which functions as a coreceptor in the Wnt-canonical pathway (Joeng et al., 2011; Korvala et al., 2012). However, recent studies have shown that Wnt3a, one of the most potent Wnt-canonical pathway activators, is needed to retain the pluripotency and multipotency of embryonic and adult stem cells as well negative regulate osteoblastogenesis in mesenchymal stem cells (Derfoul et al., 2004; Baksh et al., 2007).

In recent years, several studies have shown the contribution of Wnt-noncanonical pathways on chondrogenesis and osteogenesis (Church et al., 2002; Santos et al., 2010; Olivares-Navarrete et al., 2011). In the Wnt-Ca-dependent pathway, Wnt proteins stimulate the influx and release of calcium to activate calmodulin-dependent kinase II, protein kinase C (PKC), and calcineurin(Kohn and Moon, 2005). These proteins activate specific signaling cascades to promote osteogenic gene expression (Fig. 3). Within this noncanonical pathway, Wnt5a and Wnt5b are two of the activators, and knockout models for these proteins or their receptors have shown the importance of these proteins in chondrocytes and osteoblast differentiation as well as in cartilage and bone formation and development (Church et al., 2002; Guo et al., 2008; Witte et al., 2009; Bradley and Drissi, 2010; Bradley and Drissi, 2011).

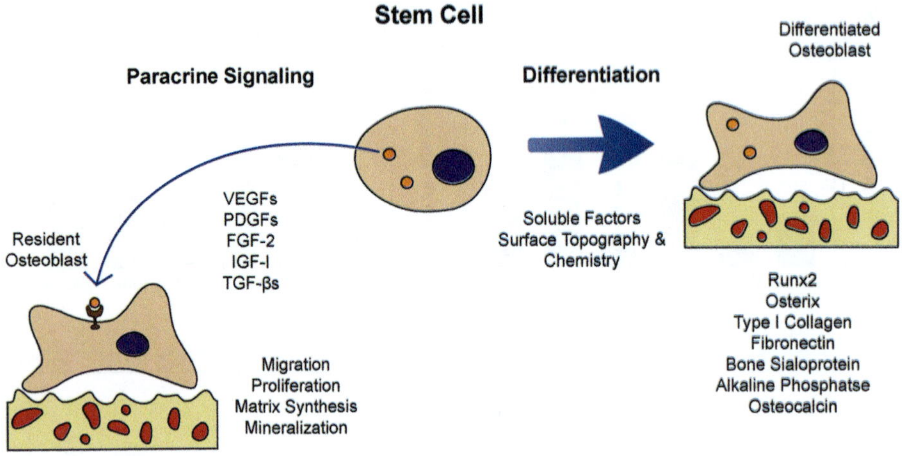

FIGURE 4 Regenerative capacities of stem cells for craniofacial bone. Stem cells can differentiate down the osteoblast lineage (osteogenesis). Different soluble factors and surface topographies and chemistries can mediate this differentiation. Stem cells can also secrete growth factors that act on resident cells (paracrine signaling) to stimulate regeneration.

4 STEM CELLS

Pluripotent and multipotent stem cells derived from different sources are used to bioengineer craniofacial bone. The vast majority of these preclinical and clinical applications use allogeneic or autologous multipotent stem cells derived from different adult tissues. Once isolated, these adult stem cells can be used in one of two ways to regenerate craniofacial bone. The first, and more common method, uses osteogenesis, or differentiation down the osteoblast lineage. Once differentiated into osteoblasts, these cells are used directly to replace damaged or diseased bone by synthesizing and mineralizing new extracellular matrix either in vitro or in situ. The second method uses these stem cells' paracrine signaling or their secreted molecules that act on neighboring cells. These cells are implanted in vivo and act as paracrine factor production sources to stimulate damaged or diseased bone tissue to regenerate itself (Fig. 4).

4.1 Tissue Origin

Mesenchymal stem cells (MSCs) are adult cells that have the capability to differentiate into bone tissue and other mesenchymal lineages. These cells have the ability to repair and regenerate tissue following injury and were first isolated and reported in bone marrow. Since their discovery, bone marrow–derived MSCs have been used in a variety of preclinical applications for craniofacial bone regeneration. Adipose stem cells (ASCs) derived from liposuction aspirates and adipose tissue reductions are an alternative progenitor cell source. Similar to MSCs, ASCs can be seeded on scaffolds and form bone de novo in subcutaneous tissues and calvarial defects in preclinical studies (Yoon et al., 2007; Ward et al., 2010). Stem cells can also be isolated from the periosteum and various craniofacial tissues, such as dental pulp, periodontal ligament, and exfoliated deciduous teeth (Ward et al., 2010; Bulgin et al., 2011).

4.2 Osteogenesis

Despite the wide range of stem cell sources and types, these cells undergo osteogenesis, or differentiation down the osteoblast lineage, in a similar manner. Different soluble factors such as chemicals, growth factors, and hormones, and different nonsoluble components such as extracellular matrix proteins and material surface topography, can facilitate or initiate osteogenesis. This process begins with increased levels of Runx2, a major transcription factor responsible for controlling osteoblast differentiation, and Osterix, a transcription factor downstream of Runx2. This upregulation is followed by three stages: proliferation, matrix secretion, and mineralization. During the proliferative stage, cells express numerous genes related to the cell cycle and extracellular matrix development. Genes for extracellular matrix proteins that are upregulated include type I collagen, fibronectin, and bone sialoprotein. Following proliferation, the gene expression and activity levels of alkaline phosphatase, an enzyme involved in bone mineralization by removing phosphate groups from molecules, increases significantly. During matrix mineralization, alkaline phosphatase activity decreases as secreted levels of osteocalcin, a noncollagenous protein that binds to bone, and other mineralization-associated proteins increase. Understanding the temporal sequence of molecule levels associated with osteogenesis allows scientists and clinicians to better bioengineer craniofacial bone.

4.3 Paracrine Signaling

Implanted stem cells secrete a wide range of paracrine factors that can elicit specific responses from resident cells. These secreted factors are responsible for modulating inflammatory and immune responses and for facilitating tissue regeneration. Specific to craniofacial bone regeneration, stem cells are known to secrete growth factors that stimulate angiogenesis and osteoconduction, such as VEGFs, PDGFs, FGFs, IGF-I, Wnt5a, and the TGFβ superfamily signaling proteins (Baraniak and McDevitt, 2010). These secreted factors stimulate resident osteoblasts and osteoprogenitor cells to migrate, proliferate, and synthesize extracellular matrix. Recent attempts to bioengineer bone in vivo try to enhance the secretion of angiogenic and osteogenic paracrine factors from stem cells by implementing different enabling technologies. These technologies include gene therapy and preconditioning with different biochemical and biophysical stimuli(Baraniak and McDevitt, 2010; Hsiao et al., 2012).

5 ENABLING TECHNOLOGIES

The combination of scaffolds, growth factors, and stem cells alone is often not enough to effectively bioengineer craniofacial bone. Enabling technologies are needed to enhance these constructs at the tissue, cell, and protein levels. In turn, bioengineered bone can better replace and integrate into surrounding craniofacial tissues.

5.1 Morphometric Analysis

The analysis of both craniofacial development and regeneration is complicated by the complex geometry and discontinuity that is seen with many craniofacial structures. Traditionally, craniofacial structures and implanted bioengineered craniofacial bone

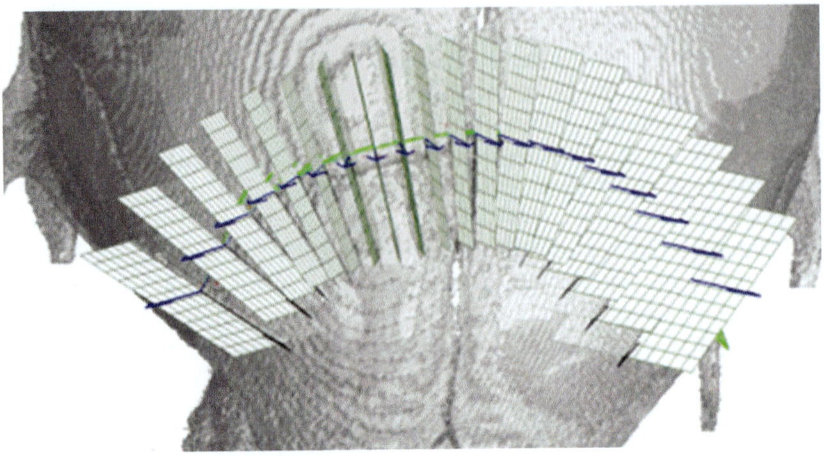

FIGURE 5 Morphometric analysis of three-dimensional computed tomography reconstruction. Advanced reconstruction and segmentation algorithms allow for thorough analysis of complex structures commonly seen in craniofacial bones. These analytical techniques can lead to more anatomically precise bioengineered bones.

have been analyzed with histology. Histological sections can be stained with different chemicals (e.g., hematoxylin and eosin) to make quantitative measurements of defect area, total bone area, marrow area, fibrous tissue area, or void areas. Additionally, sections can be stained with antibodies conjugated to fluorescent molecules to determine the presence of several markers, including angiogenic factors. If cells are implanted, an in vitro imaging system can be used to track them after surgery.

Serial histology can provide complete analysis of craniofacial structures, but it is destructive and can be prohibitively time consuming and expensive for high-throughput applications. Computed tomography for mineralized tissues and magnetic resonance imaging for soft tissues are used as alternative methods to provide high resolution, nondestructive, and complete three-dimensional imaging of craniofacial structures. To better evaluate discontinuities and changes in composition that are commonly seen in craniofacial tissues, advanced reconstruction and segmentation algorithms have recently been developed (Fig. 5). These algorithms allow for thorough analysis of the complex structures commonly seen in the cranial vault and other craniofacial bones (Hermann et al., 2012).

5.2 Three-Dimensional Printing

Once the three-dimensional shape of craniofacial bones can be imaged and analyzed, scaffolds can then be fabricated to match or to improve on the morphology of these tissues using three-dimensional printing, any process that makes three-dimensional objects from a digital file using an additive process as opposed to a subtractive process, such as carving or salt leaching. Recent advances in both computation topology design (CTD) and solid freeform fabrication (SFF), a form of three-dimensional printing, have allowed for rapid prototyping of highly reproducible engineered bone grafts with spatial resolution up to 25 μm (Ballyns and Bonassar, 2009). SFF techniques build structures by depositing materials using a layer-by-layer additive process and can be used with a wide range of polymers and hydrogels. Scaffolds for bioengineered bone can be fabricated directly or cast in a mold created with SFF. Specific strut architectures and precise porosity can also be used with this technology to create desired anisotropic elasticity and permeability (Hollister, 2005). Selective laser melting is another additive fabrication method that fuses fine metal or ceramic powders to create microarchitecture that matches that of trabecular bone (Pattanayak et al., 2011).

5.3 Nanofibrous Matrices

Native extracellular matrix is a nanofibrous network, and several scaffolds used for bioengineering craniofacial bone attempt to mimic this structure. Two common techniques for nanofibrous scaffold fabrication include phase separation and self-assembly. Electrospinning is a more recently developed process that creates nanofibers by applying a high voltage to a syringe filled with a polymer solution. This technology produces both random and oriented nanofiber polymer constructs that provide an ECM-like meshwork for cell attachment and differentiation (Lin et al., 2003; Ji et al., 2006; Ekaputra et al., 2008; Wang et al., 2012). These polymer meshes can also be made into fabric sheets that serve as a barrier, limiting epithelial cell down-growth (Yang et al., 2009) or fibrous connective tissue ingrowth (Rai et al., 2007) in bone bioengineered constructs. By combining a barrier with a biodegradable polymer scaffold, these constructs serve as an osteoconductive conduit for guiding craniofacial bone regeneration.

5.4 Biophysical Stimuli

To improve bioengineered bone osseointegration and function, biophysical stimuli can be used to recreate the mechanical, chemical, and biological properties observed in these tissues. Based on Wolff's law, mechanically stimulating viscoelastic tissue bioengineered constructs can enhance bone regeneration (Boerckel et al., 2009). For developing craniofacial bones in vitro, bioreactors are often used to mimic this dynamic environment. Perfusion and dynamic compression bioreactors have been created for anatomically shaped grafts to manipulate the local mechanical and chemical environment (Grayson et al., 2008). Additionally, spinner flasks and rotating wall vessel bioreactors are used for bone bioengineering (Rauh et al., 2011). Other stimuli, such as ultrasound (El-Bialy, 2007) and electromagnetic fields (Schwartz et al., 2008; Rauh et al., 2011) can be used to enhance bioengineering of craniofacial bones.

5.5 Biomimetic Peptides

To match or improve on the biological actions of native tissue, biomimetic peptides can be incorporated into bioengineered bone scaffolds. These biomimetic modifications are based on motifs found in different extracellular matrices such as collagen, fibronectin, and bone sialoprotein (Reyes and Garcia, 2003; Keselowsky et al., 2005; Chung et al., 2006). Arginine–glycine–aspartic acid (RGD) peptides derived from fibronectin are the most commonly used biomimetic modification for bone bioengineering scaffolds to promote cell attachment and proliferation (Rowley et al., 1999). By adjusting the presentation and surface density of these peptides, osteoblast adhesion, migration, and differentiation can be tailored. Additionally, these specific peptide sequences can provide more direct stimulatory signaling while reducing infiltration of nonosteogenic specific cells. Matrix metalloproteinase (MMP)–degradable peptides can also be incorporated into bioengineered bone scaffolds to tailor biomaterial degradation (Kim and Healy, 2003; Kim et al., 2005). This approach allows scaffolds to respond better to resident stem cells and MMPs naturally secreted during bone regeneration.

5.6 Growth Factor Delivery

To enhance osteoconductive properties while imparting osteoinductive properties to bioengineered bone scaffolds, growth factors can be incorporated into different matrices. BMPs can be released from solid constructs to enhance bone formation, and angiogenic factors can be incorporated to address mass transfer concerns (Oest et al., 2007; Guldberg, 2009). Based on different scaffold degradation properties, the release kinetics of multiple factors, such as both angiogenic and osteogenic factors, can be manipulated to provide more robust tissue formation in bioengineered bone (Richardson et al., 2001).

5.7 Gene Therapy

Gene therapy has been used for bone bioengineering to transfer genetic material into cells and stem cells using viral and nonviral vectors. Although nonviral vectors are safer than viral vectors, their efficacy in transmitting genetic material into host cells is lower. Both of these vectors can be used either to enhance the osteogenic capability of stem cells or to enhance their paracrine actions. For enhancing osteogenesis, gradients of immobilized retrovirus encoding Runx2 can be used to re-create the mineralized gradient observed at the bone–soft tissue interface (Phillips et al., 2008). To enhance stem cell paracrine actions, implanted MSCs, ASCs, and muscle-derived stem cells overexpressing BMP2, 4, 7, and 9, VEGF, PDGF-B, and IGF-I can improve endogenous MSC recruitment, cell survival, and bone regeneration (Gamradt and Lieberman, 2004; Tai et al., 2008; Waese et al., 2008; Ward et al., 2010).

6 APPLICATION OF SCAFFOLDS, GROWTH FACTORS, AND STEM CELLS

Unlike the appendicular skeleton, the cranial skeleton is derived from neural crest and paraxial mesodermal origin. Depending on their location, these different bones have distinct shapes and functions. Additionally, these tissues may undergo either endochondral or intramembranous ossification. Therefore, bone bioengineering techniques to regenerate craniofacial tissues may differ between different regions of the skull. Despite the wide array of materials and technologies available for craniofacial bone bioengineering, only a few have been used in clinical studies. The vast majority of craniofacial bone bioengineering studies are preclinical, having occurred in different animal models or in vitro (Table 2).

6.1 Preclinical

Extensive preclinical studies are essential to bring effective bone bioengineering applications to the clinic. These studies typically involve two principal design components.

The first component is the animal model. These serve as powerful tools that mimic bone deformities, injuries, and abnormal growth seen in human patients. Within these models, bioengineered craniofacial bone are exposed to several different cell types that can induce different tissue formation, encounter aggressive local and systemic environment that may enhance graft outcome, and experience a multitude of mechanical loads. These diverse factors are currently impossible to reproduce in vitro. The second component is the optimization of different materials and technologies to regenerate bone within these models while minimizing any side effects. These bioengineered bones also undergo extensive testing in vitro prior to implantation within animals.

Calvaria-defect models are the most reported preclinical studies. Mouse and rat models of calvarial defects are the most common since they are easy to breed and husbandry is simple. An advantage of using these models is the availability of athymic models, which allow for study of materials from other species while limiting host-versus-graft rejection responses. Also, these models are often used to examine the time course of bone formation and remodeling within the defect. Mice models are especially useful for recreating the genetic mutations observed in congenital diseases such as syndromic craniosynostosis, where bone growth is accelerated. Not only is functional and anatomically accurate bone regeneration important, but controlled bone regeneration to provide normal cranial development is just as critical. Larger animals, such as rabbits (Chen et al., 2009; Peters and Courtemanche, 2009; Yeom et al., 2008), sheep (Icekson et al., 2009), and goats, are also used for calvarial defect models (Yu et al., 2008).

Several combinations of scaffolds, stem cells, and growth factors have been used preclinically to regenerate calvarial defects. In mice and rats, scaffolds used range from simple versions such as gelfoam (Zhang et al., 2010), PLGA scaffolds (Cowan et al., 2007; Yoon et al., 2007), vitrigel (Zhao et al., 2007), or silk fibroin (Zhang et al., 2010), to more complex materials such as PEG hydrogels modified with biomimetic peptides (Degano et al., 2008) or apatite-coated PLGA scaffolds (Wan et al., 2007). Stem cells from various sources have also been used, including cells that overexpress Runx2 (Zhao et al., 2007), Osterix (Tu et al., 2007), or BMP7 (Zhang et al., 2010). Additionally, growth factors used in rat calvarial defects include BMP2 (Cowan et al., 2007; Schwarz et al., 2009; Young et al., 2009; Stephan et al., 2010) and VEGF (Young et al., 2009). In rabbits, sheep, and goats, scaffolds such as PCL (Lam et al., 2009), resorbable PEG hydrogels (Humber et al., 2010), silica gels (Seol et al., 2009), hyaluronic acid hydrogels (Yeom et al., 2010), TCP–chitosan–collagen hydrogels, and PLGA (Haberstroh et al., 2010) have been implanted. These scaffolds were either implanted alone or used to deliver an osteoinductive factor such as BMP2.

Bioengineering of other craniofacial bones has been investigated in larger-animal models. Maxillary sinus augmentation has been performed in rabbits (Watanabe et al., 1999; Asai et al., 2002), monkeys (Hanisch et al., 1997; Hurzeler et al., 1997a,b; Kirker-Head et al., 1997; Quinones et al., 1997a,b), minipigs (Furst et al., 2003), goats (Nevins et al., 1996), and sheep (Haas et al., 1998; Sauerbier et al., 2010). Within these sinus lift models, osteoinductive growth factors have been combined with graft materials (Sicca et al., 2008; Gruber et al., 2009). Mandible and alveolar ridge reconstructions have been investigated in canine and minipig models. Scaffolds such as porous HA, fibrin, and PCL-TCP have been used to deliver differentiated MSCs (Cancedda et al., 2007; Yeo et al., 2012). For mandible bioengineering, patient-specific anatomically shaped scaffolds are especially important because

TABLE 2 Preclinical and Clinical Applications of Scaffolds, Stem Cells, and Growth Factors to Bioengineer Craniofaical Bones

Tissue Reconstruction	Scaffolds	Stem Cells	Growth Factors	References
Preclinical				
Calvarium	PLGA, PEG, PCL, TCP, silk fibroin	Genetically modified bone marrow MSCs, ASCs	BMPs, VEGF	Degano et al., 2008; Lam et al., 2009; Zhang et al., 2010
Mandible	HA, fibrin, PCL, TCP	Bone marrow MSCs	BMPs	Cancedda et al., 2007; Yeo et al., 2012
Maxillary sinus	Collagen, HA, TCP	Bone marrow MSCs	BMP2, BMP14	Gruber et al., 2009; Sicca et al., 2008
Clinical				
Aveolar ridge	Fibrin	Bone marrow MSCs	PRP (PDGFs, TGFβs, VEGF, and IGF-I)	Hibi et al., 2006b
Hard palate	Collagen	Periosteum stem cells	BMP2	Carstens et al., 2005
Mandible	HA, fibrin	Bone marrow MSCs	BMP7, PRP (PDGFs, TGFβs, VEGF, and IGF-I)	Gronthos, 2004; Hibi et al., 2006a
Maxillary sinus	Collagen, HA, TCP, hyaluronic acid	Bone marrow MSCs	BMP2 BMP7	Shayesteh et al., 2008; Zijderveld et al., 2005; Schwartz et al., 2007; Susarla et al., 2011

of complex geometry and high variability between different patients. Three-dimensional reconstructions from CT scans can be used to create nanofibrous scaffolds to support bone formation and to restore natural appearance (Chen et al., 2006).

6.2 Clinical

Although preclinical studies provide invaluable information for designing and optimizing materials and technologies for craniofacial bone bioengineering, results from these studies do not definitively predict clinical performance. One of the few FDA-approved applications of growth factors for craniofacial reconstruction is the use of BMP2 for maxillary sinus augmentation. A randomized multicenter prospective trial compared the effectiveness of BMP2 on a collagen scaffold versus autologous bone graft for two-stage maxillary sinus floor augmentation. Success rates for both were above 75%, with significantly denser bone in the BMP2 group. Absorbable collagen sponges saturated with BMP2 can also be used to recreate the hard palate (Carstens et al., 2005).

Other maxillofacial bone bioengineering approaches without using BMP2 have been investigated. Pure TCP scaffolds alone are able to stimulate new bone formation in sinus floor augmentation after six months, but perform no better than does an autologous chin graft (Zijderveld et al., 2005). Loading autologous MSCs on a biphasic HA/TCP have the potential to provide better maxillary sinus augmentation (Shayesteh et al., 2008). These scaffolds have a mean pore size of 300 to 500 μm and are able to promote osseointegration after four months. Many clinical attempts at maxillofacial reconstruction have combined bone graft–derived material with various organic and inorganic components of bioengineered bone. Combining hyaluronic acid with demineralized freeze-dried bone allograft makes it possible to enhance new bone formation compared to each material individually (Schwartz et al., 2007). Additionally, combining allogenic graft material with different stem cells is being investigated to reconstruct massive maxillofacial defects (Susarla et al., 2011).

Several clinical cases of successful craniofacial bone bioengineering besides maxillary augmentation have also been reported. An alveolar ridge can be constructed using differentiated MSCs and PRP mixed with thrombin and calcium chloride (Hibi et al., 2006b). The PRP not only contains fibrinogen to form a fibrin matrix, but also osteoconductive growth factors that can facilitate regeneration. This approach has also been used to reconstruct the mandible (Hibi et al., 2006a). Mandible replacements consistent of HA blocks coated with BMP7 and MSCs can also be computer designed based on tomography data of a defect region (Gronthos, 2004).

Interestingly, the first clinical application of a decellularized tissue scaffold for bone tissue engineering was cleared by the FDA for use in periodontal therapy. This material, known commercially as Emdogain (Institut Straumann AG, Basel, Switzerland), consists of enamel matrix proteins derived from porcine tooth germs (EMDs) delivered with a carrier. When reconstituted with saline or other fluids, an EMD forms a matrix that facilitates the migration of progenitor cells and their differentiation into osteoblasts, cementoblasts, and ligament cells (Hammarstrom, 1997; Boyan et al., 2000; Lyngastadaas et al., 2001; He et al., 2004; Sculean et al., 2007).

7 FUTURE CONSIDERATIONS

Over the past three decades, numerous materials and enabling technologies have been developed to better bioengineer craniofacial bone. These constructs have had promising results in preclinical studies and are starting to become more common in clinical applications. Despite these advances, bone autografts are still considered the gold standard to augment or enhance craniofacial bone regeneration. Although it may just be a matter of time before bioengineered bone becomes the standard for craniofacial reconstruction, these constructs must first show a clinical improvement in osseointegration and regeneration.

To improve clinical performance of bioengineered craniofacial bone, more technologies developed preclinically must be translated into clinical studies. For example, numerous growth factors are known to facilitate bone development and regeneration, but only exogenous BMPs are prevalent clinically. It is possible that delivering more of these growth factors clinically will make bioengineered bone a more feasible option. However, effective enabling technologies must be implemented to control their delivery to optimize their performance and reduce any undesired side effects.

At the same time, engineers, biologists, and clinicians must consider the complexity of the final implant, the ability for it to be assembled quickly, and the capability for it to be incorporated into standard surgical procedures. These implants must also conform to regulatory guidelines while showing an improvement in safety and efficacy compared to autologous bone grafts. Regardless of how bioengineered craniofacial bone achieves mass adoption clinically, it is clear that optimizing scaffolds, stem cells, and growth factors will play a critical role.

Acknowledgments

The authors thank the following funding agencies for their support: the U.S. Department of Defense, the National Science Foundation, Children's Healthcare of Atlanta, IFARHU-SENACYT (Panama), the ITI Foundation (Switzerland), and the National Institutes of Health.

REFERENCES

Anselme K (2000). Osteoblast adhesion on biomaterials. *Biomaterials* 21:667–681.

Asai S, Shimizu Y, Ooya K (2002). Maxillary sinus augmentation model in rabbits: effect of occluded nasal ostium on new bone formation. *Clin Oral Implants Res* 13:405–409.

Athanasiou KA, Singhal AR, Agrawal CM, Boyan BD (1995). In vitro degradation and release characteristics of biodegradable implants containing trypsin inhibitor. *Clin Orthop Relat Res* 272–281.

Augello A, De Bari C (2010). The regulation of differentiation in mesenchymal stem cells. *Hum Gene Ther* 21:1226–1238.

Baksh D, Boland GM, Tuan RS (2007). Cross-talk between Wnt signaling pathways in human mesenchymal stem cells leads to functional antagonism during osteogenic differentiation. *J Cell Biochem* 101:1109–1124.

Ballyns JJ, Bonassar LJ (2009). Image-guided tissue engineering, *J Cell Mol Med* 13:1428–1436.

Baraniak PR, McDevitt TC (2010). Stem cell paracrine actions and tissue regeneration. *Regen Med* 5:121–143.

Bashur CA, Shaffer RD, Dahlgren LA, Guelcher SA, Goldstein AS (2009). Effect of fiber diameter and alignment of electrospun polyurethane meshes on mesenchymal progenitor cells. *Tissue Eng A* 15:2435–2445.

Boerckel JD, Dupont KM, Kolambkar YM, Lin AS, Guldberg RE (2009). In vivo model for evaluating the effects of mechanical stimulation on tissue-engineered bone repair. *J Biomech Eng* 131:084502.

Boland GM, Perkins G, Hall DJ, Tuan RS (2004). Wnt 3a promotes proliferation and suppresses osteogenic differentiation of adult human mesenchymal stem cells. *J Cell Biochem* 93:1210–1230.

Boyan BD, Weesner TC, Lohmann CH, Andreacchio D, Carnes DL, Dean DD, Cochran DL, Schartz Z (2000). Porcine fetal enamel matrix dervivative enhances bone formation induced by dermineralized freeze dried bone allograft in vivo. *J Periodontol* 71:1278–1286.

Bradley EW, Drissi MH (2010). WNT5A regulates chondrocyte differentiation through differential use of the CaN/NFAT and IKK/NF-kappaB pathways. *Mol Endocrinol* 24:1581–1593.

Bradley EW, Drissi MH (2011). Wnt5b regulates mesenchymal cell aggregation and chondrocyte differentiation through the planar cell polarity pathway. *J Cell Physiol* 226:1683–1693.

Bruens ML, Pieterman H, de Wijn JR, Vaandrager JM (2003). Porous polymethylmethacrylate as bone substitute in the craniofacial area. *J Craniofac Surg* 14:63–68.

Buechling T, Boutros M (2011). Wnt signaling signaling at and above the receptor level. *Curr Top Dev Biol* 97:21–53.

Bulgin D, Hodzic E, Komljenovic-Blitva D (2011). Advanced and prospective technologies for potential use in craniofacial tissues regeneration by stem cells and growth factors. *J Craniofac Surg* 22:342–348.

Burdick JA, Anseth KS (2002). Photoencapsulation of osteoblasts in injectable RGD-modified PEG hydrogels for bone tissue engineering. *Biomaterials* 23:4315–4323.

Cadigan KM (2012). TCFs and Wnt/beta-catenin signaling: more than one way to throw the switch. *Curr Top Dev Biol* 98:1–34.

Cancedda R, Giannoni P, Mastrogiacomo M (2007). A tissue engineering approach to bone repair in large animal models and in clinical practice. *Biomaterials* 28:4240–4250.

Carstens MH, Chin M, Ng T, Tom WK (2005). Reconstruction of #7 facial cleft with distraction-assisted in situ osteogenesis (DISO): role of recombinant human bone morphogenetic protein-2 with Helistat-activated collagen implant. *J Craniofac Surg* 16:1023–1032.

Chen D, Zhao M, and Mundy GR (2004). Bone morphogenetic proteins, *Growth Factors* 22:233–241.

Chen KY, Shyu PC, Dong GC, Chen YS, Kuo WW, Yao CH (2009). Reconstruction of calvarial defect using a tricalcium phosphate–oligomeric proanthocyanidins cross-linked gelatin composite. *Biomaterials* 30:1682–1688.

Chen VJ, Smith LA, Ma PX (2006). Bone regeneration on computer-designed nano-fibrous scaffolds. *Biomaterials* 27:3973–3979.

Choi BH, Zhu SJ, Kim BY, Huh JY, Lee SH, Jung JH (2005). Effect of platelet-rich plasma (PRP) concentration on the viability and proliferation of alveolar bone cells: an in vitro study. *Int J Oral Maxillofac Surg* 34:420–424.

Chung EH, Gilbert M, Virdi AS, Sena K, Sumner DR, Healy KE (2006). Biomimetic artificial ECMs stimulate bone regeneration. *J Biomed Mater Res A* 79:815–826.

Church V, Nohno T, Linker C, Marcelle C, Francis-West P (2002). Wnt regulation of chondrocyte differentiation. *J Cell Sci* 115:4809–4818.

Cowan CM, Aghaloo T, Chou YF, Walder B, Zhang X, Soo C, Ting K, Wu B (2007). MicroCT evaluation of three-dimensional mineralization in response to BMP-2 doses in vitro and in critical sized rat calvarial defects. *Tissue Eng* 13:501–512.

Dalby MJ, Gadegaard N, Tare R, Andar A, Riehle MO, Herzyk P, Wilkinson CD, Oreffo RO (2007). The control of human mesenchymal cell differentiation using nanoscale symmetry and disorder. *Nat Mater* 6:997–1003.

De A (2011). Wnt/Ca2$^+$ signaling pathway: a brief overview. *Acta Biochim Biophys Sin (Shanghai)* 43:745–756.

De Bruyn PP (1947). Bone formation by fresh and frozen transplants of bone, bone marrow and periosteum, Anat Rec, 99:641.

Degano IR, Vilalta M, Bago JR, Matthies AM, Hubbell JA, Dimitriou H, Bianco P, Rubio N, Blanco J (2008). Bioluminescence imaging of calvarial bone repair using bone marrow and adipose tissue-derived mesenchymal stem cells. *Biomaterials* 29:427–437.

Derfoul A, Carlberg AL, Tuan RS, Hall DJ (2004). Differential regulation of osteogenic marker gene expression by Wnt-3a in embryonic mesenchymal multipotential progenitor cells. *Differentiation* 72:209–223.

Ekaputra AK, Prestwich GD, Cool SM, Hutmacher DW (2008). Combining electrospun scaffolds with electrosprayed hydrogels leads to three-dimensional cellularization of hybrid constructs. *Biomacromolecules* 9:2097–2103.

El-Bialy T (2007). Therapeutic ultrasound applications in craniofacial growth, healing and tissue engineering. *Rejuvenation Res* 10:367–371.

Engler AJ, Sen S, Sweeney HL, Discher DE (2006). Matrix elasticity directs stem cell lineage specification. *Cell* 126:677–689.

Franz-Odendaal TA (2012). Induction and patterning of intramembranous bone. *Front Biosci* 17:3734–3746.

Furst G, Gruber R, Tangl S, Zechner W, Haas R, Mailath G, Sanroman F, Watzek G (2003). Sinus grafting with autogenous platelet-rich plasma and bovine hydroxyapatite: a histomorphometric study in minipigs. *Clin Oral Implants Res* 14:500–508.

Gamradt SC, Lieberman Jr (2004). Genetic modification of stem cells to enhance bone repair. *Ann Biomed Eng* 32:136–147.

Gautschi OP, Frey SP, and Zellweger R (2007). Bone morphogenetic proteins in clinical applications. *ANZ J Surg* 77:626–631.

Gerhardt H, Golding M, Fruttiger M, Ruhrberg C, Lundkvist A, Abramsson A, Jeltsch M, Mitchell C, Alitalo K, Shima D, Betsholtz C (2003). VEGF guides angiogenic sprouting utilizing endothelial tip cell filopodia. *J Cell Biol* 161:1163–1177.

Gittens RA, McLachlan T, Olivares-Navarrete R, Cai Y, Berner S, Tannenbaum R, Schwartz Z, Sandhage KH, Boyan BD (2011). The effects of combined micron-/submicron-scale surface roughness and nanoscale features on cell proliferation and differentiation. *Biomaterials* 32:3395–3403.

Goodkin DA, Pierce GF (1993). Role of platelet-derived growth factor in osteoblast function and bone synthesis. *Wound Repair Regen* 1:203–212.

Grayson WL, Chao PH, Marolt D, Kaplan DL, Vunjak-Novakovic G (2008). Engineering custom-designed osteochondral tissue grafts. *Trends Biotechnol* 26:181–189.

Gronthos S (2004). Reconstruction of human mandible by tissue engineering. *Lancet* 364:735–736.

Gruber RM, Ludwig A, Merten HA, Pippig S, Kramer FJ, Schliephake H (2009). Sinus floor augmentation with recombinant human growth and differentiation factor-5 (rhGDF-5): a pilot study in the Goettingen miniature pig comparing autogenous bone and rhGDF-5. *Clin Oral Implants Res* 20:175–182.

Guldberg RE (2009). Spatiotemporal delivery strategies for promoting musculoskeletal tissue regeneration. *J Bone Miner Res* 24:1507–1511.

Guldberg RE, Oest ME, Dupont K, Peister A, Deutsch E, Kolambkar Y, Mooney D (2007). Biologic augmentation of polymer scaffolds for bone repair. *J Musculoskelet Neuronal Interact* 7:333–334.

Guo J, Jin J, Cooper LF (2008). Dissection of sets of genes that control the character of wnt5a-deficient mouse calvarial cells. *Bone* 43:961–971.

Haas R, Mailath G, Dortbudak O, Watzek G (1998). Bovine hydroxyapatite for maxillary sinus augmentation: analysis of interfacial bond strength of dental implants using pull-out tests. *Clin Oral Implants Res* 9:117–122.

Haberstroh K, Ritter K, Kuschnierz J, Bormann KH, Kaps C, Carvalho C, Mulhaupt R, Sittinger M, Gellrich NC (2010). Bone repair by cell-seeded 3D-bioplotted composite scaffolds made of collagen treated tricalcium phosphate or tricalcium phosphate-chitosan-collagen hydrogel or PLGA in ovine critical-sized calvarial defects. *J Biomed Mater Res B* 93:520–530.

Hall MP, Band PA, Meislin RJ, Jazrawi LM, Cardone DA (2009). Platelet-rich plasma: current concepts and application in sports medicine. *J Am Acad Orthop Surg* 17:602–608.

Hammarstrom L (1997). Enamel matrix, cementum development and regeneration. *J Clin Periodontol* 24:658–668.

Hanisch O, Tatakis DN, Rohrer MD, Wohrle PS, Wozney JM, Wikesjo UM (1997). Bone formation and osseointegration stimulated by rhBMP-2 following subantral augmentation procedures in nonhuman primates. *Int J Oral Maxillofac Implants* 12:785–792.

Hartgerink JD, Beniash E, Stupp SI (2001). Self-assembly and mineralization of peptide-amphiphile nanofibers. *Science* 294:1684–1688.

He J, Jiang J, Safavi KE, Spangberg LS, Zhu Q (2004). Emdogain promotes osteoblast proliferation and differentiation and stimulates osteoprotegerin expression. *Oral Surg Oral Med Oral Pathol Oral Radiol Endod* 97:239–245.

Healy KE, Guldberg RE (2007). Bone tissue engineering. *J Musculoskelet Neuronal Interact* 7:328–330.

Hartley J, Tanz SS, and Schneider M (1949). Osteogenesis produced by a chemical extract of bone. *J Mt Sinai Hosp N Y* 15:383–387.

Hermann CD, Richards MA, Olivares-Navarrete R, Williams JK, Guldberg RE, Skrinjar O, Schwartz Z, Boyan BD (2012). Algorithm to assess cranial suture fusion with varying and discontinuous mineral density. *Ann Biomed Eng* 40:1597–1609.

Hibi H, Yamada Y, Kagami H, Ueda M (2006a). Distraction osteogenesis assisted by tissue engineering in an irradiated mandible: a case report. *Int J Oral Maxillofac Implants* 21:141–147.

Hibi H, Yamada Y, Ueda M, Endo Y (2006b). Alveolar cleft osteoplasty using tissue-engineered osteogenic material. *Int J Oral Maxillofac Surg* 35:551–555.

Hing KA, Best SM, Tanner KE, Revell PA, Bonfield W (1998). Histomorphological and biomechanical characterization of calcium phosphates in the osseous environment. *Proc Inst Mech Eng H* 212:437–451.

Hing KA, Best SM, Bonfield W (1999a). Characterization of porous hydroxyapatite. *J Mater Sci Mater Med* 10:135–145.

Hing KA, Best SM, Tanner KE, Bonfield W, Revell PA (1999b). Quantification of bone ingrowth within bone-derived porous hydroxyapatite implants of varying density. *J Mater Sci Mater Med* 10:663–670.

Hollister SJ (2005). Porous scaffold design for tissue engineering. *Nat Mater* 4:518–524.

Hsiao ST, Asgari A, Lokmic Z, Sinclair R, Dusting GJ, Lim SY, Dilley RJ (2012). Comparative analysis of paracrine factor expression in human adult mesenchymal stem cells derived from bone marrow, adipose, and dermal tissue. *Stem Cells Dev* 21:2189–2203.

Humber CC, Sandor GK, Davis JM, Peel SA, Brkovic BM, Kim YD, Holmes HI, and Clokie CM (2010). Bone healing with an in situ-formed bioresorbable polyethylene glycol hydrogel membrane in rabbit calvarial defects. *Oral Surg Oral Med Oral Pathol Oral Radiol Endod* 109:372–384.

Hurzeler MB, Quinones CR, Kirsch A, Gloker C, Schupbach P, Strub JR, Caffesse RG (1997a). Maxillary sinus augmentation using different grafting materials and dental implants in monkeys: Part I. Evaluation of an organic bovine-derived bone matrix. *Clin Oral Implants Res* 8:476–486.

Hurzeler MB, Quinones CR, Kirsch A, Schupbach P, Krausse A, Strub JR, Caffesse RG (1997b). Maxillary sinus augmentation using different grafting materials and dental implants in monkeys: Part III. Evaluation of autogenous bone combined with porous hydroxyapatite. *Clin Oral Implants Res* 8:401–411.

Icekson M, Regev E, Neuman R, Margulis A (2009). Transport distraction osteogenesis for closing full-thickness calvarial defects in sheep. *Cleft Palate Craniofac J* 46:363–367.

Iruela-Arispe ML, Sage EH (1993). Endothelial cells exhibiting angiogenesis in vitro proliferate in response to TGF-beta 1. *J Cell Biochem* 52:414–430.

Jakse N, Tangl S, Gilli R, Berghold A, Lorenzoni M, Eskici A, Haas R, Pertl C (2003). Influence of PRP on autogenous sinus grafts: an experimental study on sheep. *Clin Oral Implants Res* 14:578–583.

Ji Y, Ghosh K, Shu XZ, Li B, Sokolov JC, Prestwich GD, Clark RA, Rafailovich MH (2006). Electrospun three-dimensional hyaluronic acid nanofibrous scaffolds. *Biomaterials* 27:3782–3792.

Joeng KS, Schumacher CA, Zylstra-Diegel CR, Long F, Williams BO (2011). Lrp5 and Lrp6 redundantly control skeletal development in the mouse embryo. *Dev Biol* 359:222–229.

Keselowsky BG, Collard DM, Garcia AJ (2005). Integrin binding specificity regulates biomaterial surface chemistry effects on cell differentiation. *Proc Natl Acad Sci USA* 102:5953–5957.

Kim S, Healy KE (2003). Synthesis and characterization of injectable poly(*N*-isopropylacrylamide-*co*-acrylic acid) hydrogels with proteolytically degradable cross-links. *Biomacromolecules* 4:1214–1223.

Kim S, Chung EH, Gilbert M, Healy KE (2005). Synthetic MMP-13 degradable ECMs based on poly(*N*-isopropylacrylamide-*co*-acrylic acid) semi-interpenetrating polymer networks: I. Degradation and cell migration. *J Biomed Mater Res A* 75:73–88.

Kirker-Head CA, Nevins M, Palmer R, Nevins ML, Schelling SH (1997). A new animal model for maxillary sinus floor augmentation: evaluation parameters. *Int J Oral Maxillofac Implants* 12:403–411.

Kohn AD, Moon RT (2005). Wnt and calcium signaling: beta-catenin-independent pathways. *Cell Calcium* 38:439–446.

Korvala J, Juppner H, Makitie O, Sochett E, Schnabel D, Mora S, Bartels CF, Warman ML, Deraska D, Cole WG, Hartikka H, Ala-Kokko L, Mannikko M (2012). Mutations in LRP5 cause primary osteoporosis without features of OI by reducing Wnt signaling activity. *BMC Med Genet* 13:26.

Lam CX, Hutmacher DW, Schantz JT, Woodruff MA, and Teoh SH (2009). Evaluation of polycaprolactone scaffold degradation for 6 months in vitro and in vivo. *J Biomed Mater Res A* 90:906–919.

Lin AS, Barrows TH, Cartmell SH, Guldberg RE (2003). Microarchitectural and mechanical characterization of oriented porous polymer scaffolds. *Biomaterials* 24:481–489.

Lindahl P, Johansson BR, Leveen P, Betsholtz C (1997a). Pericyte loss and microaneurysm formation in PDGF-B-deficient mice. *Science* 277:242–245.

Lindahl P, Karlsson L, Hellstrom M, Gebre-Medhin S, Willetts K, Heath JK, Betsholtz C (1997b). Alveogenesis failure in PDGF-A-deficient mice is coupled to lack of distal spreading of alveolar smooth muscle cell progenitors during lung development. *Development* 124:3943–3953.

Lyngstadaas SP, Lundberg E, Ekdahl H, Andersson C, Gestrelius S (2001). Autocrine growth factors in human periodontal ligament cells cultured on enamel matrix derivative. *J Clin Periodontol* 28:181–188.

Marsell R, Sisask G, Nilsson Y, Sundgren-Andersson AK, Andersson U, Larsson S, Nilsson O, Ljunggren O, Jonsson KB (2012). GSK-3 inhibition by an orally active small molecule increases bone mass in rats. *Bone* 50:619–627.

Marx RE (2001). Platelet-rich plasma (PRP): What is PRP and what is not PRP ? *Implant Dent* 10:225–228.

McElhaney JH, Fogle JL, Melvin JW, Haynes RR, Roberts VL, Alem NM. 1970, Mechanical properties on cranial bone. *J Biomech* 3:495–511.

Miraoui H, Marie PJ (2010). Fibroblast growth factor receptor signaling crosstalk in skeletogenesis. *Sci Signal* 3:re9.

Misch CE, Qu ZM, Bidez MW (1999). Mechanical properties of trabecular bone in the human mandible: Implications for dental implant treatment planning and. Surgical placement. *J Oral Maxillofac Surg* 57:700–706.

Naski MC, Ornitz DM (1998). FGF signaling in skeletal development. *Front Biosci* 3:d781–d794.

Nevins M, Kirker-Head C, Wozney JA, Palmer R, Graham D (1996). Bone formation in the goat maxillary sinus induced by absorbable collagen sponge implants impregnated with recombinant human bone morphogenetic protein-2. *Int J Periodont Restor Dent* 16:8–19.

Nurden AT, Nurden P, Sanchez M, Andia I, Anitua E (2008). Platelets and wound healing. *Front Biosci* 13:3532–3548.

Oest ME, Dupont KM, Kong HJ, Mooney DJ, Guldberg RE (2007). Quantitative assessment of scaffold and growth factor–mediated repair of critically sized bone defects. *J Orthop Res* 25:941–950.

Olivares-Navarrete R, Raz P, Zhao G, Chen J, Wieland M, Cochran DL, Chaudhri RA, Ornoy A, Boyan BD, Schwartz Z (2008). Integrin alpha2beta1 plays a critical role in osteoblast response to micron-scale surface structure and surface energy of titanium substrates. *Proc Natl Acad Sci USA* 105:15767–15772.

Olivares-Navarrete R, Hyzy SL, Park JH, Dunn GR, Haithcock DA, Wasilewski CE, Boyan BD, Schwartz Z (2011). Mediation of osteogenic differentiation of human mesenchymal stem cells on titanium surfaces by a Wnt-integrin feedback loop. *Biomaterials* 32:6399–6411.

Pattanayak DK, Fukuda A, Matsushita T, Takemoto M, Fujibayashi S, Sasaki K, Nishida N, Nakamura T, Kokubo T (2011). Bioactive Ti metal analogous to human cancellous bone: fabrication by selective laser melting and chemical treatments. *Acta Biomater* 7:1398–1406.

Pepper MS, Vassalli JD, Orci L, Montesano R (1993). Biphasic effect of transforming growth factor-beta 1 on in vitro angiogenesis. *Exp Cell Res* 204:356–363.

Peters DA, Courtemanche DJ (2009). Role of platelet gel and calcium-coated Lactosorb membranes in healing critical calvarial defects. *J Craniofac Surg* 20:2200–2204.

Phillips JE, Burns KL, Le Doux JM, Guldberg RE, Garcia AJ (2008). Engineering graded tissue interfaces. *Proc Natl Acad Sci USA* 105:12170–12175.

Pitt CG, Jeffcoat AR, Zweidinger RA, Schindler A (1979a). Sustained drug delivery systems: 1. Permeability of poly(epsilon-caprolactone), poly(DL-lactic acid), and their copolymers. *J Biomed Mater Res* 13:497–507.

Pitt CG, Gratzl MM, Jeffcoat AR, Zweidinger R, Schindler A (1979b). Sustained drug delivery systems: 2. Factors affecting release rates from poly(epsilon-caprolactone) and related biodegradable polyesters. *J Pharm Sci* 68:1534–1538.

Quinones CR, Hurzeler MB, Schupbach P, Kirsch A, Blum P, Caffesse RG, Strub Jr (1997a). Maxillary sinus augmentation using different grafting materials and osseointegrated dental implants in monkeys: Part II. Evaluation of porous hydroxyapatite as a grafting material. *Clin Oral Implants Res* 8:487–496.

Quinones CR, Hurzeler MB, Schupbach P, Arnold DR, Strub JR, Caffesse RG (1997b). Maxillary sinus augmentation using different grafting materials and dental implants in monkeys. Part IV. Evaluation of hydroxyapatite-coated implants. *Clin Oral Implants Res* 8:497–505.

Qureshi AH, Chaoji V, Maiguel D, Faridi MH, Barth CJ, Salem SM, Singhal M, Stoub D, Krastins B, Ogihara M, Zaki MJ, Gupta V (2009). Proteomic and phospho-proteomic profile of human platelets in basal, resting state: insights into integrin signaling. *PLoS One* 4:e7627.

Raghoebar GM, Schortinghuis J, Liem RS, Ruben JL, van der Wal JE, Vissink A (2005). Does platelet-rich plasma promote remodeling of autologous bone grafts used for augmentation of the maxillary sinus floor ? *Clin Oral Implants Res* 16:349–356.

Rai B, Oest ME, Dupont KM, Ho KH, Teoh SH, Guldberg RE (2007). Combination of platelet-rich plasma with polycaprolactone–tricalcium phosphate scaffolds for segmental bone defect repair. *J Biomed Mater Res A* 81:888–899.

Ranly DM, McMillan J, Keller T, Lohmann CH, Meunch T, Cochran DL, Schwartz Z, Boyan BD (2005). Platelet-derived growth factor inhibits demineralized bone matrix-induced intramuscular cartilage and bone formation: a study of immunocompromised mice. *J Bone Joint Surg Am* 87:2052–2064.

Ranly DM, Lohmann CH, Andreacchio D, Boyan BD, Schwartz Z (2007). Platelet-rich plasma inhibits demineralized bone matrix-induced bone formation in nude mice. *J Bone Joint Surg Am* 89:139–147.

Rauh J, Milan F, Gunther KP, Stiehler M (2011). Bioreactor systems for bone tissue engineering. *Tissue Eng B* 17:263–280.

Reyes CD, Garcia AJ (2003). Engineering integrin-specific surfaces with a triple-helical collagen-mimetic peptide. *J Biomed Mater Res A* 65:511–523.

Richardson TP, Peters MC, Ennett AB, Mooney DJ (2001). Polymeric system for dual growth factor delivery. *Nat Biotechnol* 19:1029–1034.

Rohner D, Hutmacher DW, See P, Tan KC, Yeow V, Tan SY, Lee ST, Hammer B (2002). [Individually CAD-CAM technique designed, bioresorbable 3-dimensional polycaprolactone framework for experimental reconstruction of craniofacial defects in the pig]. *Mund Kiefer Gesichtschir* 6:162–167.

Rowley JA, Madlambayan G, Mooney DJ (1999). Alginate hydrogels as synthetic extracellular matrix materials. *Biomaterials* 20:45–53.

Santos A, Bakker AD, de Blieck-Hogervorst JM, Klein-Nulend J (2010). WNT5A induces osteogenic differentiation of human adipose stem cells via rho-associated kinase ROCK. *Cytotherapy* 12:924–932.

Sauerbier S, Stubbe K, Maglione M, Haberstroh J, Kuschnierz J, Oshima T, Xavier SP, Brunnberg L, Schmelzeisen R, Gutwald R (2010). Mesenchymal stem cells and bovine bone mineral in sinus lift procedures: an experimental study in sheep. *Tissue Eng C Methods* 16:1033–1039.

Schwartz Z, Kieswetter K, Dean DD, Boyan BD (1997). Underlying mechanisms at the bone-surface interface during regeneration. *J Periodont Res* 32:166–171.

Schwartz Z, Goldstein M, Raviv E, Hirsch A, Ranly DM, Boyan BD (2007). Clinical evaluation of demineralized bone allograft in a hyaluronic acid carrier for sinus lift augmentation in humans: a computed tomography and histomorphometric study. *Clin Oral Implants Res* 18:204–211.

Schwartz Z, Simon BJ, Duran MA, Barabino G, Chaudhri R, Boyan BD (2008). Pulsed electromagnetic fields enhance BMP-2 dependent osteoblastic differentiation of human mesenchymal stem cells. *J Orthop Res* 26:1250–1255.

Schwarz F, Wieland M, Schwartz Z, Zhao G, Rupp F, Geis-Gerstorfer J, Schedle A, Broggini N, Bornstein MM, Buser D, Ferguson SJ, Becker J, Boyan BD, Cochran DL (2009). Potential of chemically modified hydrophilic surface characteristics to support tissue integration of titanium dental implants. *J Biomed Mater Res B* 88:544–557.

Sculean A, Schwarz F, Becker J, Brecx M (2007). The application of an enamel matrix protein derivative (Emdogain) in regenerative periodontal therapy: a review. *Med Princ Pract* 16:167–180.

Senzel L, Gnatenko DV, Bahou WF (2009). The platelet proteome. *Curr Opin Hematol* 16:329–333.

Seol YJ, Kim KH, Kang YM, Kim IA, and Rhee SH (2009). Bioactivity, pre-osteoblastic cell responses, and osteoconductivity evaluations of the electrospun non-woven SiO_2-CaO gel fabrics. *J Biomed Mater Res B Appl Biomater* 90:679–687.

Shayesteh YS, Khojasteh A, Soleimani M, Alikhasi M, Khoshzaban A, Ahmadbeigi N (2008). Sinus augmentation using human mesenchymal stem cells loaded into a beta-tricalcium phosphate/hydroxyapatite scaffold. *Oral Surg Oral Med Oral Pathol Oral Radiol Endod* 106:203–209.

Sherwood JK, Riley SL, Palazzolo R, Brown SC, Monkhouse DC, Coates M, Griffith LG, Landeen LK, Ratcliffe A (2002). A three-dimensional osteochondral composite scaffold for articular cartilage repair. *Biomaterials* 23:4739–4751.

Shin M, Yoshimoto H, Vacanti JP (2004). In vivo bone tissue engineering using mesenchymal stem cells on a novel electrospun nanofibrous scaffold. *Tissue Eng* 10:33–41.

Sicca CM, Corotti MV, Sgarbosa SH, Cestari TM, Cavalcanti MG, Ferreira PM, Taga R, Granjeiro JM (2008). Comparative histomorphometric and tomographic analysis of maxillary sinus floor augmentation in rabbits using autografts and xenografts. *J Biomed Mater Res B* 86:188–196.

Smith KE, Hyzy SL, Sunwoo M, Gall KA, Schwartz Z, Boyan BD (2010). The dependence of MG63 osteoblast responses to (meth)acrylate-based networks on chemical structure and stiffness. *Biomaterials* 31:6131–6141.

Stephan SJ, Tholpady SS, Gross B, Petrie-Aronin CE, Botchway EA, Nair LS, Ogle RC, and Park SS (2010). Injectable tissue-engineered bone repair of a rat calvarial defect, *Laryngoscope* 120:895–901.

Stevens MM (2008). Biomaterials for bone tissue engineering. *Mater Today* 11:18–25.

Suri C, Jones PF, Patan S, Bartunkova S, Maisonpierre PC, Davis S, Sato TN, Yancopoulos GD (1996). Requisite role of angiopoietin-1, a ligand for the TIE2 receptor, during embryonic angiogenesis. *Cell* 87:1171–1180.

Susarla SM, Swanson E, Gordon CR (2011). Craniomaxillofacial reconstruction using allotransplantation and tissue engineering: challenges, opportunities, and potential synergy. *Ann Plast Surg* 67:655–661.

Tai K, Pelled G, Sheyn D, Bershteyn A, Han L, Kallai I, Zilberman Y, Ortiz C, Gazit D (2008). Nanobiomechanics of repair bone regenerated by genetically modified mesenchymal stem cells. *Tissue Eng A* 14:1709–1720.

Tamura M, Nemoto E, Sato MM, Nakashima A, Shimauchi H (2010). Role of the Wnt signaling pathway in bone and tooth. *Front Biosci (Elite Ed)* 2:1405–1413.

Tu Q, Valverde P, Li S, Zhang J, Yang P, Chen J (2007). Osterix overexpression in mesenchymal stem cells stimulates healing of critical-sized defects in murine calvarial bone. *Tissue Eng* 13:2431–2440.

Van de Putte KA, Urist MR (1965). Osteogenesis in the interior of intramuscular implants of decalcified bone matrix, *Clin Orthop Relat Res* 43:257–270.

Waese EY, Kandel RA, Stanford WL (2008). Application of stem cells in bone repair. *Skeletal Radiol* 37:601–608.

Wan DC, Pomerantz JH, Brunet LK, Kim JB, Chou YF, Wu BM, Harland R, Blau HM, Longaker MT (2007). Noggin suppression enhances in vitro osteogenesis and accelerates in vivo bone formation. *J Biol Chem* 282:26450–26459.

Wang X, Gittens RA, Song R, Tannenbaum R, Olivares-Navarrete R, Schwartz Z, Chen H, Boyan BD (2012). Effects of structural properties of electrospun TiO_2 nanofiber meshes on their osteogenic potential. *Acta Biomater* 8:878–885.

Wansleeben C, Meijlink F (2011). The planar cell polarity pathway in vertebrate development. *Dev Dyn* 240:616–626.

Ward BB, Brown SE, Krebsbach PH (2010). Bioengineering strategies for regeneration of craniofacial bone: a review of emerging technologies. *Oral Dis* 16:709–716.

Watanabe K, Niimi A, Ueda M (1999). Autogenous bone grafts in the rabbit maxillary sinus. *Oral Surg Oral Med Oral Pathol Oral Radiol Endod* 88:26–32.

Weng YL, Shang QX, Cai X, Cao YL (2001). [Experimental study for repair of cranial defects with bone marrow stromal cells and modified alginate]. *Shanghai Kou Qiang Yi Xue* 10:35–38.

Witte F, Dokas J, Neuendorf F, Mundlos S, Stricker S (2009). Comprehensive expression analysis of all Wnt genes and their major secreted antagonists during mouse limb development and cartilage differentiation. *Gene Expr Patterns* 9:215–223.

Wozney JM, Rosen V (1998). Bone morphogenetic protein and bone morphogenetic protein gene family in bone formation and repair, *Clin Orthop Relat Res* 346:26–37.

Yang F, Both SK, Yang X, Walboomers XF, Jansen JA (2009). Development of an electrospun nano-apatite/PCL composite membrane for GTR/GBR application. *Acta Biomater* 5:3295–3304.

Yeo A, Cheok C, Teoh SH, Zhang ZY, Buser D, Bosshardt DD (2012). Lateral ridge augmentation using a PCL-TCP scaffold in a clinically relevant but challenging micropig model. *Clin Oral Implants Res* 2012 Dec; 23(12):1322–32.

Yeom H, Blanchard S, Kim S, Zunt S, Chu TM (2008). Correlation between micro-computed tomography and histomorphometry for assessment of new bone formation in a calvarial experimental model. *J Craniofac Surg* 19:446–452.

Yeom J, Chang S, Park JK, Je JH, Yang DJ, Choi SK, Shin HI, Lee SJ, Shim JH, Cho DW, Hahn SK (2010). Synchrotron X-Ray Bioimaging of Bone Regeneration by Artificial Bone Substitute of MegaGen Synthetic Bone and Hyaluronate Hydrogels. *Tissue Eng Part C Methods* 16:1059–1068.

Yoon E, Dhar S, Chun DE, Gharibjanian NA, Evans GR (2007). In vivo osteogenic potential of human adipose-derived stem cells/poly lactide-*co*-glycolic acid constructs for bone regeneration in a rat critical-sized calvarial defect model. *Tissue Eng* 13:619–627.

Young S, Patel ZS, Kretlow JD, Murphy MB, Mountziaris PM, Baggett LS, Ueda H, Tabata Y, Jansen JA, Wong M, Mikos AG (2009). Dose effect of dual delivery of vascular endothelial growth factor and bone morphogenetic protein-2 on bone regeneration in a rat critical-size defect model. *Tissue Eng Part A* 15:2347–2362.

Yu D, Li Q, Mu X, Chang T, Xiong Z (2008). Bone regeneration of critical calvarial defect in goat model by PLGA/TCP/rhBMP-2 scaffolds prepared by low-temperature rapid-prototyping technology. *Int J Oral Maxillofac Surg* 37:929–934.

Zerbo IR, Bronckers AL, de Lange G, Burger EH (2005). Localisation of osteogenic and osteoclastic cells in porous beta-tricalcium phosphate particles used for human maxillary sinus floor elevation. *Biomaterials* 26:1445–1451.

Zhang Y, Fan W, Ma Z, Wu C, Fang W, Liu G, Xiao Y (2010). The effects of pore architecture in silk fibroin scaffolds on the growth and differentiation of mesenchymal stem cells expressing BMP7. *Acta Biomater* 6:3021–3028.

Zhao Z, Wang Z, Ge C, Krebsbach P, Franceschi RT (2007). Healing cranial defects with AdRunx2-transduced marrow stromal cells. *J Dent Res* 86:1207–1211.

Zhao J, Shinkai M, Takezawa T, Ohba S, Chung UI, Nagamune T (2009). Bone regeneration using collagen type I vitrigel with bone morphogenetic protein-2. *J Biosci Bioeng* 107:318–323.

Zijderveld SA, Zerbo IR, van den Bergh JP, Schulten EA, ten Bruggenkate CM (2005). Maxillary sinus floor augmentation using a beta-tricalcium phosphate (Cerasorb) alone compared to autogenous bone grafts. *Int J Oral Maxillofac Implants* 20:432–440.

20

CRANIOFACIAL TISSUE BIOENGINEERING AND REGENERATION BY ENDOGENOUS STEM CELLS

Nan Jiang, Mo Chen, Chang Hun Lee, Jian Zhou, Mildred C. Embree, Kimi Kong, Choko Cho, Avital Mendelson, Ying Zheng, Hemin Nie, and Jeremy J. Mao

Center for Craniofacial Regeneration, Columbia University Medical Center, New York, New York

1 INTRODUCTION

Directed movement of stem cells is crucial not only for organogenesis during development, but also for tissue regeneration in adult life. We present examples of stem cell migration and recruitment and the application of cell homing in orofacial regeneration. *Cell homing* refers to the recruitment of endogenous cells, including stem and progenitor cells, into a wound. Homing can be local and/or systemic. Whereas cell homing via circulation has contributed a great deal to the knowledge of pathogenesis of diverse diseases, including infections and tumor metastasis, we understand little of how to immobilize and direct cell homing in tissue regeneration. Stem and progenitor cells are highlighted to demonstrate their migration, lineage commitment, and potential roles in wound healing.

2 CELL MIGRATION AND RECRUITMENT

Cell migration is a significant and central process in the development and maintenance of organisms. Tissue formation, wound healing, and immune responses invariably

Stem Cells in Craniofacial Development and Regeneration, First Edition.
Edited by George T.-J. Huang, Irma Thesleff.

require the orchestrated movement of cells in particular directions to specific locations. In cell homing, cells migrate in response to specific external signals and chemokines (Ridley et al., 2003; Laird et al., 2008).

Homing of hematopoietic stem cells (HSCs) has long been recognized (Fig. 1). HSCs are known to migrate from circulation to their niches in bone marrow sinusoids (Laird et al., 2008). The reverse process of HSC migration from bone marrow to the periphery is also recognized (Laird et al., 2008). The adhesive mechanisms for HSC homing to bone marrow include rolling along the vascular wall, transendothelial migration, and finally, nesting in the specific niche. A cascade of cytotactic ligands and receptors are involved in HSC homing, including stromal-derived factor 1 (SDF1) and CXCR4 (Gallatin et al., 1983; Frenette et al., 1998; Quesenberry and Becker, 1998).

When infused intravenously, mesenchymal stem and progenitor cells (MSCs) can also home to injured tissue. Cultured rat and human MSCs migrate to the areas of brain injury after cerebral ischemia upon intravenous transplantation in the rat (Chen et al., 2001; Mahmood et al., 2003). Intravenously infused rat MSCs migrate to sites of allograft rejection in the heart and transform into cardiac fibroblasts and myocytes (Wu et al., 2003).

Stem cell homing appears to exhibit similarities to leukocyte trafficking in a process known as *transendothelial migration* or simply *transendotheliation* (Fig. 2). Flowing leukocytes first adhere loosely to vascular endothelium by interactions with adhesion molecules, including selectins, β1-integrins, and their respective counter ligands (Ebert et al., 2005). Leukocytes then "invade" between neighboring endothelial cells to reach the underlying tissue (Worthylake and Burridge, 2001). Chemokines that regulate the rolling and adhesion processes include integrins, selectins, and chemokine receptors,

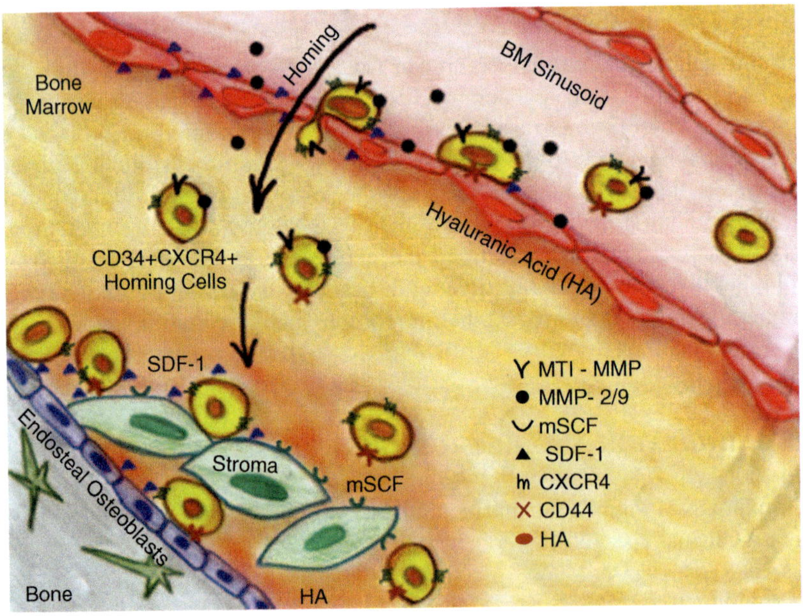

FIGURE 1 Homing of human stem cells from the blood circulation via bone marrow sinusoids to the endosteum.

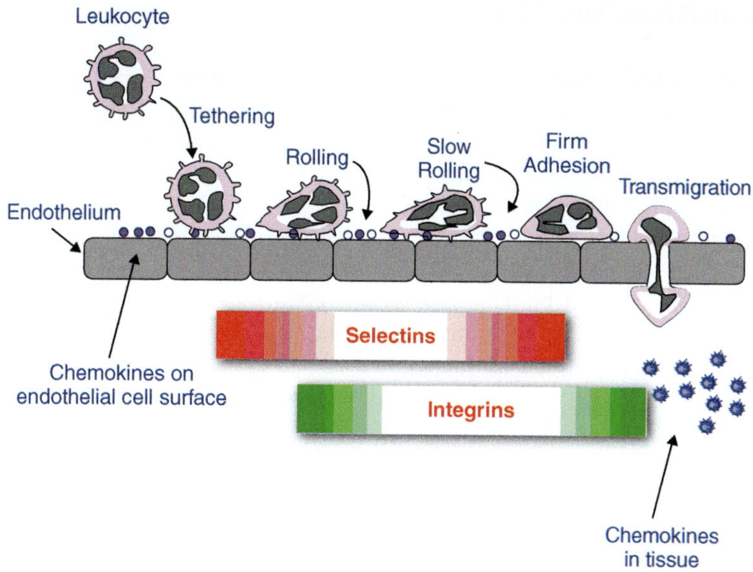

FIGURE 2 Schematic of transmigration of leukocytes across the endothelium.

especially P-selectin and VCAM1 (Lapidot et al., 2005). L- and E-selectin are involved in the initial rolling stage on leukocytes, whereas L-selectin expression is low or absent on the surface of MSCs (Ruster et al., 2006). Destabilization of the rolling cells on L-selectin ligands is also an indispensable component of the homing process (Grabovsky et al., 2002).

Chemokines are a superfamily of small glycoproteins that participate in diverse biological processes, including leukocyte trafficking, hematopoiesis, angiogenesis, and organogenesis (Baggiolini, 2001). Tissue damage stimulates the production of secreted chemokines, cytokines, and enzymes (Cottler-Fox et al., 2003). Cells migrate through subtle differences in chemokine concentrations via cytoskeletal rearrangements and adhesive interactions with the extracellular matrix (Sanchez-Madrid and del Pozo, 1999).

HSC homing involves signaling by SDF1 and stem cell factor (SCF), activation of lymphocyte function–associated antigen 1 (LFA1), very late antigen 4/5 (VLA4/5) and CD44, cytoskeleton rearrangement, membrane type 1–matrix metalloproteinase activation, and secretion of MMP2/9 (Lapidot et al., 2005). SDF1 and its receptor, CXCR4, are crucial for stem cell migration in murine bone marrow and recruitment of endothelial progenitor cells to the site of ischemic tissue (Ponomaryov et al., 2000; Cottler-Fox et al., 2003; Yamaguchi et al., 1995; Ceradini et al., 2004). Transendothelial migration is mediated further by chemokines such as PECAM1, CD99, and ICAM1 (Aurrand-Lions et al., 2002; Millan et al., 2006).

MSCs express a variety of chemokine receptors, including CCR1, CCR7, CCR9, CXCR4, CXCR5, and CXCR6 (Honczarenko et al., 2006; Ringe et al., 2007). TNFα increases CCR2, CCR3, and CCR4 expression but not CXCR4 (Ponte et al., 2007). MSC migration can be induced by PDGF, EGF, TGFβ, IGF-I, HGF, BMP2, BMP4, and HMGB1 (Fiedler et al., 2002; Dellavalle et al., 2007; Ozaki et al., 2007; Ponte et al., 2007; Meng et al., 2008).

3 CELL HOMING MODELS

Regulated migration of stem cells is critical not only in homeostasis but also in pathological conditions and wound healing. We present two models of stem cell migration to demonstrate the significance of cell trafficking in tissue regeneration. Different from classic cell migration models, what we present below accommodates extravascular guidance cues independent of blood flow.

A myogenic model incorporates the population of satellite stem and progenitor cells next to the basal lamina of muscle fibers (Mauro, 1961). Satellite cells are usually quiescent but can be activated readily upon muscle injury, and then divide, migrate, and differentiate to form myoblasts (Hawke and Garry, 2001). Activation of satellite cells depends on their proximity to a wound site and the nature of the wound (Schultz et al., 1985). Upon basal lamina injury, satellite cells migrate to myofibers (Watt et al., 1987) (Also see Chapter 13).

The rodent incisor grows continuously throughout life. Epithelium stem cells in the rodent incisor are located in the cervical loop of the developing tooth organ, known as *stellate reticulum cells*, and differentiate into ameloblasts that produce the enamel. Mesenchymal cells surround stellate reticulum cells close to the cervical loop and differentiate into odontoblasts and other pulp cells (Harada et al., 1999; Feng et al., 2011). Similar to human teeth, mesenchymal cells in rodent teeth respond to injury and differentiate into new odontoblasts (Smith et al., 1995; Feng et al., 2011). In homeostasis, few cells migrate in the incisor pulp; however, upon injury, DiI-labeled cells migrate toward the injury site (Feng et al., 2011) (Also see Chapter 18).

4 REGENERATION BY CELL HOMING

Mesenchymal cells in the orofacial region are heterogeneous and possess remarkable diversity. Orofacial mesenchymal cells include those isolated from dental pulp, periodontal ligament, apical papilla, and dental follicle (Mao et al., 2012). Orofacial mesenchymal cells regulate T-lymphocyte survival and proliferation (Yamaza et al., 2011). Despite much effort, how orofacial stem and progenitor cells contribute to the regeneration of tissue defects remains poorly understood.

4.1 Cell Delivery Approaches

Cell transplantation is the conventional approach for orofacial regeneration. The field has been searching for the "right" cell population(s) for the regeneration of various orofacial structures, including tooth or tooth elements. Epithelial and mesenchymal tissues of incisor tooth germ at the cap stage from the lower jaw at embryonic age E14.5 were reconstituted in collagen gel and cultured ex vivo (Nakao et al., 2007; Ikeda et al., 2009). After the reconstituted cells were implanted into the molar extraction sockets in 5-week-old mice, a tooth analog formed and erupted into occlusion. Cells from porcine or rat tooth buds in the biodegradable polymers formed putative dentin and enamel structures (Young et al., 2002; Duailibi et al., 2004). Similarly, tooth bud cells including dental epithelium and mesenchyme in scaffolds induced putative enamel, dentin, and pulp and connective tissue such as alveolar bone and periodontal ligament (Ohazama et al., 2004; Young et al., 2005; Kuo et al., 2008). Using a minipig model, root apical papilla stem cells and periodontal ligament stem cells were transplanted into porcine tooth sockets to generate a root–periodontal complex (Sonoyama et al., 2006) (Also refer to Chapter 27).

Tooth regeneration by cell transplantation, however, has met with hurdles in the translation and clinical application. To date, no attempts at cell-based tooth regeneration are being translated into clinical therapies (Yildirim et al., 2011). Cell source for transplantation remains problematic. Embryonic tooth germ cells are impossible to obtain autologously in patients who need tooth regeneration. Allogeneic or xenogenic embryonic tooth germ cells may elicit immunorejection, ethics problems, and mismatching tooth morphology. Postnatal tooth germ cells are scarce and appear to have lost the capacity of embryonic tooth germ cells in regenerating an entire tooth. The pluripotency of embryonic stem cells (ES cells) or induced pluripotent stem (iPS) cells may also be their "peril," with the ability to differentiate into multiple unwanted cell types in addition to cells for tooth regeneration, including ameloblasts, odontoblasts, cementoblasts, pulp cells, periodontal ligament cells, and alveolar osteoblasts. The high cost and unprecedented regulatory approval processes in association with the commercialization of ex vivo cell manipulation also act as formidable barriers against clinical translation (Yildirim et al., 2011).

4.2 Cell Homing Approaches

Cell homing refers to the recruitment of endogenous cells, including stem and progenitor cells by molecular cues. Cell homing in tissue regeneration is understudied, although several recent reports have demonstrated that proteins/peptides or small molecules are capable of recruiting cells into predefined wounds and orchestrate regeneration. The entire articular surface of the synovial joint was regenerated without cell delivery but, rather, by the homing of endogenous cells with delivery of a single biological cue, TGFβ3, spatially embedded in an anatomically correct bioscaffold (Lee et al., 2010). An overwhelming advantage of cell homing in tissue regeneration is that tissue regeneration is enabled by harnessing the host's innate capacity for regeneration and by recruiting endogenous cells, without the need for cell transplantation, thus potentially accelerating the rate of regulatory and commercialization processes.

In tooth regeneration, a recent report has demonstrated that tooth root and the periodontal ligament were regenerated, immediately following tooth extraction, by small molecules that are spatially embedded in microchannels of anatomically correct tooth roots that replaced the native tooth roots that were extracted (Kim et al., 2010b). The dimensions of the native teeth can be derived from two-dimensional images using computed tomography (CT) or magnetic resonance imaging (MRI) of a patient, or textbook averages for patients who are completely edentulous. For example, scaffolds with the shape of a human mandibular first molar were fabricated via three-dimensional layer-by-layer apposition. The scaffold is composed of 80% polycaprolactone (PCL) and 20% hydroxyapatite (HA) (wt%). PCL-HA was comolten and dispensed through a 27-gauge metal nozzle to create repeating three-dimensional microstrands and interconnecting microchannels. A blended cocktail of stromal derived factor 1 (SDF1) and bone morphogenetic protein 7 (BMP7) was adsorbed in neutralized type 1 collagen solution. SDF1, a chemokine receptor for endothelial cells and bone marrow stem and progenitor cells (Belema-Bedada et al., 2008; Kitaori et al., 2009), recruit mesenchymal and endothelial stem and progenitor cells. BMP7, which plays an important role in osteoblast differentiation and phosphorylation via SMAD pathways (Hahn et al., 1992; Itoh et al., 2001), is responsible for newly formed mineralized tissues in the regenerated tooth roots (Kim et al., 2010b; Suzuki et al., 2011; Mao et al., 2012). The potency of cell homing is substantiated not only by cell recruitment into a scaffold's microchannels, but also by regeneration of a putative periodontal ligament and newly formed alveolar bone (Fig. 3). Scaffolds in the shape of the rat mandibular incisor were integrated with surrounding tissue, showing tissue ingrowth into scaffold microchannels (Fig. 3A). Microscopically, the scaffolds within the extraction sockets clearly showed multiple tissue phenotypes, including the native alveolar bone (b), newly formed bone (nb), and a fibrous tissue interface reminiscent of the periodontal ligament (pdl) (Fig. 3A). The newly formed bone showed ingrowth into microchannel openings and interstaggered with scaffold microstrands (s). Higher magnification showed newly formed bone with bone trabeculae-like structures and embedded cells resembling osteocytes. Immediately adjacent is a structure reminiscent of the periodontal ligament, consisting of fibroblast-like cells and collagen-like structures (Fig. 3B). Von Kossa preparation showed that the newly formed bone was well mineralized, in contrast to adjacent unmineralized, putative periodontal ligament (Fig. 3C). Although host cells populated the microchannels of growth-factor-free control scaffolds (Fig. 3D), combined SDF1 and BMP7 delivery (Fig. 3E) homed significantly more cells into the microchannels of the rat incisor scaffolds ($p < 0.01$) (Fig. 3F). Angiogenesis took place in the scaffolds' microchannels with or without growth factor delivery (Fig. 3G and H). The numbers of cells and blood vessels recruited were quantified from three different locations along the entire root length of the rat mandibular incisor scaffold: the superior region of the alveolar ridge and the midpoint and inferior region of the root apex (Fig. 3J). Quantitatively, combined SDF1 and BMP7 delivery elaborated significantly more blood vessels than did the growth factor–free group ($p < 0.05$) (Fig. 3I).

In another study, dental pulp–like tissue regenerated in endodontically treated real-size native human teeth by delivery of one or several growth factors (Kim et al., 2010a). Endodontically treated human teeth were implanted in the dorsum of mice. Given that no cells were transplanted, dental pulp–like tissues regenerated in the endodontically treated root canals were orchestrated by the homing of host endogenous cells upon delivery of growth factor(s). The growth factors delivered include vascular endothelial growth factor (VEGF), basic fibroblast growth factor (bFGF), platelet-derived growth factor (PDGF), nerve growth factor (NGF), and bone morphogenetic protein-7 (BMP7). The actions of all these growth factors in other tissues have been well documented (Matsushita et al., 2000; Cooke et al., 2006; Micera et al., 2007; Ishimatsu et al., 2009). Combined delivery of bFGF, VEGF, or PDGF with basal cytokines (NGF and BMP7) yielded recellularization in the entire endodontically treated root canal of real-size human tooth, as shown by multiple reconstructed microscopic images (Fig. 4). The entire root canal was filled with regenerated dental pulp–like tissues from root apex to pulp chamber. In some but not all samples, growth factor delivery generated reparative dentin-like tissue on the surface of native dentin. Upon release of growth factor(s), host endogenous cells are homed in vivo into an anatomic compartment, in this case the endodontically treated root canals. Current root canal treatment of diseased dental pulp requires removal of substantial enamel and dentin structures because obturation with gutta-percha requires unobstructed access. Instead, it was proposed that a diseased dental pulp can be treated with a minimally invasive root canal therapy on the basis that delivery of injectable bioactive cues does not require unobstructed access to pulp chamber and root canal.

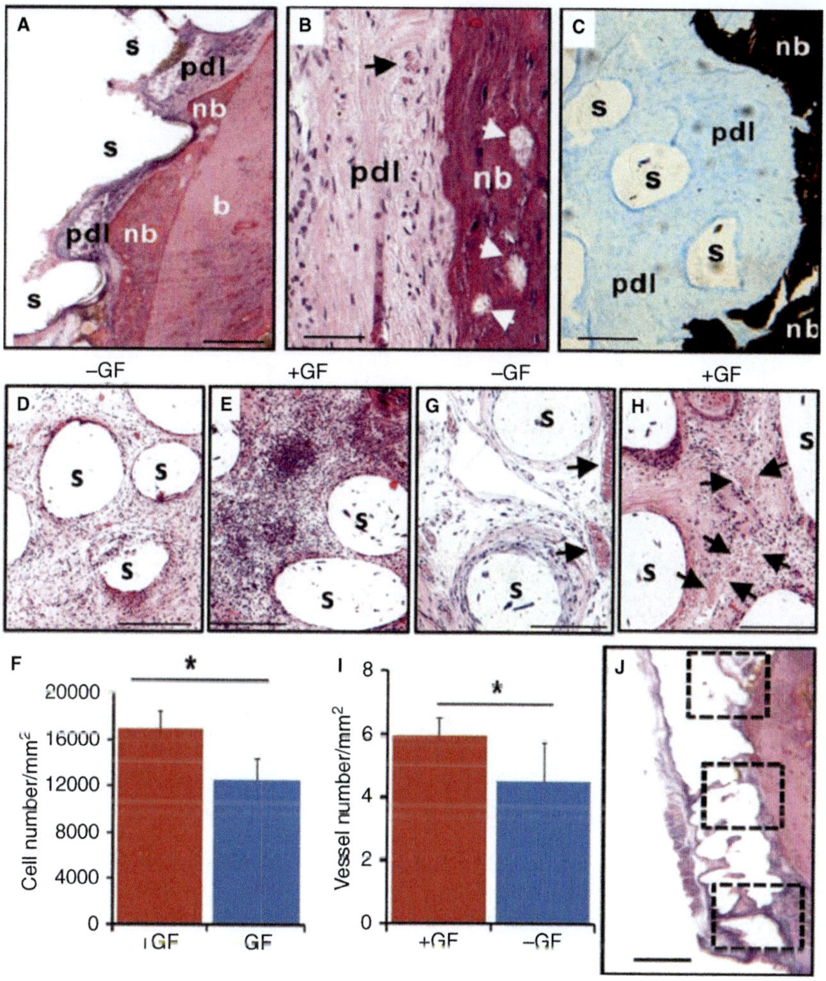

FIGURE 3 Orthotopic tooth regeneration by cell homing. (A) The rat mandibular incisor scaffold integrated with surrounding tissue, showing tissue ingrowth into scaffold microchannels and multiple tissue phenotypes, including the native alveolar bone (b), newly formed bone (nb), and a fibrous tissue interface reminiscent of the periodontal ligament (pdl). The newly formed bone (nb) showed ingrowth into microchannel openings interstaggered with scaffold microstrands (s). (B) Newly formed bone (nb) has bone trabeculae-like structures (arrows) and embedded osteocyte-like cells, immediately adjacent to a putative periodontal ligament (pdl) consisting of fibroblast-like cells and collagen bundle-like structures. (C) Newly formed bone (nb) is well mineralized (von Kossa preparation), in contrast to the adjacent unmineralized, putative periodontal ligament (pdl). (D) Cells populated the scaffold's microchannels even without growth factor delivery. Remarkably, SDF1 and BMP7 delivery yielded substantial cell homing in microchannels (E). (F) Combined SDF1 and BMP7 delivery homed significantly more cells into microchannels than without growth factor delivery ($p < 0.014$; $N = 11$). Angiogenesis took place in the scaffold microchannels without growth factor delivery (G), but was more substantial with growth factor delivery (H). (I) Combined SDF1 and BMP7 delivery elaborated significantly more blood vessels than without growth factor delivery ($p < 0.05$; $N = 11$). (J) The numbers of recruited cells and blood vessels were quantified from three different locations along the entire root length of the rat mandibular incisor scaffold: the superior region of alveolar ridge and the inferior region of root apex, with a midpoint in between. s, scaffold; GF, growth factor(s). Scale: 100 μm.

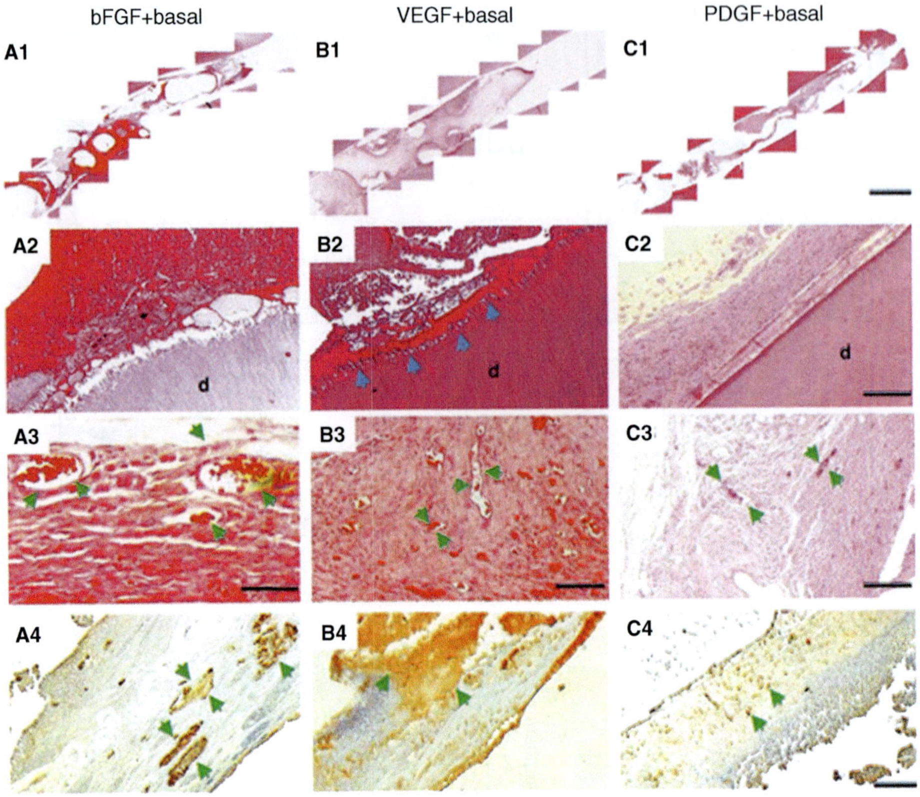

FIGURE 4 Combinatory cytokine delivery and chemotactic effects on pulp regeneration from host endogenous cells. (A1) Combined delivery of bFGF with basal cytokines of nerve growth factor (NGF) and bone morphogenetic protein 7 (BMP7) yielded re-cellularization in the entire endodontically treated root canal, as shown by reconstructed multiple microscopic images from root apex to pulp chamber. (A2) Close-up image showing abundant cells and integration with native dentin (d) in endodontically treated root canal. (A3) Higher magnification showing erythrocyte-filled blood vessels (arrows) in connective tissue. (A4) Scattered islands of positive VEGF antibody staining (arrows). (B1) Combined delivery of VEGF with basal cytokines of NGF and BMP7 induced re-cellularization in the entire endodontically treated root canal. (B2) Connective tissue with abundant cells integrated with native dentin (d), including a layer of dentin-like tissue between abundant cells and dentinal wall (arrows). (B3) Multiple blood vessel–like structures in regenerated connective tissue (arrows). (B4) Positive VEGF antibody staining. Arrows indicate positive VEGF staining. (C1) Combined delivery of platelet-derived growth factor (PDGF) with basal cytokines of NGF and BMP7 yielded fully regenerated tissues in the entire endodontically treated root canal. (C2) Abundant cells in connective tissue that integrated to native dentin (d). (C3) Blood vessel–like structures (arrows) in connective tissue. (C4) VEGF antibody staining. Arrows indicate positive VEGF staining. Scale bars (A1, B1, C1): 1 mm; (A2, B2, C2): 100 μm; (A3): 100 μm; (B3, C3): 300 μm; (A4, B4, C4): 300 μm.

Although residual inflammation in the root canal and periapical region represents challenges for pulp regeneration, chemotaxis-induced angiogenesis provides the potential for native defense mechanisms that may counteract residual infection in the root canal, leading to a vital tooth with regenerated dental pulp (Kim et al., 2010a).

5 SUMMARY

The doctrine of cells, biomaterial scaffolds, and biological signals has been the guiding principle for tissue engineering. Given the vast diversity of tissues that are being regenerated, it is difficult to imagine that one doctrine would govern all. Several reports have demonstrated that endogenous cells, probably including stem and progenitor cells, are recruited by growth factor(s) and are responsible for the regeneration of tooth roots and dental pulp. Endogenous regeneration provides a parallel strategy to cell transplantation and an important model that may be devoid of ex vivo cell manipulation. Costly procedures and uncertainty in regulatory approval are among the main hurdles that have prevented cell transplantation from clinical translation. Our continued learning of cell homing, including spontaneous migration and directed recruitment, and cell–biomaterial interactions, will further advance the field of orofacial regeneration and accelerate clinical translation.

Acknowledgments

We thank Qiongfen Guo and Jacqueline Melendez for their administrative assistance. This work was supported by National Institutes of Health grants R01DE018248, R01EB009663, and RC2DE020767 (to J.J.M.).

REFERENCES

Aurrand-Lions M, Johnson-Leger C, Imhof BA (2002). The last molecular fortress in leukocyte trans-endothelial migration. *Nat Immunol* 3:116–118.

Baggiolini M (2001). Chemokines in pathology and medicine. *J Intern Med* 250:91–104.

Belema-Bedada F, Uchida S, Martire A, Kostin S, Braun T (2008). Efficient homing of multipotent adult mesenchymal stem cells depends on FROUNT-mediated clustering of CCR2. *Cell Stem Cell* 2:566–575.

Ceradini DJ, Kulkarni AR, Callaghan MJ, Tepper OM, Bastidas N, Kleinman ME, Capla JM, Galiano RD, Levine JP, Gurtner GC (2004). Progenitor cell trafficking is regulated by hypoxic gradients through HIF-1 induction of SDF-1. *Nat Med* 10:858–864.

Chen J, Li Y, Wang L, Zhang Z, Lu D, Lu M, Chopp M (2001). Therapeutic benefit of intravenous administration of bone marrow stromal cells after cerebral ischemia in rats. *Stroke* 32:1005–1011.

Cooke JW, Sarment DP, Whitesman LA, Miller SE, Jin Q, Lynch SE, Giannobile WV (2006). Effect of rhPDGF-BB delivery on mediators of periodontal wound repair. *Tissue Eng* 12:1441–1450.

Cottler-Fox MH, Lapidot T, Petit I, Kollet O, DiPersio JF, Link D, Devine S (2003). Stem cell mobilization. *Hematol Am Soc Hematol Educ Program* 2003:419–437.

Dellavalle A, Sampaolesi M, Tonlorenzi R, Tagliafico E, Sacchetti B, Perani L, Innocenzi A, Galvez BG, Messina G, Morosetti R. et al. (2007). Pericytes of human skeletal muscle are myogenic precursors distinct from satellite cells. *Nat Cell Biol* 9:255–267.

Duailibi MT, Duailibi SE, Young CS, Bartlett JD, Vacanti JP, Yelick PC (2004). Bioengineered teeth from cultured rat tooth bud cells. *J Dent Res* 83:523–528.

Ebert LM, Schaerli P, Moser B (2005). Chemokine-mediated control of T cell traffic in lymphoid and peripheral tissues. *Mol Immunol* 42:799–809.

Feng J, Mantesso A, De Bari C, Nishiyama A, Sharpe PT (2011). Dual origin of mesenchymal stem cells contributing to organ growth and repair. *Proc Natl Acad Sci USA* 108:6503–6508.

Fiedler J, Roderer G, Gunther KP, Brenner RE (2002). BMP-2, BMP-4, and PDGF-bb stimulate chemotactic migration of primary human mesenchymal progenitor cells. *J Cell Biochem* 87:305–312.

Frenette PS, Subbarao S, Mazo IB, von Andrian UH, Wagner DD (1998). Endothelial selectins and vascular cell adhesion molecule-1 promote hematopoietic progenitor homing to bone marrow. *Proc Natl Acad Sci USA* 95:14423–14428.

Gallatin WM, Weissman IL, Butcher EC (1983). A cell-surface molecule involved in organ-specific homing of lymphocytes. *Nature* 304:30–34.

Grabovsky V, Dwir O, Alon R (2002). Endothelial chemokines destabilize L-selectin–mediated lymphocyte rolling without inducing selectin shedding. *J Biol Chem* 277:20640–20650.

Gronthos S, Mankani M, Brahim J, Robey PG, Shi S (2000). Postnatal human dental pulp stem cells (DPSCs) in vitro and in vivo. *Proc Natl Acad Sci USA* 97:13625–13630.

Hahn GV, Cohen RB, Wozney JM, Levitz CL, Shore EM, Zasloff MA, Kaplan FS (1992). A bone morphogenetic protein subfamily: chromosomal localization of human genes for BMP5, BMP6, and BMP7. *Genomics* 14:759–762.

Harada H, Kettunen P, Jung HS, Mustonen T, Wang YA, Thesleff I (1999). Localization of putative stem cells in dental epithelium and their association with Notch and FGF signaling. *J Cell Biol* 147:105–120.

Hawke TJ, Garry DJ (2001). Myogenic satellite cells: physiology to molecular biology. *J Appl Physiol* 91:534–551.

Honczarenko M, Le Y, Swierkowski M, Ghiran I, Glodek AM, Silberstein LE (2006). Human bone marrow stromal cells express a distinct set of biologically functional chemokine receptors. *Stem Cells* 24:1030–1041.

Ikeda E, Morita R, Nakao K, Ishida K, Nakamura T, Takano-Yamamoto T, Ogawa M, Mizuno M, Kasugai S, Tsuji T (2009). Fully functional bioengineered tooth replacement as an organ replacement therapy. *Proc Natl Acad Sci USA* 106:13475–13480.

Ishimatsu H, Kitamura C, Morotomi T, Tabata Y, Nishihara T, Chen KK, Terashita M (2009). Formation of dentinal bridge on surface of regenerated dental pulp in dentin defects by controlled release of fibroblast growth factor-2 from gelatin hydrogels. *J Endod* 35:858–865.

Itoh F, Asao H, Sugamura K, Heldin CH, ten Dijke P, Itoh S (2001). Promoting bone morphogenetic protein signaling through negative regulation of inhibitory Smads. *EMBO J* 20:4132–4142.

Kim JY, Xin X, Moioli EK, Chung J, Lee CH, Chen M, Fu SY, Koch PD, Mao JJ (2010a). Regeneration of dental-pulp-like tissue by chemotaxis-induced cell homing. *Tissue Eng A* 16:3023–3031.

Kim K, Lee CH, Kim BK, Mao JJ (2010b). Anatomically shaped tooth and periodontal regeneration by cell homing. *J Dent Res* 89:842–847.

Kitaori T, Ito H, Schwarz EM, Tsutsumi R, Yoshitomi H, Oishi S, Nakano M, Fujii N, Nagasawa T, Nakamura T (2009). Stromal cell–derived factor 1/CXCR4 signaling is critical for the recruitment of mesenchymal stem cells to the fracture site during skeletal repair in a mouse model. *Arthritis Rheum* 60:813–823.

Kuo TF, Huang AT, Chang HH, Lin FH, Chen ST, Chen RS, Chou CH, Lin HC, Chiang H, Chen MH (2008). Regeneration of dentin–pulp complex with cementum and periodontal ligament formation using dental bud cells in gelatin–chondroitin–hyaluronan tri-copolymer scaffold in swine. *J Biomed Mater Res A* 86:1062–1068.

Laird DJ, von Andrian UH, Wagers AJ (2008). Stem cell trafficking in tissue development, growth, and disease. *Cell* 132:612–630.

Langer R, Vacanti JP (1993). Tissue engineering. *Science* 260:920–926.

Lapidot T, Dar A, Kollet O (2005). How do stem cells find their way home? *Blood* 106:1901–1910.

Lee CH, Cook JL, Mendelson A, Moioli EK, Yao H, Mao JJ (2010). Regeneration of the articular surface of the rabbit synovial joint by cell homing: a proof of concept study. *Lancet* 376:440–448.

Mahmood A, Lu D, Lu M, Chopp M (2003). Treatment of traumatic brain injury in adult rats with intravenous administration of human bone marrow stromal cells. *Neurosurgery* 53:697–702; discussion 702–693.

Mao JJ, Stosich MS, Moioli EK, Lee CH, Fu SY, Bastian B, Eisig SB, Zemnick C, Ascherman J, Wu J, et al. (2010). Facial reconstruction by biosurgery: cell transplantation versus cell homing. *Tissue Eng B* 16:257–262.

Mao JJ, Prockop DJ (2012). Stem cells in the face: tooth regeneration and beyond. *Cell Stem Cell* 11:291–301.

Matsushita K, Motani R, Sakuta T, Yamaguchi N, Koga T, Matsuo K, Nagaoka S, Abeyama K, Maruyama I, Torii M (2000). The role of vascular endothelial growth factor in human dental pulp cells: induction of chemotaxis, proliferation, and differentiation and activation of the AP-1-dependent signaling pathway. *J Dent Res* 79:1596–1603.

Mauro A (1961). Satellite cell of skeletal muscle fibers. *J Biophys Biochem Cytol* 9:493–495.

Meng F, Guo Z, Wang H, Jin J, Wang J, Wu C, Wang L (2008). High mobility group box 1 protein inhibits the proliferation of human mesenchymal stem cells and promotes their migration and differentiation along osteoblastic pathway. *Stem Cells Dev* 17:805–813.

Micera A, Lambiase A, Stampachiacchiere B, Bonini S, Levi-Schaffer F (2007). Nerve growth factor and tissue repair remodeling: trkA (NGFR) and p75 (NTR), two receptors, one fate. *Cytokine Growth Factor Rev* 18:245–256.

Millan J, Hewlett L, Glyn M, Toomre D, Clark P, Ridley AJ (2006). Lymphocyte transcellular migration occurs through recruitment of endothelial ICAM-1 to caveola- and F-actin-rich domains. *Nat Cell Biol* 8:113–123.

Miura M, Gronthos S, Zhao M, Lu B, Fisher LW, Robey PG, Shi S (2003). SHED: stem cells from human exfoliated deciduous teeth. *Proc Natl Acad Sci USA* 100:5807–5812.

Nakao K, Morita R, Saji Y, Ishida K, Tomita Y, Ogawa M, Saitoh M, Tomooka Y, Tsuji T (2007). The development of a bioengineered organ germ method. *Nat Methods* 4:227–230.

Ohazama A, Modino SA, Miletich I, Sharpe PT (2004). Stem-cell-based tissue engineering of murine teeth. *J Dent Res* 83:518–522.

Ozaki Y, Nishimura M, Sekiya K, Suehiro F, Kanawa M, Nikawa H, Hamada T, Kato Y (2007). Comprehensive analysis of chemotactic factors for bone marrow mesenchymal stem cells. *Stem Cells Dev* 16:119–129.

Ponomaryov T, Peled A, Petit I, Taichman RS, Habler L, Sandbank J, Arenzana-Seisdedos F, Magerus A, Caruz A, Fujii N, et al. (2000). Induction of the chemokine stromal-derived factor-1 following DNA damage improves human stem cell function. *J Clin Investig* 106:1331–1339.

Ponte AL, Marais E, Gallay N, Langonne A, Delorme B, Herault O, Charbord P, Domenech J (2007a). The in vitro migration capacity of human bone marrow mesenchymal stem cells: comparison of chemokine and growth factor chemotactic activities. *Stem Cells* 25:1737–1745.

Quesenberry PJ, Becker PS (1998). Stem cell homing: rolling, crawling, and nesting. *Proc Natl Acad Sci USA* 95:15155–15157.

Ridley AJ, Schwartz MA, Burridge K, Firtel RA, Ginsberg MH, Borisy G, Parsons JT, Horwitz AR (2003). Cell migration: integrating signals from front to back. *Science* 302:1704–1709.

Ringe J, Strassburg S, Neumann K, Endres M, Notter M, Burmester GR, Kaps C, Sittinger M (2007). Towards in situ tissue repair: human mesenchymal stem cells express chemokine receptors CXCR1, CXCR2 and CCR2, and migrate upon stimulation with CXCL8 but not CCL2. *J Cell Biochem* 101:135–146.

Ruster B, Gottig S, Ludwig RJ, Bistrian R, Muller S, Seifried E, Gille J, Henschler R (2006). Mesenchymal stem cells display coordinated rolling and adhesion behavior on endothelial cells. *Blood* 108:3938–3944.

Sanchez-Madrid F, del Pozo MA (1999). Leukocyte polarization in cell migration and immune interactions. *EMBO J* 18:501–511.

Schultz E, Jaryszak DL, Valliere CR (1985). Response of satellite cells to focal skeletal muscle injury. *Muscle Nerve* 8:217–222.

Smith AJ, Cassidy N, Perry H, Begue-Kirn C, Ruch JV, Lesot H (1995). Reactionary dentinogenesis. *Int J Dev Biol* 39:273–280.

Sonoyama W, Liu Y, Fang DAJ, Yamaza T, Seo BM, Zhang CM, Liu H, Gronthos S, Wang CY, Shi ST, et al. (2006). Mesenchymal stem cell–mediated functional tooth regeneration in swine. *PLoS One* 1:e79.

Suzuki T, Lee CH, Chen M, Zhao W, Fu SY, Qi JJ, Chotkowski G, Eisig SB, Wong A, Mao JJ (2011). Induced migration of dental pulp stem cells for in vivo pulp regeneration. *J Dent Res* 90:1013–1018.

Watt DJ, Morgan JE, Clifford MA, Partridge TA (1987). The movement of muscle precursor cells between adjacent regenerating muscles in the mouse. *Anat Embryol (Berl)* 175:527–536.

Worthylake RA, Burridge K (2001). Leukocyte transendothelial migration: orchestrating the underlying molecular machinery. *Curr Opin Cell Biol* 13:569–577.

Wu GD, Nolta JA, Jin YS, Barr ML, Yu H, Starnes VA, Cramer DV (2003). Migration of mesenchymal stem cells to heart allografts during chronic rejection. *Transplantation* 75:679–685.

Yamaguchi J, Kusano KF, Masuo O, Kawamoto A, Silver M, Murasawa S, Bosch-Marce M, Masuda H, Losordo DW, Isner JM, et al. (2003). Stromal cell–derived factor-1 effects on ex vivo expanded endothelial progenitor cell recruitment for ischemic neovascularization. *Circulation* 107:1322–1328.

Yamaza T, Ren G, Akiyama K, Chen C, Shi Y, Shi, S (2011). Mouse mandible contains distinctive mesenchymal stem cells. *J Dent Res* 90:317–324.

Yildirim S, Fu SY, Kim K, Zhou H, Lee CH, Li A, Kim SG, Wang S, Mao JJ (2011). Tooth regeneration: a revolution in stomatology and evolution in regenerative medicine. *Int J Oral Sci* 3:107–116.

Young CS, Terada S, Vacanti JP, Honda M, Bartlett JD, Yelick PC (2002). Tissue engineering of complex tooth structures on biodegradable polymer scaffolds. *J Dent Res* 81:695–700.

Young CS, Kim SW, Qin C, Baba O, Butler WT, Taylor RR, Bartlett JD, Vacanti JP, Yelick PC (2005). Developmental analysis and computer modelling of bioengineered teeth. *Arch Oral Biol* 50:259–265.

21

STEM CELL–BASED BIOENGINEERING OF CRANIOFACIAL BONE

DAVID D. LO, DANIEL T. MONTORO, MONICA GROVA, JEONG S. HYUN, MICHAEL T. CHUNG, AND DERRICK C. WAN

Hagey Laboratory for Pediatric Regenerative Medicine, Department of Surgery, Plastic and Reconstructive Surgery Division, Stanford University School of Medicine, Stanford, California

MICHAEL T. LONGAKER

Hagey Laboratory for Pediatric Regenerative Medicine, Department of Surgery, Plastic and Reconstructive Surgery Division, Stanford University School of Medicine, Stanford, California, and Institute for Stem Cell Biology and Regenerative Medicine, Stanford University, Stanford, California

1 INTRODUCTION

Successful reconstruction of large skeletal defects continues to be a major challenge to craniofacial surgeons. These defects can arise from an array of etiologies, including congenital malformations, acute injuries, osteomyelitis, or tumor resection. The development of effective and consistent therapies for bone regeneration continues to prove arduous for scientists and surgeons alike. Current reconstructive strategies employ the use of autologous bone grafts, allogeneic materials, or alloplastic materials to augment bone regrowth and to restore craniofacial elements.

Approximately 600,000 bone grafts are performed annually in the United States, and nearly 6% of these procedures are in the craniofacial region (Greenwald et al., 2001). In most instances, autologous bone grafts are the preferred material for craniofacial reconstruction, as they do not elicit an immune response and possess significant osteogenic properties. Although autologous bone grafts often achieve favorable results and are

Stem Cells in Craniofacial Development and Regeneration, First Edition.
Edited by George T.-J. Huang, Irma Thesleff.

regarded as the gold standard for reconstruction, they do have major drawbacks. Bone grafts are often limited by the scarcity of donor bone graft, contour irregularities, and donor-site morbidity secondary to tissue harvesting (Laurie et al., 1984). Suitable substitutes with osteoconductive and osteoinductive properties are a high priority for stem cell scientists and surgeons alike to circumvent the drawbacks of bone graft procedures.

Commonly used allogeneic and alloplastic materials with osteoconductive properties include poly(methyl methacrylate), hydroxyapatite cement, plaster of paris, and titanium (Lew et al., 1997; Bruens et al., 2003; Taub et al., 2003). Although these bone substitutes provide an alternative to autologous bone grafts, they carry risks of disease transmission, infection, immunologic rejection, unpredictable graft resorption, poor native tissue integration, and graft-versus-host disease (Wan et al., 2006).

In the face of mounting clinical demand and suboptimal reconstructive techniques, the emergence of tissue engineering and regenerative medicine using stem cells and concurrent advancement in biotechnology have shifted the focus from available reconstructive strategies to endogenous regeneration. Such *de novo* formation of healthy craniofacial bony to replace or to reconstruct diseased or damaged tissue represents an attractive alternative. Importantly, tissue-engineered bone must have appropriate structural characteristics that approximate native bone for predictably shaped and functional reconstruction. The three-dimensional bioengineering of bone requires three essential elements—cells, growth and trophic factors, and scaffolds—each of which has already proven useful as single constituents or in combination (Zaky and Cancedda, 2009). First and foremost, cellular components must be present at the host site to give rise to new structural tissue, recruited from either local progenitor cell populations or from an exogenous cell source. Osteoinductive growth and differentiation factors must also be available to give cues to guide the appropriate development and differentiation of the cellular components. These factors may be produced by transplanted cells or derived from endogenous sources such as matrix depots. As we gain more understanding of the factors that compose bone niches, these factors can also be provided exogenously as purified proteins or biomimetic small molecules. Finally, an osteoconductive scaffolding matrix must be introduced to provide a three-dimensional substrate for cellular migration, attachment, proliferation, and differentiation. Additionally, scaffolds serve to immobilize and orient the presentation of osteoinductive signals to the responding cells (Bruder et al., 1999).

Although there are many similarities between craniofacial bone niches, there are disparate and unique regional differences that will require optimization of both osteoprogenitor cells and biomimetic scaffolds that have the ability not only to provide a temporary mechanical framework but also to facilitate cellular delivery. It is important to provide a microniche that can be comprised of cells, scaffold, and osteoinductive factors that enhance the reparative process and harness the regenerative capacity of implanted cells for robust bone healing. Achieving this level of regeneration requires the coordination of a wide array of disciplines from cell and molecular biology, developmental biology, molecular genetics, materials science, and bioengineering.

The unique biological properties of stem cells have positioned them as the ideal candidate cell source for craniofacial bone (Panetta et al., 2010). Multiple progenitor cell populations are now known to possess a capacity to undergo robust osteogenic differentiation in the presence of appropriate environmental cues: bone marrow–derived mesenchymal cells, adipose-derived stromal cells, and pluripotent stem cells (Fig. 1) (Tremoleda et al., 2008). The technical, political, and ethical

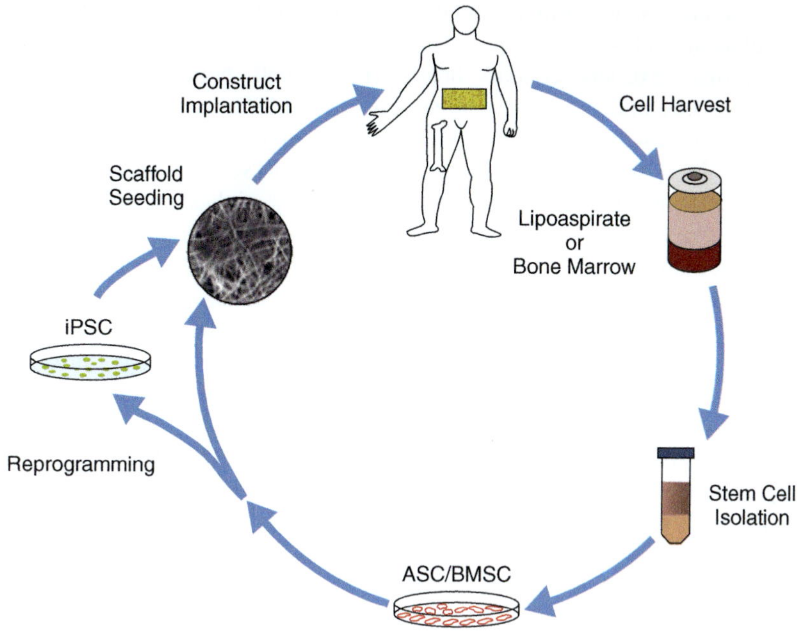

FIGURE 1 Stem cell–based tissue engineering.

implications surrounding the use of embryonic stem cells have thus far hindered their incorporation into clinical strategies; however, we will present some of the more recent advancements in the pluripotent cell field that may hold significant promise for the future use of these powerful cells.

Although osteoprogenitor cells represent the cornerstone of novel craniofacial bone engineering, we must emphasize that these cells alone are inadequate to achieve bone tissue engineering. Evidence shows that it is necessary to combine osteoprogenitor cells with a customizable biomimetic scaffold to achieve significant regeneration. We do not intend this chapter to be a rigorous discussion of scaffolds, as this topic is dealt with in depth in other chapters of the book, but we will make mention of the specific materials and designs used as applicable to our discussion of landmark studies in craniofacial bone regeneration.

2 MESENCHYMAL STROMAL CELLS

Mesenchymal stromal cells (MSCs) have been studied widely for their potential in tissue regeneration and therapeutic applicability since Friedenstein et al. (1966) identified the existence of this postnatal multipotent cell population in bone marrow. These fibroblast-like cells self-replicate in culture and have been isolated successfully from a multitude of postnatal tissues (Panetta et al., 2010). Early work by Pittenger et al. (1999) defined the *in vitro* multipotent capacity in which bone marrow–derived mesenchymal stromal cells (BMSCs) are capable of differentiating along many cell lineages when cultured under defined *in vitro* conditions with appropriate cues (Caplan, 1991). To date, BMSCs have been isolated, studied, and characterized across several vertebrate species (Alhadlaq and Mao, 2004).

Bone marrow extracts contain heterogeneous cell populations composed of hematopoietic cells and bone marrow stroma, with MSCs comprising a fraction of the total mononucleated cells within bone marrow. Given that bone marrow is a rich source of MSCs, it should come as no surprise that MSCs can readily differentiate down the osteogenic lineage. The multipotent capacity of these cells can be retained over 20 passages with no visible morphological changes (Bruder et al., 1997; Banfi et al., 2000). Although isolation and expansion of these cells can be performed readily, ongoing efforts continue to improve the culture conditions in order to increase the proliferative capacity of primary MSCs. Fibroblast growth factor-2 (FGF2), for example, has been found to enhance the proliferation rate of primary MSCs without substantially reducing their differentiation potential and delays cell senescence (Bruder et al., 1997).

Efforts have also been directed at isolation and characterization techniques that optimally identify MSCs from the milieu of isolated cells; such techniques include positive selection of the MSC surface markers STRO1, CD133, p75LNGFR, CD29, CD44, CD73, CD90, CD105, CD106, CD124, HLA-DR, and c-kit (Pittenger et al., 1999; Alhadlaq and Mao, 2004; Lee et al., 2004). Negative selection is also a helpful tool for enrichment of MSCs using antibody cocktails that sort out contaminating bone marrow cells (Alhadlaq and Mao, 2004). Although antibodies to several cell-surface antigens can be used to enrich for MSCs, monospecific and unique molecular probes still do not exist to identify these cells unequivocally in situ and obtain homogeneous populations.

Although isolation and characterization of pure MSCs populations is not trivial, their ability to expand significantly from one clone suggests that achieving a population of homogeneous cells is indeed possible. The utility of using a homogeneous versus heterogeneous population has yet to be defined, although research has been conducted with both methods. Also unclear is the degree of fate commitment versus stemness, which may be important for functional regeneration of tissue. As MSCs have multipotent potential, they can be transplanted into a strong osteoinductive environment relying on local growth factors and signals to drive differentiation. Alternatively, they can be predifferentiated into osteoblasts *in vitro* prior to transplantation. Both systems have been seen to give rise to *de novo* bone, but future work will continue to improve isolation and enrichment of pure cell populations with predictable characteristics and examine the optimal level of commitment versus plasticity of transplantable bone-producing cells.

A growing body of evidence reinforces the potential of MSCs to regenerate bone successfully in both *in vitro* and *in vivo* model systems with different scaffolding materials (Breitbart et al., 1998). Using a transgenic mouse model, Krebsbach et al. (1998) demonstrated that gelatin sponges seeded with BMSCs repaired greater than 99% of an osseous defect to full thickness within 2 weeks. Shang et al. (2001) showed augmented healing of sheep cranial defects with calcium alginate gel containing autologous MSCs. In a rabbit model, Schantz et al. (2001, 2003) found comparable *in vitro* osteogenic differentiation and *in vivo* new bone regeneration of MSCs and osteoblasts when seeded with a fibrin–glue suspension onto a polycaprolactone scaffold.

Similar MSC-based therapies can potentially be translated toward clinical practice. Early reports of cell-based TMJ grafts employed mature osteoblasts in polymer scaffolds with "painted" chondrocytes at the articular surface (Weng et al., 2006). Recent reports also suggest the possibility of generating an entire articular condyle in the same shape and dimensions of a human TMJ *in vivo*, with both cartilage and bone layers from a single population of MSCs (Grayson et al., 2010). Finally, alveolar bone regeneration and augmentation has been shown to be possible in both animals and humans utilizing BMSC-seeded constructs (Ueda et al., 2005; Weng et al., 2006).

Despite the many advantages of BMSCs, there remain several drawbacks that have limited their use in large-scale skeletal tissue engineering efforts. BMSC harvest is an invasive procedure that may impart donor site morbidity and discomfort. The number of MSCs harvested from surgical bone marrow aspiration is often limited given the heterogeneity of the marrow population. Moreover, there is an age-dependent decline in cell frequency of MSCs in bone marrow (approximately 1 : 10,000 cells in newborns to 1 : 2,000,000 in octogenarians) and a decrease in proliferative capacity and osteogenic potential (Zhou et al., 2008; Yu et al., 2011). As an alternative to BMSCs, progenitor cells isolated from the stromovascular fraction (SVF) of adipose tissue have thus emerged as an attractive alternative.

3 ADIPOSE-DERIVED STROMAL CELLS

Over the past decade, human adipose-derived stromal cells (hASCs) have been identified as potential building blocks for cell-based therapy, generating significant interest in basic science and translational research to define and harness their true potential. The initial methods to isolate cells from adipose tissue were pioneered by Rodbell in 1964, who collected the heterogeneous cell population of the stromovascular fraction (SVF) containing circulating blood cells, fibroblasts, pericytes, endothelial cells, and adipocyte progenitors. It was not until 2001, however, that hASCs were isolated from the SVF of human lipoaspirate (Zuk et al., 2001, 2002; De Ugarte et al., 2003). As an enriched source of adult multipotent cells with a proven capability to undergo osteogenic, adipogenic, chondrogenic, and myogenic differentiation, ongoing efforts seek to better define and harness ASC potential (De Ugarte et al., 2003, Thesleff et al., 2011).

As an alternative source of multipotent cells, ASCs have several advantages over BMSCs from a clinical perspective. Subcutaneous adipose tissue exists in abundance and is readily available from most individuals, translating into a larger typical yield of cells than that of bone marrow. And unlike bone marrow aspirates, lipoaspirates do not suffer from the dilutional effect of "contaminant" hematopoietic cells and peripheral blood cells. *In vitro* studies have shown no significant differences in growth kinetics and cell senescence, lineage-specific differentiation, and viral transduction efficiencies between BMSCs and ASCs (De Ugarte et al., 2003). Independent studies have also determined that cell-surface markers are relatively preserved across the two progenitor cell populations (Gronthos et al., 2001; Katz et al., 2005; Mitchell et al., 2006). Based on these observations, ASCs have spawned enthusiasm in devising clinical strategies for bone tissue engineering.

Lee et al. (2003) carried out the first published study of *in vivo* bone formation using ASCs in 2003. Significant generation of bone was observed following the subcutaneous implantation of predifferentiated rat ASCs seeded onto a poly(glycolic acid) (PGA) scaffold. Similarly, the capacity for ASCs on a hydroxyapatite–tricalcium phosphate (HA-TCP) scaffold to form osteoid subcutaneously in a severe combined immunodeficiency (SCID) mouse model was also demonstrated (Hicok et al., 2004). Subsequently, in an important study, Cowan et al. (2004) demonstrated the ability of mouse ASCs to regenerate bone in critical-sized calvarial defects using a hydroxyapatite-coated poly(lactic-*co*-glycolic acid) (PLGA) scaffold. This observation was supported by Yoon et al. (2007) and by Levi et al. (2010a) with human ASCs. Further advances have been made toward applying tissue engineering methods to other craniofacial defect models, such as the palate. In a study by Conejero et al., (2006), osteogenically induced ASC-seeded PLGA scaffolds were used to repair surgically created palatal defects in rats; the results showed significant bone regeneration in comparison to that in the control groups.

Although the literature has positively demonstrated the application of ASCs and their ostensible capacity to promote bone regeneration, it is important to point out that reproducible craniofacial bone formation has been met with varying degrees of success, suggesting that the context in which cells are placed and the method with which they are delivered is critically important. Dudas et al. (2006) reported no significant differences in bony healing in noncritical rabbit calvarial defects between empty defects and those treated with a gelfoam scaffold seeded with osteoinduced rabbit ASCs. It is possible, however, that the noncritical-sized defect played a role in this finding. The osteogenic potential of ASCs may be demonstrated more definitively with a critical-sized defect, as has been shown in studies discussed previously (Yoon et al., 2007; Levi et al., 2010a).

A number of preliminary clinical case studies using ASCs in bone regeneration have been reported in Germany and Finland. Lendeckel et al. (2004) reported a case of a large calvarial defect repair in a 7-year-old girl using ASCs in conjunction with autologous fibrin glue and bone chips from the ilium. Postoperative radiographic assessment revealed new bone formation with nearly complete calvarial continuity three months after the reconstruction. In 2009, Mesimäki et al. reported the first use of a microvascular custom-made ectopic bone flap developed from a preformed titanium cage filled with autologous ASCs, β-tricalcium phosphate (β-TCP) granules, and rhBMP-2 to reconstruct a maxillary defect. The subsequent maxillary construct appeared to have adequate rigidity to support dental implants with excellent osseointegration (Mesimäki et al., 2009). Most recently, Thesleff et al. (2011) described a novel method to reconstruct critical-sized calvarial defects of adult patients using autologous ASC and β-TCP granules, providing evidence that ASCs combined with β-TCP can be used to regenerate bone, with favorable clinical results.

Recent advances in ASC research have focused on defining homogeneous populations with more robust and specific osteogenic behavior. As with MSCs, efforts have been made to identify candidate ASC-surface markers to enrich or to isolate homogeneous subpopulations that are more capable of osteogenic differentiation than the heterogeneous mix of SVF cells. A recent study using single-cell transcriptional analysis determined a correlation between the expression of CD105, a transforming growth factor (TGF) β1 coreceptor, and osteogenic capacity (Levi et al., 2011c). FACS-sorted $CD105^{LOW}$ cells demonstrated significantly enhanced *in vitro* osteogenic differentiation and *in vivo* bone regeneration compared with unsorted or $CD105^{HIGH}$ cells.

Identification of such candidate cell-surface markers thus demonstrates the potential for subpopulation enrichment as an avenue to enhance osteogenesis in ASCs. The enhancement of osteogenic differentiation and bone regeneration of $CD105^{LOW}$ ASCs may be attributable to the isolation of a subpopulation with reduced TGFβ1/Smad2 signaling given what is already known about the inhibitory capability of TGFβ1 on osteogenesis in osteocompetent cell lines (Levi et al., 2010b).

The TGFβ superfamily is also involved in a wide range of cellular processes. In particular, bone morphogenetic protein (BMP)-2, a member of the TGFβ superfamily, has been demonstrated to potently enhance osteogenic differentiation in ASCs (Wan et al., 2006). A recent study in the authors' laboratory demonstrated increased bone formation by ASCs through manipulation of the BMP signaling pathway (Levi et al., 2011a). Using integrating viral transduction and nonintegrating minicircle transfection methods, Noggin, a BMP antagonist, was knocked down, thereby generating a population of ASCs with amplified endogenous BMP activity, which translated into more robust osteogenesis (Levi et al., 2011a). It is important to note that through the manipulation of specific signaling pathways, the cellular machinery can thus be utilized to modulate the stem cell activity.

Another important factor for successful regeneration using stem cells is the environmental niche in which they are transplanted. From the viewpoint of craniofacial reconstruction, a calvarial defect serves as a unique microenvironment that can be closely examined for bone regeneration. As with MSCs, ASCs can be engrafted into the tissue defect to stimulate bone regeneration, but understanding how the niche supports and maintains the osteogenic potential of these cells is important to optimizing their bone-forming capacity. A complex interplay of cell-to-cell signals between the surrounding calvarial osteoblasts, underlying dura mater (DM) cells, and transplanted cells has been observed. It has been shown that DM is crucial in the healing of critical-sized calvarial defects as both a participant and a producer of factors that induce ASC proliferation and enhance bone formation (Levi et al., 2011b). Additional paracrine interactions may also contribute to ASC behavior in calvarial defects. Wnt (wingless in *Drosophila*), Hedgehog, and platelet-derived growth factor (PDGF) signals have all been implicated in comprising the niche around which stem cells are implanted.

Future studies of ASCs in craniofacial bone engineering will continue to define subpopulations with specific and robust differentiation potential. The targeted delivery on a well-suited three-dimensional engineered scaffold must be paired strategically with the specific bone deficit in conjunction with appropriate osteoinductive factors to enhance bone regeneration. Finally, the components of the surrounding niche and their various modes of interaction with ASCs must be better understood to promote persistent functional bone. With progress on all fronts, ASC-mediated bone regeneration holds much promise for clinical treatment of craniofacial bone deficits.

4 INDUCED PLURIPOTENT STEM CELLS

Although BMSC and ASC possess the potential to differentiate to bone and other tissues of mesodermal lineage, pluripotent stem cells hold significant promise as an alternative candidate for craniofacial tissue engineering. Within the past five years, scientists have shown that overexpression of four key transcription factors can reprogram somatic cells into pluripotent cells, thus providing a cell source for regenerative medicine that has the flexibility to differentiate across all three embryonic germ layers. While MSCs and ASCs are a heterogeneous mix of cells with limited capacity, pluripotent cells have the potential not only to differentiate into supportive tissues such as nerve and blood vessels, but may also possess the ability to regenerate bone throughout the craniofacial skeleton.

The technology of nuclear reprogramming resets the fate of terminally differentiated somatic cells to an undifferentiated state with properties similar to that of ESCs (Okita et al., 2007, 2008). These reprogrammed cells, known as induced pluripotent stem cells (iPSCs) were first derived through viral transduction of four defined transcription factors (*Oct 3/4, Sox2, c-Myc*, and *Klf4*) into mouse embryonic fibroblasts (Takahashi et al., 2006).

iPSCs have been observed to display pluripotent properties that closely resemble those of human ESCs. Most important, they have been shown to contribute directly to germline chimeras and to possess the ability to generate live late-term embryos when injected into tetraploid blastocysts (Kang et al., 2009), proving that iPSCs can give rise not only to all three germ layers (ectoderm, mesoderm, and endoderm) but also to all of the thousands of specialized cell types that comprise a live mouse, including all of the specialized cell types important to craniofacial reconstruction.

Various cell types have now been converted to a pluripotent state: skin fibroblasts, neural stem cells, keratinocytes, peripheral blood $CD34^+$ cells, melanocytes, ASCs, and cord blood cells with varying (generally low) frequencies. Different cell lineages have also been shown to undergo reprogramming with fewer than the four key factors indicating that reprogramming may be contextually dependent and that the cell type can affect conversion efficiency (Takahashi et al., 2007; Yu et al., 2007; Aasen et al., 2008; Giorgetti et al., 2009; Kim et al., 2009; Loh et al., 2009; Sun et al., 2009, 2010; Utikal et al., 2009). Efforts thus continue to be directed toward improving the success rates of cellular conversion.

Unlike their embryonic counterparts, iPSCs can be derived from the cells of a patient, thereby enabling patient-specific autogenous tissue derivation for repair or replacement therapies. Although it was recently shown that there may be some hurdles to the widely assumed immune privileged status of autologous iPSCs, it is likely that this may be a technical and not intrinsic barrier to using iPSCs as a therapeutic cell source (Wernig et al., 2007; Zhao et al., 2011). A clear advantage of iPSCs over ESCs, however, is the avoidance of ethical debate and restrictions that are inherent with ESCs. Further work into the intrinsic versus acquired immunogenicity of iPSCs will help to define the true potential of these cells and under what set of conditions they will be useful in tissue engineering. The method of reprogramming, the epigenetic state of iPSC clones, the differentiated state of the iPSC-derived cells intended for therapy, and the method and site of cell delivery are all factors that must be researched more rigorously to be understood. Nonetheless, iPSCs still represent a powerful tool for realizing patient-specific cell therapy in bone reconstruction and regeneration.

Whereas osteogenic differentiation of MSCs has been thoroughly demonstrated, generating bone matrix–forming osteoblasts from mouse iPSCs *in vitro* has only recently been reported (Tashiro et al., 2009). In 2011, Bilousova et al. derived osteoblasts from murine iPSCs through initial formation and differentiation of an embryoid body. Mineralization and bone matrix deposits were found upon *in vivo* transplantation on an engineered gelatin scaffold after expansion and confirmation of bone lineage with osteocalcin and salioprotein expression of the osteoblasts. This method is, however, a labor-intensive process requiring an extended period of time in culture, which compounds the risks of contamination and accumulation of mutations.

A growing body of literature describes attempting to transplant iPSCs *in vivo* into osteogenic conditions and reduce the extended predifferentiation to osteoblasts *ex vivo*. *In vivo* formation of periodontal tissue has been achieved in nude mice with direct transplantation of cells predifferentiated from iPSC for 5 days. These predifferentiated cells were seeded on silk scaffolds enhanced by enamel matrix derivatives to provide differentiation factors (Duan et al, 2011). More recently, Ye et al. (2011) demonstrated *de novo* osteogenesis *in vivo* of iPSCs overexpressing a potent transcription factor SATB2 within a critical-sized calvarial bone defect. They were able to show that the iPSC did not need to be predifferentiated and can constitutively overexpress a factor that prevents transplanted cells from forming a teratoma.The possibility that *ex vivo* predifferentiation may not be necessary thus sheds new light on the balance between plasticity and lineage restriction in these cells.

Despite overwhelming enthusiasm arising from the potential benefits of iPSCs, there are inevitable barriers that need to be overcome for these cells to be used directly. Before these cells can be employed clinically, it is important to realize integration-free technologies to avoid the viral risk of insertional mutagenesis and to ensure complete fate commitment to minimize the risk of teratoma formation.

Early studies with iPSCs have utilized lentiviral infection to deliver reprogramming factors given the high efficiency of transduction. As part of the retrovirus family, genomic integration of the transgenes into host cells may result in insertional mutagenesis. Furthermore, both KLF4 and c-MYC are oncogenes, compounding the risk of tumorigenesis should incomplete silencing after reprogramming occur. Therefore, it may be advantageous to utilize technologies that bypass integration of transgenes into the genome.

As part of the efforts to circumvent this clinical conundrum, recent studies have utilized cre-excisable viral constructs, non-viral episomal plasmids, and transposon transgenesis using *piggy*Bac expression vectors to generate human iPSCs (Soldner et al., 2009; Woltjen et al., 2009; Yu et al., 2009; Sommer et al., 2010). Low reprogramming efficiencies (<0.003%) and residual vector sequences, however, have limited the application of these methods. Jia et al. (2010) reported higher transfection efficiency and longer ectopic expression using a non-viral, episomal minicircle plasmid to effectively generate transgene-free human iPSCs from adult hASCs (Narsinh et al., 2011). Improvements on this technology have included tetracycline activation, allowing temporal control of expression as well as dose–response levels of transgene expression. Alternative methods of reprogramming using nongenetic materials have been achieved, including direct application of recombinant transcription factors (Zhou et al., 2009). Stabilized RNA has also presented an alternative non-DNA method of reprogramming that shows promise. These methods allow stoichiometric control over the delivery of transcription factors, a variable that may prove important for generating iPSCs with predictable and consistent characteristics.

Another critical issue for the clinical application of iPSC is the risk of teratoma formation by residual undifferentiated stem cells. Past studies have focused on the removal of teratomas utilizing suicide genes and chemotherapy, but these techniques have not entirely eliminated the risk of a small number of unresponsive cells persisting (Schuldiner et al., 2003). Recent work by Tang et al. (2011) discovered a surface antigen, SSEA-5, targeted by a monoclonal antibody capable of binding specifically to undifferentiated pluripotent stem cells. This has allowed for the identification of undifferentiated cells for removal by FACS and has been shown to greatly reduce teratoma formation *in vitro*. This strategy shows promise for minimizing tumor risk in the generation of a wide array of iPSC-derived lineages for transplantation.

5 STEM CELLS AND GENE THERAPY

The long-term goals of stem cell–based regenerative medicine are to restore tissue structure. Although many combinations of stem cells and scaffolds have shown some promise for creating *de novo* craniofacial bone, these strategies can be optimized by coupling gene therapy with cell therapy. Gene therapy can be a powerful tool for stem cell research to enhance specific lineage differentiation; for skeletal regeneration it may be possible to direct MSCs, ASCs, or iPSCs more effectively by providing guidance cues as well as correcting genetic mutations via gene therapy.

For stem cell regeneration of craniofacial bone, a common strategy may be to enhance osteogenic differentiation through overexpression or knockdown of specific proteins in the differentiation pathway. Furthermore, transfection may be done *in vivo* directly using a vector to transfer genetic material into endogenous cell populations, or *ex vivo*, by treating cells outside the host prior to implantation. As discussed previously, viral vectors have the advantage of being highly efficient with gene transfer, but carry the risk of mutagenesis. They may also increase the immune response of the host, leading to rejection of the cells. While nonviral vectors often result in variable transfection efficiency and level of gene expression, vectors such as minicircle DNA, liposome-based particles, and nanoparticles may hold promise for more clinically relevant gene therapy.

With the goal of improving osteogenic differentiation and bone formation, gene therapy has been used to alter the expression of genes such as *Runx2, MSX2, Shh, Wnt, Osterix, Noggin*, and *BMP2*. While exogenous application of recombinant human BMP2 (rhBMP-2) has repeatedly demonstrated induction of *de novo* bone growth, its safety profile and pharmacodynamics have not been fully established. Preliminary human trials reported no to low incidence of adverse events associated with rhBMP2 use in spinal fusion procedures; however, some safety concerns have surfaced. Most recent comprehensive review of rhBMP2 use in spinal fusion procedures suggests that its use is associated with between 10 and 50% adverse events, depending on the procedure performed, which include but are not limited to cyst and ectopic bone formation, osteolysis, increased early back and leg pain, and radiculitis (Carragee et al., 2011). This may be due, in part, to uncontrolled application of a strong osteoinductive factor. Given its recent safety concerns, a question remains regarding the safety and efficacy of rhBMP's application in craniofacial skeletal engineering. To circumvent this issue, an alternative approach can focus on implanting stem cells into a skeletal defect to accelerate and restrict bone formation regionally by acting as a vehicle for providing a steady and localized release of BMP2 over time (Lieberman et al., 1998). Reducing expression of the BMP antagonist Noggin by shRNA knockdown in hASCs has also been shown to accelerate mineralization *in vitro* and bone regeneration in murine calvarial defects by amplifying the effect of BMP signaling (Levi et al., 2011a). In similar fashion, significant bone regeneration has been observed in a rabbit calvarial

TABLE 1 Summary of Strategies to Optimize Osteogenesis[a]

Isolation and enrichment of osteogenic stem cell subpopulations
FACS/MACS
Microarray analysis
Microbeads
Augmentation of key trophic factors involved in osteogenesis
Stimulatory factors
BMP (endogenous and rhBMP-2)
TGFβ[b]
SATB2
Wnt
IGF-I, II
Cellular retinol-binding protein 1 (CRBP1)
Inhibitory factors
BMP antagonists: Noggin/Gremlin/Smurf1
Wnt antagonist: Dickkopfs 1(DKK1)
TGFβ[b]
Development of biocompatible and osteoconductive scaffolds
β-Tricalcium phosphate (β-TCP)
Hydroxyapatite (HA) and composites
Poly(lactic-*co*-glycolic acid) (PLGA)/poly(L-lactic acid) (PLA)
Mesoporous bioactive glass (MBG)
Fibrin glue
Hydrogel
Gelatin sponges
Chitosan
Collagen type I (CNI)

Source: Zuk PA (2008). Tissue engineering craniofacial defects with adult stem cells? Are we ready yet? *Pediatr Res* 63(5):478–486; Deschaseaux F, Sensébé L, Heymann D (2009). Mechanisms of bone repair and regeneration. *Trends Mol Med* 15(9):417–429.

[a]FACS, fluorescence-activated cell sorting; MACS, magnetic-activated cell sorting; BMP, bone morphogenetic protein; TGFβ, transforming growth factor β; IGF, insulin-like growth factor;

[b]Both stimulatory and inhibitory.

defect model using Shh-transduced MSCs (Edwards et al., 2005). Finally, increasing focus on the Wnt signaling pathway has shown that Wnt proteins may be involved in bone growth and regeneration in MSCs (Boland et al., 2004; Chang et al., 2007). Future work will continue to clarify upstream mechanisms regulating expression of these genes and how proteins interact with each other to drive bone differentiation and bone growth.

Using similar strategies for enhancing osteogenic differentiation, gene therapy has the potential to correct or compensate for genetic bone deficiencies. Osteogenesis imperfecta (OI) is one of the few genetic bone defects caused by a well-characterized loss of function mutation and is thus an appealing model and clinical candidate for gene and cell therapy. A defect in collagen type I causes brittle bones prone to fracture, with decreased cortical mineralization. As a proof of concept, MSCs from a patient with OI were reprogrammed into iPSCs where the mutant collagen gene was targeted for replacement with a wild-type allele. These genetically altered cells were then differentiated back into MSCs for transplantation into immunocompromised mice and were found to produce *in vivo* bone (Deyle et al., 2011). This experiment lays the groundwork for the possibility of powerful patient-specific gene therapy for bone regeneration, but translating this strategy into a clinical reality depends on the continued development of a successful strategy for stem cell delivery, the persistence of these building blocks *in vivo*, and enhanced safety for the patient.

6 CONCLUSIONS

Although there have been many reports of preliminary success using stem cells both in the laboratory and clinic for craniofacial regeneration, there remain several ways to optimize their use further for the predictable, controlled, and robust bone formation necessary for routine clinical application (Table 1). MSC and ASC populations will continue as viable regenerative cell sources while methods for identification, characterization, and isolation of pure populations with defined differentiation potential are refined. Pluripotent cells are an emerging technology for tissue engineering that still require considerable effort to characterize their limitations and biological potential. However, the ability of human embryonic and induced pluripotent cells to form any tissue in the body may prove an invaluable resource for regenerative medicine. Gene therapy may be used to fine-tune desired programs of each of these cell types and may also enable correction of patient-specific mutations before transplantation. Finally, the scaffold delivery vehicle may act as the interface between the surrounding niche and the implanted stem cells and should have a synergistic effect on healing by promoting survival and appropriate differentiation of the transplanted cells. The stem cell field is thus multifaceted, but cooperation between various disciplines investigating these cells has already produced, and will continue to produce, new and exciting outcomes for craniofacial patients.

Acknowledgment

The authors acknowledge the Oak Foundation for its support.

REFERENCES

Aasen T, Raya A, Barrero MJ, Garreta E, et al. (2008). Efficient and rapid generation of induced pluripotent stem cells from human keratinocytes. *Nat Biotechnol* 26(11): 1276–1284.

Alhadlaq A, Mao JJ (2004). Mesenchymal stem cells: isolation and therapeutics. *Stem Cells Dev* 13(4):436–448.

Banfi A, Muraglia A, Dozin B, Mastrogiacomo M, et al. (2000). Proliferation kinetics and differentiation potential of ex vivo expanded human bone marrow stromal cells: implications for their use in cell therapy. *Exp Hematol* 28(6):707–715.

Bilousova G, Jun DH, King KB, De Langhe S, et al. (2011). Osteoblasts derived from induced pluripotent stem cells form calcified structures in scaffolds both in vitro and in vivo. *Stem Cells* 29(2):206–216.

Boland, G M, Perkins, G, Hall, DJ, Tuan RS (2004). Wnt 3a promotes proliferation and suppresses osteogenic differentiation of adult human mesenchymal stem cells. *J Cell Biochem* 93(6):1210–1230.

Breitbart, AS, Grande, DA, Kessler, R, Ryaby JT, et al. (1998). Tissue engineered bone repair of calvarial defects using cultured periosteal cells. 101(3):567–74.

Bruder SP, Fox BS (1999). Tissue engineering of bone: cell based strategies. *Clin Orthop Relat Res* 367S:S68–S83.

Bruder SP, Jaiswal N, Haynesworth SE (1997). Growth kinetics, self-renewal, and the osteogenic potential of purified human mesenchymal stem cells during extensive subcultivation and following cryopreservation. *J Cell Biochem* 64(2):278–294.

Bruder SP, Jaiswal N, Ricalton NS, Mosca JD, et al. (1998). Mesenchymal stem cells in osteobiology and applied bone regeneration. *Clin Orthop Relat Res* 355S:S247–S256.

Bruens ML, Pieterman H, de Wijn JR, Vaandrager JM (2003). Porous polymethylmethacrylate as bone substitute in the craniofacial area. *J Craniofac Surg* 14(1):63–68.

Caplan AI (1991). Mesenchymal stem cells. *J Orthop Res* 9(5):641–650.

Caplan AI (2007). Adult mesenchymal stem cells for tissue engineering versus regenerative medicine. *J Cell Physiol* 213(2):341–347.

Carragee EJ, Hurwitz EL, Weiner BK (2011). A critical review of recombinant human bone morphogenetic protein-2 trials in spinal surgery: emerging safety concerns and lessons learned. *Spine J* 11(6):471–491.

Chang, J, Sonoyama, W, Wang, Z, Jin Q, et al. (2007). Noncanonical Wnt-4 signaling enhances bone regeneration of mesenchymal stem cells in craniofacial defects through activation of p38 MAPK. *J Biol Chem* 282(42):30938–30948.

Conejero JA, Lee JA, Parrett BM, Terry M, et al. (2006). Repair of palatal bone defects using osteogenically differentiated fat-derived stem cells. *Plast Reconstr Surg* 117(3):857–863.

Cowan CM, Shi YY, Aalami OO, Chou YF, et al. (2004). Adipose-derived adult stromal cells heal critical-size mouse calvarial defects. *Nat Biotechnol* 22(5):560–567.

De Ugarte DA, Morizono K, Elbarbary A, Alfonso Z, et al. (2003). Comparison of multi-lineage cells from human adipose tissue and bone marrow. *Cells Tissues Organs* 174(3):101–109.

Deyle DR, Khan IF, Ren G, Wang PR, et al. (2011). Normal collagen and bone production by gene-targeted human osteogenesis imperfecta iPSCs. *Mol Ther*. Epub Oct 25, 2011.

Duan X, Tu Q, Zhang J, Ye J, et al. (2011). Application of induced pluripotent stem (iPS) cells in periodontal tissue regeneration. *J Cell Physiol* 226(1):150–157.

Dudas JR, Marra KG, Cooper GM, Penascino VM, et al. (2006). The osteogenic potential of adipose-derived stem cells for the repair of rabbit calvarial defects. *Ann Plast Surg* 56(5):543–548.

Edwards PC, Ruggiero S, Fantasia J, Burakoff R, et al. (2005). Sonic hedgehog gene-enhanced tissue engineering for bone regeneration. *Gene Ther* 12(1):75–86.

Friedenstein AJ, Piatetzky S II, Petrokova KV (1966). Osteogenesis in transplants of bone marrow cells. *J Embryol Exp Morphol* 16(3):381–390.

Giorgetti A, Montserrat N, Aasen T, Gonzalez F, et al. (2009). Generation of induced pluripotent stem cells from human cord blood using OCT4 and SOX2. *Cell Stem Cell* 5(4):353–357.

Grayson WL, Fröhlich M, Yeager K, Bhumiratana S, et al. (2010). Engineering anatomically shaped human bone grafts. *Proc Natl Acad Sci USA* 107(8):3299–3304.

Greenwald AS, Boden SD, Goldberg VM, Khan Y, et al. (2001). Bone-graft substitutes: facts, fictions, and applications. *J Bone Joint Surg Am* 83A (Suppl 2 Pt 2): 98–103.

Gronthos S, Franklin DM, Leddy HA, Robey PG, Storms RW, Gimble JM (2001). Surface protein characterization of human adipose tissue–derived stromal cells. *J Cell Physiol* 189(1):54–63.

Hicok KC, Du Laney TV, Zhou YS, Halvorsen YD, et al. (2004). Human adipose-derived adult stem cells produce osteoid in vivo. *Tissue Eng* 10(3–4):371–380.

Jia F, Wilson KD, Sun N, Gupta DM, et al. (2010). A nonviral minicircle vector for deriving human iPS cells. *Nat Methods* 7(3):197–199.

Kang L, Wang J, Zhang Y, Kou Z, et al. (2009). iPS cells can support full-term development of tetraploid blastocyst-complemented embryos. *Cell Stem Cell* 5(2):135–138.

Katz AJ, Llull R, Hedrick MH, Futrell JW (1999). Emerging approaches to the tissue engineering of fat. *Clin Plast Surg* 26(4):587–603.

Katz AJ, Tholpady A, Tholpady SS, Shang H, et al. (2005). Cell surface and transcriptional characterization of human adipose-derived adherent stromal (hADAS) cells. *Stem Cells* 23(3):412–423.

Kim JB, Greber B, Arauzo-Bravo MG, Meyer J, et al. (2009). Direct reprogramming of human neural stem cells by OCT4. *Nature* 461(7264):643–649.

Krebsbach PH, Mankani MH, Satomura K, Kuznetsov SA, et al. (1998). Repair of craniotomy defects using bone marrow stromal cells. *Transplantation* 66(10):1272–1278.

Laurie SW, Kaban LB, Mulliken JB, Murray JE (1984). Donor-site morbidity after harvesting rib and iliac bone. *Plast Reconstr Surg* 73(6):933–938.

Lee JA, Parrett BM, Conejero JA, Laser J, et al. (2003). Biological alchemy: engineering bone and fat from fat-derived stem cells. *Ann Plast Surg* 50(6):610–617.

Lee K, Majumdar MK, Buyaner D, Hendricks JK, et al. (2001). Human mesenchymal stem cells maintain transgene expression during expansion and differentiation. *Mol Ther* 3(6):857–866.

Lee RH, Kim B, Choi I, Kim H, et al. (2004). Characterization and expression analysis of mesenchymal stem cells from human bone marrow and adipose tissue. *Cell Physiol Biochem* 14(4–6):311–324.

Lendeckel S, Jödicke A, Christophis P, Heidinger K, et al. (2004). Autologous stem cells (adipose) and fibrin glue used to treat widespread traumatic calvarial defects: case report. *J Craniomaxillofac Surg* 32(6):370–373.

Levi B, James AW, Nelson ER, Vistnes D, et al. (2010a). Human adipose derived stromal cells heal critical size mouse calvarial defects. *PLoS One* 5(6):e11177.

Levi B, James AW, Xu Y, Commons GW, et al. (2010b). Divergent modulation of adipose-derived stromal cell differentiation by TGF-beta1 based on species of derivation. *Plast Reconstr Surg* 126(2):412–425.

Levi B, Hyun JS, Nelson ER, Li S, et al. (2011a). Nonintegrating knockdown and customized scaffold design enhances human adipose-derived stem cells in skeletal repair. *Stem Cells* 29(12):2018–2029.

Levi B, Nelson ER, Li S, James AW, et al. (2011b). Dura mater stimulates human adipose-derived stromal cells to undergo bone formation in mouse calvarial defects. *Stem Cells* 29(8):1241–1255.

Levi B, Wan DC, Glotzbach JP, Hyun J, et al. (2011). CD105 protein depletion enhances human adipose-derived stromal cell osteogenesis through reduction of transforming growth factor β1 (TGF-β1) signaling. *J Biol Chem* 286(45):39497–39509.

Lew D, Farrell B, Bardach J, Keller J (1997). Repair of craniofacial defects with hydroxyapatitie cement. *J Oral Maxillofac Surg* 55(12):1441–1449.

Lieberman JR, Le LQ, Wu L, Finerman GA, et al. (1998). Regional gene therapy with a BMP-2 producing murine stromal cell line induces heterotopic and orthotopic bone formation in rodents. *J Orthop Res* 16(3):330–339.

Loh YH, Agarwal S, Park IH, Urbach A, et al. (2009). Generation of induced pluripotent stem cells from human blood *Blood* 113(22):5476–5479.

Mesimäki K, Lindroos B, Törnwall J, Mauno J, et al. (2009). Novel maxillary reconstruction with ectopic bone formation by GMP adipose stem cells. *Int J Oral Maxillofac Surg* 38(3):201–209.

Mitchell JB, McIntosh K, Zvonic S, Garrett S, et al. (2006). Immunophenotype of human adipose-derived cells: temporal changes in stromal-associated and stem cell–associated markers. *Stem Cells* 24(2):376–385.

Narsinh KH, Jia F, Robbins RC, Kay MA, et al. (2011). Generation of adult human induced pluripotent stem cells using nonviral minicircle DNA vectors. *Nat Protoc* 6(1):78–88.

Okita, K, Ichisaka, T, Yamanaka, S (2007). Generation of germline-competent induced pluripotent stem cells. *Nature* 448(7151):313–317.

Okita K, Nakagawa M, Hyenjong H, Ichisaka T, et al. (2008). Generation of mouse induced pluripotent stem cells without viral vectors. *Science* 322(5903):949–953.

Panetta NJ, Gupta DM, Lee JK, Wan DC, et al. (2010). Human adipose-derived stromal cells respond to and elaborate bone morphogenetic protein-2 during in vitro osteogenic differentiation. *Plast Reconstr Surg* 125(2):483–493.

Pittenger MF, Mackay AM, Beck SC, Jaiswal RK, et al. (1999). Multilineage potential of adult human mesenchymal stem cells. *Science* 284(5411):143–147.

Rodbell M (1964). Metabolism of isolated fat cells: I. Effects of hormones on glucose metabolism and lipolysis. *J Biol Chem* 239: 375–380.

Schäffler A, Büchler C (2007). Concise review: adipose tissue–derived stromal cells—basic and clinical implications for novel cell-based therapies. *Stem Cells* 25(4):818–827.

Schantz JT, Teoh SH, Lim TC, Endres M, et al. (2001). Repair of calvarial defects with customized tissue-engineered bone grafts: I. Evaluation of osteogenesis in a three-dimensional culture system. *Tissue Eng* 9(Suppl 1): S113–S126.

Schantz JT, Hutmacher DW, Lam CX, Brinkmann M, et al. (2003). Repair of calvarial defects with customised tissue-engineered bone grafts: II. Evaluation of cellular efficiency and efficacy in vivo. *Tissue Eng* 9(Suppl 1): S127–S139.

Schuldiner M, Itskovitz-Eldor J, Benvenisty N (2003). Selective ablation of human embryonic stem cells expressing a "suicide" gene. *Stem Cells* 21(3):257–265.

Shang Q, Wang Z, Liu W, Shi Y, et al. (2001). Tissue-engineered bone repair of sheep cranial defects with autologous bone marrow stromal cells. *J Craniofac Surg* 12(6):586–593.

Soldner F, Hockemeyer D, Beard C, Gao Q, et al. (2009). Parkinson's disease patient-derived induced pluripotent stem cells free of viral reprogramming factors. *Cell* 136(5):964–977.

Sommer CA, Sommer AG, Longmire TA, Christodoulou C, et al. (2010). Excision of reprogramming transgenes improves the differentiation potential of iPS cells generated with a single excisable vector. *Stem Cells* 28(1):64–74.

Sterodimas A, de Faria J, Nicaretta B, Pitanguy I (2010). Tissue engineering with adipose-derived stem cells (ADSCs): current and future applications. *J Plast Reconstr Aesthet Surg* 63(11):1886–1892.

Sun N, Longaker MT, WU JC. (2010). Human iPS cell-based therapy: considerations before clinical applications. *Cell Cycle* 9(5):880–5.

Sun N, Panetta NJ, Gupta DM, Wilson KD, et al. (2009). Feeder-free derivation of induced pluripotent stem cells from adult human adipose stem cells. *Proc Natl Acad Sci USA* 106(37):15720–15725.

Takahashi K, Yamanaka S (2006). Induction of pluripotent stem cells from mouse embryonic and adult fibroblast cultures by defined factors. *Cell* 126(4):663–676.

Takahashi K, Tanabe K, Ohnuki M, Narita M, et al. (2007). Induction of pluripotent stem cells from adult human fibroblasts by defined factors. *Cell* 131(5):861–872.

Tang C, Lee AS, Volkmer JP, Sahoo D, et al. (2011). An antibody against SSEA-5 glycan on human pluripotent stem cells enables removal of teratoma-forming cells. *Nat Biotechnol* 29(9):829–834.

Tashiro K, Inamura M, Kawabata K, Sakurai F, et al. (2009). Efficient adipocyte and osteoblast differentiation from mouse induced pluripotent stem cells by adenoviral transduction. *Stem Cells* 27(8):1802–1811.

Taub PJ, Rudkin GH, Clearihue WJ, 3rd, Miller TA (2003). Prefabricated alloplastic implants for cranial defects. *Plast Reconstr Surg* 111(3):1233–1240.

Thesleff T, Lehtimäki K, Niskakangas T, Mannerström B, et al. (2011). Cranioplasty with adipose-derived stem cells and biomaterial: a novel method for cranial reconstruction. *Neurosurgery* 68(6):1535–1540.

Tremoleda JL, Forsyth NR, Khan NS, Wojtacha D, et al. (2008). Bone tissue formation from human embryonic stem cells in vivo. *Cloning Stem Cells* 10(1):119–132.

Ueda M, Yamada Y, Ozawa R, Okazaki Y (2005). Clinical case reports of injectable tissue-engineered bone for alveolar augmentation with simultaneous implant placement. *Int J Periodont Restor Dent* 25(2):129–137.

Utikal J, Maherali N, Kulalert W, Hochedlinger K. (2009). Sox2 is dispensable for the reprogramming of melanocytes and melanoma cells into induced pluripotent stem cells. *J Cell Sci* 122 (Pt 19): 3502–3510.

Wan DC, Shi YY, Nacamuli RP, Quarto N, et al. (2006). Osteogenic differentiation of mouse adipose-derived adult stromal cells requires retinoic acid and bone morphogenetic protein receptor type IB signaling. *Proc Natl Acad Sci USA* 103(33):12335–12340.

Weng Y, Wang M, Liu W, Hu X, et al. (2006). Repair of experimental alveolar bone defects by tissue-engineered bone. *Tissue Eng* 12(6):1503–1513.

Wernig M, Meissner A, Foreman R, et al. (2007). In vitro reprogramming of fibroblasts into a pluripotent ES-cell-like state. *Nature* 448(7151):318–324.

Woltjen K, Michael IP, Mohseni P, Desai R, et al. (2009). piggyBac transposition reprograms fibroblasts to induced pluripotent stem cells. *Nature* 458(7239):766–770.

Ye JH, Xu YJ, Gao J, Yan SG, et al. (2011). Critical-size calvarial bone defects healing in a mouse model with silk scaffolds and SATB2-modified iPSCs. *Biomaterials* 32(22):5065–5076.

Yoon E, Dhar S, Chun DE, Gharibjanian NA, et al. (2007). In vivo osteogenic potential of human adipose-derived stem cells/poly lactide-*co*-glycolic acid constructs for bone regeneration in a rat critical-sized calvarial defect model. *Tissue Eng* 13(3):619–627.

Yu J, Vodyanik MA, Smuga-Otto K, Antosiewicz-Bourget J, et al. (2007). Induced pluripotent stem cell lines derived from human somatic cells. *Science* 318(5858):1917–1920.

Antosiewicz-Bourget J, et al. (2007). Induced pluripotent stem cell lines derived from human somatic cells. *Science* 318(5858):1917–1920.

Yu JM, Wu X, Gimble JM, Guan X, et al. (2011). Age-related changes in mesenchymal stem cells derived from rhesus macaque bone marrow. *Aging Cell* 10(1):66–79.

Zaky SH, Cancedda R (2009). Engineering craniofacial structures: facing the challenge. *J Dent Res* 88(12):1077–1091.

Zhao T, Zhang ZN, Rong Z, Xu Y (2011). Immunogenicity of induced pluripotent stem cells. *Nature* 474(7350):212–215.

Zhou H, Wu S, Joo JY, Zhu S, et al. (2009). Generation of induced pluripotent stem cells using recombinant proteins. *Cell Stem Cell* 4(5):381–384.

Zhou S, Greenberger JS, Epperly MW, Goff JP, et al. (2008). Age-related intrinsic changes in human bone marrow–derived mesenchymal stem cells and their differentiation to osteoblasts. *Aging Cell* 7(3):335–343.

Zuk PA, Zhu M, Mizuno H, Huang J, et al. (2001). Multilineage cells from human adipose tissue: implications for cell-based therapies. *Tissue Eng* 7(2):211–228.

Zuk PA, Zhu M, Ashjian P, De Ugarte DA, et al. (2002). Human adipose tissue is a source of multipotent stem cells. *Mol Biol Cell* 13(12):4279–4295.

22

MUSCLE TISSUE ENGINEERING APPROACHES

JOHANNES W. VON DEN HOFF

Department of Orthodontics and Craniofacial Biology, Radboud University Nijmegen Medical Centre, Nijmegen, The Netherlands

SANDER GREFTE

Department of Biochemistry, Radboud University Nijmegen Medical Centre, Nijmegen, The Netherlands

1 INTRODUCTION

Skeletal muscle engineering is a challenging field that can contribute significantly to clinical applications for a wide variety of muscle injuries, such as strains, trauma, muscular dystrophies, and congenital malformations. Muscle strain injuries occur in sports with high-intensity sprinting, such as in football, rugby, and soccer, and have an incidence of about 30%. One of the most commonly affected muscle groups is the hamstring, and treatment is still not optimal, as shown by the recurrence rate of 30% (Garrett, 1996; Orchard and Best, 2002). Loss of muscle tissue commonly occurs in patients with large wounds, such as military personnel, victims of car accidents and gunshots, and in surgical patients. The standard treatment for these patients is the transplantation of autologous muscle tissue. However, this leads to new defects that may develop muscle fibrosis. Eventually, these patients still end up with a permanent physical handicap (Bartlett et al., 2000; Merritt et al., 2010).

Muscular dystrophies are inherited myogenic disorders and show progressive muscle wasting and weakness. Duchenne muscular dystrophy (DMD) is the most common form and is characterized by the absence of the protein dystrophin, which provides structural strength to the muscle tissue. Currently, no effective treatments are available for any of the muscular dystrophies. Gene and stem cell therapy might offer new solutions, but

Stem Cells in Craniofacial Development and Regeneration, First Edition.
Edited by George T.-J. Huang, Irma Thesleff.

major problems still exist regarding safety and the delivery into all affected muscles (Emery, 2002; Foster et al., 2006; Manzur et al., 2008).

Malformations of muscle tissue such as in cleft lip and/or palate (CLP) are characterized by disorganized muscle fibers and impaired function. About 45% of CLP patients show clefts in the soft palate (Andersson et al., 2010). These patients have difficulties with feeding and speech, and surgery is required to close the defect. However, in many patients speech and feeding problems persist after surgical closure, which is often due to the formation of fibrotic tissue in the levator veli palatini, the major muscle of the soft palate (Kogo et al., 1996; Huard et al., 2002; Mossey et al., 2009; Andersson et al., 2010).

In all of these muscle disorders, a main problem is the formation of fibrotic tissue after restoration (Huard et al., 2002). This prevents the regeneration of oriented muscle fibers and therefore impairs full functional recovery. Fibrosis is also one of the main causes of recurrent strain injuries. Present therapies are often insufficient to treat muscle injuries because of the development of fibrosis. In the field of tissue engineering and regenerative medicine, new approaches are being developed with the ultimate aim to restore muscle function by improving muscle regeneration and reducing fibrosis. In this chapter we discuss the possible strategies for the treatment of different types of muscle injuries as well as congenital and genetic muscle diseases. First, a short overview of muscle regeneration is presented.

2 SKELETAL MUSCLE REGENERATION

To design and optimize treatment strategies for muscle disorders, muscle development and regeneration have been studied extensively. Briefly, muscle regeneration occurs in three phases: inflammation, regeneration, and remodeling, which may lead to fibrosis (Huard et al., 2002; Järvinen et al., 2005). During inflammation, macrophages phagocytose necrotic debris and produce factors that, together with factors released from the extracellular matrix, start the regeneration phase (Huard et al., 2002; Chazaud et al., 2003; Chargé and Rudnicki, 2004; Järvinen et al., 2005).

Skeletal muscles regenerate by the activation of a small population of stem cells, which are associated with the myofibers (Mauro, 1961; Muir et al., 1965). These satellite cells are able to migrate to the site of injury, where they proliferate extensively and subsequently differentiate to form new muscle fibers. This process is regulated by many growth factors, such as insulin-like growth factor (IGF)-I, fibroblast growth factor (FGF)-II, and hepatocyte growth factor (HGF) (Huard et al., 2002; Chargé and Rudnicki, 2004; Järvinen et al., 2005).

A crucial event in regeneration is the self-renewal of satellite cells to replenish their numbers for future regeneration cycles, which is a key characteristic of all stem cells (Lakshmipathy and Verfaillie, 2005). Factors in the satellite cell niche are crucial for self-renewal and regeneration (see also Chapter 13). Direct contact of the satellite cell with the myofiber and the basal lamina appears to be essential for maintaining their stem cell status (Kuang et al., 2007). Loss of contact with the niche leads to proliferation and differentiation of the satellite cell.

Fibrotic tissue is often formed during the final remodeling phase, which contributes to incomplete functional recovery and recurrent muscle injuries (Huard et al., 2002; Chargé and Rudnicki, 2004; Järvinen et al., 2005). A key regulator of fibrosis in many tissues is transforming growth factor (TGF) β (Border and Noble, 1994). Thus, prevention of fibrosis might be achieved by decreasing TGFβ activity in the tissue. This basic knowledge from the field of muscle regeneration can be used to develop tissue engineering strategies for muscle defects.

3 TISSUE ENGINEERING STRATEGIES

In contrast to the extensive research on other tissues, such as skin and bone, only limited studies are available on tissue engineering of muscle. Depending on the type of muscle defect, different approaches may be followed to improve muscle function. These include the use of growth factors, specific cell populations, scaffolds, or combinations of these (Fig. 1).

3.1 Growth Factor–Based Therapy

The effects of growth factors on the activation, proliferation, and differentiation of satellite cells have been reviewed elsewhere (Chargé and Rudnicki, 2004; Wagers and Conboy, 2005; Ten Broek et al., 2010). Growth factors with stimulatory effects might be used in vivo to enhance the regeneration of muscle tissue(Fig. 1). Indeed, it appears that the administration of growth factors after muscle injury improves the healing process. In mouse models for muscle strain (Kasemkijwattana et al., 2000), contusion (Kasemkijwattana et al., 1998), and laceration injuries (Menetrey et al., 2000; Sato et al., 2003), direct injection of IGF-I and FGF2, and to a lesser extent nerve growth factor (NGF), improves muscle healing. This is indicated by an increase in the number and diameter of regenerated myofibers. Additionally, the strength of the myofibers improved.

Next to these growth factors, hepatocyte growth factor (HGF) is also present in injured muscle. HGF binds to its receptor c-met, which is expressed on satellite cells and induces their migration. Additionally, HGF influences only satellite cell proliferation in the early stages of muscle regeneration, but not their subsequent differentiation (Tatsumi et al., 1998, 2001; Suzuki et al., 2002). Recently it was found that interferon-γ and granulocyte-colony stimulating factor also stimulate satellite cell proliferation (Stratos et al., 2007; Cheng et al., 2008).

The administration of decorin, an inhibitor of TGFβ, also enhances healing, and in combination with IGF-I it seems to be the best strategy to improve muscle healing (Sato et al., 2003). However, based solely on the strength of the myofibers, the administration of decorin alone showed the best result. These conflicting data stress the importance of both histological as well as functional parameters to evaluate muscle regeneration. Since TGFβ is considered to be involved in fibrosis, its inhibition might reduce scar formation during muscle regeneration. In fact, the administration of decorin indeed inhibits muscle fibrosis (Sato et al., 2003). In contrast, the administration of FGF2 does not improve muscle regeneration at all (Mitchell et al., 1996). Therefore, it is crucial to exactly identify the growth factors that enhance muscle regeneration in vivo, and also their concentration, location, and time of administration.

Thus, growth factors seem to improve muscle healing after minor injury. However, for intrinsic muscle defects such as DMD or for larger muscle defects, growth factor–based therapy might not be appropriate. For the treatment of DMD, cell-based

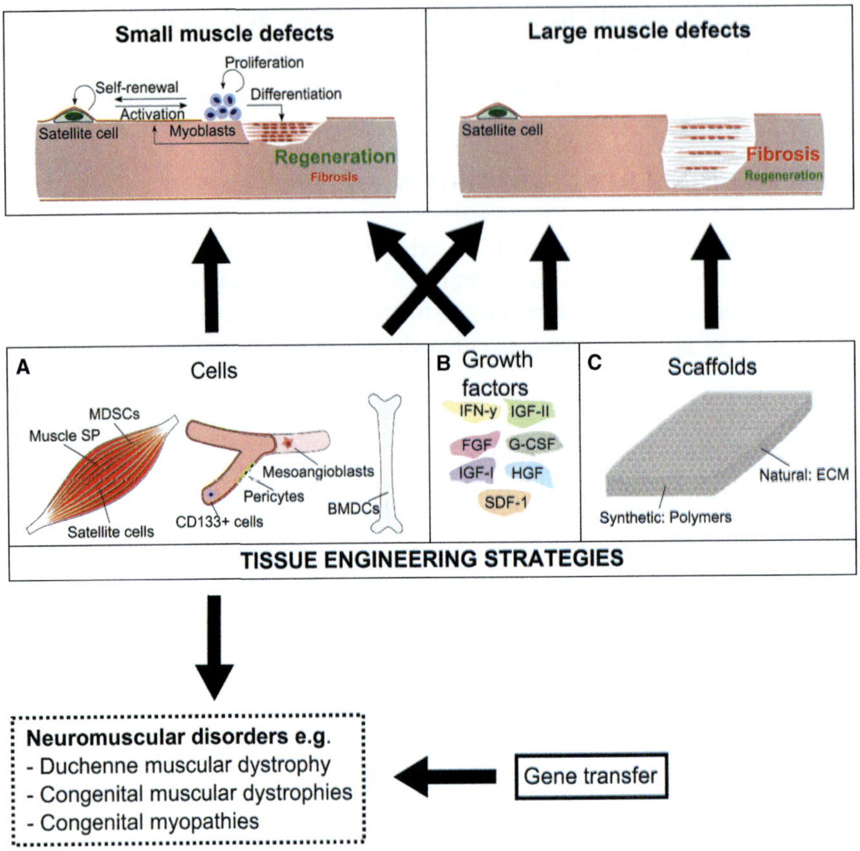

FIGURE 1 Tissue engineering strategies to restore muscle function. During muscle repair, satellite cells are activated and produce myoblasts, which in turn differentiate and fuse to regenerate muscle defects. In small muscle defects this process is efficient, although some scar tissue will be formed. In large muscle defects accompanied by extensive tissue loss, regeneration is impaired, and massive scar tissue will be formed. The scar tissue prevents full functional recovery. To restore muscle function, several tissue engineering strategies can be employed. For small muscle defects, cells with myogenic capacity can be used (A), either combined (or not) with growth factors that enhance satellite cell activation, migration, proliferation and/or differentiation (B). For large defects a scaffold (C) is required to provide structural cues for the regenerating myofibers. These scaffolds can be composed of natural extracellular matrix (ECM) components or synthetic polymers. Scaffolds can be loaded with cells and/or growth factors to improve muscle regeneration and to reduce the formation of scar tissue. Treatment of congenital muscle diseases can only be achieved using genetically corrected cells or cells isolated from healthy subjects. Preferably, these cells should be able to regenerate all muscles after systemic delivery. An alternative approach could be virus-mediated gene transfer to replace the defective gene without the use of cells.

therapy appears to be a better solution. For major muscle injuries, additional scaffolds might be required to fill up the defect and to guide the regenerating myofibers.

3.2 Cell-Based Therapy

To repair muscle through the delivery of exogenous cells such as satellite cells or intact myofibers, it is crucial that these cells not only regenerate the damaged muscle but also replenish the satellite cell pool. This allows long-term tissue maintenance and sustains the regenerative capacity of muscle. Research in this field incorporates not only isolated myoblasts and muscle-derived stem cells, but also other stem cell populations (Fig. 1).

Satellite Cells and Myoblasts. Although there is evidence that the dystrophic muscle environment is hostile for muscle regeneration (Bakay et al., 2006), the *mdx* mouse model has been used extensively to study muscle regeneration after cell transplantation. The *mdx* mouse lacks dystrophin, a structural protein that is mutated in DMD patients. Repetitive injections of notexin following the transplantation of myoblasts into irradiated *mdxnu/nu* mouse muscle resulted in new myofibers of donor origin (Gross and Morgan, 1999). The consistent formation of new muscle after repetitive muscle damage indicates that a new satellite cell population is established. Labeling studies show that the myoblasts repair the injured muscle fibers and form new satellite cells (Blaveri et al., 1999; Cousins et al., 2004). Furthermore, labeled satellite cells were detected in host muscle fibers, and isolated satellite cells from these fibers were able to become active and to proliferate (Heslop et al., 2001).

The grafting of intact myofibers that contain satellite cells results in the regeneration of damaged muscle and expansion of the satellite cell pool (Collins et al., 2005). Moreover, satellite cells that were isolated from these new myofibers generated clusters of new myofibers after grafting. This indicates that the isolation of satellite cells from the myofiber does not impair their myogenic potential (Collins et al., 2005). Additionally, a pure population of satellite cells selected by *Pax3* expression, and also expressing Pax7 and CD34, contributes to muscle regeneration and the formation of new satellite cells (Montarras et al., 2005). Satellite cell populations have also been isolated through their expression of specific cell-surface markers. Satellite cells negative for CD45, CD11b, CD31, and Sca1, but positive for integrin $\alpha 7$ and CD34 efficiently regenerate muscle injuries (Sacco et al., 2008). A different satellite cell population negative for CD45,Sca1, and Mac-1, but positive for CXCR4 and integrin $\beta 1$ also regenerates muscle injuries with high efficiency (Sherwood et al., 2004). These studies clearly show the heterogeneous nature of satellite cell populations.

Alternatively, a specific myogenic cell population was enriched based on their adhesion capabilities. These cells are able to survive after transplantation and to fuse with host myofibers, thus demonstrating their regenerative capacity (Qu et al., 1998). Two additional myogenic populations were identified by M-cadherin and CD34 expression (Lee et al., 2000). These two markers do not colocalize in skeletal muscle, but cells expressing M-cadherin or CD34 both reside between the basal lamina and the sarcolemma of the myofibers, which is also the niche for satellite cells. In contrast, another study showed that M-cadherin-positive satellite cells are also positive for CD34 (Beauchamp et al., 2000). Cloned myogenic cells from the $CD34^+$ population enhance muscle regeneration and partially restore dystrophin expression (Lee et al., 2000).

Cultured satellite cells gradually lose their myogenic potential, and the transplantation of these cells leads to less efficient muscle regeneration (Machida et al., 2004; Montarras et al., 2005). Freshly isolated satellite cells may induce muscle regeneration more efficiently (Montarras et al., 2005), but the small number of isolated cells might be a limiting factor. Therefore, new culturing methods must be developed that maintain the myogenic properties of satellite cells. Using substrates with a specific stiffness of 12 kPa, cultured satellite cells were able to self-renew, but more important, they also improved muscle regeneration after transplantation (Gilbert et al., 2010).

It has also been shown that the majority of myoblasts die after transplantation into *mdx* mice (Fan et al., 1996; Qu et al., 1998; Beauchamp et al., 1999). Inflammation may be involved since infection of the myoblasts with a retroviral vector containing the IL-1 receptor antagonist partly prevents cell death (Qu et al., 1998). Still, a small fraction of the myogenic cells survived and were able to regenerate the host muscle (Beauchamp et al., 1999).

In conclusion, most of these studies show the potential of transplanted satellite cells or myoblasts to produce new muscle fibers or to restore dystrophin expression in host muscle. Moreover, they are able to create a new satellite cell pool, which is crucial for muscle regeneration throughout life and normal tissue turnover. As cultured cells lose their myogenic capacity, they need to be freshly isolated. To allow expansion of satellite cells in culture while maintaining their myogenic capacity, new culturing systems have to be developed.

Other Stem Cells. Mice with genetically ablated satellite cells demonstrate a complete lack of muscle regeneration (Sambasivan et al., 2011). This confirms that these cells are critical for the regeneration of skeletal muscle. However, other precursor cells might also contribute to muscle regeneration, although they cannot rescue the regenerative potential of skeletal muscle. These cells also show myogenic potential and might be promising for the treatment of muscle defects or diseases. Such cell populations include muscle side population cells (SP cells), muscle-derived stem cells, mesoangioblasts (vessel-associated stem cells), pericytes, and $CD133^+$ stem cells.

Muscle SP cells are a heterogeneous and poorly defined cell population. After intramuscular injection they are able to form new myocytes and satellite cells (Gussoni et al., 1999; Asakura et al., 2002). Based on their adherence capacity, three alternative myogenic cell populations were found. Two of these represented satellite cells and showed only limited capacity to regenerate the host muscle. Excitingly, the third population strongly improved muscle regeneration. These cells are long-time proliferative, and are also called muscle derived stem cells (MDSCs) (Qu-Petersen et al., 2002). MDSCs have, in contrast to satellite cells, multilineage potential and prolonged proliferation capacity in vitro. Combined with their strong capacity to self-renew and their immune tolerance, transplantation of MDSCs into the skeletal muscle of mice gives better results than satellite cells (Qu-Petersen et al., 2002). However, it remains to be shown if they adopt a satellite cell fate and have long-term self-renewal capacity.

Two different vessel-associated (stem) cell populations, mesangioblasts and pericytes, also show myogenic potential. Systemically injected allogenic mesoangioblasts show a large potential to regenerate skeletal muscle (Sampaolesi et al., 2003). After genetical correction, transplanted autologous mesoangioblasts increase the formation of functional muscle in dystrophic dogs (Sampaolesi et al., 2006). Pericytes contribute to muscle regeneration and generate many dystrophin-rich muscle fibers in dystrophic mice after intraarterial injection (Dellavalle et al., 2007). Furthermore, pericytes might also be able to replenish the satellite cell pool (Dellavalle et al., 2007).

CD133$^+$ cells from the blood can differentiate into endothelial, hematopoietic, and muscle cells in vitro (Torrente et al., 2004). The systemic delivery of human CD133$^+$ cells in *scid/mdx* mice not only improves skeletal muscle function, but also replenishes the satellite cell pool (Torrente et al., 2004). After injection into the muscle, these cells also give better outcomes than human myoblasts regarding regenerative capacity and repopulation of the SC pool (Negroni et al., 2009). The hematopoietic stem cell (HSC) can also participate in skeletal muscle regeneration. Transplanted HSCs yield donor derived myofibers and contribute to the satellite cell pool, albeit in low numbers (LaBarge and Blau, 2002).

In summary, the satellite cell is the professional stem cell for muscle regeneration, but other stem cells may also contribute. For transplantation, satellite cells are probably not the best option, due to their inability to leave the bloodstream after injection and their limited survival. The combination of transplanted cells with suitable scaffolds might offer new strategies for the repair of major muscle damage.

3.3 Scaffold-Based Therapy

The transplantation of myogenic cells and the use of growth factors might not be sufficient for large muscle defects. The application of a three-dimensional scaffold, which fills up the defect and induces the formation of new muscle, seems to be more effective (Fig. 1).In addition, it may contain structural cues to guide the formation of new myofibers. The scaffold can be seeded with myoblasts or other myogenic cells and can be implanted into a large muscle defect to improve muscle regeneration. However, the transplantation of a collagen disk with or without myoblasts into an abdominal wall muscle defect in rats did not induce muscle formation (van Wachem et al., 1999). Similar results were obtained after the transplantation of an acellular muscle matrix into an abdominal wall defect in rabbits (Gamba et al., 2002) and rats (Conconi et al., 2005), and a dorsal muscle defect in rats (Marzaro et al., 2002). Eventually, these constructs were replaced by fibrous tissue. However, seeding the matrix with autologous satellite cells reduced the inflammation, and fibrosis occurred only at the edge of the implant. This indicates that satellite cell seeding improves the biocompatibility of the scaffold.

In vitro studies showed that an acellular muscle matrix supports the growth and differentiation of rat satellite cells, but in vivo studies showed no convincing evidence of muscle formation inside the matrix (Marzaro et al., 2002). However, electric activity was detected, indicating that skeletal muscle fibers were present (Conconi et al., 2005). This result shows the potential of using constructs seeded with myoblasts to reconstruct injured muscle. Additionally, the transplantation of a muscle matrix into full-thickness abdominal wall defects resulted in the formation of a dense capillary network, skeletal muscle fibers, and evidence of nerve formation (De Coppi et al., 2006). This matrix stained positive for FGF2 and TGFβ, which could play roles in the regeneration of injured muscle. In contrast, another study showed that muscle formation occurred only at the border of the muscle matrix after transplantation into the abdominal muscle of rats (Vindigni et al., 2004).

Thus, it seems that growth factors have a positive effect on seeded myoblasts and improve the formation of muscle fibers within the matrix. The addition of HGF and FGF2 to alginate scaffolds seeded with myoblasts increased the viability of these cells

(Hill et al., 2006a). Subsequent transplantation of such a scaffold resulted in enhanced muscle regeneration by the donor cells (Hill et al., 2006b). The need for satellite cell isolation, culture, and seeding onto scaffolds can be avoided by attracting endogenous satellite cells toward the muscle defect. This might be possible by implanting a collagen scaffold loaded with SDF1α (Grefte et al., 2010).

Other biomaterials have also been used as a three-dimensional matrix to reconstruct muscle tissue. The implantation of degradable poly(glycolic acid) (PGA) meshes (Saxena et al., 1999, 2001), alginate, and hyaluronic acid constructs (Kamelger et al., 2004) seeded with myoblasts into a nonmuscle environment resulted in vascularization and muscle formation. *In vitro* studies showed that myoblasts in a three-dimensional fibrin matrix can fuse into myotubes with physiological functions such as force generation (Huang et al., 2005). After injecting fibrinogen with myoblasts into a muscle defect in rats, the fibrin matrix was eventually resorbed, and the myoblasts fused with the host muscle (Beier et al., 2006). More important, no inflammation was observed and fibrosis was absent. However, complete integration of the scaffolds into the host muscle and full functional recovery without scarring have not yet been accomplished.

In summary, three-dimensional scaffolds might have advantages for the treatment of large muscle defects. Seeding of the scaffold with myoblasts appears to be essential to induce new muscle mass. Furthermore, the addition of growth factors might improve the proliferation, migration, and fusion of the myoblasts into new myofibers. A crucial aspect is the vascularization and innervation of the construct. It is important, however, to keep in mind that these studies were carried out in the muscles of small animals. Thus, caution must be taken in the translation of the data to humans with larger muscles. The migration of myoblasts out of the scaffold and the diffusion of signaling molecules and nutrients might be different.

3.4 Treatment of Neuromuscular Diseases

Neuromuscular diseases comprise a group of devastating inherited disorders affecting the nervous system and muscle. Two of the most common forms are Duchenne muscular dystrophy (DMD), affecting about 1 in 3500 male newborns (Emery, 2002) and the congenital muscular dystrophies (CMDs). The latter is a genetically and phenotypically heterogeneous group of inherited disorders with an incidence of 4 or 5 in 100,000 live births (Mostacciuolo et al., 1996).

DMD is characterized by the absence of dystrophin, which is crucial for the structural strength of muscle tissue. The two most frequently occurring CMDs are related to defects in the extracellular matrix (ECM) proteins laminin-$\alpha2\beta1\gamma1$ and collagen type VI (Emery, 2002). In addition to muscle weakness and joint contractions, patients show severe respiratory problems. Currently, no effective treatments are available for any of these muscular dystrophies. Several attempts have been made to replace the mutated gene. This can be reached by introducing either a full functional copy or a (functional) part of the gene, or by introducing a similar gene. The replacement of dystrophin in DMD is challenging due to the large size of the gene (13 kb) and has therefore failed up to now (Muir and Chamberlain, 2009). However, overexpression of other proteins involved in DMD and CMD, such as integrin $\beta1\alpha7$, utrophin, miniagrin, and laminin $\alpha1$ and $\alpha2$ chains, shows significant therapeutic benefits in mice (Kuang et al., 1998; Gilbert et al., 1999; Bentzinger et al., 2005; Burkin et al., 2005; Gawlik and Durbeej, 2010). Still, these strategies are not yet suitable for the clinic. Recently, use of adeno-associated viruses for gene transfer was advocated, as they transduce muscle efficiently and show almost no pathological side effects (Qiao et al., 2005).

To circumvent these problems, cell-based therapies can also be used to treat DMD. Transplanted satellite cells and myoblasts restore dystrophin expression but show only limited migration and cannot be delivered systemically (Dellavalle et al., 2007). Mesoangioblasts are able to restore muscle function in dystrophic dogs after systemic injection. A disadvantage is their limited ability to colonize the muscle, probably due to incomplete adhesion and extravasation (Sampaolesi et al., 2003; Galvez et al., 2006). Systemic delivery of pericytes into immune-deficient mice with severe muscular dystrophy (*scid-mdx* mice), results in robust regeneration of (dystrophin-rich) muscle fibers (Dellavalle et al., 2007). Next to these cells, $CD133^{+}$ cells might also be used for systemic delivery to treat DMD.

As all these strategies require complicated genetic approaches or isolated cells, downstream pathological mechanisms could also be targeted. In DMD these include modulating inflammation or regeneration through the action of TNFα, TGFβ, myostatin, and/or IGF-I (Partridge, 2011). Recent studies on mice deficient for collagen type VI and laminin α2 suggest that malfunctioning mitochondria, apoptosis, and necrosis are secondary effects of CMD. Specifically, persistent opening of the mitochondrial permeability transition pore (mPTP) and its modulator, cyclophilin D, are involved. The inhibition of cyclophilin-D by either genetic or pharmacological means prevents mPTP opening and reduces muscle pathology in DMD and several forms of CMD (Irwin et al., 2003; Angelin et al., 2007; Merlini et al., 2008; Millay et al., 2008).

4 CONCLUSIONS AND FUTURE PERSPECTIVES

Debilitating muscle defects may result from trauma, genetic muscle diseases, or congenital malformations such as cleft palate. The healing of traumatic muscle defects, either accidentally or surgically induced, often leads to fibrosis that impairs full functional recovery. The fibrotic tissue precludes the ingrowth and alignment of new myofibers. In genetic muscle diseases, an intrinsic genetic defect prevents normal muscle function. Depending on the type of muscle defect, different tissue engineering approaches are required to restore muscle function. The optimal treatment strategy should stimulate the regeneration of myofibers and prevent the formation of fibrotic tissue to facilitate their ingrowth and functional alignment.

After limited muscle trauma as in strain injuries, the local injection of growth factors such as IGF or HGF can improve regeneration. Growth factors may be combined with the antifibrotic agent decorin to prevent excessive deposition of ECM. More extensive muscle trauma or genetic muscle diseases can be treated with cell-based therapy. Freshly isolated satellite cells or myoblasts are able to restore the muscle defect and to establish a new satellite cell population. However, after expansion in vitro, they lose their myogenic capacity. This might be prevented by incorporating niche factors and mechanical cues in new culture systems (see also Chapter 13). Other precursor cell populations, such as mesangioblasts or pericytes, may also improve muscle regeneration, but it is unclear whether they can replenish the satellite cell pool.

Three-dimensional scaffolds may also improve the treatment of extensive muscle defects. Novel scaffolds should provide initial mechanical stability, induce satellite cell proliferation and differentiation, guide the alignment of muscle fibers, and provide niche factors to harbor satellite cells. Myogenic cells might be included in the scaffolds, but more ideally, resident myogenic cells are attracted by incorporated growth factors. This type of approach greatly simplifies the production and application of constructs for muscle tissue regeneration.

Up to now no cure has been available for intrinsic neuromuscular diseases such as DMD. Recent advances in genetic approaches using adeno-associated viruses or cell-mediated strategies show great promise. However, until these strategies have become available clinically, research should also aim to ameliorate the downstream pathology by pharmacological means.

In general, improvements in the tissue engineering of muscle are expected to come from the use of specific growth factors, niche components and instructive scaffolds, and new culture systems for myogenic cells.

REFERENCES

Andersson EM, Sandvik L, Abyholm F, Semb G (2010). Clefts of the secondary palate referred to the Oslo Cleft Team: epidemiology and cleft severity in 994 individuals. *Cleft Palate Craniofac J* 47:335–342.

Angelin A, Tiepolo T, Sabatelli P, Grumati P, Bergamin N, Golfieri C, et al. (2007). Mitochondrial dysfunction in the pathogenesis of Ullrich congenital muscular dystrophy and prospective therapy with cyclosporins. *Proc Natl Acad Sci USA* 104:991–996.

Asakura A, Seale P, Girgis-Gabardo A, Rudnicki MA (2002). Myogenic specification of side population cells in skeletal muscle. *J Cell Biol* 159:123–134.

Bakay M, Wang Z, Melcon G, Schiltz L, Xuan J, Zhao P, et al. (2006). Nuclear envelope dystrophies show a transcriptional fingerprint suggesting disruption of Rb-MyoD pathways in muscle regeneration. *Brain* 129:996–1013.

Bartlett CS, Helfet DL, Hausman MR, Strauss E (2000). Ballistics and gunshot wounds: effects on musculoskeletal tissues. *J Am Acad Orthop Surg* 8:21–36.

Beauchamp JR, Morgan JE, Pagel CN, Partridge TA (1999). Dynamics of myoblast transplantation reveal a discrete minority of precursors with stem cell–like properties as the myogenic source. *J Cell Biol* 144:1113–1122.

Beauchamp JR, Heslop L, Yu DS, Tajbakhsh S, Kelly RG, Wernig A, et al. (2000). Expression of CD34 and Myf5 defines the majority of quiescent adult skeletal muscle satellite cells. *J Cell Biol* 151:1221–1234.

Beier JP, Stern-Straeter J, Foerster VT, Kneser U, Stark GB, Bach AD (2006). Tissue engineering of injectable muscle: three-dimensional myoblast-fibrin injection in the syngeneic rat animal model. *Plast Reconstr Surg* 118:1113–1121; discussion 1122–1124.

Bentzinger CF, Barzaghi P, Lin S, Ruegg MA (2005). Overexpression of mini-agrin in skeletal muscle increases muscle integrity and regenerative capacity in laminin-alpha2-deficient mice. *FASEB J* 19:934–942.

Blaveri K, Heslop L, Yu DS, Rosenblatt JD, Gross JG, Partridge TA, et al. (1999). Patterns of repair of dystrophic mouse muscle: studies on isolated fibers. *Dev Dyn* 216:244–256.

Border WA and Noble NA (1994). Transforming growth factor beta in tissue fibrosis. *N Engl J Med* 331:1286–1292.

Burkin DJ, Wallace GQ, Milner DJ, Chaney EJ, Mulligan JA, Kaufman SJ (2005). Transgenic expression of {alpha}7{beta}1 integrin maintains muscle integrity, increases regenerative capacity, promotes hypertrophy, and reduces cardiomyopathy in dystrophic mice. *Am J Pathol* 166:253–263.

Chargé SB and Rudnicki MA (2004). Cellular and molecular regulation of muscle regeneration. *Physiol Rev* 84:209–238.

Chazaud B, Sonnet C, Lafuste P, Bassez G, Rimaniol AC, Poron F, et al. (2003). Satellite cells attract monocytes and use macrophages as a support to escape apoptosis and enhance muscle growth. *J Cell Biol* 163:1133–1143.

Cheng M, Nguyen MH, Fantuzzi G, Koh TJ (2008). Endogenous interferon-gamma is required for efficient skeletal muscle regeneration. *Am J Physiol Cell Physiol* 294: C1183–C1191.

Collins CA, Olsen I, Zammit PS, Heslop L, Petrie A, Partridge TA, et al. (2005). Stem cell function, self-renewal, and behavioral heterogeneity of cells from the adult muscle satellite cell niche. *Cell* 122:289–301.

Conconi MT, De Coppi P, Bellini S, Zara G, Sabatti M, Marzaro M, et al. (2005). Homologous muscle acellular matrix seeded with autologous myoblasts as a tissue-engineering approach to abdominal wall–defect repair. *Biomaterials* 26:2567–2574.

Cousins JC, Woodward KJ, Gross JG, Partridge TA, Morgan JE (2004). Regeneration of skeletal muscle from transplanted immortalised myoblasts is oligoclonal. *J Cell Sci* 117:3259–3269.

De Coppi P, Bellini S, Conconi MT, Sabatti M, Simonato E, Gamba PG, et al. (2006). Myoblast-acellular skeletal muscle matrix constructs guarantee a long-term repair of experimental full-thickness abdominal wall defects. *Tissue Eng* 12:1929–1936.

Dellavalle A, Sampaolesi M, Tonlorenzi R, Tagliafico E, Sacchetti B, Perani L, et al. (2007). Pericytes of human skeletal muscle are myogenic precursors distinct from satellite cells. *Nat Cell Biol* 9:255–267.

Emery AE (2002). The muscular dystrophies. *Lancet* 359:687–695.

Fan Y, Maley M, Beilharz M, Grounds M (1996). Rapid death of injected myoblasts in myoblast transfer therapy. *Muscle Nerve* 19:853–860.

Foster K, Foster H, Dickson JG (2006). Gene therapy progress and prospects: Duchenne muscular dystrophy. *Gene Ther* 13:1677–1685.

Galvez BG, Sampaolesi M, Brunelli S, Covarello D, Gavina M, Rossi B, et al. (2006). Complete repair of dystrophic skeletal muscle by mesoangioblasts with enhanced migration ability. *J Cell Biol* 174:231–243.

Gamba PG, Conconi MT, Lo Piccolo R, Zara G, Spinazzi R, Parnigotto PP (2002). Experimental abdominal wall defect repaired with acellular matrix. *Pediatr Surg Int* 18:327–331.

Garrett WE, Jr. (1996). Muscle strain injuries. *Am J Sports Med* 24: S2–S8.

Gawlik KI, Durbeej M (2010). Transgenic overexpression of laminin alpha1 chain in laminin alpha2 chain-deficient mice rescues the disease throughout the lifespan. *Muscle Nerve* 42:30–37.

Gilbert PM, Havenstrite KL, Magnusson KE, Sacco A, Leonardi NA, Kraft P, et al. (2010). Substrate elasticity regulates skeletal muscle stem cell self-renewal in culture. *Science* 329:1078–1081.

Gilbert R, Nalbantoglu J, Petrof BJ, Ebihara S, Guibinga GH, Tinsley JM, et al. (1999). Adenovirus-mediated utrophin gene transfer mitigates the dystrophic phenotype of *mdx* mouse muscles. *Hum Gene Ther* 10:1299–1310.

Grefte S, Kuijpers-Jagtman AM, Torensma R, Von den Hoff JW (2010). Skeletal muscle fibrosis: the effect of stromal-derived factor-1α-loaded collagen scaffolds. *Regen Med* 5:737–747.

Gross JG, Morgan JE (1999). Muscle precursor cells injected into irradiated mdx mouse muscle persist after serial injury. *Muscle Nerve* 22:174–185.

Gussoni E, Soneoka Y, Strickland CD, Buzney EA, Khan MK, Flint AF, et al. (1999). Dystrophin expression in the mdx mouse restored by stem cell transplantation. *Nature* 401:390–394.

Heslop L, Beauchamp JR, Tajbakhsh S, Buckingham ME, Partridge TA, Zammit PS (2001). Transplanted primary neonatal myoblasts can give rise to functional satellite cells as identified using the *Myf5nlacZl*$^{+}$ mouse. *Gene Ther* 8:778–783.

Hill E, Boontheekul T, Mooney DJ (2006a). Designing scaffolds to enhance transplanted myoblast survival and migration. *Tissue Eng* 12:1295–1304.

Hill E, Boontheekul T, Mooney DJ (2006b). Regulating activation of transplanted cells controls tissue regeneration. *Proc Natl Acad Sci USA* 103:2494–2499.

Huang YC, Dennis RG, Larkin L, Baar K (2005). Rapid formation of functional muscle in vitro using fibrin gels. *J Appl Physiol* 98:706–713.

Huard J, Li Y, Fu FH (2002). Muscle injuries and repair: current trends in research. *J Bone Joint Surg Am* 84:822–832.

Irwin WA, Bergamin N, Sabatelli P, Reggiani C, Megighian A, Merlini L, et al. (2003). Mitochondrial dysfunction and apoptosis in myopathic mice with collagen VI deficiency. *Nat Genet* 35:367–371.

Järvinen TAH, Järvinen TLN, Kääriäinen M, Kalimo H, Järvinen M (2005). Muscle injuries: biology and treatment. *Am J Sports Med* 33:745–764.

Kamelger FS, Marksteiner R, Margreiter E, Klima G, Wechselberger G, Hering S, et al. (2004). A comparative study of three different biomaterials in the engineering of skeletal muscle using a rat animal model. *Biomaterials* 25:1649–1655.

Kasemkijwattana C, Menetrey J, Somogyl G, Moreland MS, Fu FH, Buranapanitkit B, et al. (1998). Development of approaches to improve the healing following muscle contusion. *Cell Transplant* 7:585–598.

Kasemkijwattana C, Menetrey J, Bosch P, Somogyi G, Moreland MS, Fu FH, et al. (2000). Use of growth factors to improve muscle healing after strain injury. Clin Orthop Relat Res: 272–285.

Kogo M, Hamaguchi M, Matsuya T (1996). Observation of velopharyngeal closure patterns following isolated stimulation of levator veli palatini and pharyngeal constrictor muscles. *Cleft Palate Craniofac J* 33:273–276.

Kuang S, Kuroda K, Le Grand F, Rudnicki MA (2007). Asymmetric self-renewal and commitment of satellite stem cells in muscle. *Cell* 129:999–1010.

Kuang W, Xu H, Vachon PH, Liu L, Loechel F, Wewer UM, et al. (1998). Merosin-deficient congenital muscular dystrophy: partial genetic correction in two mouse models. *J Clin Investig* 102:844–852.

LaBarge MA, Blau HM (2002). Biological progression from adult bone marrow to mononucleate muscle stem cell to multinucleate muscle fiber in response to injury. *Cell* 111:589–601.

Lakshmipathy U, Verfaillie C (2005). Stem cell plasticity. *Blood Rev* 19:29–38.

Lee JY, Qu-Petersen Z, Cao B, Kimura S, Jankowski R, Cummins J, et al. (2000). Clonal isolation of muscle-derived cells capable of enhancing muscle regeneration and bone healing. *J Cell Biol* 150:1085–1100.

Machida S, Spangenburg EE, Booth FW (2004). Primary rat muscle progenitor cells have decreased proliferation and myotube formation during passages. *Cell Prolif* 37:267–277.

Manzur AY, Kinali M, Muntoni F (2008). Update on the management of Duchenne muscular dystrophy. *Arch Dis Child* 93:986–990.

Marzaro M, Conconi MT, Perin L, Giuliani S, Gamba P, De Coppi P, et al. (2002). Autologous satellite cell seeding improves in vivo biocompatibility of homologous muscle acellular matrix implants. *Int J Mol Med* 10:177–182.

Mauro A (1961). Satellite cell of skeletal muscle fibers. *J Biophys Biochem Cytol* 9:493–495.

Menetrey J, Kasemkijwattana C, Day CS, Bosch P, Vogt M, Fu FH, et al. (2000). Growth factors improve muscle healing in vivo. *J Bone Joint Surg Br* 82:131–137.

Merlini L, Angelin A, Tiepolo T, Braghetta P, Sabatelli P, Zamparelli A, et al. (2008). Cyclosporin A corrects mitochondrial dysfunction and muscle apoptosis in patients with collagen VI myopathies. *Proc Natl Acad Sci USA* 105:5225–5229.

Merritt EK, Hammers DW, Tierney M, Suggs LJ, Walters TJ, Farrar RP (2010). Functional assessment of skeletal muscle regeneration utilizing homologous extracellular matrix as scaffolding. *Tissue Eng A* 16:1395–1405.

Millay DP, Sargent MA, Osinska H, Baines CP, Barton ER, Vuagniaux G, et al. (2008). Genetic and pharmacologic inhibition of mitochondrial-dependent necrosis attenuates muscular dystrophy. *Nat Med* 14:442–447.

Mitchell CA, McGeachie JK, Grounds MD (1996). The exogenous administration of basic fibroblast growth factor to regenerating skeletal muscle in mice does not enhance the process of regeneration. *Growth Factors* 13:37–55.

Montarras D, Morgan J, Collins C, Relaix F, Zaffran S, Cumano A, et al. (2005). Direct isolation of satellite cells for skeletal muscle regeneration. *Science* 309:2064–2067.

Mossey PA, Little J, Munger RG, Dixon MJ, Shaw WC (2009). Cleft lip and palate. *Lancet* 374:1773–1785.

Mostacciuolo ML, Miorin M, Martinello F, Angelini C, Perini P, Trevisan CP (1996). Genetic epidemiology of congenital muscular dystrophy in a sample from north-east Italy. *Hum Genet* 97:277–279.

Muir AR, Kanji AH, Allbrook D (1965). The structure of the satellite cells in skeletal muscle. *J Anat* 99:435–444.

Muir LA, Chamberlain JS (2009). Emerging strategies for cell and gene therapy of the muscular dystrophies. *Expert Rev Mol Med* 11: e18.

Negroni E, Riederer I, Chaouch S, Belicchi M, Razini P, Di Santo J, et al. (2009). In vivo myogenic potential of human CD133(+) muscle-derived stem cells: a quantitative study. *Mol Ther* 17.1771–1778.

Orchard J, Best TM (2002). The management of muscle strain injuries: an early return versus the risk of recurrence. *Clin J Sport Med* 12:3–5.

Partridge TA (2011). Impending therapies for Duchenne muscular dystrophy. *Curr Opin Neurol* 24:415–422.

Qiao C, Li J, Zhu T, Draviam R, Watkins S, Ye X, et al. (2005). Amelioration of laminin-alpha2 deficient congenital muscular dystrophy by somatic gene transfer of miniagrin. *Proc Natl Acad Sci USA* 102:11999–12004.

Qu-Petersen Z, Deasy B, Jankowski R, Ikezawa M, Cummins J, Pruchnic R, et al. (2002). Identification of a novel population of muscle stem cells in mice: potential for muscle regeneration. *J Cell Biol* 157:851–864.

Qu Z, Balkir L, van Deutekom JC, Robbins PD, Pruchnic R, Huard J (1998). Development of approaches to improve cell survival in myoblast transfer therapy. *J Cell Biol* 142:1257–1267.

Sacco A, Doyonnas R, Kraft P, Vitorovic S, Blau HM (2008). Self-renewal and expansion of single transplanted muscle stem cells. *Nature* 456:502–506.

Sambasivan R, Yao R, Kissenpfennig A, Van Wittenberghe L, Paldi A, Gayraud-Morel B, et al. (2011). *Pax7*-expressing satellite cells are indispensable for adult skeletal muscle regeneration. *Development* 138:3647–3656.

Sampaolesi M, Torrente Y, Innocenzi A, Tonlorenzi R, D'Antona G, Pellegrino MA, et al. (2003). Cell therapy of alpha-sarcoglycan null dystrophic mice through intra-arterial delivery of mesoangioblasts. *Science* 301:487–492.

Sampaolesi M, Blot S, D'Antona G, Granger N, Tonlorenzi R, Innocenzi A, et al. (2006). Mesoangioblast stem cells ameliorate muscle function in dystrophic dogs. *Nature* 444:574–579.

Sato K, Li Y, Foster W, Fukushima K, Badlani N, Adachi N, et al. (2003). Improvement of muscle healing through enhancement of muscle regeneration and prevention of fibrosis. *Muscle Nerve* 28:365–372.

Saxena AK, Marler J, Benvenuto M, Willital GH, Vacanti JP (1999). Skeletal muscle tissue engineering using isolated myoblasts on synthetic biodegradable polymers: preliminary studies. *Tissue Eng* 5:525–532.

Saxena AK, Willital GH, Vacanti JP (2001). Vascularized three-dimensional skeletal muscle tissue-engineering. *Biomed Mater Eng* 11:275–281.

Sherwood RI, Christensen JL, Conboy IM, Conboy MJ, Rando TA, Weissman IL, et al. (2004). Isolation of adult mouse myogenic progenitors: functional heterogeneity of cells within and engrafting skeletal muscle. *Cell* 119:543–554.

Stratos I, Rotter R, Eipel C, Mittlmeier T, Vollmar B (2007). Granulocyte-colony stimulating factor enhances muscle proliferation and strength following skeletal muscle injury in rats. *J Appl Physiol* 103:1857–1863.

Suzuki S, Yamanouchi K, Soeta C, Katakai Y, Harada R, Naito K, et al. (2002). Skeletal muscle injury induces hepatocyte growth factor expression in spleen. *Biochem Biophys Res Commun* 292:709–714.

Tatsumi R, Anderson JE, Nevoret CJ, Halevy O, Allen RE (1998). HGF/SF is present in normal adult skeletal muscle and is capable of activating satellite cells. *Dev Biol* 194:114–128.

Tatsumi R, Sheehan SM, Iwasaki H, Hattori A, Allen RE (2001). Mechanical stretch induces activation of skeletal muscle satellite cells in vitro. *Exp Cell Res* 267:107–114.

Ten Broek RW, Grefte S, Von den Hoff JW (2010). Regulatory factors and cell populations involved in skeletal muscle regeneration. *J Cell Physiol* 224:7–16.

Torrente Y, Belicchi M, Sampaolesi M, Pisati F, Meregalli M, D'Antona G, et al. (2004). Human circulating AC133(+) stem cells restore dystrophin expression and ameliorate function in dystrophic skeletal muscle. *J Clin Investig* 114:182–195.

van Wachem PB, Brouwer LA, van Luyn MJ (1999). Absence of muscle regeneration after implantation of a collagen matrix seeded with myoblasts. *Biomaterials* 20:419–426.

Vindigni V, Mazzoleni F, Rossini K, Fabbian M, Zanin ME, Bassetto F, et al. (2004). Reconstruction of ablated rat rectus abdominis by muscle regeneration. *Plast Reconstr Surg* 114:1509–1515; discussion 1516–1518.

Wagers AJ, Conboy IM (2005). Cellular and molecular signatures of muscle regeneration: current concepts and controversies in adult myogenesis. *Cell* 122:659–667.

23

ENGINEERING OF DENTAL TISSUES: SCAFFOLDS AND PRECLINICAL MODELS

NA YU, ADELINA PLACHOKOVA, FANG YANG, X. FRANK WALBOOMERS, AND JOHN A. JANSEN

Department of Biomaterials, Radboud University Nijmegen Medical Centre, Nijmegen, The Netherlands

1 INTRODUCTION

The human innate ability to regenerate lost dental tissues is known to be limited. At present, due to improved understanding of the wound-healing process, together with the recent advances in materials science and molecular biology, the "engineering" of specific dental tissues or the whole tooth appears to be more and more feasible. As a result, the treatment paradigm in dentistry has been changing from "mechanical solutions" for biological problems to "biological solutions for biological problems" (Slavkin and Bartold, 2006). However, before a biological solution for a dental problem can be found in the form of a commercially available tissue engineering product, a suitable biomaterial has to be developed and tested in an appropriate animal model.

In this chapter we provide an overview of the scaffolds currently applied for dental tissue engineering and the preclinical models in which these have been evaluated. The most commonly used scaffolds and models are discussed in view of their treatment strategy: regeneration of dental pulp complex, periodontium, and alveolar bone. Based on our research experience, some guidelines for choosing the appropriate experimental setting are given, and future directions for development in this field are highlighted.

Stem Cells in Craniofacial Development and Regeneration, First Edition.
Edited by George T.-J. Huang, Irma Thesleff.

2 SCAFFOLDS IN DENTAL TISSUE ENGINEERING

Various scaffolds have been designed for dental tissue regeneration, which can basically be used with or without a combination of cells. In the first approach, cells are isolated and expanded in vitro and loaded to the scaffolds prior to implantation. In the second approach, bare scaffolds are implanted into the body and subsequently invaded by host cells. The scaffolds can also be loaded with bioactive molecules to regulate cell and tissue responses, such as growth factors. In principle, the ideal scaffold should possess the ability to:

- Facilitate cell attachment, migration, proliferation, and special organization
- Enable diffusion of cellular nutrients and oxygen
- Exert certain mechanical and biological influences to modify cell behavior
- Provide mechanical support in view of the function of the target tissue

The scaffolds can be present in various forms (e.g. sponges, gels, or fibers) and can be made from different biomaterials (e.g. polymers, ceramics, and metals). They are either biodegradable or nonbiodegradable. In view of their origin, these can be categorized as noted below.

2.1 Natural or Biosynthetic Biomaterials

Natural or biosynthetic biomaterials, mostly polymers, are derived from renewable resources widely distributed in nature: namely, from plants, animals, and microorganisms. The commonly used materials in dental tissue engineering include collagen, gelatin, alginate, and chitosan. Due to their biological origin, those materials offer the chemical and structural information on the natural extracellular matrix and possess superior biocompatibility. It has been demonstrated that they can be used to regenerate dental pulp (Mooney et al., 1996) and periodontal tissues (Rajshankar et al., 1998). For example, studies have shown that dental pulp cells and periodontal ligament cells can adhere and proliferate on chitosan-derived polymer scaffolds (Coimbra et al., 2011). In another study, a modified fibrin gel was used to carry dental pulp stem cells implanted subcutaneously, and collagenous matrix and mineral deposition were found in vivo (Galler et al., 2011a).

Furthermore, natural biomaterials can be used as growth factor carriers. In studies for periodontal tissue engineering, human bone morphogenetic protein 2 and osteogenic protein 1 were incorporated into collagen scaffolds, which resulted in beneficial cementum and bone regeneration compared to bare scaffold and empty groups (King et al., 1997).In a recent study, an injectable chitosan-based thermo-sensitive scaffold with basic fibroblast factor was tested into the periodontal defect of dog. The results showed that chitosan-based gel-like scaffolds can be used successfully as cell/growth factor carriers for periodontal regeneration (Ji et al., 2010). Despite their excellent biocompatibility, natural biomaterials hold certain limitations, such as difficulties in processing and sterilization, which can cause adverse immune reaction and pathogen transmission (Galler et al., 2011b). Therefore, synthetic biomaterials are gaining increased attention for tissue engineering applications.

2.2 Synthetic Materials

Synthetic materials can be categorized into polymers, ceramics, and metals.

Polymers. Synthetic polymers hold many benefits, as they can be prepared with controlled chemical and physical properties and made available in almost unlimited quantities. Their properties (e.g. degradation rate and mechanical strength) can be altered by simple chemical modification. They can also be generated easily into desired scaffold architectures with tailorable degradation times from several days up to years and in mechanical properties suitable for different applications (e.g. hard and soft tissue regeneration). The most commonly used synthetic polymers include poly(lactic acid) (PLA), poly(glycolic acid) (PGA), and their copolymer poly(lactide-*co*-glycolide) (PLGA), and poly(ethylene glycol) (PEG). In a study of periodontal regeneration, PLGA was fabricated into a nanosized fibrous scaffold through an electrospinning process with aligned, crossed, and random textures, which simulated the natural periodontal environment. The periodontal cells showed contact guidance effect and improved the proliferation rate when exposed to organized polymer topographies (Shang et al., 2010). In an effort to engineer pulp tissue in vitro, a research team led by Mooney and Rutherford seeded human pulp fibroblasts on PGA fibers. After 45 to 60 days of in vitro culture, they observed pulplike tissue on the PGA scaffolds (Mooney et al., 1996). These early studies are proof of principle that pulp cells can grow well and deposit matrix on suitable synthetic polymers in vitro. However, it should be noted that the accumulation of acidic degradation products from PLGA scaffolds may evoke fibrous tissue encapsulation during the wound-healing process (Gunatillake and Adhikari, 2003).

Ceramics. In general, scaffolds made from ceramics, mainly calcium phosphates (CaPs), have outstanding properties which make them suitable for clinical application, such as similarity in composition to bone mineral; ability to bond directly to bone tissue; ability to promote cellular function, which leads to formation of a uniquely strong bone–biomaterial interface; and osteoconductivity. It is evident that they are among the most frequently used biomaterials for alveolar bone regeneration. Most commonly used CaP biomaterials include hydroxyapatite (HA), tricalcium phosphate (TCP), and biphasic calcium phosphate (HA/TCP), which are available in different physical forms (e.g. particulates, blocks, cements, coatings on metal implants, and composites with polymers). Some examples of commercially available ceramic products applied successfully in the field of alveolar bone regeneration are Periograf® (Sterling Drug, Inc., Rensselaer, NY), Osteogen® (Impladent, Ltd., Holliswood, NY), Bio-Oss® (Geistlich AG, Wohusen, Switzerland), and Endobone® (Merck Biomaterialen, Darmstadt, Germany). Furthermore, ceramics can be used as carriers for growth factors [GEM 21® (β-TCP-PDGF-β, Osteohealth, Shirley, NY)] and stem cells (Pellegrini et al., 2009). Besides their applications in bone regeneration, CaP ceramics have also been used for dentin reconstruction. Gronthos et al. (2000) used HA/TCP powder as carrier materials for the dental pulp stem cells and successfully regenerated dentin-like tissue in vivo. However, there are two major limitations, which hamper the wide application of CaP ceramics: (1) their brittle mechanical behavior, which causes problems when subjected to the peak forces during mastication and; (2) the slow degradation rate, which inhibits cell and tissue ingrowth into the scaffolds. To address these problems, biodegradable

PLGA fibers and microspheres have been introduced into calcium phosphate ceramic. The in vitro and in vivo results have shown a shortened degradation time and superior bone ingrowth compared with bare ceramic scaffolds (Habraken et al., 2006; Felix Lanao et al., 2011).

Metallic Biomaterials. For several decades metallic implants (e.g. titanium and its alloys) have been installed into alveolar bone for the fixation of a crown, bridge, or full denture. Titanium is highly biocompatible, has ideal physicomechanical properties, and is superior to achieve osteointegration. Porous and fibrous titanium has also been used as scaffolding in bone regeneration and resulted in enhanced bone contact compared with a dense titanium implant (Wazen et al., 2010). In addition, titanium mesh provided with autologous bone was found to be promising in restoring a critical-sized defect in the mandible of a dog model. For dental pulp regeneration, titanium mesh can support odontoblast differentiation and mineralization in human dental pulp stem cells (Zhang et al., 2005).

2.3 Future Perspectives of Scaffolds

In view of suitable scaffolds for their target tissues, a list of scaffolds for specific dental applications is presented in Table 1. To improve the performance of currently developed tissue engineering scaffolds, researchers are focusing especially on incorporating multiple design criteria into a single material. It is thought that this approach will result in better control of biofunctionality. In particular, the specifics of the microenvironment (hard vs. soft tissue architectures) with which the material will interact with must be taken into consideration in the design process. In addition, biomolecule delivery from scaffolds still needs to be improved further in a spatially and sequentially controlled fashion (Ji et al., 2011). Therefore, significant challenges remain in integrating structural cues with biological cues to achieve optimal functional dental tissue engineering.

3 MODELS FOR DENTAL TISSUE ENGINEERING

Progress in dental tissue engineering depends strongly on preclinical research. Rigorous in vitro and in vivo tests are necessary before a new biomaterial can be used in human clinical trials. To obtain U.S. Food and Drug Administration (FDA) and European Commission (EC) regulatory approval, initial tests (e.g. cell culture), secondary tests of the local tissue reaction (e.g. animal studies), and usage tests of the final products (e.g. clinical trial) have to be completed successfully. Based on these demands, preclinical modeling appears to be the most crucial step in the evaluation of a new biomaterial before its clinical application. Therefore, there is a need for targeted preclinical studies with selection of the most suitable animal model for the biomaterial or scaffold investigated. Important considerations for the selection of an appropriate animal model are depicted in Figure 1.

It must be emphasized that selection of an appropriate animal model with a high degree of biologic similarity to the human situation is a prerequisite to transfer the experimental results to the clinic. In general, the smaller the animal model, the harder

TABLE 1 Scaffolds for Dental Tissue Engineering in View of the Target Tissue

Treatment Strategy	Scaffolds	Main Results	References
Enamel	Self-assembling peptide	Nanofibrous self-assembling peptide amphiphile supported enamel formation in the kidney of mouse.	Huang et al., 2010
Dentin–pulp	Collagen I and III, chitosan, gelatin collagen I with Dmp1	Collagen I showed superior properties in cell adhesion, proliferation, and differentiation than the rest groups when cultured with human dental pulp cells; collagen-Dmp1 scaffolds resulted in new pulp formation when cultured with dental pulp stem cells and placed in dentin block in vivo.	Kim et al., 2009; Prescott et al., 2008
	PLA, PLGA	The dental pulp stem cells together with PLGA/PLA scaffolds generated vascularized, pulp-like tissues and new tubular dentin.	Cordeiro et al., 2008; Huang et al., 2010; Sakai et al., 2010
	HA/TCP	Dental pulp stem cells together with HA/TCP powders formed dentin and dentin-pulp-like complexes in vivo; root apical papilla cells and periodontal ligament cells induced periodontal regeneration when they were loaded with HA/TCP scaffold in minipig periodontal defects; bioceramics stem cell mixture generate dentin–pulp-like complexes when they were cultured in vivo.	Gronthos et al., 2000; Miura et al., 2003; Liu et al., 2008; Sonoyama et al., 2006
Periodontal	PLGA (fiber and hydrogel)	Electrospun PLGA fibers supported periodontal cell attachment, proliferation, and differentiation; PLGA hydrogels with human growth differentiation factor5 promoted bone formation when delivered in periodontal defects in vivo.	Shang et al., 2010; Kwon et al., 2010
	Self-assembled peptide	Self-assembled peptide nanofibrous scaffolds promoted deposition of the main periodontal ligament ECM components when periodontal ligament cells were cultured in vitro.	Kumada and Zhang,

(*continued*)

TABLE 1 ***(Continued)***

Treatment Strategy	Scaffolds	Main Results	References
	Calcium phosphate cement	An injectable calcium phosphate cement provided stable wound healing and enhanced periodontal regeneration in experimental periodontitis in dogs.	Shirakata et al., 2002
Alveolar bone	HA/TCP	BMSCs combined with HA/TCP facilitated regeneration of alveolar defect around titanium implant in a dog model.	De Kok et al., 2005
	Calcium phosphate cement	Injectable composite calcium phosphate cement with and without growth factors demonstrated excellent bone regenerative capacity in a dog intrabony defect model.	Felix Lanao et al., 2011
	Calcium alginate	Calcium alginate plus cells facilitated bone formation when scaffolds were loaded with BMSCs and transplanted in vivo.	Weng et al., 2006
	PL/PGA	PL/PGA mixed with BMSC could preserve alveolar bone and reduce bone loss after dental extraction.	Marei et al., 2005
	Vegetable latex biopolymer	Scaffolds promote bone formation and decrease connective tissue formation in the dental extraction alveolar cavities in rats.	Balabanian et al., 2006
	Bio-Oss	Bio-Oss with ALP resulted in enhanced early bone formation compared to Bio-Gide alone in a rat introbony defect model.	Oortgiesen et al., 2012a
	Silk scaffold	Silk scaffold guide mineralized tissue formation when mixed with tooth bud cells and grown in the rat omentum.	Xu et al., 2008
	PGA, PLGA	Tooth bud cells seeded on PGA and PLGA scaffolds generate bioengineered tooth tissues after 12 weeks in the omenta of rat.	Duailibi et al., 2004; Young et al., 2005
	Collagen gel	Cells from epithelial and mesenchymal tissues of tooth germ generate blood vessels and nerve fibers when cultured in a collagen gel drop in vivo.	Nakao et al., 2007

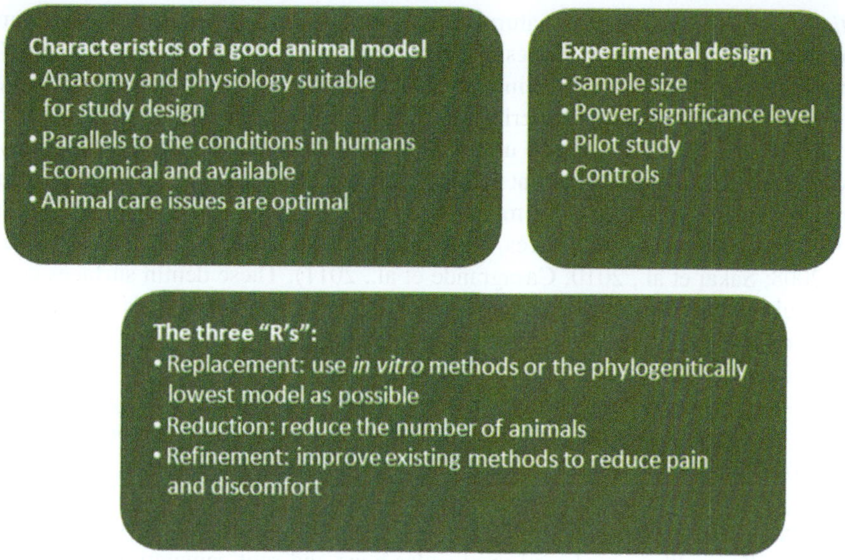

FIGURE 1 Considerations when choosing animal models.

it is to give predictable answers for clinical applicability of a biomaterial. However, for ethical and social reasons, the use of large-animal models should be reserved for the last phase of validation of a new biomaterial (i.e., safety and efficacy studies).

In the text below, emphasis is placed on frequently used animal models for dental tissue engineering. Their advantages and disadvantages, surgical protocols, and methods of evaluation are summarized briefly, and our own research experience in this field is highlighted. However, first some in vitro models for dental tissue engineering are discussed, as in vitro assays can provide an indication of the feasibility of the new approach or development.

3.1 In Vitro Models

A better understanding of cellular and molecular responses forms a prerequisite for further improvement of tissue engineering approaches in dentin–pulp, periodontal, and alveolar bone regeneration. Traditional two-dimensional cell culture systems may not always reflect the mechanical status and biochemical homeostasis of the dental environment, and thus there has been increased awareness of the need to improve in vitro models to supplement the widely used two-dimensional culture system (Pavlin and Gluhak-Heinrich, 2001). New models should be able to provide more predictability and allow the study of a number of clinical situations and biological parameters before going into an in vivo test. They should also allow the culture of tissue analogs (cell–biomaterial constructs) before transplantation use. The purpose of the following paragraphs is to provide insight into some cell culture models used in studying the effects of biomaterials and biological parameters on dental cell behavior.

Dentin–Pulp Cell Culture. In natural dental pulp, the three-dimensional environment is composed largely of several types of collagens, glycosaminoglycans, and water. The matrix provides a supporting medium for cells and for the transportation of nutrients and metabolites. In tissue engineering, such a three-dimensional structure is initially provided to the cells through the use of biodegradable and biocompatible scaffolds. They provide an environment that allows adhesion of cells and their proliferation, migration, and differentiation. In many cases, cell–scaffold constructs were cultured within natural dental circumstances, such as dentin slices, cups, or circles (Cordeiro et al., 2008; Sakai et al., 2010; Casagrande et al., 2011). These dentin surfaces contain multiple closely packed dentin tubules in which the biochemical signals (dentin matrix components) together with the surface topography may modulate the dental pulp cells to attach, align, and differentiate into the odontoblast phenotype (Heywood and Appleton, 1984). Using this model, several research groups have shown promising results using various cell types, scaffolds, and growth factors to generate dental pulp-like tissue (Cordeiro et al., 2008; Sakai et al., 2010).

Periodontal Cell Culture Models. Natural periodontal tissue is composed of highly regulated collagen fibers anchored in alveolar bone and root cementum, which allows absorption of the forces of mastication. The mechanical stimulation from a chewing load will result in adaptive responses to tooth-surrounding tissue during development, maturation, and regeneration. Since it is clear that anabolic cellular responses to mechanical stress overlap and interact with all other factors in physiological and pathological conditions, it is crucial to incorporate this factor into a tissue engineering process. However, due to the limitations of in vitro models, previous studies on mechanical regulation of periodontal cells and cementoblasts mostly utilized two-dimensional culture systems rather than three-dimensional mechanical loading circumstances (Pavlin and Gluhak-Heinrich, 2001). We have recently characterized and implemented two three-dimensional culture models for studying mechanically induced regulation of periodontal ligament cell behavior.

In the first model, a defined mechanical loading is applied to a collagen gel containing human PDL fibroblasts through a culture chamber designed to mimic a moving tooth. The culture chamber consists of a movable tapered cylinder imitating an artificial root and a surrounding wall mimicking bone surface (Fig. 2A). In between the two hard surfaces, the cell–scaffold construct is placed and attached to both sides. Controlled mechanical loading, applied for three days, resulted in magnitude-dependent changes in the expression of Cox2 and collagen I gene expression (Berendsen et al., 2009).

In the other model, PDL cells mixed with collagen gel layers were injected into silicone elastomer dishes and covered with a custom-designed porous coverslip, resulting in a 200-μm-thick collagen layer (Fig. 2B). The resulting constructs were tested up to five days in a mechanically and biochemically stimulated situation. As a mechanical stimulus, 8% longitudinal cyclic stretching was applied. As a biochemical stimulus, samples were treated with 100-μg/mL enamel matrix proteins. The morphology, proliferation, and differentiation of cells were observed and quantified. This study showed that transfer of load occurred throughout the entire sample and cells elongated perpendicular toward the loading direction. As in the normal PDL, remarkable differences occurred between the central and peripheral region of the constructs (Oortgiesen et al., 2012b).

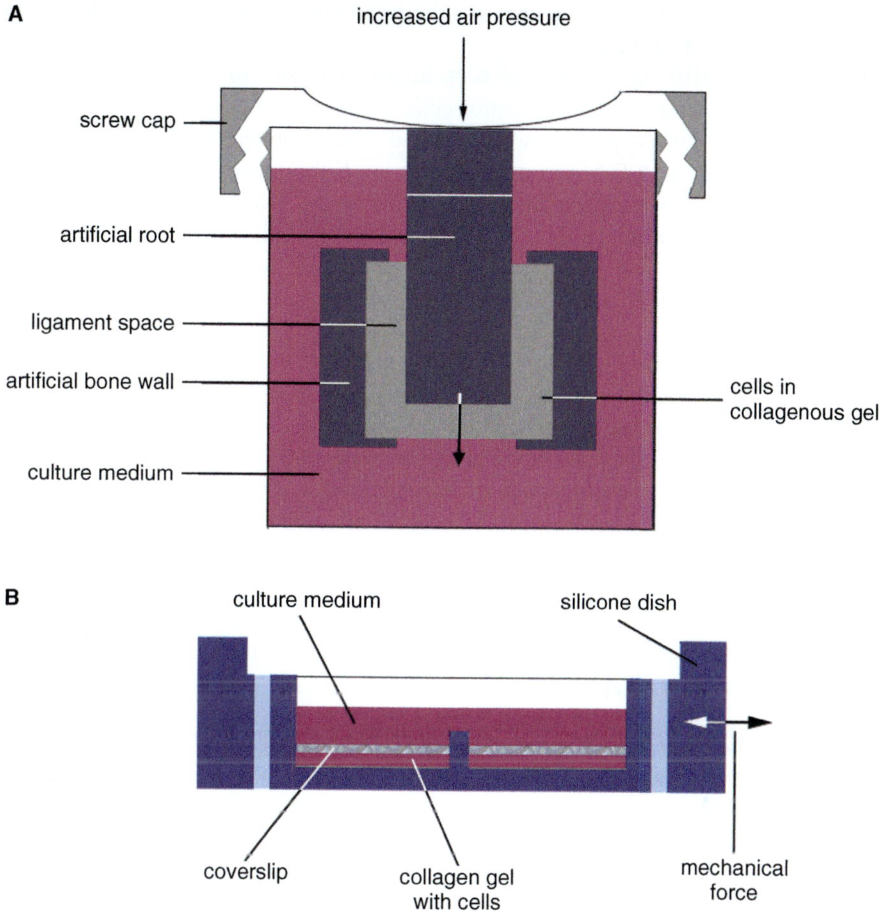

FIGURE 2 In vitro models for periodontal cell culture. (A) A three-dimensional model of functional artificial periodontal ligament by mimicking a movable dental root. (B) A model of a functional periodontal ligament by a sandwich-like mechanical loading culture system.

3.2 In Vivo Models

In the following paragraphs, frequently used animal models are discussed, especially in view of their application for dental tissue regeneration of dental–pulp complex, whole-tooth, periodontal, and bone regeneration.

Models for Pulp–Dentin Complex Regeneration. Most tissue engineering approaches to regenerate the dentin–pulp complex are based on the transplantation of cell-loaded constructs to a defect site (cell-based). Therefore, both in vitro and in vivo models must provide a suitable environment with physiological parameters to allow the cultivation of the dental pulp stem cells. Examples for suitable in vivo environments for transplants are skin, renal capsule, or omentum in various experimental animals, including mice, rats, rabbits, and pigs. After in vivo maturation, cell-based scaffolds are transplanted into their physiological anatomical region (i.e., implantation into the jaw bone), which provides biological and mechanic cues for the progenitor cells to develop into mature and functional dental tissues.

Huang et al. (2010) used a dental root fragment model to regenerate lost dental pulp and dentin from stem and progenitor cells. Stem and progenitor cells from apical papilla and dental pulp stem cells were isolated, characterized, seeded onto synthetic scaffolds consisting of PLGA, inserted into the tooth fragments, and transplanted into mice. The authors provided the first evidence showing that pulp-like tissue can be regenerated de novo in empty root canal space by stem cells from apical papilla and dental pulp stem cells that give rise to odontoblast-like cells, producing dentin-like tissue on existing dentinal walls.

Whole-Tooth Regeneration. Whole-tooth regeneration efforts consist largely of two approaches: scaffold-based, which utilizes cell-seeded scaffold to guide and support tooth formation, and organogenesis-based (or germ tissue–based), which facilitates development of a tooth from a collection of cells resembling the tooth germ (Hacking and Khademhosseini, 2009). In general, the cell-based scaffolds first need to be placed in an environment suitable for maturation, such as renal capsule and omentum, and subsequently implanted in the jaw. In the organogenesis-based approach this is not always necessary. In a study by Ikeda et al. (2009) the first bioengineered tooth was created from a tooth germ, which developed and erupted in the oral cavity of an adult mouse. The authors proposed this as a model for future replacement therapy. (More details are available in Chapter 25.)

Models for Periodontal Regeneration. To test the effects of various tissue engineering formulations on periodontal regeneration, acute and chronic periodontal defect models can be utilized. In acute defect models, lesions are created surgically and immediately treated with biomaterials. In contrast, in chronic defect models, formation of natural periodontitis lesions can be induced. Lesions are created in the course of time by placing ligatures around the teeth, and in a second procedure, biomaterials were introduced. In addition, experimental periodontal defects can be created in a combined manner (i.e., acute–chronic model): The defects are created surgically and ligatures are placed to prevent spontaneous regeneration and to ensure plaque accumulation and chronic inflammation. In this way, the advantages of acute and chronic models can be combined. Finally, periodontal lesions can be developed spontaneously in dogs and in goats. Although these models resemble to a great extent the human situation with periodontitis, the onset of the lesions is difficult to predict as well as the size and morphology of the defects.

According to the defect configuration, different types of periodontal defect models can be recognized, such as fenestration, intrabony, suprabony, and furcation (Fig. 3). These models mimic the defect morphology in humans with periodontitis and can be created in small and large animals. In general, anatomical similarities and defect size characteristics favor the selection of large animals prior to moving to clinical trials in humans.

Periodontal regeneration in animal models can be demonstrated by qualitative and quantitative analysis. The most common outcome measures include length of the new cement, new periodontal ligament (PDL) and new bone from a histological reference notch made during surgery, and histomorphometric assessment of the new bone and cementum area. In larger animals, clinical periodontal measurements (probing depth,

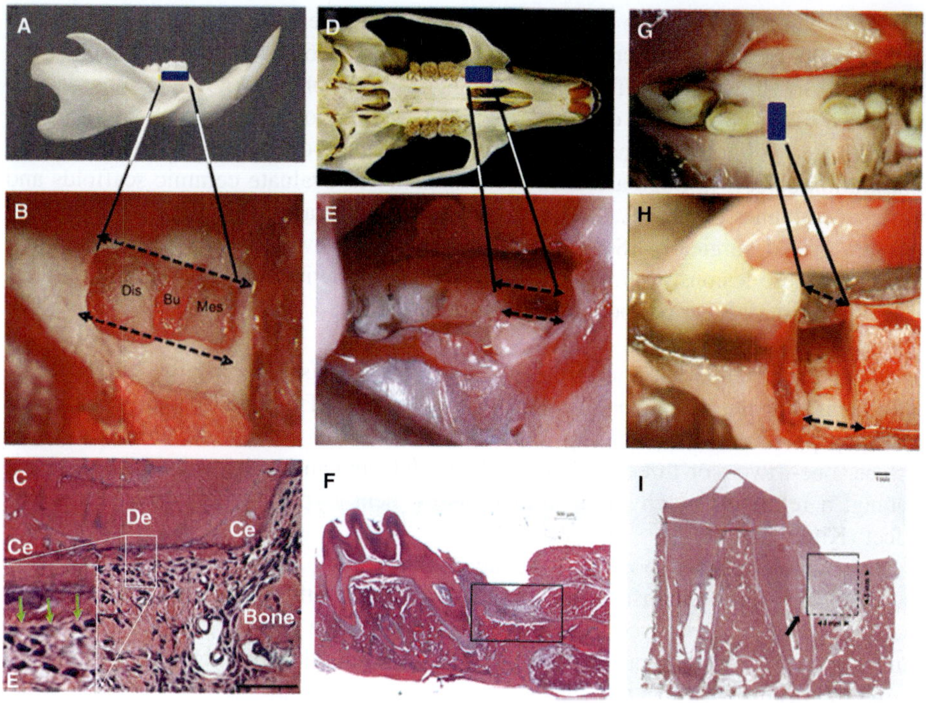

FIGURE 3 In vivo models for periodontal regeneration. Fenestration model in rat (A, B, and C) (Yang et al., 2010); intrabony model in rat (D, E, and F); intrabony defect model in dog (G, H, and I). Blue boxes (in A, D, and G) indicate surgical regions. Dashed lines (in B, E, and H) indicate direction of histological section. Black-framed boxes (in C, F, and I) indicate regions of interest in H&E staining slides.

recession, and attachment loss) can be performed. In addition, new bone formation can be evaluated radiologically and by micro-computed tomography (micro-CT). In the following paragraphs, frequently used models for periodontal regeneration are characterized briefly based on their defect configuration.

Fenestration Defect Model. The fenestration model represents a circumferential defect on the midbuccal aspect of the root surface. It can be created by extraoral (in rats and rabbits) or intraoral (in dogs and monkeys) surgical access. After flap elevation and root exposure, alveolar bone, periodontal ligament, and cementum are removed apically from the enamel–cementum junction. Subsequently, the biomaterial is placed into this well-contained defect and covered by the flap.

In rats, King et al. created standardized fenestration defects (3 × 2 × 1 mm) on the buccal surface of the first and second mandibular molars via an extraoral approach to study the effect of bone morphogenetic protein 2 in a collagen gel on periodontal wound healing. This model is used primarily to study cell-based scaffolds and growth factors for periodontal tissue engineering (King et al., 1997; Huang et al., 2005; Yang et al., 2010).

In dogs, standardized fenestration periodontal defects can be created by intraoral access in different sizes and locations (maxilla or mandible) and can be investigated for early (days) and or late time points (weeks). For example, in one of the first studies, Choi et al. (1993) created fenestration defects (6 × 4 mm) on the maxillary canine to investigate the effect of collagen on periodontal wound healing. It was concluded that the maxillary canine periodontal fenestration defect appeared a suitable model to study factors that may enhance cementum and bone regeneration. In general, fenestration defects in dogs are used predominantly to evaluate ceramic scaffolds and barrier membranes (Tal et al., 2005). In monkeys, fenestration periodontal defects can also be created surgically (Sculean et al., 2001). However, due to the spontaneous healing of these lesions, this model is not considered optimal for studying the effects of new biomaterials on periodontal regeneration.

Intrabony Defect Model. The intrabony model represents a "box-type" defect surgically created at the mesial or distal tooth root surface. The base of the box exhibits intrabony position, and according to the number of box bone walls around the root surface, one-, two-, or three-walled intrabony defects can be defined. Following root planing, a reference notch for histologic assessment is placed at the bottom of the defect (Kim et al., 2004).

This model makes it possible to analyze the healing at the junctional–epithelial and connective tissue interface, which is not possible with the fenestration model. In addition, it can be considered as being more similar to the clinical situation, due to the similar morphological characteristics and their location with microbiological influence during the healing period.

In rats, Nemcovsky et al. (2006) created intrabony defects (2 mm in diameter) bilaterally in the mesial aspect of the first maxillary molars to study the effect of enamel matrix derivative on periodontal regeneration. In contrast to the fenestration model, the intrabony model does not heal that fast and allows histological analysis of the epithelial–connective tissue interface and volumetric micro-CT analysis of the amount of the newly formed bone (Oortgiesen et al., 2012a).

In dogs, the creation of intrabony defects requires two surgical procedures with an interval of months in between. Following teeth extraction in the first surgery and a healing period, defects can be created in a second surgical procedure. As mentioned previously, healing of intrabony periodontal defects is positively correlated with the number of bone walls surrounding the defects (Kim et al., 2004), and one- and three-walled defects are preferred because more homogeneous outcomes can be expected compared to two-walled defects. The one-walled intrabony model in dogs is frequently utilized to study the effect of cell-based and cell-free scaffolds. In our recent study an injectable composite scaffold with and without growth factors was implanted for eight weeks and demonstrated excellent bone regenerative capacity three after (Felix Lanao et al., 2011).

In monkeys, intrabony defects represent an acute–chronic model (Sculean et al., 2002). First, infrabony defects have to be created surgically (depth 6 to 8 mm from the enamel–cementum junction), and these defects have to be filled with space-providing mechanical devices such as metal strips or ligatures to reduce spontaneous regeneration and to provoke plaque accumulation and chronic inflammation. After eight weeks, at surgical reentry, the biomaterials are delivered in the defects. Blumenthal et al. used this acute–chronic model to compare the efficacy of different regenerative therapies used in the clinic, such as membranes and grafts alone and in combination with two- and three-walled defects. Besides histological measurements, clinical measurements were performed. Based on their results, the authors concluded that whereas all therapies are equal in three-walled defects, the combination graft–membrane appeared to be the most appropriate therapy for two-walled defects.

Supraalveolar Defect Model. The supraalveolar periodontal defect model was developed by Wikesjo in dogs (Wikesjo and Nilveus, 1991) and is considered a nonspontaneous healing defect. In this model, a horizontal circumferential defect is generated around the premolar teeth. Biomaterials can be applied immediately or after a chronic inflammation period. The major advantages of this model are standardized anatomical characteristics, and its healing potential does not depend on the number of bony walls. Following a healing interval of 4 and 8 weeks, it has been demonstrated that periodontal regeneration is limited to the apical region (Wikesjo et al., 2004). Moreover, it demonstrates morphological similarity to class III furcation defect, one of the most challenging periodontal defects in humans (Sculean et al., 2008).

Furcation Defect Model. The furcation model represents a "box" type of defect in the furcation area of premolars and molars. They can be divided into class II and class III defects (based on their specific location). Furcation defects can only be created in large animals, such as dogs, pigs, sheep, and monkeys. In an acute model in dog, Deliberador et al. (2006) created class II furcation defects on the buccal surface of the mandibular premolars to study periodontal healing using an autogenous graft with or without a barrier membrane. After three months, a considerable amount of new bone formation resulting from spontaneous regeneration was observed in the empty defects, which could be seen as one of the drawbacks of this acute class II furcation model.

A class III furcation model appears to be more valid for the study of various approaches to periodontal regeneration. The critical–sized supraalveolar defect model can also be considered a class III furcation model. Araujo and Lindhe (1998) created class III furcation defects in dogs (4 × 3 mm) to evaluate the effect of EMD gels with and without resorbable barrier for four months.

In monkeys, class II and III furcation defects can be created in an acute–chronic model. For example, Hovey et al. (2006) created an acute–chronic model class III furcation defect (10 mm from the fornix to the reduced alveolar crest) to study the effect of enamel matrix proteins alone or in combination with bioengineered tissue on periodontal wound healing. Upon surgical induction of the defect, ligatures and cotton pellets were placed for two months. At surgical reentry the treatment modalities were placed and left for five months. The advantage of chronic defects in monkeys is that these defects do not regenerate spontaneously and therefore allow a long observation period.

Suzuki et al. (2006) evaluated the efficiency of using spontaneous periodontitis as a model in Shiba goats. They found that the clinical, histopathological, and microbiological features of spontaneous periodontitis in these goats were similar to those in human periodontitis. Moreover, the size of the oral cavity in Shiba goats is suitable for periodontal treatment, and handling and housing are relatively easy. However, the defect area of this model is difficult to standardize.

Models for Bone Regeneration. Various animal models have been developed to study bone regeneration. Although no species fulfills all the requirements for the ideal model, understanding of the differences in bone architecture and remodeling between the species can assist in selection of the most suitable model for a desired biomaterial and research question. In view of the research question, bone models can be characterized as models for (1) fundamental discovery, (2) feasibility and bioactivity testing, and (3) clinical modeling and efficacy prediction (Muschler et al., 2010). For example, to investigate basic mechanisms of action of a new tissue engineering formulation, small animals such as mice and rats are preferred. Feasibility and bioactivity testing (i.e., biocompatibility, toxicity, screening for adverse reactions, way of delivery of a biomaterial) are also usually done in mice and rats in ectopic (i.e., subcutaneous) or orthotopic (bone) implantation sites. For clinical modeling and efficacy, larger animals (dog, sheep, goat, or pig) are more appropriate, due to the close similarity with the human bone. In Table 2, a summary of the bone properties of these animals is presented.

Besides species, the following factors that can influence bone regeneration should be taken into account when choosing the most appropriate model: localization of the defect (extraoral vs. intraoral; mandibular ramus vs. alveolar ridge), size of the defect (critical-sized vs. noncritically sized), type of defect (acute vs. chronic), anatomy of the defect (containing vs. transosseous), and age of the animals. A critical-sized defect (CSD) is defined by Schmitz and Hollinger (1986) as the smallest intraosseous wound in a particular bone and species of animal which will not heal spontaneously during the lifetime of the animal. In principle, critical-sized defects are still considered the gold standard for screening of new biomaterials.

In general, bone regeneration in animal models can be assessed by qualitative (descriptive histology, immunohistochemistry radiographs) and quantitative analysis (histomorphometry and micro-CT). Although micro-CT provides three-dimensional information about the volume, distribution, and density of the mineralized matrix of the bone, a histological evaluation is necessary to characterize the cellular and

TABLE 2 Large animal models for alveolar bone regeneration (Reichert et al., 2009)

Animal	Advantages	Disadvantages
Dog	Tractable nature; bone mineral density similar to that of humans	Higher rate of solid bony fusion when compared to humans Low nonunion rates Ethical issues and negative public perception Significant interanimal variations due to breed diversity
Sheep	Docile animals; body weight similar to that of humans; dimension of long bones suitable for human implants	Age-dependant bone remodeling Haversian remodeling at 7–9 years of age (with medium-sized irregular cannals) Higher trabecular bone density than that of humans
Goat	More tolerant to ambient conditions; bone composition very similar to that of humans	Inquisitive and interactive nature
Pig	Bone mineral density, anatomy, morphology, and healing similar to those of humans	High growth rate and excessive body weight Difficult handling

tissue response: pattern, distribution, kinetics of bone formation, vascularization, inflammation, and status of the residual implant material. In some settings, mechanical testing (modulus, strength, etc.) is necessary to determine the success of bone tissue engineering.

It is beyond the scope of this chapter to provide a complete overview of all models used for bone regeneration. Moreover, numerous reviews have been published on this topic (Liebschner, 2004; Le Guehennec et al., 2005; Cancedda et al., 2007; Reichert et al., 2009). Therefore, the focus will be put on experimental defects for clinical modeling of procedures used in oral reconstructive surgery. In general, according to their aim, these procedures can be divided into (1) alveolar ridge preservation, (2) vertical and horizontal bone augmentation, (3) guided bone regeneration, and (4) sinus floor augmentation. In the following paragraphs, some examples of animal models for these procedures are given. The models are characterized based on the localization of the defect.

Alveolar Extraction Sockets. This model is used to study the effect of new biomaterials on alveolar ridge preservation or on the osteointegration of dental implants.

In rats, such defects can be created by extractions of incisors or molars and treated immediately with biomaterials. However, due to the small size of the extraction sockets and the strong healing capacity, this model is less suitable for screening of new tissue engineering formulations (Pellegrini et al., 2009).

In rabbits, Marei et al. (2005) studied cell–scaffold constructs for bone maintenance after dental extraction. It was shown that alveolar bone was preserved by filling the sockets with cell–scaffold constructs or PLGA scaffolds.

Extensive studies on ridge preservation in dogs have been carried out by Araujo's research group. In a recent study, Araujo et al. used this model to evaluate whether chips of autologous bone may allow ridge preservation following tooth extraction. For this purpose, the distal roots of the third and fourth mandibular premolars were removed and the sockets were grafted with either anorganic bovine bone or with chips of autologous bone harvested from the buccal bone plate. The results showed that after three months of healing, autologous bone chips placed in the fresh extraction socket neither stimulate nor retard new bone formation (Araujo and Lindhe, 2011).

In dogs and monkeys, alveolar sockets can be used to evaluate the process of osseointegration of dental implants placed immediately (in fresh alveolar sockets) or in a staged approach (after sockets are allowed to heal). For example, in a dog model, Wikesjo et al. (2008) found that rhBMP2 absorbed on the implant surface accelerated, in a dose-dependent manner, the bone formation process and supported osseointegration within eight weeks.

Mandibular Ramus. This location in rat can be used to study biomaterials for vertical bone augmentation. The model was developed by Kostopolous and Karring (1994) based on the principles of guided bone regeneration (GBR). Briefly, on the lateral part of the mandibular ramus of rats, a space-providing Teflon capsule filled with a biomaterial was placed and secured by sutures or screws. After 2 weeks, the space beneath the capsule was partially filled with connective tissue; after four months, up to 70% was filled with bone; and after one year the entire space was completely filled with mature lamellar bone (Kostopoulos and Karring, 1994). Using this model, Lioubavina-Hack et al. (2005) evaluated the effect of simultaneous delivery of recombinant human platelet-derived growth factor BB andinsulin-like growth factor-I (rhPDGF-BB/IGF-I) incorporated into a methyl cellulose gel on guided bone regeneration.

Inferior Border of the Mandible. The inferior border of the mandible is also a suitable location for studying bone augmentation and guided bone regeneration (GBR). Donos et al. (2005) used the inferior border of the rat mandible to compare the long-term volume stability of membranous and endochondral autogenous bone grafts with or without membrane. After 5 and 11 months of evaluation, the combined treatment of onlay grafts with GBR were found to be superior to onlay bone grafts alone. A similar model in pig was used to investigate tissue responses to resorbable and nonresorbable barrier membranes with or without bone substitute. Based on the results, the authors suggested application of bone substitutes when resorbable membranes are applied for GBR (Strietzel et al., 2006).

Bicortical Defect in the Mandible. A bicortical defect in the mandible appears to be suitable to test the ability of biomaterials to induce bone regeneration and to be applied for guided bone regeneration procedures. For example, a critical-sized bicortical defect can be created in the mandible of rabbits. In a study by Young et al. (2008), a cylindrical bone defect 10 mm in diameter was created, in full or partial thickness in premolar–molar region and left to heal for up to 16 weeks. At 8 weeks the partial thickness defect demonstrated significant bone fill, and at 16 weeks complete regeneration and bridging was found. In contrast, no bone bridging was observed in the full-thickness defect, neither difference in bone regeneration was found between 8 and 16 weeks. Based on the results, the authors concluded that 10-mm bicortical defect in the mandible of rabbits could be considered to be critical-sized.

In mongrel dogs, Sverzut et al. (2008) used unilateral 10-mm-wide segmental defects in the mandibles to test the influence of polylactide membrane permeability on the fate of iliac bone graft. The defects were created by removing both the fourth premolars and the first molars of the lower jaw. Subsequently, iliac bone grafts were placed into the defect area with different membranes (PLLA membranes or microporous laser-perforated membranes). Data from histological observation and bone labeling indicated that the bone formation process was incipient in the bone graft group at three months postoperatively, regardless of whether or not it was covered by membrane. In contrast, GBR with microporous laser-perforated membranes tended to enhance bone formation activity at three months.

Maxillary Sinus. The maxillary sinus augmentation model is developed to resemble the clinical sinus lift procedure. In this setting new biomaterials for bone augmentation can be tested. The choice of animals for this model can be rabbits, dogs, sheep, goats, pigs, or minipigs. When choosing an experimental animal, it has to be noted that only in goats can the maxillary sinus be accessed from the oral cavity (Derong et al., 2010).

In a study done by our group, a sheep model was used to evaluate the biological performance of osteoinductive microstructured tricalcium phosphate particles for sinus floor augmentation (Klijn et al., 2012). Via an extraoral approach, a window with a diameter of 10 mm was created, followed by careful removal of the resulting bone plate from the Schneiderian membrane. Subsequently, the membrane was elevated from the buccal and caudal bony wall and displaced cranially. After insertion of the material, the grafted sinus was covered by the bone plate. After 12 weeks of implantation, sinuses grafted with the particles showed an increased bone height of 6 mm and a mean total bone volume of 43%.

4 CONCLUSIONS

Since tissue engineering approaches have been introduced in dentistry, it has become possible to shift the treatment strategies from rehabilitation or restoration toward achieving real regeneration. Currently, regenerative strategies are used successfully in the fields of periodontology and oral reconstructive surgery, and the number of commercially available biomaterials for bone and periodontal tissue regeneration is constantly increasing. At the same time, the role of the biomaterial has been changing from that of a "passive" scaffold which serves mainly as a carrier vehicle, toward a "bioactive" scaffold which on its own can induce desired cellular behavior (i.e., "smart" biomaterial). Smart biomaterial should be able to mimic the properties of the target tissue, and it should be biocompatible, biodegradable in a tunable manner, and bioactive in a controllable way. Given the diversity of the tissues that are being targeted with dental tissue engineering, it is difficult to define whether this smart biomaterial should be a cell-based or a cell-free scaffold. Nevertheless, from a manufacturing and clinical point of view, cell-free scaffolds that not only target the local cells but are able to attract sufficient endogenous cells of the correct lineages from the blood circulation toward the defect site ("cell homing") appear to be more promising for the future.

With respect to the models for testing dental tissue engineering scaffolds, it is evident that no model can adequately fulfill all criteria to serve as a universal ideal experimental setting. The choice of a suitable model depends heavily on the research question, as illustrated above. Meanwhile, the results obtained must be interpreted strictly in the context of the model designed, and great care should be taken when applying the extrapolated data into humans.

REFERENCES

Araujo MG, Lindhe J (1998). GTR treatment of degree III furcation defects following application of enamel matrix proteins: an experimental study in dogs. *J Clin Periodontol* 25(6):524–530.

Araujo MG, Lindhe J (2011). Socket grafting with the use of autologous bone: an experimental study in the dog. *Clin Oral Implants Res* 22(1):9–13.

Balabanian CA, Coutinho-Netto J, Lamano-Carvalho TL, Lacerda SA, Brentegani LG (2006). Biocompatibility of natural latex implanted into dental alveolus of rats. *J Oral Sci* 48:201–205.

Benatti BB, Neto JB, Casati MZ, Sallum EA, Sallum AW, Nociti FH, Jr. (2006). Periodontal healing may be affected by aging: a histologic study in rats. *J Periodontal Res* 41(4):329–333.

Berendsen AD, Smit TH, Walboomers XF, Everts V, Jansen JA, Bronckers AL (2009). Three-dimensional loading model for periodontal ligament regeneration in vitro. *Tissue Eng C* 15(4):561–570.

Cancedda R, Giannoni P, Mastrogiacomo M (2007). A tissue engineering approach to bone repair in large animal models and in clinical practice. *Biomaterials* 28(29):4240–4250.

Casagrande L, Cordeiro MM, Nor SA, Nor JE (2011). Dental pulp stem cells in regenerative dentistry. *Odontology* 99(1):1–7.

Choi SY, Nilveus RE, Minutello RD, Zimmerman GJ, Wikesjo UM (1993). Effect of a collagen matrix on healing in periodontal fenestration defects in dogs. *J Periodontol* 64(9):878–882.

Coimbra P, Alves P, Valente TAM, Santos R, Correia IJ, Ferreira P (2011). Sodium hyaluronate/chitosan polyelectrolyte complex scaffolds for dental pulp regeneration: synthesis and characterization. *Int J Biol Macromol* 49(4):573–579.

Cordeiro MM, Dong Z, Kaneko T, Zhang Z, Miyazawa M, Shi S, et al. (2008). Dental pulp tissue engineering with stem cells from exfoliated deciduous teeth. *J Endod* 34(8):962–969.

Deliberador TM, Nagata MJ, Furlaneto FA, Melo LG, Okamoto T, Sundefeld ML, et al. (2006). Autogenous bone graft with or without a calcium sulfate barrier in the treatment of class II furcation defects: a histologic and histometric study in dogs. *J Periodontol* 77(5):780–789.

De Kok IJ, Drapeau SJ, Young R, Cooper LF (2005). Evaluation of mesenchymal stem cells following implantation in alveolar sockets: a canine safety study. *Int J Oral Maxillofac Implants* 20:511–518.

Derong Z, Lian G, Jiayu L, Xiuli Z, Zhiyuan Z, Xinquan J (2010). Anatomic and histological analysis in a goat model used for maxillary sinus floor augmentation with simultaneous implant placement. *Clin Oral Implants Res* 21(1):65–70.

Donos N, Kostopoulos L, Tonetti M, Karring T (2005). Long-term stability of autogenous bone grafts following combined application with guided bone regeneration. *Clin Oral Implants Res* 16(2):133–139.

Duailibi MT, Duailibi SE, Young CS, Bartlett JD, Vacanti JP, Yelick PC (2004). Bioengineered teeth from cultured rat tooth bud cells. *J Dent Res* 83:523–528.

Felix Lanao RP, Leeuwenburgh SC, Wolke JG, Jansen JA (2011). In vitro degradation rate of apatitic calcium phosphate cement with incorporated PLGA microspheres. *Acta Biomater* 7(9):3459–3468.

Galler KM, Cavender AC, Koeklue U, Suggs LJ, Schmalz G, D'Souza RN (2011a). Bioengineering of dental stem cells in a PEGylated fibrin gel. *Regen Med* 6(2):191–200.

Galler KM, D'Souza RN, Hartgerink JD, Schmalz G (2011b). Scaffolds for dental pulp tissue engineering. *Adv Dent Res* 23(3):333–339.

Gronthos S, Mankani M, Brahim J, Robey PG, Shi S (2000). Postnatal human dental pulp stem cells (DPSCs) in vitro and in vivo. *Proc Natl Acad Sci USA* 97(25):13625–13630.

Gunatillake PA, Adhikari R (2003). Biodegradable synthetic polymers for tissue engineering. *Eur Cell Mater* 5:1–16.

Habraken WJEM, Wolke JGC, Mikos AG, Jansen JA (2006). Injectable PLGA microsphere/calcium phosphate cements: physical properties and degradation characteristics. *J Biomater Sci Polym Ed* 17(9):1057–1074.

Hacking SA, Khademhosseini A (2009). Applications of microscale technologies for regenerative dentistry. *J Dent Res* 88(5):409–421.

Heywood BR, Appleton J (1984). The ultrastructure of the rat incisor odontoblast in organ culture. *Arch Oral Biol* 29(4):327–329.

Hovey LR, Jones AA, McGuire M, Mellonig JT, Schoolfield J, Cochran DL (2006). Application of periodontal tissue engineering using enamel matrix derivative and a human fibroblast-derived dermal substitute to stimulate periodontal wound healing in class III furcation defects. *J Periodontol* 77(5):790–799.

Huang GT, Yamaza T, Shea LD, Djouad F, Kuhn NZ, Tuan RS, et al. (2010). Stem/progenitor cell-mediated de novo regeneration of dental pulp with newly deposited continuous layer of dentin in an in vivo model. *Tissue Eng A* 16(2):605–615.

Huang KK, Shen C, Chiang CY, Hsieh YD, Fu E (2005). Effects of bone morphogenetic protein-6 on periodontal wound healing in a fenestration defect of rats. *J Periodont Res* 40(1):1–10.

Ikeda E, Morita R, Nakao K, Ishida K, Nakamura T, Takano-Yamamoto T, et al. (2009). Fully functional bioengineered tooth replacement as an organ replacement therapy. *Proc Natl Acad Sci USA* 106(32):13475–13480.

Ji QX, Deng J, Xing XM, Yuan CQ, Yu XB, Xu QC, et al. (2010). Biocompatibility of a chitosan-based injectable thermosensitive hydrogel and its effects on dog periodontal tissue regeneration. *Carbohydr Polym* 82(4):1153–1160.

Ji W, Sun Y, Yang F, van den Beucken J, Fan MW, Chen Z, et al. (2011). Bioactive electrospun scaffolds delivering growth factors and genes for tissue engineering applications. *Pharm Res* 28(6):1259–1272.

Kim CS, Choi SH, Chai JK, Cho KS, Moon IS, Wikesjo UM, et al. (2004). Periodontal repair in surgically created intrabony defects in dogs: influence of the number of bone walls on healing response. *J Periodontol* 75(2):229–235.

Kim NR, Lee DH, Chung PH, Yang HC. (2009). Distinct differentiation properties of human dental pulp cells on collagen, gelatin, and chitosan scaffolds. *Oral Surg Oral Med Oral Pathol Oral Radiol Endod* 108:e94–100.

King GN, King N, Cruchley AT, Wozney JM, Hughes FJ (1997). Recombinant human bone morphogenetic protein- 2 promotes wound healing in rat periodontal fenestration defects. *J Dent Res* 76(8):1460–1470.

Klijn RJ, Hoekstra JW, Van Den Beucken JJ, Meijer GJ, Jansen JA (2012). Maxillary sinus augmentation with microstructured tricalcium phosphate ceramic in sheep. *Clin Oral Implants Res* 23:274–280.

Kostopoulos L, Karring T (1994). Augmentation of the rat mandible using guided tissue regeneration. *Clin Oral Implants Res* 5(2):75 82.

Kumada Y, Zhang S. Significant type I and type III collagen production from human periodontal ligament fibroblasts in 3D peptide scaffolds without extra growth factors. *Plos One* 5:e10305.

Kwon DH, Bennett W, Herberg S, Bastone P, Pippig S, Rodriguez NA, Susin C, Wikesjo UM (2010). Evaluation of an injectable rhGDF-5/PLGA construct for minimally invasive periodontal regenerative procedures: a histological study in the dog. *J Clin Periodontol* 37:390–397.

Le Guehennec L, Goyenvalle E, Aguado E, Houchmand-Cuny M, Enkel B, Pilet P, et al. (2005). Small-animal models for testing macroporous ceramic bone substitutes. *J Biomed Mater Res B* 72B(1):69–78.

Liebschner MAK (2004). Biomechanical considerations of animal models used in tissue engineering of bone. *Biomaterials* 25(9):1697–1714.

Lioubavina-Hack N, Karring T, Lynch SE, Lindhe J (2005). Methyl cellulose gel obstructed bone formation by GBR: an experimental study in rats. *J Clin Periodontol* 32(12):1247–1253.

Liu Y, Zheng Y, Ding G, Fang DJ, Zhang CM, Bartold PM, Gronthos S, Shi ST, Wang SL (2008). Periodontal ligament stem cell-mediated treatment for periodontitis in miniature swine. *Stem Cells* 26:1065–1073.

Marei MK, Nouh SR, Saad MM, Ismail NS (2005). Preservation and regeneration of alveolar bone by tissue-engineered implants. *Tissue Eng* 11(5–6):751–767.

Miura M, Gronthos S, Zhao M, Lu B, Fisher LW, Robey PG, Shi S (2003). SHED: stem cells from human exfoliated deciduous teeth. *Proc Natl Acad Sci USA* 100:5807–5812.

Mooney DJ, Powell C, Piana J, Rutherford B (1996). Engineering dental pulp-like tissue in vitro. *Biotechnol Prog* 12(6):865–868.

Muschler GF, Raut VP, Patterson TE, Wenke JC, Hollinger JO (2010). The design and use of animal models for translational research in bone tissue engineering and regenerative medicine. *Tissue Eng* 16(1):123–145.

Nakao K, Morita R, Saji Y, Ishida K, Tomita Y, Ogawa M, Saitoh M, Tomooka Y, Tsuji T (2007). The development of a bioengineered organ germ method. *Nat Methods* 4:227–230.

Nemcovsky CE, Zahavi S, Moses O, Kebudi E, Artzi Z, Beny L, et al. (2006). Effect of enamel matrix protein derivative on healing of surgical supra-infrabony periodontal defects in the rat molar: a histomorphometric study. *J Periodontol* 77(6):996–1002.

Oortgiesen DA, Plachokova AS, Geenen C, Meijer GJ, Walboomers XF, van den Beucken JJ, Jansen JA (2012a). Alkaline phosphatase immobilization onto Bio-Gide(R) and Bio-Oss(R) for periodontal and bone regeneration. *J Clin Periodontol* 39:546–555.

Oortgiesen DA, Yu N, Bronckers AL, Yang F, Walboomers XF, Jansen JA. (2012b). A three-dimensional cell culture model to study the mechano-biological behavior in periodontal ligament regeneration. *Tissue Eng C Methods*. 18:81–89.

Pavlin D, Gluhak-Heinrich J (2001). Effect of mechanical loading on periodontal cells. *Crit Rev Oral Biol Med* 12(5):414–424.

Pellegrini G, Seol YJ, Gruber R, Giannobile WV (2009). Preclinical Models for oral and periodontal reconstructive therapies. *J Dent Res* 88(12):1065–1076.

Prescott RS, Alsanea R, Fayad MI, Johnson BR, Wenckus CS, Hao J, John AS, George A (2008). In vivo generation of dental pulp-like tissue by using dental pulp stem cells, a collagen scaffold, and dentin matrix protein 1 after subcutaneous transplantation in mice. *J Endod* 34:421–426.

Rajshankar D, McCulloch CA, Tenenbaum HC, Lekic PC (1998). Osteogenic inhibition by rat periodontal ligament cells: modulation of bone morphogenic protein- 7 activity in vivo. *Cell Tissue Res* 294(3):475–483.

Reichert JC, Saifzadeh S, Wullschleger ME, Epari DR, Schutz MA, Duda GN, et al. (2009). The challenge of establishing preclinical models for segmental bone defect research. *Biomaterials* 30(12):2149–2163.

Sakai VT, Zhang Z, Dong Z, Neiva KG, Machado MA, Shi S, et al. (2010). SHED differentiate into functional odontoblasts and endothelium. *J Dent Res* 89(8):791–796.

Schmitz JP, Hollinger JO (1986). The critical size defect as an experimental model for craniomandibulofacial nonunions. *Clin Orthop Relat Res* 205:299–308.

Sculean A, Berakdar M, Pahl S, Windisch P, Brecx M, Reich E, et al. (2001). Patterns of cytokeratin expression in monkey and human periodontium following regenerative and conventional periodontal surgery. *J Periodont Res* 36(4):260–268.

Sculean A, Junker R, Donos N, Berakdar M, Brecx M, Dünker N (2002). Immunohistochemical evaluation of matrix molecules associated with wound healing following regenerative periodontal treatment in monkeys. *Clin Oral Investig* 6(3):175–182.

Sculean A, Nikolidakis D, Schwarz F (2008). Regeneration of periodontal tissues: combinations of barrier membranes and grafting materials—biological foundation and preclinical evidence: a systematic review. *J Clin Periodontol* 35(8 Suppl):106–116.

Shang S, Yang F, Cheng X, Walboomers XF, Jansen JA (2010). The effect of electrospun fibre alignment on the behaviour of rat periodontal ligament cells. *Eur Cell Mater* 19:180–192.

Shirakata Y, Oda S, Kinoshita A, Kikuchi S, Tsuchioka H, Ishikawa I (2002) Histocompatible healing of periodontal defects after application of an injectable calcium phosphate bone cement. A preliminary study in dogs. *J Periodontol* 73:1043–1053.

Slavkin HC, Bartold PM (2006). Challenges and potential in tissue engineering. *Periodontol 2000* 41(1):9–15.

Sonoyama W, Liu Y, Fang D, Yamaza T, Seo BM, Zhang C, Liu H, Gronthos S, Wang CY, Wang S, Shi S (2006). Mesenchymal stem cell-mediated functional tooth regeneration in swine. Plos One 1:e79.

Strietzel FP, Khongkhunthian P, Khattiya R, Patchanee P, Reichart PA (2006). Healing pattern of bone defects covered by different membrane types: a histologic study in the porcine mandible. *J Biomed Mater Res B* 78B(1):35–46.

Suzuki S, Mitani A, Koyasu K, Oda SI, Yoshinari N, Fukuda M, et al. (2006). A model of spontaneous periodontitis in the miniature goat. *J Periodontol* 77(5):847–855.

Sverzut CE, Faria PE, Magdalena CM, Trivellato AE, Mello-Filho FV, Paccola CA, et al. (2008). Reconstruction of mandibular segmental defects using the guided-bone regeneration technique with polylactide membranes and/or autogenous bone graft: a preliminary study on the influence of membrane permeability. *J Oral Maxillofac Surg* 66(4):647–656.

Tal H, Artzi Z, Moses O, Nemcovsky C, Kozlovsky A (2005). Guided periodontal regeneration using bilayered collagen membranes and bovine bone mineral in fenestration defects in the canine. *Int J Periodont Restor Dent* 25(5):509–518.

Wazen RM, Lefebvre LP, Baril E, Nanci A (2010). Initial evaluation of bone ingrowth into a novel porous titanium coating. *J Biomed Mater Res B* 94B(1):64–71.

Weng Y, Wang M, Liu W, Hu X, Chai G, Yan Q, Zhu L, Cui L, Cao Y (2006). Repair of experimental alveolar bone defects by tissue-engineered bone. *Tissue Eng* 12:1503–1513.

Wikesjo UM, Nilveus R (1991). Periodontal repair in dogs: healing patterns in large circumferential periodontal defects. *J Clin Periodontol* 18(1):49–59.

Wikesjo UM, Qahash M, Thomson RC, Cook AD, Rohrer MD, Wozney JM, et al. (2004). rhBMP- 2 significantly enhances guided bone regeneration. *Clin Oral Implants Res* 15(2):194–204.

Wikesjo UM, Xiropaidis AV, Qahash M, Lim WH, Sorensen RG, Rohrer MD, et al. (2008). Bone formation at recombinant human bone morphogenetic protein- 2-coated titanium implants in the posterior mandible (type II bone) in dogs. *J Clin Periodontol* 35(11):985–991.

Xu WP, Zhang W, Asrican R, Kim HJ, Kaplan DL, Yelick PC (2008). Accurately shaped tooth bud cell-derived mineralized tissue formation on silk scaffolds. *Tissue Eng Part A* 14:549–557.

Yang Y, Rossi FM, Putnins EE (2010). Periodontal regeneration using engineered bone marrow mesenchymal stromal cells. *Biomaterials* 31(33):8574–8582.

Young CS, Abukawa H, Asrican R, Ravens M, Troulis MJ, Kaban LB, Vacanti JP, Yelick PC (2005). Tissue engineered hybrid tooth and bone. *Tissue Eng* 11:1599–1610.

Young S, Bashoura AG, Borden T, Baggett LS, Jansen JA, Wong M, et al. (2008). Development and characterization of a rabbit alveolar bone nonhealing defect model. *J Biomed Mater Res A* 86A(1):182–194.

Zhang W, Walboomers XF, Wolke JGC, Bian Z, Fan MW, Jansen JA (2005). Differentiation ability of rat postnatal dental pulp cells in vitro. *Tissue Eng* 11(3–4):357–368.

24

WHOLE-TOOTH ENGINEERING AND CELL SOURCES

L. KELLER, S. KUCHLER-BOPP, AND HERVÉ LESOT

INSERM UMR 1109, Team "Osteoarticular and Dental Regenerative NanoMedicine," Strasbourg, France; and Faculté de Chirurgie Dentaire, Université de Strasbourg, Strasbourg, France

1 INTRODUCTION

Tooth engineering is a recently emerged field of research combining the principles of modern tissue engineering, developmental biology, and cell biology. Initially using dental epithelial and ecto-mesenchymal cells, two distinct strategies have been developed to test the engineering and regeneration of a whole tooth. One is based on the use of preshaped scaffolds of a tooth form seeded with cells from developing teeth (Young et al., 2005; Yildirim et al., 2011). The other is not using scaffolds and is at present the method most frequently used (Ohazama et al., 2004; Hu et al., 2005a,b 2006a; Nakao et al., 2007; Honda et al., 2008; Arany et al., 2009). The absence of scaffolds allows direct interactions between dental epithelial and mesenchymal cells and the reestablishment of a continuous epithelial–mesenchymal interface, which avoids the disruption of cellular signals and their gradients. This is important for the subsequent successive steps of odontogenesis to take place in a manner similar or identical to the natural odontogenic process.

Regarding whole-tooth engineering, challenges for the successful re-formation of the epithelial and the ecto-mesenchymal compartments are quite different. For the dental epithelium–enamel organ, it is necessary to achieve complex histogenesis because at least five different cell types contribute to this tissue, some with a very transient but fundamental role in epithelial–mesenchymal interactions (Lesot and Brook, 2009). It is much more difficult to analyze and understand the organization and heterogeneity of cells in the mesenchymal compartment. Immunostaining for cell surface markers

Stem Cells in Craniofacial Development and Regeneration, First Edition.
Edited by George T.-J. Huang, Irma Thesleff.

shows the coexistence of many different cell types while having similar histological features (Fig. 1B, C, E, and F). Although it can be expected that during odontogenesis this cellular heterogeneity is specifically controlled in a temporal–spatial order, there is still very little information on how this is regulated. The origin of dental papilla cells is also difficult to determine since the contribution of the dental sac (follicle) is difficult to trace at early stages. Furthermore, "exogenous" cells are incorporated into the dental papilla, such as endothelial and neuronal cells, as well as other mesenchymal cells at later stages (Chai et al., 2000; Nait Lechguer et al., 2008; Feng et al., 2011; Rothová et al., 2011). This explains why, in contrast to epithelial histogenesis, the restoration of cell heterogeneity in the mesenchymal compartment remains a specific challenge (Nakada et al., 2011).

In this chapter we overview the specific requirements for engineering a whole organ, the search for cell sources allowing the re-formation of two very different tissues, each having either specific histogenesis or complex cell heterogeneity. Open questions concerning the initial specification of the dental epithelial–mesenchymal junction and maintenance of the developmental potential of the cells during expansion in vitro are also discussed.

2 WHOLE-TOOTH ORGAN ENGINEERING USING EMBRYONIC DENTAL CELLS

Whole-tooth organ engineering has been achieved successfully in several labs, mostly using the mouse first lower molar as a model, despite its composite origin and complex formation during embryonic development (Prochazka et al., 2010). Beyond what is generally expected from tissue engineering, organ engineering supposes the concerted development of the two dental tissues to form the crown. It is necessary to reproduce the specific geometry of the coordinated secretion of predentin–dentin by odontoblasts and enamel by ameloblasts. Maintenance of the gradient of cell differentiation, starting from cusp tips to bottoms in intercuspal regions and proceeding toward the root, is necessary to allow the crown growth in regions where functional cell differentiation is retarded (Fig. 1A and D). These differentiation gradients will also determine the formation of a correct dentin–enamel junction (Honda et al., 2008; Nait Lechguer et al., 2011). This is particularly important because dentin and enamel (the two mineralized matrices) have quite distinct biomechanical properties (Imbeni et al., 2005). The gradients of cell differentiation form reproducibly when using a two-step strategy: Dissociated cells from cap-stage molar epithelium and mesenchyme are cultured in vitro for 8 days (referred to below as cell reassociation) and then implanted under the skin of adult mice for 2 weeks (Fig. 1D; for a review, see Kuchler-Bopp et al., 2011). However, the innervation of such implants has not yet been achieved (Fig. 1G to L). Although absolutely required for tooth functionality, innervation is blocked when using dental mesenchymal cells from embryonic day (ED)14 embryos (Fig. 1I). Even when a trigeminal ganglion was implanted next to a cell reassociation, nerve fibers did not enter the dental mesenchyme, whereas they grew well into peridental tissues (Fig. 1L). This fits with what is known from tooth innervation during development. It is a late event requiring the expression of factors attracting axons (Ngf, Gdnf, Net3, Ncam, etc.) as well as decreased expression of molecules with repulsive effects on axonal growth, such as semaphorin3A (Lillesaar and Fried, 2004; Luukko et al., 2005). Apparently, these molecular requirements

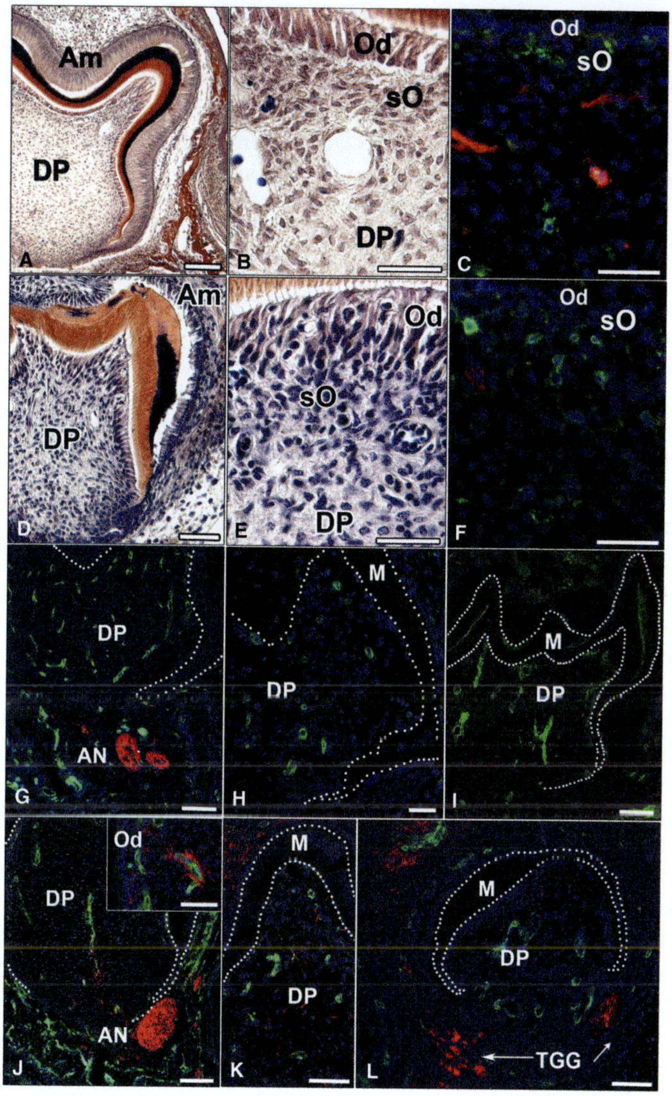

FIGURE 1 Mesenchymal cell heterogeneity and innervation in a mouse first lower molar at ED18 (G) and at postnatal day (PN)4 (A to C, H, J, K) compared to cultured reassociations of epithelial and mesenchymal cells from ED14 molar implanted for two weeks under the skin of adult mice (D to F, I, L). The cellular heterogenity in the dental mesenchyme is visible from histology (A, B, D, and E) and immunostaining for surface markers: CD146 (red) and CD45 (green) (C and F). At ED18, the molar is vascularized as seen after immunostaining for CD31 (green), but not innervated: no red staining for peripherin in dental papilla (G). At PN4, nerve fibers have entered the dental pulp (J) and reach the odontoblast layer (box in J). However, this is no longer detectable when a PN4 molar has been implanted for 2 weeks (H). When trigeminal ganglia are implanted next to a PN4 molar, fibers positive for peripherin are visible in the dental pulp (K). After 2 weeks, implanted cell reassociations of E14 teeth are vascularized but not innervated (I). The coimplantation of trigeminal ganglia next to the cell–cell reassociation does not improve the innervation (L). Am, ameloblast; AN, alveolar nerve; DP, dental pulp; M, mineralized matrix, Od, odontoblast; sO, sub-odontoblast cells; TGG, trigeminal ganglion. Scale bars: 40 μm (B, C, E, F, and H); 100 μm (A, D, G, I to L).

are not fulfilled in the implanted reassociations of cells from cap-stage tooth germs (Kuchler-Bopp et al., 2011). It will be interesting to determine why vascularization but not innervation can develop in implanted cell reassociations since the two events share common determinants (Arese et al., 2011; Rothová et al., 2011).

When mesenchymal cells were dissociated from tooth germs at different stages, their odontogenic potential decreased from the cap stage (ED14) to the early bell stage (ED16) and was lost at ED18 (Keller et al., 2011). However, it is not yet clear whether this progressive loss of the odontogenic potential reflects a change in cell heterogeneity in the dental mesenchyme during development. It has been documented that the balance between neural crest–and non-neural crest–derived cell populations changes in the dental mesenchyme during advancing development (Rothová et al., 2011). This is related to the vascularization of the dental papilla and is of major interest for tooth engineering by showing that cells with a nondental origin may participate in dental pulp formation. It was reported that tooth germs could be obtained by combining cloned cells derived from ED18 dental mesenchyme with ED14.5 dental epithelium (Arany et al., 2009). Confirmation of this work will be necessary since it raises the question of how cell heterogeneity is restored in the dental mesenchyme derived from a cell clone. One possible explanation is that exogenous stem cells participated in tooth formation since the implanted cell reassociations are vascularized. This has been observed in the case of dental pulp regeneration (Volponi et al., 2010).

One possible explanation for the decreasing odontogenic potential of dental mesenchymal cells between ED14 and ED18 may be related to tissue differentiation (Keller et al., 2011). Indeed, during the first few hours after reassociation, the epithelial and mesenchymal cells form a specific odontogenic junction allowing the formation of a bud progressing to the cap and bell stages before cell differentiation is initiated (Hu et al., 2005a, b). It might thus be easier for cells from ED14 than for cells from ED18 tooth germs to acquire earlier identity, allowing bud morphogenesis. This was supported by previous observations showing that compared to cells from ED14, reassociations performed with cells from ED13 tooth germs allowed a faster elongation of the epithelial cells to form the prospective inner dental epithelium (Hu et al., 2005b). For these reasons, it is not clear how cells cloned from ED18 dental mesenchyme can initiate tooth formation when reassociated with a cap-stage epithelium (Arany et al., 2009).

In most cases, whole-tooth organ engineering was performed using cells from embryonic tooth germs (Hu et al., 2006a; Nakao et al., 2007; Honda et al., 2008; Arany et al., 2009). Although the embryonic dental cells obviously do not represent a realistic cell source for tooth engineering in a clinical context, the research has generated valuable information about the design of a convenient methodology to achieve critical steps, such as vascularization, matrix organization, and root formation. The use of embryonic dental cells has also revealed specific endogenous constraints of the system itself.

To meet clinical requirements, attempts are being made to replace the embryonic dental cells by other cell sources more easily available and present in the adult (Fig. 2). Stem cells have already been used in attempts to engineer specific dental tissues: for example, dentin, enamel, or root. However, whole-tooth engineering continues to require the use of embryonic dental tissue or cells from at least one of the two dental components in order to induce the other nondental cells to become odontogenic (Ohazama et al., 2004; Honda et al., 2008).

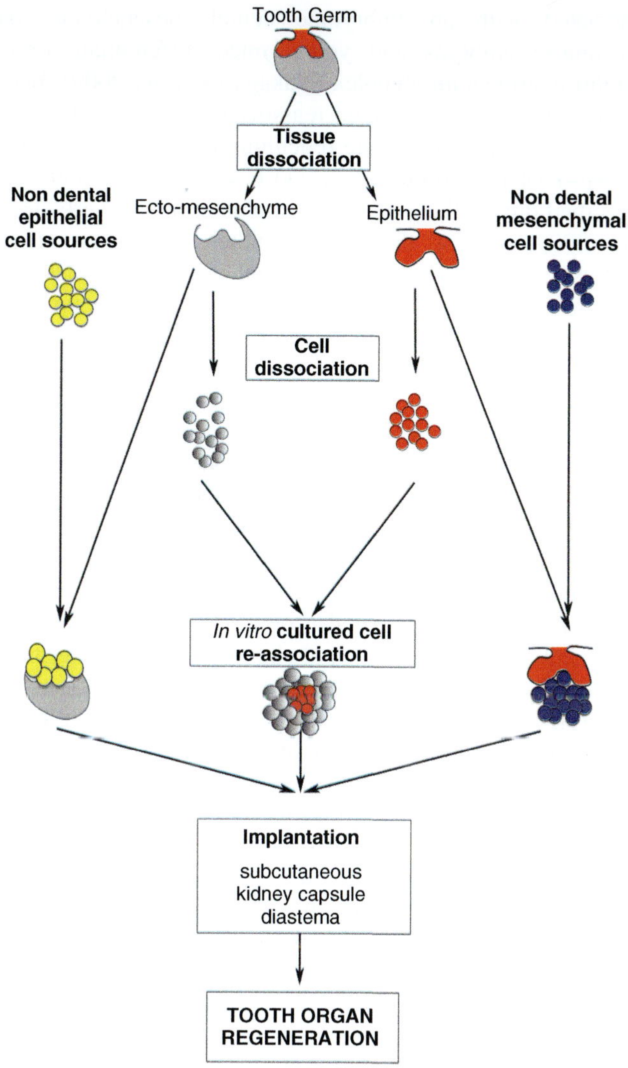

FIGURE 2 Schematic presentation of testing the potential of nondental mesenchymal and epithelial cells to form teeth in reassociation with an intact embryonic dental epithelium or mesenchyme. Reassociations are cultured in vitro and implanted. Reassociations of dental epithelium and mesenchyme are used as positive controls for whole-tooth engineering.

3 CELL SOURCES FOR ENGINEERING OF THE ENAMEL ORGAN

Unlike the dental papilla, the enamel organ disappears before tooth eruption, so that the search for alternative epithelial cell sources is still more important than for the mesenchymal cell sources. Several complementary approaches have been designed using dental and nondental cells or cell lines from embryonic as well as adult sources.

When reassociated with an embryonic dental mesenchyme, palatal epithelial cells from mouse embryos and young mice differentiate into a functional ameloblast–stratum intermedium complex (Nakagawa et al., 2009). Because the early stages of development of these implanted reassociations were not documented, many interesting questions remain: Did stellate reticulum and outer dental epithelium form; how did the crown–root junction develop; and how was the dental versus peridental mesenchyme specified during the 2 weeks of implantation? Nevertheless, the work by Nakagawa et al. (2009) suggests that palatal epithelial cells may respond to signals from the mesenchyme and also that this potential is maintained after the cells have been cultured in vitro. They detected competent epithelial cells up to postnatal (PN) day 1, but cells from later stages (up to 4 weeks PN) showed odontogenic properties when grown in vitro, prior to reassociation. This was attributed to a preferential expansion of undifferentiated epithelial cells (Nakagawa et al., 2009).

The maintenance of odontogenic potential in vitro has also been reported for epithelial cell lines derived from embryonic mouse oral epithelium (Takahashi et al., 2010). Although these cell lines may not represent a cell source for tooth bioengineering, their study may generate information about the characteristics of oral epithelial cells in vitro as well as their manipulation and expansion. Prior to their reassociation with ED16.5 molar mesenchymes, the epithelial cells were cultured on a feeder layer of Balb/c 3T3 cells for 7 to 14 days. As also observed with cultured dental epithelial cells (Honda et al., 2009), the nondental epithelial cell lines could respond to signaling from a molar mesenchyme and gave rise to ameloblasts (Takahashi et al., 2010). It thus suggests a surprising property for these cell lines. Indeed, similar to what occurs during tooth development, the differentiation of ameloblasts in these reassociations first required the induction of the differentiation of odontoblasts in the ED16.5 mesenchyme. Interestingly, the existence of a gradient of differentiation in these reassociations was suggested in one picture (Fig. 7.1 in Takahashi et al., 2010). This raises the question of how an initially homogeneous epithelial cell layer can specify or even only maintain the cell kinetic parameters in neighboring preodontoblasts required to set up gradual differentiation (Ruch et al., 1995). Experiments performed by Wang et al. (2010) report the formation of tooth structures in reassociations between human keratinocytes and molar mesenchyme from ED13.5 mouse embryos implanted under the mouse kidney capsule. Although the early stages of the implantation were not documented, the successful use of primary cultures of human keratinocytes might represent an easy way to obtain cells to engineer the epithelial compartment of a tooth.

However, although the various nondental epithelial cells can engage in the formation of dental tissues even after culture in vitro, there is no evidence that they could be used to form a complete functional tooth. When reassociated with competent dental mesenchyme, none of these cells allowed the initiation of root formation. More work will be necessary to better design the type of reassociations to be performed for that purpose and to determine the limiting parameters or conditions.

4 MESENCHYMAL CELL SOURCES

Several potential mesenchymal cell sources have been considered for tissue engineering, including dental and peridental mesenchymal stem cells, bone marrow–derived cells (BMDCs), and more recently, induced pluripotent stem cells (iPSCs) (Ohazama et al., 2004; Huang et al., 2009; Oda et al., 2010; Kuchler-Bopp et al., 2011; Otsu et al., 2011).

4.1 Dental and Peridental Mesenchymal Stem Cells

In the dental and peridental mesenchymes, several different cell types have been identified as stem cells [for a review, see Huang et al. (2009), Casagrande et al. (2011), and Kuchler-Bopp et al. (2011)]. These cells have been investigated for specific properties, including proliferation, clonogenicity, and the ability to be induced toward different types of differentiation. However, as pointed out by Bianco (2011), many stem cells have been characterized based on in vitro tests, although their real functions in vivo remain largely unknown.

Dental and peridental mesenchymes have been tested for their ability to produce dentin or cementum in different experimental conditions (Sonoyama et al., 2006; Huang et al., 2009). Dental mesenchymal stem cells can also engage in reparative processes. When reparative dentinogenesis was stimulated, long-distance cell migration was induced to finally give rise to odontoblast-like cells (Farges et al., 2003). The migrating cells were characterized as immature antigen-presenting dendritic cells, but their possible stemness was not investigated. These observations became topical recently as genetic lineage tracing showed that in response to tissue damage, pericytes as well as another population of mesenchymal stem cells migrated and differentiated into odontoblasts (Feng et al., 2011). There is evidence that stem cells may also participate in tissue repair indirectly by producing cytokines and chemokines and stimulating differentiation of endogenous cells (Prockop, 2007). The question remains whether these different stem cell populations may engage in early steps in tooth development such as the specification of a dental epithelial–mesenchymal junction or the induction of a primary enamel knot. Indeed, under physiological conditions, blood vessels enter the dental mesenchyme at the bell stage (Nait Lechguer et al., 2008; Rothová et al., 2011), and therefore pericytes and the perivascular niche appear quite late in the dental papilla.

4.2 Origin of Stem Cells in the Mesenchyme

The origin of dental mesenchymal stem cells is not clear (Huang et al., 2009; Bianco, 2011). During tooth development, changes in the heterogeneity of dental pulp cells may involve the participation of non-neural crest–derived cells (Chai et al., 2000). This may result partially from vascularization, bringing circulating progenitor cells to the forming tooth. However, it seems that the number of circulating mesenchymal progenitor cells is very low (Huang et al., 2009).

The migration of cells from peridental to dental mesenchyme during tooth development may be more important (Rothová et al., 2011). Although the two mesenchymal tissues have the same developmental origin, they have strikingly different properties at the cap and bell stages, as seen by the different chronology for both their vascularization and innervation. Stem cells are present in both dental and peridental mesenchyme (Gronthos et al., 2000; Huang, 2009; Balic et al., 2010; Kuchler-Bopp et al., 2011). However, the comparison between the two sets of stem cells is not easy since differences may either reflect intrinsic cell specificities or result from their interaction with different environments. Such an effect of the environment has been illustrated when using bone marrow cells (Hu et al., 2006b). Bone marrow cells from adult mice have been mixed with dental epithelial cells, pelleted and reassociated with a trypsin-isolated dental mesenchyme from ED14 mouse molar. At the onset of the reassociation, the basement membrane had been hydrolyzed so that bone marrow cells could either stay in the epithelium or migrate into the mesenchyme. Depending on their position on one or the other side of the new basement membrane which formed within the next few hours, bone marrow cells differentiated to give rise to either functional odontoblasts expressing DSP/DPP and DMP1 or to ameloblasts expressing amelogenin and MMP20 (Hu et al., 2006b). Stem cells can thus sense their environment and adjust to it, as also reported in the context of reparative processes (Feng et al., 2011). Altogether, these data allow reinterpreting cellular aspects of reparative dentinogenesis where the migration of progenitor cells had been suggested although they had not been characterized (Smith and Lesot, 2001). Parallel studies have shown that dental mesenchymal stem cells can be induced to give rise to odontoblast-like cells (Gronthos et al., 2000; Huang et al., 2009; Feng et al., 2011).

The existence of dental mesenchymal stem cells has been shown in a large number of publications, and their potentials to respond to specific inductions in vitro have been demonstrated. However, their functions in vivo are not clear yet and we still need evidence supporting their potential use for tooth engineering.

5 REQUIREMENTS FOR MESENCHYMAL CELLS TO BE USED FOR TOOTH ENGINEERING

The choice of cells to be used for tooth engineering is restricted by biological and technical constraints. Some of the biological constraints were identified using embryonic dental cells since their odontogenic potential depends on the stage of odontogenesis. Other constraints are related to the methodology itself. For example, maintaining odontogenic potential in vitro appears to be much more difficult in mesenchymal cells than in epithelial cells.

5.1 Specification of the Epithelial–Mesenchymal Junction

A major challenge in tooth engineering is the identification of a mesenchymal cell population that would be able to specify the formation of a dental epithelial–mesenchymal junction (i.e., to be able to mimic the initiation of odontogenesis) after recombination with competent epithelial cells. This is probably the key property to be searched for when testing the odontogenic potential of different cell sources, including dental mesenchymal stem cells. However, since the potential may be lost in vitro while expanding the cells, it will be important to investigate the reasons this happens. Possible effects of cell selection and cell adjustment to a new environment in this process are discussed below.

Embryonic mouse molars at the cap stage are most frequently used as the source of dental cells for reassociations (Hu et al., 2005b, 2006a; Nakao et al., 2007; Honda et al., 2008; Arany et al., 2009). After only 12 hours, a basement membrane is restored and the epithelial–mesenchymal junction allowing the development of an epithelial bud and its subsequent morphogenesis (Hu et al., 2005b). It would be important to understand at the molecular level how far the reassociations in vitro do "regress" and to what extent this can be compared to early stages of tooth development (Sarkar et al., 2000; Munne et al., 2009).

The cell reassociations can be cultured in vitro up to the stage of early crown formation (Hu et al., 2005b, 2006a; Nakao et al., 2007; Honda et al., 2008). Their subsequent implantation in adult mice allows their vascularization and thus the possible participation of exogenous stem cells to the formation of the dental papilla. Encouraging results were obtained using bone marrow–derived cells (BMDCs) to replace the mesenchymal compartment (Ohazama et al., 2004). When the BMDCs were reassociated with oral epithelium from ED10 mouse embryos and implanted in the kidney, tooth formation was initiated and reached an advanced stage, with functional odontoblasts and ameloblasts (Ohazama et al., 2004). The results suggest that the early-stage jaw epithelium, which has earlier been shown to have the capacity to form a tooth with nondental neural crest–derived cells (Mina and Kollar, 1987; Lumsden, 1988), is able to instruct and reprogram BMDCs to behave as dental ecto-mesenchymal cells giving rise to odontoblasts. More recent results from the same group raised the question of a possible participation of host cells since the reassociations of E10 jaw epithelium and BMDCs were cultured under the kidney capsule and thus were vascularized. However, it remains true that these conditions allowed the initiation of tooth formation, although the success rate was only 10%. It would be important to understand the reasons for the low percentage of success and what could help to increase this yield. In another context, different subpopulations of bone marrow cells were found to vary in their ability to participate in tooth development, and also to enter the tooth tissues (Hu et al., 2006b). Further experimental work will be necessary to understand the cellular and molecular characteristics of reassociations between BMDCs and oral epithelium as well as the degree of cellular heterogeneity in the mesenchyme at different stages of implantation and its similarity or potential differences with dental papilla mesenchyme. Different populations of stem cells that exist in the epidermis show differences in responsiveness to homeostatic cues and are regulated by a circadian molecular clock (Janich et al., 2011). This illustrates one further aspect of the complexity of stem cell heterogeneity.

5.2 Maintenance of the Odontogenic Potential of Cells In Vitro

This is a critical issue since in most cases cell expansion is a necessary step prior to their use for engineering. Surprisingly, embryonic dental mesenchymal cells from ED14 lose their odontogenic potential in vitro, even when cultured for only 24 hours. The reasons for this loss are still under investigation but might include cell selection due to differential adhesion, a change in cell heterogeneity, and/or a change in the phenotype when cells are grown as monolayers (Keller et al., 2011).

When dissociated dental mesenchymal cells are grown further in vitro, not all cells adhere. These differences in cell adhesion might lead to cell selection, which could explain the absence of tooth formation.

Changes in cell heterogeneity might result from cell death occurring after enzymatic separation of the dental epithelium from mesenchyme and also after dissociation of cells from each tissue. Cell death is transient and ceases rapidly after a new basement membrane has been deposited (Hu et al., 2005a). When reassociated without an intermediary culture step, the surviving cells can form teeth, indicating that they had fully retained their odontogenic potential.

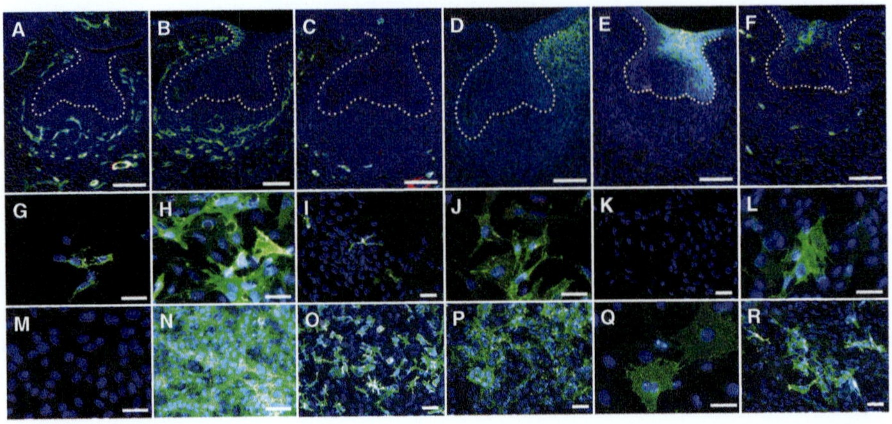

FIGURE 3 Expression of stem cell surface markers in mouse molars at ED14 (A to F) and in cultured dental mesenchymal cells dissociated from ED14 molars after 1 day (G to L) and 4 days (M to R). Immunostainings were performed for CD31 (A, G, M), CD146 (B, H, N), CD45 (C, I, O), CD90 (D, J, P), CD 73 (E, K, Q) and Sca1 (F, L, R). Scale bars: 50 μm (B to D, G to R); 100 μm (A, E, F).

A variety of changes in the phenotypes of cells after culture have been well documented. They include the expression of cell-surface markers (Fig. 3), the composition of the extracellular matrix, and/or cell proliferation. Similarly, changes in these parameters have been observed when comparing dental mesenchymal cells in situ and in primary culture. Such in vitro alterations have been reported for many different cell types, including stem cells, and represent a common technical hurdle to be overcome (Roobrouck et al., 2011; Huber and Badylak, 2011). Cells produce a variety of extracellular matrix constituents in vitro, and a large proportion of them are released to the culture medium while they are incorporated in the extracellular matrix in physiological conditions. This might have consequences on the availability of growth factors in the cultures, since most of them interact with matrix constituents. The microenvironment of the cell includes neighboring cells, extracellular matrix constituents, and diffusible molecules (cytokines, chemokines, growth factors), as well as enzymes involved in matrix remodeling and their inhibitors. The microenvironment has been investigated in detail in specific cases, such as the control of odontoblast differentiation, while much less is known about its functions in other regions of the dental mesenchyme, although it is likely that it plays a major role in the maintenance of mesenchymal cell heterogeneity and homeostasis. New methodologies such as fluorescence resonance energy transfer sensors are being developed to investigate how cells react and adjust to their environment (Zhao et al., 2011). This method might be useful to investigate how a dental mesenchymal cell leaves a specific location, such as the perivascular environment, and engages in repair (Feng et al., 2011) or how it is recruited in this environment and remains there as a progenitor (Bianco, 2011).

Mechanotransduction is one way to adjust to the microenvironment (DuFort et al., 2011). However, except for odontoblasts, very little is known on its possible role during tooth development (Magloire et al., 2010). As part of the microenvironment, specific matrix–membrane–microfilament interactions control odontoblast elongation and polarization (Lesot et al., 1990). Similar mechanisms appear to control stem cells (Guilak et al., 2009). Several lines of evidence have suggested that interfering with the cell shape might modulate the fate of stem cells (McBeath et al., 2004). In most cases, however, data were obtained by means of in vitro systems and could hardly be transposed to the physiological context of a tissue.

Since the epithelium is instructive at a very early stage of tooth development (Lumsden, 1988), it is possible that the mesenchymal stem and progenitor cells first need to be specified by such an epithelium before they can be used successfully for tooth engineering (Volponi et al., 2010). However, it remains difficult to dissociate an epithelium at a very early stage and avoid remaining mesenchymal cells. Such contaminating cells could be sufficient to induce tooth formation, as discussed by Thesleff and Tummers (2008). Since the inductive epithelial signal(s) is(are) not known, an alternative way to bypass these difficulties is to search for culture conditions that allow the maintenance of odontogenic properties by engineering a permissive microenvironment, as suggested for musculoskeletal cells (Hwang et al., 2011).

5.3 Ability to Induce the Formation of a Primary Enamel Knot

Dental mesenchymal cells at ED14 can induce the formation of a primary enamel knot when recombined with dissociated dental epithelial cells from ED14. The step of cell dissociation, although leading to a complete loss of positional information, does not impair tooth reformation in cultured cell–cell reassociations (Hu et al., 2005b). It was shown that the number of dental mesenchymal cells involved in the reassociation could be a limiting factor in the induction of the enamel knot (Hu et al., 2006a).

We performed heterotopic reassociations between an ED14 molar mesenchyme and the epithelium from an ED16 incisor (Fig. 4). To test the possibility to induce the formation of an ectopic enamel knot of the incisor, only the posterior part of the lingual portion was used, which is normally involved in root formation (Fig. 4A to D). After 4 days in vitro, an enamel knot formed, although only in 10% of the reassociations (Fig. 4E). After 10 days, the posterior part of the lingual epithelium had been transformed to an inner dental epithelium with differentiated ameloblasts (Fig. 4F). Although this never happens in physiological conditions, the differentiation of ameloblasts on the lingual side of incisors has been observed in follistatin mutant mice, probably due to enhanced BMP4 function (Wang et al., 2004). The similar presence of ectopic ameloblasts was observed in the incisors of sprouty mutant mice and presumed to result from an upregulation of FGF signaling (Boran et al., 2009). After 10 days in culture, the teeth that formed did not have a correct shape. In contrast to reassociations performed with an incisor epithelium at ED14 (Nait Lechguer et al., 2009), multicuspid teeth were never observed, showing that changes related to the stage of development also exist on the epithelial side. Furthermore, these teeth remained different from incisors in that there was no asymmetry; ameloblasts were present on the entire surface of the tooth (Fig. 4F).

In contrast to embryonic dental mesenchymal cells, BMDCs do not seem to be able to induce a primary enamel knot (Nait Lechguer et al., 2009), which might explain the absence of multicuspid teeth forming in the reassociations performed by Ohazama et al. (2004). This is one parameter that will have to be analyzed when testing the odontogenic competence of various potential mesenchymal cell sources.

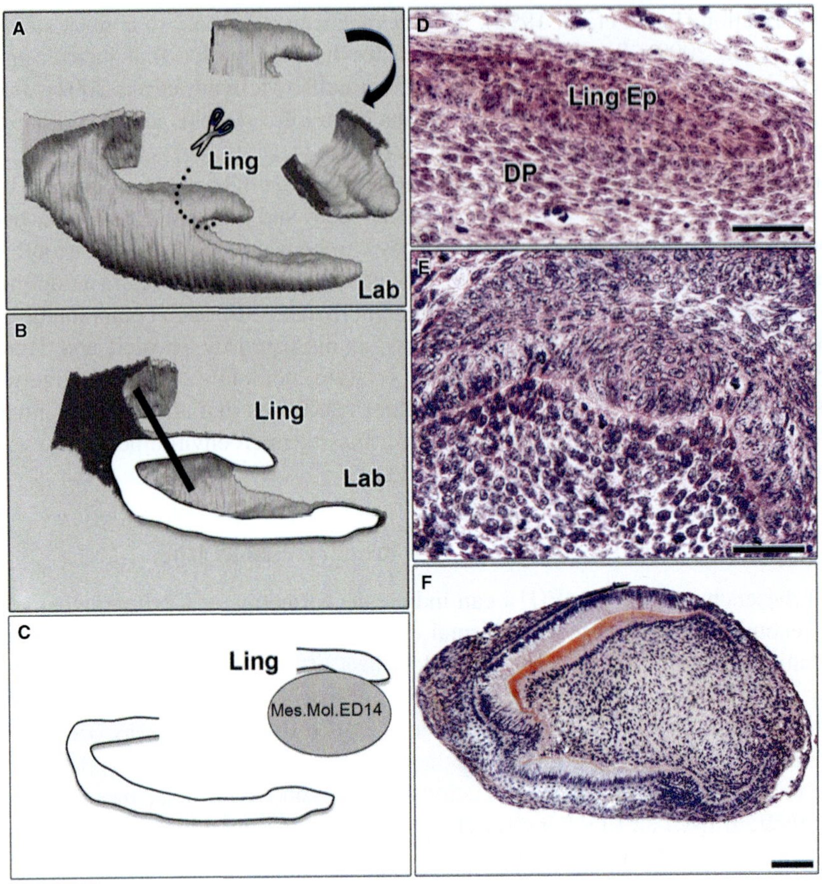

FIGURE 4 Cultures of heterotopic reassociations (E and F) using ED14 molar mesenchyme (C) and the lingual part of an incisor enamel organ at ED16 (A, B, and D). The posterior part of the epithelium was dissected (A and B) to exclude the primary enamel knot (PEK) region and reassociated with an intact molar mesenchyme (C). Before tissue dissociation, only preodontoblasts were in contact with the basement membrane on lingual part of the epithelium (D). After the reassociation has been cultured for 4 days, a PEK has been induced ectopically (E). After 10 days, odontoblasts are functional and secrete predentin which induces ameloblast cytological differentiation (F). Am, ameloblast; DP, dental pulp; IDE, inner dental epithelium; Lab, labial; Ling, lingual; Ling Ep, lingual part of the incisor epithelium; Mes, mesenchyme; Od, odontoblast; Pd, predentin; PEK, primary enamel knot; SR, stellate reticulum. Scale bars: 40 μm (D and E); 80 μm (F).

6 CONCLUSIONS

Although whole-tooth engineering is still far from clinical applications, its study may provide useful information for the engineering of other organs and tissues. The questions concerning positional information (and its nature), chronological adjustments between two interacting tissues, homeostasis, or organ size and shape remain the same in many different tissue models and regenerative processes.

Manipulating epithelial cell sources in the tissue recombination experiments appears much easier than mesenchymal cells. This is probably because experiments were performed at stages where the mesenchyme controls tooth formation, and a certain plasticity in the epithelial cells may be sufficient. Still, it is not known why mesenchymal cells, even competent embryonic ones, lose their differentiation potentials so quickly in vitro. Better knowledge of cellular microenvironment should help to better understand the molecular requirements to maintain cell identities when expanding mesenchymal cell sources. The local specificities of the extracellular environment in the dental versus peridental mesenchyme will have to be investigated not only during embryonic development, but also postnatally when innervation progresses and later when the pulp chamber becomes reduced in size. The possible mesenchymal cell selection that may occur during these late stages is still not known.

Acknowledgments

The authors thank Anne Poliard and Ann Huysseune for critical reading of the manuscript and helpful comments, Jean-Christophe Maurin for his help in dissecting trigeminal ganglions, and Hervé Gegout for the histology. This work was funded by the Dental School at Strasbourg University and by the IFRO.

REFERENCES

Arany S, Kawagoe M, Sugiyama T (2009). Application of spontaneously immortalized odontoblast cells in tooth regeneration. *Biochem Biophys Res Commun* 381(1):84–89.

Arese M, Serini G, Bussolino F (2011). Nervous vascular parallels: axon guidance and beyond. *Int J Dev Biol* 55(4–5):439–445.

Balic A, Aguila HL, Caimano MJ, Francone VP, Mina M (2010). Characterization of stem and progenitor cells in the dental pulp of erupted and unerupted murine molars. *Bone* 46(6):1639–1651.

Bianco P (2011). Back to the future: moving beyond "mesenchymal stem cells." *J Cell Biochem* 112(7):1713–1721.

Boran T, Peterkova R, Lesot H, Lyons DB, Peterka M, Klein OD (2009). Temporal analysis of ectopic enamel production in incisors from Sprouty mutant mice. *J Exp Zool B* 312B(5):473–485.

Casagrande L, Cordeiro MM, Nör SA, Nör JE (2011). Dental pulp stem cells in regenerative dentistry. *Odontol Soc Nippon Dent Univ* 99(1):1–7.

Chai Y, Jiang X, Ito Y, Bringas P Jr, Han J, Rowitch DH, et al. (2000). Fate of the mammalian cranial neural crest during tooth and mandibular morphogenesis. *Development (Camb)*, 127(8):1671–1679.

DuFort CC, Paszek MJ, Weaver VM (2011). Balancing forces: architectural control of mechanotransduction. *Nat Rev Mol Cell Biol* 12(5):308–319.

Farges JC, Romeas A, Melin M, Pin JJ, Lebecque S, Lucchini M, et al. (2003). TGF-beta1 induces accumulation of dendritic cells in the odontoblast layer. *J Dent Res* 82(8): 652–656.

Feng J, Mantesso A, De Bari C, Nishiyama A, Sharpe PT (2011). Dual origin of mesenchymal stem cells contributing to organ growth and repair. *Proc Nat Acad Sci USA* 108(16):6503–6508.

Gronthos S, Mankani M, Brahim J, Robey PG, Shi S (2000). Postnatal human dental pulp stem cells (DPSCs) in vitro and in vivo. *Proc Nat Acad Sci USA* 97(25):13625–13630.

Guilak F, Cohen DM, Estes BT, Gimble JM, Liedtke W, Chen CS (2009). Control of stem cell fate by physical interactions with the extracellular matrix. *Cell Stem Cell* 5(1):17–26.

Honda MJ, Fong H, Iwatsuki S, Sumita Y, Sarikaya M (2008). Tooth-forming potential in embryonic and postnatal tooth bud cells. *Med Mol Morphol* 41(4):183–192.

Honda MJ, Shinmura Y, Shinohara Y (2009). Enamel tissue engineering using subcultured enamel organ epithelial cells in combination with dental pulp cells. *Cells Tissues Organs* 189(1–4):261–267.

Hu B, Nadiri A, Bopp-Kuchler S, Perrin-Schmitt F, Lesot H (2005a). Dental epithelial histomorphogenesis in vitro. *J Dent Res* 84(6):521–525.

Hu B, Nadiri A, Bopp-Kuchler S, Perrin-Schmitt F, Wang S, Lesot H (2005b). Dental epithelial histo-morphogenesis in the mouse: positional information versus cell history. *Archi Oral Biol* 50(2):131–136.

Hu B, Nadiri A, Kuchler-Bopp S, Perrin-Schmitt F, Peters H, Lesot H (2006a). Tissue engineering of tooth crown, root, and periodontium. *Tissue Eng* 12(8):2069–2075.

Hu B, Unda F, Bopp-Kuchler S, Jimenez L, Wang XJ, Haïkel Y, et al. (2006b). Bone marrow cells can give rise to ameloblast-like cells. *J Dent Res* 85(5):416–421.

Huang GTJ (2009). Pulp and dentin tissue engineering and regeneration: current progress. *Regen Med* 4(5):697–707.

Huang GTJ, Gronthos S, Shi S (2009). Mesenchymal stem cells derived from dental tissues vs. those from other sources: their biology and role in regenerative medicine. *J Dent Res* 88(9):792–806.

Huber A, Badylak SF (2011). Phenotypic changes in cultured smooth muscle cells: limitation or opportunity for tissue engineering of hollow organs? *J Tissue Eng Regen Med*. Available at: http://www.ncbi.nlm.nih.gov/pubmed/21755602.

Hwang Y, Phadke A, Varghese S (2011). Engineered microenvironments for self-renewal and musculoskeletal differentiation of stem cells. *Regen Med* 6(4):505–524.

Imbeni V, Kruzic JJ, Marshall GW, Marshall SJ, Ritchie RO (2005). The dentin–enamel junction and the fracture of human teeth. *Nat Mater* 4(3):229–232.

Janich P, Pascual G, Merlos-Suárez A, Batlle E, Ripperger J, Albrecht U, et al. (2011). The circadian molecular clock creates epidermal stem cell heterogeneity. *Nature* 480(7376):209–214.

Keller L, Kuchler-Bopp S, Mendoza SA, Poliard A, Lesot H (2011). Tooth engineering: searching for dental mesenchymal cells sources. *Front Physiol* 2:7.

Kuchler-Bopp S, Keller L, Poliard A, Lesot H (2011). Tooth organ engineering: biological constraints specifying experimental approaches. In: *Tissue Engineering for Tissue and Organ Regeneration*, Eberli, D ed., InTech, New York, pp. 317–346.

Lesot H, Brook AH (2009). Epithelial histogenesis during tooth development. *Arch Oral Biol* 54 (Suppl 1): S25–S33.

Lesot H, Kubler MD, Fausser JL, Ruch JV (1990). A 165 kDa membrane antigen mediating fibronectin–vinculin interaction is involved in murine odontoblast differentiation. Differentiation 44(1):25–35.

Lillesaar C, Fried K (2004). Neurites from trigeminal ganglion explants grown in vitro are repelled or attracted by tooth-related tissues depending on developmental stage. *Neuroscience* 125(1):149–161.

Lumsden AG (1988). Spatial organization of the epithelium and the role of neural crest cells in the initiation of the mammalian tooth germ. *Development (Camb)* 103(Suppl): 155–169.

Luukko K, Kvinnsland IH, Kettunen P (2005). Tissue interactions in the regulation of axon pathfinding during tooth morphogenesis. *Dev Dyn* 234(3):482–488.

Magloire H, Maurin JC, Couble M L, Shibukawa Y, Tsumura M, Thivichon-Prince B, et al. (2010). Topical review: dental pain and odontoblasts—facts and hypotheses. *J Orofaci Pain* 24(4):335–349.

McBeath R, Pirone DM, Nelson CM, Bhadriraju K, Chen CS (2004). Cell shape, cytoskeletal tension, and RhoA regulate stem cell lineage commitment. *Dev Cell* 6(4):483–495.

Mina M, Kollar EJ (1987). The induction of odontogenesis in non-dental mesenchyme combined with early murine mandibular arch epithelium. *Arch Oral Biol*, 32(2):123–127.

Munne PM, Tummers M, Järvinen E, Thesleff I, Jernvall J (2009). Tinkering with the inductive mesenchyme: Sostdc1 uncovers the role of dental mesenchyme in limiting tooth induction. *Development (Camb)*, 136(3):393–402.

Nait Lechguer A, Kuchler-Bopp S, Hu B, Haïkel Y, Lesot H (2008). Vascularization of engineered teeth. *J Dent Res* 87(12):1138–1143.

Nait Lechguer A, Kuchler-Bopp S, Lesot H (2009). Crown formation during tooth development and tissue engineering. *J Exp Zool B* 312B(5):399–407.

Nait Lechguer A, Couble M L, Labert N, Kuchler-Bopp S, Keller L, Magloire H, et al. (2011). Cell differentiation and matrix organization in engineered teeth. *J Dent Res*, 90(5):583–589.

Nakada D, Levi BP, Morrison SJ (2011). Integrating physiological regulation with stem cell and tissue homeostasis. *Neuron* 70(4):703–718.

Nakagawa E, Itoh T, Yoshie H, Satokata I (2009). Odontogenic potential of post-natal oral mucosal epithelium. *J Dent Res* 88(3):219–223.

Nakao K, Morita R, Saji Y, Ishida K, Tomita Y, Ogawa M, et al. (2007). The development of a bioengineered organ germ method. *Nat Methods* 4(3):227–230.

Oda Y, Yoshimura Y, Ohnishi H, Tadokoro M, Katsube Y, Sasao M, et al. (2010). Induction of pluripotent stem cells from human third molar mesenchymal stromal cells. *J Biol Chem*, 285(38):29270–29278.

Ohazama A, Modino SAC, Miletich I, Sharpe P T (2004). Stem-cell-based tissue engineering of murine teeth. *J Dent Res*, 83(7):518–522.

Otsu K, Kishigami R, Oikawa-Sasaki A, Fukumoto S, Yamada A, Fujiwara N, et al. (2011). Differentiation of induced pluripotent stem cells into dental mesenchymal cells. *Stem Cells Dev*. Available at: http://www.ncbi.nlm.nih.gov/pubmed/22085204.

Prochazka J, Pantalacci S, Churava S, Rothova M, Lambert A, Lesot H, et al. (2010). Patterning by heritage in mouse molar row development. *Proc Nat Acad Sci USA* 107(35):15497–15502.

Prockop DJ (2007). "Stemness" does not explain the repair of many tissues by mesenchymal stem/multipotent stromal cells (MSCs). *Clini Pharmacol Ther* 82(3):241–243.

Roobrouck VD, Clavel C, Jacobs SA, Ulloa-Montoya F, Crippa S, Sohni A, et al. (2011). Differentiation potential of human postnatal mesenchymal stem cells, mesoangioblasts, and multipotent adult progenitor cells reflected in their transcriptome and partially influenced by the culture conditions. *Stem Cells (Dayton)*, 29(5):871–882.

Rothová M, Feng J, Sharpe Paul T, Peterková R, Tucker AS (2011). Contribution of mesoderm to the developing dental papilla. *Int J Dev Biol* 55(1):59–64.

Ruch JV, Lesot H, Bègue-Kirn C (1995). Odontoblast differentiation. *Int J Dev Biol* 39(1):51–68.

Sarkar L, Cobourne M, Naylor S, Smalley M, Dale T, Sharpe P T (2000). Wnt/Shh interactions regulate ectodermal boundary formation during mammalian tooth development. *Proc Nat Acad Sci USA* 97(9):4520–4524.

Smith AJ, Lesot H (2001). Induction and regulation of crown dentinogenesis: embryonic events as a template for dental tissue repair ? *Crit Rev Oral Biol Med* 12(5):425–437.

Sonoyama W, Liu Y, Fang D, Yamaza T, Seo B-M, Zhang C, Liu H, Gronthos S, Wang CY, Wang S, Shi S (2006). Mesenchymal stem cell–mediated functional tooth regeneration in swine. *PloS One* 1:e79.

Takahashi C, Yoshida H, Komine A, Nakao K, Tsuji T, Tomooka Y (2010). Newly established cell lines from mouse oral epithelium regenerate teeth when combined with dental mesenchyme. *In Vitro Cell Dev Biol. Anim*, 46(5):457–468.

Thesleff I, Tummers M (2008). Tooth organogenesis and regeneration. *StemBook*. Available at: http://www.stembook.org/node/535.

Volponi AA, Pang Y, Sharpe P T (2010). Stem cell-based biological tooth repair and regeneration. *Trends Cell Biol* 20(12):715–722.

Wang B, Li L, Du S, Liu C, Lin X, Chen Y, et al. (2010). Induction of human keratinocytes into enamel-secreting ameloblasts. *Dev Biol*, 344(2):795–799.

Wang XP, Suomalainen M, Jorgez CJ, Matzuk MM, Werner S, Thesleff I (2004). Follistatin regulates enamel patterning in mouse incisors by asymmetrically inhibiting BMP signaling and ameloblast differentiation. *Dev Cell* 7(5):719–730.

Yildirim S, Fu SY, Kim K, Zhou H, Lee CH, Li A, et al. (2011). Tooth regeneration: a revolution in stomatology and evolution in regenerative medicine. *Int J Oral Sci*, 3(3):107–116.

Young CS, Kim S-W, Qin C, Baba O, Butler WT, Taylor RR, et al. (2005). Developmental analysis and computer modelling of bioengineered teeth. *Arch Oral Biol* 50(2):259–265.

Zhao W, Schafer S, Choi J, Yamanaka YJ, Lombardi ML, Bose S, et al. (2011). Cell-surface sensors for real-time probing of cellular environments. *Nat Nanotechnol* 6(8):524–531.

25

BIOENGINEERING OF FUNCTIONAL TEETH

TAKASHI TSUJI
Research Institute for Science and Technology, Tokyo University of Science, and Department of Biological Science and Technology, Faculty of Industrial Science and Technology, Tokyo University of Science, Noda, Chiba, Japan, and Organ Technologies Inc., Tokyo, Japan

1 INTRODUCTION

The tooth is an ectodermal organ that is composed of unique hard tissues, including dentin, enamel, and cementum, and is associated with alveolar bone. Teeth also contain soft connective pulp tissue which functions to produce dentin and to maintain biological functions of teeth, and periodontal tissue, which sustains and constrains the teeth in the jaw bone to maintain their homeostasis (Avery, 2002). Damage, loss, and the diseases of teeth, including dental caries and periodontal disease, cause fundamental problems for oral functions and are associated with a number of health issues (Proffit et al., 2004). The restoration of tooth functions in these circumstances typically involves replacement with an artificial tooth, the use of a bridge, or with osseo-integrated dental implants (Branemark et al., 1985; Rosenstiel et al., 2001). Although these prosthetic therapies are very effective, improvements have also been made that enhance the biological functions underlying tooth movement through bone remodeling (Huang et al., 2008).

Current advances in the development of regenerative therapies have been influenced by a number of previous studies of embryonic development, stem cell biology, and tissue engineering technologies (Atala, 2005; Brockes and Kumar, 2005). Tooth tissue stem cells and the cytokine network that regulates tooth development and growth as well as dental cell differentiation have been well characterized and probably can be applied in the future for the repair of dental pulp and periodontal tissues including the alveolar bone (Thesleff, 2003; Yen and Sharpe, 2008; Mantesso and Sharpe, 2009). The ultimate goal of regenerative therapy is to develop fully functioning bioengineered

Stem Cells in Craniofacial Development and Regeneration, First Edition.
Edited by George T.-J. Huang, Irma Thesleff.

tissues that can replace lost or damaged organs following disease, injury, or aging (Atala, 2005; Ikeda and Tsuji, 2008). In the dental field, this would involve replacement of a lost or damaged tooth with a bioengineered tooth that has been built using stem cells and has the capacity to become a functional unit comprising the whole tooth, periodontal tissue, and surrounding alveolar bone (Yen and Sharpe, 2008).

In this chapter, the novel technologies used for whole-tooth replacement (i.e., tooth regenerative therapy using bioengineered tooth germs) are described. It can be argued that tooth regenerative therapy based on these technologies has the potential to provide essential functional recovery of lost teeth and therefore ultimately to replace the current artificial materials used in dental replacements.

2 TOOTH DEVELOPMENT

Organs such as hair, glands, kidney, and teeth arise from their respective germs which are induced by reciprocal interactions between epithelial and mesenchymal tissues in the developing embryo (Pispa and Thesleff, 2003; Thesleff, 2003; Tucker and Sharpe, 2004; Chapter 6, this volume). The first event in tooth development is the formation of the dental lamina (lamina stage) which is followed by thickening of the epithelium at the sites of future teeth (placode stage) and subsequent epithelial budding to the underlying neural crest-derived ectomesenchyme (bud stage). Tooth germ formation is initiated during embryonic days 10 to 11 (ED10 to 11) in mouse by epithelial signals such as FGF8, BMP4, Shh, Eda and Wnt10b. These induce the expression of transcription factors such as Barx1, Dlx1/2, Lhx6, Lhx7 Msx1, Pax9, and Gli, in the dental mesenchyme, which condenses around the developing epithelial bud (Pispa and Thesleff, 2003; Thesleff, 2003; Tucker and Sharpe, 2004). At ED14 (cap stage), a transient epithelial signaling center known as the *enamel knot*, which expresses several signaling molecules, including Shh, BMP family molecules, FGFs, and Wnts, is thought to regulate individual cell fates and crown morphogenesis. The dental mesenchymal cells form the dental papilla, which is capable of differentiating into odontoblasts and pulp cells (Sonoyama et al., 2006). The terminal differentiation of the dental mesenchymal and epithelial cells into odontoblasts and ameloblasts, respectively, occurs at the bell stage. The odontoblasts and ameloblasts secrete the mineralizing extracellular matrices for dentin and enamel, respectively, at the interface between the epithelium and mesenchyme (Fukumoto and Yamada, 2005). The dental mesenchymal cells surrounding the tooth germ form the dental follicle, generating periodontal tissues, including the periodontal ligament, cementum, and alveolar bone. Tooth root formation is initiated after the completion of crown moprhogenesis and is followed by tooth eruption in the oral cavity (Chapter 8).

3 WHOLE-TOOTH REGENERATION AS A FUTURE ORGAN-REPLACEMENT REGENERATIVE THERAPY

Current bioengineering technologies have not yet achieved their goal of reconstructing complex organs, which is in contrast to stem cell transplantation therapies used to repair damaged tissues. The reasons for this are that an organ which is comprised of

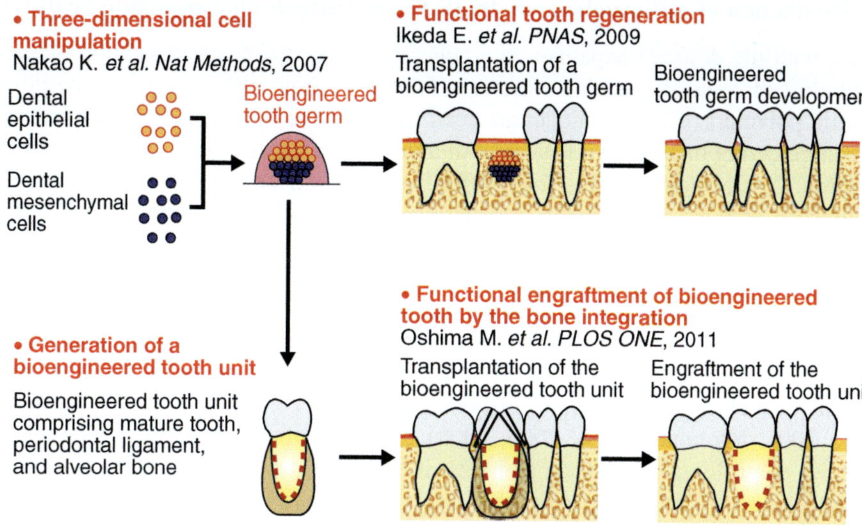

FIGURE 1 Strategies for whole-tooth replacement via regenerative therapies. Functioning teeth can be regenerated in mice from bioengineered tooth germs reconstituted from epithelial and mesenchymal cells derived from embryonic tooth germs. The bioengineered tooth germs can be transplanted to the jaw, or they can be developed to tooth units with periodontal ligament and alveolar bone before transplantation.

several tissues and cell populations performs its functions and maintains homeostasis through the coordinated interplay of the tissues and cell populations (Ikeda and Tsuji, 2008). Many attempts have been made to create fully functioning substitute organs that could replace lost or damaged tissues following disease, injury, or aging. These include fully artificial organs made from plastic or metal and containing computer chips for reproducing the physical functions of heart, eye, and kidney (Wolf, 1952; Copeland et al., 2004). They also include bioartificial organs created by tissue engineering and comprising living cells and natural or artificial polymers and reproducing the normal biochemical functions of major organs, such as the metabolic and secretory pathways of the liver and pancreas (Colton, 1995; Fort et al., 2008).

To generate ectodermal organs such as teeth (Sharpe and Young, 2005; Ikeda and Tsuji, 2008), hair follicles (Zheng et al., 2005), and salivary glands (Wei et al., 2007), the generally accepted approach is to recreate organogenesis through epithelial mesenchymal interactions and thereby develop fully functioning bioengi neered organs *in vivo* from the bioengineered organ germ (Fig. 1, upper panel; Sharpe and Young, 2005; Ikeda and Tsuji, 2008). It has been anticipated that the transplantation of a bioengineered tooth unit comprising a mature tooth, periodontal ligament (PDL), and alveolar bone into a tooth-loss region will achieve full functionality through the integration of recipient bone (Fig. 1, lower panel; Oshima et al., 2011).

In the case of whole-tooth replacement, the first major issue is to develop a three-dimensional cell manipulation technology using completely dissociated epithelial and mesenchymal cells *in vitro*.

3.1 Production of a Bioengineered Tooth Germ Using a Biodegradable Scaffold

Tissues with the desired shape can be regenerated using the scaffold method by seeding single cells onto degradable structures fabricated from either natural materials or synthetic polymers. This technique has been useful in the design of three-dimensional tissue engineering methods for tissues with a uniform cellular distribution (Griffith and Naughton, 2002; Atala, 2005). The scaffold method has been used in clinical applications such as bone and cartilage regenerative therapies (Caplan and Bruder, 2001). Poly(glycolic acid) and poly(L-lactate-*co*-glycolide) copolymer (PLA/PLGA) (Young et al., 2002; Duailibi et al., 2004) as well as a collagen sponges (Sumita et al., 2006; Honda et al., 2007) have been used as materials for tooth-shaped scaffolds onto which epithelial and mesenchymal cells isolated from porcine unerupted third molars or rat tooth buds were seeded. Small tooth structures were generated containing correct dental components including pulp, dentin, and enamel, but not periodontal tissue (Sharpe and Young, 2005; Yen and Sharpe, 2006, 2008). Although the scaffold method might be useful for controlling tooth shape and size, it has critical limitations, including, a low frequency of tooth initiation and problems in hard tissue formation. Development of the normal structures of dental hard tissues requires proper generation of the complex junctions between dentin, enamel, and cementum which result from precise spatiotemporal gradients of differentiation of the odontoblasts, ameloblasts, and cementoblasts, respectively, through epithelial–mesenchymal interactions.

3.2 Bioengineering of Tooth Germs Using the Cell Aggregation Method

The cell aggregation method is a bioengineering protocol that aims to reconstitute an organ germ by properly reproducing the epithelial and mesenchymal interactions that take place during organogenesis (Sharpe and Young, 2005; Yen and Sharpe, 2006; Ikeda and Tsuji, 2008). Previously, transplantations of bioengineered cell aggregates using hair follicle and mammary gland–derived stem cells have shown promise in the regeneration of correct organ structures with a proper arrangement of different cell types (Zheng et al., 2005; Shackleton et al., 2006). An artificial tooth germ created from precipitates of dissociated dental epithelial and mesenchymal cells after centrifugation has also been shown to generate a complete tooth (Yamamoto et al., 2003; Hu et al., 2006). Mixed cell aggregates of dissociated epithelial and mesenchymal cells isolated from molar tooth germs were also shown to develop into a tooth with the correct structure following the cell sorting of epithelial cells and the subsequent self-reorganization of epithelial and mesenchymal cells (Song et al., 2006).

We have developed an *in vitro* three-dimensional cell manipulation method, designated as a bioengineered organ germ method. This involves dissociation of the mesenchymal and epithelial cells of mouse embryonic cap-stage tooth germs, and their compartmentalization at a high cell density in type I collagen gel (Fig. 2A; Nakao et al., 2007). Multiple tooth germs with characteristic morphology of molar or incisor, depending on the origin of the cells, were induced in individual explants. The bioengineered tooth germs generated structurally correct teeth after 14 days when transplanted into a subrenal capsule *in vivo* and also when cultured in organ culture *in vitro* (Fig. 2A, right panels). In addition, the bioengineered tooth germs developed into teeth with correct structure when transplanted into a tooth socket in the oral cavity (Nakao et al., 2007). Importantly, this development occurs at a high frequency.

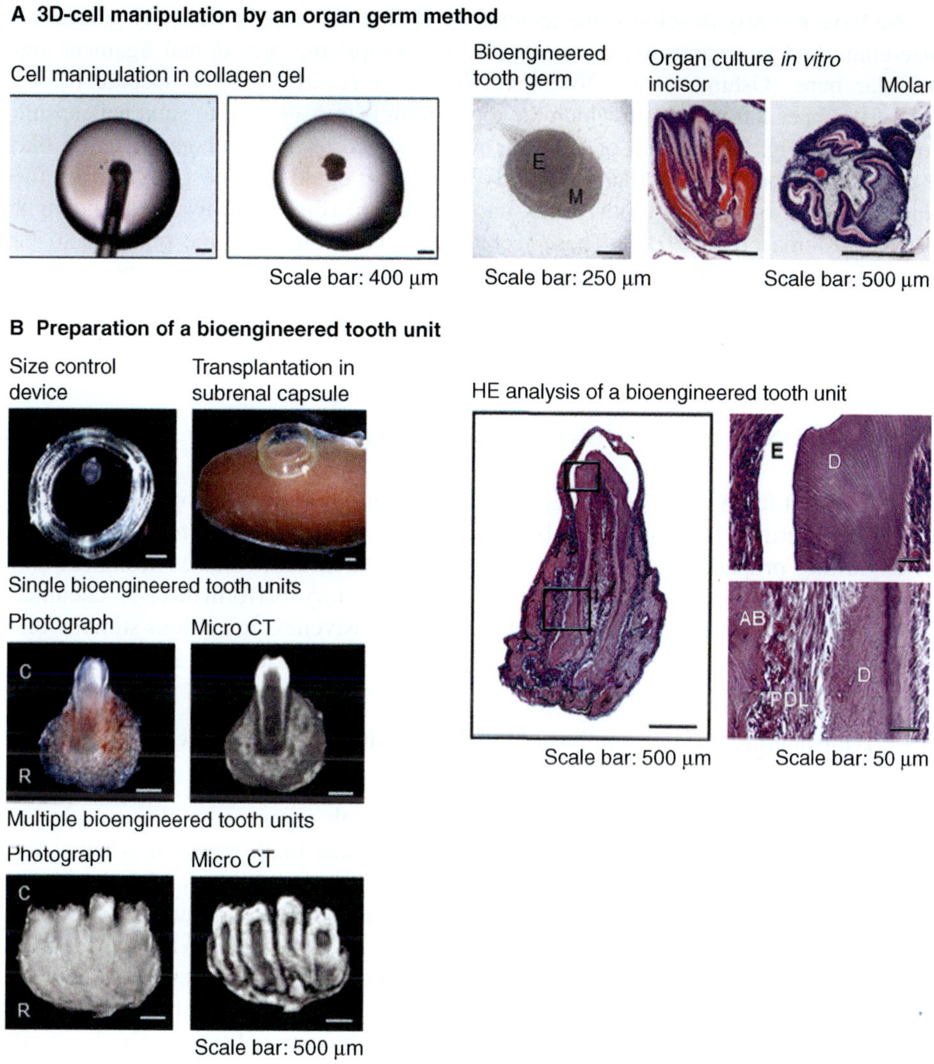

FIGURE 2 The organ germ method: a three-dimensional cell-processing system. (A) Dissociated mesenchymal cells at a high density are injected into the center of a collagen drop. Dissociated tooth germ–derived epithelial cells are subsequently injected into the drop adjacent to the mesenchymal cell aggregate (left panel). Within one day of organ culture, a bioengineered tooth germ has formed with the appropriate compartmentalization between epithelial (E) and mesenchymal (M) cells and cell-to-cell compaction (middle panels). After *in vitro* organ culture for 14 days, bioengineered incisor and molar tooth germs have developed from the dissociated cells isolated from incisor and molar tooth germs, respectively (right panels). (B) To generate the size-controlled bioengineered tooth unit, the tooth germ was inserted into a ring-shaped size-control device and transplanted into a subrenal capsule (upper left panels). By transplanting this bioengineered tooth germ into a subrenal capsule for 30 days (middle left panels), a bioengineered tooth unit can be produced comprising a mature tooth, periodontal ligament, and alveolar bone with the correct structural components, such as enamel (E), dentin (D), periodontal ligament (PDL), and alveolar bone (AB, right panels). Multiple bioengineered tooth units surrounded by alveolar bone could also be generated by the transplantation of several tooth germs into a single size-control device (lower left panels).

We have recently developed the technology further and can successfully generate a size-controlled bioengineered mature tooth unit comprising periodontal ligament and alveolar bone (Oshima et al., 2011). Tooth size is controlled by a specific device (Fig. 2B, upper left panels) in which the tooth germ is transplanted to subrenal capsule for 50 days (Fig. 2B). A unit of multiple bioengineered teeth, which can function like a denture and becomes surrounded by alveolar bone, could also be generated by the transplantation of several tooth germs into the device (Fig. 2B, lower left and right panels; Oshima et al., 2011). These technologies therefore have the potential to be adapted for successful functional tooth replacement *in vivo* and are thought to represent a substantial advance in bioengineered organ replacement for future regenerative therapies.

4 FUNCTIONAL TOOTH ORGAN REPLACEMENT *IN VIVO*

For successful tooth replacement regenerative therapy, a tooth developed from a bioengineered germ or a transplanted bioengineered mature tooth unit must be capable of engrafting properly in the region of the lost tooth in adult oral environment and achieve full functionality, including sufficient masticatory performance, biochemical cooperation with periodontal tissues and proper responsiveness to noxious stimulations via neurons in the maxillofacial region (Proffit et al., 2004; Dawson, 2006).

4.1 Transplantation of Bioengineered Tooth Germ or a Bioengineered Mature Tooth Unit as a Tooth Replacement Strategy

The most critical issue underlying the success of tooth regenerative therapy via the transplantation of bioengineered tooth germ into the lost tooth region is whether the germ can erupt and occlude properly with the opposing tooth in an adult jaw bone (Fig. 1, upper right). Tooth eruption is a complex and tightly regulated process that involves cells of the tooth organ and the surrounding alveolar bone (Wise et al., 2002). During the process of eruption, the axial movement of the tooth is achieved via the local resorption of the bone overlying the erupting tooth, and apposition of bone underlying the tooth (Wise and King, 2008). It has also been demonstrated previously that teeth can erupt in the toothless diastema region of the mouse (Ohazama et al., 2004) and that a bioengineered tooth germ can develop the correct structure when transplanted in a tooth socket (Nakao et al., 2007). We have also shown that a bioengineered tooth germ can erupt successfully at 37 days after transplantation (Fig. 3A, right panels) and that it subsequently reaches the occlusal plane and achieves and maintains occlusion with the opposing tooth from 49 days onward (Fig. 3A, middle panels; Ikeda et al., 2009). The resulting bioengineered tooth formed the correct structures, comprising enamel, dentin, odontoblasts, dental pulp, and alveolar bone (Fig. 3A, right panels), and enamel hardness was in the normal range when analyzed by the Knoop test (Ikeda et al., 2009). The bioengineered teeth generated in this way thus have the potential to successfully recover the masticatory performance of a normal mature tooth.

In the case of a transplanted bioengineered tooth unit comprising a mature tooth, periodontal ligament, and alveolar bone, the most critical consideration is whether the unit can be engrafted into a tooth-loss region through bone integration, which involves natural bone remodeling in the recipient. We demonstrated that a bioengineered tooth

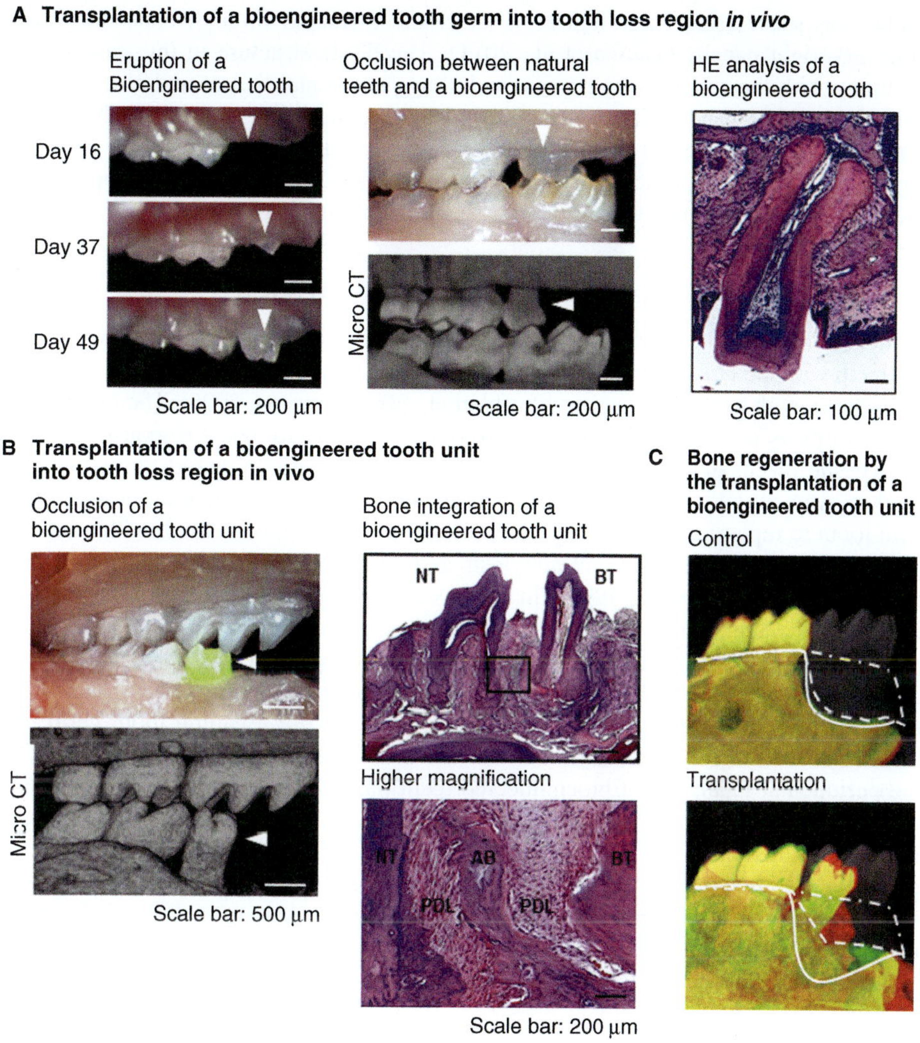

FIGURE 3 Regeneration of the bioengineered tooth by the transplantation of a bioengineered tooth germ or a bioengineered tooth unit. (A) A transplanted bioengineered tooth germ erupted (left panels) and reached the occlusal plane with the opposing lower first molar at 49 days after transplantation (middle panels). The bioengineered tooth also formed a correct structure comprising enamel, dentin, dental pulp, and alveolar bone (right panels). (B) The bioengineered tooth unit was engrafted by bone integration at a position reaching the occlusal plane with the opposing upper first molar at 40 days posttransplantation (left panels). The engrafted bioengineered tooth unit also formed a correct tooth structure. NT, natural tooth; BT, bioengineered tooth; AB, alveolar bone; PDL, periodontal ligament (middle panels). (C) Three-dimensional superposition of micro-CT images of natural dentition (gray, double dotted line), a transplanted bioengineered tooth unit (lower panel) and a no-transplantation control (upper panel) at day 0 in an extensive bone defect (red, straight line), and at 45 days after transplantation (green, dotted line) are demonstrated. The superior edges of the recipient alveolar bone are indicated by each line.

unit transplanted at a position reaching the occlusal plane with the opposing upper first molar was successfully engrafted at 40 days and thereafter through bone integration (Fig. 3B, right panels; Oshima et al., 2011). The tissue structure in this case was in harmony with the surrounding tissues, including periodontal tissues and alveolar bone (Fig. 3B, middle panels). The bioengineered tooth unit also restored enough alveolar bone in a vertical direction into an extensive bone defect of mouse lower jaw (Fig. 3A, right panels). This approach thus shows potential as a novel therapeutic approach to the recovery of masticatory performance when teeth have been lost.

4.2 Response of a Bioengineered Tooth to Mechanical Stress

Critical oral functions, such as mastication and pronunciation, require the cooperation of the teeth with the oral and maxillofacial regions through the periodontal ligament and in accordance with growth and aging (Proffit et al., 2004; Dawson, 2006). The periodontal ligament plays essential roles in the pathological and physiological responses of teeth to extreme mechanical forces: for example, bone remodelling caused by orthodontic tooth movement (Wise et al., 2008). A previous report of transplantation of a patient's own tooth to replace a tooth lost by traumatic injury showed that a tooth with a healthy periodontal ligament can prevent ankylosis and attach successfully to the surrounding alveolar tissue (Tsukiboshi, 1993). In contrast, the absence of a periodontal ligament in dental implants is associated with deficiencies in essential tooth functions and the structural relationship with the root and alveolar bone (Huang et al., 2008; Lindhe et al., 2008). A previous study also demonstrated by histochemical analysis that it is possible to reconstitute periodontal tissue after the transplantation of a natural tooth germ into the diastema region of an adult mouse (Yen and Sharpe, 2006). Our studies indicate that the periodontal ligaments of bioengineered teeth that erupted following the transplantation not only of a bioengineered tooth germ, but also the bioengineered mature tooth unit, were functionally normal. They reproduced bone remodeling successfully via the proper localization of osteoclasts and osteoblasts in response to mechanical stress, indicating that a bioengineered tooth can regenerate critical dental functions through the restoration and reestablishment of cooperation with the surrounding jawbone (Ikeda et al., 2009; Oshima et al., 2011).

4.3 Neuronal Perceptive Potential of Bioengineered Teeth

The peripheral nervous system, which is established by growing axons that navigate and form connections to their developing target organs during embryogenesis, plays an essential role in the control of organ functions and in the perception of external stimuli such as pain and mechanical stress (Guyton and Hall, 2000). It is also thought that the recovery of the nervous system following organ transplantation, which is associated with the reentry of nerve fibers, is critical for the reconstitution of organ function (Kjaer et al., 1999; Bengel et al., 2001). Teeth are peripheral target organs for the sensory trigeminal and sympathetic nerves, both of which play important roles in the function and protection of teeth (Guyton et al., 2000). The perception of mechanical forces during mastication is limited in implant patients (Hämmerle et al., 1995) and it is anticipated that tooth regenerative therapy will realize the functional recovery of the neuronal perceptive potential of teeth. Our observations indicated that nerve fibers such as sensory and sympathetic nerves innervated both the pulp and

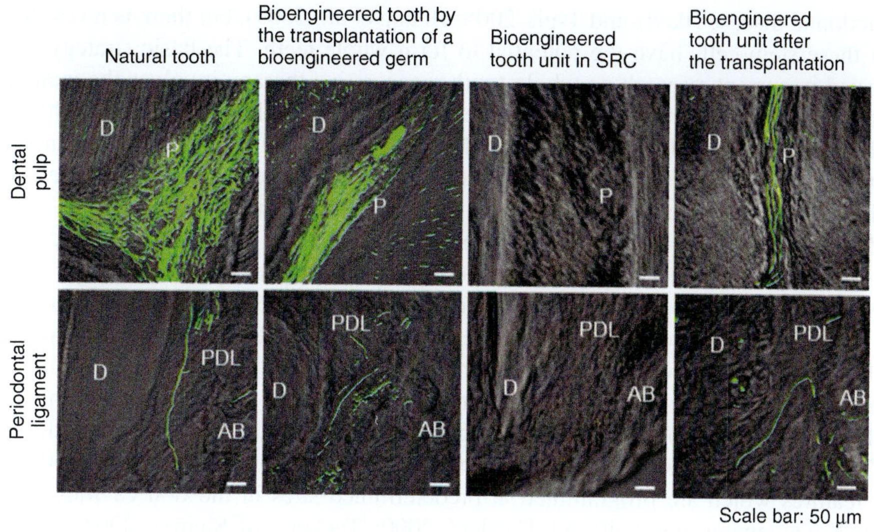

FIGURE 4 Nerve fiber innervation of a bioengineered tooth. Nerve fiber innervation of the dental pulp and periodontal ligament of natural and bioengineered teeth are demonstrated by immunostaining for neurofilament-H. D, dentin; P, pulp; PDL, periodontal ligament; AB, alveolar bone.

PDL of a regenerated tooth that had erupted successfully following transplantation of the bioengineered tooth germ as well as the bioengineered tooth unit (Fig. 4; Ikeda et al., 2009; Oshima et al., 2011). These bioengineered teeth display the appropriate perceptive potential for nociceptive pain stimulations such as pulp stimulation and orthodontic treatment, and can properly transduce these events to the central nervous system through the c-fos immune-reactive neurons. Hence, this particular whole-tooth regenerative therapy has the potential to fully restore tooth function, including not only masticatory potential and the response to mechanical stress, but also the perceptive potential for noxious stimuli (Ikeda et al., 2009; Oshima et al., 2011).

5 CONCLUSIONS AND FUTURE DIRECTIONS

To fully realize the clinical application of tooth regenerative therapy in the future, one of the major remaining research hurdles is the identification of appropriate cell sources. Tooth regenerative therapy may be optimized by using a patient's own cells to avoid immunological rejection (Ikeda and Tsuji, 2008; Mantesso and Sharpe, 2009). Recent studies of stem cells and organogenesis have provided considerable advances in our understanding of the potential of using stem cells as cell sources for regenerative therapies, including tooth regenerative therapy (Korbling and Estrov, 2003; Mantesso and Sharpe, 2009). Tooth tissue–derived stem cells, such as DPSCs, SHED, SCAP, PDLSCs, and dental follicle stem cells (Gronthos et al., 2000; Miura et al., 2003; Seo et al., 2004; Sonoyama et al., 2006, 2008), can differentiate into dental cell lineages and contribute to the turnover and supply of a wide variety of cell populations in the steady state and following tissue injury. These will be useful cell sources for stem cell transplantation regenerative therapy for dental tissue repair such as dental caries and periodontal disease (Ikeda and Tsuji, 2008; Huang et al., 2009), but there is no evidence that these stem cells have the potential to form whole teeth. The basic strategy being pursued in current research on whole-tooth regenerative therapy involves the induction of a bioengineered tooth germ and its development to a fully functioning tooth using embryonic tooth germ–derived epithelial and mesenchymal cells and the organ germ method described in this chapter (Nakao et al., 2007; Ikeda et al., 2009; Oshima et al., 2011). It will be important in the future to identify cell sources from somatic, dental and non-dental tissue–derived stem cell populations isolated from patients, which have the potency to reproduce the epithelial and mesenchymal interactions that occur during organogenesis and have tooth-forming ability, and also to identify inductive genes with the potential to initiate tooth organogenesis, reconstitute a bioengineered tooth germ, and ultimately, to develop into a functional tooth. Candidate cell sources also include multipotent stem cells, such as embryonic stem cells and induced pluripotent stem cells, which are capable of differentiating into endoderm, ectoderm, and mesoderm.

Teeth such as incisors, canines, premolars, and molars have unique morphological features which are programmed at predetermined sites in the oral cavities during tooth development (Jernvall and Thesleff, 2000; Tucker and Sharpe, 2004). Several studies have also proposed molecular mechanisms for the regulation of tooth morphology (Jernvall and Thesleff, 2000; Cai et al., 2007; Ikeda and Tsuji, 2008; Ishida et al., 2011). Methods to regulate tooth size and crown and root shape are also important considerations when generating a bioengineered tooth with the proper functional occlusion and esthetics. Tooth regenerative therapy is now regarded as a crucial study model for future replacement regenerative therapies for other more complex organs and will contribute substantially to achieving the required understanding to regenerate tissues such as hair follicles, salivary glands, liver, and kidney (Sharpe and Young, 2005; Ikeda and Tsuji, 2008).

Acknowledgments

This work was partially supported by a Grant-in-Aid for Scientific Research in Priority Areas (No. 50339131), a Grant-in-Aid for Scientific Research, a Grant-in-Aid for Young Scientists, and by an Academic Frontier Project for Private Universities from the Ministry of Education, Culture, Sports and Technology in Japan. This work was also partially supported by Health and Labour Sciences Research Grants from the Ministry of Health, Labour, and Welfare (No. 21040101).

REFERENCES

Atala A (2005). Tissue engineering, stem cells and cloning: current concepts and changing trends. *Expert Opin Biol Ther* 5:879–892.

Avery JK (2002). In: *Oral Development and Histology*, Steele PF, Avery N, eds., Thieme Publishers, New York, pp. 153–212.

Bengel FM, Ueberfuhr P, Schiepel N, Nekolla SG, Reichart B, Schwaiger M (2001). Effect of sympathetic reinnervation on cardiac performance after heart transplantation. *N Engl J Med* 345:731–738.

Branemark PI, Zarb GA, Albrektsson T (1985). Tissue-integrated prostheses. In: *Osseointegration in Clinical Dentistry*, Albrektsson T, ed., Quintessence Press, Berlin, pp. 211–232.

Brockes JP, Kumar A (2005). Appendage regeneration in adult vertebrates and implications for regenerative medicine. *Science* 310:1919–1923.

Cai J, Cho SW, Kim JY, Lee MJ, Cha YG, Jung HS (2007). Patterning the size and number of tooth and its cusps. *Dev Biol* 304:499–507.

Caplan AI, Bruder SP (2001). Mesenchymal stem cells: building blocks for molecular medicine in the 21st century. Trends Mol Med 7:259–264.

Colton CK (1995). Implantable biohybrid artificial organs. *Cell Transplant* 4:415–436.

Copeland JG, Smith RG, Arabia FA, Nolan PE, Sethi GK, Tsau PH, et al (2004). Cardiac replacement with a total artificial heart as a bridge to transplantation. *N Engl J Med* 351:859–867.

Dawson PE (2006). *Functional Occlusion: From TMJ to Smile Design*, Mosby, St. Louis, MO, pp. 18–26.

Duailibi MT, Duailibi SE, Young CS, Bartlett JD, Vacanti JP, Yelick PC (2004). Bioengineered teeth from cultured rat tooth bud cells. *J Dent Res* 83:523–528.

Fort A, Fort N, Ricordi C, Stabler CL (2008). Biohybrid devices and encapsulation technologies for engineering a bioartificial pancreas. *Cell Transplant* 17:997–1003.

Fukumoto S, Yamada Y (2005). Review: extracellular matrix regulates tooth morphogenesis. *Connect Tissue Res* 46:220–226.

Griffith LG, Naughton G (2002). Tissue engineering: current challenges and expanding opportunities. *Science* 295:1009–1014.

Gronthos S, Mankani M, Brahim J, Robey PG, Shi S (2000). Postnatal human dental pulp stem cells (DPSCs) in vitro and in vivo. *Proc Natl Acad Sci USA* 97:13625–13630.

Guyton AC, Hall JE (2000). *Textbook of Medical Physiology*, W. B. Saunders, Philadelphia.

Hämmerle CH, Wagner D, Brägger U, Lussi A, Karayiannis A, Joss A, et al. (1995). Threshold of tactile sensitivity perceived with dental endosseous implants and natural teeth. *Clin Oral Implants Res* 6:83–90.

Honda MJ, Tsuchiya S, Sumita Y, Sagara H, Ueda M (2007). The sequential seeding of epithelial and mesenchymal cells for tissue-engineered tooth regeneration. *Biomaterials* 28:680–689.

Hu B, Nadiri A, Kuchler-Bopp S, Perrin-Schmitt F, Peters H, Lesot H (2006). Tissue engineering of tooth crown, root, and periodontium. *Tissue Eng* 12:2069–2075.

Huang GT, Sonoyama W, Liu Y, Liu H, Wang S, Shi S (2008). The hidden treasure in apical papilla: the potential role in pulp/dentin regeneration and bioroot engineering. *J Endod* 34:645–651.

Huang GT, Gronthos S, Shi S (2009). Mesenchymal stem cells derived from dental tissues vs. those from other sources: their biology and role in regenerative medicine. *J Dent Res* 88:792–806.

Ikeda E, Tsuji T (2008). Growing bioengineered teeth from single cells: potential for dental regenerative medicine. *Expert Opin Biol Ther* 8:735–744.

Ikeda E, Morita R, Nakao K, Ishida K, Nakamura T, Takano-Yamamoto T et al. (2009). Fully functional bioengineered tooth replacement as an organ replacement therapy. *Proc Natl Acad Sci U S A* 106:13475–13480.

Ishida K, Murofushi M, Nakao K, Morita R, Ogawa M, Tsuji T (2011). The regulation of tooth morphologenesis is associated with epithelial cell proliferation and the expression of Sonic hedgehog through epithelial–mesenchymal interactions. *Biochem Biophys Res Commun* 405:455–461.

Jernvall J, Thesleff I (2000). Reiterative signaling and patterning during mammalian tooth morphogenesis. *Mech Dev* 92:19–29.

Kjaer M, Beyer N, Secher NH (1999). Exercise and organ transplantation. *Scand J Med Sci Sports* 9:1–14.

Korbling M, Estrov Z (2003). Adult stem cells for tissue repair: A new therapeutic concept? *N Engl J Med* 349:570–582.

Lindhe J, Lang NP, Karring T (2008). *Clinical Periodontology and Implant Dentistry*, Wiley-Blackwell, Hoboken, NJ, pp. 99–107.

Mantesso A, Sharpe P (2009). Dental stem cells for tooth regeneration and repair. *Expert Opin Biol Ther* 9:1143–1154.

Miura M, Gronthos S, Zhao M, Lu B, Fisher LW, Robey PG et al. (2003). SHED: stem cells from human exfoliated deciduous teeth. *Proc Natl Acad Sci USA* 100:5807–5812.

Nakao K, Morita R, Saji Y, Ishida K, Tomita Y, Ogawa M, et al. (2007). The development of a bioengineered organ germ method. *Nat Methods* 4:227–230.

Ohazama A, Modino SA, Miletich I, Sharpe PT (2004). Stem-cell-based tissue engineering of murine teeth. *J Dent Res* 83:518–522.

Oshima M, Mizuno M, Imamura A, Ogawa M, Yasukawa M, Yamazaki H, et al. (2011). Functional tooth regeneration using a bioengineered tooth unit as a mature organ replacement regenerative therapy. *PLoS One* 6:e21531.

Pispa J, Thesleff I (2003). Mechanisms of ectodermal organogenesis. *Dev Biol* 262:195–205.

Proffit WR, Fields HW Jr., Sarver DM (2004). *Contemporary Orthodontics*, Mosby, St. Louis, MO, pp. 78–83.

Rosenstiel SF, Land MF, Fujimoto J (2001). *Contemporary Fixed Prosthodontics*, Mosby, St. Louis, MO, pp. 209–430.

Seo BM, Miura M, Gronthos S, Bartold PM, Batouli S, Brahim J et al. (2004). Investigation of multipotent postnatal stem cells from human periodontal ligament. *Lancet* 364:149–155.

Shackleton M, Vaillant F, Simpson KJ, Stingl J, Smyth GK, Asselin-Labat ML et al. (2006). Generation of a functional mammary gland from a single stem cell. *Nature* 439:84–88.

Sharpe PT, Young CS (2005). Test-tube teeth. *Sci Am* 293:34–41.

Song Y, Zhang Z, Yu X, Yan M, Zhang X, Gu S, et al. (2006). Application of lentivirus-mediated RNAi in studying gene function in mammalian tooth development. *Dev Dyn* 235:1334–1344.

Sonoyama W, Liu Y, Fang D, Yamaza T, Seo BM, Zhang C, et al. (2006). Mesenchymal stem cell–mediated functional tooth regeneration in swine. *PLoS One* 1:e79.

Sonoyama W, Liu Y, Yamaza T, Tuan RS, Wang S, Shi S, et al. (2008). Characterization of the apical papilla and its residing stem cells from human immature permanent teeth: a pilot study. *J Endod* 34:166–171.

Sumita Y, Honda MJ, Ohara T, Tsuchiya S, Sagara H, Kagami H, et al. (2006). Performance of collagen sponge as a 3-D scaffold for tooth-tissue engineering. *Biomaterials* 27:3238–3248.

Thesleff I (2003). Epithelial–mesenchymal signalling regulating tooth morphogenesis. *J Cell Sci* 116:1647–1648.

Tsukiboshi M (1993). Autogenous tooth transplantation: a reevaluation. *Int J Periodont Restor Dent* 13:120–149.

Tucker A, Sharpe P (2004). The cutting-edge of mammalian development; how the embryo makes teeth. *Nat Rev Genet* 5:499–508.

Wei C, Larsen M, Hoffman MP, Yamada KM (2007). Self-organization and branching morphogenesis of primary salivary epithelial cells. *Tissue Eng* 13:721–735.

Wise GE, King GJ (2008). Mechanisms of tooth eruption and orthodontic tooth movement. *J Dent Res* 87:414–434.

Wise GE, Frazier-Bowers S, D'Souza RN (2002). Cellular, molecular, and genetic determinants of tooth eruption. *Crit Rev Oral Biol Med* 13:323–334.

Wolf AV (1952). The artificial kidney. *Science* 115:193–199.

Yamamoto H, Kim EJ, Cho SW, Jung HS (2003). Analysis of tooth formation by reaggregated dental mesenchyme from mouse embryo. *J Electron Microsc (Tokyo)* 52:559–566.

Yen AH, Sharpe PT (2006). Regeneration of teeth using stem cell–based tissue engineering. *Expert Opin Biol Ther* 6:9–16.

Yen AH, Sharpe PT (2008). Stem cells and tooth tissue engineering. *Cell Tissue Res* 331:359–372.

Young CS, Terada S, Vacanti JP, Honda M, Bartlett JD, Yelick PC (2002). Tissue engineering of complex tooth structures on biodegradable polymer scaffolds. *J Dent Res* 81:695–700.

Zheng Y, Du X, Wang W, Boucher M, Parimoo S, Stenn K (2005). Organogenesis from dissociated cells: generation of mature cycling hair follicles from skin-derived cells. *J Investig Dermatol* 124:867–876.

26

PULP AND DENTIN REGENERATION

MISAKO NAKASHIMA

Department of Dental Regenerative Medicine, Center of Advanced Medicine for Dental and Oral Diseases, National Center for Geriatrics and Gerontology, Research Institute, Obu, Aichi, Japan

GEORGE T.-J. HUANG

Department of Bioscience Research, College of Dentistry, The University of Tennessee Health Science Center, Memphis, Tennessee

1 INTRODUCTION

Teeth possess multiple functions. Besides mastication, they are important for taste, tactile sense, speech, and aesthetics. As such, the longevity of teeth is critical for the maintenance and enhancement of quality of life, especially relating to prevention of dementia and Alzheimer disease (Stein et al., 2010). Over 50% of extracted teeth result from caries and root fracture. Deep caries and pulp exposure have been treated by pulp capping or partial pulp amputation to preserve pulp tissue, with limited success. In case of irreversible pulpitis, the entire pulp has to be removed and the emptied root canal space is filled with artificial materials after disinfection in traditional root canal therapy. The success rate of pulpectomy is nearly 90% in three to five years, but is estimated to decrease after five years. The endodontic success rate is down to 50 to 70% for retreatment of those endodontically failed teeth with periapical lesions and/or clinical symptoms (Machtou and Reit, 2003). Enlargement and debridement of infected root canals render teeth brittle (Wilcox and Van Surksum, 1991). The postoperative procedures with postpreparation also increase the possibility of root fracture (Fuss et al., 2001). Therefore, loss of pulp vitality is widely associated with root fracture and tooth loss. There has been no superior synthetic material to replace natural pulp and dentin. Thus, regenerating pulp–dentin tissues appears to be an ideal approach to

Stem Cells in Craniofacial Development and Regeneration, First Edition.
Edited by George T.-J. Huang, Irma Thesleff.

restoring tooth functions compromised by pulp injury and/or inflammation (Nakashima and Akamine, 2005; Nakashima et al., 2009).

Under natural conditions, pulp and dentin have limited capacity to regenerate and repair when damaged. In this chapter we describe recent advances and promising approaches for pulp and dentin regeneration, including the concept of revitalization, harnessing stem and progenitor cells and/or morphogenic signaling molecules in addition to extracellular matrix scaffold for partial and d*e* novo (complete) regeneration of pulp with newly deposited regenerated dentin-like tissues.

Attempts to regenerate pulp tissue have been made in the 1960s and 1970s without success, due to a lack of understanding of stem cell biology and tissue engineering (Ostby, 1961). The advent of the modern tissue engineering concept in the 1990s prompted researchers to test pulp tissue engineering (Mooney et al., 1996). Subsequently, the discovery of dental pulp stem cells and their capability to generate an ectopic pulp–dentin complex in vivo opened a new avenue to investigate the possibility of pulp–dentin tissue regeneration (Gronthos et al., 2000).

Intervention in the progressive tooth structure damage via pulp–dentin regeneration can delay or prevent tooth loss. Before complete pulp necrosis, the remaining pulp tissue may be recovered after disinfection. To induce regeneration, engineered pulp tissues may be inserted into the pulp space to facilitate the entire recovery of pulp tissue and the generation of new dentin.

2 REVITALIZATION OF IMMATURE TOOTH: PULP HEALING AND REPAIR

2.1 Apexogenesis and Apexification

Endodontically infected immature permanent teeth have been managed via either apexogenesis or apexification, depending on the clinical conditions. These two distinct clinical situations lead to dramatically different clinical outcomes. *Apexogenesis* deals with immature teeth with vital pulp that preserves the vital tissue and allows completion of root formation and apical maturation (Bishop and Woollard, 2002). *Apexification* is performed to treat immature teeth with nonvital pulp by inducing a calcific barrier at the open apex (Rafter, 2005). After apexogenesis with vital pulp therapy, teeth develop a normal thickness of dentin, root length, and apical morphology, whereas teeth receiving apexification normally gain only an apical hard tissue bridge without further development of the root.

2.2 Revitalization–Regeneration

In the past decade, some clinicians have modified their treatment procedures for endodontically compromised immature permanent teeth based on their clinical judgment. The results have been reported and generated a paradigm shift in the endodontic management of these types of teeth (Iwaya et al., 2001; Banchs and Trope, 2004; Chueh and Huang, 2006; Huang, 2008, 2009). The term *revitalization* or, more frequently, *revascularization* has been used to describe this new approach. Huang and Lin (2008) felt that *revascularization* did not encompass the actual healing and repair process that takes place in these clinical cases. The term revitalization used by earlier studies attempting to revive tissues in the pulp space perhaps describes the phenomenon more accurately (Nevins et al., 1976). Some authors have adopted *regeneration* instead.

TABLE 1 Summary of Success using Revitalization Approach

Species	Tooth Number Cases (n)	Success	Reference
Dog	$n = 60$ teeth	64.6% no radiolucency and showing healing 43.9% hard tissue formation	Thibodeau et al., 2007
	$n = 56$ cannals	64.2% no inflammation (EndoVac) 85.7% hard tissue formation (EndoVac) 71.4% hard tissue formation (conventional)	da Silva et al., 2010
Human	$n = 14$ cases	93% resolution of radiolucency 57% canal wall thickening 71% increase root length	Shah et al., 2008
	$n = 12 \rightarrow 6$ cases	50% resolution of radiolucency, canal wall thickening, and increased root length	Ding et al., 2009

Clinical Success. Currently, the success rate of this type of approach is only available from animal study models, and some pilot clinical studies in humans as summarized in Table 1. Due to the preliminary nature of these studies, clinical success rates should be interpreted with caution.

Clinical Protocol. Management of immature teeth as described in published reports and articles has the following features: (1) minimal or no instrumentation; (2) irrigation with 1.25 to 5.25% NaOCl using regular irrigation devices or EndoVac system, and (3) intracanal medication with antimicrobial agents consisting of equal parts of metronidazol, minocycline (some authors suggest not using this compound, as it causes tooth discoloration), and ciprofloxacin in paste form at the concentration of 20 mg/mL (Professional Compounding Centers of America, Houston, TX). The gentle treatment regimen is an attempt to conserve any viable tissues that may remain in the canal system which harbor stem cells. For a detailed protocol, see the website of the American Association of Endodontists (http://www.aae.org/Dental_Professionals/Considerations_for_Regenerative_Procedures.aspx). It should be noted that ongoing research is being conducted to determine a more evidence-based protocol for revitalization procedures and the clinical outcome.

Outcome of the Revitalization–Regeneration Approach

Pulp Space Filled with Periodontal Tissues. In cases where the entire pulp, apical papilla tissues, and Hertwig's epithelial root sheath are lost, current understanding is that self-regeneration of pulp and new dentin formation is unlikely to occur. A study using a dog model indicated that severe pulp infection that destroys the pulp and apical papilla followed by disinfection results in ingrowth of periodontal tissues into the canal space (Wang et al., 2010). These ingrown periapical tissues may lead to some degree of vertical and horizontal extension of the root over time. The apposition of calcified cemental tissue increases the thickness and length of the root. A distinct feature of cementum is its connection with the periodontal ligament (PDL) by Sharpey's fibers, which can also be observed in the ingrown tissues in the pulp space. The ingrowth of periodontal tissues may reach all the way to the coronal pulp chamber.

Severely Disorganized Calcification of the Pulp Space. Following the revitalization procedures, some long-term radiographic observations demonstrated that the pulp space becomes severely narrowed or filled with radiopaque mineralized tissue over time (Jung et al., 2008).

3 PULP–DENTIN ENGINEERING AND REGENERATION: ECTOPIC MODEL

A feasible small-animal model has been utilized by researchers to regenerate ectopic human pulp–dentin. The model involves human tooth slices or root fragments inserted with scaffolds and stem cells in the canal space. This construct is then transplanted into the subcutaneous space of immunocompromised mice. The tooth slice (1-mm-thick) model was a proof-of-principle to test pulp regeneration using pulp stem cells from the primary teeth (SHED) (Cordeiro et al., 2008). Blood supply is not an issue, as the tooth slice construct is thin. Well-vascularized pulp-like tissue can be formed in the pulp space after 2 to 4 weeks of implantation with odontoblast-like cells expressing dentin sialophosphoprotien (DSP) lining against the dentin surface.

In the tooth fragment approach, a section of human tooth root about ~6 to 8 mm in length is processed and the canal content is totally removed. The canal space is enlarged to 2 to 3 mm in diameter with one end of the canal opening sealed with a cement: mineral trioxide aggregate (MTA). The condition is to simulate an immature tooth root canal size. Poly(D,L-lactide-*co*-glycolide) (PLG) scaffolds carrying a heterogeneous population of stem cells from apical papilla (SCAP) or DPSCs are packed into the canal space and the constructs transplanted subcutaneously into immunocompromised mice. After three to four months, the emptied canal space is filled with a good quality of vascularized pulp-like tissue and, more importantly, a uniform thickness of a newly generated dentin-like layer is deposited onto the canal dentin wall as well as onto the MTA cement (Fig. 1). A layer of odontoblast-like cells is lined against the mineralized dentin-like tissue. There are also few scattered cells embedded within the dentin-like structure. The odontoblast-like cells are not as well organized nor as well aligned as the natural counterparts. Nonetheless, some regions of the odontoblast-like cells are of typical odontoblast characteristics: well aligned, and each cell is spatially juxtaposed to adjacent cells with polarized morphology (Fig. 1E and 1F). The regenerated dentin-like tissue does not form observable well-organized dentinal tubules except in a few regions where odontoblast-like cells are better aligned (Fig. 1G) (Huang et al., 2010b).

4 PULP–DENTIN ENGINEERING AND REGENERATION: ORTHOTOPIC MODEL

4.1 Pulp Regeneration in Mature Teeth with Complete Apical Closure

Pulp Stem and Progenitor Cells Suitable for Pulp Regeneration. Dental pulp tissue is rich in vasculature and innervation. Pulp vasculature plays a role in nutrition and oxygen supply, functions as a conduit for the transport of metabolic waste, and regulates

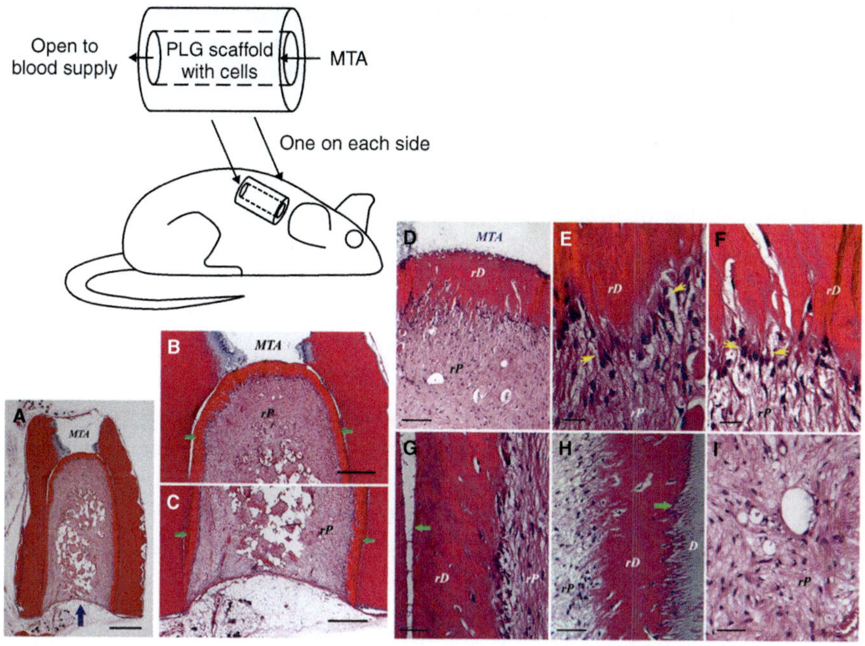

FIGURE 1 De novo regeneration of human dental pulp–dentin. Illustration at upper left depicts SCID mouse subcutaneous study model for pulp–dentin regeneration. The canal space of human tooth root fragments (about 6 to 7 mm long) was enlarged to about 2.5 mm in diameter. One end of the canal opening was sealed with MTA cement. Histological analysis of in vivo pulp–dentin regeneration using SCAP. A root fragment was prepared and the canal space inserted with SCAP/PLG and transplanted into a SCID mouse for three months. The sample was harvested and processed for H&E staining. *D*, original dentin; *rD*, regenerated dentin-like tissue; *rP*, regenerated pulp-like tissue. Blue arrow in (A) indicates the blood supply entrance; green arrows in (B) and (C) indicate a continuous layer of uniformed thickness of *rD*; yellow arrows in (E) and (F) indicate the region of well-aligned odontoblast-like cells with polarized cell bodies; green arrows in (G) and (H) indicate junctions between *D* and *rD*. Scale bars: (A) 1 mm; (B and C) 500 μm. (D) 100 μm; (E and F) 20 μm; (G to I) 50 μm. [Adapted from Huang et al. (2010), with permission].

inflammation. Pulp nerve fibers contribute to angiogenesis, extravasation of immune cells to regulate inflammation, maintenance of pulp tissue, and to strengthen pulp defense mechanisms. There is an intimate association of innervation with vasculature of the dental pulp involved in pulp homeostasis (Olgart and Kerezoudis, 1994; Nakashima and Akamine, 2005). Thus, angiogenesis and reinnervation are critical prerequisites for pulp regeneration, stimulation of migration and homing, and proliferation of stem cells and stromal matrix formation in the root canal (Nakashima and Akamine, 2005; Nakashima et al., 2009). Numerous studies have reported the isolation of adult pulp stem and progenitor cells with high proliferative and multidifferentiation potential in vitro (see Chapter 15). However, only a few studies have demonstrated angiogenesis and reinnervation in vivo after transplantation.

Porcine and human pulp $CD31^-$ side population (SP) cells and $CD105/endoglin^+$ cells, positive for CD29, CD44, CD73, and Thy1/CD90, and negative for CD31, CD45, and CD133, increase the blood flow, including a high density of capillary formation after transplantation into mouse hindlimb ischemia (Iohara et al., 2008; Nakashima et al., 2009). SP cells are enriched in the "true" or "mother" adult stem cells, which exhibited a lower level of the DNA-binding fluorescent dye Hoechst 33342 than that of the rest of the pulp cells (Iohara et al., 2006). CD105/endoglin is a component of the transforming growth factor-β receptor complex. CD105 (endoglin) is widely expressed on mesenchymal stem cells (MSCs) (Barry et al., 1999), and the positive isolation strategy based on the selection of the cells that express CD105 has been proposed for isolation of MSCs (Jarocha et al., 2008). Efficient enrichment of immature human MSCs in $CD105^{+}/CD45^{-}$ cell fraction, in terms of cloning efficiency and a high expression of OCT4 and NANOG, has been reported in bone marrow (Kastrinaki et al., 2008). Pulp $CD31^{-}$ cells and $CD105^{+}$ cells induce angiogenesis and neurogenesis in rat cerebral ischemia by releasing neurotrophic factors, such as VEGF in peri-infarct area, and promoting migration and differentiation of the endogenous neuronal progenitor cells (NPCs), and accelerate the functional recovery (Sugiyama et al., 2011). Local transplantation of these pulp stem and progenitor cells results in successful engraftment in proximity to the newly formed vasculature without being functionally incorporated into vessels, suggesting enhancement of angiogenesis and reinnervation by paracrine mediators during injury and regeneration as in other MSCs (Iohara et al., 2008; Sugiyama et al., 2011). The conditioned medium of pulp stem and progenitor cells contains a high concentration of angiogenic and neurotrophic factors, such as brain-derived neurotrophic factor (BDNF), nerve growth factor (NGF), vascular endothelial growth factor (VEGF-A), glial cell line–derived neurotrophic factor (GDNF), granulocyte–macrophage colony-stimulating factor (GM-CSF), granulocyte colony-stimulating factor (G-CSF), and matrix metalloproteinase 3 (MMP3). These cytokines and growth factors are effective on proliferation, migration, and antiapoptosis of endothelial cells and neuroprogenitor cells (Iohara et al., 2008, Sugiyama et al., 2011). On the other hand, unfractionated total pulp cells express significantly lower angiogenic and neurotrophic factors and have less effect than $CD31^{-}$ SP cells and $CD105^{+}$ cells on angiogenesis and neurogenesis after transplantation in both hindlimb ischemia and brain ischemia (Nakashima et al., 2009). Thus, some subfractions of pulp stem and progenitor cells might be useful cell sources for cell-based therapy of angiogenesis, reinnervation, and pulp regeneration.

Trophic action is a more likely therapeutic mechanism since the relatively rapid and persistent effect of the transplanted stem cells decreases, even after donor cell survival (Parr et al., 2007). MSCs release soluble cytokines and growth factors that function in a paracrine fashion, contributing to repair and regeneration by promoting cell survival, angiogenesis, neurogenesis, and neuroprotection, and by activating resident stem cells (Kassis et al., 2011).

Partial Pulp Regeneration with Pulp Stem and Progenitor Cells. Recent advances in tissue regeneration have led to the possibility of partial pulp regeneration in the case of pulp exposure or pulpotomy of partial pulptitis. The remaining healthy pulp tissue is hypothesized to be recoverable before necrosis of the whole pulp under certain conditions in immature teeth (Nakashima and Akamine, 2005). There are two methods in partial pulp regeneration even in mature teeth with complete apical closure: (1) transplantation of pulp stem and progenitor cells with scaffold or engineered pulp tissue, and (2) application of migration and homing factors with scaffolds.

It is possible to engineer pulp-like tissue by utilizing cultured pulp cells and synthetic extracellular matrix scaffolds (Mooney et al., 1996; Buurma et al., 1999). Subcutaneously implanted pulp cells seeded in PGA result in producing extracellular matrix (Buurma et al., 1999). However, their engraftment and induction of pulp regeneration has not been confirmed in vivo orthotopically after transplantation in the cavity on the remaining pulp tissue. The first method (Fig. 2A) is evidenced by an experiment in which canine autologous pulp $CD31^-$ SP cells were transplanted in the cavity on the amputated pulp using collagen scaffold. The regenerated pulp tissue filled in the cavity with well-developed vasculature and innervation, and the tubular dentin was formed along the dentinal wall (Iohara et al., 2009). Many of transplanted $CD31^-$ SP cells are close to the newly formed capillaries and expressed proangiogenic factors, implicating trophic effects on neovascularization (Iohara et al., 2009) as seen in a hindlimb (Iohara et al., 2008) and brain ischemic model (Sugiyama et al., 2011). On the other hand, after transplantation of $CD31^+$ SP cells, less newly regenerated pulp tissue and fewer capillaries are seen compared with transplantation of $CD31^-$ SP cells (Iohara et al., 2009), suggesting different potential utility in induction of pulp regeneration among subfractions of pulp stem and progenitor cells.

An experimental approach to the second method (Fig. 2A)is fibroblast growth factor 2 (FGF2) absorbed in biodegradable gelatin hydrogels, from which the release is regulated to control the final regenerated tissue, pulp or dentin (Ishimatsu et al., 2009). Another experiment related to the second method (Fig. 2A) is matrix metalloproteinase 3 (MMP3), which has a strong proliferation, migration, and antiapoptotic effect on endothelial cells (Zheng et al., 2009). Its application enhances proliferation, matrix formation, and mineralization in the cavity on the amputated pulp (Zheng et al., 2009) and regenerates the pulp tissue in case of partial pulpitis of the canine experimental model (unpublished data). The extracellular matrix proteins, particularly laminin, and chemoatractants such as sphingosine-1-phosphate (S1P) and TGFβ1, are also important promoters of pulp stem cell migration, mediating the process of pulp–dentin regeneration after tooth injury (Howard et al., 2010). When pulp stem cells seeded on the surfaces of three-dimensional collagen gel cylinders and incubated in chemically defined media with stromal cell–derived factor-1α (SDF1, CXCL12/pre-B cell growth-stimulating factor) or basic fibroblast growth factor (bFGF), more cells are recruited into collagen gel by SDF1 or bFGF than in the absence of cytokines. Upon in vivo delivery, bFGF induced recellularization and revascularization in endodontically treated human teeth implanted into the dorsum of rats (Suzuki et al., 2011). Thus, endogenous dental pulp cells, including stem and progenitor cells, may be recruited and subsequently differentiated by chemotaxis of selective cytokines in the regeneration of dental pulp. Those homing and migration factors, such as S1P, TGFβ, bFGF, and SDF1, are potentially useful as part of regenerative endodontic therapy.

Complete Pulp Regeneration with Pulp Stem and Progenitor Cells and Migration Factors. There is considerable optimism for whole-pulp regeneration induced by autologous pulp stem and progenitor cells in case of irreversible pulpitis of mature teeth with complete apical closure. It is critical for the transplanted cells to be supplied with nutrition and oxygen in the empty root canals. Thus, some subfractions of pulp stem and progenitor cells with high migration potential, such as $CD105^+$ cells or $CD31^-$ SP cells, could be more potent than unfractionated total pulp cells. As

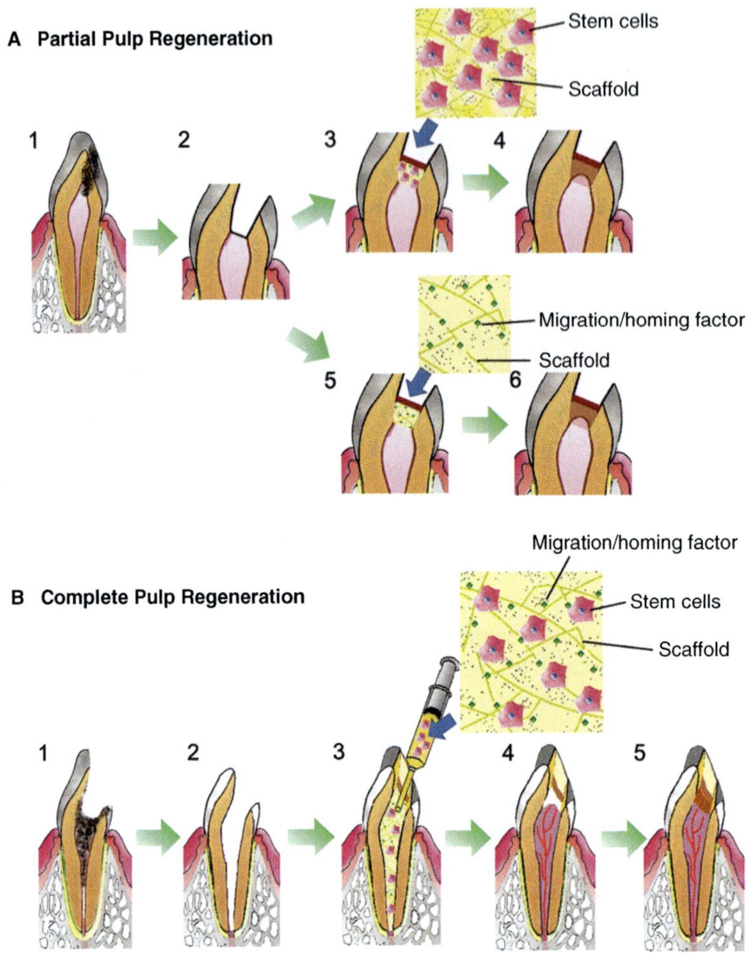

FIGURE 2 (A) Schematic diagrams for partial pulp regeneration in permanent mature teeth: (1) pulpitis model, (2) pulpetomy, (3) irrigation and filling with stem/progenitor cells with scaffold, (4) pulp and dentin regeneration, (5) irrigation and filling with migration/homing factor with scaffold, and (6) pulp and dentin regeneration. (B) Schematic diagrams for complete pulp regeneration in permanent mature teeth: (1) pulpitis model, (2) whole-pulp removal and enlargement of apical foramen, (3) irrigation and filling with stem/progenitor cells and migration/homing factor in the root canal together with scaffold, (4) complete pulp regeneration, and (5) dentin regeneration.

mentioned above, pulp stem and progenitor cells induce angiogenesis and reinnervation but not themselves forming vasculature de novo, as a result, it may be necessary to incorporate migration and homing factors for endogeneous stem and progenitor cells to move into the root canal for whole functional pulp regeneration. One candidate for migration and homing factors is SDF1, known as a chemokine for CXCR4-positive stem cells. SDF1 is involved in hematopoietic stem cells (HSCs) homing to the bone marrow niche (Chute, 2006; Kaplan et al., 2007) and in adult subventricular zone (SVZ) cells homing to areas of central nervous system injury following ischemia (Ohab et al.,

2006; Robin et al., 2006; Thored et al., 2006). Transplantation of GFP-labeled bone marrow–derived cells into irradiated wild-type mice under normal conditions resulted in more engraftment into the dental mesenchymal tissue than other organs, since expression of SDF1 was significantly higher in dental pulp than in other tissues (Zhou et al., 2011). Pulp $CD31^-$ SP cells and $CD105^+$ cells are also CXCR4-positive with high migratory and proliferative activity with SDF1 (Iohara et al., 2008; Nakashima et al., 2009). These transplanted cells can migrate to the injured site of the amputated pulp in the partial pulp regeneration (Iohara et al., 2009). Thus, it is plausible that SDF1 transplanted with pulp stem cells provides signals for homing and proliferation of endogenous stem cells originating from surrounding periodontal ligament, bone marrow, and vasculature. Other possible migration and homing factors are bFGF, VEGF, and PDGF. A recent study has shown that combinatorial delivery of bFGF, VEGF, or PDGF with NGF and BMP7 has potent cell homing effects for revascularization and pulp regeneration in endodontically treated root canals of human teeth after in vivo ectopic implantation (Kim et al., 2010).

Optimal scaffolds should also be developed for efficient and safe total pulp regeneration. It is critical to develop simple methods to inject into the empty root canal with optimal flow, devoid of bubble formation and functional mechanical properties after hardening in vivo. In addition, it has to be biocompatible, with high bioactivity to integrate trophic factors secreted by stem cells for chemotaxis and homing of progenitor cells. Finally, the scaffold has to be biodegradable, releasing these factors to replace the damaged tissue by newly formed regenerated tissue replete with extracellular matrix without immune response. In addition, the scaffold should not stimulate deposition of mineral and differentiation of odontoblasts and/or osteodentinoblasts in the root canal except along the dentinal wall for pulp regeneration (Nakashima et al., 2009). Potential scaffolds include natural polymers, such as collagen, gelatin, dextran and fibronectin, with good biocompatibility and bioactivity. Among synthetic polymers, poly(glycolic acid) (PGA) can control precisely the physicochemical properties of degradation rate, porosity, microstructure and mechanical properties (Nakashima and Akamine, 2005).

Thus, based on the concepts described above, pulp regeneration has been demonstrated by using the tissue engineering triad, stem and progenitor cells, migration and homing factors, and optimal scaffolds (Nakashima and Iohara, 2011, Iohara et al., 2011) (Fig. 2B). Pulpectomy with enlarged apical portion, 0.6 to 0.7 mm in width, was performed in mature teeth with complete apical closure in an experimental model in dogs (Iohara et al., 2011; Ishizaka et al., 2012). Current evidence shows that the root canal is filled with pulp-like loose connective tissue with vasculature and nerves after autologous transplantation of pulp stem and progenitor cells, $CD31^-$ SP cells or pulp $CD105^+$ cells, 5×10^5 cells in cell number, with SDF1 in collagen type I and type III scaffold (Fig. 3-1, 3-5, and 3-6) (Nakashima et al., 2011; Iohara et al., 2011; Ishizaka et al., 2012). The three-dimensional image of induced vascularization in the regenerated tissue (Fig. 3-8 and 3-9) is similar in vascular density and orientation to those in normal pulp. The neuronal process from regenerated pulp appears to connect to the inferior alveolar nerve (Fig. 3-10) (Iohara et al., 2011). The transplanted pulp stem cells expressed angiogenic and neurotrophic factors and are localized in the vicinity of newly formed capillaries (Iohara et al., 2011; Ishizaka et al., 2012), suggesting potential trophic effects on neovascularization. The odontoblast-like cells attached to the dentinal wall in the root canal produce dentin-like mineralized tissue extending their processes into dentinal tubules (Fig. 3-7). The enlarged apical portion following pulpectomy is filled by additional formation of dentin and cementum. Accessory and lateral canals and even the perforation site are obliterated by osteodentin, suggesting its potential utility for fractures and perforated teeth. On the other hand, transplantation of stem and progenitor cells alone (Fig. 3-2), or SDF1 alone (Fig. 22-4), yields significantly less pulp tissue. When unfractionated total pulp cells are implanted instead of fractionated pulp stem cells, there is significantly less regenerated tissue and it undergoes mineralization (Fig. 3-3) (Iohara et al., 2011). The regenerated tissue elicited by the pulp stem and progenitor cells and SDF1 is identical to normal functional pulp tissue, as demonstrated by the abundant expression of the pulp tissue markers *Syndecan* and *TRH-DE* mRNA. It is noteworthy that the qualitative and quantitative protein and mRNA expression patterns obtained by two-dimensional electrophoretic analyses and microarray analyses are virtually identical (Iohara et al., 2011; Ishizaka et al., 2012). Furthermore, in a canine experimental model of periapical disease in which the root canal is kept open for more than 3 weeks, pulp tissue has been demonstrated to be regenerated after transplantation of pulp stem cells and SDF1. The possible mechanisms of pulp regeneration may involve the CXCR4/SDF1 axis functioning as a migration or homing factor for CXCR4-positive endogenous stem cells to migrate to the coronal portion of the root canal, proliferate, and differentiate into endothelial cells, and possibly neural cells for angiogenesis and reinnervation by angiogenic and neurotrophic factors secreted by transplanted stem cells. These findings suggest potential clinical translation of complete pulp regeneration by harnessing pulp stem and progenitor cells with high angiogenic and neurogenic potential and additional migration or homing factors in endodontic treatment.

The mechanisms of recruitment and crucial molecules for cell migration are still unclear. Chemokines and their receptors, which play critical roles, need further scrutiny. The induction of MSC recruitment from surrounding tissues or from the circulation can be a helpful modality to initiate and support cell therapy for pulp regeneration.

Pulp Regeneration for Aged. In elderly patients the challenge for endodontic therapy is the need to preserve teeth and facilitate optimal oral function (Allen and Whitworth, 2004). The supply of autologous pulp tissue is very limited with age, due to shrinking pulp tissue by physiological secondary dentin formation, pathological tertiary dentin formation, and mineralization, including denticle and pulp stones (Murray et al., 2002).

In general, there are three distinct types of aging changes within MSCs: quantity, quality (differentiation and regeneration capacity), and mobilization capacity (Sethe et al., 2006). Teeth from aged patients have less pulp tissue and generate fewer DPSC colonies than those from young patients (Iida et al., 2010). Human DPSCs from the aged donors lose their differentiation capabilities after repeated replication (Takeda et al., 2008). Hypoxic cultures, grown under 3% O_2, however, succeed in recovering and expanding a small number of human pulp stem cells, suppress osteo and odontogenic differentiation, and increase the number of early stem cells, indicating the possibility of obtaining human pulp stem cells from aged patients (Iida et al., 2010). On the other hand, there are no differences in proliferative and migratory effects in pulp $CD105^+$ cells from young and aged patients. Expression of stem cell markers, mRNA expression of angionenic and neurotrophic factors, and multilineage differentiation capability are similar in both the young and the aged (Nakashima, unpublished data). Thus, the scientific literature might appear as contradictory. Many apparent discrepancies are attributable to inconsistent methods of isolating MSCs which contain various subsets of stem cells, varying in their differentiation potential and their vulnerability to senescence, and to the absence of a clear understanding of MSCs (Sethe et al., 2006;

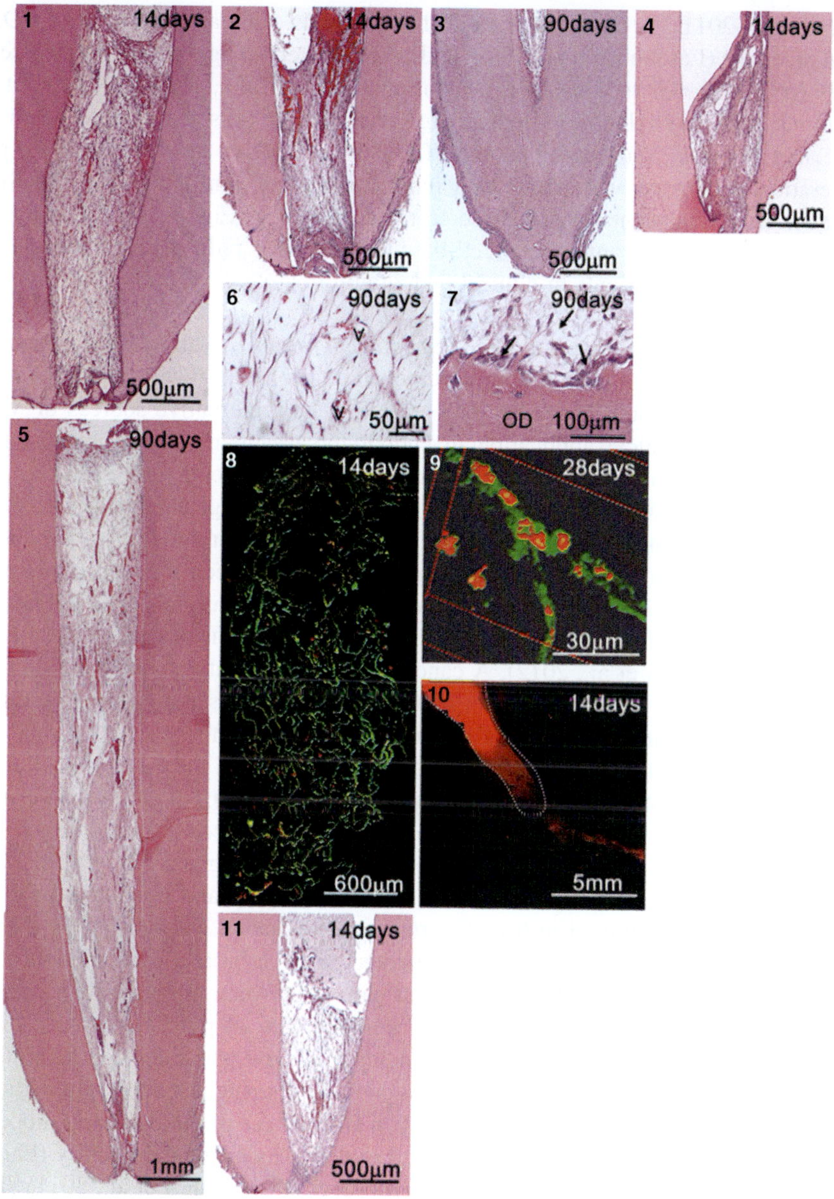

FIGURE 3 Complete regeneration of pulp tissue after autologous transplantation of pulp $CD105^+$ stem cells and SDF1 with collagen scaffold in the emptied root canal after pulpectomy in dogs. (1, 5 to 10) $CD105^+$ cells and SDF1. (2 and 3) Total pulp cells and SDF1; no mineralization in the root canal. (4) SDF1 only. (6) Newly formed blood vessels (v) in the upper part of regenerated tissue. (7) Odontoblastic cells (arrows) lining to newly formed osteodentin/tubular dentin (OD) along with the dentinal wall. (8 and 9) Three-dimensional images of new vascularization by whole-mount immunostaining with lectin (green). (8) DiI-labeled transplanted cells located throughout the tissue (red). (9) Transplanted $CD105^+$ cells in the vicinity of newly formed capillaries. (10) Reinnervation in newly formed pulp tissue (white dotted line) connecting to inferior alveolar nerve. (11) Pulp regeneration after transplantation of pulp $CD105^+$ cells with SDF1 in aged dogs. [Modified from Iohara et al. (2011)].

Lepperdinger, 2011). In the future, established methods, such as expression of CD295 (leptin receptor) (Laschober et al., 2009) and other markers of aging, such as senescence signals (p53, p16, RB tumor suppressor protein, Numb, NFk, ICAM1) (Sethe et al., 2006) and DNA methylation (Schellenberg et al., 2011), must be used to determine age-related changes. Yet autologous pulp $CD105^+$ cells from aged dogs induce pulp regeneration with good vasculature and innervation after transplantation with SDF1 as shown after transplantation of those from young dogs (Fig. 3-11). This suggests the possible efficacy of autologous cell-based therapy for pulp regeneration even in middle-aged and old patients in clinical endodontic treatment.

Other Sources of Tissue Stem Cells for Pulp Regeneration. Although the efficacy of pulp stem and progenitor cells has been demonstrated for pulp regeneration, the availability of autologous pulp tissue declines with age, and alternative sources of MSCs are necessary for clinical application in endodontic treatment. MSCs isolated from many other tissues, such as bone marrow, adipose tissue, placenta, umbilical cord, and amnion, reveal very similar profiles by transcriptional and epigenetic analysis (Aranda et al., 2009; Boeuf and Richter, 2010) Different expression profiles of trophic factors and growth factors among MSC populations has, however, been reported (Noël et al., 2008; Boeuf and Richter, 2010; Philippe et al., 2010). Therefore, it is necessary to investigate the requirements and preconditions of MSCs for effective induction of pulp regeneration (Ishizaka et al., 2012). A candidate of an alternative cell source for pulp tissue regeneration is autologous bone marrow and adipose tissue–derived MSCs, for neither ethical nor immunoreactive considerations. There are many identical characteristics between the two types MSCs (Zuk et al., 2001), although there is some difference in protein and function (Noël et al., 2008; Philippe et al., 2010). A similar level of gene expression between pulp and bone marrow MSCs is detected for more than 4000 known human genes, except a few differentially expressed genes (Shi et al., 2001). The differential proteomic expression profiles of ovine pulp and bone marrow MSCs from an individual donor identified 23 proteins upregulated in pulp MSCs (Mrozik et al., 2010). Expression patterns of gene markers and the broad differentiation potentials induced by specific methods in vitro are similar between rabbit pulp and adipose MSCs analyzed by parallel comparisons. Transplantation of pulp MSCs and adipose MSCs with BMP7 was able to grow a self-assembled new whole tooth–like structure contained mineralized dentin in rabbit extraction sockets, suggesting that they did not affect the tooth regeneration (Hung et al., 2011). Differences in the biological properties of MSC populations have previously ascribed to donor-associated variability (Phinney et al., 1999) and to cell isolation and cultivation methods (Vemuri et al., 2011). Thus, biological characteristics have been compared for pulp, bone marrow, and adipose MSCs associated with a defined protocol for isolation (SP subfraction) and cultivation and originated from a single donor (Ishizaka, 2012). Transplantation of adipose $CD31^-$ SP cells resulted in a similar amount of regenerated tissue formation compared with transplantation of pulp $CD31^-$ SP cells (Fig. 3-1 and 3-3). Although bone marrow $CD31^-$ SP cell transplantation induce significantly less regenerated tissue than do the other two (Fig. 4), those regenerated tissues are all identical to pulp tissue in respect to morphological and functional similarity, mRNA expression patterns of microarray analyses, and two-dimensional electrophoretic analyses (Ishizaka et al., 2012). Thus, it is suggested that MSCs from different sources are comparable in pulp regeneration in vivo and that their capabilities are not influenced by the niche of their origin. The local environment may provide signals driving the fate of homing endogenous stem cells originated from surrounding periodontal ligament, bone marrow, and vasculature (Ishizaka et al., 2012). However, displaying enhanced matrix formation leading to root canal obliteration in adipose cell transplantation after longer periods suggests that some trophic factors, growth and differentiation factors, and/or inhibitors of stemness may influence the pathways of differentiation in adipose stem cells. Further studies are needed to elucidate the underlying mechanisms.

Allogenic Stem Cell Transplantation. Allogeneic mesenchymal stem cells (MSCs) offer several advantages over autologous MSCs. These include an unlimited supply with little restriction in time and tremendous reduction of manufacturing costs (Vassalli and Moccetti, 2011). MSCs possess a remarkably diverse array of immunosuppressive characteristics (English and Mahon, 2011). Off-the-shelf allogeneic products have proven safe in phase I and II trials in patients with heart failure (Dib et al., 2009), acute myocardial infarction (Hare et al., 2009), multiple sclerosis (Karussis et al., 2010), and knee osteoarthritis (Davatchi et al., 2011). Little is known concerning the in vivo survival of allogeneic MSCs and immune rejection of allogeneic cells. Recent data suggest that allogeneic MSCs may switch immune states in vivo to express HLA class II, present alloantigens, and induce immune rejection (Huang et al., 2010c). If immunogenicity of human allogeneic MSCs is a concern, pharmacological immunosuppression a for limited time period after cell transplantation may be considered as an option. Furthermore, matching of donor–recipient HLA-II might be useful for prevention of alloimmune rejection. Of note, low immunogenicity and immunosuppressive

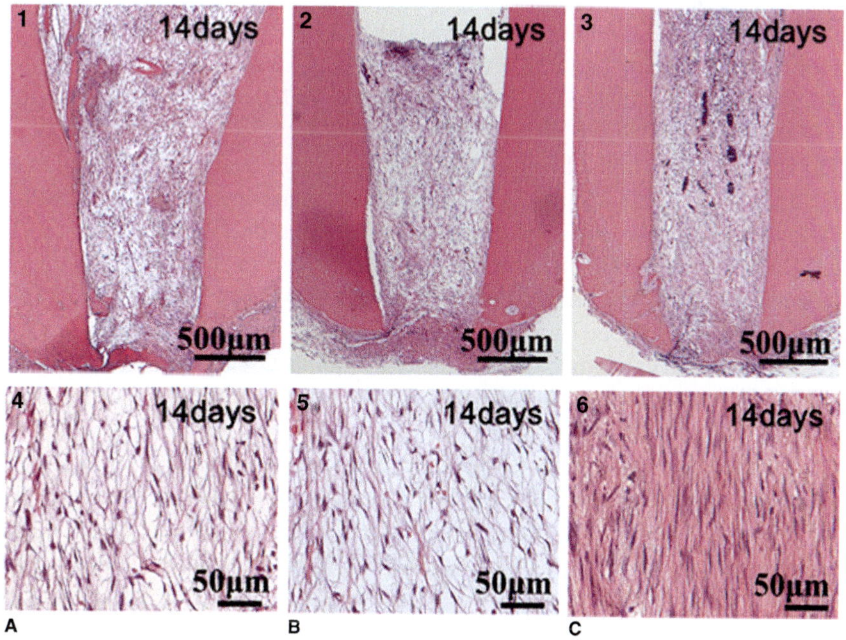

FIGURE 4 Complete regeneration of pulp tissue after autologous transplantation of CD31$^-$ SP cells from three different tissues together with SDF1 after pulpectomy in dogs. (A) Pulp CD31$^-$ SP cell transplantation. (B) Bone marrow CD31$^-$ SP cell transplantation. (C) Adipose CD31$^-$ SP cell transplantation. [Modified from Ishizaka et al. (2012)].

effects are also suggested in human pulp $CD105^+$ stem cells, due to very low expression of CD40, 80, and 86 and MHC class II and a high expression of soluble factors such as interleukins (IL) 6 and 10, TGFβ, respectively. Pulp $CD105^+$ cells inhibit immune responses of peripheral blood mononuclear cells (PBMCs) in mixed lymphocyte reactions (MLRs), suggesting their immunomodulatory effects (Nakashima, unpublished data). Similar immunomodulatory actions on lymphocytes have also been reported (Tang and Ding, 2011). In addition, paracrine mechanisms of induction of pulp regeneration suggest that long-term survival of transplanted stem cells might not be necessary to achieve sustained clinical effect. These findings suggest that allogeneic pulp stem cells might be useful for pulp regeneration.

4.2 Coronal Dentin Regeneration

Ideally, the coronal cavity should be sealed by regenerated dentin, allowing for the maintenance of pulp vitality and restoration of tooth structure and function (Nakashima and Akamine, 2005). The regenerated dentin has a greater ability to protect against physicochemical and bacterial irritation than does any restorative material and restores the mechanical strength (Smith et al., 2000). Vital pulp therapies by calcium hydroxide and mineral trioxide aggregate (MTA) considered as bioactive materials (Enkel et al., 2008; Parirokh and Torabinejad, 2010) is not sufficient, since dentin is formed toward the residual pulp tissue, constricting its size. The key elements for ideal dentin regeneration are morphogenic signaling molecules, a scaffold of extracellular matrix and stem and progenitor cells.

Morphogenic Signaling Molecules. Morphogenic signaling molecules are inductive signals that function as migration factors and/or growth and differentiation factors for homing, proliferation, and differentiation of pulp stem and progenitor cells and endothelial progenitor cells into odontoblasts and endothelial cells (Tziafas et al., 2001; Nakashima and Reddi, 2003; Smith, 2003; Goldberg et al., 2003 2006; Tziafas, 2004; Nakashima and Akamine, 2005; Komabayashi et al., 2010; Demarco et al., 2011; Simon et al., 2011; Sun et al., 2011). Recombinant human BMP2, BMP4, and BMP7 induce reparative and regenerative dentin formation (Rutherford et al., 1993, 1994; Lianjia et al., 1993; Nakashima, 1994a,b; Jepsen et al., 1997; Decup et al., 2000; Rutherford and Gu, 2000; Goldberg et al., 2001; Six et al., 2002). TGF-β1 with a collagen membrane (Hu et al., 1998), with preset calcium hydroxide (Tziafas et al., 2001) and with calcium phosphate material equipped with poly(lactic-*co*-glycolic acid) (PLGA) microspheres (Zhang et al., 2008) are also effective on the induction of dentin. Recombinant human insulin–like growth factor I with collagen membrane induces complete dentin bridging and tubular dentin formation (Lovschall et al., 2001). Adjusting the dosage of the released FGF2 from gelatin hydrogels induces dentin regeneration (Ishimatsu et al., 2009; Kitamura et al., 2012). Dentin contains numerous bioactive proteins, such as migration factors, growth and differentiation factors, enzymes, signaling molecules, and transcription factors, modulating cellular responses such as migration, proliferation, differentiation, and mineralization. The soluble proteins are released from dentin as a consequence of injury and function as regulatory signals for odontogenesis (Smith, 2003; Silva et al., 2004). A recent proteomic analysis of ethylenediaminetetraacetic acid–soluble tooth proteins (ESTPs) has detected 29 novel proteins, and ESTP treatment promotes in vivo dentin and tooth formation, as shown by xenogeneic combination and embryonic tooth-forming primordium models, suggesting that ESTPs contain dentinogenic proteins useful for tissue engineering (Chun et al., 2011).

Specific dentin extracellular matrix proteins also act as bioactive molecules, by which cells implicated in the reparative process are recruited, proliferate, and differentiate into osteoblast- and odontoblast-like cells (Goldberg et al., 2006). Those include bone sialoprotein, dentonin [a fragment from matrix extracellular phosphoglycoprotein (MERE)] and two small amelogenin gene splice products (A + 4 and A − 4) (Goldberg et al., 2009). Dentin matrix protein 1 (DMP1), a noncollagenous extracellular matrix protein of the dentin, serves as a potent growth factor. DMP1 is essential in the maturation of odontoblasts and mineralization via local and systemic mechanisms (Qin et al., 2007). Proteolytic processing of DMP1 by MMP2 can regulate the differentiation of mesenchymal cells during reparative dentin formation (Chaussain et al., 2009). These studies suggest a new therapeutic possibility of using these bioactive extracellular matrix proteins or MMPs to regenerate dentin.

Scaffolds of Extracellular Matrix. The origin and evolution of multicellular organisms was accompanied by extracellular matrix. Among the dental tissues, dentin has vast expanses of extracellular matrix. Scaffolds of extracellular matrix are native composites of biomaterials that are indispensable during dentin regeneration. The native extracellular matrix of teeth provides a scaffold for the recruitment, adhesion, proliferation, and differentiation of endogenous stem and progenitor cells. There is a growing interest in the synthetic peptides and biomimetic proteins imitating extracellular matrix that deliver morphogens to function as a scaffold with tethered morphogens and growth factors (Yuan et al., 2011). The synergistic effects of fibrin scaffolds with recombinant human BMP2 (rhBMP2) induced copious amounts of dentin formation compared to rhBMP2 alone (Ren et al., 1999). Platelet-rich fibrin stimulates proliferation and differentiation of pulp stem cells (Huang et al., 2010a). Poly(ethylene glycol) (PEG)-modified fibrin (PEGylated fibrin) that allows proliferation and differentiation of pulp stem cells can be inserted into small defects and could potentially be used for dental tissue engineering (Galler, 2011a). Allogeneic type I collagen, sponge or gel, has excellent biocompatibility and low immunogenicity. Collagen permits the binding of preodontoblasts and differentiation into odontoblasts and pulp tissue, supporting a reparative dentinogenesis (Kitasako et al., 2002). The presence of an Arg–Gly–Asp (RGD) cell attachment sequence and poly(glutamic acid) stretches in bone sialoprotein (BSP) permits binding of hydroxyapatite to BSP implantation of bone sialoprotein in the pulp tissue, which leads to upregulation of secretory activity of the extracellular matrix of newly generated odontoblasts, resulting in thick reparative dentin formation (Decup et al., 2000). Although hyaluronic acid (HA) is biocompatible and has low immunogenicity (Richter et al., 1979), its poor mechanical strength and rapid degradation rate present potential disadvantages (Jia et al., 2006). Cell attachment, spread, and proliferation are enhanced with RGD peptides (Pouyani and Prestwich, 1994). High-molecular-mass HA can provide an environment suitable for reparative dentin formation through mesenchymal cell differentiation (Sasaki and Kawamata-Kido, 1995). It is noteworthy that there is only a modest migration of granulated leukocytes into HA compared with those into collagen sponges after implantation in rat amputated pulp (Inuyama et al., 2010). Poly(ethylene glycol)–based hydrogel is a nontoxic water-soluble biocompatible polymer. PEG hydrogel has low immunogenicity and can undergo in vivo degradation. PEG hydrogel has advantages in the ability for photopolymerization, with easy control of scaffold structure and chemical composites

(Peppas et al., 1999). Alginate is a naturally derived polysaccharide and is nontoxic, biocompatible, and permeable to small-molecular-mass proteins (Lawson et al., 2004). It is difficult to control the in vivo degradation rate of alginate. Alginate hydrogels enhance dentin matrix secretion, odontoblast-like cell differentiation, and subsequent secretion of tubular dentin matrix, by hydration properties and tethering growth factors (Dobie, 2002). Dental pulp cells actively differentiate into odontoblast-like cells and induce mineralization in an alginate scaffold after subcutaneous implantation (Kumabe et al., 2006). Hydroxyapatite is biocompatible, immune tolerant, and has the ability to be osteoconductive (Furusawa et al., 1991). The performance of the porous hydroxyapatite scaffold is dictated by internal architecture, including pore size, porosity, and interconnections (Higashi and Okamoto, 1996; Zhang and Zhu, 2007). The nanocrystalline apatite coating on biphasic calcium phosphate ceramics is found to favor the adsorption of rhBMP2 while providing a means to fine tune release of the growth factor (Autefage et al., 2009).

The development of optimal scaffolds with safety and efficacy for dentin regeneration is essential for additional clinical applications. The combination of inductive scaffold with stem cells might optimize the approaches for dentin–pulp complex regeneration (Galler et al., 2011b).

Dental Pulp Stem and Progenitor Cells as a Source for Dentin Regeneration. Dental stem cells, either DPSCs or SCAPs isolated by enzymatic digestion without further fractionation, constitute a heterogeneous cell population. These unfractionated pulp cells demonstrate more mineralization in vitro and in vivo than that with fractionated $CD105^+$ cells and $CD31^-$ SP cells (Iohara et al., 2011), suggesting that unfractionated pulp cells may be a more suitable cell source if a significant amount of dentin regeneration is clinically desired or indicated. Transplantation of a three-dimensional culture of pulp cells supplemented with BMP2, with collagen as a scaffold, induces regenerated osteodentin formation on the amputated pulp in the cavity (Iohara et al., 2004). Implantation of pulp progenitors, a substitute for traditional endodontic treatments, paves the way for pulp stem cell–based therapy (Nakashima and Reddi, 2003; Nakashima, 2005; Harichane et al., 2011).

5 FUTURE PERSPECTIVES

One important goal of tissue regeneration is the regeneration of functional tissues. Pulp tissue is highly organized and complex. Revitalization procedures currently lack convincing evidence of regeneration pulp- of dentin-like tissues; instead, periodontal tissues are grown into the canal space. In contrast, regenerated pulp tissue via stem cell–based tissue engineering techniques has so far shown the following features: (1) vascularization, (2) cell density and architecture in the extracellular matrix similar to those of the natural pulp, (3) formation of new odontoblast-like cells lining against the existing dentin surface with new dentin-like tissue produced, and (4) innervation. In the ectopic pulp and dentin regeneration in the subcutaneous space, no nerve fibers were detectable by staining. However, the orthotopic regeneration showed regenerated nerve fibers in the pulp.

In the past, the vascularization in pulp was thought to be difficult for mature teeth having a small apical canal opening (<1 mm). The size of apical opening would affect the ingrowth of blood vessels into the engineered pulp tissue. The larger the opening, the more likely the angiogenesis can occur. Therefore, immature teeth with open apices are the best candidates for de novo pulp tissue regeneration. Pulp and dentin can also be completely regenerated in emptied canals of mature teeth with an apical opening enlarged to 0.7 mm in a dog study model (Iohara et al., 2011). As long as a good blood supply can be achieved, optimal cell density and a deposit of good-quality extracellular matrix should occur. New odontoblast-like cells will form against the existing dentinal wall (Huang et al., 2010b; Iohara et al., 2011). With respect to innervation, the orthotopic pulp regeneration in the dog model demonstrated reinnervations. However, the specific innervation at the pulp–dentin complex remains to be determined. It is speculated that because the newly generated dentin does not appear to have well-organized dentinal tubules and is similar to reparative dentin, if the regenerated A-δ fibers reach the pulp–dentin junction, it may not cause the dentin sensitivity normal to that of natural teeth.

The availability of cell source for cell-based tissue regeneration is an issue. For clinical success of the pulp–dentin regeneration, the stem and progenitor cells should be isolated and expanded in a safe, stable, and efficient manner. Stem cell banking and U.S. Food and Drug Administration–approved current good manufacture practice (cGMP) facilities are needed as an infrastructure to make cell-based therapy practical (Sensebé et al., 2010). Stem and progenitor cells are examined for no tumor formation after implantation subcutaneously or in testis and for freedom from chromosome aberrations before clinical application (Sensebé et al., 2010).

$CD31^-$ SP cells have to be labeled with DNA-binding Hoechst33342 and isolated by flow cytometry, which requires the confirmation of safety. $CD105^+$ cells have to be isolated by a magnetic antibody bead method if not by flow cytometry, and have to be cost-effective. In addition to stem and progenitor cells, migration and homing factors and scaffolds should consist of GMP products. Therefore, the cost involved needs to be evaluated before regenerative endodontics can replace the conventional endodontic treatment.

Although there remain challenges in regenerative endodontics, the immense clinical benefits that can be offered by this new technology warrant our further efforts into advancing this field of research.

Acknowledgments

This work was supported by Collaborative Development of Innovative Seeds, Potentiality Verification Stage, from the Japan Science and Technology Agency to M.N.; and a Research Grant for Longevity Sciences (21A-7, 23-10) from the Ministry of Health, Labour and Welfare to M.N.; a Grant-in-Aid for Scientific Research from the Ministry of Education, Science, Sports and Culture, Japan, grants 20390504 to M.N. and grants 21791872 to K.I.; and in part by National Institutes of Health/National Institute of Dental and Craniofacial Research grant R01DE019156 (to G.T.-J.H.) and an American Association of Endodontists Foundation grant (to G.T.-J.H.).

REFERENCES

Allen PF, Whitworth JM (2004). Endodontic considerations in the elderly. *Gerodontology* 21(4):185–194.

Aranda P, Agirre X, Ballestar E, Andreu EJ, Romá n-Gó mez J, Prieto I, et al. (2009). Epigenetic signatures associated with different levels of differentiation potential in human stem cells. *PLoS One* 4(11):e7809.

Autefage H, Briand-Mésange F, Cazalbou S, Drouet C, Fourmy D, Gonçalvès S, et al. (2009). Adsorption and release of BMP-2 on nanocrystalline apatite-coated and uncoated hydroxyapatite/beta-tricalcium phosphate porous ceramics. *J Biomed Mater Res B* 91(2):706–715.

Banchs F, Trope M (2004). Revascularization of immature permanent teeth with apical periodontitis: New treatment protocol? *J Endod* 30(4):196–200.

Barry FP, Boynton RE, Haynesworth S, Murphy JM, Zaia J (1999). The monoclonal antibody SH-2, raised against human mesenchymal stem cells, recognizes an epitope on endoglin (CD105). *Biochem Biophys Res Commun* 265(1):134–139.

Bishop BG, Woollard GW (2002). Modern endodontic therapy for an incompletely developed tooth. *Gen Dent* 50(3):252–256.

Boeuf S, Richter W (2010). Chondrogenesis of mesenchymal stem cells: role of tissue source and inducing factors. *Stem Cell Res Ther* 1(4):31.

Buurma B, Gu K, Rutherford RB (1999). Transplantation of human pulpal and gingival fibroblasts attached to synthetic scaffolds. *Eur J Oral Sci* 107(4):282–289.

Chaussain C, Eapen AS, Huet E, Floris C, Ravindran S, Hao J, et al. (2009). MMP2-cleavage of DMP1 generates a bioactive peptide promoting differentiation of dental pulp stem/progenitor cell. *Eur Cell Mater* 18:84–95.

Chueh L-H, Huang GTJ (2006). Immature teeth with periradicular periodontitis or abscess undergoing apexogenesis: a paradigm shift. *J Endod* 32(12):1205–1213.

Chun SY, Lee HJ, Choi YA, Kim KM, Baek SH, Park HS, et al. (2011). Analysis of the soluble human tooth proteome and its ability to induce dentin/tooth regeneration. *Tissue Eng A* 17(1–2):181–191.

Chute JP (2006). Stem cell homing. *Curr Opin Hematol* 13(6):399–406.

Cordeiro MM, Dong Z, Kaneko T, Zhang Z, Miyazawa M, Shi S, et al. (2008). Dental pulp tissue engineering with stem cells from exfoliated deciduous teeth. *J Endod* 34(8):962–969.

da Silva LAB, Nelson-Filho P, da Silva RAB, Flores DSH, Heilborn C, Johnson JD, et al. (2010). Revascularization and periapical repair after endodontic treatment using apical negative pressure irrigation versus conventional irrigation plus triantibiotic intracanal dressing in dogs' teeth with apical periodontitis. *Oral Surg Oral Med Oral Pathol Oral Radiol Endod* 109(5):779–787.

Davatchi F, Abdollahi BS, Mohyeddin M, Shahram F, Nikbin B. (2011). Mesenchymal stem cell therapy for knee osteoarthritis: preliminary report of four patients. *Int J Rheum Dis* 14(2):211–215.

Decup F, Six N, Palmier B, Buch D, Lasfargues JJ, Salih E, et al. (2000). Bone sialoprotein-induced reparative dentinogenesis in the pulp of rat's molar. *Clin Oral Investig* 4(2):110–119.

Demarco FF, Conde MC, Cavalcanti BN, Casagrande L, Sakai VT, Nör JE (2011). Dental pulp tissue engineering. *Braz Dent J* 22(1):3–13.

Dib N, Dinsmore J, Lababidi Z, White B, Moravec S, Campbell A, et al. (2009). One-year follow-up of feasibility and safety of the first U.S., randomized, controlled study using 3-dimensional guided catheter-based delivery of autologous skeletal myoblasts for ischemic cardiomyopathy (CAuSMIC study). *JACC Cardiovasc Intervint* 2(1):9–16.

Ding RY, Cheung GS-p, Chen J, Yin XZ, Wang QQ, Zhang CF (2009). Pulp revascularization of immature teeth with apical periodontitis: a clinical study. *J Endod* 35(5):745–749.

Dobie K, Smith G, Sloan AJ, Smith AJ (2002). Effects of alginate hydrogels and TGF-beta 1 on human dental pulp repair in vitro. *Connect Tissue Res* 43(2–3):387–390.

English K, Mahon BP (2011). Allogeneic mesenchymal stem cells: agents of immune modulation. *J Cell Biochem* 112(8):1963–1968.

Enkel B, Dupas C, Armengol V, Akpe Adou J, Bosco J, Daculsi G et al. (2008). Bioactive materials in endodontics. *Expert Rev Med Devices* 5(4):475–494.

Furusawa M, Nakagawa K, Asai Y (1991). Clinico-pathological studies on the tissue reactions of human pulp treated with various kinds of calcium phosphate ceramics. *Bull Tokyo Dent Coll* 32(3):111–120.

Fuss Z, Lustig J, Katz A, Tamse A (2001). An evaluation of endodontically treated vertical root fractured teeth: impact of operative procedures. *J Endod* 27(1):46–48.

Galler KM, Cavender AC, Koeklue U, Suggs LJ, Schmalz G, D'Souza RN (2011a). Bioengineering of dental stem cells in a PEGylated fibrin gel. *Regen Med* 6(2):191–200.

Galler KM, D'Souza RN, Hartgerink JD, Schmalz G (2011b). Scaffolds for dental pulp tissue engineering. *Adv Dent Res* 23(3):333–339.

Goldberg M, Six N, Decup F, Buch D, Soheili Majd E, Lasfargues JJ, et al. (2001). Application of bioactive molecules in pulpcapping situations. *Adv Dent Res* 15:91–95.

Goldberg M, Six N, Decup F, Lasfargues JJ, Salih E, Tompkins K, Veis A (2003). Bioactive molecules and the future of pulp therapy. *Am J Dent* 16(1):66–76.

Goldberg M, Lacerda-Pinheiro S, Jegat N, Six N, Septier D, Priam F, et al. (2006). The impact of bioactive molecules to stimulate tooth repair and regeneration as part of restorative dentistry. *Dent Clin N Am* 50(2):277–298.

Goldberg M, Six N, Chaussain C, DenBesten P, Veis A, Poliard A. (2009). Dentin extracellular matrix molecules implanted into exposed pulps generate reparative dentin: a novel strategy in regenerative dentistry. *J Dent Res* 88(5):396–399.

Goldstein S, Sedaghat-Zandi A, Greenberg M, Friedman S (1999). Apexification and apexogenesis. *NY State Dent J* 65(5):23–25.

Gronthos S, Mankani M, Brahim J, Robey PG, Shi S (2000). Postnatal human dental pulp stem cells (DPSCs) in vitro and in vivo. *Proc Natl Acad Sci USA* 97(25):13625–13630.

Hare JM, Traverse JH, Henry TD, Dib N, Strumpf RK, Schulman SP, et al. (2009). A randomized, double-blind, placebo-controlled, dose-escalation study of intravenous adult human mesenchymal stem cells (prochymal) after acute myocardial infarction. *J Am Coll Cardiol* 54(24):2277–2286.

Harichane Y, Hirata A, Dimitrova-Nakov S, Granja I, Goldberg A, Kellermann O, et al. (2011). Pulpal progenitors and dentin repair. *Adv Dent Res* 23(3):307–312.

Higashi T, Okamoto H (1996) Influence of particle size of calcium phosphate ceramics as a capping agent on the formation of a hard tissue barrier in amputated dental pulp. *J Endod* 22(6):281–283.

Howard C, Murray PE, Namerow KN (2010). Dental pulp stem cell migration. *J Endod* 36(12):1963–1966.

Hu CC, Zhang C, Qian Q, Tatum NB (1998). Reparative dentin formation in rat molars after direct pulp capping with growth factors. *J Endod* 24(11):744–751.

Huang FM, Yang SF, Zhao JH, Chang YC (2010a). Platelet-rich fibrin increases proliferation and differentiation of human dental pulp cells. *J Endod* 36(10):1628–1632.

Huang GTJ (2008). A paradigm shift in endodontic management of immature teeth: conservation of stem cells for regeneration. *J Dent* 36(6):379–386.

Huang GT (2009). Apexification: the beginning of its end. *Int Endod J* 42(10):855–866.

Huang GTJ, Lin LM (2008). Letter to the editor: Comments on the use of the term "revascularization." *J Endod* 34(5):511.

Huang GTJ, Yamaza T, Shea LD, Djouad F, Kuhn NZ, Tuan RS, et al. (2010b). Stem/progenitor cell–mediated de novo regeneration of dental pulp with newly deposited continuous layer of dentin in an in vivo model. *Tissue Eng A* 16(2):605–615.

Huang XP, Sun Z, Miyagi Y, McDonald Kinkaid H, Zhang L, Weisel RD, et al. (2010c). Differentiation of allogeneic mesenchymal stem cells induces immunogenicity and limits their long-term benefits for myocardial repair. *Circulation* 122(23):2419–2429.

Hung CN, Mar K, Chang HC, Chiang YL, Hu HY, Lai CC, et al. (2011). A comparison between adipose tissue and dental pulp as sources of MSCs for tooth regeneration. *Biomaterials* 32(29):6995–7005.

Iida K, Takeda-Kawaguchi T, Tezuka Y, Kunisada T, Shibata T, Tezuka K (2010). Hypoxia enhances colony formation and proliferation but inhibits differentiation of human dental pulp cells. *Arch Oral Biol* 55(9):648–654.

Inuyama Y, Kitamura C, Nishihara T, Morotomi T, Nagayoshi M, Tabata Y, et al (2010). Effects of hyaluronic acid sponge as a scaffold on odontoblastic cell line and amputated dental pulp. *J Biomed Mater Res B* 92(1):120–128.

Iohara K, Nakashima M, Ito M, Ishikawa M, Nakasima A, Akamine A. (2004). Dentin regeneration by dental pulp stem cell therapy with recombinant human bone morphogenetic protein 2. *J Dent Res* 83(8):590–595.

Iohara K, Zheng L, Ito M, Tomokiyo A, Matsushita K Nakashima M (2006). Side population cells isolated from porcine dental pulp tissue with self-renewal and multipotency for dentinogenesis, chondrogenesis, adipogenesis, and neurogenesis. *Stem Cells* 24(11):2493–2503.

Iohara K, Zheng L, Wake H, Ito M, Nabekura J, Wakita H, et al. (2008). A novel stem cell source for vasculogenesis in ischemia: subfraction of side population cells from dental pulp. *Stem Cells* 26(9):2408–2418.

Iohara K, Zheng L, Ito M, Ishizaka R, Nakamura H, Into T, et al. (2009). Regeneration of dental pulp after pulpotomy by transplantation of $CD31^-/CD146^-$ side population cells from a canine tooth. *Regen Med* 4(3):377–385.

Iohara K, Imabayashi K, Ishizaka R, Watanabe A, Nabekura J, Ito M, Matsushita K, Nakamura H, Nakashima M (2011) Complete pulp regeneration after pulpectomy by transplantation of $CD105^+$ stem cells with stromal cell-derived factor- 1. *Tissue Eng Part A* 17(15–16):1911–1920.

Ishimatsu H, Kitamura C, Morotomi T, Tabata Y, Nishihara T, Chen KK, et al. (2009). Formation of dentinal bridge on surface of regenerated dental pulp in dentin defects by controlled release of fibroblast growth factor-2 from gelatin hydrogels. *J Endod* 35(6):858–865.

Ishizaka R, Iohara K, Murakami M, Fukuta O, Nakashima M (2012). Regeneration of dental pulp following pulpectomy by fractionated stem/progenitor cells from bone marrow and adipose tissue. *Biomaterials* 33(7):2109–2118.

Iwaya SI, Ikawa M, Kubota M (2001). Revascularization of an immature permanent tooth with apical periodontitis and sinus tract. *Dent Traumatol* 17(4):185–187.

Jarocha D, Lukasiewicz E, Majka M (2008). Adventage of mesenchymal stem cells (MSC) expansion directly from purified bone marrow $CD105^+$ and $CD271^+$ cells. *Folia Histochem Cytobiol* 46(3):307–314.

Jepsen S, Albers HK, Fleiner B, Tucker M, Rueger D (1997). Recombinant human osteogenic protein-1 induces dentin formation: an experimental study in miniature swine. *J Endod* 23(6):378–382.

Ji YM, Jeon SH, Park JY, Chung JH, Choung YH, Choung PH. (2010). Dental stem cell therapy with calcium hydroxide in dental pulp capping. *Tissue Eng A* 16(6):1823–1833.

Jia X, Yeo Y, Clifton RJ, Jiao T, Kohane DS, Kobler JB, et al. (2006). Hyaluronic acid–based microgels and microgel networks for vocal fold regeneration. *Biomacromolecules* 7(12):3336–3344.

Jögren U, Hägglund B, Sundqvist G, Wing K (1990). Factors affecting the long-term results of endodontic treatment. *J Endod* 16(10):498–504.

Jung I-Y, Lee S-J and Hargreaves KM (2008). Biologically Based Treatment of Immature Permanent Teeth with Pulpal Necrosis: A Case Series.*Journal of Endodontics* 34(7):876–887.

Kaplan RN, Psaila B, Lyden D (2007). Niche-to-niche migration of bone-marrow-derived cells. *Trends Mol Med* 13(2):72–81.

Karussis D, Karageorgiou C, Vaknin-Dembinsky A, Gowda-Kurkalli B, Gomori JM, Kassis I, et al. (2010). Safety and immunological effects of mesenchymal stem cell transplantation in patients with multiple sclerosis and amyotrophic lateral sclerosis. *Arch Neurol* 67(10):1187–1194.

Kassis I, Vaknin-Dembinsky A, Karussis D. (2011). Bone marrow mesenchymal stem cells: agents of immunomodulation and neuroprotection. *Curr Stem Cell Res Ther* 6(1):63–68.

Kastrinaki MC, Andreakou I, Charbord P, Papadaki HA (2008). Isolation of human bone marrow mesenchymal stem cells using different membrane markers: comparison of colony/cloning efficiency, differentiation potential, and molecular profile. *Tissue Eng C* 14(4):333–339.

Kim JY, Xin X, Moioli EK, Chung J, Lee CH, Chen M, Fu SY, Koch PD, Mao JJ. (2010) Regeneration of dental-pulp-like tissue by chemotaxis-induced cell homing. Tissue Eng Part A. 16(10):3023–31.

Kitamura C, Nishihara T, Terashita M, Tabata Y, Washio A (2012). Local regeneration of dentin–pulp complex using controlled release of fgf-2 and naturally derived sponge-like scaffolds. *Int J Dent* 2012:190561.

Kitasako Y, Shibata S, Cox CF, Tagami J (2002). Location, arrangement and possible function of interodontoblastic collagen fibres in association with calcium hydroxide–induced hard tissue bridges. *Int Endod J* 35(12):996–1004.

Komabayashi T, Imai Y, Ahn C, Chow LC, Takagi S (2010). Dentin permeability reduction by a sequential application of calcium and fluoride phosphate solutions. *J Dent* 38(9):736–741.

Kumabe S, Nakatsuka M, Kim GS, Jue SS, Aikawa F, Shin JW, Iwai Y (2006). Human dental pulp cell culture and cell transplantation with an alginate scaffold. *Okajimas Folia Anat Jpn* 82(4):147–155.

Laschober GT, Brunauer R, Jamnig A, Fehrer C, Greiderer B, Lepperdinger G (2009). Leptin receptor/CD295 is upregulated on primary human mesenchymal stem cells of advancing biological age and distinctly marks the subpopulation of dying cells. *Exp Gerontol* 44(1–2):57–62.

Lawson MA, Barralet JE, Wang L, Shelton RM, Triffitt JT (2004). Adhesion and growth of bone marrow stromal cells on modified alginate hydrogels. *Tissue Eng* 10(9–10):1480–1491.

Lepperdinger G (2011). Inflammation and mesenchymal stem cell aging. *Curr Opin Immunol* 23(4):518–524.

Lianjia Y, Yuhao G, White FH (1993). Bovine bone morphogenetic protein-induced dentinogenesis. *Clin Orthop Relat Res* 295:305–312.

Lovschall H, Fejerskov O, Flyvbjerg A (2001). Pulp-capping with recombinant human insulin-like growth factor 1 (rhIGF-1) in rat molar. *Adv Dent Res* 15:108–112.

Machtou P, Reit C (2003). Non-surgical retreatment. In: *Textbook of Endodontology*, Bergenholtz G, Horsted-Bindslev P, Reit C, eds., 1st ed. Blackwell Munksgaad, Oxford, UK. pp. 300–310.

Messer HH (2008). To the Editor. *J Endod* 34(10):1157.

Mooney DJ, Powell C, Piana J, Rutherford B (1996). Engineering dental pulp-like tissue in vitro. *Biotechnol Prog* 12(6):865–868.

Mrozik KM, Zilm PS, Bagley CJ, Hack S, Hoffmann P, Gronthos S, et al. (2010). Proteomic characterization of mesenchymal stem cell-like populations derived from ovine periodontal ligament, dental pulp, and bone marrow: analysis of differentially expressed proteins. *Stem Cells Dev* 19(10):1485–1499.

Murray PE, Stanley HR, Matthews JB, Sloan AJ, Smith AJ (2002). Age-related odontometric changes of human teeth. *Oral Surg Oral Med Oral Pathol Oral Radiol Endod* 93(4):474–482.

Nakashima M (1994a). Induction of dentin formation on canine amputated pulp by recombinant human bone morphogenetic protein (BMP)-2 and -4. *J Dent Res* 73(9):1515–1522.

Nakashima M (1994b). Induction of dentine in amputated pulp of the dogs by recombinant human bone morphogenetic protein- 2 and - 4 with collagen matrix. *Arch Oral Biol* 39(12):1085–1089.

Nakashima M (2005). Bone morphogenetic proteins in dentin regeneration for potential use in endodontic therapy. *Cytokine Growth Factor Rev* 16(3):369–376.

Nakashima M, Akamine A (2005). The application of tissue engineering to regeneration of pulp and dentin in endodontics. *J Endod* 31(10):711–718.

Nakashima M, Iohara K (2011). Regeneration of dental pulp by stem cells. *Adv Dent Res* 23(3):313–319.

Nakashima M, Reddi AH (2003). The application of bone morphogenetic proteins to dental tissue engineering. *Nature Biotechnol* 21(9):1025–1032.

Nakashima M, Iohara K, Sugiyama M (2009). Human dental pulp stem cells with highly angiogenic and neurogenic potential for possible use in pulp regeneration. *Cytokine Growth Factor Rev* 20(5–6):435–440.

Nevins AJ, Finkelstein F, Borden BG, Laporta R (1976). Revitalization of pulpless open apex teeth in rhesus monkeys, using collagen-calcium phosphate gel. *J Endod* 2(6):159–165.

Noël D, Caton D, Roche S, Bony C, Lehmann S, Casteilla L, et al. (2008). Cell specific differences between human adipose-derived and mesenchymal–stromal cells despite similar differentiation potentials. *Exp Cell Res* 314(7):1575–1584.

Ohab JJ, Fleming S, Blesch A, Carmichael ST (2006). A neurovascular niche for neurogenesis after stroke. *J Neurosci* 26(50):13007–13016.

Olgart L, Kerezoudis NP (1994). Nerve–pulp interactions. *Arch Oral Biol* 39(Suppl): 47S–54S.

Ostby BN (1961). The role of the blood clot in endodontic therapy: an experimental histologic study. *Acta Odontol Scand* 19:324–353.

Parirokh M, Torabinejad M (2010). Mineral trioxide aggregate: a comprehensive literature review. Part I. Chemical, physical, and antibacterial properties. *J Endod* 36(1):16–27.

Parr AM, Tator CH, Keating A (2007). Bone marrow–derived mesenchymal stromal cells for the repair of central nervous system injury. *Bone Marrow Transpl* 40(7):609–619.

Peppas NA, Keys KB, Torres-Lugo M, Lowman AM (1999). Poly(ethylene glycol)–containing hydrogels in drug delivery. *J Control Release* 62(1–2):81–87.

Philippe B, Luc S, Valérie PB, Jérôme R, Alessandra BR, Louis C (2010). Culture and use of mesenchymal stromal cells in phase I and II clinical trials. *Stem Cells Int* 2010: 503593.

Phinney DG, Kopen G, Righter W, Webster S, Tremain N, Prockop DJ (1999). Donor variation in the growth properties and osteogenic potential of human marrow stromal cells. *J Cell Biochem* 75(3):424–436.

Pouyani T, Prestwich GD (1994). Biotinylated hyaluronic acid: a new tool for probing hyaluronate–receptor interactions. *Bioconjug Chem* 5(4):370–372.

Qin C, D'Souza R, Feng JQ (2007). Dentin matrix protein 1 (DMP1): new and important roles for biomineralization and phosphate homeostasis. *J Dent Res.* 86(12):1134–1141.

Rafter M (2005). Apexification: a review. *Dent Traumatol* 21(1):1–8.

Ren WH, Yang LJ, Dong SZ (1999). Induction of reparative dentin formation in dogs with combined recombinant human bone morphogenetic protein 2 and fibrin sealant. *Chin J Dent Res* 2(3–4):21–24.

Richter AW, Ryde EM, Zetterström EO (1979). Non-immunogenicity of a purified sodium hyaluronate preparation in man. *Int Arch Allergy Appl Immunol* 59(1):45–48.

Robertson A, Lundgren T, Andreasen JO, Dietz W, Hoyer I, Noren JG (1997). Pulp calcifications in traumatized primary incisors: a morphological and inductive analysis study. *Eur J Oral Sci* 105(3):196–206.

Robin AM, Zhang ZG, Wang L, Zhang RL, Katakowski M, Zhang L, et al. (2006). Stromal cell–derived factor 1alpha mediates neural progenitor cell motility after focal cerebral ischemia. *J Cereb Blood Flow Metab* 26(1):125–134.

Rutherford RB, Gu K (2000). Treatment of inflamed ferret dental pulps with recombinant bone morphogenetic protein-7. *Eur J Oral Sci* 108(3):202–206.

Rutherford RB, Wahle J, Tucker M, Rueger D, Charette M. (1993). Induction of reparative dentin formation in monkeys by recombinant human osteogenic protein-1. *Arch Oral Biol* 38(7):571–576.

Rutherford RB, Spangberg L, Tucker M, Rueger D, Charette M. (1994). he time-course of the induction of reparative dentin formation in monkeys by recombinant human osteogenic protein-1. *Arch Oral Biol* 39(10):833–838.

Sasaki T, Kawamata-Kido H (1995). Providing an environment for reparative dentine induction in amputated rat molar pulp by high-molecular-weight hyaluronic acid. *Arch Oral Biol* 40(3):209–219.

Schellenberg A, Lin Q, Schüler H, Koch CM, Joussen S, Denecke B, Walenda G, Pallua N, Suschek CV, Zenke M, Wagner W (2011). Replicative senescence of mesenchymal stem cells causes DNA-methylation changes which correlate with repressive histone marks. Aging (Albany NY) 3(9):873–888.

Sensebé L, Krampera M, Schrezenmeier H, Bourin P, Giordano R (2010). Mesenchymal stem cells for clinical application. *Vox Sang* 98(2):93–107.

Sethe S, Scutt A, Stolzing (2006). Aging of mesenchymal stem cells. *Ageing Res Rev* 5(1):91–116.

Shah N, Logani A, Bhaskar U, Aggarwal V (2008). Efficacy of revascularization to induce apexification/apexogensis in infected, nonvital, immature teeth: a pilot clinical study. *J Endod* 34(8):919–925; discussion 1157.

Shi S, Robey PG, Gronthos S (2001). Comparison of human dental pulp and bone marrow stromal stem cells by cDNA microarray analysis. *Bone* 29(6):532–539.

Silva TA, Rosa AL, Lara VS (2004). Dentin matrix proteins and soluble factors: intrinsic regulatory signals for healing and resorption of dental and periodontal tissues? *Oral Dis* 10(2):63–74.

Simon SR, Berdal A, Cooper PR, Lumley PJ, Tomson PL, Smith AJ (2011). Dentin–pulp complex regeneration: from lab to clinic. *Adv Dent Res* 23(3):340–345.

Six N, Lasfargues JJ, Goldberg M (2002). Differential repair responses in the coronal and radicular areas of the exposed rat molar pulp induced by recombinant human bone morphogenetic protein 7 (osteogenic protein 1). *Arch Oral Biol* 47(3):177–187.

Smith AJ (2003). Vitality of the dentin–pulp complex in health and disease: growth factors as key mediators. *J Dent Educ* 67(6):678–689.

Smith AJ, Sloan AJ, Matthews JB, Murray PE, Lumley PJ. (2000). Reparative processes in dentine and pulp. In: *Toothwear and Sensitivity*, Addy M, Embery G, Edgar WM, Orchardson R, eds., Martin-Dunitz, London, pp. 53–66.

Stein PS, Kryscio RJ, Desrosiers M, Donegan SJ, Gibbs MB (2010). Tooth loss, apolipoprotein E, and decline in delayed word recall. *J Dent Res* 89(5):473–477.

Sugiyama M, Iohara K, Wakita H, Hattori H, Ueda M, Matsushita K, Nakashima M (2011). Dental Pulp-Derived CD31(−)/CD146(−) side population stem/progenitor cells enhance recovery of focal cerebral ischemia in rats. *Tissue Eng A* 17(9–10):1303–1311.

Sun HH, Jin T, Yu Q, Chen FM (2011). Biological approaches toward dental pulp regeneration by tissue engineering. *J Tissue Eng Regen Med* 5(4):e1–e16.

Suzuki T, Lee CH, Chen M, Zhao W, Fu SY, Qi JJ, et al.. (2011). Induced migration of dental pulp stem cells for in vivo pulp regeneration. *J Dent Res* 90(8):1013–1018.

Takeda T, Tezuka Y, Horiuchi M, Hosono K, Iida K, Hatakeyama D, et al. (2008). Characterization of dental pulp stem cells of human tooth germs. *J Dent Res* 87(7):676–681.

Tang R, Ding G (2011). wine dental pulp stem cells inhibit T-cell proliferation. *Transplant Proc* 43(10):3955–3959.

Thibodeau B, Teixeira F, Yamauchi M, Caplan DJ, Trope M (2007). Pulp revascularization of immature dog teeth with apical periodontitis. *J Endod* 33(6):680–689.

Thored P, Arvidsson A, Cacci E, Ahlenius H, Kallur T, Darsalia V, et al. (2006). Persistent production of neurons from adult brain stem cells during recovery after stroke. *Stem Cells* 24(3):739–747.

Tziafas D (2004). The future role of a molecular approach to pulp–dentinal regeneration. *Caries Res* 38(3):314–320.

Tziafas D, Belibasakis G, Veis A, Papadimitriou S. (2001). Dentin regeneration in vital pulp therapy: design principles. *Adv Dent Res* 15:96–100.

Vassalli G, Moccetti T. (2011) Cardiac repair with allogeneic mesenchymal stem cells after myocardial infarction. Swiss Med Wkly 23; 141:w13209, 1–6.

Vemuri MC, Chase LG, Rao MS (2011). esenchymal stem cell assays and applications. *Methods Mol Biol* 698:3–8.

Wang X, Thibodeau B, Trope M, Lin LM, Huang GT (2010). Histologic characterization of regenerated tissues in canal space after the revitalization/revascularization procedure of immature dog teeth with apical periodontitis. *J Endod* 36(1):56–63.

Webber RT (1984). Apexogenesis versus apexification. *Dent Clin N Am* 28(4):669–697.

Wilcox LR, Van Surksum R (1991). Endodontic retreatment in large and small straight canals. *J Endod* 17(3):119–121.

Yuan Z, Nie H, Wang S, Lee CH, Li A, Fu SY, et al. (2011). Biomaterial selection for tooth regeneration *Tissue Eng B* 17(5):373–388.

Zhang HG, Zhu Q (2007). Preparation of porous hydroxyapatite with interconnected pore architecture. *J Mater Sci Mater Med* 18(9):1825–1829.

Zhang W, Walboomers XF, Jansen JA (2008). The formation of tertiary dentin after pulp capping with a calcium phosphate cement, loaded with PLGA microparticles containing TGF-beta1. *J Biomed Mater Res A* 85(2):439–444.

Zheng L, Amano K, Iohara K, Ito M, Imabayashi K, Into T, et al. (2009). Matrix metalloproteinase- 3 accelerates wound healing following dental pulp injury. *Am J Pathol* 175(5):1905–1914.

Zhou J, Shi S, Shi Y, Xie H, Chen L, He Y, et al. (2011). Role of bone marrow-derived progenitor cells in the maintenance and regeneration of dental mesenchymal tissues. *J Cell Physiol* 226(8):2081–2090.

Zuk PA, Zhu M, Mizuno H, Huang J, Futrell JW, Katz AJ, et al. (2001). Multilineage cells from human adipose tissue: implications for cell-based therapies. *Tissue Eng* 7(2):211–228

27

BIOENGINEERING OF ROOTS AND PERIODONTAL TISSUES

Songlin Wang, Gang Ding, Fulan Wei, and Yi Liu

Molecular Laboratory for Gene Therapy and Tooth Regeneration, Capital Medical University School of Stomatology, Beijing, China

1 INTRODUCTION

Tooth loss as a result of periodontal diseases, severe dental caries, trauma, or genetic disorders affects quality of life adversely. All current treatment protocols, including the use of removable dentures, fixed dentures, and dental implants, are relying on nonbiological restorative materials. Root is the most important part of the tooth in maintaining tooth functions (Eckert et al., 2005). Titanium dental implants have been widely used as a root substitute to support the artificial crown. However, unlike the periodontal ligament (PDL) around the roots of the teeth, the success of the implant depends on osseointegration of the alveolar bone to the implant (Baillie, 2004; Humphrey, 2006). This leads to a lack of tooth mobility and concentration of the masticatory force, which may damage the implant itself, alveolar bone, and even the temporomandibular joint (Taba et al., 2005; Earthman et al., 2006). The major drawbacks of dental implants are the absence of PDL and the need for preexisting high-quality bone structures for support, which narrow down the scope of their applications. Therefore, from a structural and functional point of view, the current implant denture is not an ideal substitute to the natural tooth.

Regenerative medicine and tissue engineering offer promising therapeutic alternatives and address the aforementioned shortcomings (Atala et al., 2006; Pham et al., 2007; Zaky and Cancedda, 2009). Recent advances in stem cell research and cell-based regeneration have encouraged researchers to explore the possibility of regenerating living functional teeth (Ikeda et al., 2009). However, owing to the complexity of the human tooth, especially in perspectives of its growth and development, regeneration

Stem Cells in Craniofacial Development and Regeneration, First Edition.
Edited by George T.-J. Huang, Irma Thesleff.

of a whole tooth, including enamel, dentin–pulp complex, and periodontal tissues, in functional form is almost impossible at present.

Periodontitis is the main cause of tooth loss and is associated with a number of systemic diseases, such as diabetes and cardiovascular disease. Thus far, several regenerative approaches, including guided tissue regeneration, application of enamel matrix derivative, and various growth factors, have been proposed to treat periodontal diseases, and favorable results have been reported. Recently, periodontal ligament stem cells (PDLSCs) have been considered as an optimal cell sources for cell-based periodontal tissue regeneration.

In this chapter we describe the current knowledge and recent development in stem cell–mediated root regeneration combined with clinical crown restoration, which may be a promising approach to restore missing teeth and to replace dental implants. Periodontal regeneration mediated by PDLSCs and their applications are also reviewed and discussed.

2 BIOROOT ENGINEERING

2.1 Stem Cells from Root Apical Papilla for Root Engineering

In 2006, identification of stem cells from root apical papilla (SCAPs) provided an opportunity to pursue root regeneration using these highly potent "young" postnatal stem cells derived from the dental tissues of 18- to 20-year-olds. Dental papilla has been considered to be the source of odontoblasts during tooth development. This developing organ evolves into dental pulp after being encased by dentin produced by odontoblasts. This apical papilla tissue is apical to the epithelial diaphragm, loosely attached to the apex of the developing root, and there is an apical cell-rich zone lying between the apical papilla and the pulp (Sonoyama et al., 2008). The discovery and isolation of SCAPs may explain in part why apexogenesis can occur in infected immature permanent teeth after disinfection.

Single colonies were formed after human SCAPs were plated at a low density (5×10^3/T25 flask) and cultured for 10 days (Sonoyama et al., 2006). Ex vivo–expanded SCAPs were also found to express the cell-surface molecule STRO1, ALP, CD24, CD29,CD73, CD90, CD105, CD106, CD146, and CD166 but were negative for CD34, CD45, CD18, and CD150 (Sonoyama et al., 2006). Isolated SCAPs grown in cultures can undergo dentinogenic differentiation when stimulated with dexamethasone supplemented with L-ascorbate-2-phosphate and inorganic phosphate. Alizarin Red–positive nodules form in SCAP cultures after 4 weeks of induction, indicating a significant calcium accumulation in vitro. Furthermore, cultured SCAPs were found to express dentinogenic markers dentin sialophosphoprotein and CBFA1/Runx2 (Sonoyama et al., 2006). In addition to their dentinogenic potential, SCAPs also exhibit adipogenic and neurogenic differentiation capabilities when treated with respective stimuli (Sonoyama et al., 2006, 2008).

SCAPs can differentiate into odontoblast-like cells, forming dentin when implanted into the subcutaneous space of immunocompromised mice using hydroxyapatite/tricalcium phosphate (HA/TCP) as a carrier vehicle (Sonoyama et al., 2006). To accomplish functional tooth regeneration, autologous SCAPs and PDLSCs were loaded onto an HA/TCP block that contained an inner post channel space and gelfoam scaffolds, respectively, and implanted into sockets of the lower jaw. Three months later, the surface of the implanted HA/SCAP-gelfoam/PDLSC structure was reopened surgically and a porcelain crown was inserted. This bioroot is different from a natural root but is encircled with PDL tissue and appears to have a natural relationship with the surrounding bone. Moreover, newly formed bioroots demonstrated a significantly improved compressive strength compared to that of the original HA/TCP carriers after three months of implantation (Sonoyama et al., 2006).

Human SCAPs were also isolated independently to regenerate vascularized human dental pulp with new dentin formation in emptied root canal space in an immunocompromised mouse model (Huang et al., 2010) (for details, see Chapter 26). It has been shown that apical tooth germ cell–conditioned medium can induce PDLSCs to exhibit several characteristics of cementoblast lineages, as indicated by morphological changes, increased proliferation, high ALP activity, and the expression of cementum-related genes and calcified nodule formation in vitro. When transplanted into immunocompromised mice, the induced PDLSCs showed tissue-regenerative capacity to produce cementum/periodontal ligament-like structures, characterized by a layer of cementum-like mineralized tissues and associated periodontal ligament-like collagen fibers connecting with the newly formed cementum-like deposits (Yang et al., 2009).

2.2 Dental Follicle Cell Sheets for Root Engineering

A recent report has shown that suitable stem cells from dental follicle and biological scaffolds have been optimized for tooth root regeneration (Yang et al., 2012). In this study, dental follicle cell sheets (DFCSs) and a natural decellularization matrix scaffold–treated dentin matrix (TDM) were used for root regeneration. Dentinal tubules were sufficiently exposed after TDM was treated with ethylenediaminetetraacetic acid. DMP1, DSPP, and BSP were highly expressed in DFCSs after being induced by the liquid extract of the TDM. Four weeks after DFCSs were seeded onto TDM in vitro, numerous fiber-like structures were observed and newly regenerated dentin formed on TDM where the DFCSs were attached. Eight weeks after being implanted subcutaneously into the dorsum of mice, distinctive dentin-pulp-like structures such as dentinal tubules, pre-dentin, polarizing odontoblast-like cells, collagen fibers, and blood vessels, were clearly observed on the TDM. The newly regenerated tissues were positive for the dentin–pulp complex markers DSP, collagen I, and bIII-tubulin, and for cementum-specific cementum attachment protein (CAP) and COLI. In another study (Guo et al., 2012), upon transplantation of DFCs combined with TDM in rats for 4 weeks, root-like tissues stained positive for markers of dental pulp, and periodontal tissues were regenerated in the alveolar fossa but not in the skull and omental pockets. This demonstrates that in combination with TDM, the microenvironment of the alveolar fossa was found to be a suitable inductive microenvironment for tooth root construction.

2.3 Allogeneic Mesenchymal Stem Cell for Root Engineering

Tooth loss is more frequent in older patients who have more limited sources of autologous dental stem cells. This largely confines their clinical uses. Interestingly, due to the low immunogenicity and immunomodulation function of mesenchymal stem cells (MSCs), allogeneic MSCs can be used instead of an autologous source (Le Blanc et al., 2004, 2008; Gonzalez-Rey et al., 2009; Sun et al., 2009; Ding et al., 2010). Previous in vitro studies have suggested that MSCs could exert a potent immunosuppressive effect in vitro (Bartholomew et al., 2002; Di Nicola et al., 2002; Krampera et al., 2003; Sudres et al., 2006). In line with this in vitro property, MSCs also display immunosuppressive capacities in vivo, evidenced by a study showing that allogeneic MSCs may prolong skin allograft survival in immunocompetent baboons (Bartholomew et al., 2002). However, other studies indicate that MSCs are not intrinsically immunoprivileged and are capable of inducing a memory T-cell response after injection in vivo in immunocompetent hosts (Nauta et al., 2006) and yielded no clinical benefit on the incidence or severity of graft-versus-host disease (GVHD) (Sudres et al., 2006). Despite the conflicting results, it has been shown that allogeneic MSCs conferred significant therapeutic effects on SLE mice and treatment-refractory patients (Sun et al., 2009). Patients with acute GVHD responded to treatment with MSCs, although little is known about the mechanisms of suppression of GVHD by MSCs (Le Blanc et al., 2004, 2008; Wu et al., 2011). These observations appear to warrant further investigation in the use of allogeneic MSCs for different therapeutic applications, including tooth root regeneration.

In a recent study, allogeneic dental pulp stem cells (DPSCs) and PDLSC sheets were used to regenerate tooth root in swine (Fig. 1) (Wei et al., 2012). Six months after transplantation, the transplanted engineered complex formed a hard root structure and a clear PDL space between the regenerated bioroot and surrounding bony tissue, as revealed by x-ray and micro-computed tomographic analysis (Fig. 2A). To test the function of the newly regenerated bioroot, a pre-made porcelain crown was cemented onto the bioroot. Six months after crown restoration, the regenerated bioroots demonstrated significantly improved compressive strength. The architecture of the PDL-like tissue also changed such that the angle of the fiber bundles turned from being parallel to the dentin-like structure to being oblique, indicating the reorganization of the ligament fibers in response to the functional performance of the bioroot (Fig. 2B and C). The clinical features of the bioroot, including probing depth, gingival recession, and attachment loss, were comparable with those of natural healthy miniature pig teeth. No significant differences were observed at 3 days, 12 weeks, and 12 months after transplantation regarding the percentage of $CD3^+$, $CD4^+$, and $CD8^+$ T-cells in whole blood between the allogeneic transplantation group and the control group for up to 12 months.

3 BIOENGINEERING OF PERIODONTAL TISSUES

The periodontium is a complex organ consisting of gingiva, PDL, cementum, and alveolar bone (Benatti et al., 2007). The tooth is held in place in the surrounding bone by a fibrocellular stratum of PDL derived from the dental follicle, which is of ectomesenchymal origin. The PDL is secured by connective tissue fibers (Sharpey's fibers) embedded between the thin mineralized outer layer of cementum and the inner wall of the alveolar bone socket. PDL not only has an important role in supporting teeth, but also contributes to tooth nutrition, homoeostasis, and repair of damaged tissue (Bartold et al., 2000; Pitaru et al., 2002; Shimono et al., 2003; Seo et al., 2004; Shi et al., 2005).

Periodontitis can affect the composition and integrity of periodontal structures, causing destruction of the connective tissue matrix and cells, loss of attachment and

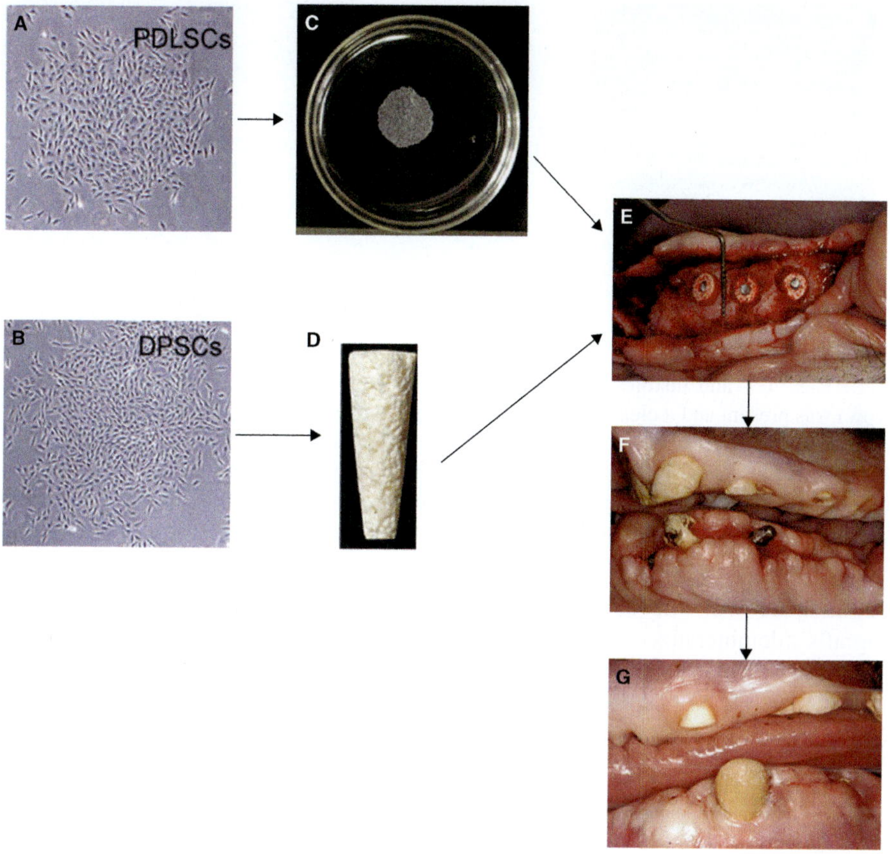

FIGURE 1 Schematic representation of the strategy for bioroot regeneration in swine. (A and B) Periodontal ligament stem cells (PDLSCs) and dental pulp stem cells (DPSCs) were isolated from an extracted tooth. (C) PDLSCs were used to prepare vitamin C–induced PDLSC sheet. (D) DPSCs were seeded on HA/TCP scaffolds. (E) PDLSC sheet wrapping the HA/TCP/DPSCs was implanted into the socket. (F and G) A porcelain crown was restored six months after bioroot implantation.

resorption of bone, and ultimately leading to tooth loss (Benatti et al., 2007). In addition, periodontitis has been associated with a number of systemic diseases, including adverse pregnancy outcomes, cardiovascular disease, stroke, pulmonary disease, and diabetes (Pihlstrom et al., 2005; Michalowicz et al., 2006; Tonetti et al., 2007). However, the periodontium has a limited capacity to regenerate. Although some minor regeneration of the periodontium may occur in the early phases of periodontal disease, once the disease becomes established, regeneration does not take place unless employing some form of therapeutic intervention.

3.1 Current Therapies

Current periodontal therapies, such as banked bone allografts, cell-occlusive barriers, or enamel matrix proteins, result in only limited bone regeneration. Attempts to regenerate periodontal tissues have focused almost exclusively on regenerating lost

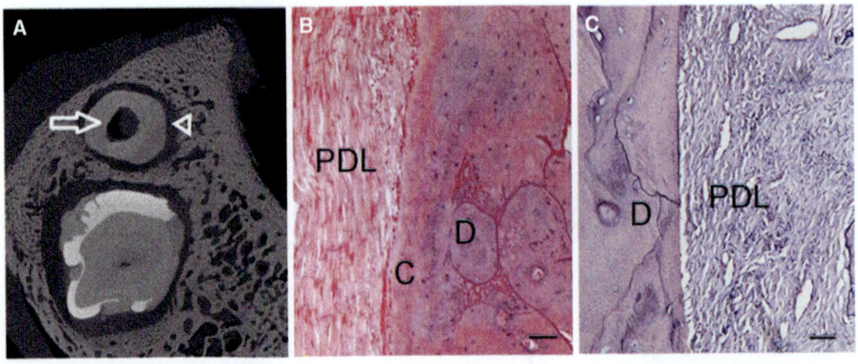

FIGURE 2 CT and histological analysis of the bioroot in swine. (A) A hard root structure (arrow) was present and a clear PDL space was found between the implant and surrounding bony tissue (arrowhead). (B) Cementum-PDL-like tissues were generated with fibers lining parallel to the long axis of dentin-like matrix structure six months after transplantation. (C) PDL-like tissue with angulated fibers in the longitudinal dentin-like structure six months after crown restoration. PDL, periodontal ligament; C, cementum; D, dentin.

alveolar bone with the use of autografts (cortical/cancellous bone), bone marrow, allografts (demineralized freeze-dried/freeze-dried bone), and alloplastic materials (ceramics, hydroxyapatite, polymers, and bioglass). However, the majority of these strategies have been plagued by variability in their safety, effectiveness, and stability over time; thus, their effectiveness as true periodontal regenerative agents has been questioned. In addition, although utilization of such grafting materials has resulted in increased periodontal attachment and radiographic evidence of bone fill, histological assessment has usually revealed that these materials had little osteoinductive capacity and appeared to generate dense fibrous connective tissue rather than organized periodontal tissue (Shi et al. 2005). To improve the therapeutic outcome, novel approaches have emerged recently, including local administration of biocompatible scaffolds with or without the presence of growth factors (for details, see Chapter 28).

In recent years, *guided tissue regeneration* (GTR) has become the gold-standard surgical approach for periodontal tissue regeneration. GTR is defined as "procedures attempting to regenerate lost periodontal structures through differential tissue responses" (Wang et al., 2005; Parrish et al., 2009). This procedure involves draping a biocompatible membrane over the periodontal defect from the root surface to the adjacent alveolar bone, often in combination with a bone graft. Barrier techniques using materials such as expanded poly(tetrafluoroethylene), polyglactin, poly(lactic acid), calcium sulfate, and collagen are employed in the GTR. The barrier membrane prevents unwanted epithelium and gingival connective tissue from entering the healing site, while promoting repopulation of the defect site by cells migrating in from the PDL. Numerous studies have confirmed that GTR is superior to open flap debridement and yields more clinical attachment (Parrish et al., 2009). However, indications for such treatments are rather limited, and the amounts of regenerated tissue are not predictable (Yamada et al., 2006; Mudda and Bajaj, 2011).

3.2 PDLSC-Mediated Periodontal Regeneration

Isolation and Identification of PDLSCs. The identification of PDLSCs has provided a new prospect and appropriate cell source for periodontal tissue regeneration (Seo et al., 2004). Multipotent stem cells have been isolated from the periodontal ligaments of humans, swine, sheep, and rats (Fig. 3) (Seo et al., 2004; Gronthos et al., 2006; Techawattanawisal et al., 2007; Liu et al., 2008). They offer a source of stem cells from an accessible tissue (Seo et al., 2004; Shi et al., 2005) (for a detailed description, see Chapter 15). PDLSCs in cultures exhibit higher rates of proliferation, approximately 30% compared to the growth of cultured bone marrow mesenchymal stem cells (BMMSCs), (Seo et al., 2004; Shi et al., 2005 Bartold et al., 2006).

Human PDLSCs. PDL is similar to the tissues in tendon, both being endowed with the ability to absorb mechanical forces of stress and strain. Given this similarity, Seo et al. (2005) examined the expression levels of scleraxis, a tendon specific transcription factor, in human cultured PDLSCs. The study showed that PDLSCs expressed measurably higher levels of scleraxis transcripts than did BMMSCs. These data imply that PDLSCs represent a unique population of postnatal stem cells, distinct from BMMSCs.

Cryopreserved human PDL can still give rise to 40% of single-colony-derived PDLSCs after thawing. Although the number of single colonies derived from cryopreserved PDL was significantly smaller than that of PDLSCs isolated from fresh PDL, they maintain a high proliferative capacity and express the cell-surface molecule STRO1, an early mesenchymal stem cell marker, along with the coexpression of cementoblastic–osteoblastic markers transforming growth factor β receptor 1 and bone sialoprotein. The PDLSCs isolated from cryopreserved PDL maintain normal stem cell characteristics, including having a normal diploid karyotype and capable of cementum/PDL-like tissue regeneration in vivo and thus utilization of frozen PDL tissues for stem cell isolation, providing a practical approach for clinical applications (Seo et al., 2005).

Ovine PDLSCs. In a study using an ovine model, Gronthos et al., (2006) generated enriched preparations of PDLSCs derived from ovine periodontal ligament using immunomagnetic bead selection, based on expression of the mesenchymal stem cell–associated antigen CD106. These $CD106^+$ ovine PDLSCs demonstrated a capacity to form adherent clonogenic clusters of fibroblast-like cells when plated at low densities in vitro. Ex vivo–expanded ovine PDLSCs exhibited a high proliferation rate in vitro and expressed a phenotype ($CD44^+$, $CD166^+$, $CBFA1^+$, $collagenI^+$, bone sialoprotein+) consistent with human-derived PDLSCs. Ex vivo–expanded ovine PDLSCs demonstrated the capacity to regenerate both cementum-like mineral and PDL when transplanted into NOD–SCID mice. Thus, ovine PDLSCs may potentially be used as a preclinical study model for periodontal regeneration prior to human clinical studies.

Rat PDLSCs. Techawattanawisal and colleagues (2007) used 6-week-old male Wistar rats to collect PDLSCs by neurosphere-forming culture system. Spheres express various markers, including GFAP (glial marker), nestin (neural precursor marker) and vimentin

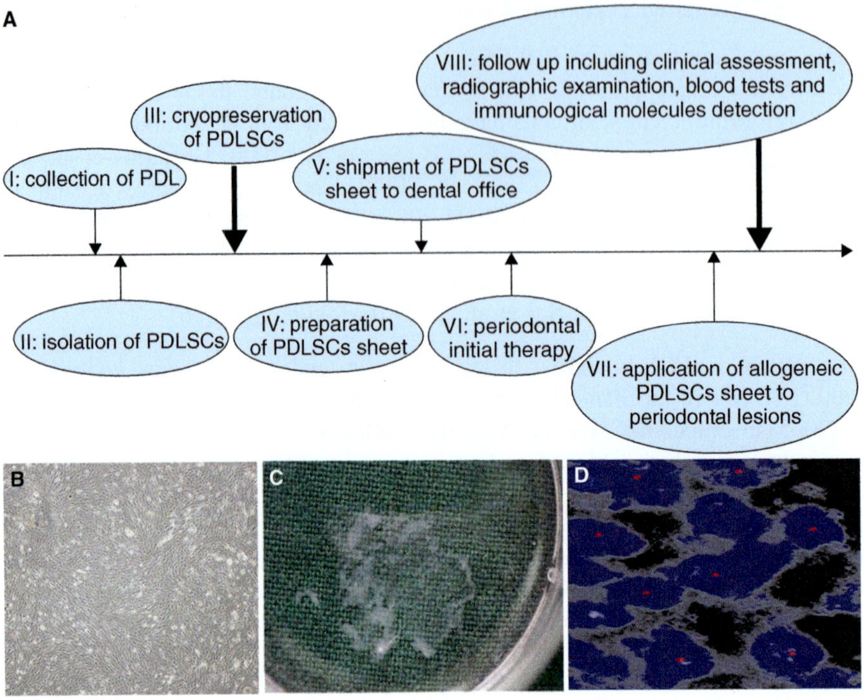

FIGURE 3 Transplantation of allogeneic periodontal ligament stem cells (PDLSCs) to treat periodontitis. (A) Steps involved in allogeneic PDLSC-mediated periodontal tissue regeneration. (B) Isolation of PDLSCs. PDLSCs displayed fibroblast-like morphology. (C) Preparation of PDLSC cell sheets. (D) Allogeneic PDLSCs used for periodontal tissue regeneration. Allogeneic PDLSC cell sheets obtained from a male miniature pig were transplanted into the periodontal bone defect area of a female miniature pig. Twelve weeks later, Y chromosome–positive signals (red) were detected in the regenerated periodontal bone tissue, indicating that regenerated periodontal tissues were derived from the transplanted allogeneic PDLSCs.

(mesenchymal marker), SSEA1 (stage-specific embryonic antigen-1), the transcription factors Twist, Slug, Sox2, and Sox9, and the multidrug-resistant transporter gene *ABCG2*. These data suggest that PDL-derived spheres possess certain characteristics of neural crest–derived stem cells. PDL-derived spheres can differentiate into multinucleated myotubes, NFM-positive neuron-like cells, GFAP-positive astrocyte-like cells, and CNPase-positive oligodendrocyte-like cells. Methylcellulose colony-forming assay revealed that a single PDL cell could form a sphere at a frequency of approximately 0.01% of total cells. These data indicate that PDL-derived spheres contained multipotent adult stem cells capable of differentiating into both neural and mesodermal progenies (Techawattanawisal et al., 2007). These findings of rat PDLSCs may help us to understand human PDLSCs and their potential applications.

Swine PDLSCs. The miniature pig is an ideal large-animal model for preclinical studies to assess the safety and efficacy of a variety of treatment regimes. The oral maxillofacial region of miniature pigs is similar to that of humans in terms of anatomy, development, physiology, pathophysiology, and disease occurrence. The miniature pig has both deciduous and permanent dentition. The morphology of the deciduous and permanent teeth and the anatomy structure of periodontal tissues are similar to those of humans (Wang et al., 2007).

Immunological Properties of PDLSCs. It is often difficult to generate enough cells from one donor source where the growth and differentiation potentials of PDLSCs vary between donors. Thus, utilizing a pool of well-characterized PDLSCs from an allogeneic source offers a solution to cell source scarcity. In terms of possible rejection of allogeneic PDLSCs, it is critical to investigate the immunological properties of these cells. Wada et al. (2009) reported that PDLSCs inhibited peripheral blood mononuclear cell (PBMCs) proliferation stimulated with mitogen or in an allogeneic mixed lymphocyte reaction. PDLSCs exhibited non-cell contact–dependent suppression of PBMCs proliferation. Furthermore, conditioned media derived from PDLSCs pretreated with IFN-γ partially suppressed PBMCs proliferation compared to conditioned media without IFN-γ stimulation. When PDLSCs were cultured with activated PBMCs, the expression of TGFβ1, hepatocyte growth factor (HGF), and indoleamine 2,3-dioxygenase (IDO) was upregulated, and IDO expression was upregulated following stimulation with IFN-γ. These results suggest that PDLSCs possess immunosuppressive properties mediated, in part, by soluble factors, produced by activated PBMCs.

Our group studied the immunological properties of human and swine PDLSCs in depth (Ding et al., 2010). In vitro study revealed that human PDLSCs expressed human leukocyte antigen (HLA)/I, but were negative for immune phenotype markers HLA-II DR, CD80, and CD86. Preirradiated PDLSCs failed to stimulate allogeneic T-cell proliferation. These results indicate that PDLSCs display low immunogenicity. PDLSCs suppress allogeneic T-cell receptor–triggered or mitogen-stimulated T-cell proliferation in a cell number–dependent and antigen-nonspecific manner. Moreover, we found that T cell proliferation was inhibited via mechanisms of soluble factor(s) involvement.

In another immunological study, canine PDLSCs showed an immunomodulatory effect on allogenic and xenogenic PBMCs, and this effect is not a product of apoptosis of PBMCs but is caused by the inhibition of cell division of PBMCs (Kim et al., 2010).

PDLSCs from Inflamed PDL Tissue. In diseased human periodontal ligament, STRO1, CD146, and CD44$^+$ cells are located primarily in the paravascular region, with small clusters of cells in the extravascular region. Moreover, wider distributions of these cells can be detected in diseased ligament (Chen et al., 2006). Inflamed human PDLSCs (ihPDLSCs) can be isolated from the inflamed PDL tissue obtained from intrabony defects. The proliferative potential of ihPDLSCs is equal to that of healthy hPDLSCs, whereas ihPDLSCs has greater migratory capacity. ihPDLSCs exhibit regenerative potential in vivo similar to that of normal hPDSCs (JC Park et al., 2011).

PDLSCs-Mediated Periodontal Regeneration.

Swine Periodontitis Model. To determine whether periodontitis-associated periodontal defects can be regenerated by PDLSC-mediated treatment, Liu et al. (2008) used miniature pigs and established a typical and stable periodontitis study model. Alveolar bone was removed using a surgical bur to create experimental periodontal defects (3 mm in width, 7 mm in length, and 5 mm in depth) in the mesial aspect of the maxilla and mandibular first molars. After the operation, 4-0 silk ligament was sutured around the cervical portion of the first molars. Twelve weeks after periodontitis formation, the inflammation of gingival tissue was distinct: swelling and redness with calculus around the gingival margin. Moreover, the inflammation was found to extend to the furcation in some samples, and the entire mesial buccal root of the first permanent molar was exposed. Radiographic analysis indicated a significant decrease in alveolar bone density in the regions of furcation and the mesial side of the first permanent molar compared with that in a preoperation sample. Furthermore, the CT images of the coronal scanning showed marked bone loss on the buccal side of the first molar. In addition, histological analysis showed marked erosion and ulcer development on the surface of sulcular epithelium, infiltration of inflammatory cells underlying connective tissue, and hyperemia and edema of blood vessels in the gingival tissues. These data indicate that a stable and typical periodontitis lesion could be successfully established in miniature pigs by surgical removal of alveolar bone in association with placement of a silk suture ligature.

PDLSCs were then harvested from the PDL extracted from the cuspids of miniature pigs. The ability of miniature pig PDLSCs to form adherent clonogenic cell clusters of fibroblast-like cells, similar to human PDLSCs, was shown by the formation of colonies in cultures. The progeny of the colony-forming cells were fusiform in shape and showed positive immunostaining for STRO1, Nestin, and ALP. Morphologically, miniature pig PDLSCs displayed many short and long branching cytoplasmic processes under a scanning electron microscope, surrounded by secreted extracellular matrix. To test whether PDLSC-mediated tissue regeneration is an ideal approach to functional periodontal tissue regeneration, one month after the generation of periodontitis lesions, approximately 2.0×10^6 of the expanded third-passage autologous PDLSCs combined with HA/TCP were transplanted into the area of bone defect, and gelatin membranes were used to cover the defects. A nontreated group and an HA/TCP group were used as controls. Before treatment, the teeth were cleaned. Then flap surgery in which a mucoperiosteal flap was raised and granulation tissue was removed from the defect was performed in all groups. In this study the authors used clinical assessment, CT scanning, and histopathology to evaluate PDLSC-mediated periodontal tissue regeneration. At 12 weeks posttransplantation, the attachment loss was 3.3 ± 0.6 mm in the PDLSC-mediated group, 5.3 ± 0.3 mm in the HA/TCP group, and 6.3 ± 0.5 mm in the untreated control group. Statistical analysis indicated that PDLSC treatment improved periodontal tissue regeneration significantly compared to the HA/TCP and control groups. CT scanning analyses showed that the height of periodontal alveolar bone in the PDLSC-mediated group recovered to approximately normal levels. In contrast, the HA/TCP group and the control group showed very limited or no bone regeneration. Histopathological photomicrographs showed that the periodontal tissue regeneration in the PDLSC-mediated group was much better than that in the HA/TCP group. The sulcular epithelium in the PDLSC-mediated group was thicker, and the epithelial pegs and dermal papillae were short and blunt, with fewer inflammatory cells in the PDLSC-mediated group. New bone, cementum, and periodontal ligament were regenerated in the periodontal defect area in the PDLSC-mediated group, and the height of the new alveolar bone was much higher than that in the HA/TCP group. Newly formed Sharpey's fibers were anchored into the newly regenerated cementum in the PDLSC-mediated group. Typical periodontitis, including marked periodontal tissue reduction and infiltration of inflammatory cells were still found in the HA/TCP group. The epithelial pegs and dermal papillae were long and slender, with infiltration of inflammatory cells underlying connective tissue. Fibers lacking the typical structure of Sharpey's fibers filled in the periodontal defect in the HA/TCP group. GFP-labeling experiments indicated that the GFP^+ cells derived from autologous ex vivo–expanded PDLSCs had differentiated into osteoblasts in vivo. This study demonstrates that miniature pig is a useful preclinical large-animal model to use to investigate the treatment of periodontitis using autologous PDLSCs (Liu et al., 2008).

Dog Periodontal Regeneration Model. Beagle dogs have been used as a large-animal model to examine PDLSCs-mediated periodontal tissue regeneration (JY Park et al., 2011). In the study, four weeks after the generation of periodontitis defect, PDLSCs were grafted to each defect. Two months after the PDLSCs graft, a 3.02-mm attachment gain was observed In histological analysis, the PDLSCs grafted defect showed regenerated alveolar bone crest and deposited cementum There were incremental lines of regenerated cementum over the root-planed dentin, and Sharpey's fibers extended from the regenerated alveolar bone. Cellular cementum was also regenerated at the root apex with embedded cementocytes. Peripheral nerve ingrowth and angiogenesis were observed inside the PDL space. Quantitative analysis of the height of new bone formation showed the significantly higher bone was regenerated in the PDLSC group (JY Park et al., 2011).

Clinical Studies. Recently, a paper reported three cases of the use of PDLSCs to treat periodontitis. Three male patients with chronic generalized periodontitis and pocket depth from 4.8 to 10 mm were selected for autologous PDLSC implantation with a total of 16 teeth. All three patients had the third molar extracted and the PDLSCs were isolated. The bone grafting material Calcitite 4060-2 was added on the surface of cultured PDLSCs. The PDLSC/Calcitite 4060-2 construct was removed as a sheet of tissue for transplantation. At several follow-up intervals, there was a consistent decrease in probing depth and attachment levels from those at presurgery. Periodontal tissue regain, specifically gingival attachment gain, was observed at 3, 6, and 42 months of posttreatment. X-ray evaluation indicated periodontal tissue regain at 72 months post-PDLSC implantation compared with pretreatment conditions. Importantly, all patients showed no inflammation in the treated area and no systemic disorder that may have been associated with autologous PDLSC transplantation, implying that this method is a safe intervention procedure for periodontitis (Feng et al., 2010).

Cell Sheets for Periodontal Regeneration. Cell sheet engineering has emerged as a novel alternative approach for periodontal tissue engineering without the use of scaffolds. It involves a cell culture system in which a temperature-responsive polymer, poly(*N*-isopropylacrylamide) (PIPAAm), is grafted covalently to surfaces of tissue culture–grade polystyrene dishes (for details, see Chapter 28, 6.2). Our team has found that vitamin C is capable of inducing telomerase activity in PDLSCs, leading to the upregulated expression of extracellular matrix type I collagen, fibronectin, and integrin b1; stem cell markers Oct4, Sox2, and Nanog; and osteogenic markers RUNX2, ALP, and OCN. Under vitamin C treatment, PDLSCs can form cell sheet structures because of increased cell matrix production in a non-temperature-responsive manner. Interestingly, PDLSC sheets demonstrated a significant improvement in periodontal tissue regeneration compared with untreated control dissociated PDLSCs and offered an effective treatment for periodontal defects in a swine model (Wei et al., unpublished data)

4 CONCLUSIONS AND FUTURE DIRECTIONS

Stem cell–mediated root regeneration offers opportunities to regenerate a bioroot and its associated periodontal tissues, which are necessary for maintaining the physiological function of teeth. Based on the findings in large animals (miniature pigs), using allogeneic MSCs to regenerate bioroot may be a viable alternative approach to replacing missing teeth. Further investigation is needed before clinical trials take place, including the establishment of growth and differentiation conditions that induce lineage specific commitment and protocols that ensure the safe and effective use of stem cells for bioroot and periodontal regeneration.

Acknowledgments

The authors thank Songtao Shi, Center for Craniofacial, Molecular Biology, Ostrow School of Dentistry, University of Southern California, for his insights and thoughtful input.

This work was supported by the National Basic Research Program of China (grants 2007CB947304 and 2010CB944801 to S.W.), the Funding Project for Academic Human Resources Development in Institutions of Higher Learning Under the Jurisdiction of Beijing Municipality (PHR20090510 to S.W.), the Funding Project to Science Facility in Institutions of Higher Learning Under the Jurisdiction of Beijing Municipality (PXM 2009-014226-074691 to S.W., PXM2011-014226-07-000066 to Z.F.), The funders had no role in study design, data collection and analysis, decision to publish, or preparation of the manuscript.

REFERENCES

Atala A, Bauer SB, Soker S, Yoo JJ, Retik AB (2006). Tissue-engineered autologous bladders for patients needing cystoplasty. *Lancet* 367:1241–1246.

Baillie B (2004). A comparative review of periodontal and peri-implant soft tissue: 1. *J West Soc Periodont Abstr* 52:69–79.

Bartholomew A, Sturgeon C, Siatskas M, Ferrer K, McIntosh K, Patil S, et al. (2002). Mesenchymal stem cells suppress lymphocyte proliferation in vitro and prolong skin graft survival in vivo. *Exp Hematol* 30:42–48.

Bartold PM, McCulloch CA, Narayanan AS, Pitaru S (2000). Tissue engineering: a new paradigm for periodontal regeneration based on molecular and cell biology. *Periodontol 2000* 24:253–269.

Bartold PM, Shi S, Gronthos S (2006). Stem cells and periodontal regeneration. *Periodontol 2000* 40:164–172.

Benatti BB, Silvério KG, Casati MZ, Sallum EA, Nociti FH Jr (2007). Physiological features of periodontal regeneration and approaches for periodontal tissue engineering utilizing periodontal ligament cells. *J Biosci Bioeng* 103:1–6.

Chen SC, Marino V, Gronthos S, Bartold PM (2006). Location of putative stem cells in human periodontal ligament. *J Periodont Res* 41:547–553.

Di Nicola M, Carlo-Stella C, Magni M, Milanesi M, Longoni PD, Matteucci P, et al. (2002). Human bone marrow stromal cells suppress T-lymphocyte proliferation induced by cellular or nonspecific mitogenic stimuli. *Blood* 99:3838–3843.

Ding G, Liu Y, Wang W, Wei FL, Liu DY, Fan ZP, et al. (2010). Allogeneic periodontal ligament stem cell therapy for periodontitis. *Stem Cells* 28:1829–1838.

Earthman JC, Li Y, VanSchoiack LR, Sheets CG, Wu JC (2006). Reconstructive materials and bone tissue engineering in implant dentistry. *Dent Clin N Am* 50:229–244.

Eckert SE, Choi YG, Sanchez AR, Koka S (2005). Comparison of dental implant systems: quality of clinical evidence and prediction of 5-year survival. *Int J Oral Maxillofac Implants* 20:406–415.

Feng F, Akiyama K, Liu Y, Yamaza T, Wang TM, Chen JH, et al. (2010). Utility of PDL progenitors for in vivo tissue regeneration: a report of 3 cases. *Oral Dis* 16:20–28.

Gonzalez-Rey E, Anderson P, González MA, Rico L, Büscher D, Delgado M (2009). Human adult stem cells derived from adipose tissue protect against experimental colitis and sepsis. *Gut* 58:929–939.

Gronthos S, Mrozik K, Shi S, Bartold PM (2006). Ovine periodontal ligament stem cells: isolation, characterization, and differentiation potential. *Calcif Tissue Int* 79:310–317.

Guo W, Gong K, Shi H, Zhu G, He Y, Ding B, et al. (2012). Dental follicle cells and treated dentin matrix scaffold for tissue engineering the tooth root. *Biomaterials* 33:1291–1302.

Huang GT, Yamaza T, Shea LD, Djouad F, Kuhn NZ, Tuan RS, et al. (2010). Stem/progenitor cell–mediated de novo regeneration of dental pulp with newly deposited continuous layer of dentin in an in vivo model. *Tissue Eng A* 16:605–615.

Humphrey S (2006). Implant maintenance. *Dent Clin N Am* 50:463–478.

Ikeda E, Morita R, Nakao K, Ishida K, Nakamura T, Takano-Yamanoto T, et al. (2009). Fully functional bioengineered tooth replacement as an organ replacement therapy. *Proc Natl Acad Sci USA* 106:13475–13480.

Kim HS, Kim KH, Kim SH, Kim YS, Koo KT, Kim TI, et al. (2010). Immunomodulatory effect of canine periodontal ligament stem cells on allogenic and xenogenic peripheral blood mononuclear cells. *J Periodont Implant Sci* 40:265–270.

Krampera M, Glennie S, Dyson J, Scott D, Laylor R, Simpson E, et al. (2003). Bone marrow mesenchymal stem cells inhibit the response of naive and memory antigen-specific T cells to their cognate peptide. *Blood* 101:3722–3729.

Le Blanc K, Rasmusson I, Sundberg B, Götherström C, Hassan M, Uzunel M, et al. (2004). Treatment of severe acute graft-versus-host disease with third party haploidenticalmesenchymal stem cells. *Lancet* 363:1439–1441.

Le Blanc K, Frassoni F, Ball L, Locatelli F, Roelofs H, Lewis I, et al. (2008). Mesenchymal stem cells for treatment of steroid-resistant, severe, acute graft-versus-host disease: a phase II study. *Lancet* 371:1579–1586.

Liu Y, Zheng Y, Ding G, Fang D, Zhang C, Bartold PM, et al. (2008). Periodontal ligament stem cell–mediated treatment for periodontitis in miniature swine. *Stem Cells* 26:1065–1073.

Michalowicz BS, Hodges JS, DiAngelis AJ, Lupo VR, Novak MJ, Ferguson JE, et al. (2006). Treatment of periodontal disease and the risk of preterm birth. *N Engl J Med* 355:1885–1894.

Mudda JA, Bajaj M (2011). Stem cell therapy: a challenge to periodontist. *Indian J Dent Res* 22:132–139.

Nauta AJ, Westerhuis G, Kruisselbrink AB, Lurvink EG, Willemze R, Fibbe WE (2006). Donor-derived mesenchymal stem cells are immunogenic in an allogeneic host and stimulate donor graft rejection in a nonmyeloablative setting. *Blood* 108:2114–2120.

Park JC, Kim JM, Jung IH, Kim JC, Choi SH, Cho KS, et al. (2011). Isolation and characterization of human periodontal ligament (PDL) stem cells (PDLSCs) from the inflamed PDL tissue: in vitro and in vivo evaluations. *J Clin Periodontol* 38:721–731.

Park JY, Jeon SH, Choung PH (2011). Efficacy of periodontal stem cell transplantation in the treatment of advanced periodontitis. *Cell Transplant* 20:271–285.

Parrish LC, Miyamoto T, Fong N, Mattson JS, Cerutis DR (2009). Non-bioabsorbable vs. bioabsorbable membrane: assessment of their clinical efficacy in guided tissue regeneration technique—A systematic review. *J Oral Sci* 51:383–400.

Pham C, Greenwood J, Cleland H, Woodruff P, Maddern G (2007). Bioengineered skin substitutes for the management of burns: a systematic review. *Burns* 33:946–957.

Pihlstrom BL, Michalowicz BS, Johnson NW (2005). Periodontal diseases. *Lancet* 366:1809–1820.

Pitaru S, Pritzki A, Bar-Kana I, Grosskopf A, Savion N, Narayanan AS (2002). Bone morphogenetic protein 2 induces the expression of cementum attachment protein in human periodontal ligament clones. *Connect Tissue Res* 43:257–264.

Seo BM, Miura M, Gronthos S, Bartold PM, Batouli S, Brahim J, et al. (2004). Investigation of multipotent postnatal stem cells from human periodontal ligament. *Lancet* 364:149–155.

Seo BM, Miura M, Sonoyama W, Coppe C, Stanyon R, Shi S (2005). Recovery of stem cells from cryopreserved periodontal ligament. *J Dent Res* 84:907–912.

Shi S, Bartold PM, Miura M, Seo BM, Robey PG, Gronthos S (2005). The efficacy of mesenchymal stem cells to regenerate and repair dental structures. *Orthod Craniofac Res* 8:191–199.

Shimono M, Ishikawa T, Ishikawa H, Matsuzaki H, Hashimoto S, Muramatsu T, et al. (2003).Regulatory mechanisms of periodontal regeneration. *Microsc Res Tech* 60:491–502.

Sonoyama W, Liu Y, Fang D, Yamaza T, Seo BM, Zhang C, et al. (2006). Mesenchymal stem cell-mediated functional tooth regeneration in swine. *PLoS One* 1:e79.

Sonoyama W, Liu Y, Yamaza T, Tuan RS, Wang S, Shi S, et al. (2008). Characterization of the apical papilla and its residing stem cells from human immature permanent teeth: a pilot study. *J Endod* 34:166–171.

Sudres M, Norol F, Trenado A, Grégoire S, Charlotte F, Levacher B, et al. (2006). Bone marrow mesenchymal stem cells suppress lymphocyte proliferation in vitro but fail to prevent graft-versus-host disease in mice. *J Immunol* 176:7761–7767.

Sun L, Akiyama K, Zhang H, Yamaza T, Hou Y, Zhao S, et al. (2009). Mesenchymal stem cell transplantation reverses multiorgan dysfunction in systemic lupus erythematosus mice and humans. *Stem Cells* 27:1421–1432.

Taba M Jr, Jin Q, Sugai JV, Giannobile WV (2005). Current concepts in periodontal bioengineering. *Orthod Craniofac Res* 8:292–302.

Techawattanawisal W, Nakahama K, Komaki M, Abe M, Takagi Y, Morita I (2007). Isolation of multipotent stem cells from adult rat periodontal ligament by neurosphere-forming culture system. *Biochem Biophys Res Commun* 357:917–923.

Tonetti MS, D'Aiuto F, Nibali L, Donald A, Storry C, Parkar M, et al. (2007). Treatment of periodontitis and endothelial function. *N Engl J Med* 356:911–920.

Wada N, Menicanin D, Shi S, Bartold PM, Gronthos S (2009). Immunomodulatory properties of human periodontal ligament stem cells. *J Cell Physiol* 219:667–676.

Wang HL, Greenwell H, Fiorellini J, Giannobile W, Offenbacher S, Salkin L, et al. (2005). Periodontal regeneration. *J Periodontol* 76:1601–1622.

Wang S, Liu Y, Fang D, Shi S (2007). The miniature pig: a useful large animal model for dental and orofacial research. *Oral Dis* 13:530–537.

Wei FL, Qu CY, Song TL, Ding G, Fan ZP, Liu DY, et al. (2012). Vitamin C treatment promotes mesenchymal stem cell sheet formation and tissue regeneration by elevating telomerase activity. *J Cell Physiol* 227:3216–3224.

Wei FL, Song TL, Ding G, Liu Y, Liu DY, Fan ZP, et al. (2012). Functional tooth restoration by allogeneic mesenchymal stem cell-based bio-root regeneration in swine. (unpublished data)

Wu KH, Chan CK, Tsai C, Chang YH, Sieber M, Chiu TH, et al. (2011). Effective treatment of severe steroid-resistant acute graft-versus-host disease with umbilical cord–derived mesenchymal stem cells. *Transplantation* 91:1412–1416.

Yamada Y, Ueda M, Hibi H, Baba S (2006). A novel approach to periodontal tissue regeneration with mesenchymal stem cells and platelet-rich plasma using tissue engineering technology: a clinical case report. *Int J Periodont Restor Dent* 26:363–369.

Yang B, Chen G, Li J, Zou Q, Xie D, Chen Y, et al. (2012). Tooth root regeneration using dental follicle cell sheets in combination with a dentin matrix–based scaffold. *Biomaterials* 33:2449–2461.

Yang ZH, Zhang XJ, Dang NN, Ma ZF, Xu L, Wu JJ, et al. (2009). Apical tooth germ cell–conditioned medium enhances the differentiation of periodontal ligament stem cells into cementum/periodontal ligament-like tissues. *J Periodont Res* 44:199–210.

Zaky SH, Cancedda R (2009). Engineering craniofacial structures: facing the challenge. *J Dent Res* 88:1077–1091.

28

PERIODONTAL BIOENGINEERING STRATEGIES: THE PRESENT STATUS AND SOME DEVELOPING TRENDS

FA-MING CHEN

Department of Periodontology and Oral Medicine, School of Stomatology, Fourth Military Medical University, Shaanxi, China

YAN JIN

Research and Development Center for Tissue Engineering, Fourth Military Medical University, Shaanxi, China

1 INTRODUCTION

Periodontitis is a globally prevalent human inflammatory disease and the most common cause of tooth loss in adults. Its advanced forms affect 10 to 15% of adults worldwide, resulting in a decreased quality of life (e.g., impairment in the abilities to eat and speak) (Al-Shammari et al., 2005). Detachment of the junctional epithelium from the tooth root surface (i.e., the formation of a periodontal pocket) and disconnection of periodontal ligament (PDL) fiber attachment to the root cementum and bone loss are the hallmarks of periodontitis. It has also been linked to many systemic disorders, such as coronary artery disease, stroke, and diabetes (Pihlstrom et al., 2005; Polimeni et al., 2006; Pizzo et al., 2010). Hence, scientists have focused much attention on developing new effective techniques to address this condition.

The key concept for periodontal therapy is to improve periodontal health and to satisfy a patient's aesthetic and functional needs. The strategies in the treatment of periodontally involved teeth have now shifted from the use of guided tissue regeneration (GTR) and various bone grafts and substitutes to cell-based therapies and tissue engineering approaches, to achieve more functional periodontal reconstruction (Ishikawa et al., 2009; Izumi et al., 2011; Chen et al., 2012b). The principle of

Stem Cells in Craniofacial Development and Regeneration, First Edition.
Edited by George T.-J. Huang, Irma Thesleff.

periodontal bioengineering is to rebuild the damaged tooth-supporting tissues, including root cementum, PDL, and alveolar bone, by using one or a combination of the following elements: cells, signals and biomaterials (Chen and Jin, 2010).

The well-accepted objectives of periodontal therapy are first, to arrest periodontal disease progression, and second, to stimulate host stem and progenitor cells to regenerate new periodontal structures (Polimeni et al., 2006; Chen et al., 2010a). Although conventional treatments such as tooth scaling, root planning, and open-flap debridement do lead to success in probing depth reduction, the healing is usually dominated by formation of an unstable long junctional epithelium instead of periodontal regeneration (Pihlstrom et al., 2005). The outcomes of introducing guided bone regeneration (GBR)/GTR and various natural and synthetic biomaterials, with or without the adjuvant use of various growth factors (GFs), into periodontal surgical approaches has been investigated (Choi et al., 2002; Sculean et al., 2008; Reynolds et al., 2010; Villar et al., 2010). The results varied greatly, and in principle, this therapeutic paradigm falls short in reaching a consistent and complete regeneration of the periodontium.

Recent advancement in tissue engineering and stem cell biology has created a new avenue for enhancing functional periodontal regeneration (Bartold et al., 2006; Ishikawa et al., 2009; Lin et al., 2009; Izumi et al., 2011). Healthy PDL harbors stem cells that are essential for periodontal regeneration. Unfortunately, in a diseased periodontal environment, tissue repair does not occur naturally because of a lack of robust stem cells, the change of extracellular milieu, and the nature of periodontal structures. Thus, exogenous regenerative "medical aid" such as stem cells, GFs, and biomaterial scaffolds will be needed for facilitating tissue regeneration (Lin et al., 2009; Izumi et al., 2011).

In this chapter we overview and discuss applied engineering strategies, including the design of biomaterials, the selection of cell types and their delivery approaches, and the necessity to replicate a favorable extracellular milieu to facilitate a predictable periodontal regeneration.

2 PERIODONTAL BIOENGINEERING

Modern dentistry relies on the use of artificial materials such as dental implants to replace missing teeth, which cannot restore the biophysiological architecture and function of the tissues. In addition, factors that interfere with osseointegration may cause implant surgery failure (Paquette et al., 2006).

The contemporary tissue engineering science is to apply the principles of cell biology, developmental biology, and biomaterials to develop strategies for regeneration of lost or damaged tissues. The same concept may be used for periodontal tissue regeneration (Bartold et al., 2006; Rios et al., 2011). The complete package of periodontal bioengineering includes the use of an appropriate cell type, a prefabricated three-dimensional scaffold (supportive matrix template), and instructive signals (e.g., GFs and matrix-attachment factors) (Scheller et al., 2009; Catón et al., 2011).

Besides cells from bone marrow and adipose tissues (Tobita et al., 2008; Tsumanuma et al., 2011), periodontal ligament stem cells (PDLSCs) from various species, including humans, have been reported to be a potent cell source for periodontal tissue regeneration (Feng et al., 2010; Liu et al., 2008; Ding et al., 2010; Park et al., 2010). These studies have, with various degree of success, attempted topical applications of various stem cells combined with GFs and/or scaffolds for the regeneration of lost periodontal structures. The combination of platelet-rich plasma (PRP) with GFs, fibrin scaffolds, and bone marrow–derived mesenchymal stem cells (BMMSCs), in particular, has been demonstrated to be a successful clinical approach (Yamada et al., 2008).

Currently, many issues need to be addressed before periodontal tissue engineering is ready for clinical practice (Izumi et al., 2011; Rios et al., 2011). It is of critical importance to understand the structure–function relationships in normal and diseased tissues and to use these insights as design criteria for engineering new tissues. Information on how tissues develop in vivo and which constituents are most critical for exerting function can be utilized to design strategies to recapitulate the processes involved during natural tissue development (Polimeni et al., 2006).

3 BIOMATERIALS FOR PERIODONTAL BIOENGINEERING

Selection of an appropriate scaffold for cell accommodation is a crucial step for any tissue engineering design. Biomaterials used in periodontal regenerative medicine include natural or synthetic polymers, ceramics, and composites. Each material offers a unique chemistry, composition and structure, degradation profile, and possibility of modification.

3.1 Natural Polymers

Natural biomaterials serve as a cornerstone in the development of matrix-based regenerative therapies that aim to accelerate clinical application with their excellent biocompatibility, biodegradability, affinity for biomolecules, and wound-healing activity (Mano et al., 2007). Collagen, hyaluronic acid, alginate, and chitosan scaffolds have been used in periodontal regenerative research for more than two decades. They can easily be modified chemically and physically to form desired structures, possess optimal properties, and perform specific functions for various applications. The use of natural polymers in the form of hydrogels allows the incorporation of biological agents during formulation. Normally, these hydrogels are degraded by enzymes and/or acid hydrolysis at a rate depending on the degree of cross-linking and/or the molecular mass (Mano et al., 2007). In addition, natural polymers can also be modified to become three-dimentional scaffolding materials, such as porous sponges, particles, films, and rods, and for the formulation of stimuli-responsive biomaterials.

Collagen. Of the many natural scaffold materials being investigated, collagen has been shown to have many advantages. Highly porous collagen lattice sponges support the in vitro growth of many types of tissues and can deliver multiple GFs. Following the clinical use of collagen carriers to deliver BMPs for tibial shaft fractures (Cancedda et al., 2007), spine fusions, and long-bone nonunions (White et al., 2007), collagen is currently being evaluated for widespread clinical periodontal regeneration. In particular, collagen is used clinically as a composite with ceramics (Bueno and Glowacki, 2009). Commercially available collagen composite scaffolds include Formagraft (Nuvasive, San Diego, CA) and ossimend (Collagen Matrix, Franklin Lakes, NJ). An absorbable collagen sponge (ACS) has been used as a carrier and/or scaffold for periodontal regeneration in animal studies (Choi et al., 2002; Kim et al., 2009a) and clinical trials (Feng et al., 2010). The ACS carrier containing recombinant human BMP-2 is now available commercially as InFuse Bone Graft (Medtronic, Minneapolis, MN) and has been used for bone augmentation during sinus lifting and dental implant placement in patients (Schliephake et al., 2005).

Alginate. Alginate hydrogels bearing cell-adhesion ligands have been used as scaffolds for cell encapsulation and transplantation, which have yielded promising results in experiments aiming at bone regeneration via growth of tissues from small numbers of implanted precursor cells (Alsberg et al., 2002).

Chitosan. This is a natural polymer from renewable resources obtained from the shells of shellfish and wastes of the seafood industry. Chitosan and its derivatives can function as carriers for GFs and functional genes encoding GFs, which can easily be integrated into the materials for a better regenerative performance. Recent studies have suggested that chitosan and its derivatives are promising candidates as a supporting material for tissue regeneration because of their porous structures, gel-forming properties, ease of chemical modification, and high affinity for in vivo macromolecules (Kim et al., 2008).

Combination. It has been proposed in recent years that a combination of several materials may offer the best clinical outcomes (Chen et al., 2009b). For example, a porous chitosan–collagen scaffold (Zhang et al., 2009) and a glycidyl methacrylate–derivatized dextran (Dex-GMA)/gelatin scaffold prepared through a freeze-drying process were developed specifically for periodontal tissue engineering (Chen et al., 2009a). Natural polymers can also be strengthened mechanically by being combined with ceramics or other polymers, such as hydroxyapatite (HA) or polylactide (blends, copolymers, and interpenetrated networks) (Hong et al., 2008). Their interactions with cells and tissues may be improved or fine-tuned by adequate chemical modifications, such as grafting with cell adhesion peptide Arg-Gly-Asp (RGD) or by surface treatments (e.g., surface plasma modification or surface chemistry modification through grafting with other polymers) (Mano et al., 2007; Kumada and Zhang, 2010). The incorporation of RGD peptides in hydrogel matrices has been found to significantly enhance the attachment, spreading, survival, and mineralization of encapsulated dental follicle progenitors, suggesting that RGD additives may promote the use of hydrogels for periodontal mineralized tissue engineering (Dangaria et al., 2009). The incorporation of peptide motifs recognized by cell receptors and the use of recombinant DNA technology has enabled the creation of tissue engineering scaffolds with new levels of biofunctionality.

3.2 Synthetic Polymers

Synthetic matrices have well-controlled and reproducible chemical and physical properties (Gunatillake et al., 2006). They can be manufactured consistently as off-the-shelf products, are biocompatible, and their biodegradation rates can be tailored for intended applications. These polymers possess predictable and reproducible mechanical and physical properties (e.g., tensile strength, elastic modulus, and degradation rate) and can be manufactured with great precision. Synthetic matrices are typically processed in the form of solid scaffolds, small particles, or hydrogels, depending on the mechanical and degradation properties for the particular application. Hydrogels or small particles can be used as injectable matrices (Gunatillake et al., 2006).

Synthetic Hydrogel. Synthetic hydrogel matrices are cross-linked water-soluble polymers that swell in water to form a gel; examples include poly(vinyl alcohol) (PVA) and poly(ethylene glycol) (PEG).

Poly(α-hydroxyester)s. Solid scaffolds are typically porous matrices fabricated by techniques such as solvent casting, gas foaming, particulate leaching, electrospinning, and rapid prototyping. Typical polymers employed in these processes include poly(α-hydroxyester)s such as poly(lactic-*co*-glycolic acid) (PLGA) or poly(L-lactic acid) (PLLA), polyanhydrides, polyorthoesters, PEG, and PVA (Gunatillake et al., 2006; Lasprilla et al., 2012). The physical properties of these polymers can be altered readily by varying the lactide/glycolide ratio, the molecular mass, and the crystallinity, or by combining them with other materials (Rezwan et al., 2006). In some cases, synthetic polymers can be used as adjunctions to natural biomaterials. For example, the rapid degradation of fibrin (a biopolymer critical to hemostasis and wound healing) can be decelerated by modification with PEG, thus creating a hybrid material for cell delivery (Galler et al., 2011).

Biodegradable polymers are used where the transient existence of materials is required; they are used as sutures, scaffolds for tissue regeneration, tissue adhesives, hemostats, transient barriers for tissue adhesion, and as drug delivery systems (DDSs) (Nair and Laurencin, 2006). Bulk biodegradable polymers have shown more promise for regeneration than have surface bioerodible polymers, due to the requirement that a tissue engineering scaffold be replaced by newly formed tissue in vivo. Amorphous poly(D, L-actic acid) (PDLLA) is one of the most popular materials considered for scaffold production and is used in combination with bioactive glasses because it can be combined with biomolecules such as GFs and antibiotics to establish a locally acting DDS (Rezwan et al., 2006). With this modification, the creation of a macroporous structure within the bioceramic materials is possible. For example, when PLGA microparticles are incorporated into calcium phosphate (CaP) cement, a macroporous CaP cement scaffold is formed after PLGA hydrolysisin vivo (Ruhé et al., 2005).

The drug delivery property of these biomaterials is determined by the polymer composition, hydrophobicity, crystallinity, and degradability, which affect the rate of drug release and tissue ingrowth (Nair and Laurencin, 2006). It is expected that a scaffold with a controlled drug delivery function will promote tissue regeneration and eliminate possible inflammatory responses upon scaffold degradation. However, many such polymers suffer from shortcomings such as eliciting persistent inflammatory reactions, eroding, or being incapable of integrating with host tissues (Sun et al., 2012). These issues need to be addressed prior to clinical application.

3.3 Bioceramics and Bioactive Glasses

Ceramics are crystalline inorganic nonmetallic minerals that are held together by ionic bonds and are usually densified by sintering, such as with calcium sulfate, HA, and CaP (Chen et al., 2006). Calcium sulfate is resorbed rapidly and is often used in dentistry and for treatment of cysts in long bones. HA and β-tricalcium phosphate (TCP) have been approved by the U.S. Food and Drug Administration (FDA) for use in dental implants since the mid-1980s and are now also approved for use in orthopedic implants or as bone substitutes (Covani et al., 2011). Clinically used forms of HA and β-TCP include injectable formulations that harden at body temperature, such as β-BSM (Etex,

Cambridge, MA) and BoneSource HA Cement (Stryker, Kalamazoo, MI). The treatment of periodontal intrabony defects with a combination of two ceramics (calcium sulfate in combination with β-TCP) leads to a favorable clinical outcome after two years (Sukumar et al., 2011). The alloplastic material Osteon (Genoss, Suwon, Korea) has an HA surface coated with β-TCP (70% HA and 30% β-TCP). The pore size of Osteon is 300 to 500 μm, and the volumetric porosity of Osteon is approximately 77%. It has recently been reported that Osteon is suitable for use in sinus graft application (Bae et al., 2010). Synthetic CaP ceramics, with their excellent biocompatibility, are common alternatives to autogenous bone, xenograft, or allograft materials when incorporated with osteogenic factors (Hallman and Thor, 2008). Because of their biocompatibility and high protein-binding affinity, which also makes them good vehicles for drug delivery, ceramics have the potential for use in periodontal bone regeneration (Jayakumar et al., 2011). These materials exhibit a range of degradation rates that span from a few weeks to permanency.

The close similarity of HA to the mineral component of bone, which has slow turnover, results in the lack of biodegradation of HA in the body—generally an undesirable feature for tissue engineering scaffold materials. For example, a recent clinical report on a six- to seven-year follow-up study has confirmed that implanted crystalline HA remains in the body for extended periods with no visible signs of biomaterial resorption (Marcacci et al., 2007). In comparison, β-TCP degradation rate can be controlled by changing the Ca/P ratio. Because of its degradation characteristics, β-TCP is regarded as an ideal material for bone substitutes for various peridontal repair (Kwon et al., 2010; Moore et al., 2010; Emerton et al., 2011; Min et al., 2011). Recently, an animal study has shown a potential for the prefabrication of vascularized bioartificial bone grafts in vivo and has introduced the principles of bone transplantation with minimal donor-site morbidity with no quantity restrictions. β-TCP scaffolds were filled intraoperatively with autogenous bone marrow for cell loading and implanted into the latissimus dorsi muscle for potential application at a later stage for segmental bone reconstruction (Kokemueller et al., 2010).

Bioactive glasses, which are osteoproductive materials, have received considerable attention as bone graft substitutes in the treatment of bony defects (Hallman and Thor, 2008). The excellent properties of bioactive glasses and their long history of applications in biomedical implants have prompted extensive research in the last 10 years regarding their use in periodontal engineering and regeneration (Lovelace et al., 1998; Yukna et al., 2001). Although bioactive glasses are mechanically weak, it has recently been discovered that 45S5 Bioglass (U.S. Biomaterials Corporation, Alachua, FL) can partially crystallize when heated to high temperatures ($>950^{\circ}C$) during scaffold fabrication and that the mechanically strong crystalline phase can transform to a biodegradable, amorphous CaP at body temperature in a biological environment (Chen et al., 2006). This transformation enables the two normally irreconcilable properties, mechanical competence and biodegradability, to be combined in a single scaffold. This discovery will promote scaffold optimization and its clinical application (Hallman and Thor, 2008). Findings indicate that Bioglass alone has the ability to support the growth of osteoblast-like cells in vitro and to promote osteoblast differentiation by stimulating the expression of major phenotypic markers. In addition, bioactive granules coated with Emdogain revealed significantly higher protein production than that of bioactive granules alone, which suggests that the use of Bioglass combined with enamel matrix proteins has the potential to achieve a predictable periodontal regeneration (Hattar et al., 2005).

3.4 Biomaterial Design

Both natural and synthetic materials have been used as regenerative biomaterials. However, in nearly every case, these materials were adopted from other areas of science and technology without substantial redesign for periodontal use (Sun et al., 2012). In terms of shape and geometry, preformed scaffolds are suitable for defects requiring mechanical strength or a predefined shape, such as one-walled bone defects, whereas injectable scaffolds may facilitate clinical use in two- or three-walled intrabony defects. Injectable types of scaffolds must solidify within a clinically acceptable time scale in situ. They have poor mechanical strength and porosity; however, pores are generated in scaffolds with rapidly degrading particles or gaseous bubbles as porogens to promote cell attachment, migration, proliferation, mass transfer, vascularization, and material resorption (Khan et al., 2008). Many biomaterials have had a long history of clinical use, but very few interact with their surrounding host environment or promote integration with host tissue in a proactive manner. The desire for more biological approaches in biomaterial design to yield materials that are more instructive to cells has led to an expansion and paradigm shift in the field of biomaterials (Scheller et al., 2009). It has been recognized for many years that the microenvironment in which a cell resides dictates many cell functions and phenotypes. Thus, it seems logical that the specifics of the ECM (e.g., the inflammatory periodontium) must be taken into account in the construction and design of a cell-seeded scaffold. The material will interact with the ECM to induce appropriate gene expression in cells that form new tissues. The complexity of this design process is exemplified by the dynamic states of the cells and tissues, which are regulated by the spatial and temporal coordination of multiple cell processes, each of which is regulated by multiple reciprocal interactions between cells and their ECM (Scheller et al., 2009). Biomaterial modification can take on different levels of complexity to produce increasing levels of physiological "mimicry" and functionality. In addition to using natural ECMs, surface and bulk chemical modifications of synthetic materials can enhance final tissue integration. For example, highly hydrophilic surfaces prevent the adsorption of proteins, or these molecules are bound very weakly. On highly hydrophobic materials, however, proteins are adsorbed in rigid and denatured forms, hampering cell adhesion (Scheller et al., 2009; Bacaková et al., 2011). Typical surface modifications include, but are not limited to, changes in hydrophilicity, functionalization with charged groups, the incorporation of insoluble ligands and peptide cell recognition sequences (e.g., RGD), the attachment of larger proteins, supramolecular self-assembly, and the development of materials that bind and release soluble factors (Langer and Tirrell, 2004). The wettability of the material surface, particularly in synthetic polymers, can be regulated effectively by physical treatments (e.g., by irradiation with ions, plasma or ultraviolet light). The irradiation-activated material surface can be functionalized by various biomolecules and nanoparticles. This characteristic further enhances its attractiveness for cells and its effectiveness in regulating cell functions (Bacaková et al., 2011). Strategies based on physical cues include the reproduction of nanoscale topology, superposition of mechanical cues, and control of degradation. Designing biological recognition into a biomaterial may also obviate the need for therapies based on the delivery of cells or recombinant GFs, which are subject to regulatory constraints (Scheller et al., 2009). To this end, scientists are creating new materials, including those with improved biocompatibility, stealth properties, responsiveness (smart materials), specificity, and other critical properties.

Optimal cell-free devices in periodontal regenerative medicine are engineered to provide mechanical stability while supporting osteogenesis, osteoconduction, and/or osteoinduction (Chen and Jin, 2010; Sun et al., 2012). The molecular and physical message that is coded within the extracellular milieu influences the development of a new generation of devices for periodontal tissue engineering (Place et al., 2009). Several powerful extracellular influences, such as BMPs and PDGFs, have already found their way into cell-instructive scaffolds in periodontal bioengineering, while others remain largely unexplored (Chen et al., 2010a). To date, biomaterials are rapidly being developed to display and deliver stem cell regulatory signals in a precise and nearly physiological fashion and to serve as powerful artificial ECMs in which to study and instruct stem cell fates both in culture and in vivo (Chen et al., 2010b). Further synergism of cell biological and biomaterials technologies promises to have a profound impact on stem cell biology and to provide insights that will advance stem cell–based clinical approaches to tissue regeneration (Lutolf et al., 2009).

4 GROWTH FACTORS AND THEIR DELIVERY

GFs modulate cell proliferation, migration, ECM formation, and other unidentified cellular functions. Some GFs may also function as cell differentiation or trafficking factors. Key GFs for periodontal tissue engineering are PDGFs, TGF-β, fibroblast growth factors (FGFs), IGFs, growth differentiation factor (GDF)-5, vascular endothelial growth factor (VEGF), parathyroid hormone (PTH), and BMPs (Chen et al., 2010a; Izumi et al., 2011). PDGF has good mitogenic and chemotactic properties for cells and is released from the clot at the start of the healing cascade. FGFs are involved in the induction of angiogenesis, the regulation of the proliferation, and the differentiation of cells such as fibroblasts, endothelial cells, osteoblasts, and PDL cells (Jin et al., 2004; Nevins et al., 2005; Peng et al., 2009). The first attempts that employed semipurified natural materials such as PRP and EMPs for periodontal regeneration have now expanded to the use of recombinant human protein products. PRP is derived from centrifuged autologous blood by drawing off the platelet-rich buffy zone. Thus, many GFs, including PDGFs, TGF-β, and IGF, are present in a well-prepared PRP formation (Chen et al., 2010a). For periodontal therapy, PRP is resuspended in a small volume of blood and placed in the periodontal or bony defect or added to a graft material. Systematic reviews of the reported effects of PRP on bone regeneration document significant improvements in clinical attachment levels but weakly beneficial effects in sinus elevation (Plachokova et al., 2008). Although PRP has been somewhat beneficial when used in periodontal or implant regeneration, it has fallen out of favor recently because of the lack of controlled clinical trials that can provide strong evidence of its efficiency (Del Fabbro et al., 2011). Enamel matrix derivative (EMD), an extract of porcine immature enamel matrix, is regarded as a candidate protein mixture that induces mesenchymal cell differentiation to form periodontal tissues (Kemoun et al., 2007). A commercial EMD, marketed as Emdogain (Institut Straumann AG, Basel, Switzerland), has received U.S. Food and Drug Administration approval and is now available for the clinical treatment of periodontal defects. EMD contains over 95%

amelogenin with small amounts of enamelin and other proteins. Initial studies showed histological evidence of regeneration in experimentally created periodontal defects in rats, pigs, and monkeys (Gestrelius et al., 1997). Subsequent clinical studies showed the material to be of benefit for cases with infrabony and angular periodontal defects (Kalpidis and Ruben, 2002). Further studies have shown that EMD also contains a number of GFs, stimulates the production of additional GFs, such as BMPs, and promotes angiogenesis (Venezia et al., 2004). However, additional well-controlled clinical trials are still needed to clarify the clinical application of EMD.

The use of recombinant GFs has the advantage over the GFs in naturally derived materials which have various concentrations. Recombinant GFs can be delivered as a single factor at desired concentrations (Tayalia and Mooney, 2009). BMPs are a family of glycoproteins that are involved in all aspects of bone formation. BMP-2 and BMP-7 have been available commercially for use primarily in orthopedics, and also in periodontics and implant dentistry. BMPs play a key role in osteoblast differentiation and bone formation and have been mixed with bone grafts and placed in collagen plugs to improve the outcome of ridge preservation (Chen et al., 2009b). To improve clinical outcomes, graft materials combined with GFs have been made into commercial products such as INFUSE Bone Graft. GDF-5, a member of the BMP family, has been used in periodontal therapy in various types of animal models and appears to be a promising for periodontal wound healing and regeneration (Kwon et al., 2010; Moore et al., 2010; Min et al., 2011; Emerton et al., 2011). Besides the INFUSE Bone Graft, recombinant human PDGF-BB homodimer has also been approved for the treatment of infrabony and furcation lesions and is available commercially as GEM-21 (Osteohealth, Shirley, NY). A controlled clinical trial with PDGF-BB homodimer and β-TCP demonstrated significant increases in clinical attachment levels and improved bone fill when compared with the results of using β-TCP alone after six months (Nevins et al., 2005). Human cementoblastoma-derived protein, which is highly expressed in cementoblasts, may regulate local cementoblast differentiation and cementum–matrix mineralization and may participate in the regeneration of alveolar bone and cementum from nonosteogenic cells (Carmona-Rodriguez et al., 2007). FGF-2 exhibits potent angiogenic activity and induces mitogenic ability in mesenchymal cells within the PDL. FGF-2 has been shown to enhance periodontal regeneration considerably in two-wall intrabony defects in dogs (Shirakata et al., 2010). These effects were also observed in a randomized double-blind placebo-controlled clinical trial conducted in 253 adult patients in Japan (Kitamura et al., 2011), which suggests that the topical application of FGF-2 can be efficacious in the regeneration of periodontal tissues.

There is still a need to devise sophisticated approaches to deliver GFs that would allow for the controlled, sustained, and localized delivery of proteins (Tayalia and Mooney, 2009). DDSs that regulate the biological presentation of GFs represent an attractive new generation of therapeutic agents for the treatment of a wide variety of diseases. The device essentially allows the loaded GFs to be released at a desired rate and concentration and to remain at injury sites for a sufficient time to recruit progenitors and stimulate the tissue-healing processes. In addition, DDSs can be formulated to have particular structures that facilitate cellular infiltration and growth (cell scaffold) (Nair and Laurencin, 2006). To improve their properties, many DDSs serve dual purposes; in addition to providing cell support, cutting-edge scaffolds interact biologically with adhering and invading cells and effectively guide cellular growth and development by releasing bioactive proteins (Chen et al., 2009a). To design controlled release systems for specific applications, it is important to understand the basic principles of protein delivery and the stability of each biomolecule applied (Sun et al., 2012).

There remain a variety of biological and engineering challenges in the use of GFs for tissue regeneration. First, it is unlikely that current delivery strategies can meet the physiological requirements of the tissue repair processes. It also seems that any treatment aiming to mimic the natural tissue regeneration processes should not be limited to the provision of a single GF but instead, should deliver multiple agents at an optimized ratio in a specific spatiotemporal pattern (Chen et al., 2010b). Owing to the complex nature of the underlying GF biology, further qualitative and quantitative understanding of the distributions and interactions of the various GFs and how the in vivo microenvironment regulates the cellular responses will probably be essential for the design of polymeric devices that lead to desired outcomes (Ramseier et al., 2006; Chen et al., 2009b). The information that is coded within the extracellular milieu is influencing the development of a new generation of biomaterials that provide the appropriate chemical and mechanical cues in a timely and spatially controlled manner to finely tune the regenerative processes (Huebsch and Mooney, 2009). The ability to modulate the response in real time will probably be important for both understanding and regulating the tissue processes, and we note that on the engineering front, it is still difficult to obtain tight "on" or "off" control over the release of proteins (Tayalia and Mooney, 2009). In addition, much is still unknown about the mechanisms by which tissues form and heal; insights from developmental biology and other biological disciplines are actively guiding the development of intelligent materials that work with nature's own mechanisms of repair (Lutolf et al., 2009). These expanding possibilities raise the question of how much extrinsic physiochemical information is required to mobilize endogenous or transplanted cells into producing a complex tissue, and specifically, what minimum level of materials complexity is required for a given task (Place et al., 2009).

5 STEM CELLS FOR PERIODONTAL BIOENGINEERING

Periodontal tissues exhibit limited endogenous repair and regeneration capacity. Adult stem cells residing in postnatal dental tissues probably provide a source of cells to regenerate lost or damaged periodontium and to facilitate periodontal repair; stem cells in the PDL and growing roots play more dynamic roles in tooth function and development (Volponi et al., 2010). Several subpopulations of stem cells have been isolated from different parts of the tooth and have demonstrated potential for future periodontal therapies (Chen et al., 2012b). These populations include PDLSCs (Seo et al., 2004), dental pulp stem cells (DPSCs) (Gronthos et al., 2000), stem cells from exfoliated deciduous teeth (SHEDs) (Miura et al., 2003), stem cells from apical papilla (SCAPs) (Sonoyama et al., 2006), dental follicle progenitor cells (DFPCs) (Handa et al., 2002; Morsczeck et al., 2005; Guo et al., 2012) (Fig. 1), and MSCs from gingival tissues (GMSCs) (Tang et al., 2011) (Fig. 2). Dental stem cells can be obtained easily, making them an attractive source of autologous stem cells for regeneration of PDL and alveolar bone lost in periodontal diseases and for the generation of complete or partial tooth structures. Nondental stem cells such as BMMSCs and adipose-derived stem cells (ASCs) have also been investigated as cell sources for periodontal regeneration (Kawaguchi et al., 2004; Tobita et al. 2008). For a detailed description of these cells, refer to Chen et al. (2012b) and Chapters 11 & 12.

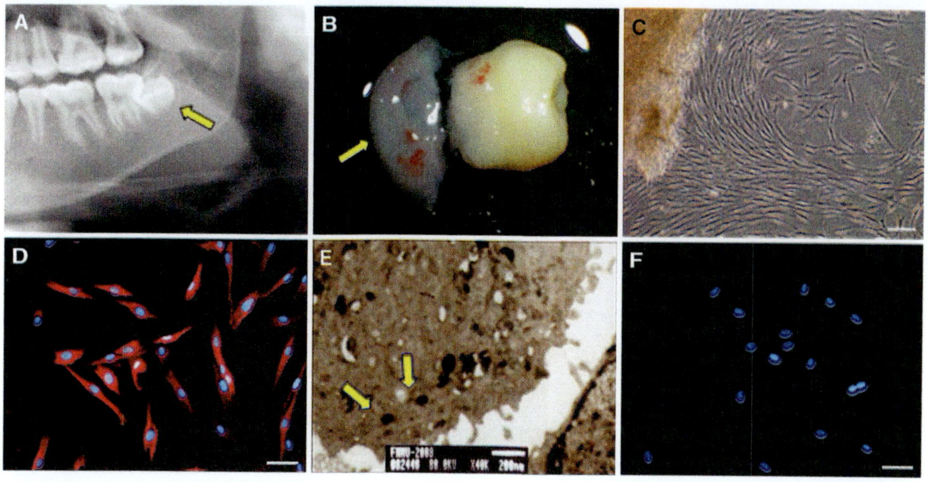

FIGURE 1 Culture and identification of dental follicle cells (DFCs) isolated from human impacted molars (HIM-DFCs) (Guo et al., 2012). (A) A representative x-ray image of an impacted mandibular third molar (indicated by a yellow arrow). (B) A representative image of tissue containing DFCs that was harvested from an impacted mandibular third molar (indicated by a yellow arrow). (C) HIM-DFCs that were cultured and passaged from a dental follicle. These cells were stained for vimentin (D) and CK14 (E), and electron-dense granular material was detected using TEM (F; indicated by a yellow arrow). Scale bars: 100 μm.

6 PERIODONTAL BIOENGINEERING STRATEGIES

6.1 Facilitating Periodontal Endogenous Regeneration

Periodontal tissue engineering strategies typically involve harvesting of suitable autologous donor tissues (e.g., the PDL), from which cells are isolated and expanded in a good manufacturing practice (GMP) facility. Expanded cells are then seeded onto a matrix and surgically implanted into the host's periodontal defects. Alternatively, the cell or matrix may be further incubated in a bioreactor or other such system prior to implantation, where the cells may be exposed to biological, chemical, or physical stimuli that promote the formation of the appropriate tissue (Chen and Jin, 2010). Considerable preclinical data and some clinical success support the applicability of these approaches (Kawaguchi et al., 2004; Tobita et al., 2008; Park et al., 2011b). However, the process is very complex and not cost-effective. Additionally, the scarcity of autologous cell source is an issue. Although allogeneic cell transplantation has shown success in animal studies (Wada et al., 2009; Ding et al., 2010), the potential risks associated with transplanting exogenous cells remain to be addressed.

These obstacles have led to recent interest in endogenous tissue engineering strategies with which the regenerative potential of a patient's own cells is harnessed to promote tissue regrowth without ex vivo cell culturing (Chen et al., 2010a). Endogenously periodontal stem or progenitor cells may be recruited via designed approaches to form new tissue. In this regard, Mao and co-workers used a simple biomaterials-based approach to regenerate dental pulp, PDL tissue and the entire articular surface of the synovial joint without the delivery of exogenous cultured cells (Kim et al., 2010a,b;

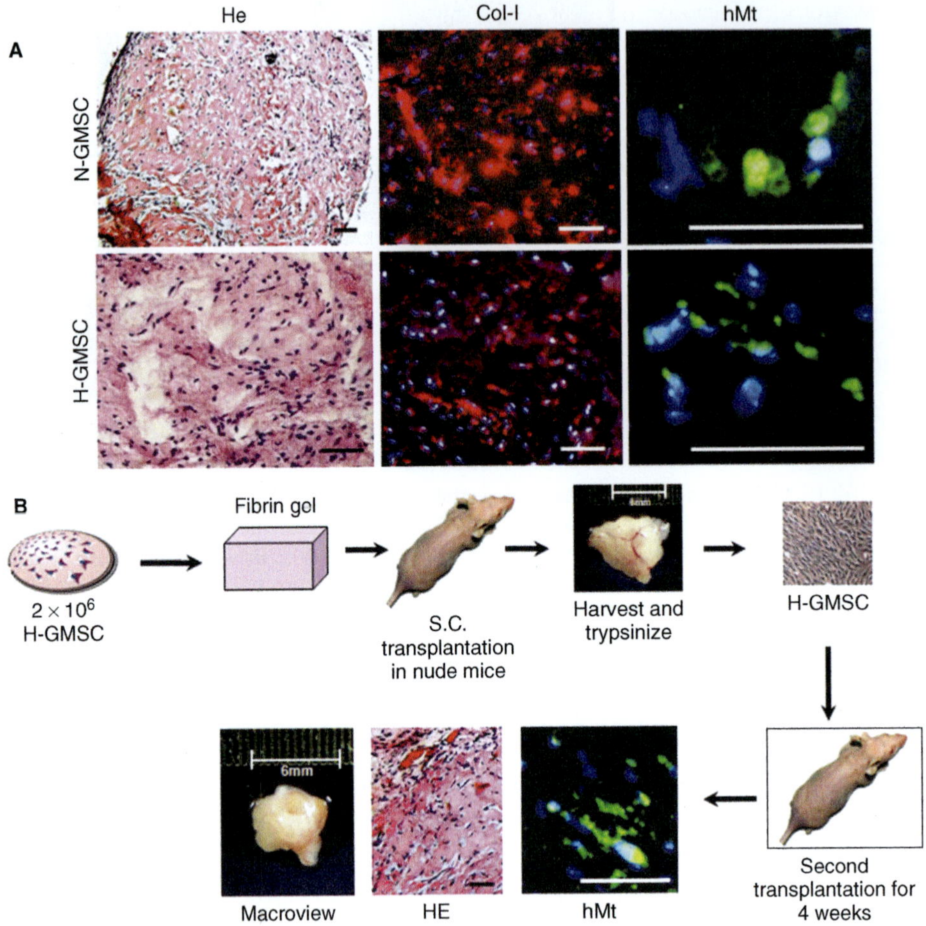

FIGURE 2 In Vivo regenerative capability of mesenchymal stem cells from human normal and hyperplastic gingival tissues (N-GMSC and H-GMSC, respectively) (Tang et al., 2011). (A) Representative image of hematoxylin–eosin (H&E) and immunofluorescence staining of H- and N-GMSC explants retrieved from immunocompromised mice after implantation for 4 weeks. (B) Schematic for strategy of serial transplantation of H-GMSC to test its genuine self-renewal and regenerative capability. H-GMSC derived from over half of primary explants (3/5) maintained the ability to form connective-like tissues in second transplantation. Due to failure of recovering single colony–derived cells from primary N-GMSC explants (1/5), no tissue formed in secondary transplantation (0/1). Scale bar: 50 μm. COL-I, type I collagen; Mt, human mitochondria; S.C., subcutaneous transplantation; hMt, human mitochondria.

Lee et al., 2010). However, such simple strategies may not offer a universal regenerative solution. Healing is restricted by an age-related decline in progenitor populations or by the intrinsically low regenerative potential of certain tissues (Place et al., 2009).

Notwithstanding such limitations, facilitated endogenous regeneration minimizes the complex procedures of cell delivery as mentioned above (Chen et al., 2011). The strategy relies on harnessing the intrinsic regenerative potential of endogenous tissues using molecular stimuli, such as GF delivery or gene transfer, to initiate reparative processes in situ. This may be achieved by forming a scaffold that leads to timed drug (small chemical or biomolecule) delivery near a stem cell niche (Chen et al., 2011).

The most challenging, but perhaps ultimate biomaterials goal is to create multicomponent injectable materials designed to act as *de novo* niches in vivo. The promise of biosynthetic approaches to biomaterials design must be weighed against the fact that very little is known about the in vivo performance of systems prepared in this way (Langer and Tirrell, 2004). Artificial niches would need to incorporate appropriate "homing" signals that could attract endogenous stem cells and localize them by means of known cell–cell or cell–matrix adhesive interactions. Once localized to these artificial niches, the cells would need to be exposed to tethered signals that control such stem cell functions as expansion by self-renewing division. Neighboring vascular cells and neural cells would need access. Upon injury, the upregulation and release of proteases would enable the newly formed stem cells to escape the niche and contribute to differentiation and tissue regeneration (Lutolf et al., 2009). The importance of the extracellular environment in determining cell behavior points to the need for regenerative materials to provide cells with biological cues. The optimization of mechanical and structural features of scaffolds and their potential to direct aspects of cell behavior illustrates that functional sophistication is synonymous with high manufacturing costs. Therefore, there is an attempt to develop novel expedited strategies for tissue self-repair and regeneration that are compatible with existing clinical modalities and cost-effective (Chen et al., 2010a).

The use of PRP and EMD in periodontal reconstructive surgery has been evaluated in various experimental and clinical models, yet the outcomes have frequently been inconsistent or conflicting between the studies (Chen et al., 2010a). If we consider these endogenous technologies for periodontal regeneration, both therapeutic products and techniques need to be improved and refined (Anitua et al., 2012). One pivotal aspect for the development of PRP technology, for example, has been the deep platelet characterization that has been developed to determine the most important GFs and proteins contained within these cells and the protocols that facilitate the manipulation and concentration of the cells safely (Anitua et al., 2010).

Endogenous tissue engineering is still at its infancy. New insights from developmental biology and other biological disciplines are actively guiding the development of biofunctonalized materials that work with nature's own mechanisms of repair, where the scaffold should be considered as both a biochemical and biophysical signaling device, modulating and orchestrating cellular activity in the defected microenvironment to harness and direct a patient's own regenerative potential.

6.2 Stem Cell–Based Periodontal Regeneration

For cases that exceed the endogenous regenerative capacity, stem cell implantation to enhance regenerative outcome has been pursued as an alternative (Wada et al., 2009; Ding et al., 2010; Feng et al., 2010; Iwata et al., 2010; Park et al., 2011b).

Scaffold-Free Cell Delivery. Without the use of scaffolds, cells can form sheets using temperature-responsive culture surfaces (e.g., UpCell, CellSeed Inc., Tokyo, Japan) to generate transplantable cell sheets for regeneration of a wide variety of cell-dense tissues. With this concept, fibroblasts or other cell types that secrete significant amounts of ECM proteins are cultured for a prolonged period to create sheet-like structures for tissue reconstruction (Yang et al., 2007). Due to the presence of deposited ECM produced during in vitro incubation, cell sheets can easily be transferred and attached to other surfaces, such as culture dishes and even host tissues. The resulting cell sheets retain their original ECM and cell–cell contacts. They can be made into multilayered to strengthen the sheet and with a higher number of cells for tissue regeneration.

Because cell sheet engineering can produce a functional layer of cells, various types of cell sheet transplantations without scaffolds have been performed with promising results (Ishikawa et al., 2009). In a rodent model, PDL cell sheets were implanted into a mesial dehiscence defect and the data showed that PDL-like tissues, including an acellular cementum-like layer and fibrils anchoring into this layer, were formed, whereas no such formation was observed in controls where no cell sheets were used (Hasegawa et al., 2005). Similar results were observed in a dog model using similar techniques. No visible adverse reactions were observed for up to 8 weeks after surgery (Akizuki et al., 2005). Recently, multilayered human PDL cell sheet was developed by Flores and co-workers (2008a,b) to regenerate periodontal tissues. Allogeneic PDLSC sheets were used to cure periodontitis in a minipig periodontitis model (Ding et al., 2010). Although the use of allogeneic cells remains inconsistent, the application of autologous PDL cell sheets for periodontal tissue regeneration may be an effective clinically strategy (Ishikawa et al., 2009). The safety and efficacy of PDL cell sheets for clinical application were evaluated both in vitro and in vivo and presented no evidence of malignant transformation (Washio et al., 2010). However, unlike scaffold-based methods, engineered tissues that are created with cell sheets harvested from temperature-responsive culture surfaces contain relatively little ECM (depending on the specific cell type), and therefore it may not be suited for regenerating tissues of large amount of ECM, such as bone or cartilage.

Yang and co-workers (2009) proposed another scaffold-free approach for cell delivery to the periodontium: a novel three-dimensional pellet cultivation system to recreate the biological microenvironment similar to that of a regenerative milieu. Ex vivo expanded dental stem cells were grafted to the periodontal defect without a scaffold, where the cellular pellets could cover the denuded root surface thoroughly and filled the defect (Park et al., 2008). Furthermore, bilayered cell pellet constructs comprising calcified bone-forming cell pellets (i.e., BMMSCs) and cementum/PDL-forming cell pellets (i.e., PDLSCs) can be fabricated in vitro via the intrinsic capacity of monodispersed cells to self-assemble into a microtissue (e.g., three-dimensional spheroid). This strategy may provide an alternative to reconstruct extensive periodontal defects and achieve a predictable and complete regeneration of the periodontium (Yang et al., 2010).

Scaffold-Based Cell Delivery. Cells, with or without co-delivery with GFs, can be carried by injectable matrices, membranes, soft scaffolds, solid load-bearing scaffolds or microencapsulation and delivered to the periodontium.

Injectable Scaffolds. Injectable materials are considered to be the ideal delivery vehicles for cells and bioactive factors because they can be delivered via a minimally invasive process and can fill irregular space of the defect. Hence, they facilitate clinical application to various types of periodontal lesions. The injectable system can offer early mechanical stabilization by in situ polymerization such as fibrin and alginate (Wei et al., 2010). The rapid degradation of fibrin can be decelerated by modification with poly(ethylene glycol) (PEG), thus creating a hybrid material for robust cell delivery (Galler et al., 2011). Knowing that PRP may enhance new bone formation and accelerate an existing wound-healing process, a combination of PRP as an autologous injectable scaffold was used to delivery BMMSCs for periodontal regeneration, resulting in successful bone formation (Yamada et al., 2008). Although cell transplantation via injectable matrices into sites of articular defects has been successful (Hu et al., 2010), it has not been tested as to whether injections of stem cells into periodontal defects can achieve significant clinical improvement.

Soft Scaffolds. Soft scaffolds such as porous membranes or fibrous structures are alternative options as cell carriers (Sittinger et al., 2004). Hydrogel sponges as vehicles for cell and drug delivery for periodontal regeneration has been tested with success in various animal models and in clinical practice (Mano et al., 2007; Nuñez et al., 2012). In terms of synthetic biomaterials, a pilot study has demonstrated that cementoblasts have a marked ability to induce mineralization in periodontal wounds when delivered via PLGA polymer sponges (Zhao et al., 2004). Clinically, collagen sponge scaffold has been shown to be an effective cell delivery vehicle for oromaxillofacial (OMF) bone tissue repair in patients (d'Aquino et al., 2009).

Solid Scaffolds. Following in vivo implantation, solid scaffold–cell transplants stimulate the repair of damaged or diseased tissue while maintaining adequate integrity (Sittinger et al., 2004). HA/TCP can be used directly as a cell delivery scaffold (Liu et al., 2008; Kim et al., 2009a; Ding et al., 2010). A major clinical challenge in the reconstruction of large oral and craniofacial defects is the neogenesis of osseous and ligamentous interfacial structures, where the natural three-dimensional shape of the tissue needs to be achieved (Sittinger et al., 2004). Recently, multiscale computational design and fabrication of composite hybrid polymeric scaffolds (PCL-PGA) to carry genetically modified human cells to regenerate human tooth dentin–ligament–bone complexes have been tested in vivo. This approach offers potential for the clinical implementation of customized periodontal scaffolds that may enable regeneration of multitissue interfaces required for oral, dental, and craniofacial engineering applications (Park et al., 2010).

Other Strategies. Cell-based research has focused on the geometric design of the scaffolds for tissue engineering. However, the orchestration of multiple tissue formation, spatial fibrous tissue organization, and endpoint functional restoration using a single in vivo scaffold system remains a significant challenge (Sun et al., 2012).

Stem cells can potentially be expanded on microcarrier beads in spin culture for direct transplantation into bone defects with significantly reduced posttransplantation apoptosis and enhanced in vivo trabecular bone formation (Yang et al., 2007, 2010). Besides, shape memory materials, although yet to be tested, are potentially useful for periodontal regeneration. These materials can deliver bulky scaffolds via minimally invasive surgery to the periodontal defects. Using these smart materials, future scaffolds may be fabricated in a condensed state before transplantation and later acquire the desired shape in vivo (Sittinger et al., 2004). The self-shaped hydrogels can be collapsed structurally into smaller, temporary shapes that permit their minimally invasive delivery in vivo. Scaffolds are rehydrated in situ with a suspension of cells or cell-free medium and delivered through the same catheter. The rapid recovery of the scaffold properties facilitates efficient cell seeding in vivo and permits neotissue formation in desired geometries (Thornton et al., 2004).

6.3 Gene Therapy for Periodontal Engineering

The major setbacks of GF delivery are the difficulty in dose control and the high costs. Gene therapy seeks to optimize the delivery of GFs to the periodontal defects so that the limitations associated with the efficacy of topical application of these factors (e.g., a short duration of action) can be overcome. This technology, which converts cells into protein-producing factories, has emerged as a promising strategy for the modulation of the host response triggered by periodontal microbes and the regeneration of periodontium during diseases (Rios et al., 2011). Gene therapy has several advantages over controlled and direct delivery of exogenous GFs for its long-term sustainability of gene expression in vivo (from weeks to years) and the potential to produce multiple endogenous agents for therapeutics. In periodontics, the strategies to deliver different genes in a spatially controlled and bioavailable manner present a great potential for regenerating tissues at the tooth–ligament–bone interface (Park et al., 2010).

Strategies for delivering therapeutic transgenes into human recipients are twofold: (1) direct infusion of the gene of interest using viral or nonviral vectors in vivo, and (2) introduction of the gene into ex vivo cultured cells (often, stem cells) followed by transferring cells back into the hosts. For periodontal tissue engineering, a therapeutic gene of interest is either injected directly into the periodontal defect via a retrovirus or incorporated into cultured stem cells, expanded and delivered into the area of interest (indirect approach). The latter is generally considered to be safer and more consistent because the target cells are isolated and the therapeutic gene is delivered ex vivo under controlled conditions (Rios et al., 2011). Adenoviral vectors have been used to enable the overexpression of growth-promoting molecules. PDGF and BMP-7 genes have been transferred successfully into cementoblasts, fibroblasts, and other periodontal cell types. The sustained release of BMP-7 and PDGF by transduced cells implanted into the experimental periodontal defects results in an enhanced regeneration of bone and cementum (Jin et al., 2003, 2004).

The application of nonviral gene therapy has also been tested for this purpose (Zhang et al., 2009). The encapsulation of DNA into polymeric depot systems can be used to spatially and temporally control DNA release, leading to a prolonged therapeutic level of the encoded proteins for tissue regeneration. Prior to encapsulation, DNA may be condensed with cationic polymers to decrease particle size, protect DNA from degradation, promote interaction with cell membranes, and facilitate endosomal release via proton sponge effect (Storrie and Mooney, 2006). The development of gene-activated matrix (GAM) is based on this concept. GAMs are potentially off-the-shelf products that are comprised of a scaffold impregnated with plasmid DNA encoding for gene products that promote tissue regeneration. The GAM is inserted into the defect site, where infiltrating progenitor cells become transfected and express the gene products (Storrie and Mooney, 2006). Peng and co-workers (2008) designed a novel GAM based on a porous chitosan–collagen composite scaffold that embeds chitosan and plasmid nanoparticles encoding platelet-derived growth factor (PDGF). The plasmid DNA entrapped in the scaffolds showed a sustained and steady release over six weeks and could be effectively protected by the chitosan nanoparticles. Recently, the gene delivery of tumor necrosis factor receptor–immunoglobulin Fc chain fusion protein was found to reverse *Porphyromonas gingivalis lipopolysaccharide*–mediated bone loss in rats, suggesting that gene delivery of anti-inflammatory factors can regulate the progression of periodontal disease (Cirelli et al., 2008). Significant efforts are still needed to develop safe and effective gene delivery platforms for periodontal tissue engineering.

7 CHALLENGES AND FUTURE DIRECTIONS

The challenges of periodontal bioengineering include development of minimally invasive techniques to harvest stem and progenitor cells, optimization of in vitro cell culture systems, controlling of cell fates, efficient delivery of cells or engineered tissue constructs to defect sites, minimization of inflammatory and/or immune responses, and seamless integration of engineered cell/tissue with the surrounding tissues in vivo (Lin et al., 2009; Oswal et al., 2011; Rios et al., 2011).

Despite evidence showing that biological regeneration can occur in the periodontium, complete and predictable regeneration is still difficult, especially in advanced periodontal defects (Polimeni et al., 2006). The use of PDLSCs for the reconstruction and repair of the periodontium is feasible (Ishikawa et al., 2009), however, little is known about the events following cell transplantation, and the risks of stem cell therapies have been underscored by several clinicians and researchers (Yoshida et al., 2012). A major obstacle is not knowing how to maximize the utility of the cells or cell constructs to be delivered into a permissive environment (Rios et al., 2011), and to replicate the key cellular events that parallel periodontal development (Lin et al., 2009). Such insights are required to improve the protocols for the isolation of pure stem cell populations and their robust ex vivo expansion in chemically defined conditions (Scheller et al., 2009; Zhang et al., 2012). Biomaterial scaffold is likely to play a critical role in cell-based therapeutics (Bartold et al., 2006).

As for preclinical research, the optimal animal model for periodontal disease has yet to be established (Lu et al., 2013). While some species (e.g., primate) are more suited than others (e.g., mouse), the research findings may not always reflect those in humans (Lin et al., 2009). Current cell culture conditions are not developed sufficiently to mimic the cell microenvironment in vivo (Rios et al., 2011). Because large numbers of cells are generally required for administration, it is often needed to perform long-term and extensive in vitro culturing. This can increase the chance of genetic or epigenetic changes in the cultured cells (Rayment and Williams, 2010). Good manufacturing practice facilities to provide clinical grade stem cell lines are still not widely available. There are challenges related to immune rejection after cell administration, the oncogenic properties of stem cells, and the functional integration of transplanted tissues into the host (Lin et al., 2009). The use of autologous stem cells should overcome immune rejection. The immunosuppressive effects of MSCs both in vitro and in vivo have also raised the possibility of using allogenic stem cells without the need for donor and recipient cross-matching (Wada et al., 2009; Ding et al., 2010). Nonetheless, careful analysis of the risks and benefits when using an allogeneic source is needed (Chen et al., 2012a). Although many issues need to be resolved before stem cell therapies become common, clinicians should continue to update themselves on the advancements (Yoshida et al., 2012). The solution to all aforementioned challenges requires multidisciplinary and interdisciplinary cooperation among material scientists, biochemical engineers, cell biologists, and clinicians to ensure development of the best possible regenerative approach (Chen et al., 2012a).

Acknowledgments

This work was supported in part by grants from the National Natural Science Foundation of China (81071253 and 31170912). The authors would like to apologize to some of the contributors of primary publications whose work is alluded to this chapter but not directly referenced because of length limitations.

REFERENCES

Akizuki T, Oda S, Komaki M, Tsuchioka H, Kawakatsu N, Kikuchi A, et al. (2005). Application of periodontal ligament cell sheet for periodontal regeneration: a pilot study in beagle dogs. *J Periodont Res* 40:245–251.

Alsberg E, Anderson KW, Albeiruti A, Rowley JA, Mooney DJ (2002). Engineering growing tissues. *Proc Natl Acad Sci USA* 99:12025–12030.

Al-Shammari KF, Al-Khabbaz AK, Al-Ansari JM, Neiva R, Wang HL (2005). Risk indicators for tooth loss due to periodontal disease. *J Periodontol* 76:1910–1918.

Anitua E, Sánchez M, Orive G (2010). Potential of endogenous regenerative technology for in situ regenerative medicine. *Adv Drug Deliv Rev* 62:741–752.

Anitua E, Alkhraisat MH, Orive G (2012). Perspectives and challenges in regenerative medicine using plasma rich in growth factors. *J Control Release* 157:29–38.

Bacaková L, Filova E, Parizek M, Ruml T, Svorcik V (2011). Modulation of cell adhesion, proliferation and differentiation on materials designed for body implants. *Biotechnol Adv* 29:739–767.

Bae JH, Kim YK, Kim SG, Yun PY, Kim JS (2010). Sinus bone graft using new alloplastic bone graft material (Osteon): II. clinical evaluation. *Oral Surg Oral Med Oral Pathol Oral Radiol Endod* 109: e14–e20.

Bartold PM, Xiao Y, Lyngstaadas SP, Paine ML, Snead ML (2006). Principles and applications of cell delivery systems for periodontal regeneration. *Periodontol 2000* 41:123–135.

Bueno EM, Glowacki J (2009). Cell-free and cell-based approaches for bone regeneration. *Nat Rev Rheumatol* 5:685–697.

Cancedda R, Giannoni P, Mastrogiacomo M (2007). A tissue engineering approach to bone repair in large animal models and in clinical practice. *Biomaterials* 28:4240–4250.

Carmona-Rodríguez B, Alvarez-Pírez MA, Narayanan AS, Zeichner-David M, Reyes-Gasga J, Molina-Guarneros J, et al. (2007). Human Cementum Protein 1 induces expression of bone and cementum proteins by human gingival fibroblasts. *Biochem Biophys Res Commun* 358:763–769.

Catón J, Bostanci N, Remboutsika E, De Bari C, Mitsiadis TA (2011). Future dentistry: cell therapy meets tooth and periodontal repair and regeneration. *J Cell Mol Med* 15:1054–1065.

Chen FM, Jin Y (2010). Periodontal tissue engineering and regeneration: current approaches and expanding opportunities. *Tissue Eng B* 16:219–255.

Chen FM, Chen R, Wang XJ, Sun HH, Wu ZF (2009a). In Vitro cellular responses to scaffolds containing two microencapulated growth factors. *Biomaterials* 30:5215–5224.

Chen FM, Shelton RM, Jin Y, Chapple IL (2009b). Localized delivery of growth factors for periodontal tissue regeneration: role, strategies, and perspectives. *Med Res Rev* 29:472–513.

Chen FM, Zhang J, Zhang M, An Y, Chen F, Wu ZF (2010a). A review on endogenous regenerative technology in periodontal regenerative medicine. *Biomaterials* 31:7892–7927.

Chen FM, Zhang M, Wu ZF (2010b). Toward delivery of multiple growth factors in tissue engineering. *Biomaterials* 31:6279–6308.

Chen FM, Wu LA, Zhang M, Zhang R, Sun HH (2011). Homing of endogenous stem/pregenitor cells for in situ tissue regeneration: promises, strategies, and translational perspectives. *Biomaterials* 32:3189–3209.

Chen FM, Zhao YM, Jin Y, Shi S (2012a). Prospects for translational regenerative medicine. *Biotech Adv* 30:658–672.

Chen FM, Sun HH, Lu H, Yu Q (2012b). Stem cell-delivery therapeutics for periodontal tissue regeneration. *Biomaterials* 33:6320–6344.

Cirelli JA, Park CH, Mackool K, Taba M Jr, Lustig KH, Burstein H, et al. (2008). AAV2/1-TNFR: Fc gene delivery prevents periodontal disease progression. *Gene Ther* 43:531–536.

Choi SH, Kim CK, Cho KS, Huh JS, Sorensen RG, Wozney JM, et al. (2002). Effect of recombinant human bone morphogenetic protein-2/absorbable collagen sponge (rhBMP-2/ACS) on healing in 3-wall intrabony defects in dogs. *J Periodontol* 73:63–72.

Covani U, Orlando B, Giacomelli L, Cornelini R, Barone A (2011). Implant survival after sinus elevation with Straumann(®) BoneCeramic in clinical practice: ad-interim results of a prospective study at a 15-month follow-up. *Clin Oral Implants Res* 22:481–484.

Dangaria SJ, Ito Y, Walker C, Druzinsky R, Luan X, Diekwisch TG (2009). Extracellular matrix-mediated differentiation of periodontal progenitor cells. *Differentiation* 78:79–90.

d'Aquino R, De Rosa A, Lanza V, Tirino V, Laino L, Graziano A, et al. (2009). Human mandible bone defect repair by the grafting of dental pulp stem/progenitor cells and collagen sponge biocomplexes. *Eur Cell Mater* 18:75–83.

Del Fabbro M, Bortolin M, Taschieri S, Weinstein R (2011). Is platelet concentrate advantageous for the surgical treatment of periodontal diseases? A systematic review and meta-analysis. *J Periodontol* 82:1100–1111.

Ding G, Liu Y, Wang W, Wei F, Liu D, Fan Z, et al. (2010). Allogeneic periodontal ligament stem cell therapy for periodontitis in swine. *Stem Cells* 28:1829–1838.

Emerton KB, Drapeau SJ, Prasad H, Rohrer M, Roffe P, Hopper K, et al (2011). Regeneration of periodontal tissues in non-human primates with rhGDF-5 and beta-tricalcium phosphate. *J Dent Res* 90:1416–1421.

Feng F, Akiyama K, Liu Y, Yamaza T, Wang TM, Chen JH, et al. (2010). Utility of PDL progenitors for in vivo tissue regeneration: a report of 3 cases. *Oral Dis* 16:20–28.

Flores MG, Hasegawa M, Yamato M, Takagi R, Okano T, Ishikawa I (2008a). Cementum–periodontal ligament complex regeneration using the cell sheet technique. *J Periodont Res* 43:364–371.

Flores MG, Yashiro R, Washio K, Yamato M, Okano T, Ishikawa I (2008b). Periodontal ligament cell sheet promotes periodontal regeneration in athymic rats. *J Clin Periodontol* 35:1066–1072.

Galler KM, Cavender AC, Koeklue U, Suggs LJ, Schmalz G, D'Souza RN (2011). Bioengineering of dental stem cells in a PEGylated fibrin gel. *Regen Med* 6:191–200.

Gestrelius S, Andersson C, Johansson AC, Persson E, Brodin A, Rydhag L, et al. (1997). Formulation of enamel matrix derivative for surface coating: kinetics and cell colonization. *J Clin Periodontol* 24:678–684.

Gronthos S, Mankani M, Brahim J, Robey PG, Shi S (2000). Postnatal human dental pulp stem cells (DPSCs) in vitro and in vivo. Proc Natl Acad Sci USA 97:13625–13630.

Gunatillake P, Mayadunne R, Adhikari R (2006). Recent developments in biodegradable synthetic polymers. *Biotechnol Annu Rev* 12:301–347.

Guo W, Chen L, Gong K, Ding B, Duan Y, Jin Y (2012). Heterogeneous dental follicle cells and the regeneration of complex periodontal tissues. *Tissue Eng A* 18:459–470.

Hallman M, Thor A (2008). Bone substitutes and growth factors as an alternative/complement to autogenous bone for grafting in implant dentistry. *Periodontol 2000* 47:172–192.

Handa K, Saito M, Tsunoda A, Yamauchi M, Hattori S, Sato S, et al. (2002). Progenitor cells from dental follicle are able to form cementum matrix in vivo. *Connect Tissue Res* 43:406–408.

Hasegawa M, Yamato M, Kikuchi A, Okano T, Ishikawa I (2005). Human periodontal ligament cell sheets can regenerate periodontal ligament tissue in an athymic rat model. *Tissue Eng* 11:469–478.

Hattar S, Asselin A, Greenspan D, Oboeuf M, Berdal A, Sautier JM (2005). Potential of biomimetic surfaces to promote in vitro osteoblast-like cell differentiation. *Biomaterials* 26:839–848.

Hong Y, Gong Y, Gao C, Shen J (2008). Collagen-coated polylactide microcarriers/chitosan hydrogel composite: injectable scaffold for cartilage regeneration. *J Biomed Mater Res A* 85:628–637.

Hu B, Ren JL, Zhang JR, Ma Q, Liu YP, Mao TQ (2010). Enhanced treatment of articular cartilage defect of the knee by intra-articular injection of Bcl-xL-engineered mesenchymal stem cells in rabbit model. *J Tissue Eng Regen Med* 4:105–114.

Huebsch N, Mooney DJ (2009). Inspiration and application in the evolution of biomaterials. *Nature* 462:426–432.

Ishikawa I, Iwata T, Washio K, Okano T, Nagasawa T, Iwasaki K, et al. (2009). Cell sheet engineering and other novel cell-based approaches to periodontal regeneration. *Periodontol 2000* 51:220–238.

Iwata T, Yamato M, Zhang Z, Mukobata S, Washio K, Ando T, et al. (2010). Validation of human periodontal ligament–derived cells as a reliable source for cytotherapeutic use. *J Clin Periodontol* 37:1088–1099.

Izumi Y, Aoki A, Yamada Y, Kobayashi H, Iwata T, Akizuki T, et al. (2011). Current and future periodontal tissue engineering. *Periodontol 2000* 56:166–187.

Jayakumar A, Rajababu P, Rohini S, Butchibabu K, Naveen A, Reddy PK, et al. (2011). Multi-centre, randomized clinical trial on the efficacy and safety of recombinant human platelet-derived growth factor with β-tricalcium phosphate in human intra-osseous periodontal defects. *J Clin Periodontol* 38:163–172.

Jin QM, Anusaksathien O, Webb SA, Rutherford RB, Giannobile WV (2003). Gene therapy of bone morphogenic protein for periodontal tissue engineering. *J Periodontol* 74:202–213.

Jin QM, Anusaksathien O, Webb SA, Printz MA, Giannobile WV (2004). Engineering of tooth-supporting structures by delivery of PDGF gene therapy vectors. *Mol Ther* 9:2716–2723.

Kalpidis CD, Ruben MP (2002). Treatment of intrabony periodontal defects with enamel matrix derivative: a literature review. *J Periodontol* 73:1360–1376.

Kawaguchi H, Hirachi A, Hasegawa N, Iwata T, Hamaguchi H, Shiba H, et al. (2004). Enhancement of periodontal tissue regeneration by transplantation of bone marrow mesenchymal stem cells. *J Periodontol* 75:1281–1287.

Kemoun P, Laurencin-Dalicieux S, Rue J, Farges JC, Gennero I, Conte-Auriol F, et al. (2007). Human dental follicle cells acquire cementoblast features under stimulation by BMP-2/-7 and enamel matrix derivatives (EMD) in vitro. *Cell Tissue Res* 329:283–294.

Khan Y, Yaszemski MJ, Mikos AG, Laurencin CT (2008). Tissue engineering of bone: material and matrix considerations. *J Bone Joint Surg Am* 90:36–42.

Kim IY, Seo SJ, Moon HS, Yoo MK, Park IY, Kim BC, et al. (2008). Chitosan and its derivatives for tissue engineering applications. *Biotechnol Adv* 26:1–21.

Kim J, Xin X, Moioli EK, Chung J, Lee CH, Chen M, et al. (2010a). Regeneration of dental pulp-like tissue by chemotaxis-induced cell homing. *Tissue Eng A* 16:3023–3031.

Kim K, Lee CH, Kim BK, Mao JJ (2010b). Anatomically shaped tooth and periodontal regeneration by cell homing. *J Dent Res* 89:842–847.

Kim SH, Kim KH, Seo BM, Koo KT, Kim TI, Seol YJ, et al. (2009b). Alveolar bone regeneration by transplantation of periodontal ligament stem cells and bone marrow stem cells in a canine peri-implant defect model: a pilot study. *J Periodontos* 80:1815–1823.

Kim TG, Wikesjö UM, Cho KS, Chai JK, Pippig SD, Siedler M, et al. (2009a). Periodontal wound healing/regeneration following implantation of recombinant human growth/differentiation factor-5 (rhGDF-5) in an absorbable collagen sponge carrier into one-wall intrabony defects in dogs: a dose-range study. *J Clin Periodontol* 36:589–597.

Kitamura M, Akamatsu M, Machigashira M, Hara Y, Sakagami R, Hirofuji T, et al. (2011). FGF-2 stimulates periodontal regeneration: results of a multi-center randomized clinical trial. *J Dent Res* 90:35–40.

Kokemueller H, Spalthoff S, Nolff M, Tavassol F, Essig H, Stuehmer C, et al. (2010) Prefabrication of vascularized bioartificial bone grafts in vivo for segmental mandibular reconstruction: experimental pilot study in sheep and first clinical application. *Int J Oral Maxillofac Surg* 39:379–387.

Kumada Y, Zhang S (2010). Significant type I and type III collagen production from human periodontal ligament fibroblasts in 3D peptide scaffolds without extra growth factors. *PLoS One* 5: e10305.

Kwon DH, Bisch FC, Herold RW, Pompe C, Bastone P, Rodriguez NA, et al. (2010). Periodontal wound healing/regeneration following the application of rhGDF-5 in a beta-TCP/PLGA carrier in critical-size supra-alveolar periodontal defects in dogs. *J Clin Periodontol* 37:667–674.

Langer R, Tirrell DA (2004). Designing materials for biology and medicine. *Nature* 428:487–492.

Lasprilla AJ, Martincz GA, Lunelli BH, Jardini AL, Filho RM (2012). Poly-lactic acid synthesis for application in biomedical devices: a review. *Biotechnol Adv* 30:321–328.

Lee CH, Cook JL, Mendelson A, Moioli EK, Yao H, Mao JJ (2010). Regeneration of the articular surface of the rabbit synovial joint by cell homing: a proof of concept study. *Lancet* 376:440–448.

Lin NH, Gronthos S, Mark Bartold P (2009). Stem cells and future periodontal regeneration. *Periodontol 2000* 51:239–251.

Liu Y, Zheng Y, Ding G, Fang D, Zhang C, Bartold PM, et al. (2008). Periodontal ligament stem cell–mediated treatment for periodontitis in miniature swine. *Stem Cells* 26:1065–1073.

Lovelace TB, Mellonig JT, Meffert RM, Jones AA, Nummikoski PV, Cochran DL (1998). Clinical evaluation of bioactive glass in the treatment of periodontal osseous defects in humans. *J Periodontol* 69:1027–1035.

Lu H, Xie C, Zhao YM, Chen FM (2013). Further research and therapeutic applications of stem cell transplantation in periodontal regenerative medicine: translational considerations. *Cell Transplant* DOI: 10.3727/096368912X656171.

Lutolf MP, Gilbert PM, Blau HM (2009). Designing materials to direct stem-cell fate. *Nature* 462:433–441.

Ma Z, Li S, Song Y, Tang L, Ma D, Liu B, et al. (2008). The biological effect of dentin noncollagenous proteins (DNCPs) on the human periodontal ligament stem cells (HPDLSCs) in vitro and in vivo. *Tissue Eng A* 14:2059–2068.

Mano JF, Silva GA, Azevedo HS, Malafaya PB, Sousa RA, Silva SS, et al. (2007). Natural origin biodegradable systems in tissue engineering and regenerative medicine: present status and some moving trends. *J R Soc Interface* 4:999–1030.

Marcacci M, Kon E, Moukhachev V, Lavroukov A, Kutepov S, Quarto R, et al. (2007). Stem cells associated with macroporous bioceramics for long bone repair: 6- to 7-year outcome of a pilot clinical study. *Tissue Eng* 13:947–955.

Min CK, Wikesjö UM, Park JC, Chae GJ, Pippig SD, Bastone P, et al. (2011). Wound healing/regeneration using recombinant human growth/differentiation factor-5 in an injectable poly-lactide-*co*-glycolide-acid composite carrier and a one-wall intra-bony defect model in dogs. *J Clin Periodontol* 38:261–268.

Miura M, Gronthos S, Zhao M, Lu B, Fisher LW, Robey PG, et al. (2003). SHED: stem cells from human exfoliated deciduous teeth. *Proc Natl Acad Sci USA* 100:5807–5812.

Moore YR, Dickinson DP, Wikesjö UM (2010). Growth/differentiation factor-5: a candidate therapeutic agent for periodontal regeneration? A review of pre-clinical data. *J Clin Periodontol* 37:288–298.

Morsczeck C, Götz W, Schierholz J, Zeilhofer F, Küuhn U, Möhl C, et al. (2005). Isolation of precursor cells (PCs) from human dental follicle of wisdom teeth. *Matrix Biol* 24:155–165.

Nair LS, Laurencin CT (2006). Polymers as biomaterials for tissue engineering and controlled drug delivery. *Adv Biochem Eng Biotechnol* 102:47–90.

Nevins M, Giannobile WV, McGuire MK, Kao RT, Mellonig JT, Hinrichs JE, et al. (2005). Platelet-derived growth factor stimulates bone fill and rate of attachment level gain: results of a large multicenter randomized controlled trial. *J Periodontol* 76:2205–2215.

Nuñez J, Sanz-Blasco S, Vignoletti F, Muñoz F, Arzate H, Villalobos C, et al. (2012). Periodontal regeneration following implantation of cementum and periodontal ligament–derived cells. *J Periodontal Res* 47:33–44.

Oswal S, Ravindra S, Sanivarapu S (2011). Stem cells in periodontics: a futuristic reality. *J Stomat Occup Med* 4:95–104.

Paquette DW, Brodala N, Williams RC (2006). Risk factors for endosseous dental implant failure. *Dent Clin North Am* 50:361–374

Park CH, Rios HF, Jin Q, Bland ME, Flanagan CL, Hollister SJ, et al. (2010) Biomimetic hybrid scaffolds for engineering human tooth–ligament interfaces. *Biomaterials* 31:5945–5952.

Park IH, Zhao R, West JA, Yabuuchi A, Huo H, Ince TA, et al. (2008). Reprogramming of human somatic cells to pluripotency with defined factors. *Nature* 451:141–146.

Park JC, Kim JM, Jung IH, Kim JC, Choi SH, Cho KS, et al. (2011a). Isolation and characterization of human periodontal ligament (PDL) stem cells (PDLSCs) from the inflamed PDL tissue: in vitro and in vivo evaluations. *J Clin Periodontol* 38:721–731.

Park JY, Jeon SH, Choung PH (2011b). Efficacy of periodontal stem cell transplantation in the treatment of advanced periodontitis. *Cell Transplant* 20:271–285.

Peng L, Cheng X, Zhuo R, Lan J, Wang Y, Shi B, et al. (2009). Novel gene-activated matrix with embedded chitosan/plasmid DNA nanoparticles encoding PDGF for periodontal tissue engineering. *J Biomed Mater Res A* 90:564–576.

Pihlstrom BL, Michalowicz BS, Johnson NW (2005). Periodontal diseases. *Lancet* 366:1809–1820.

Pizzo G, Guiglia R, Lo Russo L, Campisi G (2010). Dentistry and internal medicine: from the focal infection theory to the periodontal medicine concept. *Eur J Intern Med* 21:496–502.

Place ES, Evans ND, Stevens MM (2009). Complexity in biomaterials for tissue engineering. *Nat Mater* 8:457–470.

Plachokova AS, Nikolidakis D, Mulder J, Jansen JA, Creugers NH (2008). Effect of platelet-rich plasma on bone regeneration in dentistry: a systematic review. *Clin Oral Implants Res* 19:539–545.

Polimeni G, Xiropaidis AV, Wikesjö UM (2006). Biology and principles of periodontal wound healing/regeneration. *Periodontol 2000* 41:30–47.

Ramseier CA, Abramson ZR, Jin Q, Giannobile WV (2006). Gene therapeutics for periodontal regenerative medicine. *Dent Clin N Am* 50:245–263.

Rayment EA, Williams DJ (2010) Concise review: mind the gap: challenges in characterizing and quantifying cell- and tissue-based therapies for clinical translation. *Stem Cells* 28:996–1004.

Reynolds MA, Aichelmann-Reidy ME, Branch-Mays GL (2010). Regeneration of periodontal tissue: bone replacement grafts. *Dent Clin N Am* 54:55–71.

Rezwan K, Chen QZ, Blaker JJ, Boccaccini AR (2006). Biodegradable and bioactive porous polymer/inorganic composite scaffolds for bone tissue engineering. *Biomaterials* 27:3413–3431.

Rios HF, Lin Z, Oh B, Park CH, Giannobile WV (2011). Cell- and gene-based therapeutic strategies for periodontal regenerative medicine. *J Periodontol* 82:1223–1237.

Ruhé PQ, Hedberg EL, Padron NT, Spauwen PH, Jansen JA, Mikos AG (2005). Biocompatibility and degradation of poly(D, L-lactic-*co*-glycolic acid)/calcium phosphate cement composites. *J Biomed Mater Res A* 74:533–544.

Scheller EL, Krebsbach PH, Kohn DH (2009). Tissue engineering: state of the art in oral rehabilitation. *J Oral Rehabil* 36:368–389.

Schliephake H, Aref A, Scharnweber D, Bierbaum S, Roessler S, Sewing A (2005). Effect of immobilized bone morphogenic protein 2 coating of titanium implants on periimplant bone formation. *Clin Oral Implants Res* 16:563–569.

Sculean A, Nikolidakis D, Schwarz F (2008). Regeneration of periodontal tissues: combinations of barrier membranes and grafting materials—biological foundation and preclinical evidence: a systematic review. *J Clin Periodontol* 35:106–116.

Seo BM, Miura M, Gronthos S, Bartold PM, Batouli S, Brahim J, et al. (2004). Investigation of multipotent postnatal stem cells from human periodontal ligament. *Lancet* 364:149–155.

Shirakata Y, Taniyama K, Yoshimoto T, Miyamoto M, Takeuchi N, Matsuyama T, et al. (2010). Regenerative effect of basic fibroblast growth factor on periodontal healing in two-wall intrabony defects in dogs. *J Clin Periodontol* 37:374–381.

Sittinger M, Hutmacher DW, Risbud MV (2004). Current strategies for cell delivery in cartilage and bone regeneration. *Curr Opin Biotechnol* 15:411–418.

Sonoyama W, Liu Y, Fang D, Yamaza T, Seo BM, Zhang C, et al. (2006). Mesenchymal stem cell-mediated functional tooth regeneration in swine. *PLoS One* 1:e79.

Storrie H, Mooney DJ (2006). Sustained delivery of plasmid DNA from polymeric scaffolds for tissue engineering. *Adv Drug Deliv Rev* 58:500–514.

Sukumar S, Drí zhal I, Paulusová V, Bukac J (2011). Surgical treatment of periodontal intrabony defects with calcium sulphate in combination with beta-tricalcium phosphate: clinical observations two years post-surgery. *Acta Med* (*Hradec Kralove*) 54:13–20.

Sun HH, Qu TJ, Zhang XH, Yu Q, Chen FM (2012). Designing biomaterials for in situ periodontal tissue regeneration. *Biotechnol Prog* 28:3–20.

Tang L, Li N, Xie H, Jin Y (2011). Characterization of mesenchymal stem cells from human normal and hyperplastic gingiva. *J Cell Physiol* 226:832–842.

Tayalia P, Mooney DJ (2009). Controlled growth factor delivery for tissue engineering. *Adv Mater* 21:3269–3285.

Thornton AJ, Alsberg E, Albertelli M, Mooney DJ (2004). Shape-defining scaffolds for minimally invasive tissue engineering. *Transplantation* 77:1798–1803.

Tobita M, Uysal AC, Ogawa R, Hyakusoku H, Mizuno H (2008). Periodontal tissue regeneration with adipose-derived stem cells. *Tissue Eng A* 14:945–953.

Tsumanuma Y, Iwata T, Washio K, Yoshida T, Yamada A, Takagi R, et al. (2011). Comparison of different tissue-derived stem cell sheets for periodontal regeneration in a canine 1-wall defect model. *Biomaterials* 32:5819–5825.

Venezia E, Goldstein M, Boyan BD, Schwartz Z (2004). The use of enamel matrix derivative in the treatment of periodontal defects: a literature review and meta-analysis. *Crit Rev Oral Biol Med* 15:382–402.

Villar CC, Cochran DL (2010). Regeneration of periodontal tissues: guided tissue regeneration. *Dent Clin N Am* 54:73–92.

Volponi AA, Pang Y, Sharpe PT (2010). Stem cell–based biological tooth repair and regeneration. *Trends Cell Biol* 20:715–722.

Wada N, Menicanin D, Shi S, Bartold PM, Gronthos S (2009). Immunomodulatory properties of human periodontal ligament stem cells. *J Cell Physiol* 219:667–676.

Wang L, Shen H, Zheng W, Tang L, Yang Z, Gao Y, et al. (2011). Characterization of stem cells from alveolar periodontal ligament. *Tissue Eng A* 17:1015–1026.

Washio K, Iwata T, Mizutani M, Ando T, Yamato M, Okano T, et al. (2010). Assessment of cell sheets derived from human periodontal ligament cells: a pre-clinical study. *Cell Tissue Res* 341:397–404.

Wei N, Gong P, Liao D, Yang X, Li X, Liu Y, et al. (2010). Auto-transplanted mesenchymal stromal cell fate in periodontal tissue of beagle dogs. *Cytotherapy* 12:514–521.

White AP, Vaccaro AR, Hall JA, Whang PG, Friel BC, McKee MD (2007). Clinical applications of BMP-7/OP-1 in fractures, nonunions and spinal fusion. *Int Orthop* 31:735–741.

Yamada Y, Nakamura S, Ito K, Kohgo T, Hibi H, Nagasaka T, et al. (2008). Injectable tissue-engineered bone using autogenous bone marrow–derived stromal cells for maxillary sinus augmentation: clinical application report from a 2–6-year follow-up. *Tissue Eng A* 14:1699–1707.

Yang J, Yamato M, Shimizu T, Sekine H, Ohashi K, Kanzaki M, et al. (2007). Reconstruction of functional tissues with cell sheet engineering. *Biomaterials* 28:5033–5043.

Yang Y, Rossi FM, Putnins EE (2007). Ex vivo expansion of rat bone marrow mesenchymal stromal cells on microcarrier beads in spin culture. *Biomaterials* 28:3110–3120.

Yang Y, Rossi FM, Putnins EE (2010). Periodontal regeneration using engineered bone marrow mesenchymal stromal cells. *Biomaterials* 31:8574–8582.

Yang Z, Jin F, Zhang X, Ma D, Han C, Huo N, et al. (2009). Tissue engineering of cementum/periodontal-ligament complex using a novel three-dimensional pellet cultivation system for human periodontal ligament stem cells. *Tissue Eng C* 15:571–581.

Yang ZH, Jin F, Zhang XJ, Liu X, Zhang YF, Liu JQ, et al. (2010). A novel possible strategy based on self-assembly approach to achieve complete periodontal regeneration. *Artif Organs* 34:603–609.

Yoshida T, Washio K, Iwata T, Okano T, Ishikawa I (2012). Current status and future development of cell transplantation therapy for periodontal tissue regeneration. *Int J Dent* 2012: 307024.

Yukna RA, Evans GH, Aichelmann-Reidy MB Mayer ET (2001) Clinical comparison of bioactive glass bone replacement graft material and expanded polytetrafluoroethelyne barrier membrane in treating human mandibular molar class II furcations. *J Periodontol* 72:125–133.

Zhang J, An Y, Gao LN, Zhang YJ, Jin Y, Chen FM (2012). The effect of aging on the pluripotential capacity and regenerative potential of human periodontal ligament stem cells *Biomaterials* 33:6974–6986.

Zhang Y, Shi B, Li C, Wang Y, Chen Y, Zhang W, et al. (2009). The synergetic bone-forming effects of combinations of growth factors expressed by adenovirus vectors on chitosan/collagen scaffolds. J Control Release 136:172–178.

Zhao M, Jin Q, Berry JE, Nociti FH Jr, Giannobile WV, Somerman MJ (2004). Cementoblast delivery for periodontal tissue engineering. *J Periodontol* 75:154–161.

INDEX

Stem Cells in Craniofacial Development and Regeneration, First Edition.
Edited by George T.-J. Huang, Irma Thesleff.